Ina
Department of Dairy [illegible]ce
Virginia Tech

Mathematical Symbols (continued)

E-notation	Computer notation for a power of 10	1.1C
$E(g)$, $E(L)$	Square error of $g(x)$, $L(x)$	5.1B, 5.1C
$\mathbf{f}'(\mathbf{x})$ or J	Jacobian matrix of a nonlinear system	4.4B, 4.4C
$F[h]$	Approximation formula for Q	7.2
$F_k[h]$	Richardson's formula applied to $F[h]$	7.2E
$g(x)$	Guess function for least square fit	5.1B, 5.4B
h	Stepsize for an approximation formula	1.1A, 7.2
I, I_n	Identity matrix	3.1C
$(\text{IVP})_x$	IVP associated with $(\text{BVP})_2$	8.5A
$L]_a^b$	Left-Endpoint Rule	7.1A, 8.1D
$L(x)$	Straight-line guess function ($= mx + b$)	5.1C
$L_i(x)$	ith Lagrange polynomial for $x_k, \ldots, x_{k+n}$	6.1B
LU, $\hat{L}\hat{U}$	LU-factorization of A, $\hat{A}$ (unit L, $\hat{L}$)	3.2E, 3.3C
$L\backslash U$, $\hat{L}\backslash\hat{U}$	Compact form of LU-factorization of A, $\hat{A}$	3.2E, 3.3C
$[\hat{L}\backslash\hat{U}:\hat{\mathbf{b}}:\bar{\mathbf{c}}:\bar{\mathbf{x}}]$	Forback matrix using an LU-decomposition	3.3C, 3.3D
m_k	Slope method slope near (x_k, y_k)	2.3B
M; $M]_a^b$	Stored mantissa; Midpoint Rule	1.2A; 7.1A, 8.1D
$\\|\cdot\\|$	Norm of a vector or matrix	4.1B, 4.1D
O, $\mathbf{0}$	Zero matrix, vector	3.1B
$O(h^n)$	nth-order approximation	7.2B
$P_i(x_i, y_i)$, $P_k(x_k, y_k)$	ith knot, kth data point	6.1A, 5.1A
$p_{k,k+n}(x)$	Interpolating polynomial for $p_k, \ldots, p_{k+n}$	6.1A
$p_A(\lambda)$	Characteristic polynomial of matrix A	9.1A
$\hat{p}_n(z)$	Best nth interpolant at z	6.1E
$P_n(x)$	nth Taylor polynomial	App. II.2D, II.3
r	Stepsize ratio ($= h_{\text{larger}}/h$)	7.2E
$R_n(x)$	Remainder of nth Taylor polynomial	App. II.2D, II.3
$R(\hat{g})$	Determination index of least square $g(x)$	5.1D, 5.32A
$R_{ij}(\theta)$	Rotation matrix for Jacobi's Method	9.3B
s	Significant-digit accuracy	1.2B
$\mathbf{s}$, $\hat{\mathbf{s}}$	Vector of row scale factors	3.3C
$s(x)$	Cubic spline function	6.3B
$S_{1/3}]_a^b$, $S_{3/8}]_a^b$	Simpson rules	7.3A
$S_{1/3}[h]$	Composite Simpson's $\frac{1}{3}$-Rule	7.3B
$T]_a^b$	Trapezoidal Rule	7.1A, 8.1D
$T[h]$	Composite Trapezoidal Rule	7.3B
$T_i(\xi)$	ith Chebyshev polynomial	5.4E, 6.1F
$T_k[h]$	Romberg integration	7.3C
trid ($\mathbf{l}$, $\mathbf{d}$, $\mathbf{u}$)	Tridiagonal matrix	3.2H
u_M	Machine unit of a computer	1.2B
$y(t)$	Exact solution of a differential equation	8.1A
$y_x(t)$	Exact solution of $(\text{IVP})_x$	8.5A

NUMERICAL ANALYSIS

A Practical Approach

NUMERICAL ANALYSIS

A Practical Approach

Second Edition

M. J. Maron
University of Louisville

Macmillan Publishing Company
New York
Collier Macmillan Publishers
London

Printed in the United States of America

Macmillan Publishing Company
866 Third Avenue, New York, New York 10022

Collier Macmillan Canada, Inc.

Library of Congress Cataloging-in-Publication Data

Maron, M. J.
Numerical analysis.

Bibliography: p. 512
Includes index.
1. Numerical analysis. I. Title.
QA297.M344 1987 519.4 85-15588
ISBN 0-02-376210-1

Printing: 1 2 3 4 5 6 7 8 Year: 7 8 9 0 1 2 3 4 5

ISBN 0-02-376210-1

To my parents: Blanche and Hy

Preface

ABOUT THIS BOOK

The primary objective of *Numerical Analysis* is the same as that of the first edition: to produce intelligent users of numerical methods. An intelligent user of numerical methods has a clear understanding of how and why they work and also knows what can go wrong, so that troublesome situations can be recognized or avoided. Operationally, an intelligent user knows how to select a suitable method for a given problem and can determine from a computer printout whether the desired accuracy was achieved.

The topics covered in the book are standard for an introductory sophomore-senior-level numerical analysis or numerical methods course. A course using it can begin with roundoff error (Chapter 1), equations in one variable (Chapter 2), interpolation (Chapter 6), or linear systems (Chapter 3). Generally, only one or two methods for each type of problem is presented in detail. The selected methods are those considered closest to the current ''state of the art'' among those that can be understood by someone who has had a first calculus course.

Practical techniques for assessing whether a method worked successfully are emphasized throughout. In particular, simple indicators such as *condition numbers* of an ill-conditioned matrix (Section 4.1B) and the *index of determination* of a least square fit (Section 5.1D) and improvement formulas such as those of *Aitken* (Section 2.1F) and *Richardson* (Section 7.2E) are used more frequently than in most textbooks.

CHANGES IN THE SECOND EDITION

Numerous changes and additions were made to better achieve the primary objective. These modifications, based on my own experience using the first edition at the University of Louisville and an extensive survey of users elsewhere, are described in the following summary of the distinguishing features of the book.

The major topics covered remain essentially the same. However, all but the last chapter have been modified substantially either to proceed more effectively from the heuristic to the mathematical and computational or to incorporate new methods or real-life motivation and examples. The more signficant modifications, by topic, are

- **ERRORS IN FIXED-PRECISION ARITHMETIC Revised:** Floating-point representation is introduced using base 10 before base 2^e (Section 1.2). **Added:** Ill-conditioned calculations (Section 1.3D); designing a built-in EXP function (Section 1.4C).

- **SOLVING EQUATIONS IN ONE VARIABLE Revised:** Rates of convergence, Repeated Substitution, and Aitken's improvement formula are now discussed in Section 2.1; the convergence rate of the Newton–Raphson method is now analyzed using Repeated Substitution results (Section 2.3C). **Added:** Steffensen's method (Section 2.1F); locating roots of polynomials (Section 2.5D).

- **LINEAR SYSTEMS Completely Revised:** The solution of linear systems (Section 3.3) is now based on Gaussian Elimination rather than direct factorization but maintains the utility of an *LU*-decomposition followed by a forward/backward substitution. All ''basic'' topics, including Scaled Partial Pivoting, are introduced in Chapter 3. The instructor can then select topics of interest from Chapter 4, which now includes an improved discussion of ill conditioning (Section 4.1A) and the pivot condition number (Section 4.1B), an expanded discussion of matrix norms and cond A (Section 4.1D) and of the Choleski Decomposition (Section 4.2E), and a briefer discussion of the Crout Decomposition (Section 4.2D).

- **NONLINEAR SYSTEMS Revised:** A geometrically motivated 2×2 discussion (Sections 4.4A and 4.4B) precedes the $n \times n$ case (Sections 4.4C and 4.4D). **New:** A discussion of other methods (Section 4.4E).

- **CURVE FITTING AND FUNCTION APPROXIMATION Revised:** The determination index is introduced earlier (Section 5.1D). **New:** Practical motivation for curve fitting (Section 5.0) and function approximation (Section 5.4A); a statistical test for confidence of a straight-line fit (Section 5.1D); and orthogonal polynomials and weighted least squares (Section 5.3D).

- **POLYNOMIAL INTERPOLATION Completely Revised:** Best interpolants (Section 6.1E) and Chebyshev interpolation (Section 6.1F) are presented as soon as Lagrange interpolation is available; the discussions of h-spaced nodes (Section 6.1D) and forward difference tables for them (Section 6.2D) have been expanded. **New:** Bessel's interpolation (Section 6.2G) and inverse interpolation (Section 6.2H).

- **NUMERICAL DIFFERENTIATION AND INTEGRATION Revised:** Approximation formulas for derivatives and integrals are derived simultaneously using interpolating polynomials (Section 7.1) before using Taylor series (Section 7.2B). **New:** Strategies for tabulated functions (Sections 7.2H and 7.3D); adaptive quadrature (Section 7.4F).

- **INITIAL VALUE PROBLEMS Revised:** Second- and fourth-order Runge–Kutta methods are introduced in an intuitively appealing way based on integration formulas (Sections 8.1D and 8.2A) before Taylor series arguments are considered (Sections 8.2C and 8.2D); the discussions of stability (Section 8.2E) and stiffness (Section 8.2F) have been rewritten and now appear immediately after Runge–Kutta methods.

• **DEVELOPMENT** Each chapter begins with a description of a variety of situations in which the methods of the chapter are needed. Most methods are first described heuristically using figures and tables as needed, then illustrated in detail using a simple example and summarized as an algorithm using the pseudocode described in Appendix I. This is generally followed by a discussion of what can go wrong and how to deal with it. Proofs of theorems are given only if they either illustrate the utility of calculus or provide enhanced understanding; they are usually deferred until after the student has seen enough illustrative examples to appreciate the result being proved.

• **FORMAT AND ORGANIZATION** The book has been made more modular. Discussions relating to the effective implementation of an algorithm on a digital device are put either in a separate subsection or are clearly set off as Practical Considerations; as a result, these discussions can be emphasized, deferred, or omitted on a topic by topic basis, and in any case can easily be found when needed. Each subsection contains at most one illustrative example, algorithm, and practical consideration. This enables simple referencing such as Example 2.3F, Algorithm 4.2C, Practical Consideration 5.3B, and so on. To make these easy to find within a given subsection, practical considerations are set in a different typeface, an "end marker" (■) is used to denote the end of an example or practical consideration, and algorithms are boxed, as are important theorems and formulas. A general survey of the basic results of calculus is given in Appendix II. Summaries of several specific topics are given just before they are needed: representation of numbers in a digital device (Section 1.2), matrix algebra (Section 3.1), improper integrals (Sections 7.4D and 7.4E), existence and uniqueness of solutions of first-order initial value problems (Sections 8.1A and 8.4E), and the algebraic theory of eigenvalues (Section 9.1). These sections can be omitted and used only for reference if desired.

• **PROBLEM SETS AND SOLUTIONS** The problem sets at the end of each chapter are two to three times larger and considerably more varied than those of the first edition. They have also been separated into three categories:

Problems, which can be solved using only a hand calculator or a short program. In these problems the student is generally asked to use a method and draw conclusions from the result. To help the student learn independently, the problems are separated by subsection; and the back of the book contains answers, partial solutions, or hints to over half of them.

Miscellaneous Problems, intended for those already familiar with the use of the methods. Some probe more deeply into the mathematical or computational aspects of the methods. Others give real-world-type applications that require the methods for their solution. Still others interrelate the subsections or seek to ensure that the student clearly understands important notation and results.

Computer Problems, which vary from writing or modifying a calling program or subprogram to using existing code either to get a specific answer or to perform a "learning experiment" that would be difficult by hand.

Answers to the Miscellaneous and Computer Problems are available to instructors from the publisher.

• **COMPUTER PROGRAMS AND REALISTIC EXAMPLES** Primary emphasis is still on pseudocode rather than on any particular programming language. A preliminary example in Section 1.1A shows how pseudocode for an algorithm can be implemented with equal ease in either Fortran 77 or Pascal. All subsequent illustrative code is written in Fortran

77† and run on a DEC-1090 computer. Although significant features of this language are noted as they arise, *the discussions of computer programs are designed to describe the characteristics of well-written numerical method software in any structured programming language*. In addition, considerable attention is paid to the effective use of a general-purpose subprogram to solve a specific real-life problem (Sections 2.2C and 2.4E).

• **STYLE** The writing style is somewhat more terse than in the first edition. However, it is still conversational rather than formal and remains one of ''learning together'' from carefully selected examples. Notation is explained more carefully and used less extensively. A list of symbols and abbreviations is given inside the front and back covers for easy reference.

ACKNOWLEDGMENTS

A balanced picture of the strengths and weaknesses of the first edition was obtained as a result of reviews provided by Roger Grobe, GMI Engineering & Management Institute; Leonard M. Kahn, University of Rhode Island; Vera C. King, Prairie View A & M University; William J. Mareth, Jr., DelMar College; Loren P. Meissner, University of San Francisco; B. Prasanna, AT&T Bell Labs; and Dennis C. Smolarski, University of Santa Clara.

I am especially indebted to Gary L. Buterbaugh, Indiana University of Pennsylvania and Robert Frascatore, State University College, Buffalo, who reviewed preliminary drafts of the text, and to Christopher Hunter, Florida State University, who critically examined the text and some of the problem sets as well. Their candid assessments of the pedagogical and technical aspects of these drafts are evident throughout the book, and, I believe, improved it substantially. However, the responsibility for the selection and presentation of the topics covered is mine alone.

Thanks are due to the many colleagues at the Speed Scientific School, most notably Tom Cleaver and Dermot Collins, who offered problems from their research and undergraduate courses that were adapted to become the applied Miscellaneous Problems, and to Don Linton and Carol O'Connor who helped guide me out of some statistical impasses. Thanks are also due to many students: to Frank Nye, who helped prepare the Fortran 77 code; to Bob Costagno and Joe Bradley, whose interactive programs simplified the task of getting answers to the problems; and to the many students who earned extra credit by finding errors in drafts of the manuscript that were used as lecture notes.

Most of the book was typeset directly from diskettes provided by me. This required a substantial modification of the diskettes created by my word processor. I am grateful to Gary W. Ostedt and Robert A. Pirrung of the Macmillan Publishing Company for the courage they showed in agreeing to do this and for the advice and support they gave along the way; to Kay Downs at Waldman Graphics, Inc., who, with patience and good humor, helped me understand what had to be done; and to Kenny Hale and Danny Maron, who gallantly performed the painstaking task of actually modifying the diskettes.

Finally, my heartfelt thanks to my wife Anne and my children Danny and Melanie, who remained stoically supportive despite the fact that I transformed what once was a home into a sea of printouts, folders, and binders.

M. J. M.

†One departure from standard Fortran 77 is used freely: The exclamation point (!) is used to set off an *in-line comment*. Such comments should be omitted or suitably modified when using a Fortran 77 compiler that does not recognize them.

Contents

Chapter 2

Chapter 3

NUMERICAL ANALYSIS

A Practical Approach

1

Algorithms, Errors, and Digital Devices

1.0 Introduction

This book deals with the use of **digital devices** (calculators and computers) to obtain numerical answers to certain mathematical problems. All digital devices have one thing in common: They are incapable of recognizing *all* real numbers. However, the ''recognizable'' numbers differ from device to device and affect the *accuracy* that it is capable of achieving. Digital devices also differ with regard to *speed* (computers are generally faster), *expense* (calculators are generally cheaper to buy and use), *storage capability* and input/output (usually better on computers), and capability of being *programmed* (usually better on computers).

The purpose of this chapter is to describe how both programmable and nonprogrammable digital devices perform arithmetic, and to develop strategies for recognizing and dealing with the inaccuracy that can result.

1.1 Implementing Numerical Methods on Digital Devices

Numerical answers that are obtained using formulas will be referred to as **analytic solutions** and said to be obtained **analytically**. The formulas of algebra and calculus can be used to get answers to a remarkably diverse range of realistic problems. Unfortunately, the problems that arise in the practice of science and engineering are often either difficult or impossible to solve analytically. Consider, for example, the following four "simple" problems.

Problem 1. Find all roots between 0 and 1 of the fifth-degree equation

$$6x^5 + 20x^3 - 9 = 0$$

Problem 2. Find all $x, y > 0$ that satisfy the two nonlinear equations

$$3x^5 + 6x^3y^2 + 4xy^4 = 1 \qquad 3x^4y + 8x^2y^3 + 3y^5 = 2$$

Problem 3. Evaluate $\int_0^b \sqrt{1 + x^3}\, dx$ or $\int_0^b e^{-x^2}\, dx$ for a given $b > 0$.

Problem 4. Find $y = y(t)$ that satisfies the differential equation

$$\frac{dy}{dt} = y^2 + t^2, \quad y(0) = 1 \qquad \text{(i.e., } y = 1 \text{ at } t = 0\text{)}$$

Problem 1 would result if calculus were used to find the minimum and/or maximum values on the interval $[0, 1]$ of the function

$$F(x) = x^6 + 5x^4 - 9x + 1 \tag{1}$$

by solving the equation $F'(x) = 0$. Similarly, Problem 2 would have to be solved if calculus were used to find the minimum and/or maximum values in the first quadrant of the two-variable function

$$f(x,y) = (x^2 + y^2)^3 + x^2y^4 - 2x - 4y \tag{2}$$

by solving $\partial f/\partial x = 0$, $\partial f/\partial y = 0$ simultaneously. The integral $\int_0^b e^{-x^2}\, dx$ in Problem 3 arises frequently in statistical applications. Finally, the $y(t)$ that satisfies Problem 4 has the interesting property that the tangent line at any point $P(t, y(t))$ on its graph has slope equal to the square of the distance from P to the origin.

You are welcome to try to solve Problems 1–4 using any algebraic or calculus techniques you may know; be warned, however, that no one has yet succeeded in finding a formula involving the elementary functions for the solution of any of them (and in some cases it has been proven that no such solution exists!) Fortunately, one rarely needs *exact* numerical answers. Indeed, in the "real world" the problems themselves are generally inexact because they are posed in terms of parameters that are only approximate because they are *measured*. What one is likely to require in a realistic situation is not an exact answer, but rather one having a prescribed accuracy.

1.1A What Is a Numerical Method?

An **algorithm** for a particular problem is a step-by-step procedure that produces a solution in a finite number of steps. A **numerical method** is an algorithm for solving a problem whose solution consists of one or more numerical values. Values obtained this way are called **numerical solutions** and are said to be obtained **numerically**.

Since numerical methods are not needed if the exact answer is known, *the very first thing you must do before using a numerical method is to examine the problem and determine how close to the exact value you want your numerical solution to be*. In this book, the desired accuracy will be given with all problems. However, it is important that you do this for yourself when solving your own problems. Otherwise you will have no criterion for determining if your numerical solution is sufficiently accurate!

Perhaps the most familiar criterion for closeness of two numbers is decimal-place accuracy. We say that **x approximates $\bar{x}$ to k decimal places**, and write $\bar{x} = x\ (kd)$, if $\bar{x}$ and x differ by less than half a unit in the kth decimal place. Mathematically,

$$\bar{x} = x\ (kd) \quad \text{means} \quad |\bar{x} - x| < 0.5 \cdot 10^{-k} \qquad \textbf{(3a)}$$

When this condition holds, both x and $\bar{x}$ should yield the same number when rounded† to k decimal places. More generally, the notation

$$\bar{x} \doteq x \text{ means } x \text{ approximates } \bar{x} \text{ to all decimal places shown} \qquad \textbf{(3b)}$$

Do not confuse "$\bar{x} \doteq x$" with "$\bar{x} \approx x$," which we will use to indicate that x is *somewhere near* $\bar{x}$, although we cannot say how close. Thus we will write $y \doteq 3.8$ when $y = 3.8\ (1d)$ (i.e., $3.75 < y < 3.85$); however, when we write $y \approx 3.8$, then y might turn out to be $3.4459\ (4d)$ or $4.137\ (3d)$, and so on.

The following example is motivated by the fact that *we often do not have formulas for the functions that arise in realistic situations*. For example, suppose that the angular velocity ω of a fan [say in revolutions per minute (rpm)] is controlled by a dial whose angular setting θ can vary from 0° to 270° as shown in Figure 1.1-1(a). The ω versus θ graph will look something like the curve shown in Figure 1.1-1(b). If electrical, atmospheric, and friction conditions are not varied, then ω is uniquely determined by the dial setting θ; so $\omega = f(\theta)$. Moreover, the *derivative* of this function, $f'(\theta) = d\omega/d\theta$ (in rpm per degree change in θ), is an indicator of the **sensitivity** of the dial at a particular θ. This sensitivity would be needed if the fan, under a computer's control, is to be used to help maintain a nearly constant temperature.

Unfortunately, a formula for the function f is not likely to be known. If $f'(\theta)$ were needed for a particular dial setting θ, a rough estimate $f'(\theta) \approx \Delta\omega/\Delta\theta$ could be obtained graphically, as shown in Figure 1.1-1(b). However, if a specific accuracy were needed, say $3d$, we would have no choice but to find $f'(\theta)$ *numerically*!

†When rounding x to k decimal places, we either leave the digit in the kth decimal place unchanged or increase it by 1, *depending on which result is closer to x*. Thus -4.33492 becomes -4.33 (*not* -4.34) and 9.2451 becomes 9.25 when correctly rounded to $2d$. In the rare case when both results are equally close to x, it may be best not to round at all. If you must, use a consistent strategy such as the **even rounding rule**: *Round so that the rounded digit is even*. For example, *both* 9.235 *and* 9.245 should be rounded to 9.24 if they must be rounded to $2d$.

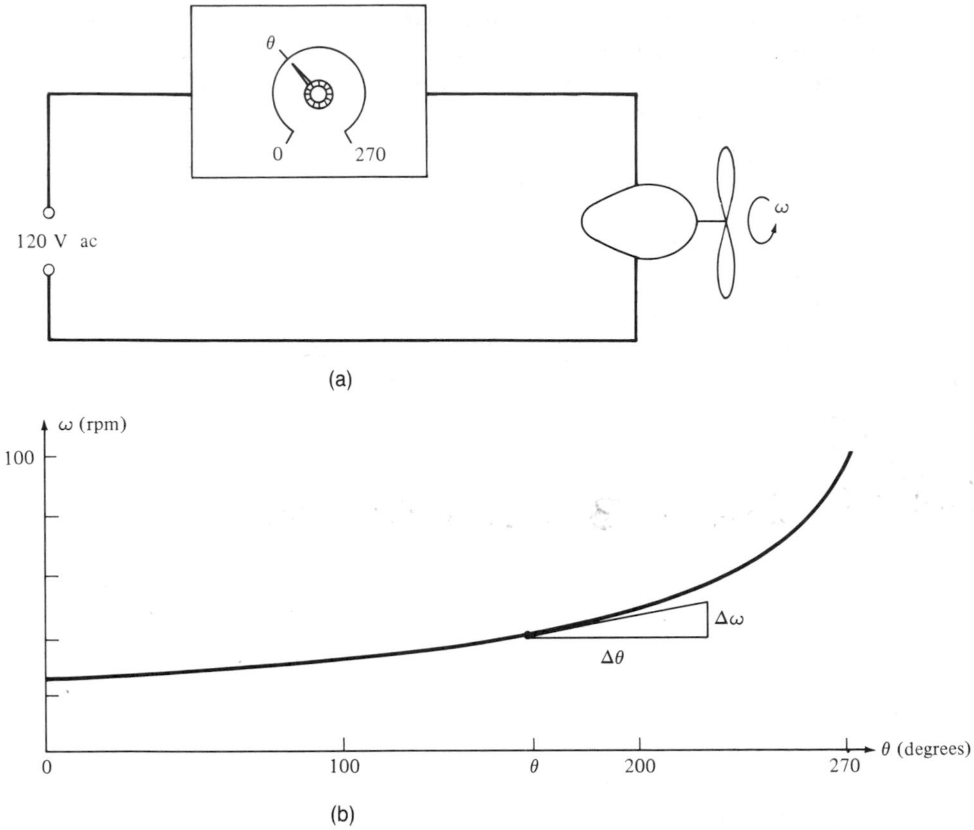

Figure 1.1-1 *(a) Using θ to control ω. (b) Graph of ω versus θ.*

EXAMPLE 1.1A Finding $f'(x)$ if you have no formula for $f(x)$

(a) Devise a numerical method for finding $f'(x)$, the derivative of a given function f at a particular x, to a prescribed number of decimal places. The method should be usable when there is no formula for $f(x)$.

(b) Test your method by using it to find to $2d$ the derivative of $f(x) = e^x$ at $x = 1$. [NOTE: $f'(x)$ is $e^1 \doteq 2.7183 = 2.72$ $(2d)$.]

SOLUTION (a): In the absence of a universal differentiation formula, we must go back to the *definition* of $f'(x)$, namely

$$f'(x) = \lim_{h \to 0} \frac{\Delta f(x)}{h}, \qquad \text{where } \frac{\Delta f(x)}{h} = \frac{f(x + h) - f(x)}{h} \tag{4}$$

This suggests that the **difference quotient**, $\Delta f(x)/h$, can be used to approximate $f'(x)$ to any desired accuracy, provided that we use a sufficiently small h. Such an h can be found empirically by finding $\Delta f(x)/h$ repeatedly for a decreasing sequence of h's until two successive difference quotients are close enough *to each other* to indicate that the desired accuracy has been achieved.

For *NumDec*-decimal-place accuracy, we can use (3a) as described using **pseudocode**† in Algorithm 1.1A. The first OUTPUT statement gives the intermediate values of h and DQ ($= \Delta f(x)/h$) while they are available (i.e., before they are updated); the second one gives the numerical solution. The rationale for the accuracy test is this: When DQ and *PrevDQ* agree *with each other* to *NumDec* decimal places, then DQ should be within *NumDec* decimal places of the (presumably unknown) exact value of $f'(x)$.

ALGORITHM 1.1A. FINDING THE DERIVATIVE OF f AT x (POSSIBLY RISKY)

PURPOSE: To find $f'(x)$ without using differentiation formulas

```
GET    x    {fixed value at which the derivative of f is desired},
       h    {initial stepsize (a small increment from x},
       r    {shrinking ratio; r > 1 causes |h| to decrease},
  NumDec    {desired number of accurate decimal places of f'(x)}
PrevDQ ← 0
Tol ← .5·10^(-NumDec)                     {tolerance for NumDec-place accuracy}

REPEAT
|  DQ ← [f(x + h) - f(x)]/h                              {current Δf(x)/h}
|  DelDQ ← DQ - PrevDQ                           {needed for accuracy test}
|  OUTPUT( h, DQ, DelDQ )                   {optional intermediate values}
|  PrevDQ ← DQ; h ← h/r               {update: prepare for next repetition}
|_ UNTIL |DeltaDQ| < Tol         {test for NumDec accurate decimal places}
OUTPUT('The derivative at 'x' to 'NumDec'-decimal places is 'DQ)
```

Algorithm 1.1A is implemented in (a) Fortran 77 and (b) Pascal in Figure 1.1-2. Notice that although the pseudocode statements of Algorithm 1.1A are not in any particular programming language, they are easily translated into syntactically correct statements in *either* Fortran 77 or Pascal.

SOLUTION (b): In both (a) and (b) of Figure 1.1-2, the input variables (x, *NumDec*, h, and r) are obtained *interactively* from a terminal during execution. Other methods (e.g., program statements, *batch* input from disk or tape) could also have been used. In anticipation of part (b), $f(x)$ was programmed as e^x. For illustrative purposes, we chose to output intermediate values of h, DQ, and *DelDQ*.

The execution of *either* program looks like the printout shown in Figure 1.1-3. (You cannot tell from the output shown whether the Fortran 77 or Pascal program was executed!) Values input from the keyboard were underlined for clarity. The choice of $r = 10$ was arbitrary. Notice that our numerical solution, $DQ = 2.718568$, becomes 2.72 when rounded to two places. So our termination criterion, $|DelDQ| < Tol = 0.005$, did indeed stop the repetitions when the desired two-place accuracy was achieved.

†See Appendix I for a detailed description of the pseudocode used in this book.

```
00100   C      Fortran program to find f'(x) to NumDec decimal place accuracy
00200          PARAMETER (C=0.5,           IR=5, IW=5) ! I/O device numbers
00300                       F(X) = EXP(X)              ! this is f(x)
00400          WRITE (IW, 1)                           ! prompt for input
00500          READ (IR, *) X, NUMDEC, H, R            ! free format READ
00600          WRITE (IW, 2)                           ! heading
00700          TOL = C * 10.**(-NUMDEC)
00800          PREVDQ = 0.0
00900   C      REPEAT UNTIL termination test is satisfied
01000     10      DQ = ( F(X + H) - F(X) ) / H
01100             DELDQ = DQ - PREVDQ
01200             IF ( PREVDQ .EQ. 0.0 ) THEN          ! first entry
01300                WRITE (IW, 3) H, DQ
01400             ELSE                                 ! other entries
01500                WRITE (IW, 3) H, DQ, DELDQ
01600             ENDIF                                ! update PREVDQ and H
01700             PREVDQ = DQ                          ! for next iteration
01800             H = H / R
01900             IF ( ABS(DELDQ) .GE. TOL ) GOTO 10   ! termination test
02000          WRITE (IW, 4) X, DQ, NUMDEC             ! final summary
02100          STOP
02200        1 FORMAT (' To get f''(x) to NumDec decimal places, input',/,
02300         &             ' x, NumDec, initial h, and shrinking ratio r')
02400        2 FORMAT ('0          h              DQ               DelDQ')
02500        3 FORMAT (E14.1, F15.6, F16.6)
02600        4 FORMAT ('0f''(',F12.6,') = ',F12.6,' to',I3,' decimal places.')
02700          END
```

(a)

> *NOTE:* The **line numbers** (00100-02700) on the left are created by the editor used to key in the programs. *They are not part of either program*, but are included for ease of reference. However, the **statement numbers** (1, 2, 10, etc.) are part of the Fortran code in (a).

```
00100   (* Pascal program to find f'(x) to NumDec decimal place accuracy *)
00200   PROGRAM DERIV     (INPUT, OUTPUT);          (* default I/O devices *)
00300   CONST C = 0.5;
00400   VAR   X, H, R, DQ, DELDQ, PREVDQ, TOL : REAL;   NUMDEC : INTEGER;
00500   FUNCTION F( X : REAL ) : REAL;
00600   BEGIN    F := EXP(X)               (* this is f(x) *)  END;
00700   BEGIN
00800      WRITELN ('To get f''(x) to NumDec decimal places, input');
00900      WRITELN (' x, NumDec, initial h, and shrinking ratio r');
01000      READ    (X, NUMDEC, H, R); WRITELN;
01100      WRITELN ('          h              DQ               DelDQ');
01200      PREVDQ := 0.0;
01300      TOL := C*EXP(LN(10)*(-NUMDEC));       (*  = C*10^(-NUMDEC)  *)
01400      REPEAT
01500         DQ := (F(X + H) - F(X))/H;  DELDQ := DQ - PREVDQ;
01600         IF PREVDQ = 0.0 THEN
01700            WRITELN('    ', H:9,'    ', DQ:12:6)  ELSE
01800            WRITELN('    ', H:9,'    ', DQ:12:6,'    ', DELDQ:12:6);
01900         PREVDQ := DQ;  H := H/R;                     (* update *)
02000      UNTIL ( ABS(DELDQ) < TOL );
02100      WRITELN; WRITELN ('f''(', X:12:6, ') = ', DQ:12:6,' to ',
02200                                 NUMDEC:2,' decimal places.')
02300      END.
```

(b)

Figure 1.1-2 *Implementations of Algorithm 1.1A in (a) Fortran 77; (b) Pascal.*

```
To get f'(x) to NumDec decimal places, input
 x, NumDec, initial h, and shrinking ratio r
 1  2  .2  1Ø

        h                 DQ               DelDQ
     Ø.2E+ØØ          3.ØØ9176
     Ø.2E-Ø1          2.745646         -Ø.263529
     Ø.2E-Ø2          2.721Ø12         -Ø.Ø24635
     Ø.2E-Ø3          2.718568         -Ø.ØØ2444

f'(     1.ØØØØØØ) =       2.718568 to  2 decimal places.
```

Figure 1.1-3 *Run of either (a) or (b) of Figure 1.1-2.* ■

1.1B Why Use a Program for a Numerical Method?

The preceding discussion illustrates the following characteristics of programmable devices (computers and programmable calculators) that make them especially well suited for numerical methods.

1. *Ease of implementation.* Algorithm 1.1A is typical of many numerical methods in that the same steps (calculate DQ and $DelDQ$, then update h and $PrevDQ$) are repeated until a **termination test** ($|DelDQ| < Tol$) is satisfied. Such algorithms are called **iterative**. An iterative algorithm is easily implemented in any programming language using a **loop** [see lines 900–1900 of Figure 1.1-2(a) and lines 1400–2000 of Figure 1.1-2(b)].
2. *Flexibility.* Once stored, either of the programs shown in Figure 1.1-2 can be used to get *any* $f'(x)$ to *any* (reasonable) number of accurate decimal places simply by changing $f(x)$ and $NumDec$. Moreover, both programs are written in accordance with the current American National Standards Institute (ANSI) standards for the language. This means that both are **transportable** in the sense that they should run with at most hardware-dependent modifications (e.g., device numbers) on *any* computer on which a "standard" compiler for the language is installed.
3. *Speed and reliability.* Once a debugged program for an algorithm is given the necessary input data, it will step through the algorithm *faster* and *more reliably* than any human being could hope to do.

The *execution time* of a program for a numerical method depends on the device executing it and the language in which it was implemented. For example, a calculator is a lot slower than a computer; BASIC code executed on an *interpreter* will take longer than *compiled* code originally written in *any* programming language. However, the *accuracy* of the numerical solution is generally *not* noticeably affected by the language used. So if execution time is not a critical factor and you have access to "canned" (i.e., ready-made) software that is *well documented* and known to be *reliable*, just use it, whether or not you are familiar with the language used to write it. But beware . . . *existing software is not always reliable!* The **Practical Consideration** discussions in this book are designed to help you create or recognize reliable software and then use it effectively.

Structured languages such as PL/I, APL, Pascal, Fortran 77, and some (nontransportable) versions of BASIC are designed to help you create code that is easy to write, document, debug, and (if necessary) modify. Unless otherwise specified, our illustrative programs will be written in Fortran 77. These programs are intended to illustrate the translation from pseudocode to transportable code and to exemplify the characteristics of "good" numerical software written in any structured language†; they should be studied carefully, along with their associated discussions, if you ever intend to write your own numerical software or modify someone else's.

1.1C Do Digital Devices Make Mistakes?

Digital devices do what they are designed to do almost flawlessly. In fact, modern computers typically perform billions of operations without making a single "mistake." Nevertheless, *digital devices often give erroneous answers!* All too frequently, they are the result of **human error** introduced by the programmer (logical error in the program), the person at the keyboard (typographical error in either the program or the input data), or the user (incorrect input data, formula, or method). *There is only one remedy for human errors: Find them and make the appropriate corrections.* The programs considered in this book are logically correct and accurately keyed in. However, in your own use of numerical methods, you should always be alert for the possibility that human error is the cause of an unreasonable computer or calculator answer.

A second source of digital device errors is a bit more subtle. There are legitimate mathematical calculations that yield correct answers when carried out using real numbers but erroneous answers when carried out on a computer or calculator. For example, we know from calculus that $\Delta f(x)/h$ will approach $f'(x)$ (if it exists) as $h \to 0$. However, this will *not* happen if the calculations are performed on a digital device. In fact, the following example shows how Algorithm 1.1A can be used to "prove" that all derivatives are zero!††

EXAMPLE 1.1C Digital device errors Table 1.1-1§ shows what happens when DQ $[= \Delta f(x)/h]$ of Figure 1.1-2 is used to try to approximate $d(e^x)/dx$ at $x = 1$ to $4d$ rather than $2d$ on (a) a DEC-10 and (b) a PDP-11. To make the errors of the calculated DQ's easier to see, we have provided the correct values of $\Delta f(x)/h$ (rounded to $7d$) in column (c). The underlined digits of DQ would differ from $\Delta f(x)/h$ after rounding. Notice first that the error in calculating DQ is device dependent. More important, in *both* (a) and (b) of Table 1.1-1, decreasing h to 0.0002 caused DQ to become closer to $f'(x) = e^1$, *but further decreases in h actually made DQ less accurate* until the termination test, $|DelDQ| <$ 0.5E−4, stopped the iteration when two successive DQ's were calculated as *zero* (which certainly does not approximate $e \doteq 2.71828$ to within four-decimal-place accuracy!)

Table 1.1-1 shows that *digital devices are capable of producing wrong answers, even if programmed correctly!* And formulas that yield accurate results for most input values (here $h > .002$) can yield highly inaccurate ones for certain others (here $0 < h < 0.00002$)! It is important to realize that the erroneous solution, $DQ = 0$, obtained in Table 1.1-1 would eventually have been obtained on *any* computer *or* calculator.

†However, no attempt was made to **optimize** the code by taking advantage of certain hardware-dependent features so as to reduce execution time, as is done by the better commercially available software.

††This is why Algorithm 1.1A was described as "possibly risky"!

§In Table 1.1-1, and throughout the book, the computer's E-notation will be used to display large or small numbers (e.g., 0.2E−3 for 0.0002).

TABLE 1.1-1 Errors in $DQ = \Delta f(x)/h$ on (a) a DEC-10 and (b) a PDP-11

h	(a) *DQ on* DEC-10	(b) *DQ on* PDP-11	(c) *Correct* $\Delta f(x)/h$
0.2	3.009176	3.009176	3.0091755
0.2E−1	2.745646	2.745652	2.7456468
0.2E−2	2.721012	2.720952	2.7210019
0.2E−3	2.718568	2.719164	2.7185537
0.2E−4	2.717972	2.729893	2.7183090
0.2E−5	2.712011	2.741814	2.7182845
0.2E−6	2.682209	3.576279	2.7182821
0.2E−7	1.490116	0.000000	2.7182819
0.2E−8	0.000000	0.000000	2.7182818
0.2E−9	0.000000		

■

Practical Consideration 1.1C (Believing computers and calculators). Most calculations performed on a digital device occur without serious error. Nevertheless, *both computers and calculators do occasionally produce highly erroneous answers; and when they do, they display them as neatly and authoritatively as correct ones!* Indeed, if either of the programs in Figure 1.1-2 were used *without printing intermediate values* to try to get $d(e^x)/dx$ at $x = 1$ to $4d$, the only output would be

```
f'(     1.ØØØØØØ) =      Ø.ØØØØØØ  to  4 decimal places
```

which is stated with a great deal of certainty but is obviously wrong, nevertheless! This is why *an intelligent user of numerical methods always scans the output of a digital device to ensure that it seems reasonable (for problem being solved) before accepting it.* ■

Serious digital device errors can often be avoided or at least minimized once we understand why they occur. This, in turn, requires an understanding of the way digital devices store real numbers and perform arithmetic on them. We consider this in the next section.

1.2 How Do Digital Devices Store Real Numbers?

Computers generally store numbers using what can be thought of as "base b cells" that can store the **base b digits** 0, 1, …, $b - 1$, where the **arithmetic base** b is typically 2 (**binary**), 8 (**octal**), or 16 (**hexadecimal**). A "base 2 cell" is called a **bit** because it stores a **b**inary dig**it** 0 or 1. The computer errors observed in the preceding section also occur on calculators, which generally store numbers using **binary-coded-decimal (BCD)** cells that can store the familiar decimal ($b = 10$) digits 0, 1, …, 9.

1.2A Normalized Floating Point Representation of Storable Numbers

Calculators use exponential notation with remarkable efficiency to store those real numbers X that can be expressed exactly as

$$X = \pm M \cdot 10^c \quad \text{where} \begin{cases} M = .d_1 d_2 \cdots d_k \ (d_1 \neq 0) \\ c = 0, \pm 1, \pm 2, \ldots, \pm 99 \end{cases} \qquad \textbf{(1)}$$

Such X's will be called **storable** because they can be stored without error on the device. The positive, k-digit fraction M in (1) is the **mantissa** of X; the two-digit integer c is its **characteristic** or **exponent**. Figure 1.2-1 shows how this X can be stored using two "sign cells" (for the signs of X and c) and $k + 2$ binary-coded-decimal cells (k for M, and two for $|c|$). A calculator that uses k BCD cells to store mantissas will be called a ***k*-significant-digit** (abbreviated ***ks***) **calculator**.

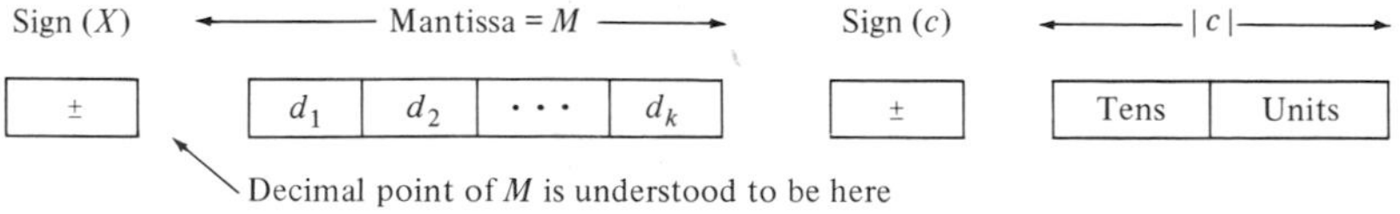

Figure 1.2-1 *Calculator storage of* $X = \pm .d_1 d_2 \cdots d_k \cdot 10^c \ (d_1 \neq 0)$.

Calculators automatically **normalize** the mantissa M of a nonzero X by "floating" the decimal point immediately to the left of the leftmost *nonzero* digit of X [d_1 in (1)] and then adjusting c accordingly. This is why the representation (1) is called the **normalized floating point decimal representation** of X. As a result of normalization, the smallest mantissa used for a nonzero X on a ks calculator is $.10\ldots00 = 10^{-1}$ *not* $.00\ldots01 = 10^{-k}$; hence $X \neq 0$ will be stored with one of

$$9 \cdot 10^{k-1} \text{ mantissas } M \quad (\text{namely, } .10\ldots00,\ .10\ldots01, \ldots,\ .99\ldots99) \qquad \textbf{(2)}$$

These correspond to $9 \cdot 10^{k-1}$ equally spaced storable X's in *each* of the intervals $[10^{c-1}, 10^c)$, where c varies over all storable characteristics as shown in Figure 1.2-2. Since there are $1 + 2 \cdot 99 = 199$ such c's (why?), there are a total of

$$199(9 \cdot 10^{k-1}) = 1791 \cdot 10^{k-1} \textit{ positive } \text{storable } X\text{'s} \qquad \textbf{(3)}$$

sprinkled over the closed interval $[s, L]$, where

$$s = \text{smallest positive storable } X = +(.10\ldots0) \cdot 10^{-99} = 10^{-100} \qquad \textbf{(4a)}$$

and

$$L = \text{largest positive storable } X = +(.99\ldots9) \cdot 10^{+99} \approx 10^{99} \qquad \textbf{(4b)}$$

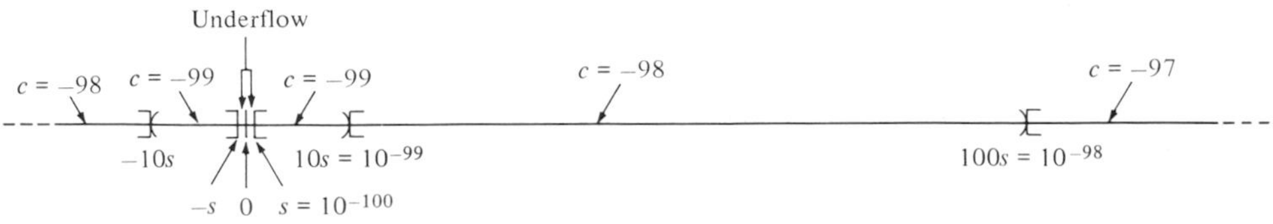

Figure 1.2-2 *Some storable X's close to zero if* $-99 \leqslant c \leqslant 99$.

with the most dense spacing near s and the most sparse spacing near L. The remaining storable X's are zero and the $1791 \cdot 10^{k-1}$ negatives of the positive storable X's, for a total of $3582 \cdot 10^{k-1} + 1$ real numbers X that can be stored exactly on a k-significant-digit calculator.

A similar analysis applies to computers, which store normalized representations of real numbers $X = \pm M \cdot 2^c$ using "base b cells" ($b = 2, 8$, or 16) for M. Thus, for *any* digital device,

> Only finitely many real numbers can be stored exactly. So there will be a smallest positive storable number s and a largest storable number L. The nonzero storable X's are sprinkled over the intervals $[-L, -s]$ and $[s, L]$, densely near s and $-s$ and sparsely near L and $-L$. **(5)**

The most important consequence of this is that most real numbers *cannot* be stored exactly on *any* given digital device! An attempt to store an X satisfying $|X| > L$ causes **floating point overflow**; an attempt to store an X satisfying $|X| < s$ causes **floating point underflow**. On calculators, underflow or overflow generally causes a blinking display. Computers experiencing underflow generally store a zero, print a warning that it has occurred, and then go on.

The **word size** of a computer is the number of bits that it uses to store a real X. Computers differ in word size and the choice of arithmetic base b ($= 2, 8$, or 16). For a given word size, computers that use a larger b generally perform arithmetic more quickly but normalize mantissas a bit less efficiently and hence are slightly less accurate. Also, those that use more bits for M can get *greater accuracy*, whereas those that allocate more bits for c have a *larger interval of storable X's*, $[s, L]$ (see Problem 1-5).

Storing the sign of c as in Figure 1.2-1 is wasteful because a zero c can be represented as either $+0$ or -0. Computers generally store a p-bit **biased characteristic** between $(00 \cdots 0)_2 = 0$ and $(11 \cdots 1)_2 = 2^p - 1$, and then get c by subtracting a **bias**, usually 2^{p-1}. For example, a 36-bit word computer that uses an 8-bit biased characteristic with a bias of $2^7 = 128$ can store characteristics c in the range

$$-128 = 0 - 128 < c < (2^8 - 1) - 128 = +127 \qquad \textbf{(6a)}$$

Hence, if nine octal cells are used to store the normalized mantissa M as shown in Figure 1.2-3(a), then all storable positive numbers X will lie between

$$s = \tfrac{1}{8} \cdot 2^{-128} \doteq .367342\text{E}-39 \quad \text{and} \quad L = (1 - 8^{-9}) \cdot 2^{127} \doteq 0.170141\text{E}39 \qquad \textbf{(6b)}$$

If all bits not shown in Figure 1.2-3(a) are 0, then the stored X has

$$M = (.640000000)_8 = \frac{6}{8} + \frac{4}{8^2} = \frac{52}{64} \quad \text{and} \quad c = (2^7 + 2^3 + 2^1) - 128 = 10 \qquad \textbf{(7a)}$$

from which $X = -(.640000000)_8 \cdot 2^{10} = -52 \cdot 2^4 = -832.0$.

When a variable is declared INTEGER in most languages† (e.g., Fortran, PL/I, and Pascal), its signed binary representation is stored in an entire word. This is illustrated for

†Standard BASIC stores *all* numbers (including integers) as $\pm M \cdot 2^c$. Some enhanced versions that do allow INTEGER variables (e.g., Applesoft BASIC) allocate only a half word.

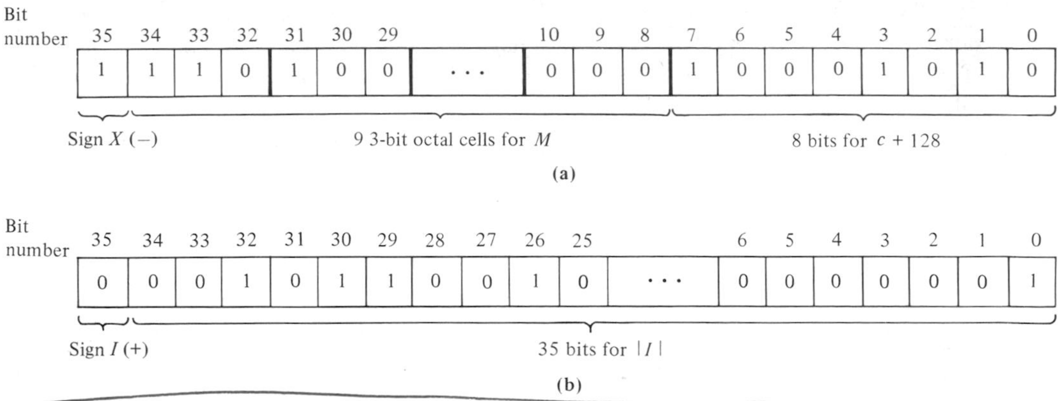

***Figure 1.2-3** Storing (a) $X = \pm M \cdot 2^c$; (b) an* INTEGER *in a 36-bit word.*

a 36-bit word in Figure 1.2-3(b), where if the bits not shown are 0, the stored INTEGER I is

$$+(2^{32} + 2^{30} + 2^{29} + 2^{26} + 1) = 5{,}972{,}688{,}897 \qquad \textbf{(7b)}$$

The largest storable integer is

$$+(11 \cdots 1)_2 = 2^{35} - 1 = 34{,}359{,}738{,}367$$

Practical Consideration 1.2A (Avoiding underflow and overflow). If overflow or underflow occurs, it can sometimes be remedied by using parentheses to *keep the values of all intermediate calculations close to* ± 1. For example, if $|x|$, $|y|$, and $|z|$ are either all very large or all very small, then

$$\frac{xy}{z} \text{ should be programmed as } \texttt{X*(Y/Z)} \text{ or } \texttt{Y*(X/Z)}, \text{ not } \texttt{X*Y/Z}$$

to remedy overflow or underflow in calculating xy. Similarly, most computers evaluating EXP(X) experience underflow when $X < -40$ and overflow when $X > 40$. So it is especially important to *keep the argument of the* EXP *function as close to zero as possible*. Thus

$$\frac{e^x}{e^y} \text{ should be programmed as } \texttt{EXP(X - Y)}, \text{ not } \texttt{EXP(X)/EXP(Y)}$$

if both x and y will be between, say, 30 and 70 or -70 and -30. ■

1.2B Significant Digit Accuracy of Fixed-Precision Devices

The **leading significant digit** of a nonzero number is the *leftmost nonzero digit* of its decimal representation. For example, both $\frac{31}{15} = 2.0666\ldots$ and $\frac{31}{15000} = 0.0020666\ldots$ have 2 as their leading *significant* digit. The abbreviation $\boldsymbol{x = \bar{x}\ (ks)}$ will be used to indicate that x and $\bar{x}$ become the same number when rounded to k significant digits. The distinction between

k-significant-digit accuracy and k-decimal-place accuracy can be seen examining the approximations

$$\tfrac{31}{15} \doteq 2.067 \qquad \text{and} \qquad \tfrac{31}{15000} \doteq 0.002067$$

Both are accurate to $4s$; however, the first is accurate to $3d$, whereas the second is accurate to $6d$. Notice that *zeros which lie to the right of the leading significant digit are significant.* The following important discussion shows why significant-digit accuracy is more natural than decimal-place accuracy for digital devices.

The **machine unit** of a digital device, denoted by $\boldsymbol{u_M}$, is the smallest possible increment in a stored mantissa. For a k-significant-digit calculator,

$$u_M = 10^{-k} \qquad \text{(a one-unit change in the last decimal digit of } M\text{)} \tag{8a}$$

Similarly, the machine unit of a computer with arithmetic base $b = 2^e$ is

$$u_M = 2^{-(\text{number of bits used to store } M)} \tag{8b}$$

with e bits used to store each "base b cell." In view of (8a), we shall call a computer a **k-significant digit** (or simply **ks**) **device** if

$$10^{-(k+1)} < u_M = \text{machine unit of the device} \leqslant 10^{-k} \tag{8c}$$

The integer k in (8c) is the **precision** of the device, that is, the maximum number of significant *decimal* digits that it can store accurately. All calculators *and* computers are thus **fixed-precision devices**. For example, a computer that uses nine octal "cells" to store mantissas, as shown in Figure 1.2-3(a), is an $8s$ device because

$$u_M = 8^{-9} = 2^{-27} \doteq 0.74 \cdot 10^{-8}, \qquad \text{hence } 10^{-9} < u_M \leqslant 10^{-8} \tag{9}$$

Practical Consideration 1.2B (Determining the precision of a device). A good calculator stores two or three **guard digits** in addition to those it displays. For example, one that displays 8 digits might store 10 (two guard digits), and hence will actually be a 10s device. To find the number of significant digits that your calculator actually stores, perform the calculation

$$\left(\frac{1}{3}\right) \cdot 10000 - 3333 = \text{displayed result} \tag{10}$$

then add 4 to the numbers of 3's in the displayed result. The machine unit of your computer can be determined as a power of 2 using Algorithm 1.2B. The idea behind it is this: If k bits are used to store the mantissa, then k is the smallest positive integer i for which $1 + 2^{-i}$ gets stored as 1. ■

Certain languages that were intended for scientific computation (e.g., Fortran and PL/I) are capable of storing numbers with **extended precision**. This option generally causes extra memory cells to be allocated to the mantissa but *not* the characteristic. For example,

ALGORITHM 1.2B. DETERMINING A COMPUTER'S MACHINE UNIT u_M

PURPOSE: To find the smallest mantissa increment as 2^{-k}.

```
u ← 1
DO FOR k = 1 TO 100
|   u ← u/2
|   test ← 1 + u
|__ UNTIL test = 1
OUTPUT('u_M is 2**-'k' = 'u)
```

declaring a variable `DOUBLE PRECISION` in Fortran at least doubles the number of storable significant digits but generally has no noticeable effect on underflow or overflow. Extended precision, if available, also reduces the speed of arithmetic calculations. Most of the practical advice offered in this book is intended to help you get the most accuracy out of the *nominal* precision (also called the **single precision**) of the device being used.

1.2C Inherent Error on a Digital Device

Most real numbers x cannot be stored exactly on any given digital device. The best one could hope to do (think about it) is the following:

$$\text{Store a storable } X \text{ that is as close as possible to the exact } x \tag{11a}$$

The error of this stored X is its **inherent error**, defined as

$$\text{inherent error} = (\text{exact value } x) - (\text{stored approximation } X) \tag{11b}$$

To get the X with the least inherent error on a k-significant-digit device, simply round x to k significant digits. For example, on a 4s device:

$$\begin{aligned}
&\text{Store } x = \tfrac{1}{15} = .066666\ldots \quad \text{as} \quad X = +.6667\cdot 10^{-1} \qquad (M = .6667,\ c = -1)\\
&\text{Store } y = -9.99972 \quad \text{as} \quad Y = -10 = -.1000\cdot 10^{2} \qquad (M = .1000,\ c = 2)\\
&\text{Store } z = 3005449 \quad \text{as} \quad Z = 300500 = +.3005\cdot 10^{6} \qquad (M = .3005,\ c = 6)
\end{aligned}$$

Thus the inherent error of storing $x = \frac{1}{15}$ as $X = +.6667\cdot 10^{-1}$ is

$$\frac{1}{15} - \frac{6667}{100000} = \frac{20000 - 3(6667)}{300000} = \frac{-1}{300000} \tag{12}$$

If $x = \frac{1}{15}$ was stored as $+.0667 \cdot 10^0$, that is, *without normalizing*, the inherent error would have been $-\frac{1}{30000}$, or ten times as large. This shows why digital devices always store normalized mantissas: *Normalization minimizes the inherent error of X by storing as many significant digits of x as possible.*

The inherent error of a stored X depends on the base b and the number of "base b cells" used to store M. For example, the "nice" number $x = .1 = \frac{1}{10}$ is exactly storable on any calculator for which $b = 10$; however, it is *not* storable on *any* computer for which $b = 2^e$. In particular, on a computer such as the IBM 370, which uses 24 bits as six hexadecimal (base $16 = 2^4$) "cells" to store normalized mantissas,

$$x = 0.1 = 0.8 \cdot 2^{-3} \quad \text{is stored as} \quad X = (.\text{CCCCCC})_{16} \cdot 2^{-3} \tag{13a}$$

The hexadecimal mantissa $(.\text{CCCCCC})_{16}$ in (13a) represents the number

$$\frac{12}{16}\left(1 + \frac{1}{16} + \frac{1}{16^2} + \frac{1}{16^3} + \frac{1}{16^4} + \frac{1}{16^5}\right) = \frac{3}{4}\left(\frac{1118481}{1048576}\right) \tag{13b}$$

which equals 0.7999999523 ... and *not* 0.8. Hence $x = 0.1$ is stored as the smaller number

$$X = (0.7999999523\ldots)\cdot\frac{1}{8} = 0.09999999404\ldots \tag{13c}$$

which is in error when rounded to the seventh significant digit.

Practical Consideration 1.2C (Chopping versus rounding). Most digital devices get X by simply **chopping** (without rounding) all digits of x that cannot be stored. For example, a 4s calculator that chops will store $\pm\frac{2}{3}$ truncated to ± 0.6666 rather than rounded to ± 0.6667. For a computer, the maximum error in a k-bit mantissa is $u_M = 2^{-k}$ if it chops, and $u_M/2 = 2^{-(k+1)}$ if it rounds. So

$$|\text{inherent error of } X = \pm M \cdot 2^c| \text{ is } \begin{cases} < u_M \cdot 2^c & \text{if } X \text{ is chopped} \\ \leq u_M \cdot 2^{c-1} & \text{if } X \text{ is rounded} \end{cases} \tag{14}$$

Thus *inherent errors get larger as c increases* (see Figure 1.2-2). ■

1.3 Errors in Fixed-Precision Arithmetic

We now describe how errors are produced when arithmetic calculations are performed on a digital device.

1.3A The Error and Relative Error of an Approximation

Let X be any approximation of the exact value x. The **error in X**, denoted by ϵ_X, is the difference†

$$\epsilon_X = x - X = (\text{exact value}) - (\text{approximation}) \tag{1}$$

So $x = X + \epsilon_X$. In words, ϵ_X *is what must be added to the approximation X to get the exact x*. A positive ϵ_X means that x lies to the right of X; a negative ϵ_X means that x lies to the left of X. The **relative error in X**, denoted by ρ_X, is the quotient

$$\rho_X = \frac{x - X}{|x|} = \frac{\epsilon_X}{|x|} = \frac{\text{error in } x}{\text{absolute value of } x} \quad \text{if } x \neq 0 \tag{2}$$

The relative error ρ_X expresses the error as a fractional part of $|x|$. It is closely related to the idea of **percent error**. For example, if it is known that $|\rho_X| < 0.02$, then by (2), $|\rho_X| < 0.02|x|$; that is, X approximates x to within 2%. More generally,

$$\text{The percent error of approximating } x \text{ by } X \text{ is } 100\rho_X \tag{3}$$

Conversely, if a value X is known to be accurate to within, say, 1.5%, then its relative error ρ_X lies between -0.015 and $+0.015$.

EXAMPLE 1.3A Find the error, relative error, and percent error of approximating $x = -43\frac{1}{3}$ $(= -\frac{130}{3})$ by $X = -43$.

Solution: The error and relative error of X are

$$\epsilon_X = \frac{-130}{3} - (-43) = \frac{-130 + 129}{3} = \frac{-1}{3}$$

and

$$\rho_X = \frac{-\frac{1}{3}}{|-\frac{130}{3}|} = \frac{-1}{130}$$

Thus the error of $-\frac{1}{3}$ unit on the number line corresponds to a relative error of -1 part in 130, or $-\frac{10}{13}\%$ of $|x|$. ■

The **absolute error** $|\epsilon_X|$ can be used to ensure a *desired decimal-place accuracy*, say *kd*,

†Before reading any numerical analysis textbook, check the index to see what is meant by "error" and "absolute error." Some authors take "error" to be $X - x$ (i.e., the negative of our ϵ_X). Others use "absolute error" (as opposed to "relative error") for what we call "error." We shall use "absolute error" when referring to $|\epsilon_X|$.

as we saw in (3) of Section 1.1A:

Absolute Error Test. If $\|x - X\| < 0.5 \cdot 10^{-k}$, then $x = X\ (kd)$.	**(4a)**

Similarly, if $x \neq 0$, $|\rho_X| = |\epsilon_X|/|x|$ can be used as follows† to ensure a desired *significant-digit accuracy*, say ks:

Relative Error Test. If $\|x - X\| < 0.5 \cdot 10^{-k}\|x\|$, then $x = X\ (ks)$.	**(4b)**

Indeed, the leading significant digit of $10^{-k}|x|$ is in the kth significant digit of x. (Why?) So $|x - X| < 0.5 \cdot 10^{-k}|x|$ assures that x and X differ by less than 0.5 in the kth significant digit, from which $x = X\ (ks)$. The following useful table is immediate from (4a) and (4b):

Change in ϵ_X or ρ_X	Corresponding Change in the Accuracy of X	
Dividing ϵ_X by 10	X has one more accurate decimal place	**(4c)**
Dividing ρ_X by 10	X has one more accurate significant digit	

The smallest (normalized) mantissa on any device that uses arithmetic base b is $M = b^{-1} = 1/b$. It follows easily from this and (14) of Section 1.2C that the *relative* inherent error of an X stored on such a device satisfies

$$|\rho_X| = \frac{|x - X|}{|x|} \text{ is } \begin{cases} < u_M b & \text{if the device chops} \\ \leq 0.5 u_M b & \text{if the device rounds} \end{cases} \qquad \textbf{(5)}$$

Thus, for a given device, the maximum magnitude of the *relative* inherent error $|\rho_X|$ is the same whether $|x|$ is large or small. This uniformity makes the Relative Error Test for significant digit accuracy, (4b), especially well suited for terminating iterative algorithms when x's of varying magnitude are to be found on a computer.

1.3B Error Propagation when Doing Fixed-Precision Arithmetic

In this section, lowercase letters $x, y, z, \ldots$ will be used to denote real numbers and uppercase letters $X, Y, Z, \ldots$ will denote the storable approximations representing them on a digital device. This means that $\epsilon_X = x - X$ *will be the inherent error of the stored* X. Note that once X is stored, there is no way for the device to know whether it is larger, smaller, or equal to the exact x!

†The test in (4b) amounts to $|\rho_x| < 0.5 \cdot 10^{-k}$, but is preferable because there is no risk of dividing by zero.

Let "$\circ$" denote one of the four arithmetic operations $+$, $-$, $*$, and $/$. Since a digital device knows only the stored values X and Y and not the exact values x and y, it cannot possibly calculate exactly $x \circ y$. The best a ks device can hope to do is calculate $X \hat{\circ} Y$, where

$$X \hat{\circ} Y \text{ means: } \begin{cases} \text{Find } X \circ Y \text{ to at least } (k+1)s, \\ \text{then store it rounded back to } ks \end{cases} \quad (6)$$

We now show how this perfectly reasonable procedure, which we call **fixed-precision arithmetic**, or more specifically ***ks* arithmetic**, can lead to a variety of errors.

Consider an *El Cheapo* calculator that is capable of storing only four (rounded) significant digits so that $u_M = 0.5 \cdot 10^{-4}$. We wish to examine the errors that can occur when this hypothetical $4s$ device does $4s$ arithmetic involving the numbers in Table 1.3-1. Note that although the inherent errors vary in size, all **relative** inherent errors are less than $0.5 \cdot 10^{-3}$ in magnitude, as predicted by (5).

TABLE 1.3-1 Storing Numbers on a 4s Device That Rounds

Exact Value	*Stored Approximation*	*Inherent Error*	*Relative Error (2s)*
$u = 122.9572$	$U = +0.1230E3$	$\epsilon_U = -0.0428$	$\rho_U = -0.35E-3$
$v = 123.1498$	$Y = +0.1231E3$	$\epsilon_Y = +0.0498$	$\rho_Y = +0.40E-3$
$w = 0.0014973$	$W = +0.1497E-2$	$\epsilon_W = +0.0000003$	$\rho_W = +0.20E-3$
$z = 457{,}932$	$Z = +0.4579E6$	$\epsilon_Z = +32$	$\rho_Z = +0.70E-4$

The examples given in Table 1.3-2 illustrate how errors in X or Y (either inherent error or the result of errors of previous arithmetic calculations) can produce worsened errors in $X \hat{\circ} Y$. We shall refer to this phenomenon as **propagated roundoff**.

TABLE 1.3-2 Error Propagation in Performing 4s Arithmetic

Exact $x \circ y$	*Rounded* $x \circ y$	$X \circ Y$	$X \hat{\circ} Y$ *(rounded* $X \circ Y$*)*
$v + z = 458055.1498$	458100	$V + Z = 458023.1$	$V \hat{+} Z = 458000$
$u - w = 122.9557027$	123.0	$U - W = 122.998503$	$U \hat{-} W = 123.0$
$v*z \doteq 0.56394234E8$	0.5639E8	$V*Z = 0.5636749E8$	$V \hat{*} Z = 0.5637E8$
$u / w \doteq 82119.2813$	82120.	$U / W \doteq 82164.3286$	$U \hat{/} W = 82160.$
$u - v = -0.1926$	-0.1926	$U - V = -0.1$	$U \hat{/} V = -0.1000$

Certain types of propagated roundoff occur frequently enough to warrant being given names. These are described in the following discussion.

Creeping Roundoff. The error in the kth significant digit of $X \hat{\circ} Y$ that can occur even though X and Y are both correctly rounded to k significant digits.

In Table 1.3-2, creeping roundoff caused $V \hat{+} Z$, $V \hat{*} Z$, and $U \hat{/} W$ to have errors in the fourth significant digit, despite the fact that U, V, W, and Z were correctly rounded to 4s.

Negligible Addition. Adding (or subtracting) two numbers whose magnitudes are so different that the resulting sum (or difference) gets rounded to the one having larger magnitude.

In Table 1.3-2, negligible addition caused $U \hat{-} W = U$. Negligible addition provides the basis for Algorithm 1.2B for finding u_M.

Error Magnification. Multiplying an erroneous number by a large (in magnitude) number, or dividing it by a number close to zero.

In Table 1.3-2, the multiplication $V \hat{*} Z$ magnified the error in V (and, to a lesser extent, the error in Z), and the division $U \hat{/} W$ magnified the error in both U and W. Although the resulting errors

$$(v*z) - (V \hat{*} Z) = 56394234.2136 - 56367490 = 26744.2136$$

and

$$(u \,/\, w) - (U \hat{/} W) = 82119.2813\ldots - 82164.3286\ldots = -45.047\ldots$$

represent inaccuracy in the least (i.e., fourth) significant digit of $V*Z$ and U/W, *they are rather large numbers*. Error magnification can thus produce unacceptably large *absolute errors* in certain situations.

Subtractive Cancellation. Subtracting two nearly equal numbers (or, equivalently, adding a number to nearly its negative).

This most insidious of roundoff errors is illustrated by the subtractions

$$\begin{aligned} u &= 122.9572 \\ v &= \underline{123.1498} \\ u - v &= -0.1926 \end{aligned} \qquad \begin{aligned} U &= 123.0 \\ V &= \underline{123.1} \\ U - V &= -0.1 = -0.1000 = U \hat{-} V \end{aligned}$$

which show how subtracting two nearly equal *rounded* numbers cancels the accurate *leading* (i.e., leftmost) significant digits and replaces them by digits that are more likely to have propagated roundoff errors. *Subtractive cancellation is also referred to as* ***loss of significance*** *or* ***catastrophic cancellation*** *because it can result in errors in the leading significant digit*, that is, complete garbage!

Further insight into propagated roundoff can be gained by examining the following formulas, which will be derived in Section 1.3E.

$$\epsilon_{X+Y} = \epsilon_X + \epsilon_Y \qquad \epsilon_{X-Y} = \epsilon_X - \epsilon_Y \tag{7a}$$

$$\rho_{X*Y} \approx (\pm)\rho_X + (\pm)\rho_Y \qquad \rho_{X/Y} \approx (\pm)\rho_X - (\pm)\rho_Y \tag{7b}$$

where $(\pm)$ in $(\pm)\rho_U$ denotes the sign of U $(= X$ or $Y)$. These formulas show how *addition and subtraction slowly erode decimal place accuracy* [see (7a)], whereas *multiplication and division slowly erode significant digit accuracy* [see (7b)]. What makes subtractive cancellation so serious is that *it can destroy precious accurate significant digits in a single operation!* The following example shows how algebraic or trigonometric identities can sometimes be used to avoid it.

EXAMPLE 1.3B   Describe the nonzero x's for which subtractive cancellation will occur in evaluating A, B, and C, where

$$A = \frac{1 - \cos x}{x^2}, \quad B = \frac{[(\sin x)/x]^2}{1 + \cos x}, \quad \text{and} \quad C = 2\left[\frac{\sin(x/2)}{x}\right]^2 \tag{8}$$

Verify your answer by evaluating A, B, and C on a computer for these x values.

SOLUTION: Since B and C are obtained from A by replacing $1 - \cos x$ by the equivalent expressions $\sin^2 x/(1 + \cos x)$ and $2\sin^2(x/2)$, respectively, A, B, and C should be equal whenever they are all defined.

For $x \approx 0$, $\cos x$ will be near 1 and hence the numerator of A will lose significance due to subtractive cancellation; this error will then get magnified when divided by x^2 (which is *smaller* than $|x|$ for $x \approx 0$!). On the other hand, for $\cos x \approx -1$ (e.g., for $x \approx \pi$) the denominator of B loses significance while itself becoming small. No loss of significance enters the calculation of C.

Figure 1.3-1 shows a Fortran 77 program that evaluates A, B, and C for x near 0 and π. In the run of this program shown in Figure 1.3-2, underlined digits are those that would be incorrect after rounding. As predicted, the only serious errors occurred when calculating A for $x \approx 0$ and B for $x \approx \pi$. The absurd results in the last line of each table resulted first

```
ØØ1ØØ   C       Program to demonstrate digital device errors
ØØ2ØØ           DATA  IW, XLIMIT /5, Ø.Ø/
ØØ3ØØ       1Ø H = Ø.2
ØØ4ØØ           WRITE (IW, 1) XLIMIT          ! heading
ØØ5ØØ           DO 2Ø K=1, 5
ØØ6ØØ               X = XLIMIT + H
ØØ7ØØ               A = ( 1. - COS(X) ) / X**2
ØØ8ØØ               B = ( SIN(X)/X )**2 / ( 1. + COS(X) )
ØØ9ØØ               C = 2. * ( SIN(X/2) / X )**2
Ø1ØØØ               WRITE (IW, 2) X, A, B , C
Ø11ØØ               H = H/1Ø.
Ø12ØØ       2Ø CONTINUE
Ø13ØØ           IF ( XLIMIT .NE. Ø.Ø ) STOP
Ø14ØØ           XLIMIT = 4. * ATAN(1.)        ! this is pi
Ø15ØØ           GOTO 1Ø
Ø16ØØ        1 FORMAT ('ØFor values of x near', F8.5, ':',//,
Ø17ØØ          &         6X, 'x', 14X, 'A', 18X, 'B', 18X, 'C')
Ø18ØØ        2 FORMAT (F1Ø.5, 3(6X, G13.6))
Ø19ØØ          END
```

Figure 1.3-1   *Program for computing A and B near 0 and π.*

```
For values of x near Ø.ØØØØØ:

     x              A                 B                 C
  Ø.2ØØØØ       Ø.498336          Ø.498336          Ø.498336
  Ø.Ø2ØØØ       Ø.499971          Ø.499983          Ø.499983
  Ø.ØØ2ØØ       Ø.5Ø1Ø52          Ø.5ØØØØØ          Ø.5ØØØØØ
  Ø.ØØØ2Ø       Ø.558794          Ø.5ØØØØØ          Ø.5ØØØØØ
  Ø.ØØØØ2       Ø.ØØØØØØE+ØØ      Ø.5ØØØØØ          Ø.5ØØØØØ

For values of x near 3.14159:

     x              A                 B                 C
  3.34159       Ø.177326          Ø.177326          Ø.177326
  3.16159       Ø.2ØØØ67          Ø.2ØØØ64          Ø.2ØØØ67
  3.14359       Ø.2Ø2384          Ø.2Ø196Ø          Ø.2Ø2384
  3.14179       Ø.2Ø2617          Ø.1813Ø1          Ø.2Ø2617
%FRSFDC FLOATING DIVIDE CHECK AT MAIN.+36 (PC 252)
  3.14161       Ø.2Ø264Ø          Ø.17Ø141E+39      Ø.2Ø264Ø
```

Figure 1.3-2 *Computer's values of A, B, and C near 0 and* π.

from the fact that the computer thought $\cos x = 1$ in the numerator of A, and then from the fact that it thought $\cos x = -1$ in the denominator of B. *Conclusion*: We should calculate C for *all* $x \neq 0$, *even if asked to evaluate* A *or* B! ■

Practical Consideration 1.3B (Division by zero). The safest thing to do when a computer divides by zero is to *ignore all values that are calculated after it occurs.* To see why, look at Figure 1.3-2, where the computer's response to the division by zero in calculating *B* was to produce the largest storable number *L* [see (6b) of Section 1.2A], output a warning *and then go on.* Since subsequent results were meaningless, it would have been better if the computer had *simply quit* after it printed the warning. ■

Summary of the Sources of Digital Device Errors

HUMAN ERROR: A *mistake* made by the programmer, keypuncher, or user.

ROUNDOFF ERROR: The general name given to errors that result from fixed-precision arithmetic. It begins with *inherent roundoff,* and then propagates by virtue of *creeping roundoff, negligible addition, error magnification, and/or subtractive cancellation.*

TRUNCATION (or DISCRETIZATION) ERROR: The error that occurs (aside from any roundoff error) as a result of using a *formula that is only approximate.* Truncation error will be discussed in Chapter 7.

1.3C The Errors of Approximating $f'(x)$ by $\Delta f(x)/h$ Explained

We can now explain what went wrong with the derivative approximation

$$e = f'(x) \approx \frac{\Delta f(x)}{h} = \frac{f(x + h) - f(x)}{h} = \frac{e^{1+h} - e}{h} \tag{9}$$

as h approached zero in Table 1.1-1 (reproduced in Table 1.3-3). For sufficiently small h, the numerator of $\Delta f(x)/h$ loses significance (subtractive cancellation), and this error in turn gets magnified when divided by h (error magnification). The combined effect rapidly increases the error as $|h|$ decreases, until h gets so small that either $x + h$ gets rounded to x (negligible addition) or the calculated values of $f(x + h)$ and $f(x)$ agree to the precision of the device, at which point $\Delta f(x)/h$ gets calculated as $0/h = 0$. It's that simple!

TABLE 1.3-3 Errors in $DQ = \Delta f(x)/h$ on (a) a DEC-10 and (b) a PDP-11

h	*(a) DQ on* DEC-10	*(b) DQ on* PDP-11	*(c) Correct* $\Delta f(x)/h$
0.2	3.009176	3.009176	3.0091755
0.2E−1	2.745646	2.745652	2.7456468
0.2E−2	2.721012	2.720952	2.7210019
0.2E−3	2.718568	2.719164	2.7185537
0.2E−4	2.717972	2.729893	2.7183090
0.2E−5	2.712011	2.741814	2.7182845
0.2E−6	2.682209	3.576279	2.7182821
0.2E−7	1.490116	0.000000	2.7182819
0.2E−8	0.000000	0.000000	2.7182818
0.2E−9	0.000000		

Column (c) of Table 1.3-3 shows the truncation error of the approximation formula $\Delta f(x)/h$ *in the absence of roundoff error*. Columns (a) (using an 7s device) and (b) (using an 8s device) show that any h small enough to give the desired 4d (i.e., 5s) accuracy produced enough *roundoff error* to *more than offset the small truncation error*! Thus, although $\Delta f(x)/h$ is a perfectly reasonable formula *analytically*, it is simply *not* a good formula for approximating $f'(x)$ *numerically*! Better ones are given in Chapter 7.

The preceding discussion points to an important difference between calculus and numerical analysis. *For numerical analysis, it is not enough to know that something converges to a limiting value* [e.g., $\Delta f(x)/h$ to $f'(x)$ as $h \to 0$]. We also need either an *estimate of the error* or some knowledge of the *rate of convergence* of the converging quantity so that we can assess *how well* it approximates the limiting value for a particular h. The following example illustrates the use of an error estimate and also shows how we got the "Correct $\Delta f(x)/h$" entries for $h \leq 0.02$ in Table 1.3-3(c).

EXAMPLE 1.3C Let $f(x) = e^x$. Devise a way to evaluate the difference quotient $\Delta f(x)/h$ accurately (regardless of x) on an 8s device for $|h| \leq 0.02$.

SOLUTION: For any fixed x, the desired approximation for $h \approx 0$ is

$$e^x = f'(x) \approx \frac{\Delta f(x)}{h} = \frac{e^{x+h} - e^x}{h} = \frac{e^h - 1}{h}\, e^x = \Phi(h)e^x \tag{10}$$

Since the device can evaluate e^x to 8s, (10) reduces the problem to that of approximating $\Phi(h) = (e^h - 1)/h$ to 8s for $|h| \leq 0.02$ [see ρ_{XY} in (7b)]. If we replace e^h by its nth *Maclaurin polynomial* plus the Lagrange form of its *remainder* [see Appendix II.2D], we get

$$\Phi(h) = \frac{1}{h}\left\{\left[\underbrace{1 + h + \frac{h^2}{2!} + \frac{h^3}{3!} + \cdots + \frac{h^n}{n!}}_{n\text{th Maclaurin polynomial for } e^h} + \underbrace{e^{\xi}\frac{h^{n+1}}{(n+1)!}}_{\text{remainder}}\right] - 1\right\} \qquad \textbf{(11a)}$$

where ξ lies somewhere between 0 and h. We can thus write $\Phi(h)$ as

$$\Phi(h) = \underbrace{1 + \frac{h}{2} + \frac{h^2}{3!} + \cdots + \frac{h^{n-1}}{n!}}_{\text{approximation of } \Phi(h)} + \underbrace{E_n(h),}_{\text{error}} \qquad \text{where } E_n(h) = \frac{e^{\xi}h^n}{(n+1)!} \qquad \textbf{(11b)}$$

We know from (10) that $\Phi(h) \approx 1$ for $h \approx 0$. So 8s accuracy of the approximation in (11b) will be assured if $|E_n(h)| < 0.5 \cdot 10^{-8}$. For $-0.02 \leq h \leq 0.02$, e^{ξ} in (11b) will be less than $e^{+0.02}$ (why?), hence

$$|E_n(h)| < \frac{e^{0.02}\,(0.02)^n}{(n+1)!} = B_n \qquad [\text{a bound for } |E_n(h)| \text{ when } |h| < 0.02] \qquad \textbf{(12)}$$

So $|E_n(h)| < 0.5 \cdot 10^{-8}$ will be assured for $|h| \leq 0.02$ if we use the smallest n for which $B_n < 0.5 \cdot 10^{-8}$. By trial and error, $n = 4$; hence by (11b),

$$1 + \frac{h}{2} + \frac{h^2}{3!} + \frac{h^3}{4!} \quad \text{approximates} \quad \Phi(h) = \frac{e^h - 1}{h} \text{ to } 8s \text{ for } |h| \leq 0.02 \qquad \textbf{(13)}$$

By (10), $(1 + h/2 + h^2/6 + h^3/24)e^x \doteq \Delta f(x)/h$ to 8s if $|h| \leq 0.02$. ■

A Maclaurin polynomial approximation can often be used as in (13) to avoid subtractive cancellation when an algebraic or trigonometric identity cannot be found to do so as in Example 1.3C or Section 1.4D.

1.3D Ill-Conditioned Calculations

Suppose that $f(x)$ is the result of a calculation involving x. How likely is it that $f(x)$ will have as many significant digits as x? Since significant digits are related to relative error [see (4b)], we can get an answer to this question if we can find a number C, called a **condition number** for the calculation, which satisfies

$$|\text{relative change in } f(x)| \approx C \cdot |\text{relative change in } x| \qquad \textbf{(14a)}$$

The condition number C can be thought of as the amount that a relative change (or error) in x ($\neq 0$) gets *magnified* in calculating a nonzero $f(x)$. For example, Table 1.3-4 shows the effect a 1% increase in $|x|$ ($\rho_x = 0.01$) has on the value of $f(x) = e^x$ when $x = 0.1$, 10, and -10. We see from Table 1.3-4 that we can take $C = 0.1$ when $x = 0$; but we need $C \approx 10$ for $x = \pm 10$. Viewed still another way [see (4c)],

$$C \approx 10^p \;\Rightarrow\; f(x) \text{ can have about } p \text{ fewer accurate significant digits than } x \qquad \textbf{(14b)}$$

TABLE 1.3-4 Relative Change in e^x Caused by a 1% Change in $|x|$

x	$dx = 0.01x$	$x + dx$	e^x	e^{x+dx}	$\rho_{e^x} = (e^{x+dx} - e^x)/e^x$
.1	0.001	0.101	1.10517	1.10628	0.001 (= 0.1%)
10	0.1	10.1	22026.4	24343.0	0.105 (= 10.5%)
−10	−0.1	−10.1	0.453993E−4	0.410796E−4	0.095 (= −9.5%)

The importance of knowing a condition number can be seen from Table 1.3-4. If we had to know e^x to say 4*s*, then x would have to be known to only 3*s* for x near 0.1, but to 5*s* for x near ± 10! Thus *the larger the condition number, the more accurate x must be to get a desired significant-digit accuracy of* $f(x)$. When C is very large, a small inherent error of x can result in large relative error in the calculated $f(x)$ *even if there were no roundoff error!* Such calculations are called **ill-conditioned**.

We can get a general formula for C in (14a) when $f(x)$ is differentiable at x. Let $x + dx$ denote a value close to x. If $|dx| << |x|$, we can use the differential of f at x [Appendix II.5C] to get

$$f(x + dx) \approx f(x) + f'(x)\,dx \quad \text{so that} \quad \frac{f(x + dx) - f(x)}{f(x)} \approx \frac{xf'(x)}{f(x)} \cdot \frac{dx}{x} \tag{15a}$$

Taking $\rho_{f(x)} = [f(x + dx) - f(x)]/|f(x)|$ and $\rho_x = dx/|x|$ in (15a) gives

$$|\rho_{f(x)}| \approx C \cdot |\rho_x|, \qquad \text{where } C = \left|\frac{xf'(x)}{f(x)}\right| \tag{15b}$$

We can test this formula using $f(x) = e^x$, in which case $f'(x) = e^x$ and so (15b) becomes simply $|\rho_{f(x)}| \approx |x| \cdot |\rho_x|$. Indeed, we saw in Table 1.3-4 (where $\rho_x = 0.01$) that $|\rho_{f(x)}| \approx 0.1|\rho_{e^x}|$ when $x = 0.1$ and $|\rho_{e^x} \approx 10|\rho_x|$ when $x = \pm 10$. So the evaluation of e^x is ill conditioned for $|x| >> 0$ and well conditioned for $x \approx 0$.

Ill-conditioned calculations generally require input variables to be stored in extended precision (to reduce $|\rho_x|$) in order to get $f(x)$ to single-precision accuracy. We will therefore note ill-conditioned calculations whenever they are encountered. It is important to note that *a calculation can still produce substantial relative error (caused by subtractive cancellation) even though it is well conditioned.* An example is $\Delta f(x)/h$ for $h \approx 0$ (see Problem 1-21).

1.3E A Quantitative Analysis of Propagated Roundoff (Optional)

We wish to show analytically how small errors in the approximations $x \approx X$ and $y \approx Y$ propagate when arithmetic is performed on X and Y. Clearly,

$$\epsilon_{X \pm Y} = (x \pm y) - (X \pm Y) = (x - X) \pm (y - Y) = \epsilon_X \pm \epsilon_Y \tag{16a}$$

In words, *the error of a sum or difference is the sum or difference of the errors in the terms*. This explains how errors can "creep" into the last significant digit of the sum or difference of two correctly rounded numbers. Since we generally do not know if ϵ_X and ϵ_Y

have the same or opposite sign, the best we can say about *absolute* error is

$$|\epsilon_{X+Y}| \le |\epsilon_X| + |\epsilon_Y| \quad \text{and} \quad |\epsilon_{X-Y}| \le |\epsilon_X| + |\epsilon_Y| \tag{16b}$$

Thus, if X and Y have absolute errors in different decimal places, the absolute error of $X \pm Y$ will be about the same as the least accurate of X and Y. In the extreme, the result is *negligible addition*.

To see how multiplication and division propagate errors, we shall view ϵ_X and ϵ_Y as increments $dX = x - X$ and $dY = y - Y$. If the relative errors of X and Y are small, the resulting errors in XY and X/Y can be approximated accurately using differentials as follows:

$$\epsilon_{XY} \approx d(XY) = Y\,dX + X\,dY = Y\epsilon_X + X\epsilon_Y \tag{17a}$$

$$\epsilon_{X/Y} \approx d\left(\frac{X}{Y}\right) = \frac{Y\,dX - X\,dY}{Y^2} = \frac{1}{Y}\epsilon_X - \frac{X}{Y^2}\epsilon_Y \tag{17b}$$

This explains quantitatively how *error magnification* occurs when multiplying by a large number or dividing by a small one.

If we divide (17a) by $|xy|$ and (17b) by $|x/y|$, we get

$$\rho_{XY} = \frac{\epsilon_{XY}}{|xy|} \approx \frac{Y}{|y|}\cdot\frac{dX}{|x|} + \frac{X}{|x|}\cdot\frac{dY}{|y|} \approx (\pm)\rho_X + (\pm)\rho_Y \tag{18a}$$

$$\rho_{X/Y} = \frac{\epsilon_{X/Y}}{|x/y|} \approx \frac{|y|}{Y}\cdot\frac{\epsilon_X}{|x|} - \frac{X}{|x|}\left(\frac{y}{Y}\right)^2\frac{\epsilon_Y}{|y|} \approx (\pm)\rho_X - (\pm)\rho_Y \tag{18b}$$

where $(\pm)$ in $(\pm)\rho_U$ denotes the sign of $U = (X \text{ or } Y)$. So if both x and y are positive, the relative error of the product XY (quotient X/Y) is approximately equal to the sum (difference) of the relative errors of X and Y. Also, if both X and Y are accurate to ks, their product and quotient will be accurate to about ks (although their absolute errors may be large).

Finally, if we divide (16a) by $|x \pm y|$, we get

$$\rho_{X\pm Y} = \frac{\epsilon_{X\pm Y}}{|x \pm y|} = \frac{\epsilon_X}{|x \pm y|} \pm \frac{\epsilon_Y}{|x \pm y|} = \left|\frac{x}{x \pm y}\right|\rho_X \pm \left|\frac{y}{x \pm y}\right|\rho_Y \tag{19}$$

Notice that when $x \pm y$ is much smaller than x and y, the factors $|x/(x \pm y)|$ and $|y/(x \pm y)|$ magnify ρ_X and ρ_Y, respectively. This *relative error* magnification is *subtractive cancellation*.

1.4

Some Practical Strategies for Minimizing Roundoff Error

Now that we understand how roundoff error occurs, we can suggest ways to avoid or at least minimize it.

1.4A Strategies for Minimizing Inherent Error

Since inherent error is the starting point for all propagated roundoff error, *an intelligent user of numerical methods knows the precision of the device being used* in order to exploit the

> **Full Mantissa Strategy.** To minimize inherent error, enter input values using as many significant digits as can be stored on the device (provided, of course, that they are known to the significant-digit accuracy of the device).

Thus, on a 7*s* device, π should be input as 3.141593, *not* as 3.14. Better yet, let the device itself calculate it as

$$\pi = 4 \tan^{-1}(1) \qquad \text{[e.g., 4.*ATAN(1.) in Fortran]} \tag{1}$$

This simple trick yields π *to whatever the device accuracy happens to be*! It was used in line 1400 of Figure 1.3-1. Similarly, e^x should be programmed as `EXP(X)`, *not* as `2.718**X` (which will result in only 4*s* accuracy *regardless of the precision of the device*!) Also, be sure to initialize `DOUBLE-PRECISION` variables in Fortran as `DOUBLE-PRECISION` constants (e.g., `0.D0`) to ensure that the entire "double mantissa" is initialized.

A related word of advice: When empirically obtained input values are known with *less precision* than that of the device, *round your final answers to the significant-digit accuracy of the **least** accurate input value*. For example, if this limiting accuracy were 3*s* and the computed solution is 23.3876, you should *record* the solution as 23.4 (i.e., rounded to 3*s*, *which is all you can trust*). If you record any more digits, you may actually believe them when you look at your answer two weeks later!

1.4B Strategies for Evaluating Expressions Accurately

On both calculators and computers, the values generated by built-in functions such as sin, cos, exp, and ln = $\log_e$ should be accurate to the precision of the device. However, they do sometimes produce slightly less accuracy for certain X's. So you should expect errors to propagate a bit faster when the calculation involves built-in functions. In particular, the evaluation of y^x on a calculator (or `Y**X` in Fortran or `Y^X` in BASIC) for any *real* exponent x uses the formula

$$y^x = e^{x(\ln y)}, \qquad \text{provided that the base } y \text{ is positive} \tag{2}$$

This propagates the error of *both* the exponential and the natural log functions. As a result, built-in routines for $\sqrt{y}$ and y^2, if available, are generally more accurate than y^x with $x = \frac{1}{2}$ and $x = 2$. More important, $y*y*\cdots*y$ *(x times)* is usually more accurate than y^x when x is a positive integer. We therefore recommend the

> **Integer Exponent Strategy.** Evaluate y^x using repeated multiplication when the exponent x is an integer less than 10.

Thus, for example, y^{-8} should be evaluated (or programmed) as $1/u$, where $u = y^8$ is obtained as, say,

$$u \leftarrow y*y; \quad u \leftarrow u*u; \quad u \leftarrow u*u \qquad \text{(three multiplications!)} \tag{3}$$

unless you know that the device actually performs repeated multiplications when executing (2) for an integer exponent such as ± 8. The calculation in (3) is an example of the†

FEWEST OPERATIONS STRATEGY. To help minimize creeping roundoff, evaluate mathematical expressions in a form that requires the fewest mathematical operations (except when this may result in subtractive cancellation).

This strategy was used in calculating B and C in (8) of Example 1.3B. For a more general application, let us compare the usual **exponent form** of the polynomial

$$p_{\text{exp}}(x) = 2x^4 - 19x^3 + 56.98x^2 - 56.834x + 5.1324 \tag{4a}$$

to its equivalent **nested form**,

$$p_{\text{nest}}(x) = (((2x - 19)x + 56.98)x - 56.834)x + 5.1324 \tag{4b}$$

Both $p_{\text{exp}}(x)$ and $p_{\text{nest}}(x)$ require four addition/subtractions and four multiplications by a power of x. However, the evaluation as $p_{\text{exp}}(x)$ requires the *additional* calculation of x^4, x^3, and x^2; the use of the x^y routine to evaluate them (as you probably would have done before reading this!) is slow, inaccurate, and *simply will not work when x is negative* [see (2)]. No matter how you evaluate x^4, x^3, and x^2, $p_{\text{nest}}(x)$ should be less vulnerable to propagated roundoff than $p_{\text{exp}}(x)$. This is why we recommend the

NESTED MULTIPLICATION STRATEGY. Evaluate polynomials in nested form.

Algorithm 1.4B describes the **Synthetic Division** algorithm (**Horner's method**) for evaluating polynomials in nested form. The reader who has seen it before may recognize it in the form shown in Figure 1.4-1.

ALGORITHM 1.4B. SYNTHETIC DIVISION (HORNER'S METHOD)

PURPOSE: To evaluate $PolValue = a_1x^n + a_2x^{n-1} + \cdots + a_nx + a_{n+1}$

GET $n, a_1, a_2, \ldots, a_n, a_{n+1}, x$
$PolValue \leftarrow 0$
DO FOR $i = 1$ TO $n + 1$
⌊ $PolValue \leftarrow PolValue \cdot x + a_i$

†The Fewest Operations Strategy takes only *accuracy* into account. If *rapid execution* is also desired, bear in mind that addition/subtractions are faster than multiplications, which are faster than divisions; and these arithmetic operations are faster than most internal function evaluations.

Find $p(x) = (((1x - 9.5)x + 28.49)x - 28.417)x + 2.5662$ when $x = -2$.

$$\begin{array}{c|ccccc}
-2 & 1 & -9.5 & 28.49 & -28.417 & 2.5662 \\
\oplus & & -2.0 & 23.0 & -102.98 & 262.794 \\
\hline
& 1 & -11.5 & 51.49 & -131.397 & 265.3602 = p(-2)
\end{array}$$

Note 1: Diagonal arrows indicate multiplication by $x = -2$.

Note 2: $\dfrac{p(x)}{x - (-2)} = x^3 - 11.5x^2 + 51.49x - 131.397 + \dfrac{265.3602}{x - (-2)}$

Figure 1.4-1 *Performing synthetic division.*

```
ØØ1ØØ            FUNCTION POLVAL (X, NP1, A)
ØØ2ØØ            DIMENSION A(NP1)
ØØ3ØØ            DOUBLE PRECISION DVALUE
ØØ4ØØ    C
ØØ5ØØ    C   POLVAL = A(1)*x^n + ... + A(n)*x + A(n+1)
ØØ6ØØ    C      NP1 = n+1 = number of coefficients
ØØ7ØØ    C
ØØ8ØØ            DVALUE = DBLE( A(1) )
ØØ9ØØ            DO 1Ø I=2, NP1
Ø1ØØØ       1Ø      DVALUE = DVALUE * X + A(I)
Ø11ØØ            POLVAL = SNGL( DVALUE )
Ø12ØØ            RETURN
Ø13ØØ            END
```

Figure 1.4-2 *Fortran* FUNCTION *subprogram for evaluating p(x).*

Practical Consideration 1.4B (Evaluating polynomials on a computer). Figure 1.4-2 shows a Fortran FUNCTION subprogram POLVAL that evaluates the nth-degree polynomial

$$p(x) = a_1 x^n + a_2 x^{n-1} + \cdots + a_n x + a_{n+1} \tag{5}$$

using Algorithm 1.4B. This subprogram need only appear *once*, no matter how many different polynomials are to be evaluated and no matter what their degree! Thus, for example, the FUNCTION calls

$$\text{P} = \text{POLVAL}(-2.3, 4, \text{COEFFP}) \quad \text{and} \quad \text{Q} = \text{POLVAL}(5.66, 3, \text{COEFFQ})$$

calculate P as a cubic at $x = -2.3$ and Q as a quadratic at $x = 5.66$. ■

1.4C Strategies for Storing Intermediate Values

> **GUARD DIGIT STRATEGY.** To minimize creeping roundoff and negligible addition, store all *intermediate* values with more accuracy than is required of the final answer.

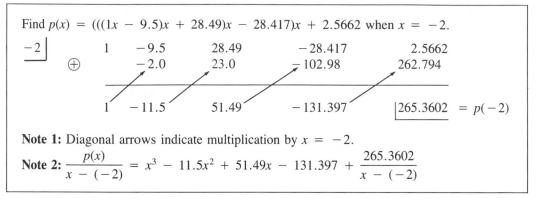

Since hand calculators generally store undisplayed guard digits, this gets done automatically if you simply *store all intermediate values in your calculator* when using it. If you must write intermediate values down, be sure to do so rounded to a couple of guard digits *more* than the desired *final* accuracy.

Similarly, it is unwise to expect the answer on a ks computer to actually be accurate to k significant digits. Thus you should only expect $7s$ (reliably) on an $8s$ computer. The risk of propagated roundoff is especially great when numbers of varying sign and size are added in a loop. The following special kind of guard digit strategy is therefore recommended whenever extended precision is available.

> **PARTIAL EXTENDED PRECISION (PEP) STRATEGY.** To minimize subtractive cancellation and negligible addition when accumulating a sum in a loop, use an extended precision variable to accumulate the sum.

This is why the DOUBLE-PRECISION variable DVALUE was used in Figure 1.4-2. The following example shows how effective it can be.

EXAMPLE 1.4C We know from calculus that the ***n*th Maclaurin polynomial**

$$P_n(x) = \sum_{k=0}^{n} \frac{x^k}{k!} = 1 + \frac{x}{1!} + \frac{x^2}{2!} + \frac{x^3}{3!} + \cdots + \frac{x^n}{n!} \qquad \textbf{(6a)}$$

converges to e^x as $n \to \infty$ for any fixed x. Use $P_{60}(x)$ to approximate e^x for $x = 12$ and $x = -12$, and discuss the results obtained.

SOLUTION: The terms added to 1 to get $P_n(x)$ satisfy

$$k\text{th term} = \frac{x^k}{k!} = \frac{x}{k}\cdot\frac{x^{k-1}}{(k-1)!} = \frac{x}{k}\cdot[(k-1)\text{st term}] \qquad \textbf{(6b)}$$

The Fortran 77 program shown in Figure 1.4-3 uses this recursive relationship to accumulate TERM as a product in the loop that is used to add these TERMs in the variable SUM. The Partial Extended Precision strategy is illustrated by accumulating the DOUBLE-PRECISION variables DTERM and DSUM in the same loop (see lines 900–1500).

The output, shown in Figure 1.4-4, displays the current TERM and DTERM and the accumulated SUM and DSUM for $n = 10, 20, \ldots, 60$. It is important to note that x itself is stored in *single precision*, which is $8s$ for the device used, so only $7s$ accuracy should be expected of the sum. For $x = 12$ the accumulation of $P_{60}(12)$ in the single-precision variable SUM yielded e^{12} to $7s$. However, for $x = -12$, SUM accumulated an obviously incorrect *negative* approximation of e^x!

The reason for the serious error when $x = -12$ can be seen from the nth TERM column of Figure 1.4-4. Since the calculations were performed on an $8s$ device, the large TERMs for $n \approx 10$ had creeping roundoff in the *third decimal place*. By the time n got to be 40, the accumulated SUM experienced a net *subtractive cancellation* of *all* significant digits of these larger TERMs, and all that was left was "roundoff noise," which happened to be negative but could just as well have been positive! On the other hand, the extended precision variable DSUM had room for about 16 accurate digits, and the use of the extended precision

```
ØØ1ØØ    C  The effect of extended precision on the series for exp(x)
ØØ2ØØ          DOUBLE PRECISION DSUM, DTERM
ØØ3ØØ          DATA IW /5/,  X /12.Ø/
ØØ4ØØ      1Ø TERM = 1.Ø      ! TERM and DTERM accumulate x^n/n (as products)
ØØ5ØØ         DTERM = 1.DØ
ØØ6ØØ         SUM = 1.Ø       ! SUM and DSUM accumulate sum of TERMs and DTERMs
ØØ7ØØ         DSUM = 1.DØ
ØØ8ØØ         WRITE(5, 1)     ! heading
ØØ9ØØ         DO 2Ø N=1,6Ø
Ø1ØØØ            TERM = TERM * X / N
Ø11ØØ            DTERM = DTERM * X / N
Ø12ØØ            SUM = SUM + TERM
Ø13ØØ            DSUM = DSUM + DTERM
Ø14ØØ            IF ( MOD(N,1Ø) .EQ. Ø ) WRITE(5, 2) N, TERM, SUM, DTERM, DSUM
Ø15ØØ      2Ø CONTINUE
Ø16ØØ         WRITE(5, 3) X, EXP(X)
Ø17ØØ         IF ( X .LT. Ø.Ø ) STOP
Ø18ØØ         X = -X                      ! X is now -12.Ø
Ø19ØØ         GOTO 1Ø
Ø2ØØØ       1 FORMAT('Ø n',4X,'nth TERM',1ØX,'SUM',11X,'nth DTERM',9X,'DSUM')
Ø21ØØ       2 FORMAT (I3, 4G16.7)
Ø22ØØ       3 FORMAT('Ø',5X,'The computer''s value of EXP(',F5.1,') is',G15.7)
Ø23ØØ         END
```

Figure 1.4-3 *Fortran program for calculating* e^{12} *and* e^{-12}.

variable DTERM ensured that they were filled accurately so that *accurate significant digits were available to replace those that subtractively canceled.* ∎

Practical Consideration 1.4C [Evaluating EXP (*X*) on a digital device]. To appreciate the challenge of designing a built-in function, consider what a *ks* device with arithmetic base *b* (= 10 or 2^e, *e* = 1, 3, or 4) must do to evaluate EXP(*X*) efficiently to *ks* for *any* representable *X*. For a given *X*, define

$$w = \frac{X}{\ln B}, \quad \text{where } B = \begin{cases} 10 & \text{if } b = 10 \\ 2 & \text{if } b = 2^e \end{cases} \text{ is stored on the device} \qquad \textbf{(7a)}$$

and let *I* be the leftmost integer within $\frac{1}{2}$ unit of *w* so that

$$w = I + z, \quad \text{where } -\tfrac{1}{2} \leqslant z < \tfrac{1}{2} \qquad \textbf{(7b)}$$

Then by (2), $e^X = e^{w(\ln B)} = e^{I(\ln B)}e^{z(\ln B)}$, that is,

$$e^X = B^I e^u, \quad \text{where } u = z(\ln B) \text{ satisfies } |u| < \tfrac{1}{2}(\ln B) \qquad \textbf{(7c)}$$

If *b* = 2 or 10 (= *B*), then B^I is exactly $(0.1)_B \cdot B^{I+1}$ [*M* = $(0.1)_B$, *c* = *I* + 1]; otherwise (i.e., if *b* = 8 or 16), b^I can be obtained to *ks* using at most one multiplication [Problem M1-15(a)]. So (7c) in effect reduces the challenge of approximating *any* e^X to *ks* to the reasonable problem of finding an efficiently calculated function $\phi(u)$ such that

n	nth TERM	SUM	nth DTERM	DSUM
1Ø	17Ø62.77	56513.25	17Ø62.77	56513.25
2Ø	1575.797	16Ø867.2	1575.797	16Ø867.2
3Ø	Ø.8949Ø58	162754.2	Ø.8949Ø58	162754.2
4Ø	Ø.18Ø1378E-Ø4	162754.8	Ø.18Ø1378E-Ø4	162754.8
5Ø	Ø.2992178E-1Ø	162754.8	Ø.2992178E-1Ø	162754.8
6Ø	Ø.6771735E-17	162754.8	Ø.6771734E-17	162754.8

The computer's value of EXP(12.Ø) is 162754.8

n	nth TERM	SUM	nth DTERM	DSUM
1Ø	17Ø62.77	91Ø9.137	17Ø62.77	91Ø9.137
2Ø	1575.797	579.3448	1575.797	579.3449
3Ø	Ø.8949Ø58	Ø.2512673	Ø.8949Ø58	Ø.2513614
4Ø	Ø.18Ø1378E-Ø4	-Ø.8392821E-Ø4	Ø.18Ø1378E-Ø4	Ø.1Ø24Ø12E-Ø4
5Ø	Ø.2992178E-1Ø	-Ø.88Ø2411E-Ø4	Ø.2992178E-1Ø	Ø.6144218E-Ø5
6Ø	Ø.6771735E-17	-Ø.88Ø2411E-Ø4	Ø.6771734E-17	Ø.6144212E-Ø5

The computer's value of EXP(-12.Ø) is Ø.6144212E-Ø5

Figure 1.4-4 *Output of the program shown in Figure 1.4-3.*

$$\phi(u) = e^u \text{ to } ks \text{ for } |u| < \begin{cases} (\ln 10)/2 \doteq 1.15129 & \text{if } b = 10 \\ (\ln 2)/2 \doteq 0.346574 & \text{if } b = 2^e \end{cases} \tag{7d}$$

A natural candidate for $\phi(u)$ is the Nth Maclaurin polynomial for e^u using the smallest N for which (7d) holds (see Example 1.3C). For 8s, N would have to be 12 if $B = 10$, and 7 if $B = 2$ [Problem M1-15(b)]. Digital devices generally take $\phi(u)$ to be a *rational function* that satisfies (7d) using fewer arithmetic operations than are required by a polynomial. Methods for finding such a $\phi(u)$ are described in Section 5.4D. ■

1.4D Strategies for Avoiding Subtractive Cancellation

It is not always possible to anticipate subtractive cancellation. But when it is, *equivalent expressions* or *Maclaurin series*, as illustrated in Examples 1.3B and 1.3C, can (*and should!*) be used to avoid it. We close this chapter with a particularly useful illustration of how subtractive cancellation can be eliminated by using an algebraically equivalent expression.

It is well known that the two roots of the general quadratic equation

$$ax^2 + bx + c = 0 \qquad (a \neq 0) \tag{8a}$$

can be obtained using the **quadratic formula** as

$$r^+ = \frac{-b + \sqrt{b^2 - 4ac}}{2a} \quad \text{and} \quad r^- = \frac{-b - \sqrt{b^2 - 4ac}}{2a} \tag{8b}$$

These two roots are calculated as RPLUS and RMINUS in lines 500 and 600, and output in

```
ØØ1ØØ    C This program demonstrates errors that can occur when using
ØØ2ØØ    C the quadratic formula to find real roots of x^2 + Bx + 1.
ØØ3ØØ          DATA IW /5/,  B /1ØØØ.ØØ1/
ØØ4ØØ    C
ØØ5ØØ       1Ø RPLUS  = ( -B + SQRT(B*B - 4.Ø) ) / 2.Ø
ØØ6ØØ          RMINUS = ( -B - SQRT(B*B - 4.Ø) ) / 2.Ø
ØØ7ØØ    C
ØØ8ØØ          WRITE (IW,1) B, RPLUS, RMINUS
ØØ9ØØ        1 FORMAT('ØFor x^2 + ',F9.3,'x + 1:  r+ = ', G13.6,
Ø1ØØØ          &                              ',  r- = ', G13.6)
Ø11ØØ          IF ( B .LT. Ø.Ø ) STOP
Ø12ØØ          B = -B                      ! B is now -1ØØØ.ØØ1
Ø13ØØ          GOTO 1Ø
Ø14ØØ          END
```

```
For x^2 +  1ØØØ.ØØ1x + 1:  r+ = -Ø.999451E-Ø3,  r- =  -1ØØØ.ØØ

For x^2 + -1ØØØ.ØØ1x + 1:  r+ =   1ØØØ.ØØ   ,  r- =  Ø.999451E-Ø3
```

Figure 1.4-5   *Finding roots of $x^2 \pm 100.01x + 1$ on a computer.*

```
ØØ1ØØ            SUBROUTINE QROOTS(A, B, C, R1, R2, COMPLX, IW, PRNT)
ØØ2ØØ            LOGICAL PRNT, COMPLX
ØØ3ØØ    C - - - - - - - - - - - - - - - - - - - - - - - - - - - - - C
ØØ4ØØ    C   This subroutine finds the two roots of the quadratic    C
ØØ5ØØ    C                    ax^2 + bx + c = Ø                       C
ØØ6ØØ    C   If PRNT = .TRUE., it prints them on output device IW.    C
ØØ7ØØ    C - - - - - - - - - - - - - - - VERSION 2  9/9/85  - - - C
ØØ8ØØ            B1 = -Ø.5*B/A
ØØ9ØØ            C1 = C/A
Ø1ØØØ            DSCR = B1*B1 - C1
Ø11ØØ            COMPLX = (DSCR .LT. Ø.Ø)
Ø12ØØ            IF (.NOT. COMPLX) THEN        ! real roots:  R1 and R2
Ø13ØØ               R1 = ABS(B1) + SQRT(DSCR)
Ø14ØØ               IF (B1 .LT. Ø.Ø) R1 = -R1
Ø15ØØ               R2 = Ø.Ø
Ø16ØØ               IF (R1 .NE. Ø.Ø) R2 = C1/R1
Ø17ØØ               IF (PRNT) WRITE (IW, 1) R1, R2
Ø18ØØ            ELSE         ! complex conjugate roots:  R1 +or- i*R2
Ø19ØØ               R1 = B1
Ø2ØØØ               R2 = SQRT(-DSCR)
Ø21ØØ               IF (PRNT) WRITE (IW, 2) R1, R2
Ø22ØØ            ENDIF
Ø23ØØ          1 FORMAT('ØReal roots:  ',G14.7,'  and  ',G14.7)
Ø24ØØ          2 FORMAT('ØComplex roots:',G14.7,' +or-  i*(',G14.7,')')
Ø25ØØ            RETURN
Ø26ØØ            END
```

Figure 1.4-6   *A Fortran* SUBROUTINE *for getting roots of $ax^2 + bx + c = 0$.*

line 800 of the Fortran program shown in Figure 1.4-5 for the following two quadratics ($a = c = 1$, $b = \pm 1000.001$):

$$x^2 + 1000.001x + 1 \qquad \text{for which } r^+ = -0.001 \text{ and } r^- = -1000$$
$$x^2 - 1000.001x + 1 \qquad \text{for which } r^+ = +1000 \text{ and } r^- = +0.001$$

The output (shown below the program listing) shows that r^+ was calculated accurately for $b = -1000.001$ but with an unexpected error in the third significant digit when $b = +1000.001$, whereas the reverse is true of r^-! A look at (8b) reveals the source of the problem:

$$\text{If } b^2 >> 4ac, \qquad \text{then } \sqrt{b^2 - 4ac} \approx \sqrt{b^2} = |b| \tag{9}$$

So r^+ can lose significance due to subtractive cancellation when b is *positive*, whereas r^- can lose significance when b is *negative*. To avoid all risk of subtractive cancellation, first observe that

$$ax^2 + bx + c = a(x - r^+)(x - r^-), \qquad \text{so that } c = ar^+r^- \tag{10}$$

Then divide through by $2a$ in (8b) to represent the desired roots as r_1 and r_2, where

```
C     Program to test the accuracy of SUBROUTINE QROOTS
      LOGICAL PRNT, COMPLX
      DATA IR /5/,  IW /5/
   1Ø WRITE (IW, 1)
    1 FORMAT ('ØFor ax^2 + bx + c = Ø, input a, b, c')
      READ (IR, *) A, B, C
      IF (A .EQ. Ø.Ø) STOP
      CALL QROOTS (A, B, C, R1, R2, COMPLX, IW, .TRUE.)
      GO TO 1Ø
      END
```

```
For ax^2 + bx + c = Ø, input a, b, c
 1  +1ØØØ.ØØ1  1

Real roots:   -1ØØØ.ØØØ       and  -Ø.1ØØØØØØE-Ø2

For ax^2 + bx + c = Ø, input a, b, c
 1  -1ØØØ.ØØ1  1

Real roots:    1ØØØ.ØØØ       and   Ø.1ØØØØØØE-Ø2

For ax^2 + bx + c = Ø, input a, b, c
 1  -2  3

Complex roots:  1.ØØØØØØ     +or-  i*(  1.414214     )
```

Figure 1.4-7 *Demonstration of the accuracy of* SUBROUTINE QROOTS.

$$r_1 = (\pm)[|b_1| + \sqrt{b_1^2 - c_1}] \quad \text{and} \quad r_2 = \frac{c_2}{r_1} \qquad \textbf{(11a)}$$

where

$$b_1 = -\frac{b}{2a}, \quad c_1 = \frac{c}{a}, \quad \text{and} \quad (\pm) \text{ is } (+) \text{ if } b_1 \geqslant 0, (-) \text{ if } b_1 < 0 \qquad \textbf{(11b)}$$

In (11a), r_1 is r^+ if $b_1 \geqslant 0$ and r^- if $b_1 < 0$. In either case there is no possibility of subtractive cancellation in calculating it because both $|b_1|$ and $\sqrt{b_1^2 - c_1}$ are positive.

Formulas (11) are implemented in the Fortran 77 SUBROUTINE QROOTS shown in Figure 1.4-6. In Figure 1.4-7, QROOTS is called to get the real roots of the equations $x^2 \pm 1000.001x + 1 = 0$ and the complex roots $(1 \pm \sqrt{2}\,i)$ of the equation $x^2 - 2x + 3 = 0$. All are correct to 8s (which is the precision of the computer used).

Practical Consideration 1.4D (Using boolean flags). Note the use of the LOGICAL **flags** PRNT and COMPLX. When SUBROUTINE QROOTS is called with PRNT = .FALSE., it prints no output, but it tells the calling program whether R1 and R2 are two real roots (COMPLX = .FALSE.) or the real and imaginary parts of a pair of complex conjugate roots (COMPLX = .TRUE.). Boolean flags with meaningful names that describe a true/false state should be used freely because they require only one bit to store but can substantially improve the readability of a program. ■

PROBLEMS

Section 1.1

1-1 Round to 2d and to 3d; use the even-rounding rule if necessary.
(a) $\frac{1000}{11}$ **(b)** $-\sqrt{6}$ **(c)** $\frac{1}{16}$ **(d)** $-\frac{3}{16}$ **(e)** 0.99999 **(f)** tan 86.5°

1-2 Evaluate $\Delta f(x)/h$ as indicated for $h = h_0$; h_0/r, h_0/r^2, and h_0/r^3. In all exercises involving a trigonometric function trig x, x is in *radians*. Be sure to set your calculator in the radian mode.

(a) $f(x) = x^2 - 3$, $x = 2$, $h_0 = -1$, $r = 10$ $[f'(x) = 4]$
(b) $f(x) = \sqrt{2 + x}$, $x = -1$, $h_0 = .64$, $r = 4$ $[f'(x) = \frac{1}{2}]$
(c) $f(x) = (x - 2)^3$, $x = 2$, $h_0 = .81$, $r = 3$ $[f'(x) = 0]$
(d) $f(x) = \ln x$, $x = 1$, $h_0 = -.5$, $r = 5$ $[f'(x) = 1]$
(e) $f(x) = \sin x$, $x = 0$, $h_0 = .81$, $r = 3$ $[f'(x) = 1]$
(f) $f(x) = \cos x$, $x = 0$, $h_0 = .64$, $r = 4$ $[f'(x) = 0]$
(g) $f(x) = e^{x^2}$, $x = 0$, $h_0 = .32$, $r = 2$ $[f'(x) = 0]$

1-3 For those of parts (a)–(g) of Problem 1-2 that you did, find a small nonzero h that makes $\Delta f(x)/h = 0$ on your calculator. Can you figure out what happened for this h?

Section 1.2

1-4 Same as Problem 1-1 but replace "to 2d and to 3d" by "to 2s and to 3s."

1-5 On a number line, sketch all *nonnegative* X's that can be stored as $M \cdot 2^c$ using a k-bit ($b = 2$) mantissa M and a p-bit biased characteristic c (with bias 2^{p-1}) for the given k and p.
(a) $k = 2; p = 1$ **(b)** $k = 2; p = 2$
(c) $k = 3; p = 2$ **(d)** $k = 2; p = 3$
In words, what is the effect of increasing k? of increasing p?

1-6 Find (i) s = smallest positive representable X, (ii) L = largest positive representable X, (iii) the number of storable real numbers X, (iv) the machine unit u_M, (v) the precision, (vi) the word size, and (vii) the largest storable INTEGER for a computer that stores real numbers as described.

(a) 32 bit word, $M = 6$ hexadecimal digits, $c = 7$ bits (bias = 64)
(b) 40 bit word, $M = 31$ bits ($b = 2$), $c = 8$ bits (bias = 128)
(c) The computer(s) available to you.

1-7 Suppose that a computer stores mantissas M using k "base $b = 2^e$ cells" and *biased* characteristics c (with bias 2^{p-1}) using p bits. Find formulas in terms of b, e, k, and p for (i)–(vii) of Problem 1-6.

1-8 How can the storage "cells" in Figure 1.2-1 be used to store $-98 \leq c \leq 100$ rather than $-99 < c < 99$? What effect will this change have on the number of *nonzero* storable numbers? on $[s, L]$?

1-9 Assuming that $-98 \leq c \leq 100$, find (i)–(v) of Problem 1-6 for a calculator that stores k (decimal) digit mantissas *including guard digits* **(a)** if $k = 10$; **(b)** if $k = 13$; **(c)** as formulas in terms of k.

1-10 Assuming that $x \approx 0$, $y >> 0$, and $z << 0$, use parentheses and/or algebraic rearrangement to minimize the possibility of overflow or underflow: **(a)** $z(e^y)^x$; **(b)** xe^ye^z; **(c)** e^{y-z}/e^{y-2z}.

1-11 For parts (a)–(f) of Problem 1-1: Find M and c of the normalized floating representation $\pm M \cdot 10^c$ that would be stored on a 4s device that rounds. What if it chopped?

1-12 Find the inherent error of your answers in Problem 1-11.

Section 1.3

1-13 Find the *relative* inherent error of your answers to Problem 1-11. Does (5) of Section 1.3A hold?

1-14 Find the error, relative error, and percent error of the approximations.
(a) $\frac{1}{11} \approx 0.1$ **(b)** $\frac{1}{11} \approx 0.09$ **(c)** $\frac{5}{9} \approx 0.56$ **(d)** $\frac{4}{9} \approx 0.44$

1-15 For the given x, describe the X's that satisfy the Relative Error Test for 3s accuracy as an inequality of the form $|x - X| < r$.
(a) $x = 10$ **(b)** $x = 0.046$ **(c)** $x = -510$ **(d)** $x = 0.99$
Note that the Relative Error Test does not get *all* X's that are correct when rounded to 3s. Describe the leading coefficients of the x's for which the test comes closest to doing so.

1-16 Let $X = 0.5289$, $Y = 0.8012$, and $Z = 0.6024$. Show that if intermediate values are stored rounded to 4s after *each* operation, then:
(a) $X*(Y + Z) \neq X*Y + X*Z$ **(b)** $(X + Y) + Z \neq X + (Y + Z)$

1-17 Make up an example for which $(X*Y)*Z \neq X*(Y*Z)$ if all intermediate values are stored rounded to 2s after each operation.

1-18 Describe the operations ($+$, $-$, $*$, $/$) that can cause negligible addition (NA), error magnification (EM), or subtractive cancellation (SC) in calculating A; give the range(s) of x (e.g., $x \approx y$, $x >> 0$, etc.) for which the errors might occur.

(a) $A = \sqrt{x^2 + 1} - 1$ **(b)** $A = \sqrt{x^2 + 1} - x$ **(c)** $A = \ln(x + y) - \ln y$
(d) $A = 1 - \cos x$ **(e)** $A = \tan^{-1}x - 1$ **(f)** $A = \sin x - \sin y$
(g) $A = \dfrac{1}{x + 1} - 1$ **(h)** $A = \dfrac{x}{x + 1} - 1$ **(i)** $A = \dfrac{e^x - 1 - x}{x^2}$

1-19 The formula $C = |xf'(x)/f(x)|$ shows that C increases as $|x|$ and $|f'(x)|$ increase but decreases as $|f(x)|$ increases. Explain in words why this is plausible based on (14a) of Section 1.3.

1-20 Is the evaluation of $f(x)$ well conditioned for the x's described? Justify.
(a) $f(x) = x^n$, any $x \neq 0$ **(b)** $f(x) = \ln x$, $x \approx 0^+$, $x \approx 1$, $x >> 0$
(c) $f(x) = \sin x - x$, $x \approx 0$ **(d)** $f(x) = x \sin(1/x)$, $x \approx 0$, $x >> 0$

1-21 Let $F(h) = \Delta f(x)/h$ for a fixed x and suppose that m is the smallest positive integer such that $f^{(m)}(x) \neq 0$. Use L'Hospital's rule to see what $hF'(h)/F(h)$ approaches as h approaches 0. When is the evaluation of $\Delta f(x)/h$ ill-conditioned for $h \approx 0$?

Section 1.4

1-22 If subtractive cancellation is likely to occur when evaluating $f(x)$ for x in an interval I, then $A \leftarrow f(x)$ should be replaced by

$$\text{IF } (x \text{ lies in } I) \text{ THEN } A \leftarrow \phi(x) \text{ ELSE } A \leftarrow f(x)$$

where $\phi(x)$ approximates $f(x)$ to the device accuracy *and* can be evaluated without subtractive cancellation for x in I. Use this strategy for A in (a)–(i) of Problem 1-18. If possible, get $\phi(x)$ using a trigonometric or algebraic identity as in Example 1.3B or Section 1.4D; otherwise, use a Maclaurin polynomial as in Example 1.3C to get $7s$ accuracy.

1-23 **(a)** If you use the Full Mantissa Strategy, how would you input $\ln \pi$ on a $7s$ device? an $8s$ device? a $10s$ device?
(b) How can you get a computer to initialize $\ln \pi$ in a language of your choice? how in extended precision (if available)?

1-24 Use the Fewest Operations Strategy without worrying about subtractive cancellation to make the evaluation most efficient.
(a) x^6 **(b)** $9e^{4x} + 6e^{2x} + 1$ **(c)** $x^3y^3(\ln x + \ln y) + x^2y^2$
(d) $\dfrac{1}{2 - \sqrt{x}}$ **(e)** $x^3 - 2x^2 + 4x - 8$ **(f)** $\cos^2(3x) - \sin^2(3x)$

1-25 Use the nested form of $p(x)$ to evaluate $p(-1.5)$.
(a) $p(x) = 9x^3 - 3x^2 - 12x + 5$ **(b)** $p(x) = 2x^5 + x^3 - 5x^2 - 10$
(c) $p(x) = x^7 - 2x^3 + 1$ **(d)** $p(x) = 3x^6 - 8x^4 + x^2 + 4$

1-26 How many multiplications are needed to evaluate an nth-degree polynomial using the Synthetic Division Algorithm 1.4B? How many addition/subtractions are needed?

MISCELLANEOUS PROBLEMS

M1-1 From the graph in Figure 1.1-1(b), estimate the sensitivity of the dial when θ is (a) 100°; (b) 200°; (c) 270°.

M1-2 Modify Algorithm 1.1A so as to stop (with an appropriate warning) when $DQ = 0$ *unless* $|DQ|$ appears to be decreasing to 0 as $h \to 0$.

M1-3 What would TOL be in line 700 of Figure 1.1-2(a) if 10. were replaced by 10 (no decimal point). How would it affect the termination test in line 1900?

M1-4 Devise a test analogous to (10) of Section 1.2B to see if your calculator chops or rounds in the least significant *guard digit*.

M1-5 Deduce from columns (a) and (b), but *not* column (c), of Table 1.1-1 (Section 1.1C) that the DEC-10 has greater precision than the PDP-11.

M1-6 **(a)** Expand $(1 + r + \cdots + r^{n-1})(1 - r)$ to obtain the **geometric series** partial sum

$$\textbf{(GS)} \quad a(1 + r + \cdots + r^{n-1}) = \frac{a(r^n - 1)}{r - 1} \qquad \text{for } r \neq 1$$

(b) Use (GS) with $r = 2$ to prove that (i) $2^m - 1$ is the largest integer that can be stored using m bits, and (ii) $1 - 2^{-k}$ is the largest k-bit mantissa.
(c) Use (GS) to show that $M = (.CCCCCC)_{16}$ in (13b) of Section 1.2C represents $\frac{3}{4} \cdot \frac{1118481}{1048576}$.
(d) Use (GS) to show that $\Sigma_{k=1}^{50} e^{6-k/5} = e^{5.8}(1 - e^{-10})/(1 - e^{-0.2})$.

M1-7 Consider the sum $S_n = \Sigma_{k=1}^{n} a_k$, where $a_1 > a_2 > \cdots > a_n > 0$ and $a_1 >> a_n \approx 0$.
(a) Why should you expect summing backward (i.e., $a_n + a_{n-1} + \cdots + a_1$) to give greater accuracy than summing forward (i.e., $a_1 + a_2 + \cdots + a_n$)?
(b) The nth partial sum of the harmonic series, that is, $S_n = \Sigma_{k=1}^{n} (1/k)$, diverges to infinity. Can a computer be used to demonstrate this by calculating S_n for $n = 100, 100^2, 100^3, 100^4$? Explain.
(c) Write a computer program that sums $S_{50} = \Sigma_{k=1}^{50} e^{6-k/5}$ both forward and backward. Do the output values verify part (a)? [See Problem M1-6(d).]

M1-8 In the 32-bit words shown in parts (a) and (b), the biased characteristic (with bias 64) is stored in the rightmost seven bits, the sign bit ($0 = +$) is the leftmost bit, and the normalized mantissa is stored between them.

(a)	0	1	0	0	1	1	0	1	0	0	0	0	0	0	0	0	0	0	0	0	0	0	0	0	0	0	0	0	1	1	0	1
(b)	0	0	1	0	1	0	1	0	1	0	1	0	1	0	1	0	1	0	1	0	1	0	1	0	1	0	1	0	1	0	1	0

Find the real number stored (if legitimate) in (a) and (b) if mantissas are stored with base b, where b is: (i) 8; (ii) 16; (iii) 2.

M1-9 If the words shown in parts (a) and (b) of Problem M1-8 store positive `INTEGERS` (leftmost bit = 0), what would they be? Use (GS) of Problem M1-6 for (b)!

M1-10 **(a)** Most computer languages have a built-in function for $\tan^{-1}x$, the principal value of arctan x (in radians, between $-\pi/2$ and $\pi/2$); but some do not have $\sin^{-1}x$ (in $[-\pi/2, \pi/2]$) or $\cos^{-1}x$ (in $[0, \pi]$). Express $\sin^{-1}x$ and $\cos^{-1}x$ for $-1 \leq x \leq 1$ using the $\tan^{-1}$ function.
(b) The cross section of a trough of length L is a semicircle of radius r (diameter on top). Write a `FUNCTION` subprogram `VFEED(H)` that returns the volume of feed when it is filled with feed to a height `H`, where $0 \leq$ `H` $\leq r$.

M1-11 Suppose that the product $p_1 p_2 \cdots p_N$ is accumulated in a loop as

$$Prod = 1; \quad \text{DO FOR } i = 1 \text{ TO } N; \quad Prod = p_i * Prod$$

If the p_i's are all positive, how can you reduce the possibility that an intermediate value of *Prod* will cause underflow or overflow?

M1-12 Verify that $a = +1$ or $a = -1$ is a root of $p(x)$, that is, $p(a) = 0$. Then use synthetic division to factor out $(x - a)$ as needed to find all roots of $p(x)$.
(a) $p(x) = 7x^3 - 3x^2 + 4x - 8$ **(b)** $p(x) = 2x^4 - 7x^3 + x^2 + 7x - 3$

M1-13 This useful exercise justifies and extends the Synthetic Division Algorithm 1.4B.
(a) Express a given $p(x) = a_1 x^n + \cdots + a_n x + a_{n+1}$ as

$$\textbf{(SD)} \quad p(x) = (x - a)Q(x) + R, \qquad \text{where } Q(x) = b_1 x^{n-1} + \cdots + b_n$$

By expanding the right-hand side in (SD), show that $b_1, \ldots, b_n$ and $R = p(a) = b_{n+1}$ can be obtained by synthetic division.
(b) Consider the following algorithm based on (SD) of part (a):

ALGORITHM. ITERATED SYNTHETIC DIVISION

PURPOSE: To expand an nth degree polynomial $p(x)$ in powers of $(x - a)$

GET $a, n, a_1, \ldots, a_{n+1}$ {coefficients of $p(x)$}

$Q_0(x) \leftarrow p(x)$
DO FOR $i = 0$ to n
    Use (SD) to get R_i and the coefficients of $Q_{i+1}(x)$, where
        $Q_i(x) = (x - a)Q_{i+1}(x) + R_i$

From the nested form of the Taylor expansion of $p(x)$ based at a,

$$p(x) = p(a) + (x - a)\left(p'(a) + (x - a)\frac{p''(a)}{2!} + (x - a)\left(\frac{p'''(a)}{3!} + \ldots\right.\right.$$

deduce that R_i of the Iterated Synthetic Division algorithm is the coefficient of $(x - a)^i$ in the Taylor expansion of $p(x)$ based at a, that is,

$$R_0 = p(a),\ R_1 = \frac{p'(a)}{1!},\ \ldots,\ R_n = \frac{p^{(n)}(a)}{n!}.$$

M1-14 Use the Iterated Synthetic Division algorithm to find the Taylor expansion of $p(x)$ based at a without differentiating $p(x)$.
(a) $p(x) = x^3 + 8x^2 - 4x + 1$; $a = 1$ and $a = -2$
(b) $p(x) = 5x^4 - 4x^3 + 3x^2 - 2x + 1$; $a = 2$ and $a = -1$
(c) $p(x) = 2x^4 + 3x^2 + 4x + 5$; $a = 1$ and $a = 0$
(d) $p(x) = x^3 - 3x^2 + 3x - 1$; $a = 1$ and $a = -1$

M1-15 **(a)** If $b = 2^e$ and I is an integer, show that $b^I = 2^i b^J$, where J is an integer and i can be $0, 1, \ldots, e - 1$. What must be stored to get b^I using at most one multiplication? Explain.
(b) Let $\phi(u)$ be the Nth Maclaurin polynomial for e^u. Use the Relative Error Test to find the smallest N for which (7d) of Practical Consideration 1.4C holds with $k = 8$ (i) if $b = 10$; (ii) if $b = 2^e$.

COMPUTER PROBLEMS

C1-1 **(a)** Implement Algorithm 1.2B on your computer to determine its machine unit, u_M.
(b) Find the machine unit for extended precision (if available). How much of a word is added to the mantissa? the characteristic?

C1-2 Write `FUNCTION` subprograms `RNDDEC` and `RNDSIG` which round any `REAL X` as indicated. Do not try to implement the even rounding rule.
(a) `RNDDEC(X, K)` returns `X` correctly rounded to `K`d; `K` can be negative.
(b) `RNDSIG(X, K)` returns `X` correctly rounded to `K`s (for any `K` smaller than the precision of the device) if `X` $\neq 0$, and zero if `X` $= 0$.

C1-3 Replace **(a)** `A = SQRT(X - 4.) - SQRT(X)` and **(b)** `A = ATAN(X) - ATAN(Y)` by program segments that compute `A` to the device accuracy for all storable `X` (see Problem 1-22).

C1-4 In exact arithmetic, adding $1/n$ to itself $n \cdot 10^k$ times gives 10^k. What does your computer get for **(a)** $n = 10$, $k = 3$; **(b)** $n = 10$, $k = 5$; **(c)** $n = 16$, $k = 3$; **(d)** $n = 16$, $k = 5$? Explain any differences.

C1-5 Write a computer program `POLTAB(NP1, ACOEFF, A, B, NSTEPS)` that calls a `FUNCTION` subprogram such as `POLVAL` of Figure 1.4-1 to form a table showing x and $p(x) = a_1x^n + \cdots + a_nx + a_{n+1}$ for $x = a$ to b in `NSTEPS` equal steps, given input values of `NP1` $(= n + 1)$, `ACOEFF` $(= [a_1 \quad \cdots \quad a_{n+1}])$, `A` $(= a)$, and `B` $(= b)$.

C1-6 Write a subprogram `EXP1(X)`, based on Problem M1-15, to evaluate e^x to $7s$ for any representable x. Compare `EXP1(X)` to `EXP(X)` (built-in) for a variety of `X`'s between -39 and $+39$, including `X` ≈ 0. See also Problem 5-41(a).

Exercises C1-7 and C1-8 are intended to give you an idea of what goes into the design of a built-in function for a calculator or computer. The actual internal subroutines generally use improved approximations such as those described in Chapter 5 together with strategies similar to those described here. Details of efficient algorithms for built-in functions are found in [7] and [25].

C1-7 Consider the two Maclaurin series (one for $+x$, the other for $-x$)

$$(\mathbf{L_1}) \quad \ln|1 \pm x| = -\left(\pm x + \frac{x^2}{2} \pm \frac{x^3}{3} + \cdots\right) = -\left[\sum_{k=1}^{\infty} \frac{(\pm x)^k}{k}\right]$$

(a) Find S_n (the sum of the first n terms) to approximate $\ln(0.7) = 1.94591$ using (L_1) for $n = 2$, 3, and 4. What is the error of S_4?

(b) From (L_1) and the identity $\ln(a/b) = \ln a - \ln b$, deduce that

$$(\mathbf{L_2}) \quad \ln\frac{|1 + x|}{|1 - x|} = -2\left(x + \frac{x^3}{3} + \frac{x^5}{5} + \cdots\right) = -2\left(\sum_{k=1}^{\infty} \frac{x^{2k-1}}{2k-1}\right)$$

(c) What must x be in (L_2) to approximate $\ln(.7)$. Use this x to get S_n with $n = 2$, 3, and 4. What is the error of S_4? How many terms of (L_1) of part (b) are needed to guarantee this accuracy?

(d) Devise an algorithm comparable to that of Practical Consideration 1.4C to evaluate `ALOG1(X)` to $7s$ accuracy. Implement it and compare `ALOG1(X)` to your language's built-in ln `X` for various positive storable `X`'s.

C1-8 **(a)** How many terms of the Maclaurin series for $\sin x$ are needed to approximate $\sin x$ to $8s$ for $|x| \le \pi/8$? How many terms are needed for $\cos x$?

(b) Use identities and your result in (a) to evaluate $\sin x$ and $\cos x$ for $\pi/8 < |x| \le \pi/4$ using only one *nontrivial* multiplication. (Multiplication by 2 is trivial on a computer. Why?)

(c) Use identities and your result in part (b) to evaluate $\sin x$ and $\cos x$ efficiently to about $7s$ for $0 \le |x| \le \pi/2$.

(d) Use identities and your results in (a)–(c) in `FUNCTION` subprograms `SIN1(X)` and `COS1(X)` that evaluate $\sin x$ and $\cos x$ for any storable x. Compare their values to those returned by the built-in `SIN(X)` and `COS(X)` functions for various storable `X`'s.

2

Numerical Methods for Solving Equations in One Variable

2.0

Introduction

The need to solve an equation of the form

$$f(x) = 0 \quad \text{or more generally} \quad g(x) = h(x)$$

arises frequently in engineering and science, either as an end in itself or as an intermediate step in solving a more complex problem. For example, the tone that can make a particular glass object shatter or the wind conditions that can make an airplane wing or suspension bridge vibrate until it cracks can often be found by solving a linear, *nth-order differential equation* of the form

$$a_1 \frac{d^n y}{dt^n} + a_2 \frac{d^{n-1} y}{dt^{n-1}} + \cdots + a_n \frac{dy}{dt} + a_{n+1} y = 0 \qquad \textbf{(1a)}$$

where n can be as high as 20 or 30. The most direct way to find a function $y = y(t)$ that satisfies (1a) is to solve its *characteristic equation*, which is the ***n*th-degree polynomial equation**

$$p(x) = 0, \qquad \text{where } p(x) = a_1 x^n + a_2 x^{n-1} + \cdots + a_n x + a_{n+1} \qquad \textbf{(1b)}$$

To be specific, suppose that the "natural response" of a structure is governed by the linear, third-order differential equation

$$y''' + 7.5y'' + 18.48y' + 15.004y = 0 \tag{2a}$$

Its characteristic equation is the cubic polynomial equation

$$p(x) = x^3 + 7.5x^2 + 18.48x + 15.004 = 0 \tag{2b}$$

which has three roots, namely $r_1 = r_2 = -2.2$ and $r_3 = -3.1$ (as you can check). One learns in differential equations that once these roots are known, the **general solution** of (2a) is (*by inspection*!)

$$y(t) = (At + B)e^{-2.2t} + Ce^{-3.1t} \tag{2c}$$

where A, B, and C are **arbitrary constants**, to be determined by three physical conditions imposed on the solution $y(t)$. Having determined that, say, $A = 5.9$, $B = 1.3$, and $C = 0$, it may then be important to know how long it takes before $y(t)$ assumes the value 1.223. The desired time, $\bar{t}$, is the smallest positive solution of the *nonpolynomial* equation $y(t) = 1.223$, that is,

$$(5.9t + 1.3)e^{-2.2t} = 1.223 \tag{2d}$$

Even if you do not understand differential equations, it should be clear that the solution of (1a) hinges on the ability to solve the polynomial equation (1b). The quadratic formula can be used to solve any quadratic equation $ax^2 + bx + c = 0$ ($n = 2$), as we saw in Section 1.4D. There are also (rather unwieldy) formulas for solving third- and fourth-degree polynomial equations. However, the brilliant French mathematician Galois (at the age of 21!) proved the impossibility of finding a formula for solving the general nth-degree polynomial equation (1b) for $n > 4$. So if (1b) is to be solved for *any* n, a numerical solution is a necessity. It is also necessary if we want to solve (2d) or equations such as

$$k \tan x = \tanh x \qquad \text{and} \qquad k \cos x + \operatorname{sech} x = 0 \tag{3}$$

which arise when describing vibrations of structures.

Equations are called **algebraic** if they involve only the four arithmetic operations and radicals [e.g., (1b)]; otherwise, they are called **transcendental** [e.g., (2d) and (3)]. The iterative methods of Sections 2.1–2.4 apply equally well to both algebraic and transcendental equations. A special method for finding real or complex roots of any polynomial equation with real coefficients is given in Section 2.5.

2.1 Iterative Algorithms

We begin with some terminology and notation that will allow us to assess the desirability of an iterative algorithm.

2.1A Accuracy and Convergence Rates of Iterative Algorithms

An **iterative algorithm** for finding a desired value $\bar{x}$ proceeds in two basic steps:

> **Initialize:** GET an **initial guess** x_0 that approximates $\bar{x}$.
>
> **Iterate:** DO FOR $k = 0, 1, 2, \ldots$
> Use x_k to obtain an improved approximation x_{k+1}
> UNTIL a **termination test** is satisfied

We shall refer to x_k as the ***k*th iterate** of the algorithm and let

$$\epsilon_k = \bar{x} - x_k = \text{the } \textbf{error} \text{ of } x_k \qquad \textbf{(1)}$$

Thus ϵ_k *is what must be added to* x_k *to get the exact desired value* $\bar{x}$.

An iterative algorithm is successful if x_k converges to $\bar{x}$, that is, if $|\epsilon_k| \to 0$ as $k \to \infty$. "Good" iterative algorithms are

Robust: x_k will converge to $\bar{x}$ even if the initial guess x_0 is not very close to $\bar{x}$.

Rapidly Convergent: Once x_k gets close to $\bar{x}$, x_{k+1} will be much closer than x_k, that is, $|\epsilon_{k+1}| << |\epsilon_k|$.

Unfortunately, ϵ_k is not a practical indicator of the accuracy of x_k because we generally do not know $\bar{x}$. (If we did, there would be no need for numerical method!) However, we do know the algorithm's iterates

$$x_0, \quad x_1, \quad x_2, \quad \ldots, \quad x_k, \quad x_{k+1}, \quad \ldots \qquad \textbf{(2a)}$$

and these, in turn, can be used to get the **increments**

$$\Delta x_0 = x_1 - x_0, \quad \Delta x_1 = x_2 - x_1, \quad \ldots, \quad \Delta x_k = x_{k+1} - x_k, \quad \ldots \qquad \textbf{(2b)}$$

If the convergence is rapid (see Figure 2.1-1), then $\Delta x_0, \Delta x_1, \Delta x_2, \ldots,$ are computable estimates of $\epsilon_0, \epsilon_1, \epsilon_2, \ldots,$ respectively; that is,

Problem:

$$\Delta x_k = x_{k+1} - x_k \quad \text{approximates} \quad \epsilon_k = \bar{x} - x_k \quad \text{if } |\epsilon_{k+1}| << |\epsilon_k| \qquad \textbf{(3)}$$

does not apply to EM, but es., to NR

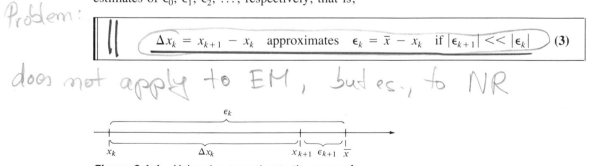

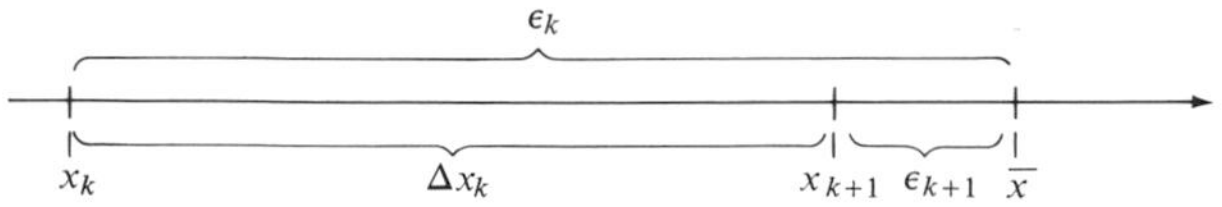

Figure 2.1-1 *Using Δx_k to estimate the error of x_k.*

Most termination tests for iterative algorithms stop the iteration when Δx_k is sufficiently small in some sense.

Before discussing any particular iterative algorithm, let us establish some guidelines for assessing how rapidly the iterates $x_0, x_1, x_2, \ldots$ converge or diverge. If there happens to be a number C_L, called the **linear convergence constant**, such that eventually

$$\Delta x_{k-1} \approx C_L \Delta x_{k-2}, \quad \Delta x_k \approx C_L \Delta x_{k-1}, \quad \Delta x_{k+1} \approx C_L \Delta x_k, \quad \ldots \tag{4a}$$

then the algorithm generating the x_k's will be called **linearly convergent** if $|C_L| < 1$, and **linearly divergent** if $|C_L| > 1$. Similarly, if there happens to be a number C_Q, called the **quadratic convergence constant**, such that eventually

$$\Delta x_{k-1} \approx C_Q(\Delta x_{k-2})^2, \quad \Delta x_k \approx C_Q(\Delta x_{k-1})^2, \quad \Delta x_{k+1} \approx C_Q(\Delta x_k)^2, \quad \ldots \tag{4b}$$

then the algorithm generating the x_k's will be called **quadratically convergent** (regardless of the size of C_Q).

It is evident from (4b) that *quadratic convergence must be* ***monotone***, that is, from only the left ($C_Q > 0$) or from only the right ($C_Q < 0$), and from (4a) that *linear convergence can be either monotone* ($0 < C_L < 1$) *or* ***alternating*** ($-1 < C_L < 0$). Also, the smaller the magnitude of the convergence constant (whether linear or quadratic), the more rapidly Δx_k will approach zero, hence the more rapid the convergence of x_k to $\bar{x}$. The following useful tests are also immediate from (4).

LINEAR CONVERGENCE/DIVERGENCE TEST. An algorithm converges or diverges *linearly* if

$$\frac{\Delta x_k}{\text{preceding } \Delta x_k} = \text{the } \textbf{linear convergence ratio} \approx \text{constant} \tag{5a}$$

(the constant being C_L) for several successive k's.

QUADRATIC CONVERGENCE TEST. An algorithm converges *quadratically* if

$$\frac{\Delta x_k}{(\text{preceding } \Delta x_k)^2} = \text{the } \textbf{quadratic convergence ratio} \approx \text{constant} \tag{5b}$$

(the constant being C_Q) for several successive k's.

The following example shows why *quadratic convergence will be our standard for rapid convergence*, whereas *linear convergence will be our standard for slow convergence*.

EXAMPLE 2.1A Consider an iterative algorithm for which

$$x_0 = 2, \quad x_1 = 2.1, \quad \text{and} \quad x_2 = 2.11$$

so that

$$\Delta x_0 = 2.1 - 2 = 0.1 \qquad \text{and} \qquad \Delta x_1 = 2.11 - 2.1 = 0.01$$

Make a table for $k = 0, 1, \ldots, 4$ showing k, x_k, Δx_k, and the appropriate convergence ratio assuming the algorithm converges *exactly* **(a)** linearly; **(b)** quadratically.

SOLUTION (a) EXACT LINEAR CONVERGENCE: By "exact" we mean that all linear convergence ratios $\Delta x_k/\Delta x_{k-1}$ are *equal* (rather than only approximately equal) to the linear convergence constant C_L. In particular, from the increments Δx_0 and Δx_1,

$$C_L = \frac{\Delta x_1}{\Delta x_0} = \frac{0.01}{0.1} = 0.1, \qquad \text{hence } \Delta x_k = 0.1\Delta x_{k-1} \quad \text{for all } k \tag{6}$$

So $\Delta x_2 = 0.1\Delta x_1 = 0.1(0.01) = 0.001$, $\Delta x_3 = 0.1\Delta x_2 = 0.0001$, and so on, hence

$$x_3 = x_2 + \Delta x_2 = 2.11 + 0.001 = 2.111$$
$$x_4 = x_3 + \Delta x_3 = 2.111 + 0.0001 = 2.1111$$

These values are summarized in Table 2.1-1(a).

TABLE 2.1-1 Convergence That Is Exactly (a) Linear and (b) Quadratic

	(a) Exact Linear Convergence			*(b) Exact Quadratic Convergence*		
k	x_k	Δx_k	$\dfrac{\Delta x_k}{\Delta x_{k-1}}$	x_k	Δx_k	$\dfrac{\Delta x_k}{(\Delta x_{k-1})^2}$
0	2.0	0.1	0.1	2.0	0.1	1
1	2.1	0.01	0.1	2.1	0.01	1
2	2.11	0.001	0.1	2.11	0.0001	1
3	2.111	0.0001	0.1	2.1101	0.000000001	1
4	2.1111	0.00001	0.1	1.11010001	1.E−16	1

SOLUTION (b) EXACT QUADRATIC CONVERGENCE: Reasoning as in part (a), we first get the quadratic convergence constant C_Q from Δx_0 and Δx_1 as

$$C_Q = \frac{\Delta x_1}{(\Delta x_0)^2} = \frac{0.01}{(0.1)^2} = 1, \qquad \text{hence } \Delta x_k = 1(\Delta x_{k-1})^2 \quad \text{for all } k \tag{7}$$

So $\Delta x_2 = 1(\Delta x_1)^2 = 0.0001$, $\Delta x_3 = 1(\Delta x_2)^2 = 0.00000001$, and so on, hence

$$x_3 = x_2 + \Delta x_2 = 2.11 + 0.0001 = 2.1101$$
$$x_4 = x_3 + \Delta x_3 = 2.1101 + 0.00000001 = 2.11010001$$

All x_k's in Table 2.1-1 are accurate to the digits shown. So Δx_k does approximate the error of x_k [see (3)]. Moreover, the values shown in Table 2.1-1 illustrate the important differences between linear and quadratic convergence summarized in Table 2.1-2. The linear convergence indicator "leading digit $\approx \alpha k$" can be viewed as: It takes roughly the same

TABLE 2.1-2 Indicators of Linear and Quadratic Convergence

Type of Convergence	*Nearly Constant Ratio*	*It Takes (Roughly) the Same Number of Iterations to:*	*The Leading Significant Digits in the Δx_k Column Lie (Roughly):*
Linear	$\dfrac{\Delta x_k}{\Delta x_{k-1}} \approx C_L$	Get *one additional* accurate decimal place of x_k	On a straight line (leading digit $\approx \alpha k$)
Quadratic	$\dfrac{\Delta x_k}{(\Delta x_{k-1})^2} \approx C_Q$	*Double* the number of accurate decimal places of x_k	On a "power" curve (leading digit $\approx \alpha^k$)

number of iterations (here one) to *add each additional leading zero* of Δx_k. The quadratic convergence indicator "leading digit $\approx \alpha^k$" can be viewed as: It takes roughly the same number of iterations (here one) to *double the number of leading zeros* of Δx_k. ■

2.1B The Repeated Substitution (RS) Algorithm

Consider the simple iterative algorithm that starts with an initial guess x_0 and then generates the succeeding iterates as

$$x_1 = g(x_0), \quad x_2 = g(x_1), \quad \ldots, \quad x_{k+1} = g(x_k), \quad \ldots \tag{8a}$$

where $g(x)$ is any continuous function of x. This algorithm is easily implemented on a programmable calculator by programming the evaluation of $g(x)$ for x in the display, putting x_0 in the display, and then repeatedly hitting the key that runs the program. We will therefore refer to (8a) as **Repeated Substitution**, abbreviated **RS**.

If the iterates in (8a) converge, say to $\bar{x}$, then

$$\bar{x} = \lim_{k\to\infty} x_{k+1} = \lim_{k\to\infty} g(x_k) = g(\lim_{k\to\infty} x_k) = g(\bar{x}) \tag{8b}$$

The assumed continuity of g at $\bar{x}$ justifies bringing the limit through g. A **fixed point** of g is a solution $\bar{x}$ of the **fixed-point equation**

$$g(x) = x \qquad \text{(in words: } g \text{ maps } x \text{ into itself)} \tag{9}$$

The analysis in (8b) shows that *if the Repeated Substitution algorithm converges to $\bar{x}$, then $\bar{x}$ is a fixed point of $g(x)$*. The RS algorithm (8a) is sometimes called **Fixed-Point Iteration** because it is a numerical method for solving the fixed-point equation (9).

EXAMPLE 2.1B Show the following for the function

$$g_p(x) = \frac{1}{p}\left[(p - 1)x + \frac{78.8}{x}\right], \quad \text{where } p \neq 0 \text{ is held fixed} \tag{10}$$

fixed point: $g(x)=x \Rightarrow \frac{1}{p}[(p-1)x+\frac{78.8}{x}]=x$
$\Rightarrow (p-1)x^2+78.8=px^2 \Rightarrow 78.8=x^2$

(a) The only fixed points of $g_p(x)$ are $\bar{x} = +\sqrt{78.8}$ and $\bar{x} = -\sqrt{78.8}$ *regardless of the (fixed) value of the parameter p.*
(b) RS converges linearly to $\bar{x} = \sqrt{78.8}$ when $p = \frac{3}{2}$.
(c) RS converges quadratically to $\bar{x} = \sqrt{78.8}$ when $p = 2$.
(d) RS diverges linearly from $\bar{x} = \sqrt{78.8}$ when $p = \frac{1}{2}$.

SOLUTION (a): We want *all* solutions $\bar{x}$ of the fixed point equation

$$g_p(x) = x, \quad \text{or equivalently,} \quad (p - 1)x^2 + 78.8 = px^2 \tag{11}$$

This simplifies to $78.8 = x^2$, from which $\bar{x} = \pm\sqrt{78.8}$.

SOLUTION (b): Taking $p = \frac{3}{2}$ in (11) gives (check this)

$$g_{3/2}(x) = \frac{2}{3}\left(\frac{x}{2} + \frac{78.8}{x}\right) \tag{12}$$

Since $\bar{x} = \sqrt{78.8} \approx \sqrt{81}$, we will take $x_0 = 9$ as an initial guess. Then

$$x_1 = g_{3/2}(x_0) = \frac{2}{3}\left(\frac{9}{2} + \frac{78.8}{9}\right) \doteq 8.837037; \qquad \text{so } \Delta x_0 = x_1 - x_0 = -0.162963$$

$$x_2 = g_{3/2}(x_1) = \frac{2}{3}\left(\frac{x_1}{2} + \frac{78.8}{x_1}\right) \doteq 8.890356; \qquad \text{so } \Delta x_1 = x_2 - x_1 = 0.053319$$

Hence $\Delta x_2/\Delta x_1 = 0.053319/(-0.162936) \doteq -0.327$. The remaining values of x_k, $\Delta x_k = x_{k+1} - x_k$, and the linear convergence ratios $\Delta x_k/\Delta x_{k-1}$ shown in Table 2.1-3 are obtained similarly. Linear convergence is indicated by the fact that the leading significant digits in the Δx_k column lie roughly on a straight line, or alternatively by the fact that it takes about two iterations to get each additional accurate digit (see Table 2.1-2). It is confirmed by the fact that the linear convergence ratio remains about -0.333 for all k so that $C_L \approx -\frac{1}{3}$. Notice that the *quadratic* convergence ratios are *not* nearly constant; instead, they increase in magnitude as x_k approaches $\bar{x}$.

TABLE 2.1-3 Linear Convergence to $\bar{x} = \sqrt{78.8}$ when $p = \frac{3}{2}$

k	$x_k = g_{3/2}(x_{k-1})$	$\Delta x_k = x_{k+1} - x_k$	$\dfrac{\Delta x_k}{\Delta x_{k-1}}$	$\dfrac{\Delta x_k}{(\Delta x_{k-1})^2}$
0	9.0	−0.162963		
1	8.837037	0.053319	−0.327	2.00
2	8.890356	−0.017880	−0.335	− 6.29
3	8.872477	0.005984	−0.333	18.7
4	8.878425	−0.001984	−0.334	− 55.4
5	8.876441	0.000661	−0.333	155.
6	8.877102			

Notice that $|\Delta x_5| = |x_6 - x_5| = 0.000661$ is just about $0.5 \cdot 10^{-3}$. This suggests that x_5 (*not* x_6) is accurate to almost $3d$ [see (3)], hence that $x_6 = 8.877102$ should be accurate to $3d$. That this reasoning is correct can be seen from the fact that $\bar{x} = \sqrt{78.8} = 8.87693641$ $(9s)$.

SOLUTION (c): Taking $p = 2$ in (10) gives (check this)

$$g_2(x) = \frac{1}{2}\left(x + \frac{78.8}{x}\right) \tag{13}$$

In anticipation of rapid convergence, let us deliberately take a poor (i.e., inaccurate) initial guess, say $x_0 = 14$. Then

$$x_1 = g_2(x_0) = \frac{1}{2}\left(14 + \frac{78.8}{14}\right) \doteq 9.81428571; \qquad \text{so } \Delta x_0 = x_1 - x_0 \doteq -4.185714$$

and so on, as in part (b). The values of x_k, $\Delta x_k = x_{k+1} - x_k$, and the *quadratic* convergence ratio $\Delta x_k/(\Delta x_{k-1})^2$ are shown in Table 2.1-4. Quadratic convergence is indicated by the fact that once $|\Delta x_k|$ becomes less than 1, the leading significant digit of Δx_k moves about twice as far from the decimal point with each iteration (see Table 2.1-2); it is confirmed by the fact that for these k's, the quadratic convergence ratio remains at about -0.056 so that $C_Q \approx -0.056$. Notice that the *linear* convergence ratios are *not* nearly constant; they decrease to zero as k increases.

TABLE 2.1-4 Quadratic Convergence to $\bar{x} = \sqrt{78.8}$ when $p = 2$

k	$x_k = g_2(x_{k-1})$	$\Delta x_k = x_{k+1} - x_k$	$\dfrac{\Delta x_k}{\Delta x_{k-1}}$	$\dfrac{\Delta x_k}{(\Delta x_{k-1})^2}$
0	14.	−4.18571429		
1	9.81428571	−0.89258682	0.213	−0.051
2	8.92169890	−0.04465020	0.050	−0.056
3	8.87704870	−0.00011229	0.003	−0.056
4	8.87693641			

The fact that $|\Delta x_3| \approx 10^{-4}$ suggests that x_3 may have a small error in the fourth decimal place. In view of the quadratic convergence, we can presume that $x_4 = 8.87693641$ may have a small error in the eighth decimal place. In fact, x_4 is accurate to all nine digits shown! The linear convergence in Table 2.1-3 would not achieve this accuracy until $k \approx 16$ (why?) even though it started with a much more accurate x_0!

SOLUTION (d): When $p = \frac{1}{2}$, $g_{1/2}(x) = 2(-x/2 + 78.8/x)$. The values of x_k, $\Delta x_k = x_{k+1} - x_k$, and $\Delta x_k/\Delta x_{k-1}$, starting with $x_0 = 9$, are shown in Table 2.1-5. Clearly, the RS algorithm is diverging; and for the x_k's near $\bar{x}$, the divergence is linear, with $C_L \approx 3$.

TABLE 2.1-5 Linear Divergence from $\bar{x} = \sqrt{78.8}$ when $p = \frac{1}{2}$

k	$x_k = g_{1/2}(x_{k-1})$	$\Delta x_k = x_{k+1} - x_k$	$\Delta x_k/\Delta x_{k-1}$
0	9.0	−0.48889	
1	8.51111	1.49475	−3.057
2	10.00586	−4.26095	−2.851
3	5.74491	15.94316	−3.742
4	21.68807	−36.10948	−2.265
5	−14.42104		

■

2.1C Finding Initial Guesses for RS Graphically

In the preceding example, we were able to solve the fixed-point equation $g(x) = x$ algebraically and then use our knowledge of the exact $\bar{x}$ to get an initial guess x_0. In a realistic situation, when a numerical method is needed because we *do not* know $\bar{x}$ exactly, the context that gives rise to the problem generally provides a rough idea of what $\bar{x}$ should be, and this understanding can be used to get x_0. If the desired $\bar{x}$'s are fixed points, say of $g(x)$, and initial guesses are not available, we can get them by solving $g(x) = x$ *graphically*, that is, as the x-coordinates of the points $\bar{P}$ where the graph of g [i.e., the curve $y = g(x)$] meets the $y = x$ line (i.e., the 45° line through the origin). The next example illustrates this.

EXAMPLE 2.1C Use Repeated Substitution to find to $4d$ all fixed points of

$$g(x) = \tfrac{1}{2}e^{x/2} \qquad \textbf{(14a)}$$

SOLUTION: The graph of g, sketched in Figure 2.1-2, meets the $y = x$ line at two points, $\bar{P}_1$ and $\bar{P}_2$, whose x-coordinates are

$$\bar{x}_1 \approx 0.7 \qquad \text{and} \qquad \bar{x}_2 \approx 4.3 \qquad \textbf{(14b)}$$

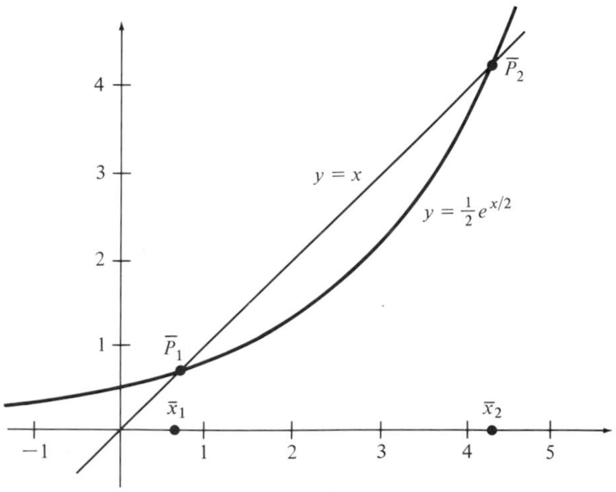

Figure 2.1-2 Estimating fixed points of $g(x) = \frac{1}{2}e^{x/2}$ graphically.

These graphically obtained initial guesses are used to get the values shown in Table 2.1-6. From the Δx_k column of Table 2.1-6(a), we see that, starting with $x_0 = 0.7$,

$$\text{RS converges linearly (with } C_L \approx 0.357\text{) to } \bar{x}_1 \approx x_6 \doteq 0.7148 \qquad \textbf{(14c)}$$

The iteration was stopped when $|\Delta x_5| = |x_6 - x_5| = 0.55 \cdot 10^{-4}$, indicating that x_5 is accurate to just about $4d$; hence x_6 should be accurate to $4d$. In fact it is, because $\bar{x}_1 = 0.714806$ ($6s$).

TABLE 2.1-6 Repeated Substitution for $g(x) = \frac{1}{2}e^{x/2}$

	(a) Seeking $\bar{x}_1 \approx 0.7$		(b) Seeking $\bar{x} \approx 4.3$	
k	$x_k = g(x_{k-1})$	$\Delta x_k = x_{k+1} - x_k$	$x_k = g(x_{k-1})$	$\Delta x_k = x_{k+1} - x_k$
0	0.7	0.009534	4.3	−0.007571
1	0.709534	0.003390	4.292429	−0.016218
2	0.712924	0.001210	4.276211	−0.034535
3	0.714134	0.000432	4.241676	−0.072615
4	0.714566	0.000154	4.169061	−0.148653
5	0.714720	0.000055	4.020408	−0.287988
6	0.714775		3.732420	

On the other hand, the increasing values of $|\Delta x_k|$ in Table 2.1-6(b) indicate that the x_k's starting with $x_0 = 4.3$ are *diverging linearly* (with $C_L \approx 2.1$). In fact, you should convince yourself that

$$\text{RS will diverge linearly from } \bar{x}_2 \doteq 4.3065847 \text{ for any } x_0 \approx \bar{x}_2 \qquad \textbf{(14d)}$$

as long as the digital device used can distinguish x_0 from $\bar{x}_2$. We will find $\bar{x}_2$ by other methods in Examples 2.1F and 2.3D. ■

2.1D Visualizing the RS Algorithm Graphically

We have seen that the RS algorithm can converge or diverge linearly (with either a positive or negative C_L) or it can converge quadratically. To visualize these cases geometrically, see Figure 2.1-3, where the kth iterates, x_k, are shown on the x-axis. The perpendicular at x_k crosses the graph of g at the point $P_k(x_k, y_k)$, where $y_k = g(x_k)$, which is x_{k+1}. To "reflect" x_{k+1} to the x-axis where it belongs, move horizontally (right or left) from P_k to the point Q_{k+1} on the $y = x$ line, then vertically (up or down) to x_{k+1} on the x-axis. Thus one iteration of the RS algorithm can be performed graphically as follows:

$$x_k \xrightarrow[\text{down to graph}]{\text{move up or}} P_k \xrightarrow[\text{left to } y\,=\,x \text{ line}]{\text{move right or}} Q_{k+1} \xrightarrow[\text{down to } x\text{-axis}]{\text{move up or}} x_{k+1} \qquad \textbf{(15)}$$

$$\sigma_u^2 = \frac{u'u + tr}{q}$$ $$\sigma_u^2 = f(\sigma_u^2)$$ fixed point equation

$$\frac{u'u + tr}{q}$$

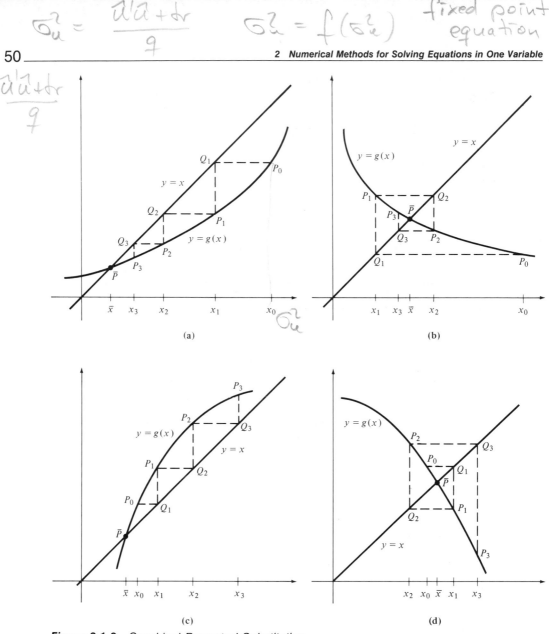

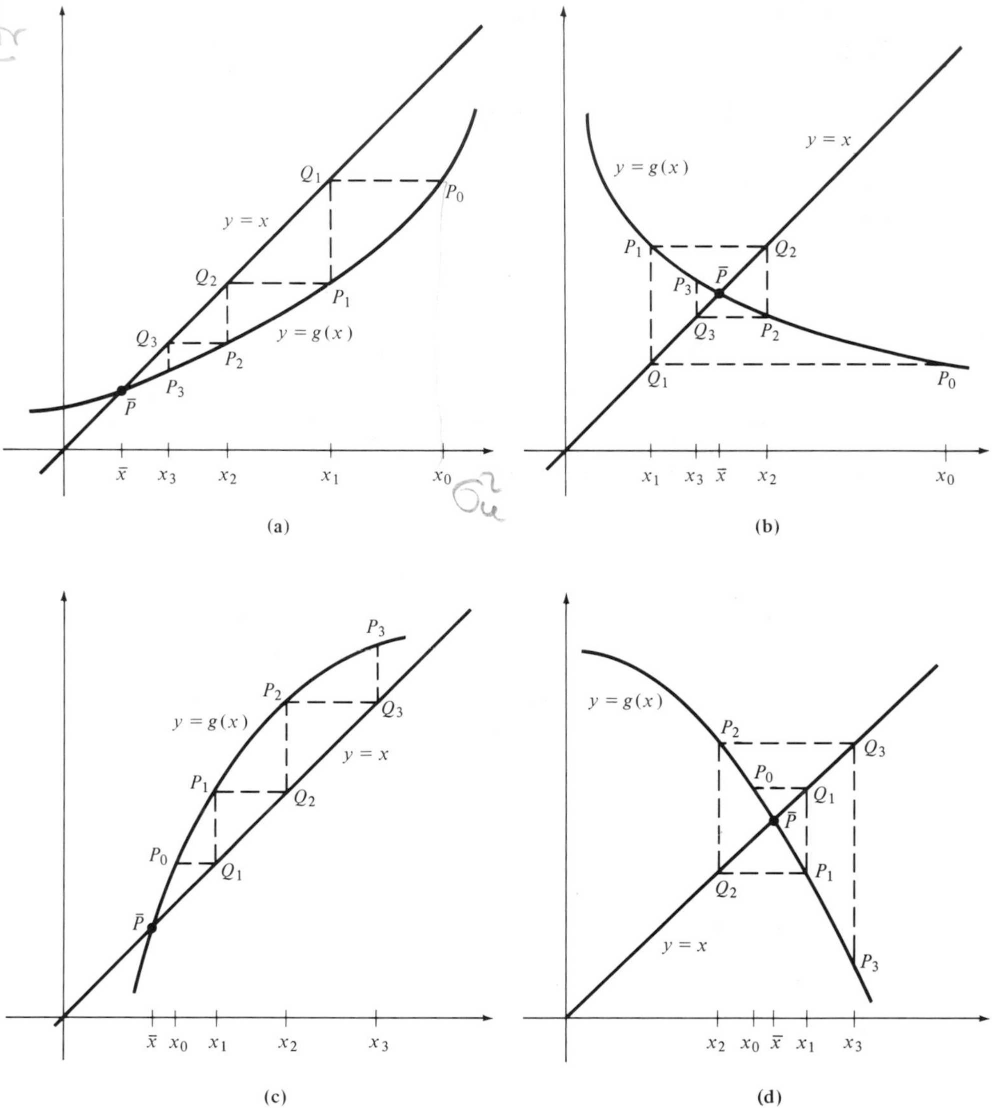

Figure 2.1-3 *Graphical Repeated Substitution.*

We see from Figure 2.1-3 that the linear convergence or divergence is monotonic if $g'(\bar{x})$ is positive, and alternating if $g'(\bar{x})$ is negative. Note how Figure 2.1-3(a) explains the results of Example 2.1C.

2.1E   When Should the RS Algorithm Converge? Theorem RS

The following theorem shows that the rate at which the RS algorithm will converge to (or diverge from) a fixed point $\bar{x}$ depends on the number $g'(\bar{x})$, that is, the slope with which the graph of g crosses the $y = x$ line. It is proved in Section 2.1F.

THEOREM RS. CONVERGENCE RATE OF REPEATED SUBSTITUTION Let $\bar{x}$ be a fixed point of a function g that has a continuous third derivative g''' near $\bar{x}$, and suppose that $x_{k+1} = g(x_k)$ for $k = 0, 1, \ldots$.

(a) If $g'(\bar{x}) \neq 0$, then $\Delta x_{k+1} \approx g'(\bar{x})\, \Delta x_k$ for $x_k \approx \bar{x}$; consequently,

$$x_k \text{ converges linearly to } \bar{x} \text{ [with } C_L = g'(\bar{x})] \text{ if } 0 < |g'(\bar{x})| < 1$$

$$x_k \text{ diverges linearly from } \bar{x} \text{ [with } C_L = g'(\bar{x})] \text{ if } |g'(\bar{x})| > 1$$

(b) If $g'(\bar{x}) = 0$, then $\Delta x_{k+1} \approx -\frac{1}{2}g''(\bar{x})\, \Delta x_k$ for $x_k \approx \bar{x}$; that is,

$$x_k \text{ converges quadratically to } \bar{x}, \text{ with } C_Q = -\tfrac{1}{2}g''(\bar{x})$$

RS may either converge (very slowly) or diverge when $g'(\bar{x}) = \pm 1$.

Theorem RS predicts the outcome of Repeated Substitution without performing any iterations! For example, if $g_p(x) = [(p - 1)x + 78.8/x]/p$ and

$$\bar{x} = \sqrt{78.8}, \qquad \text{then } g_p'(\bar{x}) = \frac{1}{p}\left[(p - 1) - \frac{78.8}{\bar{x}^2}\right] = \frac{p - 2}{p} \qquad \textbf{(16a)}$$

for *any* fixed nonzero p. In view of Theorem RS, the three values

$$g_{3/2}'(\bar{x}) = -\tfrac{1}{3}, \quad g_2'(\bar{x}) = 0, \quad \text{and} \quad g_{1/2}'(\bar{x}) = -3 \qquad \textbf{(16b)}$$

could have been used to predict the observed convergence or divergence seen in (b)–(d) of Example 2.1B. Similarly, although we only had the *estimates* $\bar{x}_1 \approx 0.7$ and $\bar{x}_2 \approx 4.3$ in Example 2.1C, the linear convergence *to* $\bar{x}_1$ and divergence *from* $\bar{x}_2$ could have been predicted from the fact that

$$|g'(0.7)| \doteq |0.3548| < 1 \qquad \text{and} \qquad |g'(4.3)| \doteq |2.146| > 1 \qquad \textbf{(17)}$$

Practical Consideration 2.1E. Since the divergence condition, $|g'(\bar{x})| > 1$, arises frequently, the RS algorithm is *not very robust*. It is also *not likely to converge rapidly* (i.e., quadratically) when it does converge because the graph of g is not likely to cross the $y = x$ line with zero slope. In summary, *Repeated Substitution is not an efficient algorithm for finding fixed points* $\bar{x}$. A remedy for its shortcomings is given next; a more efficient algorithm is given in Section 2.3D. ■

2.1F Accelerating Linear Convergence: Aitken's Formula

If an algorithm converges *exactly* linearly, its iterates x_k can be generated without the algorithm. For example, in Example 2.1A,

$$x_0 = 2, \quad x_1 = 2.1, \quad x_2 = 2.11, \quad x_3 = 2.111, \quad x_4 = 2.1111, \quad \ldots \qquad \textbf{(18)}$$

we have X_k, X_{k-1}, X_{k-2}, Δ_{k-1}, Δ_{k-2}

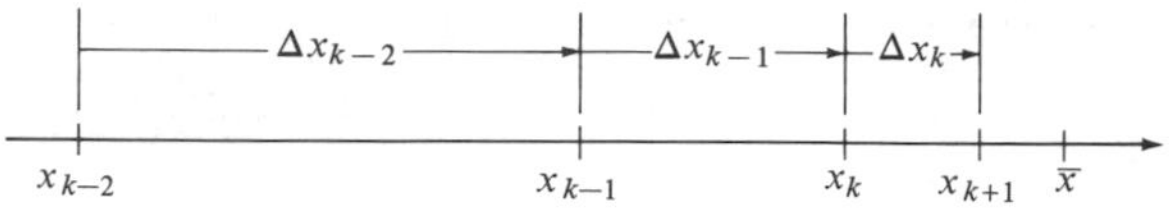

Figure 2.1-4 *Relating x_i's and Δx_i's graphically.*

Clearly, x_5 will be 2.11111, x_6 will be 2.111111, ..., and x_k will converge to $\bar{x} = 2\frac{1}{9}$ as $k \to \infty$. We now show how to do this "extrapolation to the limit" for *any* linearly convergent algorithm.

Suppose that x_i converges *exactly* linearly to $\bar{x}$, starting with x_{k-2}. Then (see Figure 2.1-4) the error of x_{k-2} can be expressed as

$$\bar{x} - x_{k-2} = \Delta x_{k-2} + \Delta x_{k-1} + \Delta x_k + \Delta x_{k+1} + \cdots \quad \textbf{(19a)}$$

$$= \Delta x_{k-2} + C_L \Delta x_{k-2} + C_L^2\,\Delta x_{k-2} + C_L^3\,\Delta x_{k-2} + \cdots \quad \textbf{(19b)}$$

where $|C_L| < 1$ = the radius of convergence of the **geometric series** (see Problem M1-6). So

$$\bar{x} = x_{k-2} + \Delta x_{k-2}(1 + C_L + C_L^2 + C_L^3 + \cdots) = x_{k-2} + \Delta x_{k-2}\frac{1}{1 - C_L} \quad \textbf{(19c)}$$

Substituting $\Delta x_{k-2}/\Delta x_{k-1}$ for C_L, and $x_k - (\Delta x_{k-2} + \Delta x_{k-1})$ for x_{k-2} and rearranging on the right-hand side of (19c) gives the following important result:

formula improving linear convergence

AITKEN'S IMPROVEMENT FORMULA. If linear convergence is indicated by the fact that

$$\frac{\Delta x_{k-1}}{\Delta x_{k-2}} \approx \frac{\Delta x_{k-2}}{\Delta x_{k-3}} \qquad \text{(say, to about 3}s\text{)} \quad \textbf{(20a)}$$

then the $\bar{x}$ to which x_{k-2}, x_{k-1}, and x_k appear to be converging is

$$\bar{x} \approx (x_k)_{\text{improved}} = x_k - \frac{(\Delta x_{k-1})^2}{\Delta x_{k-1} - \Delta x_{k-2}} = x_k - \frac{(x_k - x_{k-1})^2}{x_k - 2x_{k-1} + x_{k-2}} \quad \textbf{(20b)}$$

Formula (20b) is called **Aitken's improvement formula** or **Aitken's Δ^2-process**. It shows how to use x_{k-2} and x_{k-1} to improve the approximation x_k. In the absence of roundoff error, (20a) generally gets closer to equality as x_k gets closer to x; hence k in (20b) should be the index of the closest iterate x_k. If the x_k's are *diverging* linearly from x, then x_0 will be the closest iterate to $\bar{x}$; in this case, we can use Aitken's formula (19c) with $k = 2$ in the following form:

LINEAR DIVERGENCE FORM: $$(x_0)_{\text{improved}} = x_0 - \frac{(x_0 - x_1)^2}{x_0 - 2x_1 + x_2} \quad \textbf{(21)}$$

This formula should be viewed as showing how to use x_1 and x_2 to improve x_0. In using either (20b) or (21), keep in mind that *the more nearly constant the linear convergence ratios are in (20a), the closer $(x_k)_{\text{improved}}$ will be to $\bar{x}$.*

EXAMPLE 2.1F Improve the most accurate x_k resulting from the RS algorithm.

(a) From Example 2.1B(b): $x_4 = 8.78425$, $x_5 = 8.76441$, $x_6 = 8.77102$ $(k = 6)$
(b) From Example 2.1C(a): $x_4 = 0.714566$, $x_5 = 0.714720$, $x_6 = 0.714775$ $(k = 6)$
(c) From Example 2.1C(b): $x_0 = 4.3$, $x_1 = 4.292429$, $x_2 = 4.276211$ $(k = 2)$

SOLUTION (a): These iterates are converging linearly to $\bar{x} = \sqrt{78.8}$, so

$$(x_6)_{\text{improved}} = x_6 - \frac{(x_6 - x_5)^2}{x_6 - 2x_5 + x_4} = 8.76937 \qquad \textbf{(22)}$$

Since $\bar{x} = \sqrt{78.8} = 8.769364$ $(7s)$, the formula improved x_6 (which had only $3s$ accuracy) to just about $6s$ accuracy! This is as much accuracy as one could expect because x_4, x_5, and x_6 were themselves rounded to $6s$.

SOLUTION (b): These iterates are converging linearly to $\bar{x}_1 \doteq 0.714806$, and $(x_6)_{\text{improved}} = 0.714806$, obtained as in (22), is accurate to all digits shown.

SOLUTION (c): These iterates are *diverging* linearly *from* $\bar{x}_2 \doteq 4.3065847$. So we can use (22) to improve x_0 as follows:

$$(x_0)_{\text{improved}} = 4.3 - \frac{(4.3 - 4.292429)^2}{4.3 - 2(4.292429) + 4.276211} \doteq 4.306629 \qquad \textbf{(23a)}$$

If we apply RS with $(x_0)_{\text{improved}} = 4.306629$ as x_0 to $g(x)$ (which is $0.5e^{x/2}$ for this problem) and then reuse (21), we get

$$x_1 \doteq 4.3066801 \quad \text{and} \quad x_2 \doteq 4.3067900, \qquad \text{hence } (x_0)_{\text{improved}} \doteq 4.3065847 \qquad \textbf{(23b)}$$

This is accurate to all eight digits carried in the calculation! ■

Steffensen's method for accelerating any *linearly* convergent *or* divergent algorithm consists of forming $(x_2)_{\text{improved}}$ [using (20b) or (21)] as soon as three successive iterates are available. It will result in *quadratic convergence* when applied to *either* linear convergence *or* linear divergence [12]. So it remedies the major shortcomings of Repeated Substitution when used to accelerate it. Pseudocode for the **Modified Steffensen's method** is shown in Algorithm 2.1F. What is "modified" about it is the fact that (20b) is not used *unless* either $CL > 1$ (i.e., divergence) or two successive linear convergence ratios (CL and $CLprev$) agree to within Pct percent.

ALGORITHM 2.1F. MODIFIED STEFFENSEN'S METHOD (SUBPROGRAM)

PURPOSE: To accelerate *any* linearly convergent *or* divergent algorithm. This subprogram should be called each iteration of the algorithm it is accelerating, *after* *X* is obtained as *Xprev* + *DeltaX*.

```
GET X, DeltaX, DXprev,                              {previous iteration's DeltaX}
      Pct, CLprev                 {previous iteration's linear convergence ratio}
IF DeltaX = 0 THEN RETURN                      {X = Xprev; termination should occur}
CL ← DeltaX/DXprev                                {current linear convergence ratio}
IF UsdAit = TRUE                          {three successive iterates are not available}
|  THEN UsdAit ← FALSE                          {next iteration will be OK for Aitken}
|  ELSE          {test for ≈ linear convergence/divergence (to within Pct%)}
|  |  IF (CL ≠ 1 AND |CL − CLprev| ≤ 0.01Pct|CL|) THEN                {test OK}
|  |  |  X ← X − (DeltaX)²/(DeltaX − DXprev)                      {Aitken's formula}
|__|__|__ UsdAit ← TRUE                                   {set "Used Aitken" flag}
CLprev ← CL;  DXprev ← DeltaX                            {prepare for next iteration}
RETURN
```

2.1G Proof of Theorem RS (Section 2.1E)

Write x_k as $\bar{x} - \epsilon_k$. Then Taylor's Remainder Theorem with $n = 2$ (see Appendix II.2D) can be used to write $g(x_k)$ as

$$g(x_k) = g(\bar{x} - \epsilon_k) = g(\bar{x}) - g'(\bar{x})\epsilon_k + \tfrac{1}{2}g''(\bar{x})\epsilon_k^2 - \tfrac{1}{6}g'''(\xi)\epsilon_k^3 \tag{24}$$

where ξ lies between x_k and $\bar{x}$. But $g(x_k) = x_{k+1}$ and $g(\bar{x}) = \bar{x}$, hence

$$\epsilon_{k+1} = \bar{x} - x_{k+1} = g'(\bar{x})\epsilon_k - \tfrac{1}{2}g''(\bar{x})\epsilon_k^2 + \tfrac{1}{6}g'''(\xi)\epsilon_k^3 \tag{25}$$

When x_k gets sufficiently close to $\bar{x}$, $|\epsilon_k| >> |\epsilon_k|^2 >> |\epsilon_k|^3$ and so (by our continuity assumption on g''') $g'''(\xi)$ gets close to $g'''(\bar{x})$; hence

$$\frac{\epsilon_{k+1}}{\epsilon_k} \approx g'(\bar{x}) \text{ if } g'(\bar{x}) \neq 0 \quad \text{and} \quad \frac{\epsilon_{k+1}}{\epsilon_k^2} \approx -\frac{1}{2}g''(\bar{x}) \text{ if } g'(\bar{x}) = 0 \tag{26}$$

To complete the proof, we must replace ϵ_{k+1} by Δx_{k+1}, and ϵ_k by Δx_k in (26). We do this by proving a more general result.

CONVERGENCE LEMMA. Suppose that $x_k \to \bar{x}$ as $k \to \infty$.

Linear Convergence: If $\epsilon_{k+1}/\epsilon_k \to C_L \neq 1$, then $\Delta x_{k+1}/\Delta x_k \to C_L$.

Quadratic Convergence: If $\epsilon_{k+1}/\epsilon_k^2 \to C_Q$, then $\Delta x_{k+1}/(\Delta x_k)^2 \to C_Q$.

To prove the linear convergence part, write Δx_k as

$$\Delta x_k = (x_{k+1} - \bar{x}) + (\bar{x} - x_k) = \epsilon_k - \epsilon_{k+1} = \epsilon_k(1 - \epsilon_{k+1}/\epsilon_k) \qquad \textbf{(27a)}$$

Similarly, $\Delta x_{k+1} = \epsilon_{k+1}(1 - \epsilon_{k+2}/\epsilon_{k+1})$. So if $\epsilon_{k+1}/\epsilon_k \rightarrow C_L \neq 1$, then

$$\frac{\Delta x_{k+1}}{\Delta x_k} = \frac{\epsilon_{k+1}}{\epsilon_k} \cdot \frac{1 - \epsilon_{k+2}/\epsilon_{k+1}}{1 - \epsilon_{k+1}/\epsilon_k} \quad \rightarrow \quad C_L \frac{1 - C_L}{1 - C_L} = C_L \qquad \textbf{(27b)}$$

To prove the quadratic convergence part, assume that $\epsilon_{k+1}/\epsilon_k^2 \rightarrow C_Q$. Then

$$\frac{\epsilon_{k+1}}{\epsilon_k} = \frac{\epsilon_{k+1}}{\epsilon_k^2} \cdot \epsilon_k \quad \rightarrow \quad C_Q \cdot 0 = 0 \qquad \textbf{(28a)}$$

Similarly, $\epsilon_{k+2}/\epsilon_{k+1} \rightarrow 0$. So

$$\frac{\Delta x_{k+1}}{(\Delta x_k)^2} = \frac{\epsilon_{k+1}}{\epsilon_k^2} \cdot \frac{1 - \epsilon_{k+2}/\epsilon_{k+1}}{(1 - \epsilon_{k+1}/\epsilon_k)^2} \quad \rightarrow \quad C_Q \frac{1 - 0}{(1 - 0)^2} = C_Q \qquad \textbf{(28b)}$$

The Convergence Lemma, allows us to use Taylor series to analyze convergence rates using the mathematically convenient (*but unknown*) exact errors ϵ_k, and then apply the results using the (*known*) error estimates Δx_k.

2.2 Implementing an Iterative Algorithm on a Computer

2.2A Pseudoprogram for a General Iterative Algorithm

Although subscripts are useful for mathematical descriptions of iterative algorithms, they do not reflect the way such algorithms are actually executed (i.e., in a loop). A more realistic description of the way a general iterative algorithm for finding $\bar{x}$ proceeds is given in Algorithm 2.2A. Notice that *Xprev* and *X* (*without subscripts*) play the respective roles of x_{k-1} and x_k so that

DeltaX $= X - $ *Xprev* plays the role of $\Delta x_{k-1} = x_k - x_{k-1}$	**(1)**

Instead of incrementing the subscript of x by 1, one simply **updates** by making the current X the next *Xprev* before the **termination test**, and returning to the beginning of the loop if the test fails. The index k is used *only* to count iterations so that at most *MaxIt* iterations are performed.

ALGORITHM 2.2A. GENERAL ITERATIVE ALGORITHM

PURPOSE: To find $\bar{x}$ to a desired accuracy by repeatedly forming an improved approximation X from a current approximation *Xprev* until either X is sufficiently close to *Xprev* (**termination test**) or *MaxIt* iterations occur.

{**INITIALIZE**}
GET *MaxIt*, termination parameters, x_0　{an initial guess}
Xprev ← x_0

{**ITERATE**}
DO FOR $k = 1$ TO *MaxIt*
　| {***form* X**} X ← (formula or procedure involving *Xprev*)
　| *DeltaX* ← $X - Xprev$
　| OUTPUT(k, *Xprev*, *DeltaX*, X)　　　　　　　　　　　　{optional}
　| {**update**} *Xprev* ← X　　　　　　　{prepare for next iteration}
　|__ UNTIL $|DeltaX| \approx 0$ {**termination test**} is satisfied
IF **termination test** succeeded
　| THEN OUTPUT(X' approximates $\bar{x}$ to the desired accuracy.')
　|__ ELSE OUTPUT('Desired accuracy not evident in '*Maxit*' iterations.')
STOP

A pseudocode description of an algorithm will be referred to as a **pseudoprogram**. Pseudoprograms serve the same purpose as flowcharts but are *more compact* and can be *prepared more easily*; also, *the translation from pseudocode to an actual program is more natural than from a flowchart*. In particular, the **labels** (set in **boldface**) and indentation combine to give a quick "picture" of the algorithm and help the reader visualize its sub-program structure. We shall present algorithms as pseudoprograms, using the conventions given in Appendix I.

2.2B　Terminating Iterative Algorithms on a Computer

In view of (1) [and (3) of Section 2.1A], *DeltaX approximates the error of Xprev*, **not** X. Hence by the time $|DeltaX| < 0.5 \cdot 10^{-NumDec}$, X should approximate $\bar{x}$ to *at least NumDec decimal places*. Similarly, since $DeltaX/|X|$ approximates the relative error of *Xprev* (**not** X), X should approximate $\bar{x}$ to *at least NumSig significant digits* when $|DeltaX| < 0.5 \cdot 10^{-NumSig}|X|$. The following termination tests are based on the preceding remarks.

ABSOLUTE DIFFERENCE TEST (FOR $X \approx \bar{x}$ TO *NUMDEC* DECIMAL PLACES):

$$\text{IF } |DeltaX| \leq Const \cdot 10^{-NumDec} \text{ THEN stop the iteration.} \qquad \textbf{(2a)}$$

$\theta_k - \theta_{k-1} = \Delta_k = C \, 10^{-d}$ ： $d = $ no. of *significant decimal places*

RELATIVE DIFFERENCE TEST (FOR $X \approx \bar{x}$ TO *NUMSIG* SIGNIFICANT DIGITS):

$$\text{IF } |DeltaX| \leq Const \cdot 10^{-NumSig}|X| \text{ THEN stop the iteration.} \qquad \textbf{(2b)}$$

$$|\theta_k - \theta_{k-1}| \leq 10^{-n}|\theta_k|$$

In both (2a) and (2b), *Const* is generally taken as 1; it can be larger (say, around 9) when rapid convergence is expected, and should be smaller (say, around 0.5) when slow convergence is expected. Also, *NumSig* in (2b) should, if possible, be less than the precision of the device to allow at least one guard digit for accumulating propagated roundoff error in the ***form X*** step of Algorithm 2.2A.

If an algorithm that does not introduce much roundoff error in the ***form X*** step is implemented on an $8s$ device, then $\bar{x}$ near 0.0003 can be found to $11d$, whereas $\bar{x}$ near 30000 can only be found to $3d$. So *NumDec in (2a) must be changed each time* $|\bar{x}|$ *changes*. This can be a nuisance if $\bar{x}$'s of either unknown or varying magnitude are to be found on a single computer run. Consequently, *decimal-place accuracy is not a convenient criterion for terminating an iterative algorithm on a computer*. On the other hand, if we set $NumSig = 8$ and $Const = 2.0$ (to allow for some roundoff) in (2b), then the iteration will stop when the maximum (realistic) device accuracy has been achieved *regardless of the size of* $|\bar{x}|$, provided that $\bar{x} \neq 0$. This is why *we will use the Relative Difference Test (2b) to terminate all iterative algorithms*.

Practical Consideration 2.2B (What if $\bar{x} = 0$?). *The Relative Difference Test should not be used when $\bar{x} = 0$.* Indeed, if $X \to 0$ rapidly, then, in the absence of roundoff error, we will have

$$|X| << |Xprev| \qquad \text{hence } |DeltaX| = |X - Xprev| \approx |Xprev| \qquad \textbf{(3)}$$

So $|DeltaX| \leq Const \cdot 10^{-NumSig}|X|$ will not be satisfied unless both X and $Xprev$ are calculated as zero (which may never happen).

It is usually possible to simply "plug in" $\bar{x} = 0$ to see if zero is a desired $\bar{x}$ (root, fixed point, etc.). So automated iterative algorithms need only be used to find *nonzero* $\bar{x}$'s, for which there is no risk programming the Relative Difference Test. ■

Some final words about termination tests. There is a kind of "Murphy's law" which says that computers will get into "endless loops" in ways human beings could never dream of; and they will do so on the first run after you are finished debugging! The only reliable safeguard against this is the following:

> For any iterative algorithm, no matter how sure you are that a termination test will be met, put an upper limit (e.g., *MaxIt*) on the number of iterations (just in case you are wrong).

It is also important not to make unrealistic accuracy demands on a digital device. One must be prepared to accept less than the device accuracy when the amount of "number crunching" per iteration is large.

$$\Delta_k = \theta_k - \theta_{k-1} \approx \epsilon_{k-1}$$

2.2C   A `subprogram` for the Repeated Substitution Algorithm

The General Iterative Algorithm 2.1A becomes a pseudoprogram for Repeated Substitution by simply taking

$$X \leftarrow g(Xprev) \tag{4}$$

as the *form X* step. Rather than translate the entire pseudoprogram into a single program, we shall implement only the ITERATE phase in a Fortran 77 **subprogram**

$$\text{SUBROUTINE FIXPT(G, X, NUMSIG, MAXIT, PRNT, IW, CONVGD)} \tag{5}$$

(Figure 2.2-1). This subprogram was easy to write because it uses the "neutral" mathematical notation *X*, *Xprev*, and *DeltaX* used in Algorithm 2.2A. The variables appearing in the parentheses in (5) are the subprogram's "dummy" **parameters**†. In Fortran 77, they

```
ØØ1ØØ          SUBROUTINE FIXPT (G, X, NUMSIG, MAXIT, PRNT, IW, CONVGD)
ØØ2ØØ     C - - - - - - - - - - - - - - - - - - - - - - - - - - - - - C
ØØ3ØØ     C   This subroutine uses Repeated Substitution to solve      C
ØØ4ØØ     C             g(x) = x   (Fixed Point Equation)              C
ØØ5ØØ     C   If PRNT = TRUE, it prints the iteration on device IW.    C
ØØ6ØØ     C - - - - - - - - - - - - - -  VERSION 2  9/9/85  - - - - C
ØØ7ØØ          LOGICAL  PRNT, CONVGD
ØØ8ØØ          DATA CONST  /1.Ø/
ØØ9ØØ          RELTOL = CONST * 1Ø.**(-NUMSIG)
Ø1ØØØ          CONVGD = .FALSE.
Ø11ØØ          DO 1Ø K=1,MAXIT            ! ITERATE
Ø12ØØ             XPREV = X               ! update X
Ø13ØØ             X = G(X)                ! Form X = xk
Ø14ØØ             DELTAX = X - XPREV
Ø15ØØ             IF (PRNT) WRITE (IW,1) K, X, DELTAX
Ø16ØØ     C       Termination Test (Relative Difference Test)
Ø17ØØ             IF ( ABS(DELTAX) .LE. RELTOL*ABS(X) ) THEN
Ø18ØØ                CONVGD = .TRUE.
Ø19ØØ                IF (PRNT) WRITE (IW, 2) X, NUMSIG
Ø2ØØØ                RETURN
Ø21ØØ             ENDIF
Ø22ØØ     1Ø CONTINUE
Ø23ØØ          IF (PRNT) WRITE (IW, 3) NUMSIG
Ø24ØØ          RETURN
Ø25ØØ     1 FORMAT(' Iterate',I3,'  = ',G12.6,'  Delta  = ',E11.3)
Ø26ØØ     2 FORMAT('Ø',G15.6,' seems accurate to',I3,' digits.')
Ø27ØØ     3 FORMAT('Ø',I4,' significant digit accuracy not apparent.')
Ø28ØØ          END
```

Figure 2.2-1 *Fortran* SUBROUTINE *for Repeated Substitution.*

†The GOSUB of standard BASIC does not have the ability to pass "dummy" parameters. Many mini- and microcomputers have enhanced versions of BASIC with this capability. Unfortunately, however, these enhanced versions are generally not transportable.

can be REAL (here X), INTEGER (here MAXIT, IW), LOGICAL (here PRNT, CONVGD), COMPLEX, DOUBLE PRECISION, CHARACTER, dimensioned arrays of the foregoing types, or the name of an external subprogram (here G). The user of this (or any subprogram) must provide a **calling program** that gets the values of all the *input* parameters (here X, NUMSIG, MAXIT, PRNT, and IW). The subprogram then reports the results either by altering the values of the input parameters (here X), or by using different parameters (here CONVGD). The name of the FUNCTION subprogram used for G must be declared EXTERNAL in the calling program so that a Fortran 77 compiler will not treat it as the name of a REAL variable. Passing names of *external* subprograms makes the SUBROUTINE more flexible, and also enables the user to change the FUNCTION used *without having to edit the* SUBROUTINE.

The amount of output generated by SUBROUTINE FIXPT can be controlled *from the calling program* by setting the LOGICAL flag PRNT to .TRUE. for printed output, and .FALSE. otherwise. In the latter case, the *returned* flag CNVRGD indicates whether or not the desired accuracy was achieved in MAXIT iterations. The use of such **boolean flags** makes the subprogram flexible and allows the user to exploit this flexibility *without having to edit the subprogram*. This eliminates the risk of accidentally changing a working subprogram! Indeed, once a subprogram such as FIXPT is de-bugged, it can be stored (preferably compiled) and used with the confidence that *it will be the same every time it is called.*

An important benefit of the modularization just described is that the calling program and the external FUNCTION subprogams *for a particular application* can (and should) be written in terms of *variables that are meaningful for the application*. This enables the user, familiar with the application, to write this code quickly and accurately, as illustrated in the following example.

EXAMPLE 2.2C The steady-state temperature T of the interior of a material with embedded heat sources is known to satisfy the equation

$$e^{T/2} = \cosh(\sqrt{0.5L_{cr}}\, e^{T/2}) \qquad [\text{Reminder: } \cosh u = \tfrac{1}{2}(e^u + e^{-u})] \tag{6}$$

Find to 4*s* all possible temperatures T when $L_{cr} = 0.094$.

Solution: To simplify the analysis, we first make the substitution

$$u = e^{T/2} \qquad \text{so that} \qquad T = 2(\ln u) \tag{7a}$$

If we also set $L_{cr} = 0.094$, then (6) becomes the fixed-point equation

$$u = g_T(u), \qquad \text{where } g_T(u) = 0.5(e^{\sqrt{0.047}u} + e^{-\sqrt{0.047}u}) \tag{7b}$$

A plot of the curves $y = g_T(u)$ and $y = u$ on the same set of axes shows that $g_T(u)$ has two fixed points,

$$\bar{u}_1 \approx 1 \qquad \text{and} \qquad \bar{u}_2 \approx 16 \tag{8}$$

So (check this!) $|g_T'(\bar{u}_1)| < 1$, whereas $|g_T'(\bar{u}_2)| > 1$. In view of Theorem RS (Section 2.1D), SUBROUTINE FIXPT should be able to find $\bar{u}_1$ but not $\bar{u}_2$.

```
ØØ1ØØ   C  Calling program to use SUBROUTINE FIXPT for Example 2.2C
ØØ2ØØ         LOGICAL CONVGD
ØØ3ØØ         EXTERNAL GT
ØØ4ØØ         DATA IW /5/,  UØ /1.Ø/      ! first initial guess
ØØ5ØØ      1Ø U = UØ
ØØ6ØØ         CALL FIXPT (GT, U, 6, 12, .FALSE., IW, CONVGD)
ØØ7ØØ         IF (CONVGD) THEN
ØØ8ØØ            T = 2.Ø * ALOG(U)        ! steady state temperature
ØØ9ØØ            WRITE (IW,1) UØ, U, T  ! report convergence
Ø1ØØØ         ELSE
Ø11ØØ            WRITE (IW,2) UØ          ! report failure to converge
Ø12ØØ         ENDIF
Ø13ØØ         IF ( UØ .NE. 1.Ø ) STOP    ! ELSE
Ø14ØØ            UØ = 15.9                ! second initial guess
Ø15ØØ            GOTO 1Ø
Ø16ØØ       1 FORMAT('ØFixed pt near',F5.1,' is',F1Ø.5,' (6s).   T =',F1Ø.6)
Ø17ØØ       2 FORMAT('ØRS can''t find a fixed point near u =',F5.1)
Ø18ØØ         END
Ø19ØØ
Ø2ØØØ         FUNCTION GT(U)              ! for Example 2.2C
Ø21ØØ         X = SQRT(Ø.Ø47) * U
Ø22ØØ         GT = ( EXP(X) + EXP(-X) ) / 2.Ø   ! = cosh X
Ø23ØØ         RETURN
Ø24ØØ         END


       Fixed pt near  1.Ø is   1.Ø2478 (6s).  T =  Ø.Ø48957


       Fixed pt near 15.9 is   1.Ø2478 (6s).  T =  Ø.Ø48957
```

Figure 2.2-2 *User-provided code for* SUBROUTINE FIXPT.

Figure 2.2-2 shows a FUNCTION GT(U) and a calling program for finding $\bar{u}_1$ and $\bar{u}_2$ using SUBROUTINE FIXPT of Figure 2.2-1. Both programs were easy to write because the variables used are those of the problem at hand, namely (7). Note that the fixed points of $g_T(u)$, because they are an *intermediate* step, were found to 6s (two guard digits). As expected, SUBROUTINE FIXPT found $\bar{u}_1 \doteq 1.02478$ to get

$$T_1 = 2(\ln \bar{u}_1) = 0.04896 \quad (4s) \tag{9}$$

But it was unable to find $\bar{u}_2 \doteq 15.9743$ (6s) to get $T_2 = 5.542$ (4s). We shall find $\bar{u}_2$ by other methods in Sections 2.3D and 2.4E. ∎

Practical Consideration 2.2C (Making SUBROUTINE FIXPT **more reliable).** The preceding discussion of SUBROUTINE FIXPT was intended to illustrate the desirability of implementing a numerical method as a general-purpose subprogram. However, since the RS algorithm is not very efficient (see Section 2.1D) SUBROUTINE FIXPT *is not recommended as software for finding fixed points as it appears in Figure 2.2-1*. Problem C2-2 describes how to make it more efficient by calling a subprogram for the Modified Steffensen's method (Algorithm 2.1F) in the loop. ∎

2.3

Numerical Methods for Solving f(x) = 0

In this section we develop methods for solving the equation

$$f(x) = 0 \qquad \text{(the \textbf{root-finding equation})} \tag{1}$$

solutions $\bar{x}$ of (1) will be called **roots** of the equation $f(x) = 0$, or simply **roots of $f(x)$**. Geometrically, solutions of $f(x) = 0$ are the x-intercepts of the graph of f [Figure 2.3-1(a)]. The most general equation in the variable x can be expressed as

$$g(x) = h(x) \tag{2}$$

geometrically, solutions $\bar{x}$ of (2) are the x-coordinates of points $\bar{p}$ where the graphs of g and h intersect [Figure 2.3-1(b)]. The **fixed-point equation** $g(x) = x$ and the **inverse function equation** $g(x) = c$ (where c is a value in the range of g) are special cases of (2).

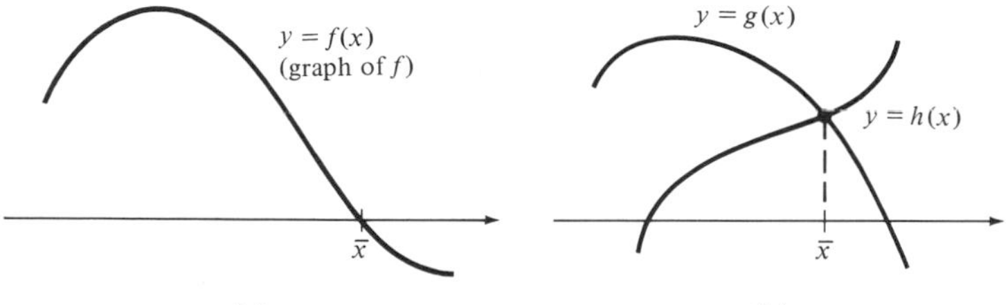

Figure 2.3-1 *(a) Roots of f(x) = 0. (b) Solutions of g(x) = h(x).*

2.3A Using Root-Finding Methods to Solve Equations

Two equations are **equivalent** if they have identical roots. A numerical method for finding roots can be used to get solutions of *any* equation because

$$g(x) = h(x) \quad \text{is equivalent to} \quad g(x) - h(x) = 0 \tag{3}$$

So *solutions of* $g(x) = h(x)$ *can be found as roots of* $g(x) - h(x)$. If desired, $g(x) - h(x)$ can be simplified† by multiplying or dividing both sides by a suitable expression before finding its roots. For example, the solutions of the equation

$$\frac{\cos x}{e^x} = 3x - x^2 \tag{4a}$$

†"Simplified" can be taken to mean "requiring the fewest keystrokes to evaluate on a calculator" unless other factors (e.g., fewest divisions or internal function evaluations, ease of differentiation) are to be considered.

can be found as the roots of either of the equivalent root-finding equations

$$e^{-x}\cos x + x^2 - 3x = 0 \qquad \text{or} \qquad \cos x + xe^x(x - 3) = 0 \tag{4b}$$

Of the many known general-purpose methods for finding real roots $\bar{x}$ of a continuous $f(x)$, we shall concentrate on two in detail: the *Newton–Raphson (NR) method* and the *Secant (SEC) Method*. These will be referred to as **slope methods** because they are based on the simple geometric idea described next.

2.3B The Slope Method Strategy

Let $\bar{x}$ be a desired root of a continuous function $f(x)$, and suppose that

- x_k is a current approximation of $\bar{x}$
- P_k is the point $(x_k, f(x_k))$ on the graph of f
- y_k is $f(x_k)$, the y-coordinate of P_k
- m_k is a nonzero number that represents the *slope* of the curve $y = f(x)$ near the point P_k
- x_{k+1} is the x-intercept of the straight line through P_k having slope m_k

as shown in Figure 2.3-2. If the straight line approximates the curve well between $P_k(x_k, y_k)$ and $I_{k+1}(x_{k+1}, 0)$, then x_{k+1} should approximate $\bar{x}$ better than x_k does. And if the procedure is repeated starting with x_{k+1}, the resulting x_{k+2} should give an even better approximation.

The computation shown in Figure 2.3-2 is a derivation of the following general formula.

$$X_{K+1} = X_k + \Delta_K$$

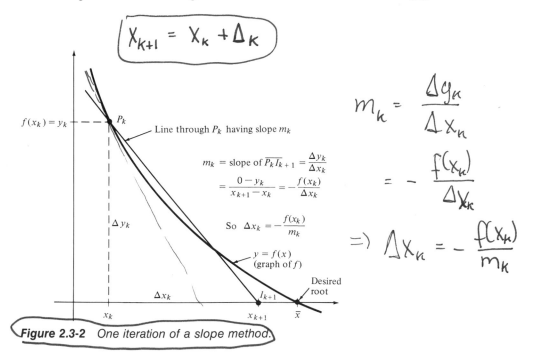

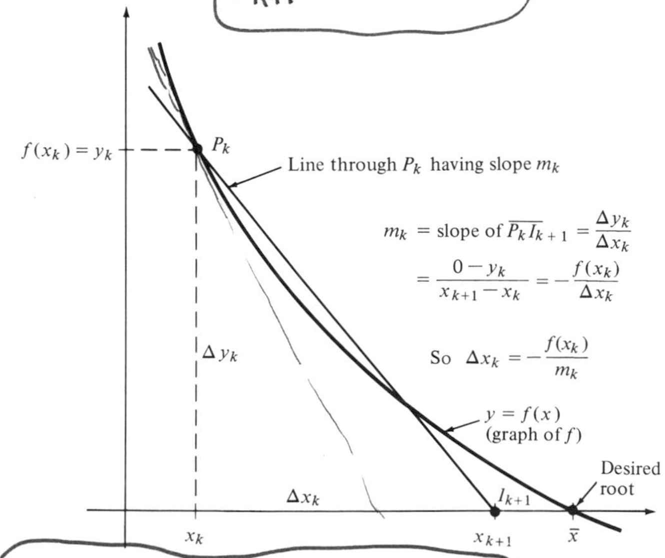

$$m_k = \frac{\Delta y_k}{\Delta x_k} = -\frac{f(x_k)}{\Delta x_k} \Rightarrow \Delta x_k = -\frac{f(x_k)}{m_k}$$

Figure 2.3-2 One iteration of a slope method.

$x_{k+1} = x_k - \frac{f(x_k)}{f'(x_k)}$

Slope Method Formula: $x_{k+1} = x_k + \Delta x_k,$

$$\text{where } \Delta x_k = -\frac{f(x_k)}{m_k} = -\frac{y_k}{m_k} \qquad \textbf{(5)}$$

Since this defines an iterative method, $\Delta x_{k-1} = DeltaX$ will be used to estimate the error of $x_{k-1} = Xprev$, as indicated in the pseudoprogram for the **Slope Method Algorithm** 2.3B. What distinguishes one slope method from another is the strategy used to form $m_k = Slope$. Two of the most effective strategies are to use the *tangent slope* at P_k (NR) or the *secant slope* determined by P_k and P_{k-1} (SEC). These are considered in Sections 2.3C and 2.3D.

ALGORITHM 2.3B. Slope Method

Purpose: To find roots $\bar{x}$ of a continuous function $f(x)$ to *NumSig* significant digits in *MaxIt* iterations or less.

```
GET x0,                                   {an initial guess of x̄}
    MaxIt, NumSig, C                      {termination parameters}
RelTol ← C·10^(-NumSig)                   {Relative Error Test tolerance}
Xprev ← x0                                {prepare for first iteration}
DO FOR k = 0, 1, ..., MaxIt
|   Yprev ← f(Xprev)
|   Form Slope using Xprev, Yprev, and perhaps other available values
|   DeltaX ← -Yprev/Slope; X ← Xprev + DeltaX
|   OUTPUT ( k, Xprev, Yprev, DeltaX, X )
|   Xprev ← X                             {update: prepare for next iteration}
|__ UNTIL |DeltaX| ≤ RelTol·|X|                       {termination test}
IFtermination test succeeded
|   THEN OUTPUT(X' approximates the root to the desired accuracy.')
|__ ELSE OUTPUT('Desired accuracy is not apparent in 'MaxIt' iterations.')
```

2.3C The Newton–Raphson (NR) Method: $m_k = m_{\tan}(x_k) = f'(x_k)$

If $f(x)$ is differentiable at x_k, then the natural candidate for m_k is

$$f'(x_k) = m_{\tan}(x_k), \quad \text{the } \textbf{tangent slope} \text{ at } P_k(x_k, y_k) \qquad \textbf{(6a)}$$

Taking m_k to be $f'(x_k)$ in (5) yields the formula for the

Newton–Raphson (NR) Method: $x_{k+1} = x_k + \Delta x_k,$

$$\text{where } \Delta x_k = -\frac{f(x_k)}{f'(x_k)} \qquad \textbf{(6b)}$$

we solve $f'(x) = 0$ in maximization/minimization problem

$$\Rightarrow \Delta x_k = -\frac{f'(x_k)}{f''(x_k)}$$

EXAMPLE 2.3C Consider the problem of finding the nth root of a given c.

(a) How can the NR method be used to find $\sqrt[n]{c}$ for any positive c and n?
(b) Use the NR method to find $\sqrt{78.8}$ to 6s.

SOLUTION (a): We can get $\sqrt[n]{c}$ by finding the unique positive solution of

$$x^n = c, \quad \text{or equivalently} \quad f(x) = x^n - c = 0 \tag{7a}$$

For this simple $f(x)$, $f'(x) = nx^{n-1}$ and the NR iteration is

$$x_{k+1} = x_k + \Delta x_k, \qquad \text{where } \Delta x_k = -\frac{x^n - c}{nx^{n-1}} = \frac{c - x^n}{nx^{n-1}} \tag{7b}$$

SOLUTION (b): When $n = 2$, (7b) becomes the square root iteration

$$x_{k+1} = x_k + \Delta x_k, \qquad \text{where } \Delta x_k = \frac{c - x_k^2}{2x_k} \tag{8}$$

Taking $c = 77.8$ and the (rather poor) initial guess $x_0 = 14$ gives

$$\Delta x_0 = \frac{78.8 - 14^2}{2(14)} \doteq -4.185714; \qquad \text{so } x_1 = 14 - 4.185714 = 9.814286$$

$$\Delta x_1 = \frac{78.8 - x_1^2}{2x_1} \doteq -0.892587; \qquad \text{so } x_2 = 9.814286 - 0.892587 = 8.921699$$

$$\Delta x_2 = \frac{78.8 - x_2^2}{2x_2} \doteq -0.044650; \qquad \text{so } x_3 = 8.921699 - 0.044650 = 8.877049$$

$$\Delta x_3 = \frac{78.8 - x_3^2}{2x_3} \doteq -0.000113; \qquad \text{so } x_4 = 8.877049 - 0.000113 = 8.876936$$

The leading zeros of Δx_1, Δx_2, and Δx_3 suggest that the convergence is quadradic. Indeed, if we eliminate Δx_k from (8) and simplify, we get

$$x_{k+1} = x_k + \frac{78.8 - x_k^2}{2x_k} = \frac{x_k^2 + 78.8}{2x_k} = g_2(x_k) \tag{9}$$

So applying NR to $f(x) = x^2 - 78.8$ can be viewed as applying RS to $g_2(x) = (x^2 + 78.8)/2x$. The latter iteration converges quadratically (see Table 2.1-4 of Section 2.1B). In fact, the convergence of NR is so rapid when $f(x) = x^2 - c$ that most calculators and computers use a formula based on a few iterations of (8) to find $\sqrt{c}$ (see Problem 2M-10). ■

More generally, the NR iteration (6) can be written as

$$\boxed{x_{k+1} = x_k - \frac{f(x_k)}{f'(x_k)} = g(x_k),} \qquad \text{where } g(x) = x - \frac{f(x)}{f'(x)} \tag{10}$$

$f(x) = 0$
$X_{k+1} = X_k + \Delta_k$
$X_{k+1} = X_k - \frac{f(x)}{f'(x)}$

$X_{k+1} = X_k$ if $\Delta_k \approx 0$
$\Leftrightarrow X_{k+1} = g(x)$ fixed point equation

Thus, using NR to find a *root* of $f(x)$ can be viewed as using Repeated Substitution to find a *fixed point* of $g(x) = x - f(x)/f'(x)$ whose derivative is

$$g'(x) = 1 - \frac{f'(x)^2 - f(x)f''(x)}{f'(x)^2} = \frac{f(x)f''(x)}{f'(x)^2} \qquad \textbf{(11)}$$

If $f(\bar{x}) = 0$ and $f'(\bar{x})$ exists and is *nonzero* (i.e., if the graph of f crosses the x-axis with *nonzero slope* at $\bar{x}$ as illustrated in Figure 2.3-2), then $\bar{x}$ is called a **simple root** of f. In this case $g'(\bar{x})$ in (11) will be zero. So we can conclude from Theorem RS of Section 2.1D that

Theorem: if $g'(\bar{x})=0 \Rightarrow$ quadratic convergence

NR will converge quadratically to a simple root **(12)**

The rapid convergence of NR once x_k gets close to $\bar{x}$ is illustrated graphically in Figure 2.3-3, which also shows why the NR method is sometimes called the **method of tangents**.

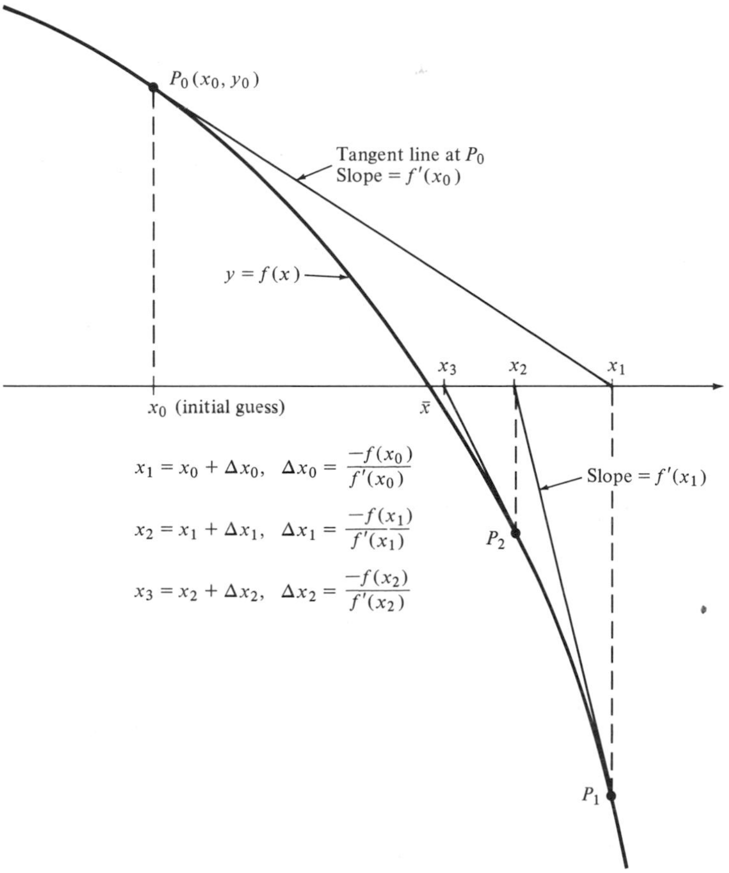

Figure 2.3-3 *Getting x_1, x_2, and x_3 by the NR method graphically.*

2.3D   The Secant (SEC) Method: $m_k = m_{sec} = \Delta y_{k-1}/\Delta x_{k-1}$

In order to use the NR method, one must first find the derivative $f'(x)$ and then evaluate $f'(x_k)$ for each iteration. The work required to find and substitute in $f'(x)$ was negligible for $f(x) = x^n - c$. But it would be substantial if, say,

$$f(x) = \sin(\sqrt{\sec x + x^3 e^{5x/\tan x}}) - e^{-x} \tag{13}$$

and *impossible* if we had no formula for $f(x)$ (see Example 1.1A). It is therefore desirable to be able to get m_k using *only* $f(x)$ itself. One such m_k is the **secant slope** determined by the *current point* $P_k(x_k, y_k)$ and the *preceding* current point $P_{k-1}(x_{k-1}, y_{k-1})$, that is,

$$\Delta x_k = -\frac{f(x_k)}{m_k}$$
$$= -\frac{y_k}{m_k}$$

$$m_{sec} = \frac{\Delta y_{k-1}}{\Delta x_{k-1}} = \frac{f(x_k) - f(x_{k-1})}{x_k - x_{k-1}} = \frac{y_k - y_{k-1}}{\Delta x_{k-1}} \tag{14a}$$

Taking m_k to be $\Delta y_{k-1}/\Delta x_{k-1}$ in the general slope method formula (5) yields the formula for the

SECANT (SEC) METHOD: $x_{k+1} = x_k + \Delta x_k$,   where $\Delta x_k = \dfrac{-y_k \Delta x_{k-1}}{y_k - y_{k-1}}$ (14b)

$$\Rightarrow \Delta x_k = -y_k \Delta x_{k-1}/\Delta y_{k-1}$$

The resulting algorithm, which requires *two* initial guesses, is called the **Secant (SEC) Method**. The use of SEC to find x_1, x_2, x_3, and x_4 from the initial guesses x_{-1} and x_0 is illustrated graphically in Figure 2.3-4. A comparison of the secant line intercepts with those of the tangent lines at P_0, P_1, P_2, and P_3 suggests that the Secant Method should converge a bit less rapidly than the Newton–Raphson method. This is confirmed in the following example.

EXAMPLE 2.3D   Use SEC to solve $f(x) = x^2 - 78.8$ for $\bar{x} = \sqrt{78.8}$ to $6s$.

SOLUTION: To compare the convergence rate with that of NR in Example 2.3C, we shall take $x_0 = 14$ and $x_{-1} \approx x_0$, say, $x_{-1} = 14.1$. Then

$$\Delta x_0 = \frac{-y_0 \Delta x_{-1}}{y_0 - y_{-1}} = \frac{-117.20(-.1)}{117.20 - 120.01} \doteq -4.170818$$

So $x_1 = x_0 + \Delta x_0 = 14.0 - 4.170818 = 9.829182$.

$$\Delta x_1 = \frac{-y_1 \Delta x_0}{y_1 - y_0} = \frac{-17.81282(-4.170818)}{17.81282 - 117.20} \doteq -0.747521$$

So $x_2 = x_1 + \Delta x_1 = 9.829182 - 0.747521 = 9.081661$.

$$\Delta x_2 = \frac{-y_2 \Delta x_1}{y_2 - y_1} = \frac{-3.676567(-0.747521)}{3.676567 - 17.81282} \doteq -0.194416$$

So $x_3 = x_2 + \Delta x_2 = 9.081661 - 0.194416 = 8.887245$.

These and the succeeding calculations are summarized in Table 2.3-1. Note that the values

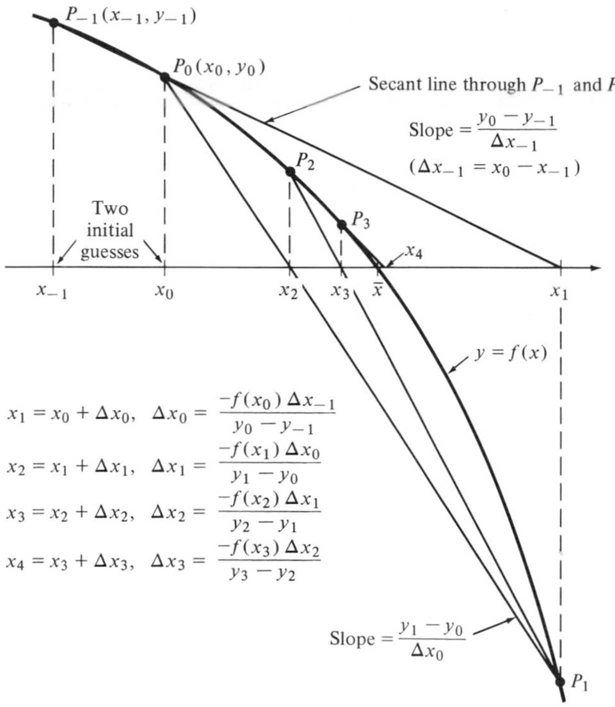

Figure 2.3-4 *Finding x_1, x_2, x_3, and x_4 using SEC graphically.*

(Δx_{k-1}, y_k, and y_{k-1}) needed to calculate any Δx_k can be found *above it and to its left*. For example, the entries used to get $\Delta x_2 = -0.194416$ in (14) are indicated in Table 2.3-1.

The iteration was stopped when $|\Delta x_4| \approx 10^{-4}$ because this indicates that x_4 is accurate to about 4d (i.e., 5s), hence that $x_5 = 8.87937$ should be (and in fact is) accurate to 6s. A more detailed analysis of the rate of convergence for this iteration is shown in Table 2.3-2, where the Δx_k's, shown to 10d, are based on a 13s calculation. The ratios shown are for $k \geq 3$ (when x_k is close to $\bar{x}$). The decreasing values down the $\Delta x_k/\Delta x_{k-1}$ column indicate that the convergence is *faster than linear*; and the increasing magnitudes down the $\Delta x_k/(\Delta x_{k-1})^2$ column indicate that the convergence is *slower than quadratic*.

TABLE 2.3-1 SEC Method Iterations for Solving $x^2 - 78.8 = 0$

k	x_k	$y_k = f(x_k)$	$\Delta x_k = \dfrac{-y_k \Delta x_{k-1}}{y_k - y_{k-1}}$	$x_{k+1} = x_k + \Delta x_k$
−1	14.1	120.01		
0	14.0	117.20	−4.170818	9.829182
1	9.829182	17.81282 $= y_1$	−0.747521 $= \Delta x_1$	9.081661
2	9.081661	3.676567 $= y_2$	−0.194416	8.887245
3	8.887245	0.183124	−0.010191	8.877054
4	8.877054	0.002088	−0.000117	8.876937

Convergence for which $|\Delta x_k| \approx C_S|\Delta x_{k-1}|^p$, where $1 < p < 2$, is said to be **superlinear** (faster than linear). It can be shown (see Problem M2-5) that when the Secant Method converges to a *simple* root,

$$\frac{|\Delta x_k|}{|\Delta x_{k-1}|^{1.618}} \approx C_S, \quad \text{the } \textbf{superlinear convergence constant} \qquad \textbf{(15a)}$$

as illustrated in Table 2.3-2. We shall refer to superlinear convergence with $p \doteq 1.618$ as **almost quadratic convergence**. In summary,

SEC will converge almost quadratically to a simple root **(15b)**

TABLE 2.3-2 Convergence Ratios for Almost Quadratic Convergence

k	Δx_k	$\Delta x_k/\Delta x_{k-1}$	$\Delta x_k/(\Delta x_{k-1})^2$	$\lvert\Delta x_k\rvert/\lvert\Delta x_{k-1}\rvert^{1.618}$
2	−0.1944155054			
3	−0.0101913285	0.052	−0.270	0.144
4	−0.0001173824	0.011	−1.126	0.196
5	−0.0000000682	0.006	−4.947	0.156

■

2.3E Finding Initial Guesses of Solutions of Equations

Iterative algorithms for solving equations must be provided with initial guesses. Very often an estimate of a desired solution $\bar{x}$ is evident from the context in which the equation arises. If such estimates are not available, it *may* be possible to rewrite the given equation as $g(x) = h(x)$, where *the graphs of **both** g **and** h can be sketched quickly*, and then find estimates of $\bar{x}$ graphically as in Figure 2.3-1(b). (This is done in the following example.) If such a rearrangement cannot be found, the equation should be written as simply as possible as $f(x) = 0$, and the desired solution(s) should be estimated as the x-intercepts of the graph of f, either drawn by hand or plotted by a computer. No matter how x_0 is obtained, the value of x_{-1} needed for the Secant Method is obtained by simply perturbing x_0 by a small amount, say, 1%.

Practical Consideration 2.3E (How accurate should a sketch be?). When initial guesses for a rapidly convergent method are obtained from hand-drawn graphs, *the graphs usually need only be accurate enough to get x_0's to 1s or maybe 2s.* This is because one additional iteration starting with a crude x_0 usually yields more accuracy (with far less trouble) than plotting more points so as to get a more accurate x_0. ■

EXAMPLE 2.3E For the transcendental equation $e^x \cos x = 1$:

(a) Estimate graphically the two smallest *positive* solutions.

(b) Use NR to find these two smallest positive solutions to 7*s*.

(c) Use SEC to find these two smallest positive solutions to 7*s*.

SOLUTION (a): The equation $e^x \cos x = 1$ has the same solutions as any of the equivalent equations

$$e^{-x}\sec x = 1 \quad \text{or} \quad e^x = \sec x \quad \text{or} \quad e^{-x} = \cos x \tag{16a}$$

Of these, the functions $g(x) = e^{-x}$ and $h(x) = \cos x$ equated last are probably the most easily sketched from memory (see Figure 2.3-5). We see from Figure 2.3-5 that the equation $e^x \cos x = 1$ has infinitely many solutions. The smallest is exactly $x_0 = 0$; the next two are

$$\bar{x}_1 \approx 1.3 \quad \text{and} \quad \bar{x}_2 \approx \frac{3\pi}{2}^+ \tag{16b}$$

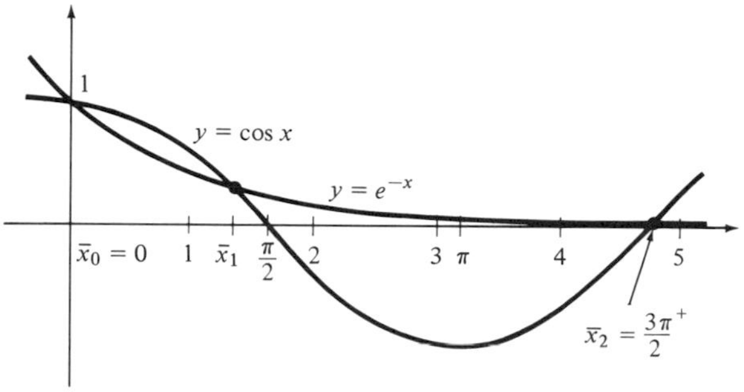

Figure 2.3-5 Estimating solutions of $e^x \cos x = 1$ graphically.

SOLUTION (b): An equation of the form $f(x) = 0$ that is equivalent to the equations (16a) and easy to differentiate is

$$f(x) = e^{-x} - \cos x = 0, \qquad \text{for which } f'(x) = \sin x - e^{-x} \tag{16c}$$

The NR iteration for this $f(x)$ (with x in *radians*) is

$$x_{k+1} = x_k + \Delta x_k, \qquad \text{where } \Delta x_k = \frac{f(x_k)}{f'(x_k)} = \frac{\cos x_k - e^{-x_k}}{\sin x_k - e^{-x_k}} \tag{16d}$$

Taking $x_0 = 1.3$ then $x_0 = 3\pi/2$ yields the iteration shown in Table 2.3-3. Both $\bar{x}_1$ and $\bar{x}_2$ were obtained to 7s (actually 8x) in only *two* iterations.

TABLE 2.3-3 NR Iterations for Two Roots of $f(x) = e^{-x} - \cos x$

k	x_k	$y_k = f(x_k)$	$\Delta x_k = \dfrac{-y_k}{f'(x_k)}$	$x_{k+1} = x_k + \Delta x_k$
0	1.3	.50329644E−2	−.0072833	1.2927167
1	1.2927167	.14403021E−4	−.0000210	1.2926957
0	4.7123890	.89832910E−2	.0089033	4.7212923
1	4.7212923	.46291601E−6	.0000005	4.7212928

Solution (c): For ease of comparison with part (b), we shall apply the Secant Method to $f(x)$ in (16c), with $x_0 = 1.3$ and $x_{-1} = 1.2$, and then with $x_0 = 3\pi/2$ and $x_{-1} = 3\pi/2 + 0.1$. The results are shown in Table 2.3-4, where 7s accuracy was evident (assuming *nearly quadratic* convergence) in two or three iterations.

TABLE 2.3-4 SEC Iterations for Two Roots of $f(x) = e^{-x} - \cos x$

k	x_k	$y_k = f(x_k)$	$\Delta x_k = \dfrac{-y_k\,\Delta x_{k-1}}{y_k - y_{k-1}}$	$x_{k+1} = x_k + \Delta x_k$
−1	1.2	−.6116354E−1		
0	1.3	.5032964E−2	−.0076031	1.2923969
1	1.2923969	−.2052559E−3	.0002980	1.2926949
2	1.2926949	−.5943326E−6	.0000008	1.2926957
−1	4.8123890	−.9170500E−1		
0	4.7123890	.8983291E−2	.0089219	4.7213109
1	4.7213109	−.1826462E−4	−.0000181	4.7212928

■

2.3F Using a Root-Finding Method to Find Fixed Points

Slope methods for finding roots are, for the most part, more robust and more rapidly convergent than the RS algorithm for finding fixed points. So from a practical point of view, equations are best solved by writing them as $f(x) = 0$ and using a root-finding method, rather than writing them as $x = g(x)$ and using Repeated Substitution. In fact, fixed points of $g(x)$ can usually be found more reliably as roots of $f(x) = g(x) - x$ than by applying the RS algorithm directly to $g(x)$. This is demonstrated in the following example. (See also Problem 2-27.)

EXAMPLE 2.3F Find the fixed points that could not be found by Repeated Substitution in **(a)** Example 2.1C (use NR); **(b)** Example 2.2C (use SEC).

Solution (a): The fixed points of $g(x) = \frac{1}{2}e^{x/2}$ are the roots of

$$f(x) = g(x) - x = \tfrac{1}{2}e^{x/2} - x \qquad \text{for which } f'(x) = \tfrac{1}{4}e^{x/2} - 1 \qquad \textbf{(17a)}$$

RS diverged from the fixed point $\bar{x}$ of $g(x)$ near 4.3. Since this $\bar{x}$ is a root of $f(x) = g(x) - x$, we can take $x_0 = 4.3$ in NR and get

$$\Delta x_0 = -\frac{0.5e^{(4.3)/2} - 4.3}{0.25e^{(4.3)/2} - 1} \doteq 0.0066; \qquad x_1 = 4.3 + 0.0066 = 4.3066 \qquad \textbf{(17b)}$$

This first iterate, x_1, is accurate to all five digits shown!

Solution (b): In Example 2.2C, SUBROUTINE FIXPT failed to find the fixed point $\bar{u}$ near 16 of $g_T(u) = (e^{\sqrt{0.047u}} + e^{-\sqrt{0.047u}})/2$. It can be found using SEC as a root of $g_T(u) - u$ or, equivalently, of $2[g_T(u) - u]$, which is

$$f(u) = e^{\sqrt{0.047u}} + e^{-\sqrt{0.047u}} - 2u, \qquad \text{with } u_0 = 16 \text{ and, say, } u_{-1} = 16.1 \tag{18a}$$

Then $\Delta u_{-1} = 16 - 16.1 = -0.1$, and one iteration of SEC gives

$$\Delta u_0 = \frac{-f(u_0)(-0.1)}{f(u_0) - f(u_{-1})} \doteq \frac{0.126705(0.1)}{0.126705 - 0.629449} \doteq -0.025203 \tag{18b}$$

Hence $u_1 = 16 - 0.025202 \doteq 15.9748$. A second SEC iteration gives

$$\Delta u_1 = \frac{-f(u_1)\,\Delta u_0}{f(u_1) - f(u_0)} \doteq -0.000486, \qquad u_2 = u_1 + \Delta u_1 \doteq 15.9743$$

From $|\Delta u_0|$, $|\Delta u_1|$, and the nearly quadratic convergence of SEC, u_2 approximates the desired $\bar{u}$ to about 6*s*. ■

2.3G Bracketing Methods

The reader should be aware of a class of methods that we shall call **bracketing methods** because they "bracket" $\bar{x}$ between the endpoints of a closed interval $[a, b]$. All bracketing methods start with an initial interval $[a, b]$ for which

$$L = f(a) \text{ and } R = f(b) \text{ have opposite sign} \tag{19}$$

If we assume f to be continuous, there must be at least one root $\bar{x}$ bracketed between a and b (Figure 2.3-6). To simplify our discussion, we assume that f has *exactly one* root $\bar{x}$ in $[a, b]$. Our objective is to find algorithms for systematically moving a and b toward each other while keeping $\bar{x}$ bracketed between them.

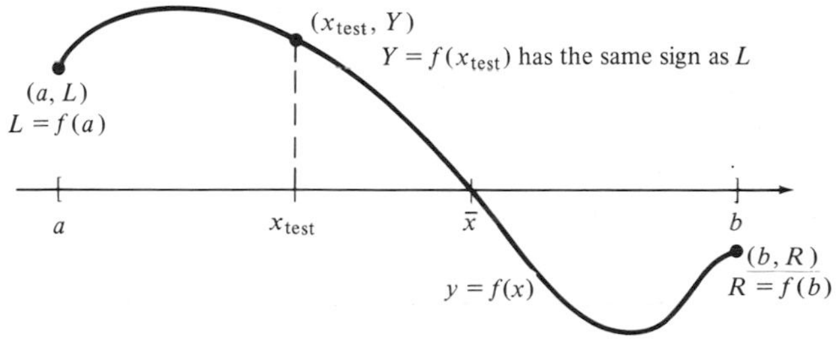

Figure 2.3-6 Bracketing method strategy.

The basic iterative step of a bracketing method is to *find a test value* x_{test} *in the open interval* (a, b) and then *replace* $[a, b]$ *by whichever of* $[a, x_{\text{test}}]$ *or* $[x_{\text{test}}, b]$ *contains* $\bar{x}$. The correct replacement is easily determined by the *sign of* $Y = f(x_{\text{test}})$:

IF sign(Y) = sign(L) THEN replace (a, L) by (x_{test}, Y)
ELSE {sign(Y) $\neq$ sign(L)} replace (b, R) by (x_{test}, Y)

In Figure 2.3-6, sign(Y) = sign(L) (both are positive); we would therefore replace (a, L)

by (x_{test}, Y). This iterative step is then repeated for the new (smaller) $[a, b]$ until either $Y = 0$ or two successive x_{test}'s are sufficiently close. Pseudocode for a bracketing method is shown in Algorithm 2.3G.

What all bracketing methods have in common is that *they can be used only if $f(x)$ has opposite sign on either side of $\bar{x}$*. What distinguishes one bracketing method from another is the strategy used to get x_{test}. The two most popular choices are†

ALGORITHM 2.3G. BRACKETING METHOD (SPECIAL CASES: BIS AND FP)

PURPOSE: To find a root $\bar{x}$ of $f(x)$ in an interval $[a, b]$ for which $L = f(a)$ and $R = f(b)$ have opposite sign.

```
{INITIALIZE}
Get a, b, MaxIt, NumSig
L ← f(a); R ← f(b); RelTol ← 10^(-NumSig)
IF sign(L) = sign(R) THEN STOP {method cannot be used}
Xprev ← b                        {this ensures that the initial DeltaX ≠ 0}

{ITERATE: Repeatedly form X (= x_test) in the open interval (a, b)}
DO FOR k = 1 TO MaxIt UNTIL termination test is satisfied
|    {form X} X← (formula for x_test involving a, b, L, R)
|    DeltaX ← X − Xprev; Y ← f(X)
|    OUTPUT (k, X, Y, DeltaX) {i.e., k, x_k, y_k, Δx_(k−1)}
|    {update}
|    IF sign(Y) = sign(L)
|    |    THEN {move a}   a ← X; L ← Y
|    |__  ELSE {move b}   b ← X; R ← Y
|    Xprev ← X
|___ {termination test: |DeltaX| ≤ RelTol·|X| OR Y = 0}
IF termination test succeeded
|    THEN OUPUT (X' approximates x̄ to 'NumSig' significant digits.')
|___ ELSE OUTPUT (MaxIt' iterations failed to yield a suitable X.')
STOP
```

$$x_{test} = x_{mid} = \frac{a + b}{2} \qquad \textbf{[Bisection (BIS) Method]} \qquad \textbf{(20)}$$

$$x_{test} = x_{FP} = \frac{bL - aR}{L - R} \qquad \textbf{[False Position (FP) Method]} \qquad \textbf{(21)}$$

†The BIS Method is also referred to as the **Interval Halving Method** or **Bolzano's method**; the FB Method is also referred to as the **Regula Falsi Method**.

In the Bisection Method, x_{test} bisects $[a, b]$. In the False Position Method, x_{test} is the x-intercept of the secant line from (a, L) to (b, R) (Figure 2.3-7).

By similar triangles ($\triangle AIJ \sim \triangle ABC$)

$$\frac{\overline{AJ}}{\overline{IJ}} = \frac{\overline{AC}}{\overline{BC}}, \text{ that is, } \frac{x_{\text{FP}} - a}{0 - L} = \frac{b - a}{R - L}$$

Solving:

$$x_{\text{FP}} = a + \frac{L}{L - R}(b - a) = \frac{bL - aR}{L - R}$$

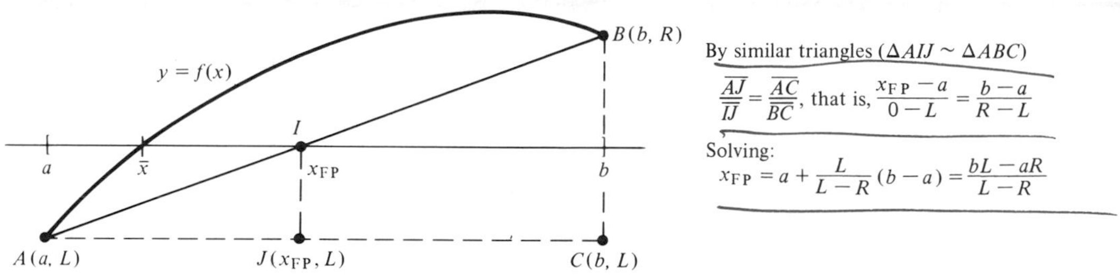

Figure 2.3-7 *Derivation of the x_{FP} formula.*

EXAMPLE 2.3G   Perform eight iterations of **(a)** BIS **(b)** FP for

$$f(x) = x^2 - 78.8, \quad a = 6, \quad \text{and} \quad b = 12 \tag{22}$$

SOLUTION: See Table 2.3-5, where x_k is the kth x_{test}. The Δx_{k-1} column reveals that:

TABLE 2.3-5   Iterations of (a) BIS and (b) FP for $f(x) = x^2 - 78.8$

$$a = 6, \quad L = f(a) = -42.8 \quad \text{(Sign of L is $\ominus$)}$$
$$b = 12, \quad R = f(b) = 65.2 \quad \text{(Sign of R is $\oplus$)}$$

(a) Bisection Method

k	a	b	$x_k = \dfrac{a + b}{2}$	$Y = f(x_k)$	$\Delta x_{k-1} = x_k - x_{k-1}$
1	6.	12.0	9.0	2.2	—
2	6.	9.0	7.5	−22.55	−1.5
3	7.5	9.0	8.25	−10.7375	0.75
4	8.25	9.0	8.625	−4.409375	0.375
5	8.625	9.0	8.8125	−1.139844	0.1875
6	8.8125	9.0	8.90625	0.5212891	0.09375
7	8.8125	8.90625	8.859375	−0.3114746	−0.046875
8	8.859375	8.90625	8.882813	0.1043579	0.023438

(b) False Position Method

k	a	b	$x_k = \dfrac{bL - aR}{L - R}$	$Y = f(x_k)$	$\Delta x_{k-1} = x_k - x_{k-1}$
1	6.0	12.0	8.377778	−8.612840	—
2	8.377778	12.0	8.800436	−1.352323	0.422658
3	8.800436	12.0	8.865450	−0.2037901	0.065014
4	8.865450	12.0	8.875217	−0.3051934E−1	0.009767
5	8.875217	12.0	8.876679	−0.4566260E−2	0.001462
6	8.876679	12.0	8.876898	−0.6831018E−3	0.000219
7	8.876898	12.0	8.876931	−0.1021884E−3	0.000033
8	8.876931	12.0	8.876936	−0.1528675E−4	0.000005

$$\text{For BIS:}\quad \frac{|\Delta x_k|}{|\Delta x_{k-1}|} = \frac{1}{2} \qquad (\text{i.e., } \Delta x_k = \pm\tfrac{1}{2}\Delta x_{k-1}) \qquad \textbf{(23a)}$$

$$\text{For FP:}\quad \frac{\Delta x_k}{\Delta x_{k-1}} \approx \text{constant} \qquad (\text{i.e., } \Delta x_k \approx C_L \Delta x_{k-1}) \qquad \textbf{(23b)}$$

where $C_L \approx .15$. This *linear* convergence to the simple root $\bar{x} = \sqrt{78.8}$ is much slower than either SEC or NR. However, Aitken's formula can be applied to the last three iterates of Table 2.3-5 to get

$$(x_8)_{\text{improved}} = x_8 - \frac{(x_8 - x_7)^2}{x_8 - 2x_7 + x_6} = 8.8769365 \qquad \textbf{(24)}$$

which approximates $\bar{x} = \sqrt{78.8} = 8.8769364$ to about 7*s*. This is about as well as NR or SEC did in the same number of iterations. ■

Note that x_{FP} always replaced a in Table 2.3-5(b). (Why?) The FP method can be modified to move both a and b in such a way that on the average

$$|\Delta x_k| \approx C_S |\Delta x_{k-1}|^{1.442} \qquad \textbf{(25)}$$

The superlinear convergence of this **Modified False Position Method** is still slower than the almost quadratic convergence of the Secant Method [see (15)]. Since it is no easier to use or program, it will not be discussed further. The interested reader is referred to [40].

2.4 Some Practical Aspects of Root Finding

In Section 2.3 we showed how slope methods and bracketing methods can be used to find roots efficiently. There are, however, roots to which these methods converge slowly, if at all, and/or are hard to find accurately even if convergence does occur. In this section we describe these roots and propose remedies for the difficulties they pose, and we present a general-purpose root-finding program.

2.4A Roots That Are Hard to Find Accurately

The problem of finding a root of $f(x)$ may be *ill conditioned* in the sense that small relative changes (errors) in the coefficients of $f(x)$ can result in disproportionately large relative changes (errors) in the *exact* value of the root. This is often the case for polynomials, as the following example shows.

EXAMPLE 2.4A Examine the changes in the values of the roots of

$$p(x) = x^6 - 21x^5 + 175x^4 - 735x^3 + 1624x^2 - 1764x + 720 \qquad (1)$$

when a_5, the coefficient of x^2, is changed from 1624 to 1624.25 to 1624.5.

Solution: The results are summarized in Table 2.4-1, which shows that $p(x)$ in (1) has six simple roots: $\bar{x}_1 = 1, \ldots, \bar{x}_6 = 6$. Adding 0.25 to 1624 changes the fifth significant digit; however, it produces changes in the *second* significant digit of $\bar{x}_3$, $\bar{x}_4$, $\bar{x}_5$, and $\bar{x}_6$! Adding a second 0.25 causes $\bar{x}_3$ and $\bar{x}_4$ to become complex! And, although not shown, increasing 1624 to 1625 (a net change of only about 0.06%) causes $\bar{x}_5$ and $\bar{x}_6$ to become complex as well.

TABLE 2.4-1 Sensitivity of Polynomial Roots to Small Changes in a Coefficient

a_5	$\bar{x}_1$	$\bar{x}_2$	$\bar{x}_3$	$\bar{x}_4$	$\bar{x}_5$	$\bar{x}_6$
1624.00	1.000000	2.000000	3.000000	4.000000	5.000000	6.000000
1624.25	1.002102	1.961491	3.268076	3.621456	5.236785	5.910090
1624.50	1.004243	1.927929	(3.407987 ±	.3728825*i*)	5.251515	5.730340

■

Practical Consideration 2.4A (Finding roots of polynomials). Let $p_{\text{stored}}(x)$ denote the polynomial whose coefficients are the *stored* coefficients of a polynomial $p(x)$. If a root-finding problem is ill conditioned, the small inherent error of the stored coefficients may move the *exact* roots of $p_{\text{stored}}(x)$ unacceptably far from the desired roots of $p(x)$. It is therefore advisable to *find roots of polynomials on a device that can store coefficients with three or four significant digits more than the desired accuracy of the roots.* ■

A second source of inaccuracy is related to multiple roots. Recall that a root $\bar{x}$ of $f(x)$ is **simple** (or **of multiplicity one**) if $f(\bar{x}) = 0$ but $f'(\bar{x}) \neq 0$. Simple roots are easily located by looking for a sign change in $f(x)$ because the graph of f crosses the x-axis with *nonzero slope* at $\bar{x}$. However, if $\bar{x}$ is a **multiple root**, that is, if both $f(\bar{x}) = 0$ *and* $f'(\bar{x}) = 0$, then

$$\text{the graph of } f \text{ will have nearly zero slope near the root } \bar{x} \qquad (2)$$

When (2) occurs, it may be difficult to recognize that a root exists and to find it accurately if you do recognize it. Some of the possible difficulties are illustrated graphically in Figure 2.4-1. The trouble in (a)–(c) arises because *the graph of f has a turning point (local maximum or minimum) on or near the x-axis.* Such roots are difficult or impossible to locate by looking for a sign change in $f(x)$. Indeed, even if you sketch the graph of f, it may not be clear from it whether the x-axis is touched (a) at a single tangency point $\bar{x}$, (b) at two

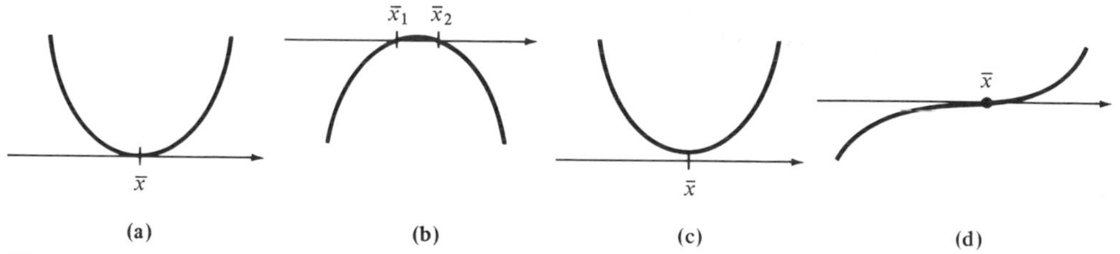

Figure 2.4-1 *Situations that present numerical difficulties.*

close points $\bar{x}_1$ and $\bar{x}_2$, or (c) not at all. In (a) and (d) there is a (multiple) root, but it is hard to say exactly where because the graph is tangent to the x-axis at $(\bar{x}, 0)$.

If a calculated value of $f(x)$ is nearly zero, it is likely that two larger but *nearly equal* values were subtracted in some intermediate step. The value of $f(x)$ is therefore likely to be contaminated by roundoff error due to subtractive cancellation, especially if error magnification occurred along the way. So if (2) holds, you should not expect to be able to find $\bar{x}$ to the accuracy of the device by *any* numerical method, simply because you are not likely to be able to evaluate $f(x)$ with sufficient precision for x near $\bar{x}$ (Problem C2-4). So *multiple roots should be found on a device that can store several digits more than the desired accuracy*. Alternatively (or in addition), one can use the results described next.

2.4B Convergence of NR and SEC to Multiple Roots

The fifth-degree polynomial $p(x)$ that can be factored as

$$p(x) = (x + 0.5)^2(x - 2.5)(x^2 - 3x + 2.3) \tag{3}$$

has $\bar{x} = 2.5$ as a simple root and $\bar{x} = -0.5$ as a double root. We extend the idea of multiplicity of roots to nonpolynomial functions as follows: $\bar{x}$ is a root of **multiplicity *m*** of $f(x)$ if $f(x)$ can be written as

$$f(x) = (x - \bar{x})^m \phi(x), \qquad \text{where } \phi \text{ is continuous at } \bar{x} \text{ and } \phi(\bar{x}) \neq 0 \tag{4a}$$

so that

$$f(x) \approx K(x - \bar{x})^m, \qquad \text{where } K = \phi(\bar{x}) \neq 0 \quad \text{for } x \approx \bar{x} \tag{4b}$$

The graphs of $K(x - \bar{x})^m$ for $m = 1, 2, 3,$ and 4 (when $K > 0$) are shown in Figure 2.4-2. All are tangent to the x-axis at $(\bar{x}, 0)$; however, they only *cross* the x-axis at $(\bar{x}, 0)$ when *m is odd*. Also, the larger the value of m, the "flatter" the graph will be near the point $(\bar{x}, 0)$.

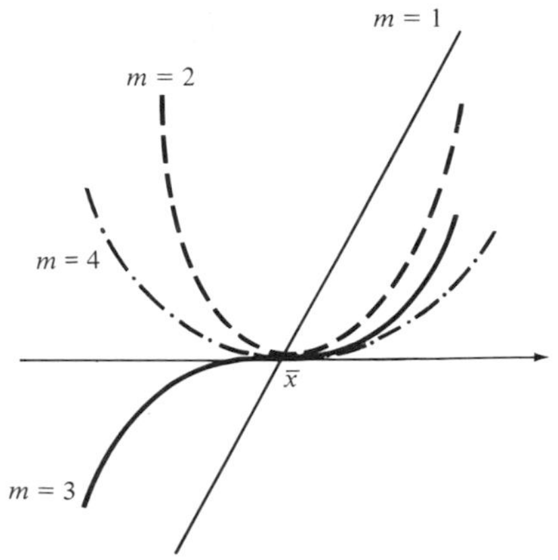

Figure 2.4-2 *$y = K(x - x)^m$ when $K > 0$, for $m = 1, 2, 3,$ and 4.*

Since multiple roots satisfy $f'(x) \approx 0$ for $x \approx \bar{x}$, we know from Section 1.4A that they are hard to find accurately. The following result [whose proof will be given in (8)] warns that they also slow the convergence of slope methods.

Both NR and SEC will converge *linearly* to a multiple root; more precisely,

$$\text{NR will satisfy } \Delta x_{k+1} \approx C_L \Delta x_k, \qquad \text{where } C_L = \frac{(m-1)}{m} \tag{5}$$

when converging to *any* root of multiplicity m, when $m > 1$.

Thus NR will satisfy $\Delta x_{k+1} \approx \frac{1}{2}\Delta x_k$ for a **double root** ($m = 2$), $\Delta x_{k+1} \approx \frac{2}{3}\Delta x_k$ for a **triple root** ($m = 3$), and so on. The multiplicity of a root can thus be deduced from the NR iteration, even if we do not know it beforehand!

EXAMPLE 2.4B Perform 12 iterations of NR and SEC with good initial guesses and deduce the multiplicity of the given root.

(a) $f(x) = x^5 - 4.5x^4 + 4.55x^3 + 2.675x^2 - 3.3x - 1.4375, \quad \bar{x} = -0.5$

(b) $g(x) = \tan x - xe^{-x^2}, \quad \bar{x} = 0$

SOLUTION (a): This $f(x)$ is the polynomial $p(x)$ shown in factored form in (3). So it has $\bar{x} = -0.5$ as a double root [$\phi(x)$ in (4a) is $(x - 2.5)(x^2 - 3x + 2.3)$]. Table 2.4-2 shows the iteration that results from using NR with

$$x_0 = -0.51 \qquad \text{and} \qquad f'(x) = 5x^4 - 18x^3 + 13.65x^2 + 5.35x - 3.3 \tag{6a}$$

and SEC with $x_0 = -0.51$ and $x_{-1} = -0.52$. It is clear from the $\Delta x_{k-1}/\Delta x_{k-2}$ columns that both NR and SEC are converging linearly, hence that $-.5$ is a multiple root. And since $C_L = \frac{1}{2}$ for the NR iteration, we could have deduced from (5) that -0.5 is actually a double root ($m = 2$), even if we did not know it from (3).

SOLUTION (b): Although $\bar{x} = 0$ is clearly a root of $g(x) = \tan x - xe^{-x^2}$, its multiplicity is not immediately apparent. Table 2.4-3 shows the iteration that results from using NR with

$$x_0 = 0.05 \qquad \text{and} \qquad g'(x) = \sec^2 x - e^{-x^2}(1 - 2x^2) \tag{6b}$$

and SEC with $x_0 = 0.05$ with $x_{-1} = 0.10$. The nearly constant linear convergence ratios for both NR and SEC show that $\bar{x} = 0$ is a multiple root. But this time the NR linear convergence constant $C_L \doteq \frac{2}{3}$ indicates that it is actually a triple root ($m = 3$) (see Problem 2-31). ■

Practical Consideration 2.4B (Using Aitken's formula to get multiple roots). Notice how the linear convergence ratios become less nearly constant for the last iterations shown in Tables 2.4-2 and 2.4-3. This will happen for *any* root because the Δx_k's will suffer subtractive can-

TABLE 2.4-2 NR and SEC Iteration for a Double Root

(a) Newton–Raphson (NR) method

k	x_k	$y_k = f(x_k)$	$\Delta x_{k-1} = x_k - x_{k-1}$	$\Delta x_{k-1}/\Delta x_{k-2}$
0	−0.5100000	−0.12311E−02		
1	−0.5050327	−0.30978E−03	0.49673E−02	
2	−0.5025248	−0.77695E−04	0.25080E−02	0.50
3	−0.5012647	−0.19446E−04	0.12601E−02	0.50
4	−0.5006335	−0.48727E−05	0.63118E−03	0.50
5	−0.5003174	−0.12219E−05	0.31613E−03	0.50
6	−0.5001590	−0.29802E−06	0.15834E−03	0.50
7	−0.5000819	−0.74506E−07	0.77099E−04	0.49
8	−0.5000445	−0.29802E−07	0.37419E−04	0.49
9	−0.5000170	0.00000E+00	0.27552E−04	0.74

Computed $f(-0.5000170)$ is zero; iteration discontinued.

(b) Secant (SEC) Method

k	x_k	$y_k = f(x_k)$	$\Delta x_{k-1} = x_k - x_{k-1}$	$\Delta x_{k-1}/\Delta x_{k-2}$
−1	−0.5200000	−0.49895E−02		
0	−0.5100000	−0.12311E−02		
1	−0.5067244	−0.55428E−03	0.32756E−02	
2	−0.5040419	−0.19954E−03	0.26825E−02	0.82
3	−0.5025330	−0.78216E−04	0.15089E−02	0.56
4	−0.5015602	−0.29624E−04	0.97278E−03	0.64
5	−0.5009671	−0.11370E−04	0.59303E−03	0.61
6	−0.5005978	−0.43362E−05	0.36938E−03	0.62
7	−0.5003700	−0.16689E−05	0.22773E−03	0.62
8	−0.5002276	−0.62585E−06	0.14249E−03	0.63
9	−0.5001421	−0.25332E−06	0.85491E−04	0.60
10	−0.5000839	−0.89407E−07	0.58132E−04	0.68
11	−0.5000522	−0.29802E−07	0.31707E−04	0.55
12	−0.5000364	−0.14901E−07	0.15855E−04	0.50

cellation once the x_k's approximate $\bar{x}$ to nearly the device accuracy. However, it happens much sooner for multiple roots for the reasons given in Section 2.4A. As a result, Aitken's formula generally gives better improvement if you *use the three closest iterates for which the linear convergence ratios $\Delta x_{k-1}/\Delta x_{k-2}$ remain nearly constant*, rather than the three closest iterates. For example, from the SEC iteration in Table 2.4-2,

$$(x_7)_{\text{improved}} = -0.5000033 \quad \text{whereas} \quad (x_{12})_{\text{improved}} = -0.5000207 \tag{7a}$$

Similarly, for the NR iteration in Table 2.4-2,

$$(x_6)_{\text{improved}} = -0.4999999 \quad \text{whereas} \quad (x_9)_{\text{improved}} = -0.4999406 \tag{7b}$$

As a rule, Aitken's formula yields multiple roots more accurately when applied this way to NR than when applied to SEC (Problem 2-32).

TABLE 2.4-3 NR and SEC Iteration for a Triple Root

(a) Newton–Raphson (NR) method

k	x_k	$y_k = g(x_k)$	$\Delta x_{k-1} = x_k - x_{k-1}$	$\Delta x_{k-1}/\Delta x_{k-2}$
0	0.0500000	0.16655E−03		
1	0.0333257	0.49334E−04	0.16674E−01	
2	0.0222151	0.14616E−04	0.11111E−01	0.67E+00
3	0.0148094	0.43305E−05	0.74057E−02	0.67E+00
4	0.0098727	0.12830E−05	0.49367E−02	0.67E+00
5	0.0065819	0.38015E−06	0.32908E−02	0.67E+00
6	0.0043882	0.11263E−06	0.21937E−02	0.67E+00
7	0.0029262	0.33382E−07	0.14620E−02	0.67E+00
8	0.0019524	0.11161E−07	0.97380E−03	0.67E+00
9	0.0012199	0.27212E−08	0.73254E−03	0.75E+00
10	0.0007616	0.66939E−09	0.45826E−03	0.63E+00
11	0.0004755	0.16735E−09	0.28613E−03	0.62E+00
12	0.0002943	0.40018E−10	0.18114E−03	0.63E+00

(b) Secant (SEC) Method

k	x_k	$y_k = g(x_k)$	$\Delta x_{k-1} = x_k - x_{k-1}$	$\Delta x_{k-1}/\Delta x_{k-2}$
−1	0.1000000	0.13297E−02		
0	0.0500000	0.16655E−03		
1	0.0428404	0.10478E−03	0.71596E−02	
2	0.0306959	0.38554E−04	0.12144E−01	0.17E+01
3	0.0236261	0.17581E−04	0.70699E−02	0.58E+00
4	0.0176994	0.73924E−05	0.59267E−02	0.84E+00
5	0.0133994	0.32074E−05	0.43000E−02	0.73E+00
6	0.0101039	0.13752E−05	0.32955E−02	0.77E+00
7	0.0076303	0.59238E−06	0.24736E−02	0.75E+00
8	0.0057585	0.25466E−06	0.18718E−02	0.76E+00
9	0.0043471	0.10949E−06	0.14114E−02	0.75E+00
10	0.0032826	0.47119E−07	0.10645E−02	0.75E+00
11	0.0024783	0.20256E−07	0.80422E−03	0.76E+00
12	0.0018719	0.98516E−08	0.60644E−03	0.75E+00

■

Proof That NR Satisfies $\Delta x_{k+1} \approx [(m - 1)/m]\Delta x_k$ when $m > 1$. Let us assume that $\phi(x)$ in (4) is twice differentiable and write (4) in the more compact form

$$f = h^m\phi, \quad \text{where } f = f(x), \quad \phi = \phi(x), \quad \text{and} \quad h = x - \bar{x} \qquad \textbf{(8a)}$$

Then $d(h^m)/dx = mh^{m-1}\cdot dh/dx = mh^{m-1}\cdot 1$, hence by the product rule,

$$f' = h^m\phi' + mh^{m-1}\phi \quad \text{and} \quad f'' = h^m\phi'' + 2mh^{m-1}\phi' + m(m - 1)h^{m-2}\phi \qquad \textbf{(8b)}$$

We saw in (10a) of Section 2.3C that the NR iteration can be viewed as Repeated Substitution applied to $g(x) = x - f(x)/f'(x)$ for which

$$g'(x) = \frac{f(x)f''(x)}{f'(x)^2} = \frac{\phi h^m[h^m\phi'' + 2mh^{m-1}\phi' + m(m-1)h^{m-2}\phi]}{(h^m\phi' + mh^{m-1}\phi)^2} \qquad \textbf{(8c)}$$

Now $g'(\bar{x})$ is not defined because $f'(\bar{x}) = 0$. However, we can divide the numerator and denominator by h^{2m-2} in (8c), to determine that

$$\lim_{x \to \bar{x}} g'(x) = \frac{\phi(\bar{x})[0 \cdot \phi''(\bar{x}) + 0 \cdot \phi'(\bar{x}) + m(m-1)\phi(\bar{x})]}{[0 \cdot \phi'(\bar{x}) + m\phi(\bar{x})]^2} = \frac{m-1}{m} \qquad \textbf{(8d)}$$

The desired conclusion (5) now follows from the fact that the RS algorithm satisfies $\Delta x_{k+1} \approx g'(\bar{x})\Delta x_k$ when $g'(\bar{x}) \neq 0$ [see (26) of Section 2.1G]. A proof that the Secant Method converges linearly to a multiple root is given in [40].

2.4C Overshoot, Cycling, and Wandering of Slope Methods

In certain situations, slope methods will either converge to a different root than the one intended or will fail to converge at all. The reasons for this are easily understood graphically and will be illustrated for NR. Figure 2.4-3 shows how NR can experience **cycling** if the graph of f has an inflection point near $(\bar{x}, 0)$, and **wandering** if it has a turning point near the x-axis but does not actually cross or touch it. The Secant Method can also experience wandering but not cycling. In fact, it is easy to see graphically that inflection points near $\bar{x}$ generally accelerate the convergence of SEC!

Figure 2.4-4 shows some possible adverse effects of a poor initial guess on the Newton–Raphson method. In (a) x_k will converge to a root but not the desired one; in (b) x_k will diverge to $-\infty$. The trouble in both (a) and (b) occurred because the tangent line at $P_0(x_0, y_0)$ was nearly horizontal [i.e., $m_0 = f'(x_0) \approx 0$] and consequently its x-intercept x_1 was far from both x_0 and the desired $\bar{x}$. This phenomenon is called **overshoot**. The NR slope $m_k = m_{\tan}$ makes no use of prior iterates. So if overshoot puts x_k "on the other side of the hill" as in Figure 2.2-4, the subsequent iteration is not likely to find its way back to $\bar{x}$. Overshoot is not quite as serious for the Secant Method because the SEC slope $m_k = m_{\sec}$ uses P_{k-1} as well as P_k and so "remembers where it came from"; so if x_{k-1} is near $\bar{x}$, but overshoot causes x_k to move far from $\bar{x}$, then x_{k+1} is likely to be a lot closer to $\bar{x}$ than x_k, as the following example shows.

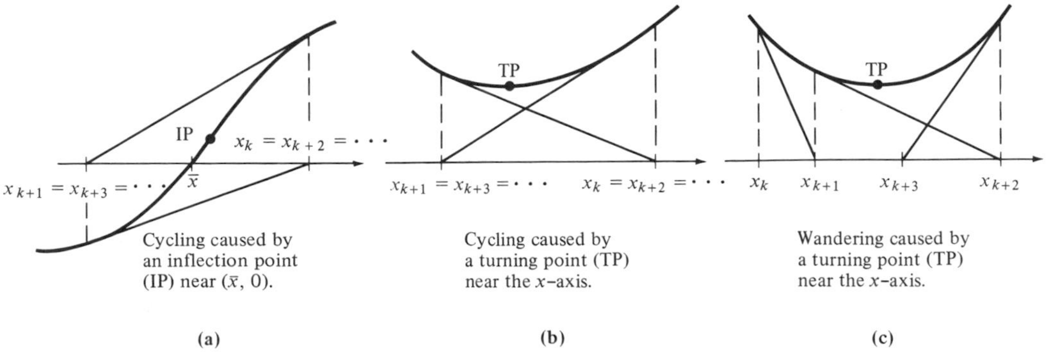

Figure 2.4-3 *Cycling and wandering of NR iterates.*

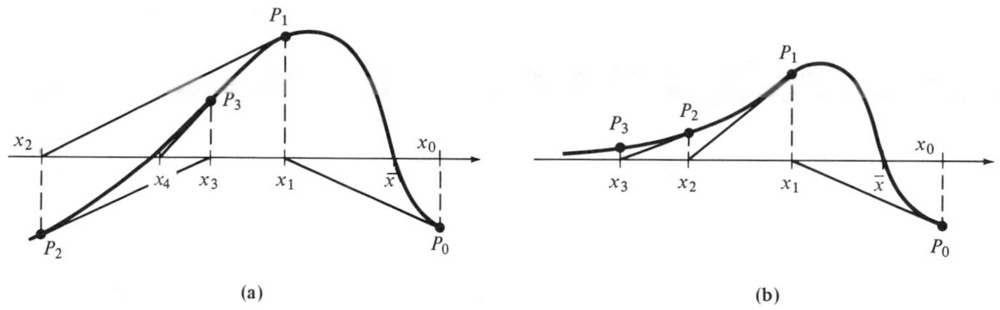

Figure 2.4-4 *Possible consequences of a poor initial guess on NR.*

EXAMPLE 2.4C Figure 2.4-5 shows the graph of the fifth-degree polynomial

$$p(x) = x^5 - 4.5x^4 + 4.55x^3 + 2.675x^2 - 3.3x - 1.3375 \quad \textbf{(9)}$$

Try to find to 6*s* the rightmost of the two roots near $\bar{x} = -0.5$ using **(a)** NR with $x_0 = -0.4975$ and **(b)** SEC with $x_0 = -0.4975$ and $x_{-1} = -0.5$.

SOLUTION: The results are shown in Table 2.4-4. Note first that the fact that $p'(x_0) \approx 0$ caused NR to overshoot to $x_1 = 1.555202$, where it ''wandered'' near the turning point at $x = 1.5$ until it overshot again, finally converging to the simple root 2.4891163. The nearly zero initial slope caused SEC to overshoot even *farther* to $x_1 = 2.8026036$, but $x_2 = -0.5546964$ *returned* to near x_0! Unfortunately, it actually passed x_0 and converged to the *leftmost* of the two close roots. Thus both methods eventually converged, but to different wrong roots! ■

Practical Consideration 2.4C (Preventing overshoot). One can prevent overshoot by modifying the slope method algorithm slightly as follows:

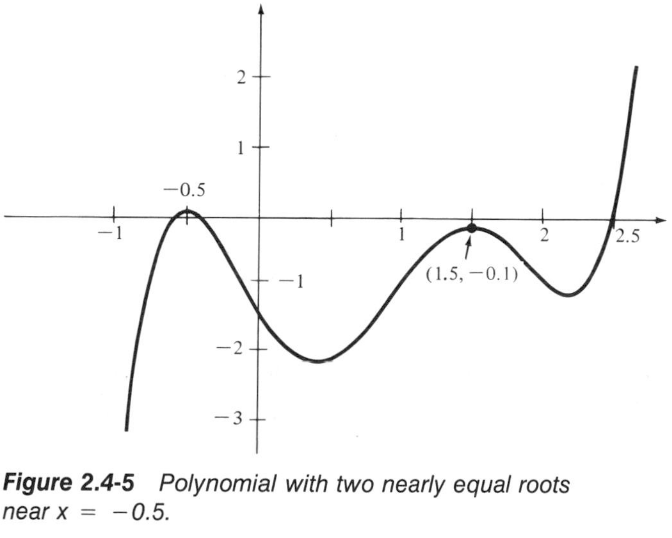

Figure 2.4-5 *Polynomial with two nearly equal roots near x = −0.5.*

TABLE 2.4-4 Consequences of an Initial Guess Between Two Close Roots

(a) Newton–Raphson (NR) method

k	x_k	$y_k = p(x_k)$	$\Delta x_{k-1} = x_k - x_{k-1}$
0	−0.4975000	0.99924E−01	
1	1.1555202	−0.52152E+00	0.16530E+01
2	1.3872877	−0.14852E+00	0.23177E+00
3	1.5613136	−0.11442E+00	0.17403E+00
4	1.3172022	−0.22581E+00	−0.24411E+00
5	1.4851089	−0.10085E+00	0.16791E+00
6	2.3647369	−0.78558E+00	0.87963E+00
7	2.5606911	0.76808E+00	0.19595E+00
8	2.4999759	0.99773E−01	−0.60715E−01
9	2.4894166	0.26827E−02	−0.10559E−01
10	2.4891166	0.23991E−05	−0.29998E−03
11	2.4891163	0.17881E−06	−0.26869E−06

$x_{11} = 2.4891163$ appears accurate to 6*s*.

(b) Secant (SEC) Method

k	x_k	$y_k = p(x_k)$	$\Delta x_{k-1} = x_k - x_{k-1}$
−1	−0.5000000	0.10000E+00	
0	−0.4975000	0.99924E−01	
1	2.8026036	0.58653E+01	0.33001E+01
2	−0.5546964	0.60961E−01	−0.33573E+01
3	−0.5899570	−0.10469E−01	−0.35261E−01
4	−0.5847891	0.25015E−02	0.51679E−02
5	−0.5857858	0.69231E−04	−0.99670E−03
6	−0.5858142	−0.49174E−06	−0.28369E−04
7	−0.5858140	0.14901E−07	0.20010E−06

$x_7 = -0.5858140$ appears accurate to 6*s*.

$$\begin{aligned}&\text{Form } m_k;\quad \Delta x_k \leftarrow -\frac{y_k}{m_k};\quad DxMax \leftarrow Fract\cdot|x_k|\\ &\text{IF } |\Delta x_k| > DxMax \text{ THEN } \Delta x_k \leftarrow (\pm)DxMax \qquad \{(\pm) = \text{sign of } \Delta x_k\}\end{aligned} \tag{10}$$

Thus if, say, *Fract* is taken as 0.1, then $|\Delta x_k|$ will be no larger than 10% of $|x_k|$. This will keep $x_1, x_2, \ldots$ near x_0 until $|\Delta x_k| \le DxMax$. However, once rapid (quadratic or superlinear) convergence begins, (10) will have no effect on the iteration. ■

The following conclusions about slope methods can be drawn from the preceding discussion.

1. The Secant Method is more robust than the Newton–Raphson method.
2. When using any slope method, try to avoid initial guesses where the graph has nearly zero slope. In particular, take an initial guess on either side of (*not* between) two close roots.
3. As with any iterative method, put an upper limit on the number of iterations that can be performed.

2.4D Root-Finding Methods Compared

The following discussion compares the slope and bracketing methods considered thus far.

Rate of Convergence. For simple roots, NR will converge quadratically, and the SEC method will converge almost quadradically (i.e., superlinearly with $p \doteq 1.618$) as described in Section 2.3. Both NR and SEC will converge linearly to roots of multiplicity $m > 1$, with $C_L = (m - 1)/m$ for NR. The BIS and FP methods will converge linearly to simple or multiple roots. However, the FP method can be modified by Steffensen's method to become quadratically convergent.

Robustness. Both NR and SEC can experience overshoot or wandering, but the effects are generally more serious for NR, which can also experience cycling. So SEC is more robust than NR. Both methods become considerably more robust if a limit is placed on the percent change in x_k using (10). Bracketing methods require initial guesses on either side of the root; but once found, they will converge to a root. However, they cannot be used to find multiple roots of even multiplicity m.

Accuracy. Simple roots can generally be found to the device accuracy by any method. However, Aitken's formula should be used with NR, SEC, or FP (but not BIS) as described in Practical Consideration 2.4B if multiple roots are needed to this accuracy.

Hand Calculation. If $f(x)$ and its derivative are available and not too complicated, NR [with $\Delta x = -f(x)/f'(x)$ *simplified* before being used to get Δx_k] is advised; otherwise, SEC is probably best.

Ease of Writing and Using a Program. NR is slightly easier to implement than SEC; but it is much less convenient to use because one must first find and then write code for $f'(x)$ as well as $f(x)$. A program for BIS can be written so as to require very little memory. (This is why calculators that have "built-in" root-finding routines often use it.) FP is about as easy to write as SEC but requires more work to find suitable initial guesses. A program for NR or SEC should have a safeguard against overshoot. A program for FP should incorporate Steffensen's method to accelerate its linear convergence.

Most commercial software uses a slow, reliable method initially, then changes to a more rapidly convergent method when there is some certainty that x_k is near a root. One of the better known such methods is Brent's ZEROIN program [6].

2.4E A Subroutine for the Secant Method

In view of the preceding discussion, we shall use the Secant (SEC) Method to illustrate the use of a computer to find roots. The Fortran 77 `SUBROUTINE SEC` shown in Figure 2.4-6 is based on Algorithm 2.3B, with *Slope* calculated as m_{sec} in line 2300. The input parameters `NUMSIG`, `MAXIT`, and `IOUT` allow the termination criteria and the amount of output to be controlled *from the calling program*, without having to modify the subroutine itself. When `SUBROUTINE SEC` is used with `IOUT` = 0 to get roots of $f(x)$ as an *intermediate step* of a more complex problem, the returned values of `X` ($\doteq \bar{x}$) and `IEXIT` provide the information needed by the calling program. `SUBROUTINE SEC` was used with `IOUT` = 2 to get the SEC results of Tables 2.4-2 and 2.4-3. In line 2600, it implements the strategy given in (10) to avoid the kind of overshoot that occurred in Table 2.4-4.

```
00100          SUBROUTINE SEC (F, NUMSIG, MAXIT, XPREV, X, Y, IOUT, IEXIT, IW)
00200  C - - - - - - - - - - - - - - - - - - - - - - - - - - - - - - - - C
00300  C   SUBROUTINE TO FIND ROOTS OF f(x) = 0 USING THE SECANT METHOD    C
00400  C   In the parameter list, the user must provide:                  C
00500  C          F = name of FUNCTION subprogram for f(x)                 C
00600  C      NUMSIG = desired number of significant digits                C
00700  C       MAXIT = maximum number of iterations                        C
00800  C   XPREV, X = two initial guesses                                  C
00900  C        IOUT = 0 to suppress all output (to device IW)             C
01000  C               1 to output final results only                     C
01100  C               2 to output details for each iteration             C
01200  C   The subroutine returns:                                        C
01300  C       X, Y = current x, f(x) when termination occurred           C
01400  C       IEXIT = 1, 2, 3, or 4 (see FORMAT statements 1 - 4)        C
01500  C Subprogram name F must be declared EXTERNAL in calling program.  C
01600  C - - - - - - - - - - - - - - - - - - - VERSION 2  9/9/85  - C
01700          RELTOL = 10.**(-NUMSIG)
01800          YPREV = F(XPREV)
01900          Y = F(X)
02000          IF (IOUT .GT. 1) WRITE(IW, 5) XPREV, YPREV, X, Y
02100          IEXIT = 1
02200          DO 10 K=1, MAXIT                              ! ITERATE
02300             SLOPE = (Y - YPREV) / (X - XPREV)          ! secant slope
02400             IF (SLOPE .EQ. 0.) GOTO 20
02500             DELTAX = -Y / SLOPE                        ! slope method
02600             ABSDX = AMIN1(ABS(DELTAX), ABS(0.1*X))     ! overshoot?
02700             DELTAX = SIGN(ABSDX, DELTAX)
02800             XPREV = X                                  ! update X
02900             X = XPREV + DELTAX
03000             YPREV = Y                                  ! update Y
03100             Y = F(X)
03200             IF (IOUT .GT. 1) WRITE (IW, 6) K, X, Y, DELTAX
03300             IF (Y .EQ. 0.0) IEXIT = 2                  ! terminate?
03400             IF (ABSDX .LE. RELTOL*ABS(X)) IEXIT = 3
03500             IF (IEXIT .GT. 1) GOTO 20
03600      10 CONTINUE
03700          IEXIT = 4
03800      20 IF (IOUT .EQ. 0) RETURN
03900          IF (IEXIT .EQ. 1) WRITE (IW, 1) X
04000          IF (IEXIT .EQ. 2) WRITE (IW, 2) X
04100          IF (IEXIT .EQ. 3) WRITE (IW, 3) X, NUMSIG
04200          IF (IEXIT .EQ. 4) WRITE (IW, 4) MAXIT
04300          RETURN
04400      1 FORMAT('0Slope = 0 when X =',G12.7,'.  Iteration discontinued')
04500      2 FORMAT('0Computed F( ',G12.7,') is 0.  Iteration discontinued')
04600      3 FORMAT('0ROOT:   ',G12.7,' appears to be accurate to ',I1,'s')
04700      4 FORMAT('0Desired accuracy is not evident in ',I3,' iterations')
04800      5 FORMAT('0SECANT METHOD:  X[-1], X[0] are initial guesses',/,
04900     &         '0 k',4X,'X = Xk',7X,'Y = F(X)',7X,'X - XPREV',/,
05000     &         ' -1 ', G12.7, E12.5, /,' 0  ', G12.7, E12.5)
05100      6 FORMAT(I3, 3X, G12.7, E12.5, E15.5)
05200          END
```

Figure 2.4-6 *A Fortran* SUBROUTINE *for the Secant Method.*

```
ØØ1ØØ  C      Using SUBROUTINE SEC to find a fixed point of Example 2.2C
ØØ2ØØ         PARAMETER (IR=5, IW=5)
ØØ3ØØ         EXTERNAL GTMU              ! gT(u) - u
ØØ4ØØ         WRITE (IW, 1)
ØØ5ØØ       1 FORMAT ('ØEnter an initial guess')
ØØ6ØØ         READ(IR, *) U
ØØ7ØØ         UPREV = 1.Ø1 * U           ! vary initial guess by 1%
ØØ8ØØ  C
ØØ9ØØ         CALL SEC (GTMU, 6, 2Ø, UPREV, U, FOFU, 1, IEXIT, IW)
Ø1ØØØ  C
Ø11ØØ         STOP
Ø12ØØ         END
Ø13ØØ
Ø14ØØ         FUNCTION GTMU(U)
Ø15ØØ         X = SQRT(Ø.Ø47) * U
Ø16ØØ         GT = ( EXP(X) + EXP(-X) ) / 2.Ø  ! = GT(U)
Ø17ØØ         GTMU = GT - U
Ø18ØØ         RETURN
Ø19ØØ         END

              Enter an initial guess
              15.9

              ROOT:   15.97431      appears to be accurate to 6s

              Enter an initial guess
              1.Ø

              ROOT:    1.Ø2478      appears to be accurate to 6s
```

Figure 2.4-7 *Using* SUBROUTINE SEC *to solve* $g_T(u) = u$.

Figure 2.4-7 shows how SUBROUTINE SEC can be used to find the fixed point of $g_T(u)$ that could not be found by SUBROUTINE FIXPT in Example 2.2C. Since SEC is a root finding-method, the fixed-point equation $g_T(u) = u$ must first be changed to the equivalent equation

$$f(u) = 0, \text{ where } f(u) = g_T(u) - u \ [= \text{GTMU in line 1700}] \qquad \textbf{(11)}$$

The parameters NUMSIG = 6, MAXIT = 20 and IOUT = 1 were specified in the calling statement in line 900. The answer, $\bar{u} \doteq 15.9743$ ($6s$), agrees with that obtained using NR in Example 2.3F(b).

2.5 Finding Roots of Real Polynomials; Bairstow's Method

In many situations (e.g., when analyzing the stability of mechanical structures or electrical systems) one must know the *complex* roots of polynomials whose degree can be as high as

30. It has been proved that general formulas for roots of polynomials cannot be found for polynomials of degree ≥ 5. So we generally have no choice but to use a numerical method to find such roots. A Fortran program for the Newton–Raphson method can be adapted to find complex roots by changing all `REAL` variables to `COMPLEX` (and suitably modifying the input/output statements). See Problem C2-7. **Bairstow's method** of this section accomplishes the same thing using only *real* arithmetic, and so is especially desirable when using a language that does not have built-in complex arithmetic. This suggests that when $n \geq 3$, all roots of the *real* polynomial $p(x)$ can be found using the **Quadratic Deflation** Algorithm 2.5A.

The ***solve*** step of the Quadratic Deflation algorithm can be performed by a call to a subroutine such as `QROOTS` in Figure 1.4-6. The method we will use to find a quadratic factor in the ***factor*** step is due to Bairstow. It uses real arithmetic and converges quadratically when the roots of $q(x)$ are *not* also roots of $Q(x) = p(x)/q(x)$. The basis for Bairstow's method is the synthetic division algorithm described next.

2.5A Deflating Real Polynomials Using Quadratic Factors

Even if available, complex arithmetic should not be necessary when $p(x)$ is a **real polynomial** [i.e., when all coefficients of $p(x)$ are real]. This is because the *complex roots of real polynomials occur in conjugate pairs.* For any such pair, say, $\alpha \pm i\beta$, the real quadratic

$$q(x) = (x - \alpha - i\beta)(x - \alpha + i\beta) = (x - \alpha)^2 + \beta^2$$

is a *real* factor of $p(x)$; that is, there is a *real* polynomial $Q(x)$ such that

$$p(x) = q(x)Q(x), \qquad \text{where degree } Q(x) = \text{degree } p(x) - 2$$

ALGORITHM 2.5A. QUADRATIC DEFLATION

PURPOSE: To find all roots of a real polynomial $p(x) = a_1x^n + \cdots + a_nx + a_{n+1}$.

GET $n, a_1, \ldots, a_{n+1}$ $\{a_1 \neq 0\}$

DO WHILE $n \geq 2$
  {***factor***} Find a real quadratic $q(x)$ such that $p(x) = q(x)Q(x)$
  {***solve***} Find the two roots r_1, r_2 of $q(x)$ by the quadratic formula
  OUTPUT ('Roots: 'r_1, r_2)
  {***deflate***} $p(x) \leftarrow Q(x)$; $n \leftarrow n - 2$

IF $n = 1$ THEN *LastRoot* $\leftarrow -a_2/a_1$; OUTPUT ('Root: '*LastRoot*)
STOP

2.5B Synthetic Division by $q(x)$

If the real nth-degree polynomial

$$p(x) = a_1x^n + a_2x^{n-1} + \cdots + a_nx + a_{n+1} \quad (a_1 \neq 0) \tag{1}$$

is divided by a real quadratic

$$q(x) = x^2 - rx - s \quad \textit{(notice the minus signs!)} \tag{2}$$

the result will be a *quotient* $Q(x)$ of degree $n - 2$ and a *remainder* $R(x)$ of degree ≤ 1. This division can be displayed as

$$\frac{p(x)}{q(x)} = Q(x) + \frac{R(x)}{q(x)} \quad \text{or} \quad p(x) = q(x)Q(x) + R(x) \tag{3}$$

Our notation for this division will be as follows (notice the b_n term):

$$p(x) = \underbrace{(x^2 - rx - s)}_{q(x)}\underbrace{(b_1x^{n-2} + b_2x^{n-3} + \cdots + b_{n-2}x + b_{n-1})}_{Q(x)} + \underbrace{b_n(x - r) + b_{n+1}}_{R(x)} \tag{4}$$

Expanding (4) and equating coefficients with (1) gives the following table.

Power of x	**Equated Coefficients**	**Solving for b_i**
x^n	$a_1 = b_1$	$b_1 = a_1$
x^{n-1}	$a_2 = b_2 - rb_1$	$b_2 = a_2 + rb_1$
x^{n-2}	$a_3 = b_3 - rb_2 - sb_1$	$b_3 = a_3 + rb_2 + sb_1$
x^{n-3}	$a_4 = b_4 - rb_3 - sb_2$	$b_4 = a_4 + rb_3 + sb_2$
$\vdots$	$\vdots$	$\vdots$
x	$a_n = b_n - rb_{n-1} - sb_{n-2}$	$b_n = a_n + rb_{n-1} + sb_{n-2}$
constant	$a_{n+1} = b_{n+1} - rb_n - sb_{n-1}$	$b_{n+1} = a_{n+1} + rb_n + sb_{n-1}$

(5)

The right-hand column shows that the b_i's that define $Q(x)$ and $R(x)$ can be obtained **synthetically** [i.e., using only r, s, and the coefficients of $p(x)$, as shown in Figure 2.5-1(a)]. This quadratic **synthetic division** algorithm is illustrated in Figure 2.5-1(b) for the division

$$\frac{2x^6 - 4x^5 + x^3 - 40x}{x^2 - x + 4} = (2x^4 - 2x^3 - 10x^2 - x + 39) + \frac{3(x - 1) - 153}{x^2 - x + 4} \tag{6}$$

(a)

	a_1	a_2	a_3	a_4	$\cdots$	a_{n-1}	a_n	a_{n+1}
$\times r>$		$+b_1r$	$+b_2r$	$+b_3r$	$\cdots$	$+b_{n-2}r$	$+b_{n-1}r$	$+b_nr$
$\times s>$			$+b_1s$	$+b_2s$	$\cdots$	$+b_{n-3}s$	$+b_{n-2}s$	$+b_{n-1}s$
	b_1	b_2	b_3	b_4	$\cdots$	b_{n-1}	b_n	b_{n+1}

(b)

	2	-4	0	1	0	-40	0
$\times 1>$		2	-2	-10	-1	39	3
$\times(-4)>$			-8	8	40	4	-156
	2	-2	-10	-1	$+39$	$+3 = b_6$	$-153 = b_7$

Figure 2.5-1 *Quadratic Synthetic Division: (a) $p(x)/(x^2 - rx - s)$; (b) $(2x^6 - 4x^5 + x^3 - 40x)/(x^2 - x + 4)$.*

2.5C Bairstow's Method

Our objective is to find quadratics $q(x)$ that are factors of $p(x)$, that is, which make $R(x) = b_n(x - r) + b_{n+1} = 0$. To do this, we must find r and s that make

$$b_n = 0 \qquad \text{and} \qquad b_{n+1} = 0 \tag{7}$$

This is not a trivial problem because, as can be seen from (5), each of $b_2, \ldots, b_{n+1}$ depends on the preceding b_i's, which are themselves functions of r and s.

Let $\bar{r}$ and $\bar{s}$ be values that satisfy (7), and suppose that r and s are approximations of $\bar{r}$ and $\bar{s}$. If the increments

$$dr = \bar{r} - r \qquad \text{and} \qquad ds = \bar{s} - s \tag{8}$$

are small and we view b_n and b_{n+1} as functions of (r, s), then we can use the total differentials of b_n and b_{n+1} to get the approximations

$$0 = b_n(\bar{r}, \bar{s}) = b_n(r + dr, s + ds) \approx b_n(r, s) + \frac{\partial b_n}{\partial r}dr + \frac{\partial b_n}{\partial s}ds \tag{9a}$$

$$0 = b_{n+1}(\bar{r}, \bar{s}) = b_{n+1}(r + dr, s + ds) \approx b_{n+1}(r, s) + \frac{\partial b_{n+1}}{\partial r}dr + \frac{\partial b_{n+1}}{\partial s}ds \tag{9b}$$

where the partial derivatives are evaluated at (r, s) (see Appendix II.5C). Hence, if r and s are kth iterates (i.e., $r = r_k$ and $s = s_k$) and we use dr_k and ds_k to denote the solution of (9) with "$\approx$" replaced by "$=$", then in view of (8), r_{k+1} and s_{k+1} defined by

$$r_{k+1} = r_k + dr_k \qquad \text{and} \qquad s_{k+1} = s_k + ds_k \tag{10}$$

should approximate $\bar{r}$ and $\bar{s}$ better than r_k and s_k.

In order to be able to use iteration (10), we must know the values of the four partial derivatives in (9). What Bairstow astutely observed is that if synthetic division by

$$q(x) = x^2 - rx - s$$

is performed *with the b_i's replacing the a_i's* to get $c_1, \ldots, c_n$ as follows:

$$\begin{array}{l|ccccc|c}
 & b_1 & b_2 & b_3 & \cdots & b_n & b_{n+1} \\
\times r> & & +c_1 r & +c_2 r & & +c_{n-1} r & \\
\times s> & & & +c_1 s & & +c_{n-2} s & \\
\hline
 & c_1 & c_2 & c_3 & \cdots & c_n & \text{not needed}
\end{array} \tag{11}$$

then the required partial derivatives in (9) can be obtained as

$$\frac{\partial b_n}{\partial r} = c_{n-1}, \quad \frac{\partial b_n}{\partial s} = c_{n-2}, \quad \frac{\partial b_{n+1}}{\partial r} = c_n, \quad \text{and} \quad \frac{\partial b_{n+1}}{\partial s} = c_{n-1} \tag{12}$$

A proof of (12) will be given in Section 2.5E. From (12) and (9) we see that dr_k and ds_k in (10) are obtained by solving the *linear* equations

$$\begin{aligned} c_{n-1}dr + c_{n-2}ds &= -b_n \\ c_n\,dr + c_{n-1}ds &= -b_{n+1} \end{aligned} \tag{13}$$

Using Cramer's rule [see (8) of Section 3.5B], we get

$$dr_k = \frac{b_n c_{n-1} - b_{n+1} c_{n-2}}{c_n c_{n-2} - c_{n-1}^2} \quad \text{and} \quad ds_k = \frac{b_{n+1} c_{n-1} - b_n c_n}{c_n c_{n-2} - c_{n-1}^2} \tag{14}$$

The resulting iterative method for finding $q(x)$ is called **Bairstow's method**. Pseudocode for Bairstow's method is shown in Algorithm 2.5C.

ALGORITHM 2.5C. BAIRSTOW'S METHOD

PURPOSE: To find a quadratic factor $q(x) = x^2 - rx - s$ of an nth-degree polynomial $p(x) = a_1x^n + \cdots + a_nx + a_{n+1}$ $(a_1 \neq 0)$.

```
{INITIALIZE}
GET n, a_1, a_2, ..., a_(n+1),      {parameters of p(x)}
    MaxIt, NumSig,                  {termination parameters}
    r_0, s_0                        {initial guesses of r, s}
b_1 ← a_1;  c_1 ← b_1;  r ← r_0;  s ← s_0;  Tol ← 10^(-NumSig)

{ITERATE}
DO FOR k = 1 TO MaxIt UNTIL termination test is satisfied
  | {get remaining b's and c's by quadratic synthetic division}
  | b_2 ← a_2 + b_1 r;  c_2 ← b_2 + c_1 r
```

```
DO FOR i = 3 TO n + 1
    b_1 ← a_i + r b_{i-1} + s b_{i-2}                    {This is (5)}
    c_i ← b_i + r c_{i-1} + s c_{i-2}                    {This is (11)}

{solve}
Det ← c_n c_{n-2} - c_{n-1}^2
dr ← (b_n c_{n-1} - b_{n+1} c_{n-2})/Det       {This is (14)}
ds ← (b_{n+1} c_{n-1} - b_n c_n)/Det
r ← r + dr;   s ← s + ds                       {This is (10)}
OUTPUT (r, dr, s, ds)
{termination test: |dr| ≤ Tol·max(1, |r|) and |ds| ≤ Tol·max(1, |s|)}

IF termination test succeeded {i.e., (|ρ_r| or |ε_r| ≤ Tol) and (|ρ_s| or |ε_s| ≤ Tol)}
  THEN OUTPUT ('q(x) = 'x^2 - rx - s' is a factor of p(x), and
               Q(x) = 'b_1 x^{n-2} + ··· + b_{n-2} x + b_{n-1} 'is p(x)/q(x).')
     {In this case the p(x) ← Q(x) step of Quadratic Deflation is simply}
     {        DO FOR i = 2 TO n - 1;   a_i ← b_i                          }
  ELSE OUTPUT ('Convergence is not apparent in 'MaxIt' iterations'.)
STOP
```

Initial guesses r_0 and s_0 are usually taken to be 0. Alternatively:

For large roots, try $r_0 = -\dfrac{a_2}{a_1}$ and $s_0 = -\dfrac{a_3}{a_1}$ **(15a)**

For small roots (if $a_{n-1} \neq 0$), try $r_0 = -\dfrac{a_n}{a_{n-1}}$ and $s_0 = -\dfrac{a_{n+1}}{a_{n-1}}$ **(15b)**

This is because very large roots $\bar{x}$ will satisfy

$$0 = p(\bar{x}) \approx (a_1\bar{x}^2 + a_2\bar{x} + a_3)\bar{x}^{n-2} \qquad \text{so that} \qquad \bar{x}^2 + \frac{a_2}{a_1}\bar{x} + \frac{a_3}{a_1} \approx 0$$

and, similarly, very small roots $\bar{x}$ satisfy $a_{n-1}\bar{x}^2 + a_n\bar{x} + a_{n+1} \approx 0$.

EXAMPLE 2.5C Find all roots of the fifth-degree polynomial

$$p(x) = x^5 - 4.5x^4 + 4.55x^3 + 2.675x^2 - 3.3x - 1.4375 \tag{16}$$

using Bairstow's method. Try to find $\bar{x}$'s of smallest magnitude first.

Solution: Using (15b), we try as initial guesses for r and s,

$$r_0 = -\frac{-3.3}{2.675} \doteq 1.233645 \qquad \text{and} \qquad s_0 = -\frac{-1.4375}{2.675} \doteq 0.5373832$$

The c_i values required to get dr_0 and ds_0 are obtained using the "double" synthetic division as follows:

a_i's:	1	−4.5	+4.55	+2.675	−3.3	−1.4375
×1.233645>		+1.233645	−4.029522	+1.305025	+2.744540	+0.016057
×0.5373832>			+0.537383	−1.755284	+0.568477	+1.195538
b_i's:	1	−3.266355	+1.057861	+2.224740	+0.013016	−0.225904
×1.233645>		+1.233645	−2.507643	−1.125576	+0.008414	
×0.5373832>			+0.537383	−1.092344	−0.490308	
c_i's:	1	−2.032710	−0.912398	+0.006820	−0.468877	

The desired increments are then obtained using (14):

$$dr_0 = \frac{(0.013016)(0.006820) - (-0.225904)(-0.912398)}{(-0.468877)(-0.912398) - (0.006820)^2}$$

$$= \frac{-0.2060261}{0.4277564} = -0.4816435$$

$$ds_0 = \frac{(0.006820)(-0.225904) - (-0.468877)(0.013016)}{-0.4277564} = +0.01066556$$

Adding these increments to the initial values, we get

$$r_1 = r_0 + dr_0 = 1.233645 - 0.4816435 = 0.752001$$
$$s_1 = s_0 + ds_0 = 0.5373832 + 0.01066556 = 0.5480488$$

If we continue the iteration, we get Table 2.5-1.

We see from the Δr and Δs columns of Table 2.5-1 that *the convergence was quadratic once r_k and s_k were near $\bar{r} = -1$ and $\bar{s} = -0.25$* (i.e., for $k \geqslant 7$). However, *both r_k and s_k "wandered" for the first six iterations.* As it turned out, the roots of the quadratic factor obtained, namely

$$q(x) = x^2 - (-1)x - (-0.25) = (x + 0.5)^2 \qquad (-0.5 \text{ is a double root}) \qquad \textbf{(17)}$$

are the two roots of $p(x)$ of smallest magnitude [see (21)].

TABLE 2.5-1 Convergence of Bairstow's Method

k	Δr	Δs	r	s
0	—	—	1.233645	0.5373832
1	−0.4816435	0.0106656	0.752001	0.5480488
2	0.1478756	0.2232268	0.899877	0.7712756
3	0.5234931	0.1512589	1.423370	0.9225345
4	−0.0828045	−0.7647468	1.340565	0.1577877
5	0.6070840	0.3324581	1.948274	0.4902458
6	−2.8041133	−0.6506271	−0.855840	−0.1603809
7	−0.1514431	−0.0790732	−1.007283	−0.2394512
8	0.766207E−2	−0.990590E−2	−0.999620	−0.2493600
9	0.379778E−3	−0.640298E−3	−1.000000	−0.2500003
10	0.262E−6	0.3218E−6	−1.000000	−0.2500000

The initial "wandering" can go on for a long time before converging. In fact, had $r_0 = s_0 = 0$ been used as initial guesses, then over 30 iterations (for which some of the increments are larger than 100!) would have been needed before finally finding the $q(x)$ in (17). However, it could just as well have found a different quadratic factor or not converged at all. In view of the amount of work needed for a single iteration, *this method is not recommended for hand calculation unless good initial guesses are known.*

Dividing $p(x)$ synthetically by $q(x) = x^2 + x + 0.25$, we get

a_i's:	1	−4.5	+4.55	+2.675	−3.3	−1.4375
×(−1)>		−1.0	+5.5	−9.8	+5.75	0
×(−0.25)>			−0.25	+1.375	−2.45	+1.4375
b_i's:	1	−5.5	+9.8	−5.75	0	0

so that

$$p(x) = (x + 0.5)^2 Q(x), \qquad \text{where } Q(x) = x^3 - 5.5x^2 + 9.8x - 5.75 \tag{18}$$

We can now seek a quadratic factor of $Q(x)$ by taking either

$$r_0 = \frac{-9.8}{-5.5} \doteq 1.78 \qquad \text{and} \qquad s_0 = -\frac{-5.75}{-5.5} \doteq -1.05 \tag{19a}$$

or $r_0 = s_0 = 0$. Either choice will yield $r = 3$, $s = -2.3$, so that

$$q(x) = x^2 - 3x + 2.3 = (x - 1.5)^2 + 0.05 \tag{19b}$$

This irreducible quadratic has a pair of complex conjugate roots, namely $1.5 \pm \sqrt{0.05}i$. Dividing the deflated $p(x)$ [i.e., $Q(x)$ in (18)] by it leaves $(x - 2.5)$ as the remaining factor. So the five roots of $p(x)$ in (16) are

$$-0.5, \quad -0.5. \quad 1.5 + \sqrt{0.05}i, \quad 1.5 - \sqrt{0.05}i, \quad 2.5 \tag{20}$$

In fact, the preceding discussion could have been used to get the factorization given in (3) of Section 2.4B. ■

2.5D Some Practical Aspects of Polynomial Root Finding

We first remind the reader that the possibility of ill conditioning makes it advisable to *use extended precision when finding roots of polynomials* (Section 2.4A). For the same reason, any $\bar{x}_{\text{calc}}$ found as a root of a *deflated* polynomial should be used as an *initial guess* of a root of the *original* $p(x)$. This will eliminate the errors that occur when deflation is performed with a $q(x)$ that is not exact.

The approximate location of roots can be expedited by using the following two useful results.

ROOT BOUND THEOREM. All (real and complex) roots $\bar{x}$ of

$$p(x) = a_1x^n + a_2x^{n-1} + \cdots + a_nx + a_{n+1} \qquad (a_1 \neq 0) \tag{21a}$$

lie inside the circle $|z| = R$ in the complex plane, that is,

$$|\bar{x}| \leq R, \qquad \text{where } R = 1 + \max\left\{\frac{|a_2|}{|a_1|}, \frac{|a_3|}{|a_1|}, \ldots, \frac{|a_{n+1}|}{|a_1|}\right\} \tag{21b}$$

Descartes' Rule of Signs. Let $S(+)$ be the number of sign changes that occur when the *nonzero* coefficients of $p(x)$ are listed in order, and let $R(+)$ be the number of *positive* (real) roots of $p(x)$. Then

$$R(+) \leq S(+); \qquad \text{and} \qquad R(+) \text{ is even/odd if } S(+) \text{ is even/odd} \tag{22}$$

The number of (real) *negative* roots, $R(-)$, is related to $S(-)$, the number of sign changes in the list of coefficients of $p(-x)$. Also, the number of roots on either side of a fixed x, say, $x = c$ (where, of course, $|c| < R$ of the Root Bound Theorem) can be found by applying Descartes' rule of signs to the coefficients of $p_c(x)$, where $p_c(x)$ is $p(x)$ written as a polynomial in $(x - c)$. These strategies are illustrated in the following example.

EXAMPLE 2.5D Discuss the location and number of real roots of

$$p(x) = x^5 + x^4 - 14.8x^3 + 23.4x^2 - 12.6x + 2 \tag{23}$$

Solution: The Root Bound Theorem assures us that all roots satisfy $|\bar{x}| \leq 1 + |23.4|/|1| = 24.4$. To apply Descartes' rule of signs, we list the coefficients:

$$\textit{For } p(x): \quad 1, \quad 1, \quad -14.8, \quad 23.4, \quad -12.6, \quad 2; \qquad \text{so } S(+) = 4 \tag{24a}$$

$$\textit{For } p(-x): \quad -1, \quad 1, \quad 14.8, \quad 23.4, \quad 12.6, \quad 2; \qquad \text{so } S(-) = 1 \tag{24b}$$

It follows from (24a) that $p(x)$ has zero, two, or four positive roots [in (0, 24.4)], and from (24b) that it has exactly one negative root [in $(-24.2, 0)$]. To further pinpoint the location of the positive roots, let us see where they are with respect to $c = 3$. We first rewrite $p(x)$ as

$$p_3(x) = (x - 3)^5 + 16(x - 3)^4 + 87.2(x - 3)^3 + 214.2(x - 3)^2 + 241.2(x - 3) + 97.2 \tag{25}$$

(see Problem 2-37). The coefficient lists for this polynomial are

$$\textit{For } p_3(x): \quad 1, \quad 16, \quad 87.2, \quad 214.2, \quad 241.2, \quad 99.2; \qquad \text{so } S(+) = 0 \tag{26a}$$

$$\textit{For } p_3(-x): \quad -1, \quad 16, \quad -87.2, \quad 214.2, \quad -241.2, \quad 99.2; \qquad \text{so } S(-) = 5 \tag{26b}$$

We see from (26a) that all real roots actually lie to the left of $c = 3$. Nothing new is learned from (26b). (Why?) However, we do have enough information to try, say, $q(x) =$

$(x - 1)(x - 2)$ (i.e., $r_0 = 3$, $s_0 = -2$) to approximate a pair of real roots for Bairstow's method and $x_0 = -3$ as a reliable initial guess for the Newton–Raphson method. ■

The preceding example shows that R of the Root Bound Theorem is a rather crude bound. However, it can be useful in a computer program that searches for real roots (see Problem C2-3).

2.5E Derivation of the Bairstow's Method Formulas

We wish to show by induction that if the b_j's are obtained synthetically from $a_1, \ldots, a_{n+1}$, that is, if

$$b_1 = a_1; \quad b_2 = a_2 + b_2 r; \quad b_j = a_j + b_{j-1}r + b_{j-2}s \text{ for } j = 3, \ldots, n+1 \tag{27a}$$

and the c_j's are obtained synthetically from $b_1, \ldots, b_n$, that is, if

$$c_1 = b_1; \quad c_2 = b_2 + c_1 r; \quad c_j = b_j + c_{j-1}r + c_{j-2}s \quad \text{for } j = 3, \ldots, n \tag{27b}$$

then the partial derivatives of b_i satisfy

$$\frac{\partial b_i}{\partial r} = c_{i-1} \quad \text{and} \quad \frac{\partial b_i}{\partial s} = c_{i-2} \qquad \text{for } i = 3, 4, \ldots, n \tag{28}$$

When $i = 3$ (to begin the induction) we first note from (27a) that

$$\begin{aligned} \frac{\partial b_1}{\partial r} = \frac{\partial b_1}{\partial s} = 0, \quad \text{so } \frac{\partial b_2}{\partial r} &= 0 + \left[b_1 \cdot 1 + \frac{\partial b_1}{\partial r} r \right] = b_1 \\ \text{and } \frac{\partial b_2}{\partial s} &= 0 + \frac{\partial b_1}{\partial s} r = 0 \end{aligned} \tag{29}$$

(Recall that $\partial/\partial r$ is with s fixed, and $\partial/\partial s$ is with r fixed.) Hence

$$\frac{\partial b_3}{\partial r} = 0 + \left[b_2 \cdot 1 + \frac{\partial b_2}{\partial r} r \right] + \frac{\partial b_1}{\partial r} s = b_2 + b_1 r = b_2 + c_1 r = c_2$$

$$\frac{\partial b_3}{\partial s} = 0 + \frac{\partial b_2}{\partial s} r + \left[\frac{\partial b_1}{\partial s} s + b_1 \cdot 1 \right] = b_1 = c_1$$

by (29) and (27). Similarly, assuming that (28) holds for $j = 3, \ldots, i - 1$, we get

$$\frac{\partial b_i}{\partial r} = 0 + \left[b_{i-1} \cdot 1 + \frac{\partial b_{i-1}}{\partial r} r \right] + \frac{\partial b_{i-2}}{\partial r} s = b_{i-1} + c_{i-2}r + c_{i-3}s = c_{i-1}$$

$$\frac{\partial b_i}{\partial s} = 0 + \frac{\partial b_{i-1}}{\partial s} r + \left[\frac{\partial b_{i-2}}{\partial s} s + b_{i-2} \cdot 1 \right] = b_{i-2} + c_{i-3}r + c_{i-4}s = c_{i-2}$$

This completes the induction and hence the proof of (28).

PROBLEMS

Section 2.1

2-1 For parts (a)–(d): If x_{k-2}, x_{k-1}, and x_k are converging *exactly linearly*, find x_{k+1} and the $\bar{x}$ to which they are converging.
(a) $x_2 = \frac{5}{2}, x_3 = 3, x_4 = \frac{13}{4}$ **(b)** $x_1 = -2, x_2 = -\frac{4}{3}, x_3 = -1$
(c) $x_0 = 2.5, x_1 = 2.7, x_2 = 2.8$ **(d)** $x_2 = -.5, x_3 = -1, x_4 = -1.2$

2-2 For parts (a)–(d) of Problem 2-1: If the convergence is *exactly quadratic*, find x_{k+1} exactly, and $\bar{x}$ to $8s$.

2-3 Is the convergence described most likely to be linear, quadratic, or almost quadratic? Justify, using convergence ratios. If linear, use Aitken's formula to estimate $\epsilon_4 = \bar{x} - x_4$.
(a) $\Delta x_0 \doteq .274, \Delta x_1 \doteq .182, \Delta x_2 \doteq .121, \Delta x_3 \doteq .080$
(b) $\Delta x_0 \doteq .274, \Delta x_1 \doteq .233, \Delta x_2 \doteq .168, \Delta x_3 \doteq .087$
(c) $\Delta x_0 \doteq .274, \Delta x_1 \doteq .219, \Delta x_2 \doteq .175, \Delta x_3 \doteq .140$
(d) $\Delta x_0 \doteq .274, \Delta x_1 \doteq .209, \Delta x_2 \doteq .135, \Delta x_3 \doteq .067$

Does the most rapid convergence give the smallest Δx_3? Explain.

2-4 For each given x_0, find x_1, x_2, x_3, and x_4 using the RS algorithm. Guess the rate of convergence or divergence by inspection of $\Delta x_0, \ldots, \Delta x_3$; confirm your guess by showing that the appropriate convergence ratio is nearly constant.
(a) $g(x) = \frac{1}{4}(x - 1)^2;\ x_0 = 1, 6$ **(b)** $g(x) = x/2 + 2/x;\ x_0 = 1, -3$
(c) $g(x) = e^{x/2} - 2;\ x_0 = -4, 3$ **(d)** $g(x) = -\frac{1}{2}x^2 + 4x - 4;\ x_0 = \frac{9}{5}, \frac{7}{2}$
(e) $g(x) = x/3 + 18/x^2;\ x_0 = \frac{5}{2}, \frac{7}{2}$ **(f)** $g(x) = 5 \ln x;\ x_0 = 2, 20$

2-5 For parts (a)–(f) of Problem 2-4: Sketch the graph of $g(x)$. Use it to get reasonable approximations of *all* fixed points $\bar{x}$, and to get x_1 and x_2 *graphically* as in Figure 2.1-2 for each x_0 given in Problem 2-4. Are x_1 and x_2 roughly the same as x_1 and x_2 obtained analytically in Problem 2-4?

2-6 For parts (a)–(f) of Problem 2-4: Use the approximations of $\bar{x}$ obtained in Problem 2-5 to estimate $g'(\bar{x})$ for each $\bar{x}$. Does Theorem RS (Section 2.1E) predict the convergence/divergence seen graphically in Problem 2-5?

2-7 For parts (a)–(f) of Problem 2-4: If Steffensen's method applies, use it (unmodified) starting with the given x_0, until Aitken's formula is used twice; compare the accuracy to that of x_4 of Problem 2-4.

2-8 Verify that $x = a$ is a fixed point of the given $g(x)$. Then, *without taking any iterations*, find all values of a for which RS will converge to $x = a$ linearly; quadratically.
(a) $g(x) = \dfrac{x + a}{2}$ **(b)** $g(x) = -x^2 + 3ax + a - 2a^2$ **(c)** $g(x) = \dfrac{a^3}{x^2} + ax - x^2$

2-9 For $g_1(x) = 78.8/x$ ($p = 1$ in Example 2.1A) and $x_0 = 9$:
(a) Find x_1, x_2, and x_3 using Repeated Substitution. Does what happens contradict Theorem RS of Section 2.1E? Explain.
(b) Can Aitken's formula be used in part (a)? Explain.

2-10 Do parts (a) and (b) of Problem 2-9 for $g(x) = x^{1/3}$ and $x_0 = \frac{1}{256}$ ($\bar{x} = 0$).

2-11 If $|\Delta x_0| = 0.4$, how many iterations are needed to get $4d$ accuracy if the convergence is *exactly* **(a)** linear, $C_L = 0.2$? **(b)** linear, $C_L = 0.5$? **(c)** linear, $C_L = 0.8$? **(d)** quadratic, $C_Q = 1$? **(e)** quadratic, $C_Q = 2$? **(f)** almost quadratic, $C_S = 0.5$? **(g)** almost quadratic, $C_S = 1$?

Section 2.2

2-12 How would you modify the General Iterative Algorithm 2.2A so as to use **(a)** Steffensen's method? **(b)** the Modified Steffensen's method (Algorithm 2.1F)?

2-13 In a programming language of your choice, write a *syntactically correct* statement that sets CONVGD to TRUE if X and XPREV agree **(a)** to 4*d*; **(b)** to 6*s*.

2-14 The following termination tests are sometimes used. Describe in words (in terms of significant digits or decimal places) when they stop the iteration.
(a) IF $|X - Xprev| \leq 0.5 \cdot 10^{-p} \cdot \max(1, |X|)$ THEN stop the iteration.
(b) IF $|X - Xprev| \leq 0.5 \cdot 10^{-p} \cdot \min(|Xprev|, |X|)$ THEN stop the iteration.

2-15 Form $\Delta x_0, \ldots, \Delta x_3$ for $x_0 = 0.1101000100000001$, $x_1 = 0.0101000100000001$, $x_2 = 0.0001000100000001$, $x_3 = 0.0000000100000001$, $x_4 = 0.0000000000000001$ to show that x_k is converging to $\bar{x} = 0$ quadratically. Would the Relative Difference Test for 4*s* stop the iteration? Explain.

2-16 In Example 2.2C, was it actually necessary to get $\bar{u}_1$ and $\bar{u}_2$ to 6*s* (two guard digits)? Explain [see Problem 1-20(b)].

Section 2.3

2-17 For a given $c > 0$, find an equation of the form $f(x) = 0$ whose solution is **(a)** $\sqrt[5]{c}$; **(b)** $\tan^{-1}c$; **(c)** a fixed point of $1 + xe^{-cx}$.

2-18 Rewrite the given equation as $g = h$, where *both* g and h are functions having *familiar graphs* that can be sketched quickly using *at most* one or two plotted points. Graph g and h on the same set of axes to get estimates of *all* solutions of the given equation.
(a) $x^5 + x^2 = 9$ **(b)** $2/v^4 - 3/(v + 1) = 0$
(c) $(u - 1)e^u - 2 = 0$ **(d)** $3 - t - \sqrt{t} = 0$
(e) $e^x - x^2 = 2$ **(f)** $\theta^2 \sin\theta = \cos\theta$

2-19 For (a)–(f) of Problem 2-18: Find the smallest positive root to 5*s* by the Newton–Raphson (NR) method. Use your graphical estimate as your initial guess, x_0.

2-20 Same as Problem 2-19 but using the Secant (SEC) Method with $x_{-1} = 1.01x_0$.

2-21 Same as Problem 2-19 but using the Bisection (BIS) Method. Be sure that $f(a)f(b) < 0$.

2-22 Same as Problem 2-19 but using the False Position (FP) Method. Be sure that $f(a)f(b) < 0$.

2-23 Same as Problem 2-19 but using the False Position (FP) Method with Steffensen's method (unmodified) used to accelerate it.

2-24 Use the appropriate convergence ratio to verify that the rate of convergence was as expected in those of Problems 2-19–2-23 that you did.

2-25 Let $f(x) = x^4 - 3x^3 - 2x^2 + 12x - 8$. Note that $f(1) = f(2) = 0$.
(a) Find x_1, x_2, x_3, and x_4 using the Bisection (BIS) Method with $a = 0.8$ and $b = 1.3$.
(b) Same as part (a) but using the False Position (FP) Method.
(c) Which of BIS or FP will find $\bar{x} = 2$ starting with $a = 2 - h$ and $b = 2 + 2h$, where $h > 0$? Explain. [Consider $f'(2)$ and $f''(2)$.]

2-26 Let n be the number of iterations of the Bisection Method needed to ensure that a root $\bar{x}$ is bracketed in an interval of length less than ϵ. Express n in terms of ϵ and the initial a and b.

2-27 Use NR with the given x_0 to find to 6*s* the *fixed point* that could not be found by Repeated Substitution
(a) in Problem 2-9; **(b)** in Problem 2-5(c).

2-28 Same as Problem 2-27 but using SEC with $x_{-1} = 1.01x_0$.

2-29 Let $f(x) = x^3 - 7x$. This f has three simple roots, 0 and $\pm\sqrt{7}$.
(a) Find x_1, x_2, and x_3 using NR if x_0 is (i) 1.19; (ii)1.5.
(b) Use a neat sketch of $y = f(x)$ to explain why x_k will converge but not to the root closest to x_0 in both (i) and (ii) of part (a).

Section 2.4

2-30 In (3) of Section 2.4B, what is $\phi(x)$ for $\bar{x} = -0.5$? for $\bar{x} = 2.5$?

2-31 Prove that $\bar{x} = 0$ is a triple root of $g(x) = \tan x - xe^{-x^2}$ two ways.
(a) Show that $g(0) = g'(0) = g''(0) = 0$, but $g'''(0) \neq 0$.
(b) Using the Maclaurin series for tan and exp (see Appendix II.2D) to show that definition (4) of Section 4.2B holds with $m = 3$.

2-32 Use Aitken's formula to get the best estimate of $\bar{x}$ from **(a)** the NR iteration; **(b)** the SEC iteration of Table 2.4-3.

2-33 Will a polynomial with integer coefficients be stored without inherent error on a computer? Explain.

Section 2.5

2-34 Perform two iterations of Bairstow's method starting with r_0, s_0.
(a) $p(x) = x^4 - 5x^3 + 8.5x^2 - 6x + 2$; $r_0 = 1$, $s_0 = 0$; $c = 1$
(b) $p(x) = 2x^4 + 5x^3 - 5x^2 - 5x + 3$; $r_0 = -3$, $s_0 = -3$; $c = -2$
(c) $p(x) = 1x^4 - x^3 + 1x^2 + 2$; $r_0 = 2$, $s_0 = -1$; $c = 1$

2-35 For parts (a)–(c) of Problem 2-34, do (i)–(iii) as in Example 2.5D.
(i) Use the Root Bound Theorem of Section 2.5D to find a crude bound for the magnitude of the real or complex roots.
(ii) What does Descartes' rule of signs tell you about the real roots? Apply it to $p_c(x)$ for the given c. What can you say now about the real roots?
(iii) Based on your results in (ii), describe reliable initial guesses for Bairstow's method.
(iv) Same as (iii) but for the Newton–Raphson method.

2-36 Let $p(x) = x^5 + x^4 - 14.8x^3 + 23.4x^2 - 12.6x + 2$. Use Bairstow's method, starting with $r_0 = 3$, $s_0 = \frac{9}{4}$, to find to 6s a pair of real roots in the interval (0, 3) (see Example 2.5D). Deflate $p(x)$ and apply Descartes' Rule of Signs to the deflated cubic. Can you deduce the nature and location of the three remaining roots?

2-37 Use the Iterated Synthetic Division Algorithm of Problem 1-13 to get (25) of Example 2.5D from the given $p(x)$ in (23).

MISCELLANEOUS PROBLEMS

M2-1 You can tell *by inspection* of the Δx_k column in Table 2.1-3 [Example 2.1B(b)] that the convergence is not quadratic. Why?

M2-2 Illustrate graphically a fixed point $\bar{x}$ of $g(x)$ for which RS will result in **(a)** quadratic convergence **(b)** $x_2 = x_0$ ("cycling").

M2-3 **(a)** Obtain Aitken's formula (20b) from (19c) (Section 2.1F).
(b) Show that $(x_0)_{\text{improved}}$ in (21) is simply a rearrangement of $(x_2)_{\text{improved}}$ in (20b).

M2-4 The SEC formula can be rearranged as $x_{n+1} = (x_{n-1}y_n - x_ny_{n-1})/(x_n - x_{n-1})$. Why is this form less desirable than (13b) of Section 2.3D? Does the similar FP method formula (21) of Section 2.3G have the same shortcoming? Explain.

M2-5 **(a)** It can be shown [40] that for simple roots, the SEC Method satisfies $|\Delta x_{k+1}| \approx K|\Delta x_k|\,|\Delta x_{k-1}|$. Verify this for Table 2.3-1.
(b) Suppose that $|\Delta x_{k+1}| \approx K|\Delta x_k||\Delta x_{k-1}|$ and that $|\Delta x_{k+1}| \approx C_S|\Delta x_k|^p$ and $|\Delta x_k| \approx C_S|\Delta x_{k-1}|^p$, where $1 < p < 2$ (superlinear convergence). Show that $p = 1 + 1/p$, hence $p = (1 + \sqrt{5})/2 \doteq 1.618$.

M2-6 Of the two equivalent equations of the form $f(x) = 0$ in (4b) of Section 2.3A, which is best for NR? for SEC? Explain.

M2-7 Taking $m_k = [f(x_k + h_k) - f(x_k)]/h_k$ where $h_k = f(x_k)$ (**Steffensen's stepsize**) in the Slope Method Algorithm 2.3B results in quadratic convergence [12].
(a) Verify this result for $f(x) = x^2 - 78.8$, using $x_0 = 14$.
(b) Describe the advantages and disadvantages of this m_k, as compared to that of NR and SEC.

M2-8 **(a)** Show that $f(x) = x^5 + 20x^3 - 9 = 0$ has exactly one root $\bar{x}$, and that it lies between 0 and 1. Find $\bar{x}$ to $6s$ by any method.
(b) Solve Problem 1 at the beginning of Section 1.1 to $5s$. (Check endpoints!)

M2-9 Use any root-finding method to find the following to $6s$:
(a) The desired $\bar{t} \approx 1$ that satisfies (2d) of Section 2.0.
(b) All solutions on (0, 1] of $\sin(\sqrt{\sec x + x^3 e^{5x/(\tan x)}}) = e^{-x}$.
(c) The smallest positive values of y defined implicitly by the equation $x \cos y = y^3 e^x - 5 y/x$ when $x = 1$.
(d) $5 \cos u + \operatorname{sech} u = 0$. NOTE: $\operatorname{sech} u = 2/(e^u + e^{-u})$.

M2-10 Suppose that NR is applied to $f(x) = x^2 - c$ to find $\bar{x} = \sqrt{c}$ for $c > 0$.
(a) Find a formula for x_2 as a rational function of x_0.
(b) Use your formula with $x_0 = (c + 1)/2$ if c is (i) 0.01; (ii) 4; (iii) 16; (iv) 900. What is the largest relative error (in magnitude) of x_2?
(c) Repeat (b), but with $x_0 = (5 + c)/4 - 1/(1 + c)$. Compare the accuracy to that of part (b). NOTE: Many devices use a rational function similar to this x_0 for their built-in square root routines.

M2-11 The only root of $f(x) = \tan^{-1}x$ is $\bar{x} = 0$. Take three iterations of NR **(a)** with $x_0 = \sqrt{8}$; **(b)** with $x_0 = 2$; **(c)** with $x_0 = \sqrt{2}$. Explain the different results in words, referring to a graph if desired.

M2-12 The cube root function $f(x) = x^{1/3}$ has only root, namely $x = 0$.
(a) Use the NR iteration formula for $f(x) = x^{1/3}$ to express x_{k+1} in terms of only x_k. *Simplify algebraically as much as you can!*
(b) Your formula in part (a) should tell you that NR will not converge to $\bar{x} = 0$ unless x_0 is *exactly* zero (to the accuracy of the device)! Does this contradict (11) of Section 2.3C? Explain. Refer to an accurate sketch of the graph of f.

M2-13 Under what conditions on $f'(x)$ and $f''(x)$ near $\bar{x}$ will the FP Method always move a? Repeat for b.

M2-14 The polynomial $p(x) = x^4 - 10x^3 + 35x^2 - 50x + 24$ has roots 1, 2, 3, 4. To see the effects of roundoff, do the following.
(a) Use Bairstow's method with $r_0 = 3.2$, $s_0 = -2.2$. Stop when both $|dr_k|$ and $|ds_k|$ are less than 0.005 (i.e., have $2d$ accuracy). Then deflate and find the two remaining roots exactly using the quadratic formula.
(b) Same as part (a) but starting with $r_0 = 7.3$, $s_0 = -12.3$. Based on your results, which roots should you seek first when using deflation, the smallest (in magnitude) or largest?

M2-15 To see how Bairstow's method behaves near multiple roots, consider the cubic $p(x) = x^3 - 3x^2 + 3x - 1$ which is $(x - 1)^3$.
(a) Perform four iterations of Bairstow's method with $r_0 = 2.1$, $s_0 = -0.9$. Does the convergence to $\bar{x} = 1$ appear to be linear or quadratic?
(b) Same as part (a) but with $r_0 = s_0 = 0$. Is convergence evident?

M2-16 For the nth degree polynomial $a_1x^n + a_2x^{n-1} + \cdots + a_nx + a_{n+1}$, show that the product and sum of the n roots are, respectively, a_{n+1}/a_1 and $-a_2/a_1$.

M2-17 The current i (in microamps, μa) through a diode is related to the voltage v (in volts) across it by the **diode equation** $i = I_s(e^{v/\theta} - 1)$, where I_s is the saturation current (in μa) and θ is the diode variable. A junction diode for which $I_s = 20$ and $\theta = 0.052$ is connected to a circuit for which v and i must also satisfy $v + 10^4 i = 4$. Find the current (in μa) through the diode. Get an initial guess by approximating the diode equation by $i = I_s v/\theta$.

M2-18 A feeding trough is L meters (m) long. Its cross section is a semicircle of radius r meters, diameter on top. When filled with feed to within x meters of the top, the volume V (in cubic meters, m^3) of the feed is

$$V = L\left[\frac{1}{2\pi r^2} - \frac{x}{r^2} - x^2 + r^2\sin^{-1}\left(\frac{x}{r}\right)\right]$$

If $L = 25$ m and $r = 2$ m, find x when V is **(a)** 10 m^3; **(b)** 50 m^3; **(c)** 100 m^3. For an initial guess, try $(V/V_{max})r$, where $V_{max} = \frac{1}{2}\pi r^2 L$.

M2-19 **Archimedes' law** states that when a solid of density δ_S is placed in a liquid of density δ_L, it will sink to a depth h that displaces an amount of liquid whose weight equals that of the solid. For a sphere of radius r, Archimedes' law becomes

$$\frac{\pi}{3}(3rh^2 - h^3)\delta_L = \frac{4}{3}\pi r^3 \delta_S$$

If $r = 5$ ft and $\delta_L = 62.5$ lb/ft^3 (density of water), find h if δ_S in lb/ft^3 is **(a)** 20; **(b)** 45; **(c)** 70. Usc $2r\delta_S/\delta_L$ as an initial guess of the desired h.

M2-20 The current I in amperes when an E volt voltage is applied across a varistor is $I = 0.0001^5 - 0.1273^2 + E$. What voltage E is needed to produce a current of (a) 0.75 amperes? (b) 1.5 amperes? Note that $I \approx E$ for small E; use this to get an initial guess.

M2-21 A widely used generalization of the **ideal gas law** $PV = nRT$ is the **van der Waals equation**

$$\left(P + \frac{a}{V^2}\right)(V - b) = nRT, \qquad \text{where } R = \text{gas constant} = 0.08205\,\frac{\text{l}\cdot\text{atm}}{°K\cdot\text{mole}}$$

and n is the number of moles of gas. For isobutane, $a = 12.87$ l^{2}atm/mole^2 and $b = 0.1142$ l/mole. Find the volume of 1 mole of isobutane at a temperature $T = 313°K$ and a pressure $P = 2$ atm. Use the ideal gas law to get an initial guess. SUGGESTION: You may want to rewrite the equation as a polynomial equation in V.

M2-22 A w pound (lb) object is released with zero velocity at $t = 0$. If it encounters air resistance with a coefficient of friction r lb·sec/ft, then the distance s in feet that it has fallen after t seconds is

$$s = v_{ss}\left[t - v_{ss}\frac{1 - e^{-gt/v_{ss}}}{g}\right], \qquad \text{where } v_{ss} = \frac{w}{r}$$

and $g = 32$ ft/sec^2, the gravitational acceleration constant. If $w = 4$ and $r = 0.1$, how long does it take for the object to fall **(a)** 16 ft; **(b)** 256 ft? Get initial guesses from the fact that $s = \frac{1}{2}gt^2$ when $r = 0$.

M2-23 The torsion T (Kip·in) and maximum shear stress τ_{max} (Kip/in.^{2}) for a pipe of inner radius R_i (in.) and outer radius R_o (in.) are related by the equation $TR_o = \pi(R_o^4 - R_i^4)/2$. What should R_o be to get $\tau_{max} = 36$ (Kip/in.^{2}) when T is 0.9 Kip·in.? Take $R_o = 2R_i$ as an initial guess.

M2-24 In analyzing the antisymmetric buckling of beams, one must find the **stability factor** ϕ that satisfies $0 < \phi \leq \pi/2$, by solving the equation $6\cos\phi = \gamma\phi\sin\phi$, where γ depends on the geometry of the beams and the critical stresses on them. Find ϕ when γ is **(a)** 0.1; **(b)** 0.5; **(c)** 1; **(d)** 5.

COMPUTER PROBLEMS

C2-1 Modify `SUBROUTINE FIXPT` of Section 2.2C so as to find $(x_2)_{\text{improved}}$ and use it as a new x_0 as soon as x_2 is found (*unmodified* Steffensen's method.) Test it using $g_p(x)$ of Example 2.1B with $p = \frac{3}{2}, 2, \frac{1}{2}$ and 1 (see Problem 2-9). [SUGGESTION: In Fortran, pass p to `FUNCTION G(X)` using a block common: `COMMON /PARAM/ P`.] Which shortcomings of the RS algorithm still remain?

C2-2 **(a)** Write a subprogram `STEFF(X, XPREV, DELTAX, DXPREV, PCT, CLPREV)` for the Modified Steffensen's method Algorithm 2.1F.

(b) Modify `SUBROUTINE FIXPT` of Section 2.2C so that it calls `STEFF` each iteration. Test it as described in Problem C2-1. Which shortcomings of the RS algorithm still remain?

C2-3 ***An Initial Guess Finder*** Write a computer program `FINDX0` that will evaluate $f(x)$ at $x = a, a + h, a + 2h, \ldots$ until $x > b$, where a, h, and b are input by the user. It should print a message only if it detects either a sign change (definite root) or a turning point less than h units from the x-axis (possible root). (BE WARNED: This is not as easy as it might appear.) Test `FINDX0` with both $p(x)$ and $-p(x)$, where $p(x)$ is given in (3) of Section 2.4B.

C2-4 Use `FINDX0` of Problem C2-3 with $a = -0.0001$ and $b = 0.0002$ for $g(x)$ of Example 2.4B(b). Take 20 steps to get from a to b. Explain the results obtained.

C2-5 Refer to the Iterated Synthetic Division Algorithm (Problem M1-13).

(a) Propose an efficient algorithm that performs two synthetic divisions *in the same loop* to get both $p(x_k)$ and $p'(x_k)$ for the NR algorithm.

(b) Write a subprogram `NRPOL(NP1, A, x, NUMSIG, MAXIT, PRNT, IW, CNVRGD)` similar to `SUBROUTINE SEC` of Section 2.4E to find roots of a polynomial $p(x)$ of degree n. It should use an array `A` of `NP1` $= n+1$ coefficients of $p(x)$ (in order of descending powers of x, including zeros for missing terms) instead of an external `FUNCTION` subprogram for the polynomial.

(c) Suppose that the nth-degree polynomial $p(x)$ satisfies $p(\bar{x}) = 0$ so that $p(x) = (x - \bar{x})Q(x)$. If `A` stores the `NP1` ($= n + 1$) coefficients of $p(x)$ and `B` stores the n coefficients of $Q(x)$, write a subprogram `DFLAT(N, NP1, A, B)` that **deflate**s $p(x)$ by replacing it by $Q(x)$ as follows:

$$NP1 \leftarrow n; \quad n \leftarrow n - 1$$
$$\text{DO FOR } i = 1 \text{ TO } NP1; \quad b_i \leftarrow a_i$$

Modify `NRPOL` so as to call `DFLAT` repeatedly after each root is found, until $n = 1$, at which point the remaining root is $-a_2/a_1$.

(d) Test your `NRPOL` subprogram (including `DFLAT` if you wrote it) using $p(x)$'s in Problem 2-34.

C2-6 Write a subprogram `PROOTS` to implement Algorithm 2.5A for quadratic deflation of an nth-degree polynomial $p(x)$. It should call subprograms `BAIRST` (which implements Algorithm 2.5C) for the **factor** step and `DEFLAT` for the **deflate** step). If in Fortran, `QROOTS` (Figure 1.4-6) should be used to get the roots of $q(x)$ obtained in the **solve** step. Test `PROOTS` using $p(x)$'s in Problem 2-34.

C2-7 Adapt your program in Problem C2-5 so as to find complex roots of real or complex polynomials. Use it to find all roots of **(a)** $p(x)$ of Problem M2-8; **(b)** $p(x)$ of Problem M2-14; **(c)** $F(x)$ in (1) of Problem 1 of Section 1.1.

C2-8 Thin bars of length c and a are pinned to the x-axis at $x = 0$ and $x = d$. Their (counterclockwise) angles with the positive x-axis are, respectively, ϕ and θ. If their free ends are pinned to the ends of a bar of length b, where $a < d < a + b + c$, then ϕ and θ of the resulting *linkage* will satisfy the **Freudenstein equation**

$$\cos(\theta - \phi) = R_1 \cos\theta - R_2 \cos\phi + R_3$$

where $R_1 = d/c$, $R_2 = d/a$, and $R_3 = (a^2 - b^2 + c^2 + d^2)/(2ac)$. If $a : b : c : d$ are in the ratio $1 : 2 : 2 : 2$, find ϕ when θ is **(a)** 0°; **(b)** 30°; **(c)** 60°; **(d)** 90°.

3

Direct Methods for Solving Linear Systems

3.0 Introduction

A **linear combination** of the variables $x_1, x_2, \ldots, x_n$ is a weighted sum

$$a_1x_1 + a_2x_2 + \cdots + a_nx_n$$

A **linear equation** in $x_1, \ldots, x_n$ is obtained by requiring a linear combination to assume a prescribed value b, that is,

$$a_1x_1 + a_2x_2 + \cdots + a_nx_n = b$$

Such equations arise frequently, generally n at a time, that is, as $\boldsymbol{n \times n}$ **linear systems**. For example, the $n \times n$ linear system

$$\begin{array}{ccc} a_{11}x_1 + a_{12}x_2 + \cdots + a_{1n}x_n & = & b_1 \\ \vdots & & \vdots \\ a_{n1}x_1 + a_{n2}x_2 + \cdots + a_{nn}x_n & = & b_n \end{array}$$

TABLE 3.0-1 Situations Described by $n \times n$ Linear Systems

Discipline	*Context*	*Units of $x_1, \ldots, x_n$*
Electrical engineering	Passive networks	Voltage at certain nodes, or current in certain branches
Mechanics	Static equilibrium	Component of force at a point
Transportation, economics, engineering	Efficient allocation among several points	Net flow from one point to another
Biology/sociology	Population growth	Number of individuals in a particular age group at a particular time

Genetics — Estimation of breeding value — eg. kg milk

can arise in the contexts shown in Table 3.0-1 as well as many others that occur in the natural, social, and engineering sciences. A readable reference that presents details of a variety of these occurrences is [37]. Typically, in these contexts, $2 \leqslant n \leqslant 20$. However, n as large as 100 to 1000 can occur.

The reader no doubt has had experience solving 2×2 or 3×3 linear systems. Our primary purpose in this chapter is to describe efficient methods that yield the exact solution (if one exists) of an $n \times n$ *linear* system in a finite number of steps if exact arithmetic (i.e., no rounding) is used. Such methods, generally termed **direct methods**, fall into two categories: **elimination methods** and **factorization methods**.

To discuss linear systems, it will be necessary to introduce the basic algebraic properties of *matrices*. The time spent doing this will be well rewarded because matrices make it possible to describe linear systems in a concise way that makes solving $n \times n$ linear systems seem like solving the familiar 1×1 linear system as follows:

$$ax = b \quad \Rightarrow \quad \bar{x} = \frac{b}{a} = a^{-1}b \qquad \text{if } a \neq 0$$

Moreover, matrices can be used to generalize the Newton–Raphson method to solve *nonlinear* $n \times n$ systems.

3.1 Basic Properties of Matrices

3.1A Terminology and Notation

A **matrix** is a rectangular array of numbers, usually enclosed in square brackets. A matrix having m rows and n columns is called an $\boldsymbol{m \times n}$ (read "m by n") **matrix** or an $\boldsymbol{m \times n}$ **array** and will be denoted by any of the notations

$$A = (a_{ij})_{m\times n} = \begin{bmatrix} a_{11} & a_{12} & \cdots & a_{1j} & \cdots & a_{1n} \\ a_{21} & a_{22} & \cdots & a_{2j} & \cdots & a_{2n} \\ \vdots & \vdots & & \vdots & & \vdots \\ a_{i1} & a_{i2} & \cdots & a_{ij} & \cdots & a_{in} \\ \vdots & \vdots & & \vdots & & \vdots \\ a_{m1} & a_{m2} & \cdots & \underbrace{a_{mj}}_{\text{col}_j A\ (1 \le j \le n\ =\ \text{number of columns})} & \cdots & a_{mn} \end{bmatrix} \begin{matrix} \\ \\ \\ \}\ \text{row}_i A \\ (1 \le i \le m = \text{number of rows}) \\ \\ \end{matrix} \tag{1}$$

The number a_{ij} in the ith row ($\text{row}_i A$) and jth column ($\text{col}_j A$) is called the ijth **entry** of the matrix A. The **transpose of $A_{m\times n}$** is the $n \times m$ matrix A^T obtained by interchanging the rows and columns of A, that is, $A^T = (a_{ji})_{n\times m}$. A is **symmetric** if it is **square** (i.e., $m = n$) and $A^T = A$ so that $\text{col}_j A^T = \text{row}_j A$ and $\text{row}_i A^T = \text{col}_i A$. A **diagonal matrix**, which we shall denote by $\text{diag}(d_{11}, \ldots, d_{nn})$, is a square matrix whose **main diagonal** entries are $d_{11}, d_{22}, \ldots, d_{nn}$ (some or all of which can be zero), and whose off-diagonal entries are all zero; diagonal matrices are always symmetric.

Matrices for which both $m > 1$ and $n > 1$ will be denoted by uppercase letters $A, B, C, \ldots$ as in (1). Matrices with $m = 1$ (respectively, $n = 1$) will be called **row** (respectively, **column**) **vectors** and denoted by lowercase **boldface** letters **a**, **b**, **x** .... Entries of vectors will be called **components** and written with *single* subscripts: row vectors as $\mathbf{y} = [y_1 \ \cdots \ y_n]$, and column vectors as $\mathbf{c} = [c_1 \ \cdots \ c_n]^T$. Numbers will be referred to as **scalars** and denoted by lowercase letters $a, b, x, y, \alpha, \beta, \ldots$ to distinguish them from matrices and vectors. We will not make a distinction between a 1×1 matrix, say $[a]$, and the scalar a. To illustrate some of these definitions, consider

$$A = \begin{bmatrix} 1 & 4 \\ 0 & 5 \\ 0 & 6 \end{bmatrix}, \quad \mathbf{b} = [1 \quad 0 \quad 3], \quad C = \begin{bmatrix} 3 & 0 \\ \sqrt{2} & 0 \\ \pi & -1 \end{bmatrix}, \quad D = \begin{bmatrix} 5 & 0 \\ 0 & 8 \end{bmatrix}, \quad \mathbf{e} = \begin{bmatrix} 3 \\ -7 \\ 0 \end{bmatrix} \tag{2}$$

A and C are 3×2 matrices with $\text{row}_3 A = [0 \quad 6]$, $\text{col}_2 C = [0 \quad 0 \quad -1]^T$, $a_{12} = 4$, and $c_{31} = \pi$; $D = \text{diag}(5, 8)$; **b** is a row vector with $b_2 = 0$; and **e** is a column vector with $e_1 = 3$. Column vectors will often be written in one line either as $\mathbf{e} = [3 \quad -7 \quad 0]^T$ or by writing $\mathbf{e}^T = [3 \quad -7 \quad 0]$.

3.1B Addition, Subtraction, and Scalar Multiplication

Two matrices $A = B$ are **equal**, written $A = B$, if they are the same size and have identical entries (i.e., $a_{ij} = b_{ij}$ for all i, j). If A and B are equisized, say $A = (a_{ij})_{m\times n}$ and $B = (b_{ij})_{m\times n}$, and s is any scalar, then the **sum** $A + B$, **difference** $A - B$, and **scalar multiple** sA are the $m \times n$ matrices defined in the natural way by

$$A \pm B = (a_{ij} \pm b_{ij})_{m\times n} \qquad \text{and} \qquad sA = (sa_{ij})_{m\times n} \tag{3}$$

Of special importance for these algebraic operations is the matrix

$$\mathrm{O}_{m\times n} = (0)_{m\times n} \qquad \text{(all zeros)} \tag{4}$$

called the **zero matrix**. It will be written simply as O if its size is clear. For example, for A, $\mathbf{b}$, C, D, and $\mathbf{e}$ given in (2),

$$(-2)\mathbf{e} = \begin{bmatrix} -6 \\ 14 \\ 0 \end{bmatrix}, \quad 2A - 3C = \begin{bmatrix} -7 & 8 \\ -3\sqrt{2} & 10 \\ -3\pi & 15 \end{bmatrix}, \quad \text{and} \quad \mathbf{b}^T + \mathbf{e} = \begin{bmatrix} 4 \\ -7 \\ 3 \end{bmatrix} \qquad \textbf{(5)}$$

whereas $A \pm \mathbf{b}$, $\mathbf{b} \pm C$, $C \pm D$, and $\mathbf{b} \pm \mathbf{e}$ are not defined.

It follows from definitions (3)–(4) that for any equisized matrices A, B, and C, and any scalar s:

A1:	$A + B = B + A$	(Matrix addition is *commutative*)
A2:	$A + \mathrm{O} = \mathrm{O} + A = A$	(O is the *zero* for matrix addition)
A3:	$-A$, defined as $(-1)A$, satisfies $A + (-A) = (-A) + A = \mathrm{O}$	($-A$ is the *negative of* A)
A4:	$(A + B) + C = A + (B + C)$	(Matrix addition is *associative*)
A5:	$s(A + B) = sA + sB$	(Scalar *distributive law*)

The associative law **A4** assures us that sums such as $A + B + C + D$, when defined, can be written unambiguously *without parentheses*. In particular,

$$A + A + \cdots + A \ (k \text{ times}) = kA, \qquad k = 0, 1, 2, \ldots$$

that is, adding A to itself k times has the same effect as scaling it by the nonnegative integer k; in particular $0A = \mathrm{O}$, and $1A = A$.

Just as the sum of the two $m \times n$ matrices A and B was defined "entrywise," we could define their product AB as $(a_{ij}b_{ij})_{m\times n}$. However, there are not enough mathematical applications of this definition to make it worth considering. On the other hand, the less "natural" definition of AB given next is so useful that it is the key to the importance of matrices in studying linear processes.

3.1C Matrix Multiplication

The **inner product** (or **scalar product**) of a row vector $\boldsymbol{r} = [r_1 \ \ r_2 \ \ \cdots \ \ r_n]$ and a column vector $\mathbf{c} = [c_1 \ \ c_2 \ \ \cdots \ \ c_n]^T$ is the scalar $\mathbf{rc}$ obtained as the sum

$$\mathbf{rc} = r_1c_1 + r_2c_2 + \cdots + r_nc_n = \sum_{k=1}^{n} r_kc_k \qquad \textbf{(6a)}$$

If $A = (a_{ij})_{m\times n}$ and $B = (b_{ij})_{n\times p}$ so that *A has as many columns (n) as B has rows*, then the **product** AB is defined as the $m \times p$ matrix whose ijth entry is the inner product $\text{row}_i A$

col_jB. To get this ijth entry of AB, scan *across* the ith row of the left factor A and *down* the jth column of the right factor B, summing the products $a_{ik}b_{kj}$ as you go:

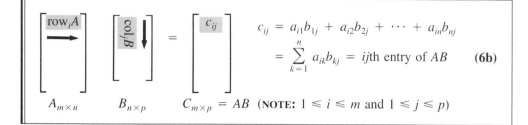

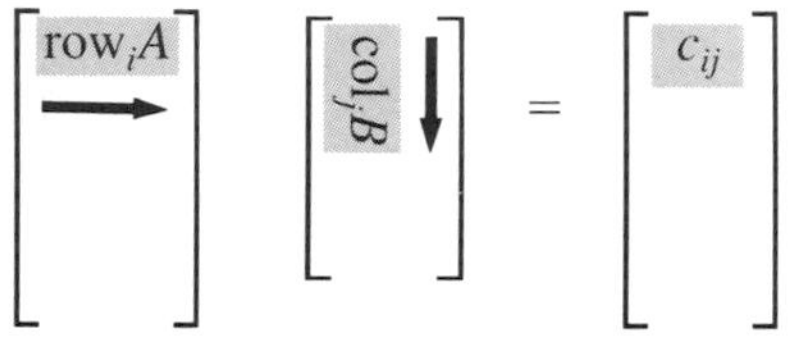

$$c_{ij} = a_{i1}b_{1j} + a_{i2}b_{2j} + \cdots + a_{in}b_{nj} = \sum_{k=1}^{n} a_{ik}b_{kj} = ij\text{th entry of } AB \qquad \textbf{(6b)}$$

$A_{m\times n}$ $\quad B_{n\times p}$ $\quad C_{m\times p} = AB$ (NOTE: $1 \leq i \leq m$ and $1 \leq j \leq p$)

Notice that, when defined, the product AB has as many rows (m) as the left factor A, and as many columns (p) as the right factor B. It is immediate from definition (6b) that for $1 \leq i \leq m$ and $1 \leq j \leq p$:

$$\text{row}_i(AB) = (\text{row}_iA)B \qquad \text{and} \qquad \text{col}_j(AB) = A(\text{col}_jB) \qquad \textbf{(7)}$$

The following example illustrates multiplying $A_{3\times 2}$ by $B_{2\times 2}$.

$$\begin{bmatrix} 1 & 2 \\ 3 & 4 \\ 5 & 6 \end{bmatrix}\begin{bmatrix} 7 & 9 \\ 8 & 10 \end{bmatrix} = \begin{bmatrix} (1\cdot 7 + 2\cdot 8) & (1\cdot 9 + 2\cdot 10) \\ (3\cdot 7 + 4\cdot 8) & (3\cdot 9 + 4\cdot 10) \\ (5\cdot 7 + 6\cdot 8) & (5\cdot 9 + 6\cdot 10) \end{bmatrix} = \begin{bmatrix} 23 & 29 \\ 53 & 67 \\ 83 & 105 \end{bmatrix}$$

Since the product $B_{2\times 2}A_{3\times 2}$ is not defined, it makes no sense to ask if $BA = AB$. In fact, the question makes sense only if both A and B are square and of the same size (i.e., both $n \times n$). If it does happen that $AB = BA$, we say that A and B **commute**. The following example shows that *we should not expect two $n \times n$ matrices to commute*.

EXAMPLE 3.1C Suppose that $A = \begin{bmatrix} 1 & -2 \\ -2 & 4 \end{bmatrix}$ and $B = \begin{bmatrix} 2 & 0 \\ 1 & 0 \end{bmatrix}$. Then

$$AB = \begin{bmatrix} 0 & 0 \\ 0 & 0 \end{bmatrix} = \mathrm{O}_{2\times 2} \quad \text{whereas} \quad BA = \begin{bmatrix} 2 & -4 \\ 1 & -2 \end{bmatrix} \neq \mathrm{O}_{2\times 2}$$

Notice that unlike the *scalar* zero, *a zero matrix can have* ***proper divisors***, that is, *nonzero* factors A and B such that $AB = \mathrm{O}$. ■

Of special importance for matrix multiplication is the $n \times n$ **identity matrix**

$$I_n = \begin{bmatrix} 1 & 0 & 0 & \cdots & 0 \\ 0 & 1 & 0 & \cdots & 0 \\ 0 & 0 & 1 & \cdots & 0 \\ \vdots & \vdots & \vdots & & \vdots \\ 0 & 0 & 0 & \cdots & 1 \end{bmatrix} \quad \left(\begin{array}{l} n \times n \text{ matrix with} \\ a_{ii} = 1 \text{ for } 1 \leq i \leq n \\ \text{and } a_{ij} = 0 \text{ for } i \neq j \end{array}\right) \qquad \textbf{(8)}$$

written as simply I when its size is clear. It follows from definitions (6) and (8) that whenever all sums and products are defined:

M1:	$(sA) = A(sB) = s(AB)$	(Scalars can be moved through products)
M2:	For $A_{m \times n}$, $I_m A = AI_n = A$	(I is the *identity for matrix multiplication*)
M3a:	$(A + B)C = AC + BC$	(*Right distributive law*)
M3b:	$A(B + C) = AB + AC$	(*Left distributive law*)
M4:	$A(BC) = (AB)C$	(Matrix multiplication is *associative*)

In view of **M2**, I plays the role for matrices that the number 1 does for scalars. It follows from the associative law **M4** that products such as $ABCD$ (when defined) can be written *without parentheses* with the assurance that the answer will come out the same whether calculated as

$$(AB)(CD) \quad \text{or} \quad A[B(CD)] \quad \text{or} \quad A[(BC)D] \quad \text{or} \quad [(AB)C]D \quad \text{or} \quad [A(BC)]D$$

as long as the *order* A, B, C, D is maintained. **From this point on we shall restrict our discussion to *column vectors* and *square matrices*.** This will ensure that the linear combinations $sA + tB$ and $s\mathbf{x} + t\mathbf{y}$ and the products $A\mathbf{x}$, AB, and BA are defined (although AB and BA may not be equal). It then follows from **A1–A4** of Section 3.1B and **M1–M4** above that *the usual algebraic rules governing addition, subtraction, and multiplication apply to (square) matrices, as long we do not reverse the order in which noncommutative matrices are multiplied*. For example,

$$(A+2B)(A - 2B) = AA + 2BA - 2AB - 4B^2 = (A^2 - 4B^2) + 2(BA - AB)$$

This will *not* equal $A^2 - 4B^2$ unless A and B commute (i.e., $BA = AB$).

We shall not define division by a matrix A; however, there may exist a matrix A^{-1} such that multiplication by A^{-1} has the effect of dividing by A.

3.1D The Inverse A^{-1} of a Nonsingular Matrix A

An $n \times n$ matrix A is called **nonsingular** or **invertible** if an $n \times n$ matrix B can be found such that

$$AB = BA = I_n \tag{9}$$

If no such B exists, then A is called **singular**.

Let us first show that *at most one matrix B can be found to satisfy* (9). Indeed, if also $A\hat{B} = \hat{B}A = I_n$, then, using **M2** and **M4** of Section 3.1C,

$$B = I_n B = (\hat{B}A)B = \hat{B}(AB) = \hat{B}I_n = \hat{B}, \qquad \text{that is, } B = \hat{B}$$

If A is invertible, then the *unique* matrix B in (9) is called the **inverse** of A and is denoted by A^{-1}. Thus, by definition,

$$AA^{-1} = A^{-1}A = I \qquad \text{if } A \text{ is nonsingular} \qquad \textbf{(10)}$$

Since $I_n I_n = I_n$, *I_n is invertible and is in fact its own inverse*, that is, $I_n = I_n^{-1}$ for any n. Also, since $O_{n\times n}B = BO_{n\times n} = O_{n\times n}$, there is no way to satisfy (9) when $A = O_{n\times n}$. So *$O_{n\times n}$ is singular for every n.*

I1: If A and B are invertible, so is AB; in fact, $(AB)^{-1} = B^{-1}A^{-1}$, that is, *the inverse of a product is the product of the inverses in reverse order.*

I2: If A is invertible, so is A^k for any k; in fact, $(A^k)^{-1} = (A^{-1})^k$, that is, *the inverse of A^k is the* kth *power of the inverse of A.*

I3: If A is invertible, so is sA for any nonzero s; in fact, $(sA)^{-1} = \dfrac{1}{s}A^{-1}$.

I4: If $AB = O$ (but $A \neq O$ and $B \neq O$), then A and B must be singular.

To prove **I1**, we must show that $B^{-1}A^{-1}$ is the inverse of AB, that is,

$$(AB)(B^{-1}A^{-1}) = (B^{-1}A^{-1})(AB) = I$$

But this is immediate from (10) and the associative law because

$$(AB)(B^{-1}A^{-1}) = A(BB^{-1})A^{-1} = AIA^{-1} = AA^{-1} = I$$

and similarly $(B^{-1}A^{-1})(AB) = I$. The proofs of **I2–I4** are left as exercises. It follows from **I4** that neither A nor B of Example 3.1C has an inverse.

Although, in general, the value of AB tells us nothing about the value of BA, the following remarkable result assures us that if either AB or BA yields the identity matrix, so will the other!

I5: If a matrix B can be found that is *either* a right inverse of A (i.e., $AB = I$) *or* a left inverse of A (i.e., $BA = I$), then A is invertible and $B = A^{-1}$.

Thus verifying *either* that $AB = I$ *or* that $BA = I$ is sufficient to show that $B = A^{-1}$. The proof of **I5** is not trivial. It is given in linear algebra.

The inverse of a 2×2 matrix A can be obtained *by inspection* using the *formula*

$$\begin{bmatrix} a & b \\ c & d \end{bmatrix}^{-1} = \frac{1}{\det A}\begin{bmatrix} d & -b \\ -c & a \end{bmatrix} \qquad \text{provided that } \det A = ad - cb \neq 0 \qquad \textbf{(11)}$$

For example, if $A = \begin{bmatrix} 1 & 2 \\ 3 & 4 \end{bmatrix}$, then $\det A = 1\cdot 4 - 3\cdot 2 = -2 \neq 0$, so A is nonsingular and

$$\begin{bmatrix} 1 & 2 \\ 3 & 4 \end{bmatrix}^{-1} = \frac{1}{-2}\begin{bmatrix} 4 & -2 \\ -3 & 1 \end{bmatrix} = \begin{bmatrix} -2 & 1 \\ \frac{3}{2} & -\frac{1}{2} \end{bmatrix}$$

To prove (11), let

$$A = \begin{bmatrix} a & b \\ c & d \end{bmatrix} \quad \text{and} \quad C = \begin{bmatrix} d & -b \\ -c & a \end{bmatrix}$$

Then

$$AC = \begin{bmatrix} a & b \\ c & d \end{bmatrix}\begin{bmatrix} d & -b \\ -c & a \end{bmatrix} = \begin{bmatrix} ad - bc & 0 \\ 0 & -cd + da \end{bmatrix} = (\det A)\begin{bmatrix} 1 & 0 \\ 0 & 1 \end{bmatrix} \qquad \textbf{(12)}$$

If $\det A \neq 0$, then $A \cdot (1/\det A)C = (1/\det A)AC = I_2$ by **M1** of Section 3.1F; so $(1/\det A)C$ is a *right* inverse of A. By **I5**, it must be A^{-1}. Note that if $\det A = 0$, then $AC = O_{2\times 2}$; hence A is singular by **I4**.

3.1E The Matrix Form of an $n \times n$ Linear System

An $n \times n$ **linear system** is a system of n *linear* equations in n unknowns. Our notation for a general linear system will be

$$\begin{aligned} (E_1)\ & a_{11}x_1 + a_{12}x_2 + \cdots + a_{1n}x_n = b_1 \\ (E_2)\ & a_{21}x_1 + a_{22}x_2 + \cdots + a_{2n}x_n = b_2 \\ & \vdots \qquad\qquad\qquad\qquad\qquad\quad \vdots \\ (E_n)\ & a_{n1}x_1 + a_{n2}x_2 + \cdots + a_{nn}x_n = b_n \end{aligned} \qquad \textbf{(13)}$$

where the a_{ij}'s and b_i's are scalars and $x_1, \ldots, x_n$ are the unknowns. The definitions of matrix multiplication and matrix equality make it possible to write this linear system as a single matrix equation

$$A\mathbf{x} = \mathbf{b} \qquad \textbf{(14a)}$$

where the **coefficient matrix** A, **vector of unknowns** $\mathbf{x}$, and **target vector** $\mathbf{b}$ are

$$A = (a_{ij})_{n\times n} = \begin{bmatrix} a_{11} & a_{12} & \cdots & a_{1n} \\ a_{21} & a_{22} & \cdots & a_{2n} \\ \vdots & \vdots & & \vdots \\ a_{n1} & a_{n2} & \cdots & a_{nn} \end{bmatrix}, \quad \mathbf{x} = \begin{bmatrix} x_1 \\ x_2 \\ \vdots \\ x_n \end{bmatrix}, \quad \text{and} \quad \mathbf{b} = \begin{bmatrix} b_1 \\ b_2 \\ \vdots \\ b_n \end{bmatrix} \qquad \textbf{(14b)}$$

We shall identify the matrix equation (14) with the linear system (13) and speak of the *linear system* $A\mathbf{x} = \mathbf{b}$.

Solutions of linear systems will be denoted by putting bars over the variables as we did in Chapter 2. Thus a **solution** of (13) is a set of numerical values $\bar{x}_1, \ldots, \bar{x}_n$ which, when substituted for $x_1, \ldots, x_n$, satisfy $(E_1), \ldots, (E_n)$. In view of (2), a solution can be viewed as a **solution vector** $\bar{\mathbf{x}} = [\bar{x}_1 \cdots \bar{x}_n]^T$ which the coefficient matrix multiplies into the target vector $\mathbf{b}$ (i.e., $A\bar{\mathbf{x}} = \mathbf{b}$). If A is nonsingular, then the vector $A^{-1}\mathbf{b}$ satisfies (2) because $A(A^{-1}\mathbf{b}) = I\mathbf{b} = \mathbf{b}$ (by the associative law **M4**); hence it is a solution vector. Conversely, any vector $\bar{\mathbf{x}}$ that satisfies $A\bar{\mathbf{x}} = \mathbf{b}$ must satisfy $\bar{\mathbf{x}} = A^{-1}(A\mathbf{x}) = A^{-1}\mathbf{b}$. Thus:

> If A is nonsingular, then the linear system $A\mathbf{x} = \mathbf{b}$ has a *unique* solution given by the *formula* $\bar{\mathbf{x}} = A^{-1}\mathbf{b}$ for *any* target vector $\mathbf{b}$. **(15)**

This formula makes it easy to solve $A\mathbf{x} = \mathbf{b}$ when A^{-1} is readily available, in particular when $n = 2$. For example, the 2×2 linear system

$$\begin{aligned} -8x - 3y &= 6 \\ 7x + 2y &= -1 \end{aligned} \qquad \left(\mathbf{x} = \begin{bmatrix} x \\ y \end{bmatrix},\ A = \begin{bmatrix} -8 & -3 \\ 7 & 2 \end{bmatrix},\ \mathbf{b} = \begin{bmatrix} 6 \\ -1 \end{bmatrix}\right) \tag{16a}$$

is easily solved using formula (11) as follows:

$$\begin{bmatrix} \bar{x} \\ \bar{y} \end{bmatrix} = \bar{\mathbf{x}} = A^{-1}\mathbf{b} = \tfrac{1}{5}\begin{bmatrix} 2 & 3 \\ -7 & -8 \end{bmatrix}\begin{bmatrix} 6 \\ -1 \end{bmatrix} = \tfrac{1}{5}\begin{bmatrix} 9 \\ -34 \end{bmatrix} = \begin{bmatrix} \frac{9}{5} \\ -\frac{34}{5} \end{bmatrix} \tag{16b}$$

Thus $\bar{x} = \frac{9}{5}$ and $\bar{y} = -\frac{34}{5}$, as is easily verified.

Practical Consideration 3.1E (How useful is the formula $\bar{\mathbf{x}} = A^{-1}\mathbf{b}$?). There is no better way to solve a 2×2 linear system $A\mathbf{x} = \mathbf{b}$ than to use the formula $\bar{\mathbf{x}} = A^{-1}\mathbf{b}$ as we did in (15). Unfortunately, when $n > 2$, it is not easy to determine whether a given $n \times n$ matrix A is nonsingular, and if so, to find A^{-1}. Consequently, as we shall see in Section 3.4C, *finding A^{-1} and then multiplying it by* $\mathbf{b}$ *is generally* **not** *the best way to solve $A\mathbf{x} = \mathbf{b}$ when $n > 2$.* ■

We will sometimes write "$\bar{\mathbf{x}} = A^{-1}\mathbf{b}$" as an abbreviated way of saying "$\bar{\mathbf{x}}$ is a solution of $A\mathbf{x} = \mathbf{b}$" *even if $\bar{\mathbf{x}}$ is not actually obtained as* $A^{-1}\mathbf{b}$. This will allow us to say simply "$\bar{\mathbf{x}} = A^{-1}\mathbf{b} = U^{-1}\mathbf{c}$" to indicate that $\bar{\mathbf{x}}$ is a solution of *both* linear systems $A\mathbf{x} = \mathbf{b}$ and $U\mathbf{x} = \mathbf{c}$.

One final remark before going on to describe efficient methods for solving $A\mathbf{x} = \mathbf{b}$. *The unknowns of a linear system are just placeholders.* Changing $\mathbf{x}$ in the linear system $A\mathbf{x} = \mathbf{b}$ to $\boldsymbol{\gamma}$ or $\mathbf{c}$ merely has the effect of changing the *name* of the numerical solution vector (which will be $A^{-1}\mathbf{b}$ in any case!) to $\bar{\boldsymbol{\gamma}}$ or $\bar{\mathbf{c}}$. In fact, all the information needed to find a solution of $A\mathbf{x} = \mathbf{b}$ (*or* $A\boldsymbol{\gamma} = \mathbf{b}$, *or* $A\mathbf{c} = \mathbf{b}$) can be stored efficiently in an $n \times (n + 1)$ **augmented matrix**

$$[A\!:\!\mathbf{b}] = \begin{bmatrix} a_{11} & a_{12} & \cdots & a_{1n} & : & b_1 \\ a_{21} & a_{22} & \cdots & a_{2n} & : & b_2 \\ \vdots & \vdots & & \vdots & & \vdots \\ a_{n1} & a_{n2} & \cdots & a_{nn} & : & b_n \end{bmatrix} \tag{17}$$

3.2

Solving Linear Systems Directly

A **direct method** for solving a linear system is one that yields the exact solution in a finite number of steps if all calculations are performed exactly, that is, without roundoff error. Most direct methods are variants of the *elimination* or *factorization* methods introduced in this section.

3.2A Triangular Systems

An $n \times n$ matrix L is called **lower triangular** if all entries *above* the main diagonal are zero. Similarly, an $n \times n$ matrix U is called **upper triangular** if all entries *below* the main diagonal are zero. For example, of the five triangular matrices

$$A = \begin{bmatrix} 2 & -2 \\ 0 & 6 \end{bmatrix}, \quad B = \begin{bmatrix} 8 & 0 \\ 4 & -2 \end{bmatrix}, \quad C = \begin{bmatrix} 0 & 1 \\ 0 & 0 \end{bmatrix}, \quad D = \begin{bmatrix} 3 & 0 \\ 0 & -5 \end{bmatrix}, \quad E = \begin{bmatrix} 2 & 0 \\ 5 & 0 \end{bmatrix} \qquad \textbf{(1)}$$

A, C, and D are upper triangular and B, D, and E are lower triangular.

An $n \times n$ linear system is called **upper** (**lower**) **triangular** if it has an upper (lower) triangular coefficient matrix. Triangular systems are particularly easy to solve, as the following example shows.

EXAMPLE 3.2A Solve the following 3×3 linear systems†.

(a) Lower triangular system $L\mathbf{c} = \mathbf{b}$

$$\begin{aligned} (E_1) \quad & -2c_1 & &= 8 \\ (E_2) \quad & c_1 + 3c_2 & &= 11 \\ (E_3) \quad & -2c_1 - c_2 + 6c_3 & &= 9 \end{aligned}$$

(b) Upper triangular system $U\mathbf{x} = \mathbf{c}$

$$\begin{aligned} (E_1) \quad & 2x_1 - 5x_2 - 8x_3 &&= -3 \\ (E_2) \quad & -x_2 + 3x_3 &&= -6 \\ (E_3) \quad & -4x_3 &&= 4 \end{aligned}$$

Solution (a): We solve (E_1) for $\bar{c}_1$, then (E_2) for $\bar{c}_2$, then (E_3) for $\bar{c}_3$:

$$\bar{c}_1 = \frac{8}{-2} = -4$$

$$\bar{c}_2 = \frac{11 - \bar{c}_1}{3} = \frac{15}{3} = 5$$

$$\bar{c}_3 = \frac{9 + 2(-4) + 5}{6} = 1$$

Solution (b): We solve (E_3) for $\bar{x}_3$, then (E_2) for $\bar{x}_2$, then (E_1) for $\bar{x}_1$:

$$\bar{x}_3 = \frac{4}{-4} = -1$$

$$\bar{x}_2 = \frac{-6 - 3\bar{x}_3}{-1} = 3$$

$$\bar{x}_1 = \frac{-3 + 5(3) + 8(-1)}{2} = 2$$

Thus $\bar{\mathbf{c}} = [-4 \quad 5 \quad 1]^T$ and $\bar{\mathbf{x}} = [-1 \quad 3 \quad 2]^T$ are the desired solutions. ■

†The reason for using $\mathbf{c}$ for the vector of unknowns in part (a), but for the given target vector in part (b), will become clear in Section 3.2E.

The method used in part (a) of Example 3.2A is called **forward substitution**; it can be used to get the unknowns of any $n \times n$ *lower triangular* system $L\mathbf{c} = \mathbf{b}$ for which all $l_{ii} \neq 0$ in the *forward* order $\bar{c}_1, \bar{c}_2, \ldots, \bar{c}_n$, as shown in Algorithm 3.2A(a). The method used in part (b) of Example 3.2A is called **backward substitution**; it can be used to get the unknowns of any $n \times n$ *upper triangular* system $U\mathbf{x} = \mathbf{c}$ for which all $u_{ii} \neq 0$ in the *backward* order $\bar{x}_n, \bar{x}_{n-1}, \ldots, \bar{x}_1$, as shown in Algorithm 3.2A(b).

ALGORITHM 3.2A. FORWARD AND BACKWARD SUBSTITUTION

(a) FORWARD SUBSTITUTION $\{l_{ii} \neq 0\}$

PURPOSE: To solve $L\mathbf{c} = \mathbf{b}$ for $\bar{\mathbf{c}}$

GET n, L {lower triangular}, $\mathbf{b}$
$\bar{c}_1 = b_1/l_{11}$
DO FOR $i = 2$ TO n

$$\bar{c}_i = \frac{1}{l_{ii}}\left(b_i - \sum_{j=1}^{i-1} l_{ij}\bar{c}_j\right)$$

OUTPUT ('$[\bar{c}_1 \quad \cdots \quad \bar{c}_n]^T$ is $L^{-1}\mathbf{b}$')

(b) BACKWARD SUBSTITUTION $\{u_{ii} \neq 0\}$

PURPOSE: To solve $U\mathbf{x} = \mathbf{c}$ for $\bar{\mathbf{x}}$

GET n, U {upper triangular}, $\mathbf{c}$
$\bar{x}_n = c_n/u_{nn}$
DO FOR $i = n - 1$ DOWNTO 1

$$\bar{x}_i = \frac{1}{u_{ii}}\left(c_i - \sum_{j=i+1}^{n} u_{ij}\bar{x}_j\right)$$

OUTPUT ('$[\bar{x}_1 \quad \cdots \quad \bar{x}_n]^T$ is $U^{-1}\mathbf{c}$')

3.2B Efficient Forward and Backward Substitution

The summations in the formulas for $\bar{c}_i$ and $\bar{x}_i$ in Algorithm 3.2A can be viewed as *inner products* as follows:

FORWARD SUBSTITUTION:
$$\bar{c}_i = \frac{1}{l_{ii}}\left\{b_i - [l_{i1} \quad \cdots \quad l_{i,i-1}]\begin{bmatrix}\bar{c}_1\\ \vdots \\ \bar{c}_{i-1}\end{bmatrix}\right\} \tag{2a}$$

BACKWARD SUBSTITUTION:
$$\bar{x}_i = \frac{1}{u_{ii}}\left\{c_i - [u_{i,i+1} \quad \cdots \quad u_{i,n}]\begin{bmatrix}\bar{x}_{i+1}\\ \vdots \\ \bar{x}_{n}\end{bmatrix}\right\} \tag{2b}$$

When performing forward substitution by hand, the $\bar{c}_i$'s should be entered *down* the $\bar{\mathbf{c}}$ column of a "doubly augmented" matrix $[L : \mathbf{b} : \bar{\mathbf{c}}]$ *as they are obtained*. The values needed to calculate $\bar{c}_i$ by (2a) can then be found *by inspection* of the ith row of $[L : \mathbf{b}]$ (*to the left of* $\bar{c}_i$) and the *current* $\bar{\mathbf{c}}$ column (*above* $\bar{c}_i$), as shown shaded in Figure 3.2-1(a). Similarly, when performing backward substitution by hand, the $\bar{x}_i$'s should be entered *up* the $\bar{\mathbf{x}}$ column of a "doubly augmented" matrix $[U : \mathbf{c} : \bar{\mathbf{x}}]$ as they are obtained. The values needed to calculate $\bar{x}_i$ by (2b) can then be found *by inspection* of the ith row of $[U : \mathbf{c}]$ (*to the left of* $\bar{x}_i$) and the *current* $\bar{\mathbf{x}}$ column (*below* $\bar{x}_i$), as shown shaded in Figure 3.2-1(b).

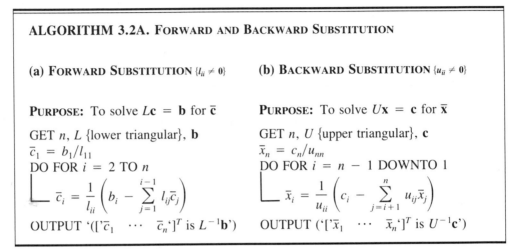

(a) FORWARD SUBSTITUTION Solve $L\mathbf{c} = \mathbf{b}$ for $\bar{c}_1, \bar{c}_2, \ldots, \bar{c}_n$:

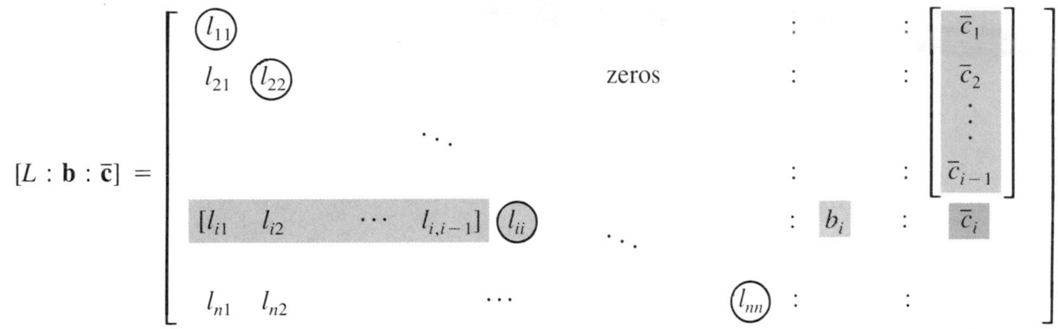

(b) BACKWARD SUBSTITUTION Solve $U\mathbf{x} = \mathbf{c}$ for $\bar{x}_n, \ldots, \bar{x}_2, \bar{x}_1$:

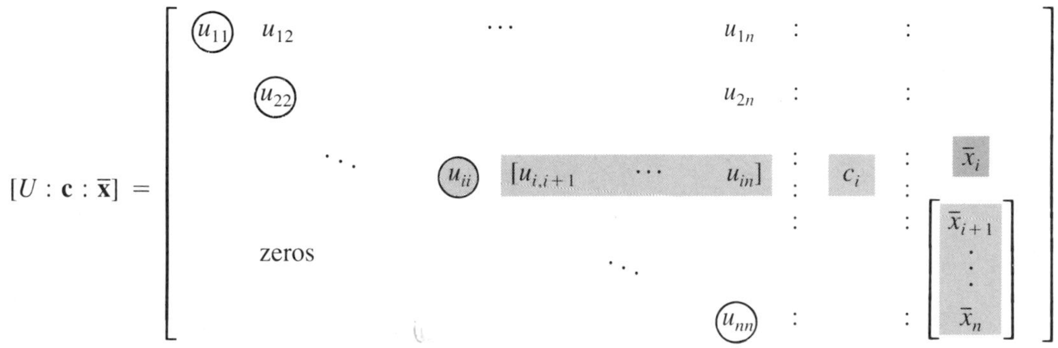

NOTE: We will circle the diagonal entries (l_{ii} and u_{ii}) to separate them from the row vectors used in the inner products in (2) and as a reminder to divide by l_{ii} in (2a) and by u_{ii} in (2b).

Figure 3.2-1 *Getting (a)* $\bar{\mathbf{c}}$ *down* $[L:\mathbf{b}:\bar{\mathbf{c}}]$; *(b)* $\bar{\mathbf{x}}$ *up* $[U:\mathbf{c}:\bar{\mathbf{x}}]$.

EXAMPLE 3.2B Solve the following triangular systems using (2) as just described.

$$\textbf{(a)}\quad \begin{aligned} (E_1)&\; -2c_1 &&= 8\\ (E_2)&\; c_1 + 3c_2 &&= 11\\ (E_3)&\; -2c_1 - c_2 + 6c_3 &&= 9 \end{aligned} \qquad \textbf{(b)}\quad \begin{aligned} (E_1)&\; x_1 - x_2 - 3x_3 &&= 4\\ (E_2)&\; x_2 + 2x_3 &&= 0\\ (E_3)&\; x_3 &&= 3 \end{aligned} \tag{3}$$

SOLUTION: We calculate $\bar{\mathbf{c}}$ and $\bar{\mathbf{x}}$ in the doubly augmented matrices

$$\left[\begin{array}{ccc:c:c} \textcircled{-2} & & & 8 & -4\\ 1 & \textcircled{3} & & 11 & 5\\ -2 & -1 & \textcircled{6} & 9 & 1 \end{array}\right] \quad\text{and}\quad \left[\begin{array}{ccc:c:c} 1 & -1 & -3 & 4 & 7\\ & 1 & 2 & 0 & -6\\ & & 1 & 3 & 3 \end{array}\right] \tag{4a}$$

$$\qquad L \qquad \mathbf{b} \qquad \bar{\mathbf{c}} \qquad\qquad\qquad \text{unit } U \qquad \mathbf{c} \qquad \bar{\mathbf{x}}$$

The forward substitution used to get $\bar{\mathbf{c}}$ using (2a) [see Example 3.2A(a)] and the backward substitution used to get $\bar{\mathbf{x}}$ using (2b) proceed as follows:

$$\bar{c}_1 = \frac{1}{(-2)}\{8\} = -4 \qquad \bar{x}_3 = 3\ (= c_3) \tag{4b}$$

$$\bar{c}_2 = \frac{1}{(3)}\{11 - [1][-4]\} = \frac{15}{3} = 5 \qquad \bar{x}_2 = 0 - [2][3] = -6 \tag{4c}$$

$$\bar{c}_3 = \frac{1}{(6)}\left\{9 - \underbrace{[-2 \quad -1]\begin{bmatrix}-4\\5\end{bmatrix}}_{(-2)(-4)+(-1)(5)}\right\} = 1 \qquad \bar{x}_1 = 4 - \underbrace{[-1 \quad -3]\begin{bmatrix}-6\\3\end{bmatrix}}_{(-1)(-6)+(-3)(3)} = 7 \tag{4d}$$

So the desired solutions are $\bar{\mathbf{c}} = [-4 \quad 5 \quad 1]^T$ and $\bar{\mathbf{x}} = [7 \quad -6 \quad 3]^T$. ■

A **unit** triangular matrix is a triangular L or U that has all 1's on its main diagonal. For such matrices, the formulas (2) simplify to

(a) FORWARD SUBSTITUTION (UNIT L): **(b) BACKWARD SUBSTITUTION (UNIT U):**

$$\bar{c}_i = b_i - [l_{i1} \quad \cdots \quad l_{i,i-1}]\begin{bmatrix}\bar{c}_1\\ \vdots \\ \bar{c}_{i-1}\end{bmatrix} \qquad \bar{x}_i = c_i - [u_{i,i+1} \quad \cdots \quad u_{in}]\begin{bmatrix}\bar{x}_{i-1}\\ \vdots \\ \bar{x}_n\end{bmatrix} \tag{5}$$

Since the 1's on the diagonal of a *unit* L or U do not enter into the formulas (5), they need not be circled. [See Example 3.2B(b), where (5b) was used.]

In Section 3.2E we will show how equations (2) and (5) can be used *together* to solve *general* (not necessarily triangular) linear systems efficiently. The idea used there is based on the important strategy described next.

3.2C Basic Gaussian Elimination

Two systems of equations are called **equivalent** if they have identical solutions. The mth and ith equations of a general (not necessarily triangular) $n \times n$ linear system $A\mathbf{x} = \mathbf{b}$ are, respectively,

$$(\mathrm{E}_m)\ a_{m1}x_1 + a_{m2}x_2 + \cdots + a_{mm}x_m + \cdots + a_{mn}x_n = b_m \tag{6a}$$

$$(\mathrm{E}_i)\ a_{i1}x_1 + a_{i2}x_2 + \cdots + a_{im}x_m + \cdots + a_{in}x_n = b_i \tag{6b}$$

If $a_{mm} \neq 0$ and both sides of (E_m) are multiplied by a_{im}/a_{mm}, then the coefficient of x_m in (E_m) becomes a_{im}. As a result,

$$(\mathrm{E}_i) - \frac{a_{im}}{a_{mm}}(\mathrm{E}_m) \text{ has zero as its coefficient of } x_m \tag{7}$$

In words, subtracting $(a_{im}/a_{mm})(\mathrm{E}_m)$ from (E_i) "eliminates" x_m from the ith equation. The equivalent linear system resulting from such a subtraction will be denoted by $\tilde{A}\mathbf{x} = \tilde{\mathbf{b}}$; and the ith row of its augmented matrix $[\tilde{A} : \tilde{\mathbf{b}}]$ will be abbreviated† as ρ_i. Thus, for $i = 1, 2, \ldots, n$,

$$\rho_i = [\tilde{a}_{i1} \quad \tilde{a}_{i2} \quad \cdots \quad \tilde{a}_{in} : \tilde{b}_i] = \text{row}_i[\tilde{A} : \tilde{\mathbf{b}}] \tag{8}$$

In this notation, the "a_{im} zeroing" operation (7), when performed on the augmented matrix $[\tilde{A} : \tilde{\mathbf{b}}]$, will be described using the following notation:

0_{im}-SUBTRACT. Replace ρ_i by $\rho_i - l_{im}\rho_m$ to get $\tilde{a}_{im} = 0$ by taking

$$l_{im} = \textit{\textbf{im}}\textbf{th multiplier} = \frac{\tilde{a}_{im}}{\tilde{a}_{mm}} = \frac{\text{entry being ``zeroed'' in } \rho_i}{\text{diagonal entry of } \rho_m} \tag{9}$$

0_{im}-subtracts can be used as follows to transform $[A : \mathbf{b}]$ into the augmented matrix of an equivalent *upper triangular* system $U\mathbf{x} = \bar{\mathbf{c}}$ *one column at a time from left to right*:

```
DO FOR m = 1 TO n − 1   {reduce the mth column of Ã}
|  DO FOR i = m + 1 TO n   {below the mmth entry of Ã}        (10a)
L  L  Perform a 0_im-subtract to get ã_im = 0
```

This fundamental algorithm will be referred to as **upper triangulating** $[A : \mathbf{b}]$, or **reducing** it to **upper triangular form** and abbreviated as

$$[A:\mathbf{b}] \quad \rightarrow \quad \{\, [\tilde{A}:\tilde{\mathbf{b}}] \,\} \quad \rightarrow \quad [U:\bar{\mathbf{c}}] \tag{10b}$$

where $\{[\tilde{A}:\tilde{\mathbf{b}}]\}$ denotes the sequence of *intermediate* augmented matrices obtained in (10a). Initially, we take $[\tilde{A}:\tilde{\mathbf{b}}]$ to be $[A:\mathbf{b}]$ (the augmented matrix of the *given* system). Then, at any intermediate step, $[\tilde{A}:\tilde{\mathbf{b}}]$ is the augmented matrix of the *current* equivalent system whose ith equation

$$(\text{E}_i)\ \tilde{a}_{i1}x_1 + \tilde{a}_{i2}x_2 + \cdots + \tilde{a}_{in}x_n = \tilde{b}_i \tag{11}$$

is represented by ρ_i in (8) for $i = 1, 2, \ldots, n$. The final $[\tilde{A}:\tilde{\mathbf{b}}]$ is the desired $[U:\bar{\mathbf{c}}]$.

†If you *add* $(a_{im}/a_{mm})(\text{E}_m)$ to the (E_i) that results from (7), you get the *original* (E_i). So the operation (7) is reversible, hence the resulting system must be equivalent to the original one. Also, the symbol "ρ" is the lowercase Greek rho; so read "ρ_i" as "row i."

Basic† Gaussian Elimination is the method of solving the system $A\mathbf{x} = \mathbf{b}$ by first *upper triangulating* its augmented matrix, and then using *backward substitution* to solve the resulting equivalent upper triangular system $U\mathbf{x} = \bar{\mathbf{c}}$.

EXAMPLE 3.2C Use Basic Gaussian Elimination to solve

$$\begin{aligned} (\mathrm{E}_1)&\quad -x + y - 4z = 0 \\ (\mathrm{E}_2)&\quad 2x + 2y = 1 \\ (\mathrm{E}_3)&\quad 3x + 3y + 2z = \tfrac{1}{2} \end{aligned} \qquad [A\!:\!\mathbf{b}] = \begin{bmatrix} -1 & 1 & -4 & : & 0 \\ 2 & 2 & 0 & : & 1 \\ 3 & 3 & 2 & : & \tfrac{1}{2} \end{bmatrix} \tag{12a}$$

Solution: The (E_i) and ρ_i notations will be illustrated side by side. We begin the **upper triangulation** by taking $[A\!:\!\mathbf{b}]$ as the initial $[\tilde{A}\!:\!\tilde{\mathbf{b}}]$.

***m* = 1.** Get zeros below $\tilde{a}_{11}$ $(= a_{11})$ using the following two $0_{i,1}$-subtracts:

$$\rho_2 - \frac{\tilde{a}_{21}}{\tilde{a}_{11}}\rho_1 = [2 \quad 2 \quad 0 : 1] - \frac{2}{-1}[-1 \quad 1 \quad -4 : 0] = [0 \quad 4 \quad -8 : 1] \tag{12b}$$

$$\rho_3 - \frac{\tilde{a}_{31}}{\tilde{a}_{11}}\rho_1 = [3 \quad 3 \quad 2 : \tfrac{1}{2}] - \frac{3}{-1}[-1 \quad 1 \quad -4 : 0] = [0 \quad 6 \quad -10 : \tfrac{1}{2}] \tag{12c}$$

New $\tilde{A}\mathbf{x} = \tilde{\mathbf{b}}$ | New $[\tilde{A}\!:\!\tilde{\mathbf{b}}]$

$$\begin{aligned} (\mathrm{E}_1)&\quad -x + y - 4z = 0 \\ (\mathrm{E}_2) - \frac{2}{-1}(\mathrm{E}_1)&\quad 4y - 8z = 1 \\ (\mathrm{E}_3) - \frac{3}{-1}(\mathrm{E}_1)&\quad 6y - 10z = \tfrac{1}{2} \end{aligned} \qquad \begin{matrix} \rho_1 \\ \rho_2 - \frac{2}{-1}\rho_1 \\ \rho_3 - \frac{3}{-1}\rho_1 \end{matrix} \begin{bmatrix} -1 & 1 & -4 & : & 0 \\ & 4 & -8 & : & 1 \\ & 6 & -10 & : & \tfrac{1}{2} \end{bmatrix} \tag{12d}$$

***m* = 2.** Get a zero below the current $\tilde{a}_{22}$ $(= 4)$ using the 0_{32}-subtract.

$$\rho_3 - \frac{\tilde{a}_{32}}{\tilde{a}_{22}}\rho_2 = [0 \quad 6 \quad -10 : \tfrac{1}{2}] - \tfrac{6}{4}[0 \quad 4 \quad -8 : 1] = [0 \quad 0 \quad 2 : -1] \tag{12e}$$

Desired $U\mathbf{x} = \bar{\mathbf{c}}$ | Desired $[U\!:\!\bar{\mathbf{c}}]$

$$\begin{aligned} (\mathrm{E}_1)&\quad -x + y - 4z = 0 \\ (\mathrm{E}_2)&\quad 4y - 8z = 1 \\ (\mathrm{E}_3) - \tfrac{6}{4}(\mathrm{E}_2)&\quad 2z = -1 \end{aligned} \qquad \begin{matrix} \rho_1 \\ \rho_2 \\ \rho_3 - \frac{6}{4}\rho_2 \end{matrix} \begin{bmatrix} -1 & 1 & -4 & : & 0 \\ & 4 & -8 & : & 1 \\ & & 2 & : & -1 \end{bmatrix} \tag{12f}$$

The equivalent upper triangular system $U\mathbf{x} = \bar{\mathbf{c}}$ is easily solved by **backward substitution:**

†Here "basic" refers to the fact that the algorithm uses only 0_{im}- subtracts, so it fails if $\tilde{a}_{mm} = 0$ is encountered. To deal with this possibility, one must also interchange equations of $\tilde{A}\mathbf{x} = \tilde{\mathbf{b}}$ or, equivalently, rows of $[\tilde{A}\!:\!\tilde{\mathbf{b}}]$. This will be done in Section 3.3. The reason for the bar over the $\mathbf{c}$ in $U\mathbf{x} = \bar{\mathbf{c}}$ will become clear in Section 3.2E.

$$\begin{bmatrix} \textcircled{-1} & 1 & -4 & : & 0 & : & \frac{5}{4} \\ & \textcircled{4} & -8 & : & 1 & : & -\frac{3}{4} \\ & & \textcircled{2} & : & -1 & : & -\frac{1}{2} \end{bmatrix} \quad \begin{aligned} \bar{x}_1 &= \frac{1}{-1}\{0 - [(1)(-\tfrac{3}{4}) + (-4)(-\tfrac{1}{2})]\} = \tfrac{5}{4} \\ \bar{x}_2 &= \frac{1}{4}\{1 - (-8)(-\tfrac{1}{2})\} = -\tfrac{3}{4} \\ \bar{x}_3 &= \frac{-1}{2} = -\tfrac{1}{2} \end{aligned} \qquad \textbf{(12g)}$$

U $\qquad$ $\bar{\mathbf{c}}$ $\qquad$ $\bar{\mathbf{x}}$ [obtained using (2b)]

Verify that $\bar{\mathbf{x}} = [\frac{5}{4} \quad -\frac{3}{4} \quad -\frac{1}{2}]^T$ satisfies the given system (12a). ■

3.2D Performing Basic Gaussian Elimination in Place

When performing Basic Gaussian Elimination, the upper triangulation is most efficiently done **in place**, that is, by storing the current $[A:\mathbf{b}]$ in the memory originally allocated to $[A:\mathbf{b}]$. Also, since we know that a_{im} (below $\tilde{a}_{mm}$) becomes 0 after an 0_{im}-subtract, *the* ***i,m*- multiplier**, l_{im}, *should be stored in its place* and then used to carry out the 0_{im}-subtract *to its right* in $\text{row}_i[\tilde{A}:\tilde{\mathbf{b}}]$. In this way, no locations of A are wasted, and the upper triangulation is more accurately represented as

$$[\tilde{A}:\tilde{\mathbf{b}}] \rightarrow \{\, [\tilde{A}:\tilde{\mathbf{b}}] \,\} \rightarrow [L\backslash U:\bar{\mathbf{c}}] \qquad \textbf{(13)}$$

where L represents the multipliers l_{im} stored below the diagonal of U. For example, the upper triangulation (12) of Example 3.2C would be

$$[A:\mathbf{b}] \rightarrow \begin{bmatrix} \textcircled{-1} & 1 & -4 & : & 0 \\ \underline{-2} & 4 & -8 & : & 1 \\ \underline{-3} & 6 & -10 & : & \frac{1}{2} \end{bmatrix} \rightarrow \begin{bmatrix} \textcircled{-1} & 1 & -4 & : & 0 \\ \underline{-2} & \textcircled{4} & -8 & : & 1 \\ \underline{-3} & \underline{\frac{3}{2}} & \textcircled{2} & : & -1 \end{bmatrix} = [L\backslash U:\bar{\mathbf{c}}] \qquad \textbf{(14a)}$$

if performed in place. The diagonal entry u_{mm} is called the ***m*th pivot entry** of $L\backslash U$ (the final $\tilde{A}$). Each pivot entry will be circled just *before* reducing its column. These circles will provide a ''barrier'' between the U and L entries, and will remind us to divide by u_{mm} first when getting $l_{im} = \tilde{a}_{im}/\tilde{a}_{mm}$ stored below it, and later when getting $\bar{x}_m$ by backward substitution. We will also underline the *stored* multipliers l_{im} to distinguish them from the *actual* entries of $[\tilde{A}:\tilde{\mathbf{b}}]$ during the reduction, as we did in (14a). This will remind us to use the actual zeros (*not* the underlined multipliers) in subsequent 0_{im}-subtracts [see (12b), (12c), and (12e)].

Pseudocode for **Basic Gaussian Elimination in Place** is given in Algorithm 3.2D. Notice that the backward substitution is also carried out in place; that is, $\mathbf{b}$, then $\bar{\mathbf{c}}$, then $\bar{\mathbf{x}}$ is stored in the memory originally allocated to $\mathbf{b}$. It is important to realize that *Algorithm 3.2D is not a practical algorithm for solving linear systems*. First of all, it fails if $\tilde{a}_{mm} = 0$ is encountered; and even when it succeeds, the solution $\bar{\mathbf{x}}$ obtained on a digital device may be considerably less accurate than it could be. We will remedy this (serious) deficiency in Section 3.3. Second, it is inefficient when $A\mathbf{x} = \mathbf{b}$ must be solved with *several different* $\mathbf{b}$'s. The remedy for this deficiency is given next.

ALGORITHM 3.2D. BASIC GAUSSIAN ELIMINATION IN PLACE

PURPOSE: To solve a linear system $A\mathbf{x} = \mathbf{b}$ by reducing $[A:\mathbf{b}]$ one column at a time to $[U:\bar{\mathbf{c}}]$, where $Ux = \bar{\mathbf{c}}$ is an equivalent upper triangular system which is then solved by backward substitution. The given augmented matrix $[A:\mathbf{b}]$ is returned as $[L\backslash U:\bar{\mathbf{x}}]$, where $\bar{\mathbf{x}}$ is the desired solution and l_{im} is the multiplier of the 0_{im}-subtract used to "zero" $\tilde{a}_{im}$ below $\tilde{a}_{mm}$ in $\text{row}_i\tilde{A}$.

Get n, A, $\mathbf{b}$ {$[A:\mathbf{b}]$ stores the current $[\tilde{A}:\tilde{\mathbf{b}}]$}

{**UPPER TRIANGULATION:** Reduce $[A:\mathbf{b}]$ to $[L\backslash U:\bar{\mathbf{c}}]$ in place}

DO FOR $m = 1$ TO n {a_{mm} stores the mth pivot entry u_{mm}}
  IF $a_{mm} = 0$ THEN STOP {algorithm fails because $\tilde{a}_{mm} = 0$}
  {$\text{row}_m[A:\mathbf{b}]$ now stores $\text{row}_m[L\backslash U:\bar{\mathbf{c}}]$}
  DO FOR $i = m + 1$ TO n {perform a 0_{im}-subtract below $\tilde{a}_{mm}$}
    $a_{im} \leftarrow a_{im}/a_{mm}$ {a_{im} now stores l_{im}}
    DO FOR $j = m + 1$ TO n {to the right of a_{im}}
      $a_{ij} \leftarrow a_{ij} - a_{im}a_{mj}$ {a_{ij} now stores the current $\tilde{a}_{ij}$}
    $b_i \leftarrow b_i - a_{im}b_m$ {b_i now stores the current $\tilde{b}_i$}
  {$\text{col}_m[A:\mathbf{b}]$ now stores $\text{col}_m[L\backslash U:\bar{\mathbf{c}}]$}

{**BACKWARD SUBSTITUTION:** Solve $U\mathbf{x} = \bar{\mathbf{c}}$ in place}

$b_n \leftarrow b_n/a_{nn}$ {b_n now stores $\bar{x}_n$}
DO FOR $i = n - 1$ DOWNTO 1
  DO FOR $j = i + 1$ TO n {form $\bar{c}_i - \sum_{j=i+1}^{n} u_{ij}\bar{x}_j$ in storage for b_i}
    $b_i \leftarrow b_i - a_{ij}b_j$
  $b_i \leftarrow b_i/a_{ii}$ {b_i now stores $\bar{x}_i$}

OUTPUT ('$\bar{\mathbf{x}} = [$'$b_1, \ldots, b_n$'$]^T$ is the solution of the given system.')
STOP

3.2E Upper Triangulation = *LU*-Factorization + Forward Substitution

When the upper triangulation phase of Basic Gaussian Elimination is performed in place, the circled pivot entries of $L\backslash U$ should be thought of as "covering" the 1's on the diagonal of a *unit* L. We will therefore read "$L\backslash U$" as "**L under U**." A rather remarkable thing about L and U obtained from $L\backslash U$ this way (see Problem M3-8) is that *the matrix product LU will always be the given coefficient matrix A*! For example, from $L\backslash U$ obtained of Example 3.2D [see (14a)],

$$\begin{bmatrix} \textcircled{-1} & 1 & -4 \\ -2 & \textcircled{4} & -8 \\ -3 & \frac{3}{2} & \textcircled{2} \end{bmatrix} \Rightarrow \begin{bmatrix} 1 & & \\ -2 & 1 & \\ -3 & \frac{3}{2} & 1 \end{bmatrix}\begin{bmatrix} -1 & 1 & -4 \\ & 4 & -8 \\ & & 2 \end{bmatrix} = \begin{bmatrix} -1 & 1 & -4 \\ 2 & 2 & 0 \\ 3 & 3 & 2 \end{bmatrix} \quad \textbf{(14b)}$$

L under U unit L U A of Example 3.2D

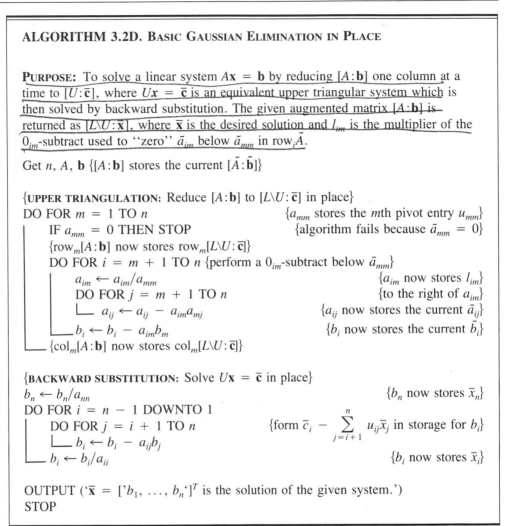

$L U \bar{x} = b$

$L \bar{c} = b \Rightarrow \bar{c} = L^{-1} b \Rightarrow U \bar{x} = L^{-1} b$

$\Rightarrow \bar{x} = U^{-1} L^{-1} b$

An ***LU*-factorization** of A is a representation of an $n \times n$ matrix A as a product LU, where L (the left factor) is lower triangular and U (the right factor) is upper triangular. Since the "L under U" matrix $L\backslash U$ obtained in (14a) stores both L and U in the same $n \times n$ array, we shall refer to it as the **compact form** of the LU-factorization of $A = LU$.

If $A = LU$ and all diagonal entries of both L and U are nonzero, then for any target vector $\mathbf{b}$, the linear system $A\mathbf{x} = \mathbf{b}$ can be written as $L(U\mathbf{x}) = \mathbf{b}$, that is, as $L(\mathbf{c}) = \mathbf{b}$, where $\mathbf{c} = U\mathbf{x}$. We can therefore solve the system in *two simple* steps:

$$\text{First solve } L\mathbf{c} = \mathbf{b} \text{ for } \bar{\mathbf{c}}; \text{ then solve } U\mathbf{x} = \bar{\mathbf{c}} \text{ for the desired } \bar{\mathbf{x}} \qquad \textbf{(15)}$$

We have just seen that the reduction $[A:\mathbf{b}] \rightarrow [L\backslash U:\bar{\mathbf{c}}]$, if successful, provides such a factorization $A = LU$ (with at unit L). So (15) says that rather than reduce $\mathbf{b}$ to $\bar{\mathbf{c}}$ in the **UPPER TRIANGULATION** phase of Basic Gaussian Elimination, we can reduce only A to $L\backslash U$, and then get $\bar{\mathbf{c}}$ by solving the unit lower triangular system $L\mathbf{c} = \mathbf{b}$. The algorithm that results when the **UPPER TRIANGULATION** phase of Basic Gaussian Elimination is separated into an ***LU*-FACTORIZATION** (reduce *only* A to get $L\backslash U$) followed by a **FORWARD SUBSTITUTION** (solve $L\mathbf{c} = \mathbf{b}$ to get $\bar{\mathbf{c}}$) will be called the ***LU*-Factorization Algorithm**. It is outlined in Algorithm 3.2E.

ALGORITHM 3.2E. *LU*-FACTORIZATION

PURPOSE: To solve an $n \times n$ linear system $A\mathbf{x} = \mathbf{b}$.

***LU*-FACTORIZATION:** Reduce A in place to get $L\backslash U$, where $LU = A$.

FORWARD SUBSTITUTION: Form [unit $L:\mathbf{b}:\bar{\mathbf{c}}$] to solve $L\mathbf{c} = \mathbf{b}$ for $\bar{\mathbf{c}}$.

BACKWARD SUBSTITUTION: Form $[U:\bar{\mathbf{c}}:\bar{\mathbf{x}}]$ to solve $U\mathbf{x} = \bar{\mathbf{c}}$ for $\bar{\mathbf{x}} = A^{-1}\mathbf{b}$.

Separating the **UPPER TRIANGULATION** $[A:\mathbf{b}] \rightarrow [L\backslash U:\bar{\mathbf{c}}]$ into an ***LU*-FACTORIZATION** ($A \rightarrow L\backslash U$) followed by a **FORWARD SUBSTITUTION** (solve $L\mathbf{c} = \mathbf{b}$) enables you to avoid needlessly repeating the $A \rightarrow L\backslash U$ part of the upper triangulation if *several* systems having the *same coefficient matrix* must be solved. This is illustrated in the following example.

EXAMPLE 3.2E Solve the following 3×3 linear system.

$$\begin{aligned} (E_1)\ & -x_1 + x_2 - 4x_3 = 5 \\ (E_2)\ & 2x_1 + 2x_2 = 14 \\ (E_3)\ & 3x_1 + 3x_2 + 2x_3 = 15 \end{aligned} \qquad \left([A:\mathbf{b}] = \begin{bmatrix} -1 & 1 & -4 & : & 5 \\ 2 & 2 & 0 & : & 14 \\ 3 & 3 & 2 & : & 15 \end{bmatrix} \right) \qquad \textbf{(16a)}$$

SOLUTION: The compact form $L\backslash U$ of the LU-factorization of the coefficient matrix A was obtained in Section 3.2D. We shall therefore use the LU-Factorization Algorithm 3.2E, but *reusing this previously obtained* $L\backslash U$ [see (14)] rather than repeating the ***LU*-FACTORIZATION** phase. The remaining steps are

FORWARD SUBSTITUTION

$$\left[\begin{array}{ccc:c:c} 1 & & & 5 & 5 \\ -2 & 1 & & 14 & 24 \\ -3 & \frac{3}{2} & 1 & 15 & -6 \end{array}\right]$$

unit L $\quad$ $\mathbf{b}$ $\quad$ $\overline{\mathbf{c}}$

BACKWARD SUBSTITUTION

$$\left[\begin{array}{ccc:c:c} \textcircled{-1} & 1 & -4 & 5 & 7 \\ & \textcircled{4} & -8 & 24 & 0 \\ & & \textcircled{2} & -6 & -3 \end{array}\right] \qquad \textbf{(16b)}$$

U $\quad$ $\overline{\mathbf{c}}$ $\quad$ $\overline{\mathbf{x}}$

For this *unit* L, the forward substitution [using (5a)] is simply

$$\overline{c}_1 = 5; \qquad \overline{c}_2 = 14 - (-2)5 = 24; \qquad \overline{c}_3 = 15 - [(-3)5 + \tfrac{3}{2}(24)] = -6 \qquad \textbf{(16c)}$$

It is easy to see that $\overline{\mathbf{x}} = [7 \quad 0 \quad -3]^T$, obtained by backward substitution as in (12g), satisfies the given linear system (16a). ■

Practical Consideration 3.2E (Why factor *A* first?). The Basic Gaussian Elimination and *LU*-Factorization Algorithms 3.2D and 3.2E will solve a *single* linear system with the same amount of computation. However, getting $\overline{\mathbf{c}}$ from $\mathbf{b}$ using just a forward substitution as in (16c) requires less work than performing an entire upper triangulation $[A{:}\mathbf{b}] \rightarrow [L\backslash U{:}\overline{\mathbf{c}}]$. In fact, for large n, it takes about $2n^3/3$ arithmetic operations to do an entire upper triangulation, compared to only about n^2 for a forward substitution (see Section 3.2G). So the use of the *LU*-Factorization Algorithm 3.2E as in Example 3.2E can save considerable computing time when n is large and *several* linear systems with the *same* coefficient matrix must be solved. We shall pursue this further in Section 3.4. ■

3.2F Compact Forward/Backward Substitution

Once a compact form $L\backslash U$ has been obtained for A, the forward substitution $[L{:}\mathbf{b}{:}\overline{\mathbf{c}}]$, then backward substitution $[U{:}\overline{\mathbf{c}}{:}\overline{\mathbf{x}}]$ needed to solve $A\mathbf{x} = \mathbf{b}$ can be carried out efficiently in a single *triply augmented* matrix $[L\backslash U{:}\mathbf{b}{:}\overline{\mathbf{c}}{:}\overline{\mathbf{x}}]$, which we will refer to as a **forback matrix**. For example, the calculations in (16b) can be carried out in the forback matrix

$$[L\backslash U{:}\mathbf{b}{:}\overline{\mathbf{c}}{:}\overline{\mathbf{x}}] = \left[\begin{array}{ccc:c:c:c} \textcircled{-1} & 1 & -4 & 5 & 5 & 7 \\ \underline{-2} & \textcircled{4} & -8 & 14 & 24 & 0 \\ \underline{-3} & \underline{\tfrac{3}{2}} & \textcircled{2} & 15 & -6 & -3 \end{array}\right] \qquad \textbf{(16d)}$$

The calculations of $\overline{\mathbf{c}}$ then $\overline{\mathbf{x}}$ in (16d) use the "inspection" formulas (5a) and (2b) *exactly* as they were used to get $\overline{\mathbf{c}}$ in $[L{:}\mathbf{b}{:}\overline{\mathbf{c}}]$, then $\overline{\mathbf{x}}$ in $[U{:}\overline{\mathbf{c}}{:}\overline{\mathbf{x}}]$ in (16b). First $[L\backslash{:}\mathbf{b} : \; : \;]$, shaded in (16d), is used to get the $\overline{c}_i$'s *down* the $\overline{\mathbf{c}}$ column. Then $[\backslash U : \; : \overline{\mathbf{c}} : \;]$, bounded by dashed lines in (16d), is used to get the $\overline{x}_i$'s *up* the $\overline{\mathbf{x}}$ column. When completing a forback matrix by hand, remember that *the underlined l_{im} entries are used to get* $\overline{\mathbf{c}}$, and that divisions by the (circled) pivot entries of $L\backslash U$ are needed to get $\overline{\mathbf{x}}$, *but not* $\overline{\mathbf{c}}$.

When $L\backslash U$ is obtained in place, its entries replace those of A. If desired, A can be reconstructed from $L\backslash U$ as the product LU. For example,

$$L\backslash U = \begin{bmatrix} \textcircled{4} & 3 \\ 2 & \textcircled{-1} \end{bmatrix} \Rightarrow A = \begin{bmatrix} 1 & 0 \\ 2 & 1 \end{bmatrix}\begin{bmatrix} 4 & 3 \\ 0 & -1 \end{bmatrix} = \begin{bmatrix} 4 & 3 \\ 8 & 5 \end{bmatrix} \qquad \textbf{(17)}$$

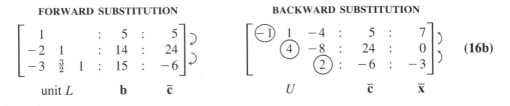

Practical Consideration 3.2F (Checking a hand calculation). When obtained by hand, the entries of first $L\backslash U$, then $\bar{\mathbf{c}}$, then $\bar{\mathbf{x}}$, should be entered in $[L\backslash U:\mathbf{b}:\bar{\mathbf{c}}:\bar{\mathbf{x}}]$ (and then used) as they are obtained. *You can check your work along the way* by verifying the products

$$LU = A, \qquad \text{then } L\bar{\mathbf{c}} = \mathbf{b}, \qquad \text{then } U\bar{\mathbf{x}} = \bar{\mathbf{c}} \text{ (or better, } A\bar{\mathbf{x}} = \mathbf{b}) \tag{18}$$

after calculating $L\backslash U$, then $\bar{\mathbf{c}}$, then $\bar{\mathbf{x}}$, respectively. ■

EXAMPLE 3.2F Use the LU-Factorization Algorithm 3.2E to solve the 4×4 system

$$\begin{array}{lrcrcrcrcr} (\mathrm{E}_1) & 2x_1 & + & 6x_2 & & & - & 2x_4 & = & -2 \\ (\mathrm{E}_2) & -2x_1 & - & 5x_2 & - & 3x_3 & + & 4x_4 & = & 83 \\ (\mathrm{E}_3) & -x_1 & - & 2x_2 & & & - & 1x_4 & = & 20 \\ (\mathrm{E}_4) & & & 2x_2 & + & 10x_3 & + & 5x_4 & = & 10 \end{array} \tag{19}$$

SOLUTION: Initialize $\tilde{A}$ as A. The reduction of $\mathrm{col}_1\tilde{A}$ is

$$\tilde{A} = \begin{bmatrix} 2 & 6 & 0 & -2 \\ -2 & -5 & -3 & 4 \\ -1 & -2 & 0 & -1 \\ 0 & 2 & 10 & 5 \end{bmatrix} \begin{array}{l} \longrightarrow \\ \rho_2 - \frac{-2}{2}\rho_1 \\ \rho_3 - \frac{-1}{2}\rho_1 \\ \rho_4 - \frac{0}{2}\rho_1 \end{array} \begin{bmatrix} \textcircled{2} & 6 & 0 & -2 \\ \underline{-1} & 1 & -3 & 2 \\ \underline{\frac{1}{2}} & 1 & 0 & -2 \\ \underline{0} & 2 & 10 & 5 \end{bmatrix} \tag{20a}$$

The reduction of column 2 of

$$\rho_3 - 1\rho_2 = [0 \quad 1 \quad 0 \quad -2] - 1[0 \quad 1 \quad -3 \quad 2] = [0 \quad 0 \quad 3 \quad -4] \tag{20b}$$

$$\rho_4 - 2\rho_2 = [0 \quad 2 \quad 10 \quad 5] - 2[0 \quad 1 \quad -3 \quad 2] = [0 \quad 0 \quad 16 \quad 1] \tag{20c}$$

To complete the ***LU*-FACTORIZATION**, we need only reduce column 3 by using these two resulting rows in the 0_{34}-subtract:

$$\rho_4 - \tfrac{16}{3}\rho_3 = [0 \quad 0 \quad 16 \quad 1] - \tfrac{16}{3}[0 \quad 0 \quad 3 \quad -4] = [0 \quad 0 \quad 0 \quad \tfrac{67}{3}] \tag{20d}$$

Note that the actual zeros, *not* the stored (underlined) multipliers, were used in the $0_{i,3}$- and $0_{i,4}$-subtracts. The safest way to ensure this is to *perform the* 0_{im}*-subtracts only to the right of the entry being "zeroed."* The final $L\backslash U$ (which you can check by verifying that $LU = A$) is then put in the forback matrix

$$[L\backslash U:\mathbf{b}:\bar{\mathbf{c}}:\bar{\mathbf{x}}] = \left[\begin{array}{cccc:c:c:c} \textcircled{2} & 6 & 0 & -2 & -2 & -2 & -98 \\ \underline{-1} & \textcircled{1} & -3 & 2 & 83 & 81 & 35 \\ \underline{-\frac{1}{2}} & \underline{1} & \textcircled{3} & -4 & 20 & -62 & -10 \\ \underline{0} & \underline{2} & \underline{\frac{16}{3}} & \textcircled{\frac{67}{3}} & 10 & \frac{536}{3} & 8 \end{array}\right] \tag{20e}$$

The final calculation in getting $\bar{\mathbf{c}}$ [using (5a)]

$$\bar{c}_4 = 10 - [0 \quad -2 \quad \tfrac{16}{3}]\begin{bmatrix} -2 \\ 81 \\ -62 \end{bmatrix} = 10 - (0 + 162 - \tfrac{992}{3}) = \tfrac{536}{3} \qquad \textbf{(20f)}$$

At this point you can check that $\bar{\mathbf{c}} = [-2 \quad 81 \quad -62 \quad \frac{536}{3}]^T$ satisfies $L\bar{\mathbf{c}} = \mathbf{b}$. The final calculation in getting $\bar{\mathbf{x}}$ [using (2b)] is

$$\bar{x}_1 = \frac{1}{2}\left(-2 - [6 \quad 0 \quad -2]\begin{bmatrix} 35 \\ -10 \\ 8 \end{bmatrix}\right) = \frac{1}{2}\{-2 - (210 - 0 - 16)\} = -98 \qquad \textbf{(20g)}$$

At this point you should check that $\bar{\mathbf{x}} = [-98 \quad 35 \quad -10 \quad 8]^T$ is the desired solution of (19), by showing that $A\bar{\mathbf{x}} = \mathbf{b}$. ■

3.2G Counting Arithmetic Operations

We wish to determine how many arithmetic operations it takes to solve $A\mathbf{x} = \mathbf{b}$ using the *LU*-Factorization Algorithm 3.2E. The abbreviations $N(\pm)$, $N(*)$, and $N(/)$ will be used to denote the number of additions or subtractions, multiplications, and divisions respectively. To begin, the algorithm for a single 0_{im}-subtract $\rho_i - l_{im}\rho_m$ is

$l_{im} \leftarrow \tilde{a}_{im}/\tilde{a}_{mm}$	{one division}
DO FOR $j = m+1$ TO n	$\{(n - m)$ entries to the right of $\tilde{a}_{im}\}$
└ $\tilde{a}_{ij} = \tilde{a}_{ij} - l_{im}\tilde{a}_{mj}$	{one subtraction and one multiplication}

for which $N(/) = 1$, and $N(\pm) = N(*) = (n - m)$. The reduction of the mth column requires $n - m$ such subtracts (for $i = m + 1, \ldots, n$) for which

$$N(/) = (n - m) \quad \text{and} \quad N(\pm) = N(*) = (n - m)^2 \qquad \textbf{(21a)}$$

Summing these from $m = 1$ to $n - 1$, and using the familiar closed-form formulas for the sum of the first $n - 1$ integers and of their squares gives the total number of operations needed to reduce A to $L\backslash U$:

$$N(/) = (n - 1) + (n - 2) + \cdots + 2 + 1 = \frac{n(n - 1)}{2} \qquad \textbf{(21b)}$$

$$N(\pm) = N(*) = (n - 1)^2 + (n - 2)^2 + \cdots + 2^2 + 1^2 = \frac{n(n - 1)(2n - 1)}{6} \qquad \textbf{(21c)}$$

We get the other entries of Table 3.2-1 from a similar analysis of the formulas

$$\bar{c}_i = b_i - \sum_{j=1}^{i-1} l_{ij}\bar{c}_j \quad \text{and} \quad \bar{x}_i = \frac{1}{u_{ii}}\left(\bar{c}_i - \sum_{j=i+1}^{n} u_{ij}\bar{x}_j\right) \qquad \textbf{(21d)}$$

Basic Gaussian Elimination requires just as many arithmetic operations because its **UPPER TRIANGULATION** phase requires the same arithmetic steps that it takes to get $L\backslash U$ then $\bar{\mathbf{c}}$ by *LU*-factorization.

Thus the total number of arithmetic operations needed to solve an $n \times n$ linear system by elimination *or* factorization is 28 if $n = 3$; 62 if $n = 4$; 805 if $n = 10$; 87,025 if $n = 50$; and 681,550 if $n = 100$. Note that over half of the operations are multiplications or divisions (which typically take two to seven times as long as an addition/subtraction), and that most of the "number crunching" is needed to get $L\backslash U$.

For large n, the number of arithmetic operations needed to reduce an $n \times n$ matrix A to $L\backslash U$ (i.e., to find an LU-factorization of A) is roughly proportional to n^3. This means that doubling n multiplies the amount of computation by about $2^3 = 8$, tripling n multiplies it by $3^3 = 27$, and so on. We describe this cubic "growth rate" by saying that the calculation is of **order n^3**, abbreviated $\boldsymbol{O(n^3)}$, for large n. On the other hand, forward and backward substitution (to get $\bar{\mathbf{c}}$ and $\bar{\mathbf{x}}$) each require about n^2 arithmetic operations, hence are only $O(n^2)$ for large n. So when several linear systems with the same coefficient matrix are to be solved, the larger the value of n, the more desirable it is to *reuse* $L\backslash U$ in the LU-Factorization Algorithm 3.2E, as compared to *recalculating* it in the Basic Gaussian Elimination Algorithm 3.2D.

TABLE 3.2-1 Arithmetic Operations Needed to Solve $A_{n\times n}\mathbf{x} = \mathbf{b}$ Using *LU*-Factorization or Gaussian Elimination

Matrix	*N(±)*	*N(∗)*	*N(/)*	*Total Operations*
$L\backslash U$	$\frac{n(n-1)(2n-1)}{6}$	$\frac{n(n-1)(2n-1)}{6}$	$\frac{n(n-1)}{2}$	$\frac{n(n-1)(4n+1)}{6}$
$\bar{\mathbf{c}}$	$\frac{n(n-1)}{2}$	$\frac{n(n-1)}{2}$	0	$n(n-1)$
$\bar{\mathbf{x}}$	$\frac{n(n-1)}{2}$	$\frac{n(n-1)}{2}$	n	n^2
$[L\backslash U:\mathbf{b}:\bar{\mathbf{c}}:\bar{\mathbf{x}}]$	$\frac{n(n-1)(2n+5)}{6}$	$\frac{n(n-1)(2n+5)}{6}$	$\frac{n(n+1)}{2}$	$\frac{n(4n^2+9n-7)}{6}$

Table 3.2-1 makes it clear that the danger of large errors in $\bar{\mathbf{x}}$ due to propagated roundoff error increases rapidly as n gets larger. We shall confront this fact of life in Section 3.3.

3.2H Gaussian Elimination

If the *nonzero* entries of a square matrix lie at most p entries above the main diagonal and at most q entries below it, and p and q are *both* less than n, then the matrix is called **banded**, with **bandwidth** $p + q + 1$. For example, the 6×6 matrix

$$\begin{bmatrix} 1 & 1 & & & & \\ 2 & 2 & 0 & & & \\ 3 & 8 & 8 & 3 & & \\ & 0 & 4 & 4 & 0 & \\ & & 5 & 0 & 3 & 2 \\ & & & 1 & 6 & 0 \end{bmatrix} \quad \text{(entries not shown are zero)} \qquad \textbf{(22)}$$

is banded, with $p = 1$ and $q = 2$, so its bandwidth is 4.

Of special importance are matrices T for which $p = q = 1$ (bandwidth 3); such matrices are called **tridiagonal**. To emphasize the fact that all nonzero entries of a tridiagonal matrix lie either on the main diagonal or one entry above or below it, we will use the notation

$$T = \text{trid}(\mathbf{l}, \mathbf{d}, \mathbf{u}) = \begin{bmatrix} d_1 & u_1 & & & & \\ l_2 & d_2 & u_2 & & & \\ & l_3 & d_3 & u_3 & & \\ & & \ddots & \ddots & \ddots & \\ & & & l_{n-1} & d_{n-1} & u_{n-1} \\ & & & & l_n & d_n \end{bmatrix} \tag{23a}$$

where l_1 and u_n (not shown) can be left undefined. The vectors

$$\mathbf{l} = [l_1 \quad l_2 \quad \cdots \quad l_n], \quad \mathbf{d} = [d_1 \quad d_2 \quad \cdots \quad d_n], \quad \text{and} \quad \mathbf{u} = [u_1 \quad u_2 \quad \cdots \quad u_n] \tag{23b}$$

contain, respectively, the **subdiagonal** (lower), **diagonal**, and **superdiagonal** (upper) entries of T. A **tridiagonal system** is a linear system whose matrix form is $T\mathbf{x} = \mathbf{b}$, where T is a tridiagonal matrix. For such systems, both the Gaussian Elimination and LU-Factorization Algorithms are particularly simple, as the following example shows.

EXAMPLE 3.2H Use Basic Gaussian Elimination to solve the 4×4 system

$$\begin{aligned} 5x_1 - 3x_2 \qquad\qquad\quad &= 7 \\ x_1 + 4x_2 - 2x_3 \qquad\quad &= 6 \\ -x_2 + 3x_3 + x_4 &= -4 \\ 2x_3 + 5x_4 &= -15 \end{aligned} \qquad \left([T{:}\mathbf{b}] = \begin{bmatrix} 5 & -3 & & & : & 7 \\ 1 & 4 & -2 & & : & 6 \\ & -1 & 3 & 1 & : & -4 \\ & & 2 & 5 & : & -15 \end{bmatrix} \right) \tag{24}$$

SOLUTION: The **UPPER TRIANGULATION** reduces $[T{:}\mathbf{b}]$ to the $[L\backslash U{:}\bar{\mathbf{c}}]$ part of

$$\begin{bmatrix} \textcircled{5} & -3 & & & : & 7 & : & 2 \\ \frac{1}{5} & \frac{23}{5} & -2 & & : & \frac{23}{5} & : & 1 \\ & -\frac{5}{23} & \frac{59}{23} & 1 & : & -3 & : & 0 \\ & & \frac{46}{59} & \frac{249}{59} & : & -\frac{747}{59} & : & -3 \end{bmatrix} \tag{25}$$

$$L\backslash U \qquad\qquad \bar{\mathbf{c}} \qquad\qquad \bar{\mathbf{x}}$$

The reduction $[T{:}\mathbf{b}] \rightarrow [L\backslash U{:}\bar{\mathbf{c}}]$ proceeds as follows:

$$\rho_2 - \frac{1}{5}\rho_1{:} \quad [1 \quad 4 \quad -2 \quad 0{:}6] - \tfrac{1}{5}[5 \quad -3 \quad 0 \quad 0{:}7] = [0 \quad \tfrac{23}{5} \quad -2 \quad 0{:}\tfrac{23}{5}]$$

$$\rho_3 - \frac{-1}{\frac{23}{5}}\rho_2{:} \quad [0 \quad -1 \quad 3 \quad 1{:}-4] - \tfrac{-5}{23}[0 \quad \tfrac{23}{5} \quad -2 \quad 0{:}\tfrac{23}{5}] = [0 \quad 0 \quad \tfrac{59}{23} \quad 1{:}-3]$$

$$\rho_4 - \frac{2}{\frac{59}{23}}\rho_3{:} \quad [0 \quad 0 \quad 2 \quad 5{:}-15] - \tfrac{46}{59}[0 \quad 0 \quad \tfrac{59}{23} \quad 1{:}-3] = [0 \quad 0 \quad 0 \quad \tfrac{249}{59}{:}-\tfrac{747}{59}]$$

The **BACKWARD SUBSTITUTION** then proceeds as follows:

$$\bar{x}_4 = \frac{-\frac{747}{59}}{\frac{249}{59}} = -3; \quad \bar{x}_3 = \frac{-3 - 1\cdot(-3)}{\frac{59}{23}} = 0;$$

$$\bar{x}_2 = \frac{\frac{14}{5} - (-2)\cdot 0}{\frac{14}{5}} = 1; \quad \bar{x}_1 = \frac{7 - (-3)\cdot 1}{5} = 2$$

Notice that in the reduction $[T\!:\!\mathbf{b}] \to [L\backslash U\!:\!\bar{\mathbf{c}}]$ a single $0_{m+1,m}$-subtract, which did not alter the superdiagonal ($\mathbf{u}$) values, was all that was needed to reduce the mth column ($m = 1, 2, 3$). Also, the inner products used in the backward substitution were just single multiplications. ■

The simplifications just noted will always occur, as can be seen by examining the pseudocode for the **Gaussian Elimination Algorithm for Tridiagonal Systems** shown in Algorithm 3.2H. To get familiar with the trid($\mathbf{l}$, $\mathbf{d}$, $\mathbf{u}$) notation, you should verify that the formulas of Algorithm 3.2H yield the solution (25) for Example 3.2H.

ALGORITHM 3.2H. GAUSSIAN ELIMINATION (IN PLACE) FOR TRIDIAGONAL SYSTEMS

PURPOSE: To solve in place a given $n \times n$ tridiagonal system $T\mathbf{x} = \mathbf{b}$, where $T =$ trid($\mathbf{l}$, $\mathbf{d}$, $\mathbf{u}$). $[T\!:\!\mathbf{b}]$ is returned as $[L\backslash U\!:\!\bar{\mathbf{x}}\,]$.

```
GET n, l, d, u, b                          {l_1 and u_n can be left undefined}

{UPPER TRIANGULATION: Reduce [T:b] → [L\U:c̄] in place}
DO FOR m = 1 TO n
|   IF d_m = 0 THEN                                       {the algorithm fails}
|       OUTPUT ('T appears to be singular.')
|       STOP
|   IF m < n THEN             {perform an m,m+1-subtract to reduce col_m T}
|   |   l_{m+1} ← l_{m+1}/d_m              {l_{m+1} is now the (m+1,m)th multiplier}
|   |   d_{m+1} ← d_{m+1} − l_{m+1}u_m     {d_{m+1} is now the (m+1)st pivot entry)}
|___|__ b_{m+1} ← b_{m+1} − l_{m+1}b_m                        {b now stores c̄_i}

BACKWARD SUBSTITUTION: Solve Ux = c̄ for x̄ (over c̄ in b)
b_n ← b_n/d_n                                              {b_n now stores x̄_n}
DO FOR i = n − 1 DOWNTO 1;
|___  b_i ← (1/d_i)(b_i − u_i b_{i+1})                     {b_i now stores x̄_i}
OUTPUT('x̄ = ['b_1, ..., b_n']^{-1} is the desired solution of Tx = b.')
STOP
```

Practical Consideration 3.2H (The need for a separate algorithm). Algorithm 3.2H exploits the fact that T is tridiagonal, and so uses only

$$3n - 3 \text{ subtractions}, \quad 3n - 3 \text{ multiplications}, \quad \text{and} \quad 2n - 1 \text{ divisions} \tag{26}$$

So the number of arithmetic operations needed to solve $T\mathbf{x} = \mathbf{b}$ is proportional to n, that is, only $O(h)$ for large n. Tridiagonal systems with n as large as 100,000 arise in the solution of differential equations. For this n, Algorithm 3.2H requires about $8n = 8 \cdot 10^5$ arithmetic operations; the general LU-Factorization Algorithm 3.2E would require about $2n^3/3 \approx 0.7 \cdot 10^{15}$ (about 10 billion times as many, most being *unnecessary* multiplications by, and addition/subtractions of, zero!). This would take prohibitively long, even on the fastest computers. Consequently, a software library should have a special tridiagonal system solver such as the Fortran SUBROUTINE TRIDAG shown in Figure 3.2-2 *in addition to* a general linear system solver such as the one we shall describe in Section 4.2B. ■

A square matrix for which each diagonal element is larger in magnitude than the *sum* of

```
      SUBROUTINE TRIDAG (N, L, D, U, B, OK)
      REAL L(N), D(N), U(N), B(N)
      LOGICAL OK
C - - - - - - - - - - - - - - - - - - - - - - - - - - - - - - - C
C  Subroutine to solve an n by n tridiagonal system TX = B in   C
C  place.  L, D, and U are the sub-, on-, and super- diagonal   C
C  entries of T.  L(1) and U(n) are not used.  The solution is  C
C  returned in B (if OK); L and D are also altered by the call. C
C - - - - - - - - - - - - - - - - - - - - - VERSION 2  9/9/85 - C
      OK = .TRUE.        ! UPPER TRIANGULATE:
      DO 10 M=1, N
         IF (D(M) .EQ. 0.0) THEN              ! algorithm fails
            OK = .FALSE.
            RETURN
         ENDIF
         IF (M .LT. N) THEN
            MP1 = M + 1
            L(MP1) = L(MP1) / D(M)            ! (m+1,m)th multiplier
            D(MP1) = D(MP1) - L(MP1)*U(M) ! reduce mth row of T
            B(MP1) = B(MP1) - L(MP1)*B(M) ! reduce mth row of B
         ENDIF
   10 CONTINUE           ! BACKWARD SUBSTITUTE:
      B(N) = B(N)/D(N)
      IF (N .GT. 1) THEN
         DO 20 I=N-1, 1, -1
            B(I) = ( B(I) - U(I)*B(I+1) ) / D(I)
   20    CONTINUE
      ENDIF              ! B now stores the solution
      RETURN
      END
```

Figure 3.2-2 TRIDAG: *A Fortran* SUBROUTINE *for solving* $T\mathbf{x} = \mathbf{b}$.

the other entries in its row is called **strictly diagonally dominant**. An $n \times n$ tridiagonal matrix $T = \text{trid}(\mathbf{l}, \mathbf{d}, \mathbf{u})$ is strictly diagonally dominant if and only if

$$|d_1| > |u_1|, \quad |d_n| > |l_n|, \quad \text{and} \quad |d_i| > |l_i| + |u_i| \quad \text{for } 1 < i < n \tag{27}$$

The matrix T in (24) is diagonally dominant. In fact, the tridiagonal coefficient matrices that occur in practice are almost always diagonally dominant. It can be shown [37] that $d_m = 0$ will never occur when T is strictly diagonally dominant. However, we test for it in Algorithm 3.2H (and Figure 3.2-2) just in case T is not.

3.2I Gauss–Jordan Elimination

A popular variant of Basic Gaussian Elimination is **Basic Gauss–Jordan Elimination**. It consists of using 0_{im}-subtracts to get zeros *above as well as below* the mth pivot entry a_{mm} for $m = 1, 2, \ldots, n$. Its use in solving an $n \times n$ linear system $A\mathbf{x} = \mathbf{b}$ involves two phases:

DIAGONALIZATION: $[A{:}\mathbf{b}] \to [D{:}\mathbf{e}]$, where $D = \text{diag}(d_{11}, \ldots, d_{nn})$ **(28a)**

SCALING: $[D{:}\mathbf{e}] \to [I_n{:}\bar{\mathbf{x}}]$, where $\bar{\mathbf{x}}$ is the desired solution **(28b)**

The ith equation of the equivalent system $D\mathbf{x} = \mathbf{e}$ is simply $d_{ii}x_i = e_i$. So the SCALING phase consists of simply dividing the ith equation by d_{ii} to get $\bar{x}_i = e_i/d_{ii}$, for $i = 1, \ldots, n$.

EXAMPLE 3.2I Use Basic Gauss–Jordan Elimination to solve

$$\begin{array}{ll} (E_1) & -x_1 + x_2 - 4x_3 = 0 \\ (E_2) & 2x_1 + 2x_2 = 1 \\ (E_3) & 3x_1 + 3x_2 + 2x_3 = \frac{1}{2} \end{array} \qquad \left([A{:}\mathbf{b}] = \begin{bmatrix} -1 & 1 & -4 & : & 0 \\ 2 & 2 & 0 & : & 1 \\ 3 & 3 & 2 & : & \frac{1}{2} \end{bmatrix} \right) \tag{29}$$

SOLUTION: The DIAGONALIZATION phase can be described as follows:

$$[A{:}\mathbf{b}] \xrightarrow[\substack{\rho_2 - (-2)\rho_1 \\ \rho_3 - (-3)\rho_1}]{\rho_1} \begin{bmatrix} -1 & 1 & -4 & : & 0 \\ 0 & 4 & -8 & : & 1 \\ 0 & 6 & -10 & : & \frac{1}{2} \end{bmatrix} \xrightarrow[\substack{\rho_2 \\ \rho_3 - (\frac{3}{2})\rho_2}]{\rho_1 - (\frac{1}{4})\rho_2} \begin{bmatrix} -1 & 0 & -2 & : & -\frac{1}{4} \\ 0 & 4 & -8 & : & 1 \\ 0 & 0 & 2 & : & -1 \end{bmatrix} \tag{30a}$$

column 1 reduced column 2 reduced

$$\xrightarrow[\substack{\rho_2 - (-4)\rho_3 \\ \rho_3}]{\rho_1 - (-1)\rho_3} \begin{bmatrix} -1 & 0 & 0 & : & -\frac{5}{4} \\ 0 & 4 & 0 & : & -3 \\ 0 & 0 & 2 & : & -1 \end{bmatrix} = [D{:}\mathbf{e}] \tag{30b}$$

column 3 reduced

The SCALING phase is simply: $\bar{x}_1 = (-\frac{5}{4})/(-1) = \frac{5}{4}$, $\bar{x}_2 = (-3)/4 = -\frac{3}{4}$, and $\bar{x}_3 = (-1)/2 = -\frac{1}{2}$. The solution $\mathbf{x} = [\frac{5}{4} \;\; -\frac{3}{4} \;\; -\frac{1}{2}]^T$ agrees with that of Example 3.2C.

Practical Consideration 3.2I (Gauss or Gauss–Jordan?). A computer program for Gauss–Jordan Elimination is shorter than one for Gaussian Elimination because code for the SCALING phase is shorter than that for **BACKWARD SUBSTITUTION**. However, Gauss–Jordan Elimination uses more arithmetic operations to execute this shorter program [Problem 3-18(e)]. The resulting possibililty of a longer execution time and extra roundoff makes Gauss–Jordan Elimination somewhat less desirable as a method for solving a *single* $n \times n$ linear system. We shall therefore continue to concentrate on Gaussian Elimination. ■

3.3

LU-Decomposition = *LU*-Factorization + Pivoting

In this important section we demonstrate the need to rearrange the order of the equations when using the *LU*-Factorization Algorithm 3.2E and we develop computational techniques for doing this so as to reduce the propagation of roundoff error.

3.3A The Need for Pivoting (Exact Arithmetic)

We saw in Example 3.2C that $\bar{\mathbf{x}} = [\frac{5}{4} \;\; -\frac{3}{4} \;\; -\frac{1}{2}]^T$ is the solution of the linear system $A\mathbf{x} = \mathbf{b}$ given by

$$\begin{array}{ll} (\mathrm{E}_1) & -x_1 + x_2 - 4x_3 = 0 \\ (\mathrm{E}_2) & 2x_1 + 2x_2 \phantom{{}-4x_3} = 1 \\ (\mathrm{E}_3) & 3x_1 + 3x_2 + 2x_3 = \frac{1}{2} \end{array} \qquad [A\!:\!\mathbf{b}] = \begin{bmatrix} -1 & 1 & -4 & : & 0 \\ 2 & 2 & 0 & : & 1 \\ 3 & 3 & 2 & : & \frac{1}{2} \end{bmatrix} \tag{1}$$

Suppose that these equations were given with the first and third equations reversed, that is, as

$$\begin{array}{ll} (\mathrm{E}_1) & 3x_1 + 3x_2 + 2x_3 = \frac{1}{2} \\ (\mathrm{E}_2) & 2x_1 + 2x_2 \phantom{{}+2x_3} = 1 \\ (\mathrm{E}_3) & -x_1 + x_2 - 4x_3 = 0 \end{array} \qquad [A\!:\!\mathbf{b}] = \begin{bmatrix} 3 & 3 & 2 & : & \frac{1}{2} \\ 2 & 2 & 0 & : & 1 \\ -1 & 1 & -4 & : & 0 \end{bmatrix} \tag{2}$$

Since the equations in (2) are simply a rearrangement of the equations in (1), we know that $\bar{\mathbf{x}} = [\frac{5}{4} \;\; -\frac{3}{4} \;\; -\frac{1}{2}]^T$ is also the solution of (2). However, if we attempt to use the *LU*-Factorization Algorithm 3.2E to solve (2), then the result of reducing the first column of A would be

$$A \to \tilde{A} = \begin{bmatrix} \textcircled{3} & 3 & 2 \\ \frac{2}{3} & 0 & -\frac{4}{3} \\ -\frac{1}{3} & 2 & -\frac{10}{3} \end{bmatrix} \qquad \begin{array}{l} \\ (\text{after } \rho_2 - l_{21}\rho_1;\; l_{21} = \frac{2}{3}) \\ (\text{after } \rho_3 - l_{31}\rho_1;\; l_{31} = -\frac{1}{3}) \end{array} \tag{3}$$

At this point the *LU*-Factorization Algorithm 3.2E fails because the 0_{32}-subtract, $\rho_3 - (\tilde{a}_{32}/\tilde{a}_{22})\rho_2$, would require division by zero. Thus, although the system (2) has a solution, it cannot be found using the *LU*-Factorization Algorithm 3.2E. However, we have seen enough to conclude the following about a given linear system $A\mathbf{x} = \mathbf{b}$:

If the *LU*-Factorization Algorithm 3.2E fails because $\tilde{a}_{mm} = 0$ for some m, it may be possible to find the solution by solving an *equivalent* system obtained by rearranging its equations. **(4)**

In fact, there may be *several* such rearrangements. For example, (2) can be solved by interchanging equations (E_1) and (E_3) to get back to (1). Alternatively, interchanging equations (E_2) and (E_3) in (2) yields a *different* equivalent system, $\hat{A}\mathbf{x} = \hat{\mathbf{b}}$, namely

$$\begin{array}{ll} (E_1) & 3x_1 + 3x_2 + 2x_3 = \tfrac{1}{2} \\ (E_2) & -x_1 + x_2 - 4x_3 = 0 \\ (E_3) & 2x_1 + 2x_2 \qquad = 1 \end{array} \qquad [\hat{A}:\hat{\mathbf{b}}] = \begin{bmatrix} 3 & 3 & 2 & : & \frac{1}{2} \\ -1 & 1 & -4 & : & 0 \\ 2 & 2 & 0 & : & 1 \end{bmatrix} \qquad \textbf{(5)}$$

for which the *LU*-Factorization Algorithm 3.2E succeeds, as summarized in the following forback matrix.

$$[\hat{L}\backslash\hat{U}:\hat{\mathbf{b}}:\bar{\mathbf{c}}:\bar{\mathbf{x}}] = \begin{bmatrix} \textcircled{3} & 3 & 2 & : & \frac{1}{2} & : & \frac{1}{2} & : & \frac{5}{4} \\ -\frac{1}{3} & \textcircled{2} & -\frac{10}{3} & : & 0 & : & \frac{1}{6} & : & -\frac{3}{4} \\ \frac{2}{3} & 0 & \textcircled{-\frac{4}{3}} & : & 1 & : & \frac{2}{3} & : & -\frac{1}{2} \end{bmatrix} \qquad \textbf{(6)}$$

Since $\hat{A}\mathbf{x} = \hat{\mathbf{b}}$ in (5) is equivalent to $A\mathbf{x} = \mathbf{b}$ in (1), it should be no surprise that $\bar{\mathbf{x}} = [\frac{5}{4} \;\; -\frac{3}{4} \;\; -\frac{1}{2}]^T$ obtained in (6) is its solution. Notice that $\hat{L}\backslash\hat{U}$ in (6) could be obtained by interchanging the second and third rows of $\tilde{A}$ in (3) and then continuing the reduction *starting with* $m = 2$. Note, too, that $\hat{L}\backslash\hat{U}$ in (6) satisfies $\hat{L}\hat{U} = \hat{A}$ in (5), *not* $\tilde{A}$ in (3).

If *exact arithmetic* (without roundoff) is used, then *the LU-Factorization Algorithm 3.2E will yield the exact* $\bar{\mathbf{x}}$ *regardless of the order in which the equations are given*, as long as $\tilde{a}_{mm} \neq 0$ for $m = 1, 2, \ldots, n-1$. In this case, there is no reason to rearrange the equations *unless* $\tilde{a}_{mm} = 0$ is encountered during the reduction of A to $L\backslash U$. Unfortunately, as we now show, this simple strategy is not suitable when the reduction is performed on a digital device.

3.3B The Need for Pivoting (Fixed-Precision Arithmetic)

The following example shows that when calculations are performed in *fixed-precision arithmetic* (see Section 1.3B), the accuracy of the calculated $\bar{\mathbf{x}}$ depends on the order in which the equations are given!

EXAMPLE 3.3B Solve using the *LU*-Factorization Algorithm 3.2E:

$$\text{First System: } (E_1)\; 3x + 2y = 13 \quad \text{and} \quad (E_2)\; 7x + 90y = 713 \qquad \textbf{(7a)}$$

$$\text{Second System: } (E_1)\; 7x + 90y = 713 \quad \text{and} \quad (E_2)\; 3x + 2y = 13 \qquad \textbf{(7b)}$$

To simulate the best a 4*s* device can do, calculate all entries in extended precision, but store them rounded to 4*s*.

NOTE: The equations in (7b) are the same as those of (7a) but in the reverse order. Both systems have $\bar{\mathbf{x}} = [-1 \;\; 8]^{\mathrm{T}}$ as their exact solution.

SOLUTION: The *LU*-Factorization Algorithm 3.2E yields the following forback matrices for the systems (7a) and (7b), respectively:

$$\begin{bmatrix} \textcircled{3} & 2 & : & 13 & : & 13 & : & -1.000 \\ 2.333 & \textcircled{85.33} & : & 713 & : & 682.7 & : & 8.001 \end{bmatrix}; \qquad \text{so } \bar{\mathbf{x}}_{\text{calc}} = \begin{bmatrix} -1.000 \\ 8.001 \end{bmatrix} \qquad \textbf{(8a)}$$

$$\left[\begin{array}{cc:c:c:c} \textcircled{7} & 90 & 713 & 713 & -1.013 \\ \underline{0.4286} & \textcircled{-36.59} & 13 & -292.6 & 8.001 \end{array}\right]; \qquad \text{so } \bar{\mathbf{x}}_{\text{calc}} = \begin{bmatrix} -1.013 \\ 8.001 \end{bmatrix} \tag{8b}$$

The small errors in $\bar{\mathbf{x}}_{\text{calc}}$ are the result of creeping roundoff. It just "crept" faster in (8b)! Both calculations yielded comparable accuracy through the calculation of $\bar{x}_2$ at the beginning of the backward substitution. The difference occurred in calculating $\bar{x}_1$:

$$\text{For (7a): } \bar{x}_1 = \tfrac{1}{3}\{13 - 2(8.001)\} = \tfrac{1}{3}\{-3.002\} \doteq -100 \ (4s) \tag{9a}$$

$$\text{For (7b): } \bar{x}_1 = \tfrac{1}{7}\{713 - 90(8.001)\} = \tfrac{1}{7}\{-7.090\} \doteq -1.013 \ (4s) \tag{9b}$$

At first glance, it might appear that the larger error in (9b) occurred when the error in $\bar{x}_2 = 8.001$ got magnified when multiplied by the relatively large number $u_{12} = 90$. However, if the systems (7) were given with one of the equations divided by 100†, say as

$$\text{First System: } (\mathrm{E}_1) \quad 3x + 2y = 13 \quad \text{and} \quad (\mathrm{E}_2)\ 0.07x + 0.9y = 7.13 \tag{10a}$$

$$\text{Second System: } (\mathrm{E}_1)\ 0.07x + 0.9y = 7.13 \quad \text{and} \quad (\mathrm{E}_2) \quad 3x + 2y = 13 \tag{10b}$$

then the resulting forback matrices would have been, respectively,

$$\left[\begin{array}{cc:c:c:c} \textcircled{3} & 2 & 13 & 13 & -1.000 \\ \underline{0.02333} & \textcircled{0.8533} & 7.13 & 6.827 & 8.001 \end{array}\right]; \text{ so } \bar{\mathbf{x}}_{\text{calc}} = \begin{bmatrix} -1.000 \\ 8.001 \end{bmatrix} \tag{11a}$$

$$\left[\begin{array}{cc:c:c:c} \textcircled{0.07} & 0.9 & 7.13 & 7.13 & -1.013 \\ \underline{42.86} & \textcircled{-36.59} & 13 & -292.6 & 8.001 \end{array}\right]; \text{ so } \bar{\mathbf{x}}_{\text{calc}} = \begin{bmatrix} -1.013 \\ 8.001 \end{bmatrix} \tag{11b}$$

which are neither better nor worse than the results obtained in (8)! ■

To see what did cause the differences observed in (9), let us examine the calculation of $\bar{x}_1$ in the last (backward substitution) step of the *LU*-Factorization Algorithm 3.2E for *any* $n \times n$ system. We shall assume that $L\backslash U$ was calculated with insignificant error but that subsequent propagated roundoff caused $\bar{x}_2, \ldots, \bar{x}_n$ to be calculated with errors $\bar{\epsilon}_2, \ldots, \bar{\epsilon}_n$, respectively. Then by (5b) of Section 3.2B,

$$\bar{x}_{1,\text{calc}} = \frac{1}{u_{11}}\left\{\bar{c}_1 - [u_{12} \ \cdots \ u_{1n}]\begin{bmatrix} \bar{x}_2 - \bar{\epsilon}_2 \\ \vdots \\ \bar{x}_n - \bar{\epsilon}_n \end{bmatrix}\right\}$$

$$= \underbrace{\frac{1}{u_{11}}\left\{\bar{c}_1 - [u_{12} \ \cdots \ u_{1n}]\begin{bmatrix} \bar{x}_2 \\ \vdots \\ \bar{x}_n \end{bmatrix}\right\}}_{\text{formula for exact } \bar{x}_1} - \underbrace{\begin{bmatrix} \dfrac{u_{12}}{u_{11}} & \dfrac{u_{13}}{u_{11}} & \cdots & \dfrac{u_{1n}}{u_{11}} \end{bmatrix}\begin{bmatrix} \bar{\epsilon}_2 \\ \vdots \\ \bar{\epsilon}_n \end{bmatrix}}_{\bar{\epsilon}_1 = \text{ error of } \bar{x}_{1,\text{calc}}} \tag{12}$$

†This might occur if the unscaled equations in (7) equated measurements made in centimeters, whereas the corresponding scaled equations in (10) equated the *same* measurements but in *meters*.

We see from (12) that $\bar{\epsilon}_1$ *is most likely to be small when* $|u_{12}|, \ldots, |u_{1n}|$ *are as small as possible compared to* $|u_{11}|$, that is, when the equations are ordered so that the entries of $\text{row}_1 A$ (= $\text{row}_1 U$) satisfy

$$\frac{|a_{12}|}{|a_{11}|}, \quad \frac{|a_{13}|}{|a_{11}|}, \quad \ldots, \quad \frac{|a_{1n}|}{|a_{11}|} \quad \text{are as small as possible} \qquad \textbf{(13)}$$

When $n = 2$, there is only one ratio in (13), namely $|a_{12}|/|a_{11}|$ (= u_{12}). For the systems just solved, it is

$$\tfrac{2}{3} \text{ for (7a) and (10a),} \quad \text{but} \quad \tfrac{90}{7} \text{ for (7b) and (10b)}$$

So criterion (13) would yield the more accurate solution (9a) for *either* system in (7), and the more accurate solution (11a) for *either* system in (10). Note that the ratios in (13) do not change if equations are rescaled as in (10b).

3.3C Pivoting Strategies for the *LU*-Decomposition Algorithm

It is clear from Sections 3.3A and 3.3B that some provision must be made for rearranging the order of the equations when performing Gaussian Elimination or *LU*-factorization. The following terminology and notation will be used to do this. It will be used extensively for the rest of this chapter and also in Chapter 4. So *learn it well.*

Terminology and Notation. The interchange of the mth row of a matrix with a *lower*† row, say the pth, will be called an ***m,p*-pivot**. The use of m,p-pivots in solving $A\mathbf{x} = \mathbf{b}$ will be referred to as **row pivoting**, or simply **pivoting**. Reducing A using pivoting will be referred to as **decomposing** A and abbreviated as

$$\hat{A} \;\to\; \{\tilde{A}\} \;\to\; \hat{L}\backslash\hat{U} \qquad \textbf{(14a)}$$

where $\hat{A}$ (read as "A roof" or "A hat") and $\tilde{A}$ (read as "A squiggle" or "A tilde") are related to the given coefficient matrix A as follows:

$$\hat{A}\mathbf{x} = \hat{\mathbf{b}} \text{ (initially, } A\mathbf{x} = \mathbf{b}\text{) is the current system being solved} \qquad \textbf{(14b)}$$

$$\tilde{A} \text{ is the partial reduction of the current coefficient matrix } \hat{A} \qquad \textbf{(14c)}$$

Any m,p-pivots performed on $\tilde{A}$ in (14a) must also be performed on $[\hat{A}:\hat{\mathbf{b}}]$. So the final $[\hat{A}:\hat{\mathbf{b}}]$ will *not* be the given $[A:\mathbf{b}]$ *unless* it turns out that no m,p-pivots were used. In any case, the final reduced $\tilde{A}$ will be $\hat{L}\backslash\hat{U}$ (unit $\hat{L}$), where $\hat{L}\hat{U} = \hat{A}$. In words, $\hat{L}\backslash\hat{U}$ will be the

†If the first $m - 1$ columns of $\tilde{A}$ have zeros *below* the main diagonal, then interchanging the mth row of $\tilde{A}$ with a row *above* it will bring the *nonzero* diagonal entry of that row *below* the main diagonal. This will undo the previous reduction to an upper triangular matrix!

compact form of the LU-factorization of the *final* $\hat{A}$; hence the *final* equivalent system $\hat{A}\mathbf{x} = \hat{\mathbf{b}}$ can be solved for the desired $\bar{\mathbf{x}}$ using forward/backward substitution.

An m,p-pivot on $\tilde{A}$ moves $\tilde{a}_{pm}$ up to the (circled) mth pivot entry of $\hat{L}\backslash\hat{U}$. A **pivoting strategy** is a strategy for selecting $\tilde{a}_{pm}$. This selection must be made (and the m,p-pivot performed) *before reducing the mth column of* $\tilde{A}$. We saw in the last paragraph of Section 3.3A that the natural pivoting strategy when calculations are performed *exactly* is the

BASIC PIVOTING (BP) STRATEGY. Take $\tilde{a}_{pm}$ to be *the topmost nonzero entry* among $\tilde{a}_{mm}, \tilde{a}_{m+1,m}, \ldots, \tilde{a}_{nm}$ (at and below the current $\tilde{a}_{mm}$).

Basic Pivoting is also referred to as **trivial pivoting** or **naive pivoting**. It will result in the LU-Factorization Algorithm 3.2E (i.e., no m,p-pivots) unless a zero $\tilde{a}_{mm}$ occurs.

On the other hand, the discussion in the last paragraph of Section 3.3B suggests that in the presence of roundoff error, $\tilde{a}_{pm}$ should be taken to be whichever of $\tilde{a}_{mm}, \tilde{a}_{m+1,m}, \ldots, \tilde{a}_{nm}$ is "relatively largest" (compared to the other entries in its row in $\tilde{A}$). A simple strategy for selecting a "relatively large" $\tilde{a}_{pm}$ is the

SCALED PARTIAL PIVOTING (SPP) STRATEGY. Take $\tilde{a}_{pm}$ to be the topmost entry having the largest $|\tilde{a}_{im}|/\hat{s}_i$ ratio among

$$\frac{|\tilde{a}_{mm}|}{\hat{s}_m}, \quad \frac{|\tilde{a}_{m+1,m}|}{\hat{s}_{m+1}}, \quad \ldots, \quad \frac{|\tilde{a}_{nm}|}{\hat{s}_n} \qquad \textbf{(15a)}$$

where $\tilde{a}_{mm}, \tilde{a}_{m+1}, \ldots, \tilde{a}_{nm}$ lie at and below the mth pivot entry, and

$$\hat{s}_i = \text{the largest absolute value in row}_i(\text{current } \hat{A}) \text{ for } i = m, \ldots, n \qquad \textbf{(15b)}$$

The SPP strategy is **scale independent** in the sense that the values of the ratios in (15a) do not change when one or more equations of the system are rescaled as in (10). In implementing the SPP strategy, the **row scale factors** $s_1, \ldots, s_n$ are calculated *for the given* A (the initial $\hat{A}$) and used to initialize the **row scale vector** $\hat{\mathbf{s}}$ *before* beginning the decomposition $\hat{A} \to \{\tilde{A}\} \to \hat{L}\backslash\hat{U}$. Then, any m,p-pivot performed on $\tilde{A}$ is also performed on the vector $\hat{\mathbf{s}}$ as illustrated in the following example.

EXAMPLE 3.3C Decompose the coefficient matrix of the given system using the Scaled Partial Pivoting strategy; then use the LU-factorization of the final $\hat{A}$ to find $\bar{\mathbf{x}}$.

$$\begin{aligned} 2x_1 + 6x_2 - 2x_4 &= -2 \\ -2x_1 - 5x_2 - 3x_3 + 4x_4 &= 83 \\ -x_1 - 2x_2 - 1x_4 &= 20 \\ 2x_2 + 10x_3 + 5x_4 &= 10 \end{aligned} \qquad A = \begin{bmatrix} 2 & 6 & 0 & -2 \\ -2 & -5 & -3 & 4 \\ -1 & -2 & 0 & -1 \\ 0 & 2 & 10 & 5 \end{bmatrix}; \quad \mathbf{s} = \begin{bmatrix} 6 \\ 5 \\ 2 \\ 10 \end{bmatrix} \qquad \textbf{(16)}$$

SOLUTION: We begin by using A and $\mathbf{s}$ in (16) to initialize $\tilde{A}$ and $\hat{\mathbf{s}}$, respectively. The reduction of the mth column of $\tilde{A}$ for $m = 1, 2, 3$ then proceeds as follows.

$\boldsymbol{m = 1.}$ All entries of $\mathrm{col}_1\tilde{A}$ are candidates for the first pivot entry. Their $|\tilde{a}_{i1}|/\hat{s}_i$ ratios (15a) are

$$\frac{|2|}{6};\ \frac{|-2|}{5};\ \frac{|-1|}{2};\ \frac{|0|}{10} \qquad \left(\text{largest is } \frac{|-1|}{2}, \text{ in row 3}\right) \tag{17a}$$

We therefore perform a 1,3-pivot in $\hat{\mathbf{s}}$ and $\tilde{A}$, and then reduce $\mathrm{col}_2\tilde{A}$

$$\hat{\mathbf{s}} = \begin{bmatrix} 2 \\ \\ 5 \\ 6 \\ 10 \end{bmatrix}; \quad \tilde{A} = \begin{bmatrix} \enclose{circle}{-1} & -2 & 0 & -1 \\ \\ -2 & -5 & -3 & 4 \\ 2 & 6 & 0 & -2 \\ 0 & 2 & 10 & 5 \end{bmatrix} \begin{matrix} \}\ \mathrm{row}_1\hat{L}\backslash\hat{U} \\ \xrightarrow{\hspace{5em}} \\ \rho_2 - (2)\rho_1 \\ \rho_3 - (-2)\rho_1 \\ \rho_4 - (0)\rho_1 \end{matrix} \begin{bmatrix} \enclose{circle}{-1} & -2 & 0 & -1 \\ \\ \underline{2} & -1 & -3 & 6 \\ \underline{-2} & 2 & 0 & -4 \\ \underline{0} & 2 & 10 & 5 \end{bmatrix} \tag{17b}$$

$\boldsymbol{m = 2.}$ The $|\tilde{a}_{i2}|/\hat{s}_i$ ratios (15a) for the three candidates (at and below $\tilde{a}_{22}$) for the second pivot entry are

$$\frac{|-1|}{5};\ \frac{|2|}{6};\ \frac{|2|}{10} \qquad \left(\text{largest is } \frac{|2|}{6}, \text{ in row 3}\right) \tag{17c}$$

We therefore perform a 2,3-pivot in $\hat{\mathbf{s}}$ and $\tilde{A}$, and reduce $\mathrm{col}_3\tilde{A}$:

$$\hat{\mathbf{s}} = \begin{bmatrix} 2 \\ 6 \\ 5 \\ 10 \end{bmatrix}; \quad \tilde{A} = \begin{bmatrix} \enclose{circle}{-1} & -2 & 0 & -1 \\ \underline{-2} & \enclose{circle}{2} & 0 & -4 \\ \underline{2} & -1 & -3 & 6 \\ \underline{0} & 2 & 10 & 5 \end{bmatrix} \begin{matrix} \}\ \mathrm{row}_1\hat{L}\backslash\hat{U} \\ \}\ \mathrm{row}_2\hat{L}\backslash\hat{U} \\ \rho_3 - (-\frac{1}{2})\rho_2 \\ \rho_4 - (1)\rho_2 \end{matrix} \begin{bmatrix} \enclose{circle}{-1} & -2 & 0 & -1 \\ \underline{-2} & \enclose{circle}{2} & 0 & -4 \\ \underline{2} & \underline{-\frac{1}{2}} & -3 & 4 \\ \underline{0} & \underline{1} & 10 & 9 \end{bmatrix} \tag{17d}$$

$\boldsymbol{m = 3.}$ Since $|\tilde{a}_{43}|/\hat{s}_4 = |10|/10$ (in row 4) is larger than $|\tilde{a}_{33}|/\hat{s}_3 = |-3|/5$, the SPP strategy requires that we perform a 3,4-pivot in $\hat{\mathbf{s}}$ and $\tilde{A}$. To complete the reduction, we then perform the 0_{43}-subtract $\rho_4 - (-3/10)\rho_3$ on the resulting $\tilde{A}$ to get $\hat{L}\backslash\hat{U}$, the compact form of the LU-factorization of the coefficient matrix $\hat{A}$ of the final equivalent system $\hat{A}\mathbf{x} = \hat{\mathbf{b}}$, where

$$\underset{\hat{L}\backslash\hat{U}\ (\text{unit } \hat{L})}{\begin{bmatrix} \enclose{circle}{-1} & -2 & 0 & -1 \\ \underline{-2} & \enclose{circle}{2} & 0 & -4 \\ \underline{0} & \underline{1} & \enclose{circle}{10} & 9 \\ \underline{2} & \underline{-\frac{1}{2}} & \underline{\frac{3}{10}} & \enclose{circle}{\frac{67}{10}} \end{bmatrix}}; \quad \underset{\text{final } [\hat{A}:\hat{\mathbf{b}}]}{\begin{bmatrix} -1 & -2 & 0 & -1 & : & 20 \\ 2 & 6 & 0 & -2 & : & -2 \\ 0 & 2 & 10 & 5 & : & 10 \\ -2 & -5 & -3 & 4 & : & 83 \end{bmatrix}}; \quad \underset{\text{final } \hat{\mathbf{s}}}{\begin{bmatrix} 2 \\ 6 \\ 10 \\ 5 \end{bmatrix}} \tag{17e}$$

At this point you might wish to verify that the product $\hat{L}\hat{U}$ equals the final $\hat{A}$, and that the final $[\hat{A}:\hat{\mathbf{b}}]$ is what results when the three pivots used (i.e., 1,3-, then 2,3-, then 3,4-) are performed on the given $[A:\mathbf{b}]$ in (16). We can now get the desired solution by solving $\hat{L}(\hat{U}\mathbf{x}) = \hat{\mathbf{b}}$ as usual in a forback matrix:

$$[\hat{L}\backslash\hat{U} = \hat{\mathbf{b}}:\bar{\mathbf{c}}:\bar{\mathbf{x}}] = \begin{bmatrix} \enclose{circle}{-1} & -2 & 0 & -1 & : & 20 & : & 20 & : & -98 \\ \underline{-2} & \enclose{circle}{2} & 0 & -4 & : & -2 & : & 38 & : & 35 \\ \underline{0} & \underline{1} & \enclose{circle}{10} & 9 & : & 10 & : & -28 & : & -10 \\ \underline{2} & \underline{-\frac{1}{2}} & \underline{-\frac{3}{10}} & \enclose{circle}{\frac{67}{10}} & : & 83 & : & \frac{268}{5} & : & 8 \end{bmatrix} \tag{17f}$$

Since exact arithmetic was used, the $\overline{\mathbf{x}}$ obtained as the solution of the *equivalent* system $\hat{A}\mathbf{x} = \hat{\mathbf{b}}$ in (17f) is identical to $\overline{\mathbf{x}}$ obtained by solving the *given* system (using Basic Pivoting) in (20e) of Example 3.2F. ■

In Example 3.3C, the SPP strategy called for an m,p-pivot for each $m = 1, 2$, and 3. This will not always happen. In fact, *no m,p- pivots would have been used if the system* had been *given* as in (17e)! However, *the SPP strategy would result in the forback computation in (17f) no matter how the equations of the given system were arranged;* and if fixed precision were used, the calculated $\overline{\mathbf{x}}$ resulting from this computation would be about as accurate as could be obtained on the device used.

The ***LU*-Decomposition Algorithm** shown in Algorithm 3.3C shows how to introduce a **pivot** step in the LU-Factorization Algorithm 3.2E in order to use a pivoting strategy before reducing the mth column of $\tilde{A}$ for $m = 1, \ldots, n - 1$. Example 3.3C used Algorithm 3.3C, with the **FACTOR $\hat{A}$** phase performed in (17a)–(17e), and the **FORBACK** phase performed in (17f).

ALGORITHM 3.3C. *LU*-DECOMPOSITION with SPP

PURPOSE: To solve an $n \times n$ linear system $A\mathbf{x} = \mathbf{b}$ by finding an equivalent rearranged system $\hat{A}\mathbf{x} = \hat{\mathbf{b}}$ for which $\hat{A} = \hat{L}\hat{U}$, then solving $\hat{L}(\hat{U}\mathbf{x}) = \hat{\mathbf{b}}$ by forward/backward substitution.

GET n, A, $\mathbf{b}$ {A stores the current reduction $\tilde{A}$ (initially A)}
Choose a pivoting strategy for selecting a_{pm} in the **pivot** step.

{**FACTOR $\hat{A}$**: Get $\hat{L}\backslash\hat{U}$ for which the product $\hat{L}\hat{U}$ equals the final $\hat{A}$.}
DO FOR $m = 1$ TO n
  IF $a_{mm}, \ldots, a_{nm}$ are all zero THEN {cannot find a nonzero mth pivot}
    OUTPUT('*A is singular*'); STOP
  IF $m < n$ THEN {reduce the mth column of A}
    {**pivot**} Select a nonzero a_{pm} from among $a_{mm}, \ldots, a_{nm}$.
    IF $p \neq m$ THEN perform an m,p-pivot in A
    {row$_m A$ now stores the mth row of $\hat{L}\backslash\hat{U}$}
    DO FOR $i = m + 1$ TO n {perform a 0_{im}-subtract in A}
      $a_{im} \leftarrow a_{im}/a_{mm}$ {a_{im} now stores $\hat{l}_{im}$}
      DO FOR $j = m + 1$ TO n {to the right of a_{im}}
        $a_{ij} \leftarrow a_{ij} - a_{im}a_{mj}$ {a_{ij} now stores $\tilde{a}_{ij}$}
{A now stores $\hat{L}\backslash\hat{U}$ (unit $\hat{L}$) for which $\hat{L}\hat{U} = \hat{A}$}

{**FORBACK:** Solve the equivalent system $\hat{L}(\hat{U}\mathbf{x}) = \hat{\mathbf{b}}$ for $\overline{\mathbf{x}}$.}
Perform the m,p-pivots of the **FACTOR $\hat{A}$** phase in $\mathbf{b}$ to get $\hat{\mathbf{b}}$.
Complete the forback matrix $[\hat{L}\backslash\hat{U} : \hat{\mathbf{b}} : \overline{\mathbf{c}} : \overline{\mathbf{x}}]$ to get $\overline{\mathbf{x}}$

OUTPUT('The calculated solution is $\overline{\mathbf{x}} = [$'$\overline{x}_1, \ldots, \overline{x}_n$'$]^T$')
STOP

Practical Consideration 3.3C (When should the SPP strategy be used?). It is easy to see that when the given coefficient matrix A satisfies $s_1 = s_2 = \cdots = s_n$, the SPP strategy results in the same m,p-pivots (hence yields the same solution) as the simpler

> **PARTIAL PIVOTING (PP) STRATEGY.** Take $\tilde{a}_{pm}$ to be the topmost entry having the largest $|\tilde{a}_{im}|$ among $|\tilde{a}_{mm}|, |\tilde{a}_{m+1,m}|, \ldots, |\tilde{a}_{nm}|$.

When all s_i's are *nearly* equal, the calculated $\bar{\mathbf{x}}$'s obtained using SPP and PP generally have *comparable* accuracy, even if different m,p-pivots are used, especially for small values of n. However, *the m,p-pivots resulting from the PP strategy are affected by row scaling and might actually result in an inaccurate solution*. For example, the PP strategy would result in the worse solution (8b) if *either* of the systems (7) were given, but the better solution (11a) if either of the systems (10) were given! We therefore offer the following advice:

When using the *LU*-Decomposition Algorithm 3.3C with fixed-precision arithmetic to solve a linear system $A\mathbf{x} = \mathbf{b}$, use Scaled Partial Pivoting as the pivoting strategy. **(18)**

As with any strategy for reducing propagated roundoff error, SPP is not guaranteed to work best for every system (see Problem 3-23). However, it generally produces solutions that are *about as accurate as is possible* for the device used; this assurance more that offsets the slight extra effort needed to implement it. There is a *Scaled Full Pivoting* strategy that can get slightly better accuracy than SPP, but it is much more complicated to implement (see Section 3.3F).

To set a good example, we shall adhere to the advice given in (18) from now on, with *two important exceptions*: It has been shown [50] that when the coefficient matrix is either *positive definite*† or *strictly diagonally dominant*, then the *LU-Factorization* Algorithm 3.2E (i.e., with *Basic Pivoting*) will succeed and yield solutions as accurate as those obtained using the *LU-Decomposition* Algorithm 3.3C with *any* pivoting strategy! Since tridiagonal systems usually have diagonally dominant coefficient matrices, `SUBROUTINE TRIDAG` shown in Figure 3.2-2 of Section 3.2H will generally be reliable as written, that is, without a **pivot** step. ■

3.3D LU-Decompositions of a Given $A_{n\times n}$

The FACTOR $\hat{A}$ phase of the LU-decomposition algorithm (Algorithm 3.3C) can be displayed as an ***LU*-decomposition** $< \hat{L}\backslash\hat{U},\ \boldsymbol{\rho}_{m,p} >$, in which $\hat{L}\backslash\hat{U}$ is the compact form that satisfies $\hat{L}\hat{U} = \hat{A}$, and $\boldsymbol{\rho}_{m,p}$ is a description of the m,p-pivots used to convert the *given* coefficient matrix A to the final (factored) $\hat{A}$. An m,p-pivot will be indicated by writing "$(\rho_m \leftrightarrow \rho_p)$" to the right of $\text{row}_m\hat{L}\backslash\hat{U}$. For example, the reductions performed in Section 3.3A and Example 3.2C can be summarized by saying that

†A matrix A is **positive definite** if the scalar $\mathbf{x}^T A\mathbf{x}$ is positive for all nonzero vectors $\mathbf{x}$. Such matrices are considered in Section 9.5C.

$$\begin{bmatrix} -1 & 1 & -4 \\ 2 & 2 & 0 \\ 3 & 3 & 2 \end{bmatrix} \text{ has } \begin{bmatrix} \textcircled{3} & 3 & 2 \\ \underline{-\frac{1}{3}} & \textcircled{2} & -\frac{10}{3} \\ \underline{\frac{2}{3}} & \underline{0} & \textcircled{-\frac{4}{3}} \end{bmatrix} \begin{matrix} (\rho_1 \leftrightarrow \rho_3) \\ (\rho_2 \leftrightarrow \rho_3) \\ \; \end{matrix} \text{ and } \begin{bmatrix} \textcircled{-1} & 1 & -4 \\ \underline{-2} & \textcircled{4} & -8 \\ \underline{-3} & \underline{\frac{3}{2}} & \textcircled{2} \end{bmatrix} = L\backslash U \qquad \textbf{(19)}$$

as LU-decompositions. The first would result from the Partial Pivoting strategy because (underlined) multipliers l_{im} are $\leqslant 1$ (Problem M3-7). The second, obtained using Basic Pivoting, illustrates the way we will denote an LU-decomposition when it turns out that *no m,p-pivots are used*. Similarly, we saw in Example 3.3C that

$$A = \begin{bmatrix} 2 & 6 & 0 & -2 \\ -2 & -5 & -3 & 4 \\ -1 & -2 & 0 & -1 \\ 0 & 2 & 10 & 5 \end{bmatrix} \text{ has } \begin{bmatrix} \textcircled{-1} & -2 & 0 & -1 \\ \underline{-2} & \textcircled{2} & 0 & -4 \\ \underline{0} & \underline{1} & \textcircled{10} & 9 \\ \underline{0} & \underline{-\frac{1}{2}} & \underline{-\frac{3}{10}} & \textcircled{\frac{67}{10}} \end{bmatrix} \begin{matrix} (\rho_1 \leftrightarrow \rho_3) \\ (\rho_2 \leftrightarrow \rho_3) \\ (\rho_3 \leftrightarrow \rho_4) \\ \; \end{matrix} \qquad \textbf{(20)}$$

as the LU-decomposition that results when Scaled Partial Pivoting is used. Note that the transformation from the given $[A:\mathbf{b}]$ to the final $[\hat{A}:\hat{\mathbf{b}}]$ is described by reading *down* the $\boldsymbol{\rho}_{m,p}$ list.

For any given matrix A, there are as many different LU-decompositions as there are sequences of m,p-pivots that result in nonzero pivot entries $\hat{u}_{11}, \ldots, \hat{u}_{nn}$ in $\hat{L}\backslash\hat{U}$. Any one of them can be used to get the solution of a linear system $A\mathbf{x} = \mathbf{b}$ by first using $\boldsymbol{\rho}_{m,p}$ to rearrange the given $\mathbf{b}$ to get the $\hat{\mathbf{b}}$ of the equivalent system $\hat{A}\mathbf{x} = \hat{\mathbf{b}}$, and then solving $\hat{A}\mathbf{x} = \hat{\mathbf{b}}$ as $\hat{L}(\hat{U}\mathbf{x}) = \hat{\mathbf{b}}$ by completing the forback matrix $[\hat{L}\backslash\hat{U}:\hat{\mathbf{b}}:\bar{\mathbf{c}}:\bar{\mathbf{x}}]$. In the LU-decomposition in (20), for example,

$$\boldsymbol{\rho}_{m,p} \text{ is } \begin{matrix} (\rho_1 \leftrightarrow \rho_3) \\ (\rho_2 \leftrightarrow \rho_3); \\ (\rho_3 \leftrightarrow \rho_4) \end{matrix} \quad \text{so} \quad \mathbf{b} = \begin{bmatrix} -2 \\ 83 \\ 20 \\ 10 \end{bmatrix} \quad \text{becomes} \quad \hat{\mathbf{b}} = \begin{bmatrix} 20 \\ -2 \\ 10 \\ 83 \end{bmatrix} \begin{matrix} \{\hat{b}_1 \text{ is } b_3\} \\ \{\hat{b}_2 \text{ is } b_1\} \\ \{\hat{b}_3 \text{ is } b_4\} \\ \{\hat{b}_4 \text{ is } b_2\} \end{matrix} \qquad \textbf{(21)}$$

To solve $\hat{A}\mathbf{x} = \hat{\mathbf{b}}$, you must use the *rearranged* $\hat{\mathbf{b}}$ (*not* the given $\mathbf{b}$) to solve the *equivalent* system $\hat{A}\mathbf{x} = \hat{\mathbf{b}}$ in the forback matrix $[\hat{L}\backslash\hat{U}:\hat{\mathbf{b}}:\bar{\mathbf{c}}:\bar{\mathbf{x}}]$. If you use $\mathbf{b}$, you would be solving $\hat{A}\mathbf{x} = \mathbf{b}$ whose solution is *not* the desired solution $A^{-1}\mathbf{b}$ unless $\hat{A} = A$!

Practical Consideration 3.3D (Storing $\rho_{m,p}$ efficiently on a computer). If the *LU*-Decomposition Algorithm 3.3C is carried out in place with the SPP strategy, then each m,p-pivot must be performed on the current $\tilde{A}$, $\hat{\mathbf{s}}$, and $\hat{\mathbf{b}}$, as follows:

FACTOR $\hat{A}$: DO FOR $j = 1$ *TO* n
  └ *Temp* $\leftarrow \tilde{a}_{mj}$; $\tilde{a}_{mj} \leftarrow \tilde{a}_{pj}$; $\tilde{a}_{pj} \leftarrow$ *Temp*
  Temp $\leftarrow \hat{s}_m$; $\hat{s}_m \leftarrow \hat{s}_p$; $\hat{s}_p \leftarrow$ *Temp*

FORBACK: *Temp* $\leftarrow \hat{b}_m$; $\hat{b}_m \leftarrow \hat{b}_p$; $\hat{b}_p \leftarrow$ *Temp*

An efficient alternative to all this rearranging is to *leave* A, $\mathbf{s}$, and $\mathbf{b}$ *in the row order of the given system* $A\mathbf{x} = \mathbf{b}$ and, instead, perform all required m,p-pivots in a one-dimensional INTEGER array that stores *the current rearrangement of the row subscripts*. Such an array, called a **row pointer vector**, will be *NewRow*. It is initialized as $[1 \;\; 2 \;\; \cdots \;\; n]^T$ (no rearrangement initially). Then any required m,p-pivots are performed only in *NewRow*, and *not* in A, $\mathbf{s}$, or **b**. For example, if a pointer array *NewRow* was used in Example 3.3C to get the *LU*-decomposition (20), *NewRow* would become

$$\begin{bmatrix}1\\2\\3\\4\end{bmatrix} \xrightarrow{\rho_1 \leftrightarrow \rho_3} \begin{bmatrix}3\\2\\1\\4\end{bmatrix} \xrightarrow{\rho_2 \leftrightarrow \rho_3} \begin{bmatrix}3\\1\\2\\4\end{bmatrix} \xrightarrow{\rho_3 \leftrightarrow \rho_4} \begin{bmatrix}3\\1\\4\\2\end{bmatrix} = \begin{bmatrix}NewRow\ (1)\\NewRow\ (2)\\NewRow\ (3)\\NewRow\ (4)\end{bmatrix} \qquad \textbf{(22a)}$$

After the **FACTOR** $\hat{A}$ phase, rows 1, 2, 3, and 4 of $\hat{A}$, $\hat{\mathbf{s}}$, and $\hat{\mathbf{b}}$, would be stored in rows 3, 1, 4, and 2 of A, **s**, and **b**, as follows:

$$\begin{bmatrix} \underline{-2} & \textcircled{2} & 0 & -4 \\ \underline{2} & \underline{-\frac{1}{2}} & \underline{-\frac{3}{10}} & \textcircled{\tfrac{67}{10}} \\ \textcircled{-1} & -2 & 0 & -1 \\ \underline{0} & \underline{1} & \textcircled{10} & 9 \end{bmatrix} \begin{matrix} \}\text{row}_2\hat{L}\backslash\hat{U} \\ \}\text{row}_4\hat{L}\backslash\hat{U} \\ \}\text{row}_1\hat{L}\backslash\hat{U} \\ \}\text{row}_3\hat{L}\backslash\hat{U} \end{matrix} \qquad \begin{bmatrix} 6 \\ 5 \\ 2 \\ 10 \end{bmatrix} \begin{matrix} \}\hat{s}_2 \\ \}\hat{s}_4 \\ \}\hat{s}_1 \\ \}\hat{s}_3 \end{matrix} \qquad \begin{bmatrix} -2 \\ 83 \\ 20 \\ 10 \end{bmatrix} \begin{matrix} \}\hat{b}_2 \\ \}\hat{b}_4 \\ \}\hat{b}_1 \\ \}\hat{b}_3 \end{matrix} \qquad \textbf{(22b)}$$

final storage in A final **s** final **b**

Note that **s** and **b** are exactly as they were for the *given* system in Example 3.3C [see (20) and (21)]. At any time *during or after* the **FACTOR** $\hat{A}$ phase,

$$\tilde{a}_{i,j} \text{ is stored in } A_{NewRow(i),j}, \text{ and } \hat{s}_i \text{ is stored in } \mathbf{s}_{NewRow(i)} \qquad \textbf{(23)}$$

Then in the **FORBACK** phase, *it is not necessary to actually form* $\hat{\mathbf{b}}$ *from* **b**; instead, $\hat{b}_i$ is simply found in $\mathbf{b}_{NewRow(i)}$! Thus, using a pointer vector *NewRow* amounts to simply replacing the row index i by *NewRow*(i). See Section 4.2B. ■

3.3E What if $\tilde{a}_{mm}, \tilde{a}_{m+1,m}, \ldots, \tilde{a}_{nm}$ Are All Zero?

The only way the *LU*-Decomposition Algorithm 3.3C can fail is if $\tilde{a}_{mm}, \tilde{a}_{m+1,m}, \ldots, \tilde{a}_{nm}$ all turn out to be zero for some m in one of the **pivot** steps of the decomposition $\hat{A} \to \{\tilde{A}\} \to \hat{L}\backslash\hat{U}$. For example, if Basic Pivoting is used, then

$$A = \begin{bmatrix} 2 & -2 & -1 \\ -2 & 3 & 3 \\ 0 & 2 & 4 \end{bmatrix} = \hat{A} \to \tilde{A} = \begin{bmatrix} \textcircled{2} & -2 & -2 \\ \underline{-1} & \textcircled{1} & 2 \\ \underline{0} & \underline{1} & 0 \end{bmatrix} \quad \tilde{a}_{33} = 0 \qquad \textbf{(24)}$$

If such a failure occurs using one pivoting strategy, it will occur for the same m using any other pivoting strategy because the problem is with the matrix A, *not* the strategy. Specifically,

$$A \text{ is singular} \Leftrightarrow \tilde{a}_{mm} = \cdots = \tilde{a}_{nm} = 0 \text{ occurs in decomposing } A \qquad \textbf{(25)}$$

To appreciate the consequences of (25), consider the linear system

$$A = \begin{bmatrix} 2 & -2 & -1 \\ 0 & 2 & 4 \\ -2 & 3 & 3 \end{bmatrix} \Rightarrow \tilde{A} = \begin{bmatrix} 2 & -2 & -1 \\ 0 & 2 & 4 \\ 0 & 1 & 2 \end{bmatrix} \Rightarrow \tilde{A} = \begin{bmatrix} 2 & -2 & -1 \\ 0 & 2 & 4 \\ 0 & 0 & 0 \end{bmatrix}$$

$$\begin{array}{ll} (\mathrm{E}_1) & 2x - 2y - 1z = 2 \\ (\mathrm{E}_2) & -2x + 3y + 3z = -1 \\ (\mathrm{E}_3) & 0x + 2y + 4z = b_3 \end{array} \qquad \left([A:\mathbf{b}] = \begin{bmatrix} 2 & -2 & -1 & : & 2 \\ -2 & 3 & 3 & : & -1 \\ 0 & 2 & 4 & : & b_3 \end{bmatrix} \right) \quad \textbf{(26a)}$$

whose coefficient matrix is the singular matrix A in (24). If we try to solve (26a) using Basic Gaussian Elimination, we get the equivalent upper triangular system $U\mathbf{x} = \mathbf{c}$, where

$$\begin{array}{l} (\mathrm{E}_1)\ 2x - 2y - 1z = 2 \\ (\mathrm{E}_2)\ 0x + 1y + 2z = 1 \\ (\mathrm{E}_3)\ 0x + 0y + 0z = b_3 - 2 \end{array} \qquad \left([U:\bar{\mathbf{c}}] = \begin{bmatrix} 2 & -2 & -1 & : & 2 \\ 0 & 1 & 2 & : & 1 \\ 0 & 0 & 0 & : & b_3 - 2 \end{bmatrix} \right) \quad \textbf{(26b)}$$

It is evident from (26b) that the nature of the solution depends on the value of b_3.

Case 1 ($b_3 = 2$): In this case (E_3) *puts no restriction on* x, y, and z; so (26b) in effect has only *two* equations, namely (E_1) and (E_2), in these *three* unknowns. As a result, one of x, y, z, say $z = z_0$, can be chosen *arbitrarily*, and the other two can then be found in terms of z_0 by backward substitution:

$$y = 1 - 2z_0, \qquad \text{then } x = \tfrac{1}{2}(2 + 2y + z) = \tfrac{1}{2}[2 + 2(1 - 2z_0) + z_0] \quad \textbf{(27a)}$$

Thus $U\mathbf{x} = \bar{\mathbf{c}}$ (and hence $A\mathbf{x} = \mathbf{b}$) has *infinitely many solutions*, namely

$$\bar{\mathbf{x}} = [(2 - \tfrac{3}{2}z_0) \quad (1 - 2z_0) \quad z_0]^T \qquad (z_0 \text{ can be any real number}) \quad \textbf{(27b)}$$

Case 2 ($b_3 \neq 2$): In this case (E_3) in (26b) cannot be satisfied. So *neither* $A\mathbf{x} = \mathbf{b}$ *nor* $U\mathbf{x} = \bar{\mathbf{c}}$ *has any solutions.*

The preceding discussion illustrates the following important result. Its proof can be found in any linear algebra text.

ALTERNATIVE THEOREM: If A is nonsingular, then $A\mathbf{x} = \mathbf{b}$ has a unique solution $\bar{\mathbf{x}} = A^{-1}\mathbf{b}$ for any given $\mathbf{b}$. If A is singular, then $A\mathbf{x} = \mathbf{b}$ can *never* have a unique solution; it will have either infinitely many solutions (the system is **underdetermined**) or no solutions (the equations are **inconsistent**). **(28)**

Intuitively, a system is underdetermined if one of the equations imposes a constraint that is already imposed by the other equations; it is inconsistent if one of the equations imposes a constraint that is incompatible with the constraints imposed by the other equations.

Practical Consideration 3.3E (Looking for zero $\tilde{a}_{im}$'s on a computer). It is unrealistic to expect a computer to determine that $\tilde{a}_{im} = 0$ for $i = m, \ldots, n$. What is likely to be found are *very small entries* consisting of the accumulated roundoff error that preceded the calculation of $\tilde{a}_{mm}, \ldots, \tilde{a}_{nm}$. Unfortunately, there is simply no way that a computer can be made to distinguish this roundoff "noise" from legitimately small entries. This is why canned software for solving linear systems warns of a `NEARLY SINGULAR MATRIX` when these entries are all "nearly zero." We shall see in Section 4.1 that such a warning might be needed even if A is not actually singular. ■

In what follows, we shall only consider linear systems with nonsingular coefficient matrices. However, singular coefficient matrices do occur as a result of either an ill-posed problem or an error in finding or entering $[A\!:\!\mathbf{b}]$. So *you should always be alert for the possibility that a linear system to be solved has a singular coefficient matrix.*

3.3F Scaled Full Pivoting (Optional)

For $m = 1, 2, \ldots, n$, let p_m denote the jth selected pivot entry, and let $\tilde{A}^{(m)}$ denote the *unreduced* part of the current $\tilde{A}$, that is, $\tilde{A}$ with all rows and columns containing a *previously* selected (circled) pivot entries removed. The discussion that motivated the SPP strategy [see (13) of Section 3.3B] suggests that the best choice for p_m is the entry of $\tilde{A}^{(m)}$ that is "relatively largest in its row" in the sense that the ratio

$$\frac{|p_m|}{\text{largest absolute value of } \textit{other} \text{ entries of its row in } \tilde{A}^{(m)}} \qquad \textbf{(29)}$$

is larger than the corresponding ratio of any other row of $\tilde{A}^{(m)}$. This is the **Scaled Full Pivoting (SFP)** strategy. Note that the p_m selected by SFP need not lie in $\mathrm{col}_m\tilde{A}$.

EXAMPLE 3.3F Use *LU*-factorization with Scaled Full Pivoting to solve

$$\begin{array}{ll} (\mathrm{E}_1) & -x_1 + x_2 - 4x_3 = 0 \\ (\mathrm{E}_2) & 2x_1 + 2x_2 \qquad\;\; = 1 \\ (\mathrm{E}_3) & 3x_1 + 3x_2 + 2x_3 = \tfrac{1}{2} \end{array} \qquad [A\!:\!\mathbf{b}] = \begin{bmatrix} -1 & 1 & -4 & : & 0 \\ 2 & 2 & 0 & : & 1 \\ 3 & 3 & 2 & : & \tfrac{1}{2} \end{bmatrix} \qquad \textbf{(30)}$$

SOLUTION: The **FACTOR** $\hat{A}$ phase, with $\tilde{A}^{(1)}$ $(= A)$, $\tilde{A}^{(2)}$, and $\tilde{A}^{(3)}$ shown shaded, proceeds as follows:

$$\begin{array}{ccccc}
p_1 = -4 & & p_1 \text{ column reduced} & & p_2 \text{ column reduced} \\[1ex]
\begin{bmatrix} -1 & 1 & \enclose{circle}{-4} \\ 2 & 2 & 0 \\ 3 & 3 & 2 \end{bmatrix} &
\begin{array}{l} \underline{(m = 1)} \\ \rho_2 - 0\rho_1 \\ \rho_3 + \tfrac{1}{2}\rho_1 \end{array} &
\begin{bmatrix} -1 & 1 & \enclose{circle}{-4} \\ 2 & 2 & \underline{0} \\ \tfrac{5}{2} & \enclose{circle}{\tfrac{7}{2}} & \underline{-\tfrac{1}{2}} \end{bmatrix} &
\begin{array}{l} \underline{(m = 2)} \\ \rho_2 \leftrightarrow \rho_3 \\ \rho_3 - \tfrac{4}{7}\rho_2 \end{array} &
\begin{bmatrix} -1 & 1 & \enclose{circle}{-4} \\ \tfrac{5}{2} & \enclose{circle}{\tfrac{7}{2}} & \underline{-\tfrac{1}{2}} \\ \enclose{circle}{\tfrac{4}{7}} & \underline{\tfrac{4}{7}} & \underline{0} \end{bmatrix} \\[1ex]
\dfrac{|-4|}{|\pm 1|} > \dfrac{2}{2} = \dfrac{3}{3} & & \dfrac{\tfrac{7}{2}}{\tfrac{5}{2}} > \dfrac{2}{2} \;\; (p_2 = \tfrac{7}{2}) & & p_3 = \tfrac{4}{7}
\end{array} \qquad \textbf{(31a)}$$

The 0_{im}-subtracts used to get zeros below p_m are performed in the current $\tilde{A}^{(m)}$, and do not alter the (underlined) stored multipliers. (Why?)

For the **FORBACK** phase, we first perform a 2,3-pivot in the given $\mathbf{b}$ to get $\hat{\mathbf{b}} = [0 \;\; \tfrac{1}{2} \;\; 1]^T$. Since the pivot entries p_1, p_2, and p_3 multiply x_3, x_2, and x_1 *in that order*, we *temporarily* rearrange the components of $\mathbf{x}$ so that p_i multiplies x_i by taking

$$\mathbf{x}_r = [x_3 \quad x_2 \quad x_1]^T \qquad \text{(rearranged } \mathbf{x}) \qquad \textbf{(31b)}$$

and we correspondingly rearrange the *columns* of the final reduction in (31a) so that p_i is in the ith column of $\hat{L}\backslash\hat{U}$. The forback matrix to solve $\hat{L}(\hat{U}\mathbf{x}) = \hat{\mathbf{b}}$ then looks familiar:

$$[\hat{L}\backslash\hat{U}:\hat{\mathbf{b}}:\bar{\mathbf{c}}:\bar{\mathbf{x}}_r] = \left[\begin{array}{ccc:c:c:c} \textcircled{-4} & 1 & -1 & 0 & 0 & -\frac{1}{2} \\ -\frac{1}{2} & \textcircled{\frac{7}{2}} & \frac{5}{2} & \frac{1}{2} & \frac{1}{2} & -\frac{3}{4} \\ \underline{0} & \underline{\frac{4}{7}} & \textcircled{\frac{4}{7}} & 1 & \frac{5}{7} & \frac{5}{4} \end{array}\right] \quad \textbf{(31c)}$$

So the desired solution of (30) is $\bar{x}_1 = \frac{5}{4}, \bar{x}_2 = -\frac{3}{4}, \bar{x}_3 = -\frac{1}{2}$. This agrees with the solution obtained previously in Sections 3.2C and 3.3A. ■

Practical Consideration 3.3F (Is SFP worth the trouble?). A column pointer vector *NewCol* can be used as in Practical Consideration 3.3D to avoid actually having to rearrange **x** and $\hat{L}\backslash\hat{U}$ as in (31b). Nevertheless, searching for the "relatively largest" p_m and keeping track of the column order of the selected *m*th pivots complicates a program for the *LU*-Factorization Algorithm 3.3E and increases its execution time. Experience has shown that the improvement in accuracy over the SPP strategy is generally not great enough to justify this extra computational overhead. Consequently, Scaled Full Pivoting is not recommended unless the maximum safeguard against propagated roundoff error is needed and the increased execution time causes no problem. ■

3.4 Reusing an LU-Decomposition

In this section we examine the problem of solving several linear systems, *all having the same coefficient matrix*, say

$$A\mathbf{x}_1 = \mathbf{b}_1, \quad A\mathbf{x}_2 = \mathbf{b}_2, \quad \ldots, \quad A\mathbf{x}_k = \mathbf{b}_k \quad \textbf{(1)}$$

An efficient algorithm for solving these k linear systems efficiently is presented in Section 3.4A. Then, in Sections 3.4B, we show how to use this algorithm to find the inverse of $A_{n\times n}$ if you have to do so.

3.4A Efficient Solution of $A\mathbf{x}_1 = \mathbf{b}_1, \ldots, A\mathbf{x}_k = \mathbf{b}_k$

Since all k linear systems in (1) have the same coefficient matrix A, an efficient approach to solving them all is to find *one LU-* decomposition of A and then use it k times. Pseudocode for this strategy, which we will call the ***LU*-Decomposition Algorithm for Several Linear Systems** is given in Algorithm 3.4A.

ALGORITHM 3.4A. ***LU*-DECOMPOSITION FOR SEVERAL LINEAR SYSTEMS**

PURPOSE: To solve k linear systems $A\mathbf{x}_1 = \mathbf{b}_1, \ldots, A\mathbf{x}_k = \mathbf{b}_k$

GET $n, A, k, \mathbf{b}_1, \ldots, \mathbf{b}_k$
Find a suitable *LU*-decomposition $< \hat{L}\backslash\hat{U}, \boldsymbol{\rho}_{m,p} >$ for A
DO FOR $j = 1$ TO k {Solve $\hat{A}\mathbf{x}_j = \hat{\mathbf{b}}_j$ (equivalent to $A\mathbf{x}_j = \mathbf{b}_j$)}
  $\mathbf{b} \leftarrow \mathbf{b}_j$; Use $\boldsymbol{\rho}_{m,p}$ to form $\hat{\mathbf{b}}$ from $\mathbf{b}$ as in Section 3.3D
  Complete the forback matrix $[\hat{L}\backslash\hat{U}:\hat{\mathbf{b}}:\bar{\mathbf{c}}:\bar{\mathbf{x}}]$ {$\bar{\mathbf{x}}$ is $\bar{\mathbf{x}}_j$}
  OUTPUT('The solution of system 'j' is ['$\bar{x}_1, \ldots, \bar{x}_n$']^T')

The steps of Algorithm 3.4A will be displayed in a **composite forback matrix**

$$[\hat{L}\backslash\hat{U} : \underbrace{\hat{\mathbf{b}}_1 : \hat{\mathbf{b}}_2 : \cdots : \hat{\mathbf{b}}_k}_{\hat{B}} : \underbrace{\bar{\mathbf{c}}_1 : \bar{\mathbf{c}}_2 : \cdots : \bar{\mathbf{c}}_k}_{\bar{C}} : \underbrace{\bar{\mathbf{x}}_1 : \bar{\mathbf{x}}_2 : \cdots : \bar{\mathbf{x}}_k}_{\bar{X}}] \tag{2}$$

which will be denoted as $[\hat{L}\backslash\hat{U} : \hat{B} : \bar{C} : \bar{X}]$. It is formed as follows: First find and enter $\hat{L}\backslash\hat{U}$ for an LU-decomposition of A. Then use $\boldsymbol{\rho}_{m,p}$ to form the columns $\hat{\mathbf{b}}_j$ of $\hat{B}$ from the given target vectors $\mathbf{b}_j$ as in Section 3.3D. The columns of $\bar{C}$ then $\bar{X}$ are then obtained one at a time in the usual way (i.e., forward substitution using the unit $\hat{L}$ and $\hat{\mathbf{b}}_j$, then backward substitution using U and $\bar{\mathbf{c}}_j$).

EXAMPLE 3.4A Solve the 4×4 linear systems $A\mathbf{x}_1 = \mathbf{b}_1$ and $A\mathbf{x}_2 = \mathbf{b}_2$ if

$$A = \begin{bmatrix} 2 & -2 & 0 & 4 \\ 3 & -3 & 0 & -1 \\ -1 & 6 & 5 & -7 \\ -5 & 1 & 0 & -6 \end{bmatrix}, \quad \mathbf{b}_1 = \begin{bmatrix} 2 \\ 10 \\ -11 \\ 3 \end{bmatrix}, \quad \text{and} \quad \mathbf{b}_2 = \begin{bmatrix} 4 \\ 13 \\ 13 \\ -14 \end{bmatrix} \tag{3}$$

Solution: The LU-decomposition of A using Basic Pivoting is

$$\begin{bmatrix} \textcircled{2} & -2 & 0 & 4 \\ \underline{-\frac{1}{2}} & \textcircled{5} & 5 & -5 \\ \underline{-\frac{5}{2}} & \underline{-\frac{4}{5}} & \textcircled{4} & 0 \\ \underline{\frac{3}{2}} & \underline{0} & \underline{0} & \textcircled{-7} \end{bmatrix} \begin{matrix} \\ (\rho_2 \leftrightarrow \rho_3) \\ (\rho_3 \leftrightarrow \rho_4) \\ \\ \end{matrix} \tag{4a}$$

The 2,3- and 3,4-pivots are needed because the calculated values of a_{22} and a_{33} are zero (Problem 3-20). As a result of these m,p-pivots,

$$B = [\mathbf{b}_1 : \mathbf{b}_2] = \begin{bmatrix} 2 & : & 4 \\ 10 & : & 13 \\ -11 & : & 13 \\ 3 & : & -14 \end{bmatrix} \quad \text{becomes} \quad \hat{B} = [\hat{\mathbf{b}}_1 : \hat{\mathbf{b}}_2] = \begin{bmatrix} 2 & : & 4 \\ -11 & : & 13 \\ 3 & : & -14 \\ 10 & : & 13 \end{bmatrix} \tag{4b}$$

The resulting composite forback matrix $[\hat{L}\backslash\hat{U} : \hat{B} : \bar{C} : \bar{X}]$ is

$$\left[\begin{array}{cccccccccccccccc} \textcircled{2} & -2 & 0 & 4 & : & 2 & : & 4 & : & 2 & : & 4 & : & 0 & : & 4 \\ \underline{-\frac{1}{2}} & \textcircled{5} & 5 & -5 & : & -11 & : & 13 & : & -10 & : & 15 & : & -3 & : & 0 \\ \underline{-\frac{5}{2}} & \underline{-\frac{4}{5}} & \textcircled{4} & 0 & : & 3 & : & -14 & : & 0 & : & 8 & : & 0 & : & 2 \\ \underline{\frac{3}{2}} & \underline{0} & \underline{0} & \textcircled{-7} & : & 10 & : & 13 & : & -7 & : & -7 & : & -1 & : & -1 \end{array}\right] \tag{4c}$$

$$\underbrace{\qquad\qquad}_{\hat{L}\backslash\hat{U}} \qquad \underbrace{\qquad}_{\hat{B}} \qquad \underbrace{\qquad}_{\bar{C}} \qquad \underbrace{\qquad}_{\bar{X} = [\bar{\mathbf{x}}_1 : \bar{\mathbf{x}}_2]}$$

So $\bar{\mathbf{x}}_1 = [0 \quad -3 \quad 0 \quad -1]^T$ and $\bar{\mathbf{x}}_2 = [4 \quad 0 \quad 2 \quad -1]^T$ are the desired solutions. ■

Two subroutines are needed to implement Algorithm 3.4A on a computer: one to form an LU-decomposition of A, the other (used in the loop) to complete the $\hat{\mathbf{b}}$, $\bar{\mathbf{c}}$ and $\bar{\mathbf{x}}$ columns of the forback matrix $[\hat{L}\backslash\hat{U} : \hat{\mathbf{b}} : \bar{\mathbf{c}} : \bar{\mathbf{x}}]$ for $\hat{\mathbf{b}} = \hat{\mathbf{b}}_j$. Such subroutines will be described in detail in Section 4.2B.

The k systems in (1) are related to the solution of the single *matrix* equation $AX = B$. To understand this relationship, consider the $n \times k$ matrices

$$X = [\mathbf{x}_1 : \mathbf{x}_2 : \cdots : \mathbf{x}_k] \qquad \text{and} \qquad B = [\mathbf{b}_1 : \mathbf{b}_2 : \cdots : \mathbf{b}_k] \tag{5a}$$

for which $\mathbf{x}_j = \text{col}_j X$ and $\mathbf{b}_j = \text{col}_j B$ for $j = 1, 2, \ldots, k$. We saw in (7a) of Section 3.1C that

$$\text{the } j\text{th column of } AX \text{ is the product } A(\text{col}_j X), \text{ that is, } A\mathbf{x}_j \tag{5b}$$

It follows that if $\overline{X}$ denotes the numerical matrix $[\overline{\mathbf{x}}_1 : \cdots : \overline{\mathbf{x}}_k]$, then

$$A\overline{X} = B \Leftrightarrow \overline{\mathbf{x}}_j = \text{col}_j \overline{X} \text{ satisfies } A\overline{\mathbf{x}}_j = \mathbf{b}_j \qquad \text{for } j = 1, \ldots, k \tag{6}$$

In words, *$\overline{X}_{n \times k}$ is a solution of the matrix equation $AX = B$ if and only if its* jth *column, $\overline{\mathbf{x}}_j$, is the solution of the linear system $A\mathbf{x}_j = \mathbf{b}_j$ for $j = 1, \ldots, k$.* So a matrix equation $AX = B$ can be solved *one column at a time* using Algorithm 3.4A. For example, for A and B given in (3) and (4b), $\overline{X}_{4 \times 2} = [\overline{\mathbf{x}}_1 : \overline{\mathbf{x}}_2]$ obtained in (4c) is the solution of the system $AX = B$.

3.4B Finding A^{-1}

Electrical circuits are often described in terms of n *voltages* $v_1, \ldots, v_n$ and n *currents* $i_1, \ldots, i_n$. In some situations one can get an $n \times n$ **impedance matrix** Z that relates the voltages and currents as follows

$$\mathbf{v} = Z\mathbf{i}, \qquad \text{where } \mathbf{v} = \begin{bmatrix} v_1 \\ \vdots \\ v_n \end{bmatrix} \quad \text{and} \quad \mathbf{i} = \begin{bmatrix} i_1 \\ \vdots \\ i_n \end{bmatrix} \tag{7a}$$

In other situations one can get an $n \times n$ **admittance matrix** that satisfies

$$\mathbf{i} = Y\mathbf{v}, \qquad \text{so that } Y = Z^{-1}, \quad \text{or, equivalently,} \quad Z = Y^{-1} \tag{7b}$$

When one of Z and Y is convenient to get but the other is needed, it becomes necessary to invert the more easily obtained matrix. We now show how this can be done.

Given an $n \times n$ matrix A, the problem of finding A^{-1} can be viewed as:

$$\boxed{\text{Solve the matrix equation } AX = I_n \text{ for } \overline{X} = A^{-1}} \tag{8a}$$

This is of the form $AX = B$, with $B = I_n$. In view of (6), we can find $\overline{X} = A^{-1}$ by solving the n linear systems

$$A\mathbf{x}_j = \mathbf{u}_j, \qquad \text{where } \mathbf{u}_j \text{ is the } j\text{th column of } I_n, \quad j = 1, \ldots, n \tag{8b}$$

So once an LU-decomposition of A has been found, we need only form $\hat{I}_n$ for the equivalent (rearranged) matrix equation $\hat{A}X = \hat{I}_n$, and then

$AX = I$ $\quad X \triangleq A^{-1}$

like n systems with equal coefficient matri...

complete the composite forback matrix $[\hat{L}\backslash\hat{U}:\hat{I}_n:C:\overline{X} = A^{-1}]$ **(8c)**

Alternatively, one can use Gauss–Jordan Elimination (Section 3.2I) to get A^{-1} as

DIAGONALIZATION SCALING
$$[A:I_n] \;\to\; [D:E] \;\to\; [I_n:A^{-1}]$$ **(9)**

When pivoting is used with Gauss–Jordan Elimination, m,p-pivots must still be performed with $p > m$ (why?); so they will be identical to those used for Gaussian Elimination or LU-factorization with the same pivoting strategy.

EXAMPLE 3.4B We observed in (20) of Example 3.3D that the 3×3 matrix

$$A = \begin{bmatrix} -1 & 1 & -4 \\ 2 & 2 & 1 \\ 3 & 3 & 2 \end{bmatrix} \quad \text{has} \quad \begin{bmatrix} ③ & 3 & 2 \\ -\frac{1}{3} & ② & -\frac{10}{3} \\ \frac{2}{3} & 0 & \boxed{-\frac{4}{3}} \end{bmatrix} \begin{matrix} (\rho_1 \leftrightarrow \rho_3) \\ (\rho_2 \leftrightarrow \rho_3) \\ \; \end{matrix}$$ **(10)**

as the LU-decomposition that results from the use of Partial Pivoting.
  (a) Use Algorithm 3.4A with this decomposition to find A^{-1}.
  (b) Find A^{-1} using Gauss–Jordan Elimination with Partial Pivoting.

SOLUTION (a): We must first use a 1,3-pivot followed by a 2,3-pivot to get $\hat{I}_3$. The resulting forback matrix for solving $\hat{A}X = \hat{I}_3$ is

$$\begin{bmatrix} ③ & 3 & 2 & : & 0 & 0 & 1 & : & 0 & 0 & 1 & : & -\frac{1}{2} & \frac{7}{4} & -1 \\ -\frac{1}{3} & ② & -\frac{10}{3} & : & 1 & 0 & 0 & : & 1 & 0 & \frac{1}{3} & : & \frac{1}{2} & -\frac{5}{4} & 1 \\ \frac{2}{3} & 0 & \boxed{-\frac{4}{3}} & : & 0 & 1 & 0 & : & 0 & 1 & -\frac{3}{8} & : & 0 & -\frac{3}{4} & \frac{1}{2} \end{bmatrix}$$ **(11)**

$$\underbrace{\hat{L}\backslash\hat{U}} \qquad \underbrace{\hat{I}_3} \qquad \underbrace{\overline{C}} \qquad \underbrace{\overline{X} = A^{-1}}$$

You might wish to verify that $A\overline{X} = I_3$, hence that $\overline{X}$ is in fact A^{-1}.

SOLUTION (b): The reduction $[A:I_3] \to [D:E]$ using Partial Pivoting is

column 1 reduced

$$[A:I_3] = \begin{bmatrix} -1 & 1 & -4 & : & 1 & 0 & 0 \\ 2 & 2 & 0 & : & 0 & 1 & 0 \\ ③ & 3 & 2 & : & 0 & 0 & 1 \end{bmatrix} \begin{matrix} \rho_1 \rightleftarrows \rho_3 \\ \xrightarrow{\rho_2 - \frac{2}{3}\rho_1} \\ \rho_3 + \frac{1}{3}\rho_1 \end{matrix} \begin{bmatrix} ③ & 2 & 2 & : & 0 & 0 & 1 \\ 0 & 0 & -\frac{4}{3} & : & 0 & 1 & -\frac{2}{3} \\ 0 & ② & -\frac{10}{3} & : & 1 & 0 & \frac{1}{3} \end{bmatrix}$$

$|3|$ is largest of $|-1|, |2|, |3|$ $|2|$ is largest of $|0|, |2|$

$$\begin{matrix} \rho_1 - \frac{3}{2}\rho_2 \\ \xrightarrow{\hspace{1cm}} \\ \rho_2 \rightleftarrows \rho_3 \end{matrix} \begin{bmatrix} ③ & 0 & 7 & : & -\frac{3}{2} & 0 & \frac{1}{2} \\ 0 & ② & -\frac{10}{3} & : & 1 & 0 & \frac{1}{3} \\ 0 & 0 & \boxed{-\frac{4}{3}} & : & 0 & 1 & -\frac{2}{3} \end{bmatrix} \begin{matrix} \rho_1 + \frac{21}{4}\rho_3 \\ \xrightarrow{\hspace{1cm}} \\ \rho_2 - \frac{5}{2}\rho_3 \end{matrix} \begin{bmatrix} ③ & 0 & 0 & : & -\frac{3}{2} & \frac{21}{4} & -3 \\ 0 & ② & 0 & : & 1 & -\frac{5}{2} & 2 \\ 0 & 0 & \boxed{-\frac{4}{3}} & : & 0 & 1 & -\frac{2}{3} \end{bmatrix}$$

column 2 reduced   column 3 reduced

In the 0_{i1}-subtracts used to reduce column 1, ρ_1 denotes the first row of $[A:I_2]$ *after* the 1,3-pivot, and similarly for ρ_2 in the 0_{12}-subtract $(\rho_1 - \frac{3}{2}\rho_2)$ needed to reduce column 2. It is easy to see that SCALING (i.e., dividing rows 1, 2, and 3 of $[D:E]$ by 3, 2, and $-\frac{4}{3}$, respectively) yields the same A^{-1} as part (a). Note that the two m,p-pivots used are the same as those given in (10). ■

Practical Consideration 3.4B (Finding A^{-1} on a computer). If you need the inverse of an $n \times n$ matrix, then a program for doing so using Gauss–Jordan Elimination can be written more easily and will find A^{-1} just as accurately as Algorithm 3.4A with the same pivoting strategy (see Problem C3-6). Such a program need not store multipliers because once A^{-1} has been found, a linear system $A\mathbf{x} = \mathbf{b}$ can (and should) be solved as the product $A^{-1}\mathbf{b}$! Alternatively, if reliable software for solving a single linear system is available, it can be used as in (8b) to get A^{-1} one column at a time, without having to write a new program [see Problem C4-2(d)]. ■

3.4C Counting Arithmetic Operations

The solution of k linear systems using Algorithm 3.4a requires one LU-decomposition and k forward/backward substitutions. Table 3.4-1, which follows directly from Table 3.2-1 of Section 3.2G, summarizes the number of arithmetic operations required.

TABLE 3.4-1 Number of Arithmetic Operations for Algorithm 3.4A

Problem (A is $n \times n$)	*Number of Operations Required*
Solve $A\mathbf{x} = \mathbf{b}_1, \ldots, A\mathbf{x} = \mathbf{b}_k$	$\dfrac{n(n-1)(4n+1)}{6} + k(2n^2 - n)$
Find A^{-1} (solve $AX = I_n$)	$\dfrac{n(16n^2 - 9n - 1)}{6}$

Practical Consideration 3.4C (Should $A\mathbf{x} = \mathbf{b}$ be solved as $A^{-1}\mathbf{b}$?). For large n, it takes about $2n^3/3$ operations to solve one linear system $a\mathbf{x} = \mathbf{b}$ by completing $[\hat{L}\backslash\hat{U}:\hat{\mathbf{b}}:\overline{\mathbf{c}}:\overline{\mathbf{x}}]$, but about $8n^3/3$ to find A^{-1} (by completing $[\hat{L}\backslash\hat{U}:\hat{I}_n:\overline{C}:\overline{X} = A^{-1}]$). So the amount of calculation needed to find A^{-1} and *then* form $A^{-1}\mathbf{b}$ is more than four times that needed to complete $[\hat{L}\backslash\hat{U}:\hat{\mathbf{b}}:\overline{\mathbf{c}}:\overline{\mathbf{x}}]$ (with corresponding increases in the amount of propagated roundoff error). This gives quantitative justification for our earlier statements that *the formula* $\overline{\mathbf{x}} = A^{-1}\mathbf{b}$ is not a computationally desirable way to solve a single linear system for $n > 2$. There is one possible exception to this: If software for solving a linear system is unavailable, but you do have access to an enhanced version of `BASIC` that has `MAT` commands, then a short `BASIC` program can be written quickly and used to solve a linear system $A\mathbf{x} = \mathbf{b}$ as $A^{-1}\mathbf{b}$ (see Problem C3-3). ■

3.5 The Determinant of an $n \times n$ Matrix

For $n > 1$, an $n \times n$ matrix A is an *array* of numbers and as such has no numerical "value." However, the *determinant* of square matrix A is a number that is formed from the entries of A and gives some useful information about A. In this section we survey the basic properties of determinants and develop an efficient method for calculating them.

3.5A Definition and Basic Properties

The reader no doubt has evaluated a 2×2 determinant as

$$\det A_{2 \times 2} = \begin{vmatrix} a_{11} & a_{12} \\ a_{21} & a_{22} \end{vmatrix} = a_{11}a_{22} - a_{12}a_{21} \qquad \textbf{(1a)}$$

where the downward arrow gives the positive term and the upward arrow gives the negative one. A similar "arrow rule" is usually used to get the sign of the six terms of the 3×3 determinant

$$\det A_{3 \times 3} = \begin{vmatrix} a_{11} & a_{12} & a_{13} \\ a_{21} & a_{22} & a_{23} \\ a_{31} & a_{32} & a_{33} \end{vmatrix} = \begin{cases} +a_{11}a_{22}a_{33} + a_{12}a_{23}a_{31} + a_{13}a_{21}a_{32} \\ -a_{11}a_{23}a_{32} - a_{12}a_{21}a_{33} - a_{13}a_{22}a_{31} \end{cases} \qquad \textbf{(1b)}$$

More generally, the **determinant of an $n \times n$** matrix A is defined as

$$\det A = \begin{vmatrix} a_{11} & \cdots & a_{1n} \\ \vdots & & \vdots \\ a_{n1} & \cdots & a_{nn} \end{vmatrix} = \Sigma \, (-1)^{N(\pi)} a_{1\alpha} a_{2\beta} \cdots a_{n\nu} \qquad \textbf{(2)}$$

where the sum Σ is taken over the $n!$ **permutations** (i.e., rearrangements) $\pi = \{\alpha, \beta, \ldots, \nu\}$ of the list $\{1, 2, \ldots, n\}$, and $N(\pi)$ denotes the number of **inversions** (i.e., order reversals) in the list π. In (1b), " $+$ " precedes $a_{12}a_{23}a_{31}$ and " $-$ " precedes $a_{12}a_{21}a_{33}$ because

$$a_{12}a_{23}a_{31} \text{ has } N(\pi) = N(\{2, 3, 1\}) = 2 \qquad (2 > 1 \text{ and } 3 > 1) \qquad \textbf{(3a)}$$
$$a_{12}a_{21}a_{33} \text{ has } N(\pi) = N(\{2, 1, 3\}) = 1 \qquad (2 > 1) \qquad \textbf{(3b)}$$

Unfortunately, *there are no "arrow rules" for $n > 3$*. Indeed, even for $n = 4$ there is no way to draw arrows to describe the signs of the $4! = 24$ terms in definition (2). An efficient alternative procedure for evaluating $n \times n$ determinants for $n > 3$ is based on the following properties that hold for any $A_{n \times n}$ and $B_{n \times n}$.

D1: If A is upper or lower triangular, then $\det A = a_{11} \cdot a_{22} \cdot \cdots \cdot a_{nn}$.

interchange of row m with lower row p

D2: If B is obtained from A by an m,p-pivot, then $\det B = -(\det A)$.

D3: If B is obtained from A by multiplying a single row of A by a scalar s, then $\det B = s(\det A)$.

D4. (THE PRODUCT THEOREM): $\det(AB) = (\det A)(\det B)$

Properties **D1**–**D3** follow directly from definition (2). The Product Theorem **D4** says that the number obtained as the determinant of the product matrix AB is the same as that obtained as the product of the determinants of A and B! The proof of this remarkable result can be found in any linear algebra book. If we apply it, along with **D1**, to the identity $AA^{-1} = I_n$, we get $(\det A)(\det A^{-1}) = 1^n$, from which we can conclude:

> **D5:** If A is nonsingular, then $\det A \neq 0$ and $\det(A^{-1}) = 1/(\det A)$.

Thus the determinant of the inverse of A can be found as the *reciprocal* of the (nonzero) determinant of A, *without finding* A^{-1}.

3.5B Getting det(*A*) from an *LU*-Decomposition

Let $< \hat{L}\hat{U}, \boldsymbol{\rho}_{m,p} >$ be an LU-decomposition of $A_{n\times n}$. Then $\hat{L}\hat{U} = \hat{A}$, where $\hat{A}$ is obtained from A using the m,p-pivots indicated in $\boldsymbol{\rho}_{m,p}$. If $N(\boldsymbol{\rho}_{m,p})$ denotes the number of m,p-pivots used, then by **D4** and **D2**,

$$\det \hat{A} = (\det \hat{L})(\det \hat{U}) = (-1)^{N(\boldsymbol{\rho}_{m,p})} \det A \tag{4}$$

By **D1**, $\det \hat{L} = 1^n$ and $\det \hat{U} = \hat{u}_{11}\cdot\hat{u}_{22}\cdot \cdots \cdot\hat{u}_{nn}$. So solving (4) gives

> **D6:** $\det A = (-1)^{N(\boldsymbol{\rho}_{m,p})}$ {product of the (circled) pivot entries of $\hat{L}\backslash\hat{U}$}

This tells us how to get det A by inspection of *any* LU-decomposition of A! For example, we observed in (20) of Section 3.3D that

$$\begin{bmatrix} -1 & 1 & -4 \\ 2 & 2 & 0 \\ 3 & 3 & 2 \end{bmatrix} \quad \text{has} \quad \begin{bmatrix} \textcircled{3} & 3 & 2 \\ \underline{-\frac{1}{3}} & \textcircled{2} & -\frac{10}{3} \\ \underline{\frac{2}{3}} & \underline{0} & \textcircled{$-\frac{4}{3}$} \end{bmatrix} \begin{matrix} (\rho_1 \leftrightarrow \rho_3) \\ (\rho_2 \leftrightarrow \rho_3) \\ \end{matrix} \quad \text{and} \quad \begin{bmatrix} \textcircled{-1} & 1 & -4 \\ \underline{-2} & \textcircled{4} & -8 \\ \underline{-2} & \underline{\frac{3}{2}} & \textcircled{2} \end{bmatrix} = L\backslash U \tag{5}$$

as LU-decompositions for which $N(\boldsymbol{\rho}_{m,p})$ is 2 and 0, respectively. Applying **D6** to either of these yields $\det A = -8$, in agreement with (1b), as you can check. Hence by **D5**, $\det(A^{-1}) = -\frac{1}{8}$; this can be verified by using (1b) to evaluate $\det(A^{-1})$ using A^{-1} obtained in Example 3.4B. Similarly, we saw in (20) of Section 3.3D that

$$\begin{bmatrix} 2 & 6 & 0 & -2 \\ -2 & -5 & -3 & 4 \\ -1 & -2 & 0 & -1 \\ 0 & 2 & 10 & 5 \end{bmatrix} \quad \text{has} \quad \begin{bmatrix} \textcircled{-1} & -2 & 0 & -1 \\ \underline{-2} & \textcircled{2} & 0 & -4 \\ \underline{0} & \underline{1} & \textcircled{10} & 9 \\ \underline{2} & \underline{-\frac{1}{2}} & \underline{-\frac{3}{10}} & \textcircled{$\frac{67}{10}$} \end{bmatrix} \begin{matrix} (\rho_1 \leftrightarrow \rho_3) \\ (\rho_2 \leftrightarrow \rho_3) \\ (\rho_3 \leftrightarrow \rho_4) \\ \end{matrix} \tag{6}$$

as an LU-decomposition for which $N(\boldsymbol{\rho}_{m,p})$ is 1. We can therefore use **D6** and **D5** to determine that $\det A = (-1)^3(-1)(2)(10)(\frac{67}{10}) = 134$ and $\det(A^{-1}) = \frac{1}{134}$. To fully appreciate **D6** and **D5**, try to get these determinants using definition (2) or any other method you may know.

It follows from **D6** that det A can be zero if and only if a diagonal entry of $\hat{L}\backslash\hat{U}$ is zero. This can occur only if an A having zeros *on and below* the *mm*th entry is encountered in decomposing A. Consequently (see Section 3.3E),

$$A_{n\times n} \text{ is singular if and only if } \det A = 0 \tag{7}$$

This result accounts for much of the *theoretical* importance of determinants. However, since an *LU*-decomposition of A is likely to be needed to calculate det A (using **D6**), (7) is of no more *practical* value than simply looking for $\tilde{a}_{mm} = \cdots = \tilde{a}_{nm} = 0$ in decomposing A.

Practial Consideration 3.5B (Cramer's rule). If the linear system $A\mathbf{x} = \mathbf{b}$ has a nonsingular coefficient matrix A, the components of its unique solution $\overline{\mathbf{x}} = [\overline{x}_1 \ \ \overline{x}_2 \ \ \cdots \ \ \overline{x}_n]^T$ can be expressed as the formula

$$\overline{x}_j = \frac{\det A_j}{\det A} \quad \text{for } j = 1, 2, \ldots, n \qquad \textbf{(Cramer's rule)} \tag{8}$$

where A_j is the $n \times n$ matrix obtained by replacing $\text{col}_j A$ by $\mathbf{b}$. The arrow rules make Cramer's rule about as easy to use as the formula $A^{-1}\mathbf{b}$ when $n = 2$, and about as easy to use as the *LU*-Factorization Algorithm when $n = 3$. However for $n \geq 4$, the evaluation of det A *alone* (using **D6**) requires an $O(n^3)$ *LU*-decomposition of A. Once this is done, it would be foolish to even consider finding the *n additional* $n \times n$ determinants det A_1, … , det A_n needed in (8), when with a lot less effort, $\overline{\mathbf{x}}$ can be found by simply completing the $\overline{\mathbf{c}}$ and $\overline{\mathbf{x}}$ columns of the forback matrix $[\hat{L}\hat{U}:\hat{\mathbf{b}}:\overline{\mathbf{c}}:\overline{\mathbf{x}}]$ which is only $O(h^2)$! Thus, although Cramer's rule is often useful when *algebraic expressions* for solutions are needed, it is not efficient *numerically* when $n > 3$. ■

3.5C Geometric Interpretation of det(A)

The determinant is related to the concept of volume. To see this, consider the *row vectors*

$$\text{row}_i A = [a_{i1} \ \ a_{i2} \ \ \cdots \ \ a_{in}] \qquad \text{for } i = 1, 2, \ldots, n \tag{9}$$

For $n = 2$ we can identify $\text{row}_i\, A$ with the geometric vector $\overrightarrow{OP}_i$, where P_i is the point in the plane (2-space) having coordinates (a_{i1}, a_{i2}); similarly, for $n = 3$ we can identify $\text{row}_i A$ with OP_i, where P_i is the point in 3-space having coordinates (a_{i1}, a_{i2}, a_{i3})(Figure 3.5-1).

With these identifications, one can show (Figure 3.5-2) that

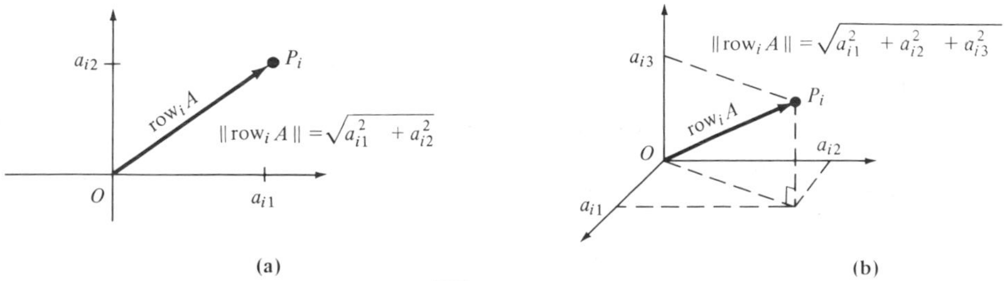

Figure 3.5-1 *Identifying $\text{row}_i A$ with $\overline{OP}_i$: (a) 2-space; (b) 3-space.*

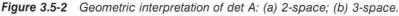

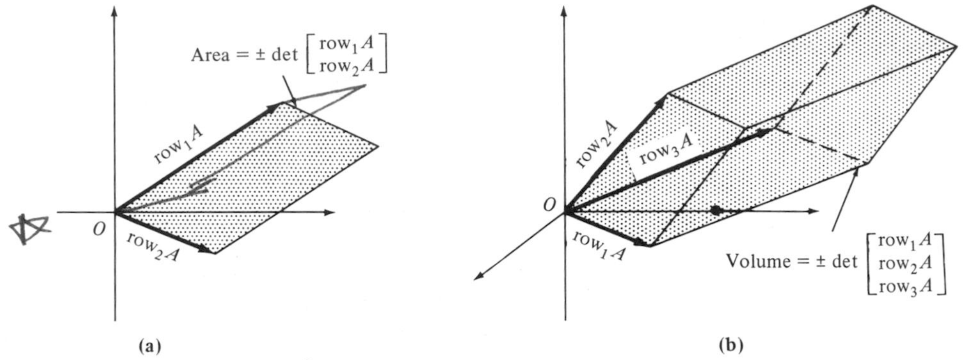

Figure 3.5-2 *Geometric interpretation of det A: (a) 2-space; (b) 3-space.*

If $n = 2$, $\det A = \pm$(area of parallelogram with sides $\text{row}_1 A$, $\text{row}_2 A$) **(10a)**

If $n = 3$, $\det A = \pm$(volume of parallelepiped with sides $\text{row}_i A$, $i = 1, 2, 3$) **(10b)**

Although it is impossible to draw n-space for $n > 3$, it is useful to think of the nonnegative number $|\det A|$ as the volume of the "parallelepiped" in n-space having sides $\text{row}_1 A, \ldots, \text{row}_n A$, where $\text{row}_i A$ is viewed as the geometric vector from the origin of n-space to the "point" having coordinates $(a_{i1}, a_{i2}, \ldots, a_{in})$. With this interpretation, if we define

$$\|\text{row}_i A\| = \textbf{Euclidean length of } \text{row}_i A = \sqrt{a_{i1}^2 + a_{i2}^2 + \cdots + a_{in}^2} \qquad \textbf{(11)}$$

then the volume $|\det A|$ can be no larger than the product of the lengths of its sides; this occurs when the vectors $\text{row}_1 A, \ldots, \text{row}_n A$ are mutually orthogonal. Stated symbolically (this is **Hadamard's inequality**),

$$|\det A| \leq \|\text{row}_1 A\| \, \|\text{row}_2 A\| \cdots \|\text{row}_n A\| \qquad \textbf{(12)}$$

Moreover, the volume $|\det A|$ will be "small" if either (i) one or more of the lengths $\|\text{row}_1 A\|, \|\text{row}_2 A\|, \ldots, \|\text{row}_n A\|$ is "small," or (ii) there is a side $\text{row}_i A$ that is almost parallel to the "plane" spanned by some of the other sides. In the limiting cases, we get the following result:

D7: $\det A = 0$ *if and only if there is an i such that* (i) $\text{row}_i A = \mathbf{0}$ *or* (ii) $\text{row}_i A \neq \mathbf{0}$ *but can be expressed as a linear combination of the remaining rows; that is, there exist scalars* w_{ik} *such that*

$$\text{row}_i A = \sum_{k \neq i} w_{ik} \, \text{row}_k A \qquad \textbf{(13)}$$

The preceding intuitive presentation can be justified rigorously using the results of linear algebra or vector analysis. The characterization of singular matrices given in **D7** will be helpful in explaining why some coefficient matrices are more sensitive to roundoff errors than others (see Section 4.1B).

PROBLEMS

Section 3.1

3-1 Evaluate $\mathbf{x}^T\mathbf{y}$ and $\mathbf{x}\mathbf{y}^T$: **(a)** $\mathbf{x} = [3 \quad 1]^T$, $\mathbf{y} = [2 \quad -1]^T$; **(b)** $\mathbf{x} = y = [2 \quad -1 \quad 1]^T$

3-2 Let $A = \begin{bmatrix} 1 & 2 \\ 3 & 4 \end{bmatrix}$, $B = \begin{bmatrix} 2 & 1 \\ 1 & 1 \end{bmatrix}$, $\mathbf{c} = [4 \quad 1]$, $\mathbf{d} = \begin{bmatrix} 2 \\ 3 \end{bmatrix}$, $E = \begin{bmatrix} 1 & 0 & 2 \\ 0 & 1 & 0 \end{bmatrix}$.

Form AB and BA. Then use them as needed to find (*if defined*):
(a) $2A - 3B$ **(b)** ABA **(c)** A^2B **(d)** BA^2 **(e)** EBA
(f) ABE **(g)** BAE **(h)** $\mathbf{c}B$ **(i)** $E\mathbf{d}$ **(j)** $\mathbf{d}A$

3-3 Verify using or A, B, $\mathbf{c}$, and $\mathbf{d}$ of Problem 3-2.
(a) $(\mathbf{c}A)B = \mathbf{c}(AB)$ **(b)** $(A + B)\mathbf{d} = A\mathbf{d} + B\mathbf{d}$ **(c)** $A(BA) = (AB)A$

3-4 For A and B of Problem 3-2, use formula (11) of Section 3.1D to find A^{-1} and B^{-1}. Then verify that $(AB)^{-1} = B^{-1}A^{-1}$.

3-5 Prove that if A is $m \times n$ and B is $n \times p$, so that AB is defined, then
(a) $\text{row}_i(AB) = a_{i1}\text{row}_1B + a_{i2}\text{row}_2B + \cdots + a_{in}\text{row}_nB \quad (1 \leq i \leq m)$
(b) $\text{col}_j(AB) = b_{1j}\text{col}_1A + b_{2j}\text{col}_2A + \cdots + b_{nj}\text{col}_nA \quad (1 \leq j \leq p)$

3-6 What does Problem 3-5 say about the products AD and DA when $D = \text{diag}(d_{11}, \ldots, d_{nn})$? Verify for $D = \text{diag}(a, b)$ using A of Problem 3-2.

3-7 Prove that if AB is defined, then $(AB)^T = B^TA^T$.

3-8 For the given 2×2 linear system, form the augmented matrix and use the inverse of the coefficient matrix to solve it.

(a) $\begin{aligned} 1x + 2y &= -3 \\ 4x + 3y &= 18 \end{aligned}$ **(b)** $\begin{aligned} -2s + 1t &= -1 \\ 3s + -1t &= 3 \end{aligned}$ **(c)** $\begin{aligned} 2x_1 - x_2 &= 0 \\ 2x_1 + x_2 &= 1 \end{aligned}$

3-9 How many arithmetic operations are needed to solve $A_{2\times 2}\mathbf{x} = \mathbf{b}$ as $A^{-1}\mathbf{b}$?

3-10 Deduce formulas for $\bar{x}_1$ and $\bar{x}_2$ from the formula $\bar{\mathbf{x}} = A^{-1}\mathbf{b}$ when $n = 2$.

Section 3.2

3-11 Solve the given triangular system by using (2a), (2b), (5a), or (5b) of Section 3.2B as appropriate to complete a "doubly augmented" matrix as in Example 3.2B.

(a)
$$\begin{aligned} 1x &= 3 \\ 2x + 3y &= 0 \\ 4x + 5y + 6z &= 4 \end{aligned}$$

(b)
$$\begin{aligned} 6x + 5y + 4z &= 4 \\ 3y + 2z &= 0 \\ 1z &= 3 \end{aligned}$$

(c)
$$\begin{aligned} 2s &= -2 \\ 0s + 1t &= -1 \\ -2s + 0t + 3u &= 2 \\ 0s + 1t + 1u + 2v &= -1 \end{aligned}$$

(d)
$$\begin{aligned} 2s + 1t + 1u + 0v &= -1 \\ 3t + 0u - 2v &= 2 \\ 1u + 0v &= -1 \\ 2v &= -2 \end{aligned}$$

(e)
$$\begin{aligned} c_1 &= -1 \\ 7c_1 + c_2 &= -5 \\ 6c_1 + 5c_2 + c_3 &= 4 \\ 4c_1 + 3c_2 + 2c_3 + 4c_4 &= -5 \end{aligned}$$

(f)
$$\begin{aligned} x_1 + 2x_2 + 1x_3 + 1x_4 &= 6 \\ x_2 + 5x_3 - 3x_4 &= -10 \\ x_3 - 2x_4 &= -5 \\ x_4 &= 2 \end{aligned}$$

3-12 Express the following matrix equality as four scalar equations:

$$\begin{bmatrix} a & 0 \\ b & c \end{bmatrix}\begin{bmatrix} d & e \\ 0 & f \end{bmatrix} = \begin{bmatrix} a_{11} & a_{12} \\ a_{21} & a_{22} \end{bmatrix} \qquad \text{(this is } LU = A\text{)}$$

Deduce that a given A can have infinitely many LU-factorizations.

3-13 For the given matrix A, use the four scalar equations of Problem 3-12 to factor A as LU (if possible) three different ways:

(i) Assume that L is unit lower triangular (i.e., $a = c = 1$).
(ii) Assume that U is unit upper triangular (i.e., $d = f = 1$).
(iii) Assume that $a = d > 0$ and $c = f > 0$.

(a) $A = \begin{bmatrix} 4 & 3 \\ 2 & 2 \end{bmatrix}$ **(b)** $A = \begin{bmatrix} 2 & 1 \\ 1 & 0 \end{bmatrix}$ **(c)** $A = \begin{bmatrix} 4 & -2 \\ -2 & 2 \end{bmatrix}$

3-14 For the linear systems in parts (a)–(f): Solve using Basic Gaussian Elimination as in Example 3.2C; stop if $\tilde{a}_{mm} = 0$ is encountered.

(a)
$$\begin{aligned} x + 3y + z &= 3 \\ 2x + 8y + 2z &= -2 \\ 2y + 2z &= -6 \end{aligned}$$

(b)
$$\begin{aligned} 2x + 2y - 2z &= -4 \\ 2x + 2y - 4z &= 8 \\ -x + y + 3z &= 4 \end{aligned}$$

(c)
$$\begin{aligned} x_1 + 2x_2 + x_3 &= 1 \\ 2x_1 - x_2 + 2x_3 &= -3 \\ x_2 + 3x_3 &= 1 \end{aligned}$$

(d)
$$\begin{aligned} 2x_2 + x_3 &= 3 \\ -x_1 + x_2 &= 2 \\ 2x_1 - x_2 + 3x_3 &= -5 \end{aligned}$$

(e)
$$\begin{aligned} 2t + 1u + 4v &= -3 \\ s - t - 2u &= 0 \\ 2s + 6v &= -2 \\ -s + 3t + 2u &= 0 \end{aligned}$$

(f)
$$\begin{aligned} s - t + v &= -3 \\ 2s - 3t - u + 3v &= -6 \\ 3s - 3t + 2u + 5v &= 9 \\ 4s - 2t + 2u + 0v &= -14 \end{aligned}$$

3-15 Same as Problem 3-14, but using the LU-Factorization Algorithm 3.2E as in Example 3.2E.

3-16 Same as Problem 3-14, but using Basic Gauss–Jordan Elimination as in Example 3.2I.

3-17 For the given augmented matrix $[T:\mathbf{b}]$, solve $T\mathbf{x} = \mathbf{b}$ using the LU-Factorization Algorithm for Tridiagonal Systems (Algorithm 3.2H).

(a) $\left[\begin{array}{ccc:c} 3 & 2 & & 5 \\ 1 & 2 & 1 & 2 \\ & 2 & 6 & -4 \end{array}\right]$ **(b)** $\left[\begin{array}{ccc:c} 4 & 1 & & 6 \\ 1 & 3 & 1 & 6 \\ & 2 & -3 & 10 \end{array}\right]$

(c) $\left[\begin{array}{cccc:c} 4 & 1 & & & 6 \\ 1 & 4 & 1 & & -4 \\ & 1 & 4 & 1 & 5 \\ & & 1 & 4 & 2 \end{array}\right]$ **(d)** $\left[\begin{array}{cccc:c} 3 & -1 & & & -1 \\ -1 & 3 & -1 & & 3 \\ & -1 & 3 & -1 & -2 \\ & & -1 & 3 & 3 \end{array}\right]$

3-18 For this problem, refer to Table 3.2-1 of Section 3.2G.

(a) Derive the formulas for the $\bar{\mathbf{c}}$ and $\bar{\mathbf{x}}$ rows.

(b) Find the table's values when $n = 2, 3, 5, 20, 100$, and 1000. Do the values when $n = 100$ and 1000 verify the $O(h^2)$ and $O(h^3)$ growth described in Sections 3.2G?

(c) From your values in part (b), determine the minimum time it will take to solve $A\mathbf{x} = \mathbf{b}$ on a computer that can do arithmetic at the following rate (1 microsecond is 10^{-6} second):

Arithmetic operation:	$\pm$	$*$	$/$
Time required (microseconds):	0.4	1	4

NOTE: You may wish to replace these times by those of the computer(s) available to you.

(d) Make a similar table for the UPPER TRIANGULATION and BACKWARD SUBSTITUTION phases of the Gaussian Elimination Algorithm for Tridiagonal Systems (Algorithm 3.2H). Then find the table values and execution times as in parts (b) and (c) for $n = 2, 3, 5, 20, 100$, and 1000. Do your results confirm the need for a separate subprogram for solving tridiagonal systems? Why?

(e) Make a comparable table for the DIAGONALIZATION and SCALING phases of Gauss–Jordan Elimination (Section 3.2I). Then find the table values and execution times as in parts (b) and (c) for $n = 2, 3, 5, 20, 100$, and 1000. For what n does Gauss–Jordan Elimination require fewer arithmetic operations than Gaussian Elimination or LU-factorization?

Section 3.3

3-19 Decompose A in (1) of Section 3.3A using Scaled Partial Pivoting.

3-20 Show that Basic Pivoting yields the LU-decomposition in (4a) of Example 3.4A.

3-21 For parts (a)–(f) of Problem 3-14: Solve using the LU-Decomposition Algorithm 3.3C with Basic Pivoting.

3-22 For parts (a)–(f) of Problem 3-14: Solve using the LU-Decomposition Algorithm 3.3C with Scaled Partial Pivoting.

3-23 Let $A = \begin{bmatrix} 13 & 22 \\ 3 & 5 \end{bmatrix}$ and $\mathbf{b} = \begin{bmatrix} 1 \\ 0 \end{bmatrix}$. NOTE $A^{-1} = \begin{bmatrix} -5 & 22 \\ 3 & -13 \end{bmatrix}$.

Decompose A using BP and using SPP. Store $\hat{L}\backslash\hat{U}$ entries rounded to $2s$ as they are obtained, and use the rounded $\hat{l}_{12}$ to get $\hat{u}_{22}$. Use these two decompositions to do parts (a)–(c).

(a) Which LU-decomposition would be obtained using PP?

(b) Form the product LU for both decompositions. Which has the smallest (absolute) errors? Which has the smallest relative errors?

(c) Solve $A\mathbf{x} = \mathbf{b}$ by completing a forback matrix (with values entered rounded to $2s$) for both decompositions. Which solution is most accurate?

What conclusions (if any) can you draw from parts (a)–(c)?

3-24 Repeat Problem 3-23 but round to $3s$ rather than $2s$. Reexamine your conclusions in Problem 3-23 in light of the results of this problem.

3-25 The following LU-decompositions result from the FACTOR $\hat{A}$ phase of the LU-Decomposition Algorithm 3.3C. Find first $\hat{A}$, then the *given* A.

(a) $\begin{bmatrix} \textcircled{1} & 2 & 1 \\ \underline{2} & \textcircled{-5} & 0 \\ \underline{0} & \underline{-\frac{1}{5}} & \textcircled{3} \end{bmatrix} = L\backslash U$

(b) $\begin{bmatrix} \textcircled{2} & 2 & -2 \\ \underline{-\frac{1}{2}} & \textcircled{2} & 2 \\ \underline{1} & \underline{0} & \textcircled{-2} \end{bmatrix}$ $(\rho_2 \leftrightarrow \rho_3)$

(c) $\begin{bmatrix} \textcircled{1} & -1 & 0 \\ \underline{0} & \textcircled{2} & 4 \\ \underline{2} & \underline{2} & \textcircled{1} \end{bmatrix}$ $(\rho_1 \leftrightarrow \rho_3)$ $(\rho_2 \leftrightarrow \rho_3)$

(d) $\begin{bmatrix} \textcircled{2} & 8 & 2 \\ \underline{0} & \textcircled{2} & 2 \\ \underline{\frac{1}{2}} & \underline{-\frac{1}{2}} & \textcircled{1} \end{bmatrix}$ $(\rho_1 \leftrightarrow \rho_2)$ $(\rho_2 \leftrightarrow \rho_3)$

(e) $\begin{bmatrix} \textcircled{2} & 0 & 0 & 6 \\ \underline{-\frac{1}{2}} & \textcircled{3} & 2 & 3 \\ \underline{\frac{1}{2}} & \underline{-\frac{1}{3}} & \textcircled{-\frac{4}{3}} & -2 \\ \underline{0} & \underline{\frac{2}{3}} & \underline{\frac{1}{4}} & \textcircled{\frac{5}{2}} \end{bmatrix}$ $(\rho_1 \leftrightarrow \rho_3)$ $(\rho_2 \leftrightarrow \rho_4)$ $(\rho_3 \leftrightarrow \rho_4)$

(f) $\begin{bmatrix} \textcircled{3} & 0 & -1 & 0 \\ \underline{0} & \textcircled{-1} & -1 & 0 \\ \underline{\frac{1}{3}} & \underline{0} & \textcircled{-\frac{5}{3}} & 1 \\ \underline{0} & \underline{-2} & \underline{\frac{6}{5}} & \textcircled{\frac{14}{5}} \end{bmatrix}$ $(\rho_1 \leftrightarrow \rho_3)$ $(\rho_2 \leftrightarrow \rho_4)$

3-26 Use the indicated LU-decomposition to solve $A\mathbf{x} = \mathbf{b}$ for the given $\mathbf{b}$ by forming $\hat{\mathbf{b}}$, and then completing the forback matrix $[\hat{L}\backslash\hat{U} : \hat{\mathbf{b}} : \bar{\mathbf{c}} : \bar{\mathbf{x}}]$.

(a) Problem 3-25(c); $\mathbf{b} = [2 \quad -3 \quad -4]^T$ **(b)** Problem 3-25(d); $\mathbf{b} = [1 \quad 4 \quad 2]^T$

(c) Problem 3-25(e); $\mathbf{b} = [6 \quad -2 \quad 16 \quad 2]^T$ **(d)** Problem 3-25(f); $\mathbf{b} = [-1 \quad 8 \quad 3 \quad -1]^T$

3-27 From the details of your work in (a)–(f) of Problem 3-25: Form the $|\tilde{a}_{im}|/\hat{s}_i$ ratios *by inspection* of the given $\hat{L}\backslash\hat{U}$ and $\hat{A}$ $(= \hat{L}\hat{U})$ for $m = 1, \ldots, n-1$ to determine if the decomposition would result from the Scaled Partial Pivoting strategy. Do *not* actually decompose A using SPP!

3-28 For the decompositions given in (a)–(f) of Problem 3-25: Find the final pointer vector *NewRow* as in Practical Consideration 3.3D.

3-29 For parts (a)–(f) of Problem 3-14: Solve the system using the LU-Decomposition Algorithm 3.3C with Scaled Full Pivoting (Section 3.3F).

3-30 Show that the following matrices are singular (see Section 3.3E). Express one equation as a linear combination of the others.

(a) $\begin{bmatrix} 1 & 2 & 3 \\ 4 & 5 & 6 \\ 7 & 8 & 9 \end{bmatrix}$ **(b)** $\begin{bmatrix} -1 & 2 & 3 \\ 0 & 1 & 4 \\ 2 & -4 & -6 \end{bmatrix}$ **(c)** $\begin{bmatrix} 1 & -1 & -1 & -2 \\ 2 & -3 & 2 & -1 \\ 3 & -5 & 5 & 0 \\ 1 & -2 & 3 & 1 \end{bmatrix}$

Section 3.4

3-31 Use Algorithm 3.4A with the indicated LU-decomposition to solve the given systems $A\mathbf{x}_j = \mathbf{b}_j$ as in Example 3.4A(a).
(a) Problem 3-25(a); $\mathbf{b}_1 = [11 \quad 7 \quad 6]^T$, $\mathbf{b}_2 = [13 \quad 1 \quad 11]^T$
(b) Problem 3-25(c); $\mathbf{b}_1 = [2 \quad 7 \quad 1]^T$, $\mathbf{b}_2 = [6 \quad 7 \quad -3]^T$, $\mathbf{b}_3 = [0 \quad 2 \quad 1]^T$
(c) Problem 3-25(d); $\mathbf{b}_1 = [1 \quad 4 \quad 2]^T$, $\mathbf{b}_2 = [1 \quad 4 \quad -2]^T$

3-32 Invert the matrix using Algorithm 3.4A as in Example 3.4B(a).
(a) A of Problem 3-25(b) **(b)** A of Problem 3-25(c)
(c) A of Problem 3-25(e) **(d)** T of Problem 3-17(c)

3-33 Same as Problem 3-32, but using Gauss–Jordan Elimination as in Example 3.2B(b). Use PP in part (a), SPP in part (b), and BP in parts (c) and (d).

3-34 Find the values of Table 3.4-1 for $n = 2, 3, 5, 20, 100$, and 200.

3-35 Make a table comparable to Table 3.4-1 (Section 3.4C) for Gauss–Jordan Elimination for finding A^{-1} [(9) of Section 3.4B]. Find the values of your table for $n = 2, 3, 5, 20, 100$, and 200.

Section 3.5

3-36 Deduce from **D3** of Section 3.5A that $\det(sA) = s^n(\det A)$ for any scalar s.

3-37 If $a = \det(A_{n\times n})$ and $b = \det(B_{n\times n}) \neq 0$, express in terms of a, b, and n:
(a) $\det(-A)$ **(b)** $\det(AB)$ **(c)** $\det(3AB^2)$
(d) $\det(AB^{-1})$ **(e)** $\det(A + A)$ **(f)** $\det[(B^{-1}A^2)^3]$

3-38 From the LU-decompositions in parts (a)–(f) of Problem 3-25: Use **D6** and **D5** of Section 3.5A to find $\det A$ and $\det A^{-1}$.

3-39 **(a)** Modify Algorithm 3.3C so that it outputs $\det A$ along with $\bar{\mathbf{x}}$.
(b) Modify your solution in part (a) to use the result of Problem M1-11 to avoid the possibility of over/underflow.

3-40 Find a formula in terms of n for the total number of arithmetic operations needed to solve an $n \times n$ linear system $A\mathbf{x} = \mathbf{b}$ by Cramer's rule (8) of Section 3.5B, assuming that **D6** of Section 3.5C is used to evaluate all determinants. Evaluate your formula for $n = 2, 3, 5, 20, 100$, and 1000? For what values of n is it less work than the LU-Decomposition Algorithm?

MISCELLANEOUS PROBLEMS

M3-1 It is immediate from the definition of the product AB that if the product AB is defined and A and B are partitioned as shown then

$$AB = \left[\begin{array}{c|c} A_{11} & A_{12} \\ \hline A_{21} & A_{22} \end{array}\right]\left[\begin{array}{c|c} B_{11} & B_{12} \\ \hline B_{21} & B_{22} \end{array}\right] = \left[\begin{array}{c|c} A_{11}B_{11} + A_{12}B_{12} & A_{11}B_{12} + A_{12}B_{22} \\ \hline A_{21}B_{11} + A_{22}B_{12} & A_{21}B_{12} + A_{22}B_{22} \end{array}\right]$$

provided, of course, that the products $A_{ik}B_{kj}$ are defined. For AB given in (a) and (b), partition B as needed to verify this.

(a) $\left[\begin{array}{cc|c} 1 & 2 & 3 \\ \hline 4 & 5 & 6 \end{array}\right]\begin{bmatrix} 9 & 8 & 7 \\ 6 & 5 & 4 \\ 3 & 2 & 1 \end{bmatrix}$ **(b)** $\left[\begin{array}{c|cc} 2 & 1 & 0 \\ 3 & 0 & 1 \\ \hline 4 & 5 & 6 \end{array}\right]\begin{bmatrix} 3 & 2 & 8 & 7 \\ 1 & 0 & 1 & 0 \\ 0 & 1 & 0 & 1 \end{bmatrix}$

M3-2 Prove: For any θ, the **rotation matrix** $R(\theta) = \begin{bmatrix} \cos\theta & -\sin\theta \\ \sin\theta & \cos\theta \end{bmatrix}$ satisfies

(a) $R(\theta)^{-1} = R(\theta)^T = R(-\theta)$; **(b)** $R(\theta + \phi) = R(\theta) + R(\phi)$.

M3-3 A symmetric matrix P is **positive definite** if $\mathbf{x}^T P\mathbf{x} > 0$ for all $\mathbf{x} \neq \mathbf{0}$.

(a) Show that for any (not necessarily square) matrix K, both KK^T and K^TK are (symmetric and) positive definite if they are nonsingular.

(b) For any positive definite matrix P, show that $\|\mathbf{x}\|$ defined as $\mathbf{x}^T P\mathbf{x}$ satisfies (i) $\|\mathbf{x}\| \geqslant 0$, with equality if and only if $\mathbf{x} = \mathbf{0}$; (ii) $\|a\mathbf{x}\| = |a|\ \|\mathbf{x}\|$ for all scalars a.

M3-4 Verify for $A = \begin{bmatrix} a & b \\ c & d \end{bmatrix}$, $B = \begin{bmatrix} e & f \\ g & h \end{bmatrix}$, and $C = \begin{bmatrix} x & y \\ z & w \end{bmatrix}$.

(a) $(AB)C = A(BC)$ **(b)** $\det(AB) = (\det A)(\det B)$

(c) $\det A^T = \det A$ **(d)** $\det A^{-1} = 1/\det A$

M3-5 Will a coefficient matrix that has all integer entries be stored without inherent roundoff error on a computer? Explain.

M3-6 You may have learned to do Gaussian Elimination by **scaling** (E_m) by dividing both sides of (E_m) by the current $\tilde{a}_{mm}$ (if it is nonzero) *before* reducing $\mathrm{col}_m\tilde{A}$. This makes the *scaled* $\tilde{a}_{mm} = 1$ so that the subsequent 0_{im}-subtracts become simply $\rho_i - \tilde{a}_{im}\rho_m$.

(a) Reduce A of Example 3.2C *in place* using scaling as just described, but storing (circling) the $\tilde{a}_{mm}$ *before* scaling instead of the resulting 1. For the resulting $L\backslash U$, how are the off-diagonal column entries of L and row entries of U related to those of $L\backslash U$ in Example 3.2C? If this $L\backslash U$ is viewed as "L over (unit) U," does $LU = A$?

(b) Why should Algorithm 3.2C be implemented on digital devices *without* scaling?

M3-7 Show that unscaled Partial Pivoting (PP) ensures that all $|\hat{l}_{ij}| \leqslant 1$, whereas Scaled Full Pivoting (SPP) ensures that all $|\hat{u}_{ij}| \leqslant 1$; hence PP limits propagated roundoff during the *forward* substitution, whereas SFP limits propagated roundoff during the *backward* substitution. Which alternative is more desirable? Explain.

M3-8 ***(Proof That $A = LU$ when the Reduction $A \to \{\tilde{A}\} \to L\backslash U$ Succeeds)***

(a) Prove that the product of (unit) lower triangular matrices is (unit) lower triangular. SUGGESTION: Use Problem 3-5(a).

(b) Suppose that the reduction $A \to \{\tilde{A}\} \to L\backslash U$ succeeds for $A_{n\times n}$. For any $m = 1, \ldots, n-1$, let M_m be the matrix obtained by writing the l_{im}'s used to reduce $\mathrm{col}_m\tilde{A}$ in place in the identity matrix I_n. Use Problem 3-5(a) to show that:

(i) M_m is unit lower triangular and nonsingular; in fact, M_m^{-1} is obtained by replacing each stored l_{im} by $-l_{im}$ in M_m.

(ii) The product $M_m\tilde{A}$ performs on $\tilde{A}$ the 0_{im}-subtracts used to reduce $\mathrm{col}_m\tilde{A}$. Hence $\tilde{L} = M_{n-1}\cdots M_2M_1$ is a unit lower triangular matrix for which the product $\tilde{L}A$ is an upper triangular matrix U.

(c) Deduce from part (b) that $L = M_1^{-1}M_2^{-1}\cdots M_{n-1}^{-1}$ $(= \tilde{L}^{-1})$ is a unit lower triangular matrix that satisfies $A = LU$.

(d) For the upper triangulation $[A\,\vdots\,\mathbf{b}] \to [U\,\vdots\,\bar{\mathbf{c}}]$, derive a formula for $\bar{\mathbf{c}}$ in terms of $\mathbf{b}$ and $\tilde{L}$ or L of part (c).

M3-9 For the indicated linear system, form the L and $\bar{\mathbf{c}}$ of parts (c) and (d) of Problem M3-8.

Compare to $[L\backslash U : \bar{\mathbf{c}}]$ obtained using the LU-Factorization Algorithm 3.2E.
(a) Example 3.2C **(b)** Example 3.2D **(c)** Problem 3-15(a)

M3-10 ***(Using a Permutation Matrix P to Represent A as $P\hat{L}\hat{U}$)*** A **permutation matrix** is any matrix obtained by performing m,p-pivots on the identity matrix. In particular, $P(m, p)$ is obtained by performing just an m,p-pivot. Note that $P(m, m) = I$. Show that:
(a) $P(m, p)^2 = I$; i.e., $P(m, p)$ is its own inverse for any m and p.
(b) The product of permutation matrices is a permutation matrix.
(c) Suppose that the decomposition $\hat{A} \to \{\tilde{A}\} \to \hat{L}\backslash\hat{U}$ succeeds. If $\boldsymbol{\rho}_{m,p} = (m_1 \leftrightarrow p_1, \ldots, m_k \leftrightarrow p_k)$ is used to decompose A, then the permutation matrix $\tilde{P} = P(m_k, p_k) \cdots P(m_2, p_2)P(m_1, p_1)$ satisfies $\tilde{P}A = \hat{A}$; hence $A = P\hat{L}\hat{U}$, where $P = P(m_1, p_1) P(m_2, p_2) \cdots P(m_k, p_k)\ (= \tilde{P}^{-1})$.

M3-11 For the LU-decompositions given in parts (a)–(f) of Problem 3-25: Find P of Problem M3-10 for which $A = P\hat{L}\hat{U}$. How is P related to the pointer vector *NewRow* in Problem 3-28?

M3-12 Write two subprograms, one for Basic Gaussian Elimination (Section 3.2C) and one for Gauss–Jordan elimination (Section 3.2I). Use as few statements as possible and do not worry about $a_{mm} = 0$. Which subprograms is shorter? When $n = 20$, which will take longer to run? Which will probably give a more accurate answer? *Justify!* NOTE: Neither should actually be implemented as written. Why?

M3-13 Describe how a given LU-decomposition of A can be used to solve $A^T\mathbf{x} = \mathbf{b}$. SUGGESTION: See Problem 3-7.

M3-14 Use your algorithm of Problem M3-13 with the indicated LU-decomposition of A to solve $A^T\mathbf{x} = \mathbf{b}$.
(a) Problem 3-25(a), $\mathbf{b} = [3 \quad 4 \quad 0]^T$ **(b)** Problem 3-25(c), $\mathbf{b} = [2 \quad -4 \quad -3]^T$

M3-15 Let $P = \text{pent}(\mathbf{o}, \mathbf{l}, \mathbf{d}, \mathbf{u}, \mathbf{p})$ denote a band matrix of bandwidth 5 consisting of $\text{trid}(\mathbf{l}, \mathbf{d}, \mathbf{u})$ with possible nonzero entries in $\mathbf{p}$ (one entry above $\mathbf{u}$) or in $\mathbf{o}$ (one entry below $\mathbf{l}$).
(a) Write pseudocode using Basic Pivoting as in Algorithm 3.2H for solving $P\mathbf{x} = \mathbf{b}$.
(b) Find formulas analogous to those in (26) of Section 3.2H for the number of arithmetic operations that your algorithm in part (a) requires.

M3-16 Deduce from the Gauss–Jordan algorithm (9) of Section 3.4B that if a matrix A is lower (respectively, upper) triangular and all diagonal entries are nonzero, then A is nonsingular and A^{-1} is also lower (respectively, upper) triangular.

M3-17 Prove **I2–I4** of Section 3.1D.

M3-18 **(a)** Use the Product Theorem (**D4** of Section 3.5A) to prove that if $A_{n \times n}$ satisfies $A^2 = I$, then $\det A = \pm 1$.
(b) Find an $A_{2 \times 2}$ with all a_{ij}'s nonzero such that $A^2 = I_2$ and $\det A = -1$.

M3-19 When a force F acts at angles α, β, and γ with the x-, y-, and z- axes, then its components along the coordinate axes are, respectively, $F \cos \alpha$, $F \cos \beta$, and $F \cos \gamma$. If three wires with tensions F_A, F_B, F_C (in Newtons [N]) act at the origin to suspend an m kilogram [kg] mass, then the equilibrium equations are

$$\begin{aligned}
F_A\cos \alpha_A + F_B\cos \alpha_B + F_C\cos \alpha_C &= 0 \\
F_A\cos \beta_A + F_B\cos \beta_B + F_C\cos \beta_C &= 0 \\
F_A\cos \gamma_A + F_B\cos \gamma_B + F_C\cos \gamma_C &= 981m
\end{aligned}$$

Find F_A, F_B, and F_C if wire A lies on the positive x-axis; wire B is fastened at the point $P_B(-1, 2, 2)$; wire C has $\alpha_C = 120°$, $\beta_C = 135°$, and $\gamma_C = 60°$; and $m = 100$ kg. NOTE: For wire B, the **direction cosines** $\cos \alpha_B$, $\cos \beta_B$, and $\cos \gamma_B$ are the components of the unit vector in the direction from the origin to P_B.

M3-20 A flexible horizontal beam, fixed at end A and movable at end B, is viewed as having four translational degrees of freedom $u_1, \ldots, u_4$, where u_i occurs $i/5$ of the way from A to B. If a unit load is applied at u_3, then $\mathbf{u} = [u_1 \ \ u_2 \ \ u_3 \ \ u_4]^T$ satisfies the banded system $K\mathbf{u} = \mathbf{r}$, where the **stiffness matrix** K and **load vector r** are given by

$$K = \begin{bmatrix} 5 & -4 & 1 & 0 \\ -4 & 6 & -4 & 1 \\ 1 & -4 & 6 & -4 \\ 0 & 1 & -4 & 5 \end{bmatrix}, \quad \text{and} \quad \mathbf{r} = \begin{bmatrix} 0 \\ 0 \\ EI \\ 0 \end{bmatrix}$$

EI depends on the beam's material and geometry. Find $\mathbf{u}$ when $EI = 1$.

COMPUTER PROBLEMS

C3-1 **(a)** Write a subprogram `ADD(M, N, A, S, B, C)` that returns $C = A_{m\times n} + sB_{m\times n}$, where s is any `REAL` scalar.

(b) Write a subprogram `MULT(M, N, P, A, B, C)` that returns $C = A_{m\times n}B_{n\times p}$. If extended precision is available, use Partial Extended Precision (Section 1.4C) to accumulate the summations. Test it on $\mathbf{d}^T\mathbf{d}$, $\mathbf{d}\mathbf{d}^T$, EE^T, and E^TE for $\mathbf{d}$ and E of Problem 3-2.

C3-2 Write a subprogram to multiply two *complex* $n \times n$ matrices

$$A = \text{Re}(A) + i\text{Im}(A) \qquad \text{and} \qquad B = \text{Re}(B) + i\text{Im}(\text{B})$$

[Re(A), Im(A), Re(B), and Im(B) are stored REAL $n \times n$ matrices] as

$$AB = \text{Re}(AB) + i\text{Im}(AB), \qquad \text{where} \begin{cases} \text{Re}(AB) = \text{Re}(A)\text{Re}(B) - \text{Im}(A)\text{Im}(B) \\ \text{Im}(AB) = \text{Re}(A)\text{Im}(B) + \text{Im}(A)\text{Re}(B) \end{cases}$$

Use the subprograms you wrote in Problem C3-1 to get Re(AB) and Im(AB).

C3-3 If a version of BASIC that supports MAT (matrix) statements is available, do parts (a) and (b).

(a) Write a short program to solve an $n \times n$ linear system $A\mathbf{x} = \mathbf{b}$ as $A^{-1}\mathbf{b}$. Test it on Example 3-2D. What are the advantages and disadvantages of such a program compared to one that uses the LU-Decomposition Algorithm 3.3C as in Figures 4.2-1–4.2-3?

(b) Insert the statement: `PRINT DET` after the `MAT INV` statement of your program in part (a). If it is recognized, the output value will be the determinant of the inverted matrix.

C3-4 **(a)** Write a calling program for `SUBROUTINE TRIDAG` (Figure 3.2-2). Test it on Example 3.2H.

(b) Separate `SUBROUTINE TRIDAG` into two `SUBROUTINE`s, `FACTOR` and `FWDBAK`, which implement the LU-Factorization Algorithm 3.2E. Test on Example 3.2H.

(c) Write a program for finding T^{-1} when $T = \text{trid}(\mathbf{l}, \mathbf{d}, \mathbf{u})$. Use `SUBROUTINE`s `FACTOR` and `FWDBAK` of part (b) as indicated in Section 3.4B. Test it on Problem 3-32(d).

C3-5 Consider the the 100×100 tridiagonal system $T\mathbf{x} = \mathbf{b}$, where all $\mathbf{l}$ and $\mathbf{u}$ entries are 1, all $\mathbf{d}$ entries are d, and all $\mathbf{b}$ entries are $d + 2$ except b_1 and b_n, which are $d + 1$. (The exact solution is $\bar{x}_i = 1$ for all i.) Use a computer to solve it with $d = 4$ and with $d = 5$. Explain any differences in the accuracy obtained.

C3-6 Write a program that uses Gauss–Jordan Elimination to find the inverse of $A_{n\times n}$ as descibed in (9) of Section 3.4B. Use SPP. Test using Example 3-32(c).

NOTE: Other computer problems relating to solving linear systems are given at the end of Chapter 4.

4

Solving Systems of Equations in Fixed-Precision Arithmetic

4.0 Introduction

In this chapter we focus our attention on systems of equations that are difficult to solve, either because they are sensitive to roundoff error (i.e., ill conditioned), very large, or nonlinear. The reader is assumed to be familiar with the *LU*-Decomposition Algorithm as described in Section 3.3.

4.1 Recognizing and Dealing with Ill Conditioning

Consider the linear system $A\mathbf{x} = \mathbf{b}$. Intuitively, we expect small changes in A or $\mathbf{b}$ to produce correspondingly small changes in the exact solution $\overline{\mathbf{x}} = A^{-1}\mathbf{b}$. However, even for the trivial $n = 1$ case of solving $ax = b$, small variations in a or b can get magnified and so produce disproportionately large changes in $\overline{x} = a^{-1}b$ when $a \approx 0$. The generalization of $a \approx 0$ is described next.

4.1A Ill-Conditioned Linear Systems

For a 2×2 linear system $A\mathbf{x} = \mathbf{b}$, the formula $\overline{\mathbf{x}} = A^{-1}\mathbf{b}$ can be used to examine how changes in either the coefficient matrix A or the target vector $\mathbf{b}$ affect the components of the solution vector $\overline{\mathbf{x}}$. For example,

$$\begin{array}{ll}(E_1) & x - 11y = b_1 \\ (E_2) & -9x + 100y = b_2\end{array} \Rightarrow \begin{bmatrix}\overline{x}\\ \overline{y}\end{bmatrix} = \overline{\mathbf{x}} = \frac{1}{1}\begin{bmatrix}100 & 11\\ 9 & 1\end{bmatrix}\begin{bmatrix}b_1\\ b_2\end{bmatrix} = \begin{bmatrix}100b_1 + 11b_2\\ 9b_1 + b_2\end{bmatrix} \quad (1)$$

If $\mathbf{b} = [10 \quad -1]^T$ is changed to $\mathbf{b}_1 = [9 \quad -1]^T$ ($\Delta b_1 = -10\%$), then

$$\begin{bmatrix}\overline{x}\\ \overline{y}\end{bmatrix} = A^{-1}\mathbf{b} = \begin{bmatrix}989\\ 89\end{bmatrix} \text{ is changed to } \begin{bmatrix}\overline{x}_1\\ \overline{y}_1\end{bmatrix} = A^{-1}\mathbf{b}_1 = \begin{bmatrix}889\\ 80\end{bmatrix} \quad \begin{array}{l}(\Delta\overline{x} \doteq -10\%)\\ (\Delta\overline{y} \doteq -10\%)\end{array} \quad (2)$$

whereas if $\mathbf{b} = [-1 \quad 10]^T$ is changed to $\mathbf{b}_1 = [-1 \quad 9]^T$ ($\Delta b_2 = -10\%$), then

$$\begin{bmatrix}\overline{x}\\ \overline{y}\end{bmatrix} = A^{-1}\mathbf{b} = \begin{bmatrix}10\\ 1\end{bmatrix} \text{ is changed to } \begin{bmatrix}\overline{x}_1\\ \overline{y}_1\end{bmatrix} = A^{-1}\mathbf{b}_1 = \begin{bmatrix}-1\\ 0\end{bmatrix} \quad \begin{array}{l}(\Delta\overline{x} = -110\%)\\ (\Delta\overline{y} = -100\%)\end{array} \quad (3)$$

The changes in $\overline{\mathbf{x}}$ in (2) are perfectly reasonable; however, those in (3) are larger than one might expect from a -10% change in b_2.

To see what happened in (3), we have sketched the graphs of (E_1) and (E_2) for $A\mathbf{x} = \mathbf{b} = [-1 \quad 10]^T$ in Figure 4.1-1(a) and for $A\mathbf{x} = \mathbf{b}_1 = [-1 \quad 9]^T$ in Figure 4.1-1(b).

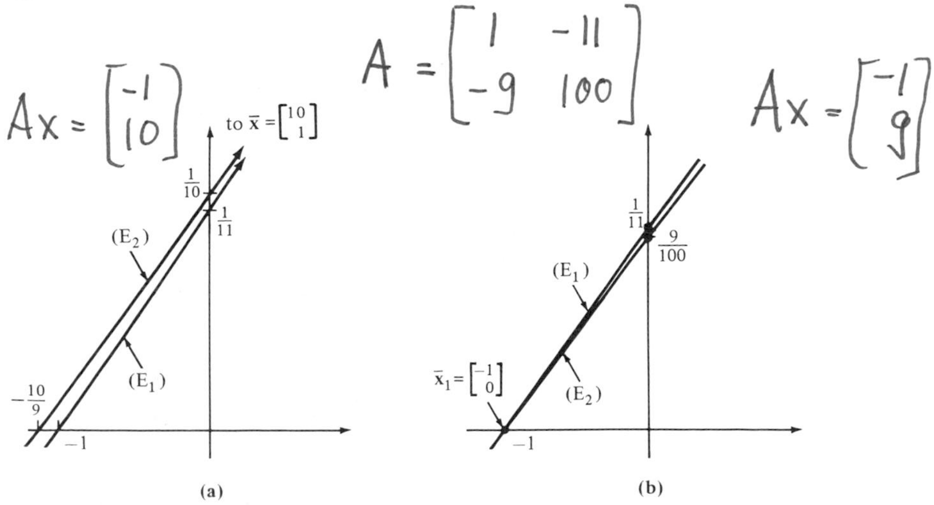

Figure 4.1-1 *Graphical visualization of (a)* $A\mathbf{x} = \mathbf{b}$ *and (b)* $A\mathbf{x} = \mathbf{b}_1$.

These graphs are straight lines which are *nearly parallel* because their slopes, determined solely by A, are nearly equal. Changing $\mathbf{b}_2$ from 10 to 9 moves the graph of (E_2) parallel to itself in such a way that its x-intercept moves from $-\frac{10}{9}$ to -1; this small movement ($\frac{1}{9}$ of a unit to the right) moves the point of intersection of these nearly parallel lines from $\overline{\mathbf{x}} = [10 \quad 1]^T$ all the way to $\overline{\mathbf{x}}_1 = [-1 \quad 0]^T$. Similar *absolute* changes in $\overline{x}$ and $\overline{y}$ occurred in (2), but they were not as large as *percent* changes.

For further evidence of how ''volatile'' the linear system (1) is, let us change $a_{12} = -11$ to -11.1 ($\Delta a_{12} \approx -0.9\%$), thus changing A to

$$A_1 = \begin{bmatrix} 1 & -11.1 \\ -9 & 100 \end{bmatrix}; \qquad \text{hence} \begin{bmatrix} \bar{x}_1 \\ \bar{y}_1 \end{bmatrix} = \bar{\mathbf{x}}_1 = (A_1)^{-1}\mathbf{b} = 10 \begin{bmatrix} 100b_1 + 11.1b_2 \\ 9b_1 + b_2 \end{bmatrix} \quad \textbf{(4)}$$

Upon comparing $\bar{\mathbf{x}}_1$ in (4) to $\bar{\mathbf{x}}$ in (1), we see that $\bar{\mathbf{x}}_1 \approx 10\bar{\mathbf{x}}$. Thus a change of less than 1% in a_{12} produces a change of about 900% in the solution, regardless of the target vector **b**! And if, instead, $a_{22} = 100$ was changed to 99 ($\Delta a_{22} = -1\%$), the matrix A would change to

$$A_2 = \begin{bmatrix} 1 & -11 \\ -9 & 99 \end{bmatrix}, \qquad \text{for which } \det(A_2) = 0; \text{ so } A_2 \text{ is } \textit{singular}! \quad \textbf{(5)}$$

A matrix that becomes singular if one or more of its entries is given a small relative perturbation will be called **nearly singular**. What we have learned from (2) through (4) is that

> When the coefficient matrix of a linear system $A\mathbf{x} = \mathbf{b}$ is nearly singular, the problem of finding its solution $\bar{\mathbf{x}}$ is ill conditioned, because small changes in A or $\mathbf{b}$ can (*but need not*) cause disproportionately large changes in $\bar{\mathbf{x}}$.

When this occurs, we will simply refer to $A\mathbf{x} = \mathbf{b}$ as an **ill-conditioned linear system** and to A itself as an **ill-conditioned matrix**.

Practical Consideration 4.1A (The need to know if *A* is nearly singular). The augmented matrix of a given linear system $A\mathbf{x} = \mathbf{b}$ generally is stored with inherent roundoff error, and so is actually that of a slightly different system $A_{\text{stored}}\mathbf{x} = \mathbf{b}_{\text{stored}}$. If A is *not* ill conditioned and the system is solved using error-reducing strategies (e.g., SPP and Partial Extended Precision), the solution of this *stored* system, that is, $\bar{\mathbf{x}}_{\text{calc}} = (A_{\text{stored}})^{-1}\mathbf{b}_{\text{stored}}$, will generally give $\bar{\mathbf{x}}$ to about the precision of the device as long as n is not too large. However, if A is severely ill conditioned, $\bar{\mathbf{x}}_{\text{calc}} = (A_{\text{stored}})^{-1}\mathbf{b}_{\text{stored}}$ may be seriously in error even if calculated exactly and $n = 2$! *It is therefore important to know if a coefficient matrix is nearly singular*. We now show how this can be done. ■

4.1B Using Condition Numbers to Indicate Ill Conditioning

An $n \times n$ matrix is singular if and only if its determinant is zero. This suggests that a matrix will be nearly singular if its determinant is nearly zero. We must be careful about what we mean by ''nearly zero'' here, because if $\tilde{A}\mathbf{x} = \tilde{\mathbf{b}}$ is obtained from $A\mathbf{x} = \mathbf{b}$ by multiplying one of its equations by a nonzero scalar s, then both systems should be equally sensitive to errors in A or $\mathbf{b}$. However, $\det \tilde{A} = s(\det A)$ (see **D3** of Section 3.5A), so it can be made ''large'' or ''small'' simply by adjusting s! A useful rule of thumb is

> A is ill conditioned iff $|\det A|$ is small compared to the row scale factors $s_i = \max\{|a_{i1}|, \ldots, |a_{in}|\}$, for $i = 1, \ldots, n$. **(6a)**

In view of **D7** of Section 3.5C, this can occur only if one row of A is nearly a linear combination of the others, that is, if scale factor s: largest absolute number in a row

$$\text{row}_i A \approx \sum_{k \neq i} w_k \text{row}_k A \qquad \text{for some } i \tag{6b}$$

When $n = 2$, this means that the rows of A are nearly proportional (see Figure 4.1-1); so ill conditioning can be recognized by inspection.

However, it is usually impossible to detect (6b) by inspection of A when $n > 2$. What we need for $n > 2$ is a **condition number** C that tells how large the "relative error magnification ratio"

$$\frac{|\text{relative error}| \text{ in components of } \bar{\mathbf{x}} = A^{-1}\mathbf{b}}{|\text{relative error}| \text{ in entries of } A \text{ or components of } \mathbf{b}} \tag{7a}$$

might be, with possible large ratios corresponding to ill conditioning. To do this mathematically, one must first select a **norm** $\|\cdot\|$ (which gives the "size" of a vector or matrix just as absolute value $|\cdot|$ gives the "size" a number) and then find a C satisfying

$$\frac{\|\delta\bar{\mathbf{x}}\|}{\|\bar{\mathbf{x}}\|} \leq C\frac{\|\delta A\|}{\|A\|} \qquad \text{and} \qquad \frac{\|\delta\bar{\mathbf{x}}\|}{\|\bar{\mathbf{x}}\|} \leq C\frac{\|\delta \mathbf{b}\|}{\|\mathbf{b}\|} \tag{7b}$$

[where δ denotes change (or error)] for all possible $\delta\mathbf{b}$ and δA.

For any particular choice of norm, the *smallest* possible C in (7b) is

$$\text{cond } A = \|A\| \cdot \|A^{-1}\| = \text{the **condition number** of } A \tag{8}$$

(see Section 4.1D). Unfortunately, finding cond A exactly requires a lot more work than is needed to solve $A\mathbf{x} = \mathbf{b}$. A more useful C is an approximation of cond A that can be determined directly from an LU-decomposition of A. One, due to Conte and motivated by (6a), is

Aprorox. 1

$$C_C(A) = \frac{\hat{s}_1\hat{s}_2 \cdots \hat{s}_n}{|\hat{u}_{11}\hat{u}_{22} \cdots \hat{u}_{nn}|} = \frac{\Pi\{\text{row scale factors of } \hat{A}\}}{|\Pi\{\text{circled entries of } \hat{L}\backslash\hat{U}\}|} \tag{9}$$

An even simpler indicator is the **pivot condition number** $C_p(A)$ given by

Aprrox 2

$$C_p(A) = \frac{\hat{s}_n}{|\hat{u}_{nn}|} = \frac{\text{scale factor of row}_n\hat{A}}{|\text{circled entry of col}_n\hat{L}\backslash\hat{U}|} \tag{10}$$

Practical Consideration 4.1B (Using a condition number C). There is no "threshold" value of C in (7) that separates coefficient matrices that are ill conditioned from those that are not. What can be said is

If $C > 10^c$, then errors in the kth significant digit of $[A\!:\!\mathbf{b}]$ can (but need not) affect the $(k - c)$th significant digit of $\bar{\mathbf{x}}$. **(11)**

e.g. $k=3$
$c=2$
$10^2 = 100 \Rightarrow k-c=1$

Thus, if $C > 1000 = 10^3$ ($c = 3$), then the components of $\bar{\mathbf{x}}$ calculated on an 7s device may well have errors in their *fourth* significant digit, *regardless of n*. In addition to this, there will be propagated roundoff error that increases with *n*. ■

It should be noted that $C_p(A)$ in (10) depends on the pivoting strategy used to decompose A, just as cond A in (8) depends on norm used to get $\|A\|$; however, if SPP is used (as it should), $C_p(A)$ will be as reliable as any C when used in (11). In view of the simplicity with which $C_p(A)$ can be obtained, we shall use it exclusively to determine how ill conditioned a coefficient matrix A is.

EXAMPLE 4.1B Show how the *LU*-Factorization Algorithm 3.2E would solve the following 3×3 linear system on a 3s device. Assume that intermediate values are calculated in extended precision, but stored rounded to 3s.

$$\begin{aligned} &(E_1) & -0.0720x_1 + 0.2997x_2 - 0.2103x_3 &= 1.7667 \\ &(E_2) & 0.8736x_1 - 0.2674x_2 + 0.1327x_3 &= -1.7411 \\ &(E_3) & -0.5008x_1 - 0.1226x_2 + 0.1254x_3 &= -0.6046 \end{aligned} \qquad \text{for which } \bar{\mathbf{x}} = \begin{bmatrix} 0 \\ 8 \\ 3 \end{bmatrix} \tag{12}$$

Find $C_p(A)$. Does it predict the worst possible relative error in $\bar{\mathbf{x}}$?

SOLUTION: At the outset, the augmented matrix $[A : \mathbf{b}]$ is stored with inherent error (i.e., rounded to 3s) as

$$[A_{\text{stored}} : \mathbf{b}_{\text{stored}}] = \begin{bmatrix} -0.072 & 0.300 & -0.210 & : & 1.77 \\ 0.874 & -0.267 & 0.133 & : & -1.74 \\ -0.501 & -0.123 & 0.125 & : & -0.605 \end{bmatrix} \tag{13}$$

Since we are using fixed-precision arithmetic, we shall use Scaled Partial Pivoting. For $\text{col}_1\tilde{A}$ ($= \text{col}_1 A$ when $m = 1$), the $|\tilde{a}_{i1}|/\hat{s}_i$ ratios are

$$\frac{|\tilde{a}_{11}|}{\hat{s}_1} = \frac{0.072}{0.300} = 0.024, \qquad \frac{|\tilde{a}_{21}|}{\hat{s}_2} = \frac{0.874}{0.874} = 1, \qquad \frac{|\tilde{a}_{31}|}{\hat{s}_3} = \frac{0.501}{0.501} = 1 \tag{14a}$$

Since $\tilde{a}_{21}$ is the topmost of the two entries having the largest ratio in (14a), the SPP strategy [see (15a) of Section 3.3C] calls for a 1,2-pivot. The resulting *LU*-decomposition, which does not need a 2,3-pivot, is

$$\begin{bmatrix} \boxed{0.874} & -0.267 & 0.133 \\ \underline{-0.082} & \boxed{0.278} & -0.199 \\ \underline{-0.573} & \underline{-0.993} & \boxed{0.00339} \end{bmatrix} (\rho_1 \leftrightarrow \rho_2) \qquad \begin{aligned} &\underline{|\tilde{a}_{i2}|/\hat{s}_i \text{ ratios } (m = 2)} \\ &|\tilde{a}_{22}|/\hat{s}_2 = 0.278/0.300 \doteq 0.93 \\ &|\tilde{a}_{32}|/\hat{s}_3 = 0.276/0.501 < 0.93 \end{aligned} \tag{14b}$$

Here $\text{row}_3\hat{A} = \text{row}_3 A$. So $\hat{s}_3$ is 0.501 and the pivot condition number of $A_{3\times 3}$ is

$$C_p(A) = \frac{\hat{s}_3}{|\hat{u}_{33}|} = \frac{0.501}{0.00339} \doteq 148 \tag{15}$$

Since $C_p(A) > 10^2$, the small relative (inherent) errors in the third significant digit of A_{stored}

and $\mathbf{b}_{\text{stored}}$ can possibly get multiplied by more than 10^2 and thus produce errors in the *leading* significant digits of $\overline{\mathbf{x}}_{\text{calc}}$! To get $\overline{\mathbf{x}}_{\text{calc}}$, we complete the forback matrix $[\hat{L}\backslash\hat{U}\colon \mathbf{b}:\overline{\mathbf{c}}:\overline{\mathbf{x}}_{\text{calc}}]$, rounding entries to $3s$ as they are stored.

$$\left[\begin{array}{ccc:c:c:c} 0.874 & 0.267 & 0.133 & -1.74 & -1.74 & 0.126 \\ \underline{-0.082} & 0.278 & -0.199 & 1.77 & 1.63 & 9.37 \\ \underline{-0.573} & \underline{-0.993} & 0.00339 & -0.605 & 0.0166 & 4.90 \end{array}\right] \qquad \textbf{(16)}$$

The exact solution of (12) is $\overline{\mathbf{x}} = [0 \quad 8 \quad 3]^T$. So the calculated values of $\overline{x}_2$ and $\overline{x}_3$ do in fact have errors in their leading significant digit. The erroneously calculated nonzero $\overline{x}_1$ is the error that accumulated as a result of storing the entries of $\hat{L}\backslash\hat{U}$, $\overline{\mathbf{c}}$, and $\overline{\mathbf{x}}$ rounded to $3s$ and then using these *rounded* values in all subsequent calculations. Even greater errors would result from chopping. ■

To see that no serious mistakes were made in (16), you might wish to show that $\hat{L}\hat{U} = \hat{A}$ (i.e., A with a 1,2-pivot) to $3s$. Also, the $\overline{\mathbf{x}}_{\text{calc}}$ obtained in (16) satisfies

$$A_{\text{stored}}\overline{\mathbf{x}}_{\text{calc}} = \begin{bmatrix} -0.072 & 0.300 & -0.210 \\ 0.874 & -0.267 & 0.133 \\ -0.573 & -0.123 & 0.125 \end{bmatrix}\begin{bmatrix} 0.126 \\ 9.37 \\ 4.90 \end{bmatrix} \doteq \begin{bmatrix} 1.77 \\ -1.74 \\ -0.603 \end{bmatrix} \qquad \textbf{(17)}$$

Since $\mathbf{b}_{\text{stored}} = [1.77 \quad -1.74 \quad -0.605]^T$, our $3s$ calculation did a reasonable job of solving the *stored* system $A_{\text{stored}}\mathbf{x} = \mathbf{b}_{\text{stored}}$ (see Problem M4-1). Unfortunately, as a result of the ill conditioning of A, the exact solution of the *stored* system differs from the exact solution of the *given* system in the leading significant digit. Thus any chance of finding $\overline{\mathbf{x}} = [0 \quad 8 \quad 3]^T$ to $3s$ was lost as soon as $[A:\mathbf{b}]$ was stored with inherent error in (13)!

4.1C The Residual and Iterative Improvement

Suppose that $\overline{\mathbf{x}}_{\text{calc}}$ is the solution of $A\mathbf{x} = \mathbf{b}$ obtained on a *ks* device using the *LU*-Decomposition Algorithm 3.3C, and that $C_p(A)$ for this decomposition indicates that A is ill conditioned. Since the entries of the stored augmented matrix $[A_{\text{stored}}:\mathbf{b}_{\text{stored}}]$ have relative errors of as much as $0.5 \cdot 10^{-k}$ (see Section 1.2C), the components of $\overline{\mathbf{x}}_{\text{calc}}$ can (but need not) have relative errors of about $0.5 \cdot 10^{-k} C_p(A)$ [see (7)]. The natural check on the accuracy of $\overline{\mathbf{x}}_{\text{calc}}$ is to "plug it in," that is, to form the product $A\overline{\mathbf{x}}_{\text{calc}}$ and see how close it comes to the given target vector $\mathbf{b}$. A useful way to do this is to form

$$\mathbf{r} = \mathbf{b} - A\overline{\mathbf{x}}_{\text{calc}}, \qquad \text{the } \textbf{residual vector} \text{ for } \overline{\mathbf{x}}_{\text{calc}} \qquad \textbf{(18a)}$$

This residual vector can be used to get

$$\overline{\mathbf{e}} = \overline{\mathbf{x}} - \overline{\mathbf{x}}_{\text{calc}}, \qquad \text{the } \textbf{error vector} \text{ for } \overline{\mathbf{x}}_{\text{calc}} \qquad \textbf{(18b)}$$

because $\mathbf{r} = A\overline{\mathbf{x}} - A\overline{\mathbf{x}}_{\text{calc}} = A(\overline{\mathbf{x}} - \overline{\mathbf{x}}_{\text{calc}}) = A\overline{\mathbf{e}}$. Consequently,

$\bar{\mathbf{e}}$ is the solution of the linear system $A\mathbf{e} = \mathbf{r}$ **(18c)**

Practical Consideration 4.1C (Calculating residuals). It has been shown [50] that if SPP is used to decompose A, then $A\bar{\mathbf{x}}_{\text{calc}}$ will be close to **b** whether $\bar{\mathbf{x}}_{\text{calc}}$ is accurate or not! [See (17).] Consequently, *small* **r** *entries cannot be taken as an indication that* $\bar{\mathbf{x}}_{\text{calc}}$ *is accurate if A is ill conditioned*. Also, to avoid the inevitable subtractive cancellation in calculating **r**,

$A\bar{\mathbf{x}}_{\text{calc}}$ should be calculated and subtracted from **b** in extended precision, if possible with enough digits to allow **r** to be rounded accurately to the significant digit accuracy of the device used to obtain $\bar{\mathbf{x}}_{\text{calc}}$. **(18d)**

If the device used to obtain $\bar{\mathbf{x}}_{\text{calc}}$ does not have extended precision, **r** should be calculated on a more accurate device (perhaps, a hand calculator!) *The iterative improvement algorithm to be described next will be reliable only if* **r** *is calculated in extended precision.* ■

The algorithm suggested by (18) is described in Algorithm 4.1C. The name **iterative improvement** comes from the fact that it can be reapplied with $\bar{\mathbf{x}}_{\text{improved}}$ taken as $\bar{\mathbf{x}}_{\text{calc}}$. It should be noted that *this algorithm should not be needed unless A is known to be ill conditioned or n is very large* and a significant amount of accumulated roundoff is likely to have occurred in calculating $\bar{\mathbf{x}}_{\text{calc}}$.

ALGORITHM 4.1C. ITERATIVE IMPROVEMENT

PURPOSE: To improve the accuracy of a solution $\bar{\mathbf{x}}_{\text{calc}}$ of $A\mathbf{x} = \mathbf{b}$ obtained using an LU-decomposition $< \hat{L}\backslash\hat{U},\ \boldsymbol{\rho}_{m,p} >$ on a fixed precision device, when $\bar{\mathbf{x}}_{\text{calc}}$ is suspected of having an unacceptable amount of roundoff error.

Calculate $\mathbf{r} = \mathbf{b} - A\bar{\mathbf{x}}_{\text{calc}}$ in extended precision, then round **r** to the nominal accuracy of the device.

Reuse $< \hat{L}\backslash\hat{U},\ \boldsymbol{\rho}_{m,p} >$ to solve $A\mathbf{e} = \mathbf{r}$ as $[\hat{L}\backslash\hat{U}:\hat{\mathbf{r}}:\bar{\mathbf{c}}:\bar{\mathbf{e}}_{\text{calc}}]$.

Add $\bar{\mathbf{e}}_{\text{calc}}$ to $\bar{\mathbf{x}}_{\text{calc}}$ to get $\bar{\mathbf{x}}_{\text{improved}} = \bar{\mathbf{x}}_{\text{calc}} + \bar{\mathbf{e}}_{\text{calc}}$.

EXAMPLE 4.1C Improve the accuracy of $\bar{\mathbf{x}}_{\text{calc}}$ obtained in Example 4.1B.

SOLUTION: To get the residual vector $\mathbf{r} = \mathbf{b} - A\bar{\mathbf{x}}_{\text{calc}}$ accurately, we shall evaluate it using the *given A* and **b** in (12) rather than A_{stored} and $\mathbf{b}_{\text{stored}}$ in (13):

$$\mathbf{r} = \begin{bmatrix} 1.7667 \\ -1.7411 \\ -0.6046 \end{bmatrix} - \begin{bmatrix} -0.0720 & 0.2997 & -0.2103 \\ 0.8736 & -0.2674 & 0.1327 \\ -0.5008 & -0.1226 & 0.1254 \end{bmatrix} \begin{bmatrix} 0.126 \\ 9.37 \\ 4.90 \end{bmatrix} \qquad \textbf{(19a)}$$

$$= \begin{bmatrix} 1.7667 \\ -1.7411 \\ -0.6046 \end{bmatrix} - \begin{bmatrix} 1.7686470 \\ -1.7452344 \\ -0.5974028 \end{bmatrix} \doteq \begin{bmatrix} -0.00195 \\ 0.00413 \\ -0.00720 \end{bmatrix} \ (3s) \qquad \textbf{(19b)}$$

(Note that if we did not know better, these rather small residual entries might have led us to believe that $\overline{\mathbf{x}}_{\text{calc}}$ was reasonably accurate!) The LU-decomposition of A obtained in (14) will now be reused to solve $A\overline{\mathbf{e}} = \mathbf{r}$. We first form $\hat{\mathbf{r}}$ (in this case using $\rho_1 \leftrightarrow \rho_2$), and then find $\overline{\mathbf{c}}$ and $\overline{\mathbf{e}}_{\text{calc}}$ in the forback matrix $[\hat{L}\backslash\hat{U}:\hat{\mathbf{r}}:\overline{\mathbf{c}}:\overline{\mathbf{e}}_{\text{calc}}]$, which is

$$\begin{bmatrix} 0.874 & -0.267 & 0.133 & : & 0.00413 & : & 0.00413 & : & -0.125 \\ -0.082 & 0.278 & -0.199 & : & -0.00195 & : & -0.00161 & : & -1.37 \\ -0.573 & -0.993 & 0.00339 & : & -0.00643 & : & -0.00643 & : & -1.90 \end{bmatrix} \tag{20}$$

The $\overline{\mathbf{c}}$ and $\overline{\mathbf{e}}_{\text{calc}}$ entries were stored rounded to $3s$ as in Example 4.1B. Adding $\overline{\mathbf{e}}_{\text{calc}}$ to $\overline{\mathbf{x}}_{\text{calc}}$ gives the improved solution vector

$$\overline{\mathbf{x}}_{\text{improved}} = \begin{bmatrix} 0.126 \\ 9.37 \\ 4.90 \end{bmatrix} + \begin{bmatrix} -0.125 \\ -1.37 \\ -1.90 \end{bmatrix} \doteq \begin{bmatrix} 0.001 \\ 8.00 \\ 3.00 \end{bmatrix} \quad (3s) \tag{21}$$

This is as accurate a solution as one might expect from a $3s$ calculation. ■

One improvement of an $\overline{\mathbf{x}}_{\text{calc}}$ as inaccurate as that obtained in (20) generally does not yield as much accuracy as obtained in (21). The roundoff errors just happened to work in our favor. To show this, let us redo *only* the forward and backward substitution in (20), but storing values *chopped* rather than rounded. The result, as you can verify, is

$$\overline{\mathbf{c}} = \begin{bmatrix} 0.00413 \\ -0.00160 \\ -0.00642 \end{bmatrix}, \qquad \overline{\mathbf{e}}_{\text{calc}} = \begin{bmatrix} -0.120 \\ -1.35 \\ -1.89 \end{bmatrix}, \qquad \overline{\mathbf{x}}_{\text{improved}} = \begin{bmatrix} 0.005 \\ 8.02 \\ 3.01 \end{bmatrix} \tag{22}$$

Still less accuracy would have resulted if chopping were used for the *entire* calculation (as is done on most digital devices).

The number of improvements needed to get nearly the device accuracy depends on how close the product $\hat{L}\hat{U}$ is to the *actual* $\hat{A}$, how ill conditioned A is, and the accuracy of the original $\overline{\mathbf{x}}_{\text{calc}}$. As with any iterative process, iterative improvement should be continued until the components of $\overline{\mathbf{e}}_{\text{calc}}$ are sufficiently small compared to those of $\overline{\mathbf{x}}_{\text{calc}}$. Unless A is extremely ill conditioned, one or two iterations should suffice.

4.1D Matrix Norms and the Condition Number cond A

It is customary to use the word *norm* and the symbol "$\|\cdot\|$" to indicate the "size" or "length" of a member of a vector space. A natural indicator of the size of $\mathbf{x} = [x_1 \cdots x_n]^T$ in n-space is the **Euclidean norm** (or **2-norm**)

$$\|\mathbf{x}\|_2 = \sqrt{\mathbf{x}^T\mathbf{x}} = \sqrt{x_1^2 + x_2^2 + \cdots + x_n^2} \tag{23a}$$

Unfortunately, $\|\mathbf{x}\|_2$ is expensive to compute on a digital device. More frequently used are $\|\mathbf{x}\|_1$ and $\|\mathbf{x}\|_\infty$ defined as follows:†

†In this notation, the row scale factor $\hat{s}_n$ used to get $C_p(A)$ is $\|\text{row}_n\hat{A}\|_\infty$.

$$\|\mathbf{x}\|_1 = \sum_{i=1}^{n} |x_i| \quad \text{(the 1-norm of } \mathbf{x}) \tag{23b}$$

and

$$\|\mathbf{x}\|_\infty = \max\{|x_1|, \ldots, |x_n|\} \quad \text{(the max-norm of } \mathbf{x}) \tag{23c}$$

Let us select one of $\|\cdot\|_1$, $\|\cdot\|_2$, or $\|\cdot\|_\infty$ as a norm on n-space and denote it by simply $\|\cdot\|$. Having done this, we wish to introduce a $\|\cdot\|$ to describe the "size" of an $n \times n$ matrix. In order to be called a **norm**, $\|\cdot\|$ must satisfy

$$\|A\| \geq 0 \quad \text{and} \quad \|A\| = 0 \Leftrightarrow A = O_{n \times n} \tag{24a}$$

$$\|sA\| = |s| \, \|A\| \quad \text{for any scalar } s \tag{24b}$$

$$\|A + B\| \leq \|A\| + \|B\| \quad \text{(Triangle Inequality)} \tag{24c}$$

$$\|AB\| \leq \|A\| \, \|B\| \tag{24d}$$

for any $n \times n$ matrices A and B. There are a variety of definitions of $\|A\|$ that satisfy these conditions; however, the one that has been found to be especially useful is

$$\|A\| = \max_{\mathbf{x} \neq 0} \frac{\|A\mathbf{x}\|}{\|\mathbf{x}\|}, \quad \text{the norm of } A_{n \times n} \tag{25a}$$

It is immediate from (25a) that

$$\|A\mathbf{x}\| \leq \|A\| \cdot \|\mathbf{x}\| \quad \text{for any } \mathbf{x} \tag{25b}$$

Thus $\|A\|$ is the most $\mathbf{x}$ can be "stretched" when multiplied by A.

For a given A and a given norm on n-space, $\|A\|$ is generally not easily found using definition (25a). The following theorem provides formulas for finding $\|A\|$ when $\|\mathbf{x}\|$ is one of the norms in (26).

MATRIX NORM REPRESENTATION THEOREM. Let A be any $n \times n$ matrix.

$$\text{If } \|\mathbf{x}\| = \|\mathbf{x}\|_2, \text{ then } \|A\| = \sqrt{\hat{\lambda}}, \quad \text{where } \hat{\lambda} = \text{largest eigenvalue of } AA^T \tag{26a}$$

$$\text{If } \|\mathbf{x}\| = \|\mathbf{x}\|_1, \text{ then } \|A\| = \max\{\|\text{col}_1 A\|_1, \|\text{col}_2 A\|_1, \ldots, \|\text{col}_n A\|_1\} \tag{26b}$$

$$\text{If } \|\mathbf{x}\| = \|\mathbf{x}\|_\infty, \text{ then } \|A\| = \max\{\|\text{row}_1 A\|_1, \|\text{row}_2 A\|_1, \ldots, \|\text{row}_n A\|_1\} \tag{26c}$$

The formula in (26a) is explained and derived in Section 9.5. Derivations of (26b) and (26c) are outlined in Problems M4-6 and M4-7. In view of (24b), they can be remembered, respectively, as the maximum absolute column sum and maximum absolute row sum of A.

The following example uses the matrix A whose inverse was obtained in Example 3.4B.

EXAMPLE 4.1D For $\mathbf{x} = \begin{bmatrix} 2 \\ -5 \\ 3 \end{bmatrix}$, $A = \begin{bmatrix} -1 & 1 & -4 \\ 2 & 2 & 0 \\ 3 & 3 & 2 \end{bmatrix}$, and $A^{-1} = \frac{1}{4}\begin{bmatrix} -2 & 7 & -4 \\ 2 & -5 & 4 \\ 0 & -3 & 2 \end{bmatrix}$, find $\|\mathbf{x}\|$, $\|A\|$, and $\|A^{-1}\|$ **(a)** if $\|\mathbf{x}\| = \|\mathbf{x}\|_1$; **(b)** if $\|\mathbf{x}\| = \|\mathbf{x}\|_\infty$.

SOLUTION (a): By (24b), $\|\mathbf{x}\| = \|\mathbf{x}\|_1 = 2 + 5 + 3 = 10$. And by (26b),

$$\|A\| = \max\{6, 6, 6\} = 6 \qquad \text{and} \qquad \|A^{-1}\| = \tfrac{1}{4}\max\{4, 15, 10\} = \tfrac{15}{4}$$

SOLUTION (b): By (24c), $\|\mathbf{x}\| = \|\mathbf{x}\|_\infty = \max\{2, 5, 3\} = 5$. And by (26c),

$$\|A\| = \max\{6, 4, 8\} = 8 \qquad \text{and} \qquad \|A^{-1}\| = \tfrac{1}{4}\max\{13, 11, 5\} = \tfrac{13}{4}$$

Note that $\|A\|$ depends on the particular choice of $\|\mathbf{x}\|$ and that, unlike determinants, the norm of A^{-1} is *not* $1/\|A\|$. ■

To understand the role of matrix norms in assessing ill conditioning, consider the linear system $A\mathbf{x} = \mathbf{b}$, where A is nonsingular and $\mathbf{b} \neq \mathbf{0}$. We wish to get a quantitative description of the effect that small changes in A and $\mathbf{b}$ can have on the solution $\bar{\mathbf{x}} = A^{-1}\mathbf{b}$. To begin, suppose that A is exact but $\mathbf{b}$ has an error $\delta\mathbf{b}$. If the resulting error in $\bar{\mathbf{x}}$ is denoted by $\delta\bar{\mathbf{x}}$, then

$$A(\bar{\mathbf{x}} + \delta\bar{\mathbf{x}}) = \mathbf{b} + \delta\mathbf{b} \tag{27a}$$

Since $A(\bar{\mathbf{x}} + \delta\bar{\mathbf{x}}) = A\bar{\mathbf{x}} + A\delta\bar{\mathbf{x}} = \mathbf{b} + A\delta\bar{\mathbf{x}}$, the errors $\delta\mathbf{b}$ and $\delta\bar{\mathbf{x}}$ satisfy

$$A\delta\bar{\mathbf{x}} = \delta\mathbf{b}, \qquad \text{or equivalently,} \qquad \delta\bar{\mathbf{x}} = A^{-1}\delta\mathbf{b} \tag{27b}$$

Applying (25b) to the equations $\mathbf{b} = A\bar{\mathbf{x}}$ and $\delta\bar{\mathbf{x}} = A^{-1}\delta\mathbf{b}$, we get

$$\|\mathbf{b}\| \leq \|A\|\,\|\bar{\mathbf{x}}\| \qquad \text{and} \qquad \|\delta\bar{\mathbf{x}}\| \leq \|A^{-1}\|\,\|\delta\mathbf{b}\|$$

Rearranging these inequalities (in which $\|\mathbf{b}\| > 0$ and $\|\bar{\mathbf{x}}\| > 0$) gives

$$\frac{\|\delta\bar{\mathbf{x}}\|}{\|\bar{\mathbf{x}}\|} \leq \|A\|\,\|A^{-1}\|\,\frac{\|\delta\mathbf{b}\|}{\|\mathbf{b}\|} \tag{28a}$$

Since $\|\delta\bar{\mathbf{x}}\|/\|\bar{\mathbf{x}}\|$ and $\|\delta\mathbf{b}\|/\|\mathbf{b}\|$ are the relative errors of $\bar{\mathbf{x}} + \delta\bar{\mathbf{x}}$ and $\mathbf{b} + \delta\mathbf{b}$ (see Section 1.3A), (28a) can be interpreted as follows:

> The relative error of $\bar{\mathbf{x}} + \delta\bar{\mathbf{x}}$ can be at most $\|A\|\,\|A^{-1}\|$ times the relative error of $\mathbf{b} + \delta\mathbf{b}$.

Similarly, if A has an error δA with $\mathbf{b}$ exact, the inequality

$$\frac{\|\delta \overline{\mathbf{x}}\|}{\|\overline{\mathbf{x}} + \delta \overline{\mathbf{x}}\|} \leq \|A\| \, \|A^{-1}\| \frac{\|\delta A\|}{\|A\|} \tag{28b}$$

can be obtained and given an analogous interpretation.

There is no reason to expect the relative error in $\overline{\mathbf{x}} + \delta\overline{\mathbf{x}}$ to be any smaller than that of $A + \delta A$ or $\mathbf{b} + \delta \mathbf{b}$; that is, the bound $\|A\| \, \|A^{-1}\|$ in (28) should be at least 1. Indeed, this follows from the fact that, by (25b),

$$\|\mathbf{x}\| = \|I\mathbf{x}\| = \|A(A^{-1}\mathbf{x})\| \leq \|A\| \, \|A^{-1}\mathbf{x}\| \leq \|A\| \, \|A^{-1}\| \, \|\mathbf{x}\|$$

Thus the **condition number** of A defined by

$$\text{cond } A = \|A\| \, \|A^{-1}\| \text{ satisfies cond } A \geq 1 \text{ for any } A \tag{29}$$

We can therefore conclude the following about A in Example 4.1D.

$$\text{If } \|\mathbf{x}\| = \|\mathbf{x}\|_1, \text{ then cond } A = \|A\| \cdot \|A^{-1}\| = 6(\tfrac{15}{4}) = \tfrac{45}{2} = 22.5 \tag{30a}$$

$$\text{If } \|\mathbf{x}\| = \|\mathbf{x}\|_\infty, \text{ then cond } A = \|A\| \cdot \|A^{-1}\| = 8(\tfrac{13}{4}) = 26 \tag{30b}$$

If $\|\mathbf{x}\| = \|\mathbf{x}\|_2$ were used, cond A would have been 15.9 (3s) [this is Problem 9-39(c)]. Thus the condition number depends on the norm used for n-space. In general, *$\|A\|$ resulting from the euclidean norm $\|\mathbf{x}\|_2$ is the smallest possible value of $\|A\|$* (see Section 9.5D); hence it yields the smallest value of cond $A = \|A\|\|A^{-1}\|$. This smallest value is the "best" bound that can be used with certainty as $\|A\|\|A^{-1}\|$ in the fundamental inequalities (28a) and (28b). Unfortunately, it is difficult to find.

For A of Example 4.1D, the LU-decomposition obtained with (unscaled) Partial Pivoting has $\hat{u}_{33} = -\frac{4}{3}$ and $\hat{s}_3 = 4$ [see (19) of Section 3.3D]. From this decomposition and the one obtained using SPP in Problem 3-19, we can conclude that:

$$\text{If PP is used to decompose } A, \quad \text{then } C_p(A) = \frac{4}{|-\frac{4}{3}|} = 3 \tag{31a}$$

$$\text{If SPP is used to decompose } A, \quad \text{then } C_p(A) = \tfrac{3}{2} = 1.5 \tag{31b}$$

Thus the pivot condition number depends on the pivoting strategy and might actually be smaller than cond $A = \|A\| \cdot \|A^{-1}\|$. So the accuracy of $\overline{\mathbf{x}}_{\text{calc}}$ may actually be *worse* than indicated by $C_p(A)$.

4.2 Solving Linear Systems Directly on a Computer

No software library would be complete without at least one reliable subroutine for solving linear systems directly. In this section we describe the considerations that should be taken into account when selecting, using, or writing such a subroutine.

4.2A Suggestions for Selecting and Using a Canned Program

To begin with, try to avoid using a program for which no documentation is available, especially if it is in object code (i.e., already compiled) and you cannot see the source code. Whoever wrote it may have known a lot less than you do about solving linear systems!

Before using an available program, check to see what pivoting strategy is used. If Basic Pivoting (or no pivoting strategy) is used, the code should be modified to incorporate at least Partial Pivoting (see Practical Consideration 3.3C). If Partial Pivoting is used and the row scale factors of A (i.e., $s_1, \ldots, s_n$) differ by several orders of magnitude, you should **row equilibrate** the linear system by multiplying the equations as necessary (preferably by suitable powers† of the device's arithmetic base) to ensure that they are nearly equal before using the program. Similarly, if the column scale factors are substantially different, **column equilibration** might reduce roundoff when used with *any* pivoting strategy. Be careful, though; if $\text{col}_j A$ is multiplied by α_j, then $\bar{x}_j$ of the calculated solution must be multiplied by α_j to get the desired $\bar{x}_j$ for the given system.

If iterative improvement is used, the coefficient matrix A should be stored in extended precision so that residuals $\mathbf{r}$ can be obtained to the nominal (single) precision of the device. And if ample active memory is available or systems with even moderately large n (say, $n > 10$) are to be solved, it might be best to simply store all real variables in extended precision.

for direct methods

Finally, and perhaps most important, the program should output a condition number such as cond(A) or at least $|\det A|$ or the nth pivot entry [so that $C_C(A)$ or $C_p(A)$ can be formed] to indicate how many significant digits of the calculated solution should be accurate in the absence of propagated roundoff (see Section 4.1).

Reliable, well-documented subprograms in Fortran are included in the IMSL library, the LINPAK library developed at the Sandia Laboratory, and the ITPACK library developed at the University of Texas.

4.2B Programming the *LU*-Decomposition Algorithm

Pseudocode for a reliable general-purpose algorithm for solving linear systems is given in Algorithm 4.2B. It implements the *LU*-Decomposition Algorithm 3.3C in place using Scaled Partial Pivoting, and returns the calculated solution $\bar{\mathbf{x}}$ along with the pivot condition number $C_p(A)$ and the resulting *LU*-decomposition in the form of $< \hat{L} \backslash \hat{U}, NewRow >$ where *NewRow* is a *row pointer vector*. *NewRow* is used to avoid actually performing m,p-pivots in $\hat{A}$, $\hat{\mathbf{s}}$, and $\hat{\mathbf{b}}$; instead, it "points to" them in A, $\mathbf{s}$, and $\mathbf{b}$ according to the prescription

$$\hat{a}_{ij},\ \hat{s}_i, \text{ and } \hat{b}_i \quad \text{are stored in} \quad A(ii, j),\ \mathbf{s}(ii), \text{ and } \mathbf{b}(ii) \tag{1}$$

where $ii = NewRow(i)$ as described in Practical Consideration 3.3D.

†Multiplying (E_i) by a power of the device's arithmetic base changes the characteristic, but *not the mantissa*, of the stored entries of $\text{row}_i[A:\mathbf{b}]$. In this way, no additional inherent error is introduced by row or column equilibration.

ALGORITHM 4.2B. *LU*-DECOMPOSITION USING SCALED PARTIAL PIVOTING

PURPOSE: To solve an $n \times n$ linear system $A\mathbf{x} = \mathbf{b}$ using the Scaled Partial Pivoting Strategy to find an equivalent rearranged system $\hat{A}\mathbf{x} = \hat{\mathbf{b}}$ for which $\hat{A} = \hat{L}\hat{U}$, then solving $\hat{L}(\hat{U}\mathbf{x}) = \hat{\mathbf{b}}$ for $\bar{\mathbf{x}}$ by forward/backward substitution. A is returned as $\hat{L}\backslash\hat{U}$ along with a row pointer vector *NewRow* and $C_p(A)$.

GET n, A, $\hat{\mathbf{b}}$, {$\text{row}_{NewRow(i)}[A:\mathbf{b}:\mathbf{s}]$ will store $\text{row}_i[\hat{A}:\hat{\mathbf{b}}:\hat{\mathbf{s}}]$}
    EPS {tolerance for determining if A is nearly singular}
DO FOR $i = 1$ TO n; $NewRow(i) \leftarrow i$; $\mathbf{s}(i) \leftarrow$ largest absolute value in $\text{row}_i A$

{**FACTOR** $\hat{A}$: Reduce A in place to $\hat{L}\backslash\hat{U}$, where $\hat{L}\hat{U}$ = final $\hat{A}$}
DO FOR $m = 1$ TO n {get $\text{row}_m \hat{L}\backslash\hat{U}$ using SPP}
  MaxRat $\leftarrow 0$ {*MaxRat* will store $|\tilde{a}_{pm}|/\hat{s}_p$ = largest ratio}
  DO FOR $i = m$ TO n {get $|\tilde{a}_{im}|/\hat{s}_i$; update if largest}
    Ratio $\leftarrow |A(NewRow(i), m)|/\mathbf{s}(NewRow(i))$
    IF Ratio $>$ *MaxRat* THEN {***update***} $MaxRat \leftarrow Ratio$; $p \leftarrow i$
  IF *MaxRat* $\leqslant$ EPS THEN OUTPUT('**A* is singular*'); STOP
  IF $p \neq m$ THEN {perform an m,p-pivot in *NewRow*}
    $mp \leftarrow NewRow(p)$; $NewRow(p) \leftarrow NewRow(m)$; $NewRow(m) \leftarrow mp$
  {Reduce $\text{col}_m\tilde{A}$ in place (if $m < n$)}
  DO FOR $i = m + 1$ TO n; $ii \leftarrow NewRow(i)$
    $A(ii, m) \leftarrow A(ii, m)/A(mp, m)$ {stored *im*th multiplier}
    DO FOR $j = m + 1$ TO n {to the right of $\tilde{a}_{im}$}
      $A(ii, j) \leftarrow A(ii, j) - A(ii, m)A(mp, j)$ {0_{im}-subtract}
{A now stores $\hat{L}\backslash\hat{U}$, where $\hat{L}\hat{U} = \hat{A}$; *MaxRat* stores $|\hat{a}_{nn}|/\hat{s}_n = 1/C_p(A)$}

{**FORBACK**: Solve the equivalent system $\hat{A}\mathbf{x} = \hat{\mathbf{b}}$ for the desired $\bar{\mathbf{x}}$}
{***Forward Substitution***: Solve $\hat{L}\mathbf{c} = \hat{\mathbf{b}}$ for $\bar{\mathbf{c}}$ in storage for $\bar{\mathbf{x}}$}
DO FOR $i = 1$ TO n; $ii \leftarrow NewRow(i)$; $\mathbf{x}(i) \leftarrow \mathbf{b}(ii)$
  DO FOR $j = 1$ TO $i - 1$; $\mathbf{x}(i) \leftarrow \mathbf{x}(i) - A(ii, j)\mathbf{x}(j)$ {$\mathbf{x}(i)$ now stores $\bar{c}_i$}
{***Backward Substitution***: Solve $\hat{U}\mathbf{x} = \bar{\mathbf{c}}$ (stored in $\mathbf{x}$) in place}
DO FOR $i = n$ DOWNTO 1; $ii \leftarrow NewRow(i)$
  DO FOR $j = i + 1$ TO n; $\mathbf{x}(i) \leftarrow \mathbf{x}(i) - A(ii, j)\mathbf{x}(j)$
  $\mathbf{x}(i) \leftarrow \mathbf{x}(i)/A(ii, i)$ {$\mathbf{x}(i)$ now stores $\bar{x}_i$}

OUTPUT('The calculated $\bar{\mathbf{x}}$ is ['$\mathbf{x}(1), \ldots, \mathbf{x}(n)$']^T; $C_p(A)$ is '$1/MaxRat$)
STOP

The Fortran 77 SUBROUTINE DECOMP shown in Figure 4.2-1 implements the **FACTOR** $\hat{A}$ phase of Algorithm 4.2B. First, the arrays NEWROW and S are initialized in lines 1400–1900. The SPP strategy is then implemented in lines 2000–2800 , where (for M = 1, . . . , N) MAXRAT becomes the largest of the ratios

$$\frac{|\mathtt{A(NEWROW(I), M)}|}{\mathtt{S(NEWROW(I))}} \left(= \frac{|\tilde{a}_{im}|}{\hat{s}_i} \right), \qquad \mathtt{I = M, \ldots, N} \tag{2}$$

and P becomes the smallest index I for which A(NEWROW(I), M) ($= \tilde{a}_{im}$) has this maximum ratio. If P $\neq$ M, an M,P-pivot is made in NEWROW (lines 3100–3300) but *not* in A or S. The

```
ØØ1ØØ            SUBROUTINE DECOMP(N, MAXN, EPS, A, S, NEWROW, SNGULR, MP, CPVT)
ØØ2ØØ            INTEGER   NEWROW(N), P
ØØ3ØØ            REAL      A(MAXN,N), S(N), MAXRAT
ØØ4ØØ            LOGICAL   SNGULR
ØØ5ØØ      C - - - - - - - - - - - - - - - - - - - - - - - - - - - - - - - - - - C
ØØ6ØØ      C Given A(n,n) and EPS, this subroutine uses SPP to reduce A        C
ØØ7ØØ      C in place to (unit L^)\U^, where L^*U^ = A^.  It returns:          C
ØØ8ØØ      C        S = array of row scale factors of A                        C
ØØ9ØØ      C NEWROW = row pointer: a^(i,j)=A(NewRow(i),j); s^(i)=S(NewRow(i)) C
Ø1ØØØ      C SNGULR = .TRUE. if for some m, |a^(i,m)|/s^(i) <= EPS for i >= m C
Ø11ØØ      C     MP = (nearly?) dependent equation (when SNGULR = .TRUE.)      C
Ø12ØØ      C   CPVT = pivot condition number Cp(A) (when SNGULR = .FALSE.)     C
Ø13ØØ      C - - - - - - - - - - - - - - - - - - - - - -  VERSION 2  9/9/85  - C
Ø14ØØ            DO 1Ø I=1,N          ! initialize NEWROW and S
Ø15ØØ               NEWROW(I) = I     ! current [s^ : A^] starts as given [s : A]
Ø16ØØ               S(I) = Ø.Ø        ! becomes largest of |a(i,1)|, ..., |a(i,n)|
Ø17ØØ               DO 1Ø J=1,N
Ø18ØØ                  S(I) = AMAX1( S(I), ABS(A(I,J)) )
Ø19ØØ      1Ø CONTINUE
Ø2ØØØ            DO 4Ø M=1,N          ! A -> L^/U^; use Scaled Partial Pivoting
Ø21ØØ               MAXRAT = Ø.Ø      ! becomes largest |a^(i,m)|/s^(i), i >= m
Ø22ØØ               P = M             ! becomes row of A^ having largest ratio
Ø23ØØ               DO 2Ø I=M,N
Ø24ØØ                  II = NEWROW(I)
Ø25ØØ                  IF (ABS(A(II,M)) .GT. MAXRAT*S(II)) THEN
Ø26ØØ                     MAXRAT = ABS(A(II,M)) / S(II)
Ø27ØØ                     P = I
Ø28ØØ                  ENDIF
Ø29ØØ      2Ø      CONTINUE                       ! now MAXRAT = |a^(p,m)|/s^(p)
Ø3ØØØ               MP = NEWROW(P)
Ø31ØØ               IF (P .GT. M) THEN            ! perform an m,p-pivot in NEWROW
Ø32ØØ                  NEWROW(P) = NEWROW(M)
Ø33ØØ                  NEWROW(M) = MP
Ø34ØØ               ENDIF                         ! rowMP[A] now stores rowM[L^\U^]
Ø35ØØ               SNGULR = (MAXRAT .LT. EPS)
Ø36ØØ               IF (SNGULR) RETURN            ! A is (nearly?) singular
Ø37ØØ               IF (M .EQ. N) THEN            ! get pivot condition number:
Ø38ØØ                  CPVT = 1./MAXRAT           ! = s^(n)/|a^(n,n)| = Cp(A)
Ø39ØØ               ELSE                          ! get zeros below a^(m,m):
Ø4ØØØ                  DO 3Ø I=M+1,N              ! perform a Øi,m-subtract in A^
Ø41ØØ                     II = NEWROW(I)
Ø42ØØ                     A(II,M) = A(II,M)/A(MP,M)  ! store i,m-multiplier
Ø43ØØ                     DO 3Ø J=M+1,N
Ø44ØØ                        A(II,J) = A(II,J) - A(II,M)*A(MP,J)
Ø45ØØ      3Ø         CONTINUE
Ø46ØØ               ENDIF                         ! colM[A] now stores colM[L^\U^]
Ø47ØØ      4Ø CONTINUE
Ø48ØØ            RETURN
Ø49ØØ            END
```

Figure 4.2-1 SUBROUTINE DECOMP *Decomposes A using SPP.*

passed parameter EPS controls how small MAXRAT must be for A to be considered "nearly singular." If MAXRAT < EPS, then SNGULR is set to .TRUE. (line 3500) and MP then tells which row of the *given* A appears to be linearly dependent. Otherwise, the pivot condition number $C_p(A)$ (= 1/MAXRAT when M = N) is returned as CPVT (line 3800).

Note that the reduction $\hat{A} \to \{\tilde{A}\} \to \hat{L}\backslash\hat{U}$ in the **FACTOR** $\hat{A}$ phase is performed in place, with $\hat{L}\backslash\hat{U}$ stored over the coefficient matrix in A. So a copy of A should be made before invoking SUBROUTINE DECOMP if iterative improvement is wanted [Problem C4-1(b)].

The **FORBACK** phase of Algorithm 4.2B is implemented in SUBROUTINE FORBAK shown in Figure 4.2-2. The desired $\bar{\mathbf{x}} = A^{-1}\mathbf{b}$ is returned in the array XBAR. Notice that the array B, which stores the target vector **b**, is *not* altered by the FORBAK call. (If the forward/backward substitution were done in place in B, then the components of $\bar{\mathbf{x}}$ would

```
      SUBROUTINE FORBAK (N, NMAX, LU, NEWROW, B, XBAR)
      REAL    LU(NMAX,N), XBAR(N), B(N)
      INTEGER NEWROW(N)
      DOUBLE PRECISION DSUM     ! for Partial Extended Precision
C - - - - - - - - - - - - - - - - - - - - - - - - - - - - - - - - C
C This subroutine solves A(n,n)*X = B for XBAR by solving first  C
C L^C = B^ (forward subst'n) then U^X = CBAR (backward subst'n)  C
C using LU (= L^\U^) and NEWROW returned by SUBROUTINE DECOMP.   C
C - - - - - - - - - - - - - - - - - - - - - - - VERSION 2  9/9/85 - C
      IF (N .EQ. 1) THEN
         XBAR(1) = B(1)/LU(1,1)            ! solution of a 1x1 system
         RETURN
      ENDIF
C    *Forward Substitution                (Note that B is not altered.)
      XBAR(1) = B(NEWROW(1))
      DO 20 I = 2, N
         II = NEWROW(I)
         DSUM = DBLE(B(II))                             ! PEP strategy
         DO 10 J = 1, I-1
            DSUM = DSUM - DBLE(LU(II,J))*XBAR(J)
   10    CONTINUE
         XBAR(I) = SNGL(DSUM)
   20 CONTINUE                    ! XBAR now stores CBAR (= reduced B^)
C    *Backward Substitution
      XBAR(N) = XBAR(N)/LU(NEWROW(N),N)
      DO 40 I = N-1, 1, -1
         II = NEWROW(I)
         DSUM = DBLE(XBAR(I))                           ! PEP strategy
         DO 30 J = I+1, N
            DSUM = DSUM - DBLE(LU(II,J))*XBAR(J)
   30    CONTINUE
         XBAR(I) = SNGL(DSUM/LU(II,I))
   40 CONTINUE
      RETURN                                 ! solution returned in XBAR
      END
```

Figure 4.2-2 SUBROUTINE FORBAK *Forward/backward substitution.*

necessarily be returned in the permuted order of NEWROW rather than their natural order. Why?)

Figure 4.2-3 shows a minimal calling program for solving a single system $A\mathbf{x} = \mathbf{b}$ by Algorithm 4.2B using SUBROUTINEs DECOMP and FORBAK. SUBROUTINE GET (not shown) gets N (< 20), A(N, N), B(N), and EPS from the keyboard, a tape, a file on disk, or whatever is desired. The augmented matrix [A:B] is then printed so that the user will know the system that was solved if you look at the output a week after it was created. If $n > 20$ is needed, you need only change NMAX in the PARAMETER statement of the calling program, *without modifying* DECOMP *or* FORBAK.

If k (> 1) linear systems $A\mathbf{x}_j = \mathbf{b}_j$ are to be solved, the FORBAK call can be put in a loop, with B used to store $\mathbf{b}_j$ for $j = 1, \ldots, k$ as described in Algorithm 3.4A. And if A^{-1} is needed, it can be found by taking B to be $\text{col}_j I_n$ for $j = 1, \ldots, n$ as described in Section 3.4B [Problem C4-7]. It is often the case that a matrix to be inverted has entries that are complex numbers; if so, all appropriate REAL variables in DECOMP, FORBAK, and the calling program will have to be declared COMPLEX *explicitly* (or see Problem C4-6).

```
C - - - - - - - - - - - - - - - - - - - - - - - - - - - - - - - - - - - C
! Program to solve an nxn linear system AX = B by calling SUBROUTINEs !
!    DECOMP  to get L^\U^ (in place), NewRow, and CPVT (= Cp(A))       !
!    FORBAK  to do forward/backward substitution                       !
C - - - - - - - - - - - - - - - - - - - - - - - - - VERSION 2  9/9/85  - C
      PARAMETER (MAXN=20, IR=5, IW=5)
      REAL A(MAXN,MAXN), B(MAXN), X(MAXN)     ! MAXN is maximum size n
      LOGICAL SNGULR
      INTEGER NEWROW(MAXN), DEPROW            ! dependent row if SNGULR
      CALL GET(N, MAXN, NUMDEC, A, B)         ! get user-provided input
      EPS = 10.**(-NUMDEC)                    ! singularity tolerance
      WRITE(IW,1)                             ! echo back [A : B ]
      DO 10 I=1,N
         WRITE (IW,2) (A(I,J), J=1,N), B(I)
10    CONTINUE
      CALL DECOMP (N, MAXN, EPS, A, S, NEWROW, SNGULR, DEPROW, CPVT)
      IF (SNGULR) THEN
         WRITE (IW,3) DEPROW, EPS             ! singularity warning
         STOP
      ENDIF                                   ! A now stores L^\U^
      CALL FORBAK(N, MAXN, A, NEWROW, B, X)
      WRITE(IW,4) EPS, CPVT, (X(I), I=1,N)    ! desired output
      STOP
1     FORMAT ('0Solving the linear system AX = B, where [A : B] is'/)
2     FORMAT (2(1X, 11G12.6))
3     FORMAT ('0Row',I3' seems to be dependent.   EPS =',E8.1//)
4     FORMAT ('0EPS is 'E7.1,'    Pivot Condition Number is ',G9.3,
     &       '    Solution is:',//2(1X,10G13.7))
      END
```

Figure 4.2-3 *Simple calling program for* DECOMP *and* FORBAK.

4.2C Formulas for LU-Factorizations of A

$$\underbrace{\begin{bmatrix} l_{11} & & & \\ l_{21} & l_{22} & & \\ \vdots & \vdots & \ddots & \\ l_{n1} & l_{n2} & \cdots & l_{nn} \end{bmatrix}}_{l_{ij}=0 \text{ for } j>i} \underbrace{\begin{bmatrix} u_{11} & u_{12} & \cdots & u_{1n} \\ & u_{22} & \cdots & u_{2n} \\ & & \ddots & \vdots \\ & & & u_{nn} \end{bmatrix}}_{u_{ij}=0 \text{ for } i>j} = \begin{bmatrix} a_{11} & a_{12} & \cdots & a_{1n} \\ a_{21} & a_{22} & \cdots & a_{2n} \\ \vdots & \vdots & & \vdots \\ a_{n1} & a_{n2} & \cdots & a_{nn} \end{bmatrix} \quad \textbf{(3)}$$

This one matrix equality represents n^2 scalar equations, which can be grouped as follows:

$$\textbf{COL}_m\text{: } [l_{i1}u_{1m} + \cdots + l_{i,m-1}u_{m-1,m}] + l_{im}u_{mm} = a_{im} \qquad (\text{for } i \geqslant m) \qquad \textbf{(4a)}$$

$$\textbf{ROW}_m\text{: } [l_{m1}u_{1j} + \cdots + l_{m,j-1}u_{j-1,j}] + l_{mm}u_{mj} = a_{mj} \qquad (\text{for } j \geqslant m) \qquad \textbf{(4b)}$$

Note that the diagonal entry a_{mm} appears twice, when $i = m$ in (4a), and when $j = m$ in (4b). The sum in square brackets should be taken as zero when $m = 1$.

Since the n^2 scalar equations in (3) are in terms of the $n^2 + n$ unknown entries of L and U, we must impose n additional conditions in order to specify these unknowns uniquely. If we set all l_{mm}'s to 1, then L will be a *unit* matrix, which we shall denote by L_1. In this case we can rearrange (4) as follows:

$$\textbf{COL}_m\boldsymbol{L}_1\text{: } l_{im} = \frac{1}{u_{mm}}[a_{im} - (l_{i1}u_{1m} + \cdots + l_{i,m-1}u_{m-1,m})] \qquad (\text{for } i > m) \qquad \textbf{(5a)}$$

$$\textbf{ROW}_m\boldsymbol{U}\text{: } u_{mj} = a_{mj} - (l_{m1}u_{1j} + \cdots + l_{m,j-1}u_{j-1,j}) \qquad (\text{for } j \geqslant m) \qquad \textbf{(5b)}$$

Note that (5a) cannot be used unless u_{mm} [$j = m$ in (5b)] is known. This will be the case if the formulas (5) are used in the following order:

$$\underbrace{\textbf{row}_1\boldsymbol{U}, \textbf{col}_1\boldsymbol{L}_1}_{m=1}, \quad \underbrace{\textbf{row}_2\boldsymbol{U}, \textbf{col}_2\boldsymbol{L}_1}_{m=2}, \quad \ldots, \quad \underbrace{\textbf{row}_{n-1}\boldsymbol{U}, \textbf{col}_{n-1}\boldsymbol{L}_1}_{m=n-1}, \quad \underbrace{\textbf{row}_n\boldsymbol{U}}_{m=n} \qquad \textbf{(5c)}$$

The resulting LU-factorization, $L_1U = A$, is called the **Doolittle factorization** of A.

Alternatively, if we set all u_{mm}'s to 1, then U will be a unit matrix, which we shall denote by U_1. In this case we can rearrange (4) as follows:

$$\textbf{COL}_m\boldsymbol{L}\text{: } l_{im} = a_{im} - [l_{i1}u_{1m} + \cdots + l_{i,m-1}u_{m-1,m}] \qquad (\text{for } i \geqslant m) \qquad \textbf{(6a)}$$

$$\textbf{ROW}_m\boldsymbol{U}_1\text{: } u_{mj} = \frac{1}{l_{mm}}[a_{mj} - (l_{m1}u_{1j} + \cdots + l_{m,j-1}u_{j-1,j})] \qquad (\text{for } j > m) \qquad \textbf{(6b)}$$

In order to use (6b), we must *first* know l_{mm} [$i = m$ in (6a)]. To ensure this, the formulas (6) must be used in the order

$$\underbrace{\mathbf{col}_1 L,\ \mathbf{row}_1 U_1}_{m\,=\,1},\quad \underbrace{\mathbf{col}_2 L,\ \mathbf{row}_2 U_1}_{m\,=\,2},\quad \ldots,\quad \underbrace{\mathbf{col}_{n-1} L,\ \mathbf{row}_{n-1} U_1}_{m\,=\,n-1},\quad \underbrace{\mathbf{col}_n L}_{m\,=\,n} \qquad \textbf{(6c)}$$

The resulting LU-factorization, $LU_1 = A$, is called the **Crout factorization** of A.

The entries obtained using (5) can be stored in a "compact form" $L_1\backslash U$ (read as "$\mathbf{L_1}$ **under** $\mathbf{U}$") with the u_{mm} entries "covering" the 1's on the diagonal of L_1. Similarly, the entries obtained using (6) can be stored in a "compact form" $L\backslash U_1$ (read as "$\mathbf{L}$ **over** $\mathbf{U_1}$") with the l_{mm} entries "covering" the 1's on the diagonal of U_1. For example,

$$A = \begin{bmatrix} -1 & 1 & -4 \\ 2 & 2 & 0 \\ 3 & 3 & 2 \end{bmatrix} \quad \text{has} \quad L_1\backslash U = \begin{bmatrix} \boxed{-1} & 1 & -4 \\ -2 & \boxed{4} & -8 \\ -3 & \frac{3}{2} & \boxed{2} \end{bmatrix}, \qquad L\backslash U_1 = \begin{bmatrix} \boxed{-1} & -1 & 4 \\ 2 & \boxed{4} & -2 \\ 3 & 6 & \boxed{2} \end{bmatrix} \qquad \textbf{(7)}$$

as the compact forms of its Doolittle (unit L) and Crout (unit U) factorizations; you can check this by forming the products L_1U and LU_1, respectively. The dashed lines in $L_1\backslash U$ and $L\backslash U_1$ indicate the order in (5c) and (6c), respectively. In general, the compact form "$L\backslash U$" obtained by reducing A in the FACTOR $\hat{A}$ phase of the LU-factorization Algorithm 3.2E will be $L_1\backslash U$. In fact, $L_1\backslash U$ in (7) was obtained this way in Example 3.2D.

4.2D The Crout Decomposition Algorithm with SPP (Optional)

The FACTOR $\hat{A}$ phase LU-Decomposition Algorithm 3.3C with any pivoting strategy can be modified so as to use the Crout Factorization† formulas (6) to get the entries of the compact form

$$\hat{L}\backslash\hat{U}_1 = \begin{bmatrix} \boxed{\hat{l}_{11}} & \hat{u}_{12} & \hat{u}_{13} & \cdots & \hat{u}_{1n} \\ \hat{l}_{21} & \boxed{\hat{l}_{22}} & \hat{u}_{23} & \cdots & \hat{u}_{2n} \\ \hat{l}_{31} & \hat{l}_{32} & \boxed{\hat{l}_{33}} & \cdots & \hat{u}_{3n} \\ \vdots & \vdots & \vdots & & \vdots \\ \hat{l}_{n-1,1} & \hat{l}_{n-1,2} & \hat{l}_{n-1,3} & \cdots & \hat{u}_{n-1,n} \\ \hat{l}_{n1} & \hat{l}_{n2} & \hat{l}_{n3} & \cdots & \boxed{\hat{l}_{nn}} \end{bmatrix} \begin{matrix} \{\mathrm{row}_1 \hat{U}_1 \\ \{\mathrm{row}_2 \hat{U}_1 \\ \{\mathrm{row}_3 \hat{U}_1 \\ \vdots \\ \{\mathrm{row}_{n-1} \hat{U}_1 \\ \\ \end{matrix} \qquad \textbf{(8)}$$

$$\qquad\quad \mathrm{col}_1\hat{L} \qquad \mathrm{col}_2\hat{L} \qquad \mathrm{col}_3\hat{L} \qquad \cdots \qquad \mathrm{col}_n\hat{L}$$

Pseudocode for the resulting **Crout Decomposition Algorithm** (with Scaled Partial Pivoting) is given in Algorithm 4.2D. For ease of reading, the entries of $L\backslash U_1$ are referred to as l_{im} and u_{mj}, although they are usually stored over a_{im} ($i \geqslant m$) and a_{mj} ($j > m$), respectively; also, a row pointer vector is not implemented, and the **FORBACK** phase is not done in

†The ordering in (6c) (i.e., get $\mathrm{col}_m L$ *before* $\mathrm{row}_m U$) makes it possible to use *any* pivoting strategy of Section 3.3C to select the mth pivot of $\hat{L}\backslash\hat{U}_1$ from among $\hat{l}_{mm}, \ldots, \hat{l}_{nm}$ exactly as it was used there to select the mth pivot $\hat{L}\backslash\hat{U}$ ($= \hat{L}_1\backslash\hat{U}$) from among $\tilde{a}_{mm}, \ldots, \tilde{a}_{nm}$. The ordering (5c) for Doolittle's factorization cannot get $\mathrm{col}_m\hat{L}_1$ until *after* $\mathrm{row}_m\hat{U}$ is obtained; this defeats the purpose of row pivoting, which is to get "relatively small" $\hat{U}$ entries (see Section 3.3C).

place (as in Algorithm 4.2B). The forward and backward substitution formulas used with $\hat{L}\backslash\hat{U}_1$ are slightly different than those used (with $\hat{L}_1/\hat{U}$) in Algorithm 4.2B. However, both algorithms require the same number of arithmetic operations to solve $A_{n\times n}\mathbf{x} = \mathbf{b}$.

ALGORITHM 4.2D. CROUT DECOMPOSITION (WITH SCALED PARTIAL PIVOTING)

PURPOSE: To solve an $n \times n$ linear system $A\mathbf{x} = \mathbf{b}$ by using the Scaled Partial Pivoting Strategy to find the compact form $\hat{L}\backslash\hat{U}_1$ for which $\hat{L}\backslash\hat{U}_1 = \hat{A}$, then solving the equivalent (rearranged) system $\hat{L}(\hat{U}_1\mathbf{x}) = \hat{\mathbf{b}}$ by forward/backward substitution.

```
GET n, A, b;   Â ← A                    {L̂\Û₁ will be formed in place over Â}
DO FOR i = 1 TO n;   ŝᵢ ← largest absolute value in rowᵢÂ

{FACTOR Â: Get L̂ and Û₁ for which L̂\Û₁ equals the final Â.}
DO FOR m = 1 TO n                     {get colₘL̂ using SPP, then get rowₘÛ₁}
    MaxRat ← 0                       {MaxRat will store |l̂ₚₘ|/ŝₚ = largest ratio}
    {Get colₘL̂ using (10a), and check for largest ratio}
    DO FOR i = m TO n;   l̂ᵢₘ ← âᵢₘ − Σ_{j<i} l̂ᵢⱼûⱼₘ
    |    IF |l̂ᵢₘ|/ŝᵢ > MaxRat THEN                       {update MaxRat and p}
    |_   |_   MaxRat ← |l̂ᵢₘ|/ŝᵢ;   p ← i
    IF MaxRat = 0 THEN {colₘL̂ = 0} OUTPUT('A is singular');   STOP
    IF p ≠ m THEN perform an m,p-pivot in [ŝ: Â]
    {Get rowₘÛ₁ using (10b) (if m < n)}
    DO FOR j = m + 1 TO n;   ûₘⱼ ← (âₘⱼ − Σ_{k<m} l̂ₘₖûₖⱼ)/l̂ₘₘ
{Now Â stores L̂\Û₁, and MaxRat stores |l̂ₙₙ|/ŝₙ = 1/Cₚ(A).}

{FORBACK: Solve the equivalent system Âx = b̂ for the desired x̄.}
{Forward substitution: Solve L̂c = b̂ using (5a) of Section 3.2B
Perform the m,p-pivots of the FACTOR Â phase on b to get b̂.
DO FOR i = 1 TO n;   c̄ᵢ ← (b̂ᵢ − Σ_{j<i} l̂ᵢⱼc̄ⱼ)/l̂ᵢᵢ
{Backward substitution: Solve Û₁x = c̄ using (2b) of Section 3.2B.}
DO FOR i = n DOWNTO 1;   x̄ᵢ ← c̄ᵢ − Σ_{j>i} ûᵢⱼx̄ⱼ
OUTPUT('The calculated x̄ is ['x̄₁, ..., x̄ₙ']ᵀ, and Cₚ(A) is '1/MaxRat)
STOP
```

EXAMPLE 4.2D Use Algorithm 4.2D to solve the system $A\mathbf{x} = \mathbf{b}$, where

$$\begin{aligned} (E_1)\ & -x_1 + x_2 - 4x_3 = 0 \\ (E_2)\ & 2x_1 + 2x_2 = 1 \\ (E_3)\ & 3x_1 + 3x_2 + 2x_3 = \tfrac{1}{2} \end{aligned} \qquad \left([s\!:\!A\!:\!\mathbf{b}] = \begin{bmatrix} 4 & : & -1 & 1 & -4 & : & 0 \\ 2 & : & 2 & 2 & 0 & : & 1 \\ 3 & : & 3 & 3 & 2 & : & \tfrac{1}{2} \end{bmatrix}\right) \tag{9}$$

SOLUTION: We begin by initializing $[\hat{A}:\hat{\mathbf{b}}]$ as $[A:\mathbf{b}]$. For $m = 1$, formula (6a) reduces to $\text{col}_1\hat{L} = \text{col}_1\hat{A}$. Hence

$$\text{col}_1\hat{L} = \begin{bmatrix} -1 \\ 2 \\ 3 \end{bmatrix}, \qquad \text{for which} \quad \begin{aligned} |\hat{l}_{11}|/\hat{s}_1 &= |-1|/4 = \tfrac{1}{4} \\ |\hat{l}_{21}|/\hat{s}_2 &= |2|/2 = 1 \\ |\hat{l}_{31}|/\hat{s}_3 &= |3|/3 = 1 \end{aligned} \tag{10a}$$

In view of the "tie" for largest $|\hat{l}_{i1}|/\hat{s}_i$ ratio, the SPP strategy calls for a 1,2-pivot (in *both* $[\hat{s}:\hat{A}]$ and $\hat{L}\backslash\hat{U}_1$). The result is

$$[\hat{s}:\hat{A}] = \begin{bmatrix} 2 & : & 2 & 2 & 0 \\ 4 & : & -1 & 1 & -4 \\ 3 & : & 3 & 3 & 2 \end{bmatrix} \quad \text{and} \quad \hat{L}\backslash\hat{U}_1 = \begin{bmatrix} 2 & \hat{u}_{12} & \hat{u}_{13} \\ -1 & \hat{l}_{22} & \hat{}_{23} \\ 3 & \underbrace{\hat{l}_{32}}_{\text{col}_2\hat{L}} & \underbrace{\hat{l}_{33}}_{\text{col}_3\hat{L}} \end{bmatrix} \begin{matrix} \{\text{row}_1\hat{U}_1 \\ \{\text{row}_2\hat{U}_1 \\ \\ \end{matrix} \qquad \textbf{(10b)}$$

$\hat{L}\backslash\hat{U}_1$ in (10b) is the partial compact form of the Crout factorization of the current $\hat{A}$ in (10b). The rest of the $m = 1$ calculation using (6b)] is

$$\textbf{ROW}_1\hat{\boldsymbol{U}}_1\textbf{:} \quad \hat{u}_{12} = \frac{\hat{a}_{12}}{\hat{l}_{11}} = \frac{2}{2} = 1 \quad \text{and} \quad \hat{u}_{13} = \frac{\hat{a}_{13}}{\hat{l}_{11}} = 0 \qquad \textbf{(10c)}$$

For $m = 2$, we begin by finding $\text{col}_2\hat{L}$ [using (6a)]. The details are

$$\hat{l}_{22} = \hat{a}_{22} - \hat{l}_{21}\hat{u}_{12} = 1 - (-1)1 = 2; \qquad \hat{l}_{32} = \hat{a}_{32} - \hat{l}_{31}\hat{u}_{12} = 3 - 3\cdot 1 = 0 \qquad \textbf{(10d)}$$

For the SPP strategy, we must compare the $|\hat{l}_{i2}|/\hat{s}_i$ ratios

$$\frac{|\hat{l}_{22}|}{\hat{s}_2} = \frac{|2|}{4} \quad \text{and} \quad \frac{|\hat{l}_{32}|}{\hat{s}_3} = \frac{|0|}{3} \qquad \textbf{(10e)}$$

Since $\frac{1}{2} > 0$, we take $\hat{l}_{22} = 2$ as the second pivot entry (a $2,p$-pivot is not necessary). The remaining $m = 2$ calculation [using (6b)] is

$$\textbf{ROW}_2\hat{\boldsymbol{U}}_1\textbf{:} \quad \hat{u}_{23} = \frac{\hat{a}_{23} - (\hat{l}_{21}\hat{u}_{13})}{\hat{l}_{22}} = \frac{-4 - (-1)\cdot 0}{2} = -2 \qquad \textbf{(10f)}$$

Finally, when $m = n = 3$, the only calculation needed is

$$\textbf{COL}_3\hat{\boldsymbol{L}}\textbf{:} \quad \hat{l}_{33} = \hat{a}_{33} - (\hat{l}_{31}\hat{u}_{13} + \hat{l}_{32}\hat{u}_{23}) = 2 - [3\cdot 0 + 0(-2)] = 2 \qquad \textbf{(10g)}$$

This concludes the **FACTOR** $\hat{A}$ phase. The **FORBACK** phase is summarized in the forback matrix

$$[\hat{L}\backslash\hat{U}_1:\hat{\mathbf{b}}:\bar{\mathbf{c}}:\bar{\mathbf{x}}] = \begin{bmatrix} \textcircled{2} & 1 & 0 & : & 1 & : & \frac{1}{2} & : & \frac{5}{4} \\ -1 & \textcircled{2} & -2 & : & 0 & : & \frac{1}{4} & : & -\frac{3}{4} \\ 3 & 0 & \textcircled{2} & : & \frac{1}{2} & : & -\frac{1}{2} & : & -\frac{1}{2} \end{bmatrix} \qquad \textbf{(11)}$$

First $\hat{\mathbf{b}}$ was obtained by performing a 1,2-pivot on $\mathbf{b}$ in (9); then $\bar{\mathbf{c}}$ and $\bar{\mathbf{x}}$ were obtained using (2a) and (5b) of Section 3.2C, respectively. The calculation in (11) solved the equivalent system $\hat{A}\mathbf{x} = \hat{\mathbf{b}}$, where $\hat{A}$ is the "final" coefficient matrix shown in (10b). You might wish to compare this solution to that of Example 3.3C, where Algorithm 3.3C was used (with SPP) to solve the same problem. ∎

Practical Consideration 4.2D (Error-reducing strategies). If extended precision is available, the summations in the formulas for the entries of $\hat{L}\backslash\hat{U}_1$, $\bar{\mathbf{c}}$ and $\bar{\mathbf{x}}$ in Algorithm 4.2D can be *accumulated in extended precision*, and then *stored in single precision*. This is the Partial Extended Precision (PEP) strategy of Section 1.4C. It can be implemented using *one* extended precision variable, say DSUM, to accumulate the summations of *all* entries of $\hat{L}\backslash\hat{U}_1$, $\bar{\mathbf{c}}$, *and* $\bar{\mathbf{x}}$! The solution obtained using Algorithm 4.2D with this strategy is generally about as accurate as that obtained by performing the *LU*-Decomposition Algorithm 4.2B *entirely* in extended precision and then rounding the solution to the single precision.

Unfortunately, the reduction of A to $\hat{L}_1\backslash\hat{U}$ in Algorithm 4.2B requires that the A entries be *restored after each* 0_{im}*-subtract*. So all n^2 entries of A would have to be stored in extended precision to implement the PEP strategy. This makes Algorithm 4.2D with PEP more accurate than Algorithm 4.2B when extended precision is available but there is not enough storage to use it to store A. On the other hand, *Scaled Full Pivoting* (Section 3.3F) is not possible in performing the Crout Decomposition Algorithm 4.2D. (Why?)

Roundoff error of either Algorithm 4.2B or 4.2D can be further reduced by storing *both* the $\hat{l}_{im}$ *and* $\hat{u}_{mj}$ entries **without** dividing by the circled pivot entry. This **symmetric compact form** of the *LU*-factorization of A will be denoted by $L\backslash D\backslash U$ (D for "diagonal"). For example, for the 3×3 matrix A in (7):

$$\underset{A}{\begin{bmatrix} -1 & 1 & -4 \\ 2 & 2 & 0 \\ 3 & 3 & 2 \end{bmatrix}} \quad \underset{L_1\backslash U}{\begin{bmatrix} \textcircled{-1} & 1 & -4 \\ -2 & \textcircled{4} & -8 \\ -3 & \frac{3}{2} & \textcircled{2} \end{bmatrix}} \quad \underset{L\backslash U_1}{\begin{bmatrix} \textcircled{-1} & -1 & 4 \\ 2 & \textcircled{4} & -2 \\ 3 & 6 & \textcircled{2} \end{bmatrix}} \quad \underset{L\backslash D\backslash U}{\begin{bmatrix} \textcircled{-1} & 1 & -4 \\ 2 & \textcircled{4} & -8 \\ 3 & 6 & \textcircled{2} \end{bmatrix}} \qquad \textbf{(12)}$$

If we view the (circled) D entries as $l_{11}, \ldots, l_{nn}$ (i.e., the diagonal entries of L), then the formulas for getting $L\backslash D\backslash U$ are

$$\textsc{col}_m \boldsymbol{L\backslash D}\textbf{:} \quad l_{im} \leftarrow a_{im} - \Sigma_{j<i}\, l_{ij}u_{jm}/l_{jj} \qquad (i \geqslant m) \qquad \textbf{(13a)}$$

$$\textsc{row}_m \boldsymbol{U}\textbf{:} \quad u_{mj} \leftarrow a_{mj} - \Sigma_{k<m}\, l_{mk}u_{kj}/l_{kk} \qquad (j > m) \qquad \textbf{(13b)}$$

The reason these formulas propagate less error is this: When a computer executes an arithmetic expression in a single statement, the calculations are performed in extended precision on its accumulator. *Roundoff error does not actually propagate until intermediate answers are stored rounded (or more likely chopped) to the devices nominal precision*. So a term such as $l_{ij}u_{jm}/l_{jj}$ in (13a) will be calculated with less roundoff than the comparable term $l_{ij}u_{jm}$ in (6a) because the u_{jm} in (6a) is the *stored* value of u_{jm}/l_{mm} in (13a). However, the price you pay for the extra bit of accuracy is a much slower execution time [because each u-term in (13) must be divided by a pivot entry]. So $L\backslash D\backslash U$ should be considered only if every bit of accuracy is desired *and* a relatively long execution time can be tolerated. ■

4.2E Choleski's Factorization for Positive Definite Matrices

If A is symmetric (i.e., if $A^T = A$), then substituting the n conditions

$$l_{ii} = u_{ii}, \qquad i = 1, 2, \ldots, n \qquad \textbf{(14)}$$

along with the symmetry conditions $a_{ji} = a_{ij}$ in the general LU-factorization equations (4) yields the **Choleski factorization** $A = LU$, where $U = L^T$. The L entries l_{im} $(= u_{mi})$ are obtained as follows for $m = 1, \ldots, n$:

$$\textbf{DIAG}_m\textbf{L:}\quad l_{mm} = \sqrt{a_{mm} - \Sigma_{i<m}\, l_{mj}^2} \qquad \textbf{(15a)}$$

$$\textbf{COL}_m\textbf{L:}\quad l_{im} = \frac{1}{l_{mm}}(a_{im} - \Sigma_{j<i}\, l_{ij}l_{mj}), \qquad i < m \qquad \textbf{(15b)}$$

Practical Consideration 4.2E (Using Choleski's factorization). The Choleski factorization will succeed only if A is *positive definite*, in which case the expression under the radical in (15a) will be positive for all m. Positive definite coefficient matrices arise frequently in engineering and statistics. Since the Choleski factorization can be obtained with about half the computation needed to get the Doolittle or Crout factorization, it is worthwhile to have a separate subprogram for it. In view of Practical Consideration 3.3C, a program implementing (15) can use Basic Pivoting as shown in Algorithm 4.2E. ■

ALGORITHM 4.2E. CHOLESKI FACTORIZATION

PURPOSE: To solve an $n \times n$ linear system $A\mathbf{x} = \mathbf{b}$ when A is positive definite by finding its Choleski decomposition $A = LL^T$ and then solving $L(L^T\mathbf{x}) = \mathbf{b}$ by forward/backward substitution.

GET n, A, $\mathbf{b}$ {**NOTE:** You only need the lower triangular part of A.}

{**FACTOR** A: Get L for which $LL^T = A$.}
DO FOR $m = 1$ TO n {get col$_m L$ entries using (15)}
  Under $\leftarrow a_{mm} - \Sigma_{i<m}\, l_{mj}^2$
  IF *Under* $\leqslant 0$ THEN OUTPUT('A is not positive definite.'); STOP
  $l_{mm} \leftarrow \sqrt{Under}$
  DO FOR $i = m + 1$ TO n; $l_{im} \leftarrow \dfrac{1}{l_{mm}}(a_{im} - \Sigma_{j<i}\, l_{ij}l_{mj})$

{**FORBACK:** Solve $LL^T\mathbf{x} = \mathbf{b}$ for the desired $\bar{\mathbf{x}}$.}
{***Forward substitution:*** Solve $L\mathbf{c} = \mathbf{b}$ using (5a) of Section 3.2B.}
DO FOR $i = 1$ TO n; $\bar{c}_i \leftarrow \dfrac{1}{l_{ii}}(b_i - \Sigma_{j<i} l_{ij}\bar{c}_j)$
{***Backward substitution:*** Solve $L^T\mathbf{x} = \bar{\mathbf{c}}$ using (2a) of Section 4.2B.}
DO FOR $i = n$ DOWNTO 1; $\bar{x}_i \leftarrow \bar{c}_i - \Sigma_{j>i}\, l_{ij}\bar{x}_j$

OUTPUT('The calculated $\bar{\mathbf{x}}$ is ['$\bar{x}_1, \ldots, \bar{x}_n$']^T and $C_p(A)$ is '$1/MaxRat$)
STOP

4.3 Iterative Methods for Solving Linear Systems

Recall that a direct (i.e., factorization or elimination) method requires about $2n^3/3$ arithmetic operations to solve an $n \times n$ linear system $A\mathbf{x} = \mathbf{b}$. This limits the size of the systems for which direct methods can realistically be used. For example, it would take the fastest known computer days to perform the $0.7 \cdot 10^{12}$ operations needed to solve a 10,000 variable system; and the roundoff error that would accumulate is more than likely to make the result meaningless anyway!

Linear systems with n between 1000 and 100,000 do actually arise in the solution of differential equations. The coefficient matrices for these systems are generally **sparse** (i.e., most of the entries are zero). This desirable property (which makes it feasible to even store A for such large n) is usually destroyed by direct methods unless A is banded. Therefore, an alternative approach must be found to solve large sparse systems. The *iterative* methods discussed in this section provide the desired alternative. They preserve sparseness and can achieve a high degree of accuracy, even for large n.

4.3A The Jacobi and Gauss–Seidel Methods

The vector equation $A\mathbf{x} = \mathbf{b}$ represents the n linear equations

$$(\mathrm{E}_i)\ a_{i1}x_1 + a_{i2}x_2 + \cdots + a_{in}x_n = b_i, \qquad i = 1, 2, \ldots, n \tag{1a}$$

If all diagonal entries of A are nonzero, we can solve (E_i) for

$$x_i = \frac{1}{a_{ii}}\left[b_i - \sum_{j\neq i} a_{ij}x_j\right], \qquad i = 1, \ldots, n \tag{1b}$$

It follows that $\bar{\mathbf{x}} = [\bar{x}_1 \ \cdots \ \bar{x}_n]^T$ satisfies $A\mathbf{x} = \mathbf{b}$ if and only if it satisfies

$$\mathbf{x} = \mathbf{g}(\mathbf{x}) \qquad \text{(the \textbf{fixed-point equation})} \tag{2a}$$

where the vector function $\mathbf{g}(\mathbf{x}) = [g_1(\mathbf{x}) \ \cdots \ g_n(\mathbf{x})]^T$ is defined by

$$g_i(\mathbf{x}) = \frac{1}{a_{ii}}\left(b_i - \sum_{j\neq i} a_{ij}x_j\right), \qquad i = 1, 2, \ldots, n \tag{2b}$$

Jacobi's method (also called the **Method of Simultaneous Displacements**) for solving $A\mathbf{x} = \mathbf{b}$ consists of solving (2a) by *Repeated Substitution*, that is, by starting with an initial guess $\mathbf{x}$, then repeatedly replacing $\mathbf{x}$ by $\mathbf{g}(\mathbf{x})$, as follows:

$$\mathbf{x} \leftarrow \mathbf{g}(\mathbf{x}) \qquad \{\text{simultaneous displacement}\} \tag{3a}$$

until $\mathbf{x}$ and $\mathbf{g}(\mathbf{x})$ are sufficiently close. As might be expected from our experience in Section 2.1D, this iteration may not converge, even with a good initial guess, and when it does, it does so only linearly.

A slight improvement of Jacobi's method is **Gauss–Seidel iteration** (also called the **Method of Successive Displacements**), which replaces x_i by $g_i(\mathbf{x})$ in (2b) *as soon as it is obtained*, as follows:

DO FOR $i = 1$ TO n

$$x_i \leftarrow \frac{1}{a_{ii}}\left(b_i - \sum_{j\neq i} a_{ij}x_j\right) \qquad \{\text{successive displacement}\} \tag{3b}$$

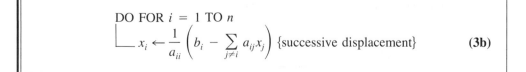

Under which conditions does GS converge?

The x_i that results in (3b) causes $\mathbf{x} = [x_1 \ \ x_2 \ \ \cdots \ \ x_n]^T$ to satisfy (E_i) by construction; so this $\mathbf{x}$ might be expected to be closer to $\bar{\mathbf{x}}$ than the $\mathbf{x}$ used to get x_i (i.e., *before* the replacement). In fact, it has been shown [47] that *Gauss–Seidel iteration generally converges whenever Jacobi's method does, and faster*. Since Gauss–Seidel iteration is also easier to implement on a computer (why?), our discussion of iterative methods will be restricted to Gauss–Seidel iteration.

If Gauss–Seidel iteration converges, it usually does so only linearly. Moreover, whether or not convergence will occur depends on the order in which the equations are solved and the variables for which thay are solved, as the following example shows.

EXAMPLE 4.3A Use Gauss–Seidel iteration to solve the linear system

$$\begin{aligned} (E_1)\ 1x_1 + 0x_2 + 3x_3 &= 2 \\ (E_2)\ 5x_1 + 1x_2 + 2x_3 &= -5 \\ (E_3)\ 1x_1 + 6x_2 + 2x_3 &= -11 \end{aligned} \qquad \left(\text{exact solution is } \bar{\mathbf{x}} = \begin{bmatrix} -1 \\ -2 \\ 1 \end{bmatrix}\right) \tag{4}$$

SOLUTION: Since all diagonal coefficients are nonzero, we can solve (E_i) for x_i, $i = 1, 2, 3$, to get the following iterative step:

$$\begin{aligned} x_1 &\leftarrow \tfrac{1}{1}[2 - 0x_2 - 3x_3] \quad \{\text{displace } x_1\} \\ \text{then } x_2 &\leftarrow \tfrac{1}{1}[-5 - 5x_1 - 2x_3] \quad \{\text{displace } x_2\} \\ \text{then } x_3 &\leftarrow \tfrac{1}{2}[-11 - 1x_1 - 6x_2] \quad \{\text{displace } x_3\} \end{aligned} \tag{5}$$

Starting with $\mathbf{x} = \mathbf{x}_0 = [0 \ \ 0 \ \ 0]^T$, we find that one iteration of (5) gives

$$\left.\begin{aligned} x_1 &= \tfrac{1}{1}[2 - 0\cdot 0 - 3\cdot 0] = 2 \\ x_2 &= \tfrac{1}{1}[-5 - 5\cdot 2 - 2\cdot 0] = -15 \\ x_3 &= \tfrac{1}{2}[-11 - 1\cdot 2 - 6(-15)] = 38.5 \end{aligned}\right\} \Rightarrow \quad \bar{\mathbf{x}}_1 = \begin{bmatrix} 2 \\ -15 \\ 38.5 \end{bmatrix}$$

The second iteration, starting with $\mathbf{x} = \mathbf{x}_1$, gives

$$\left.\begin{aligned} x_1 &= \tfrac{1}{1}[2 - 0(-15) - 3(38.5)] = -113.5 \\ x_2 &= \tfrac{1}{1}[-5 - 5(-113.5) - 2(38.5)] = 485.5 \\ x_3 &= \tfrac{1}{2}[-11 - 1(-113.5) - 6(485.5)] = 1405.25 \end{aligned}\right\} \Rightarrow \quad \mathbf{x}_2 = \begin{bmatrix} -113.5 \\ 485.5 \\ 1405.25 \end{bmatrix}$$

Clearly, $\mathbf{x}_k$ is diverging. Suppose, however, that instead of simply solving (E_i) for $\mathbf{x}_i$, we solve (E_3) for x_2, then (E_1) for x_3, then (E_2) for (x_1) to get the following iterative step:

$$\begin{aligned} x_2 &\leftarrow \tfrac{1}{6}[-11 - 1x_1 - 2x_3] \quad \{\text{displace } x_2\} \\ \text{then} \quad x_3 &\leftarrow \tfrac{1}{3}[\ 2 - 1x_1 - 0x_2] \quad \{\text{displace } x_3\} \\ \text{then} \quad x_1 &\leftarrow \tfrac{1}{5}[\ -5 - 1x_2 - 2x_3] \quad \{\text{displace } x_1\} \end{aligned} \tag{6}$$

Starting with $\mathbf{x} = \mathbf{x}_0 = \mathbf{0}$ again, one iteration of (6) gives

$$\left.\begin{aligned} x_2 &= \tfrac{1}{6}[-11 - 1\cdot 0 - 2\cdot 0] = -\tfrac{11}{6} \\ x_3 &= \tfrac{1}{3}[\ 2 - 1\cdot 0 - 0(-\tfrac{11}{6})] = \tfrac{2}{3} \\ x_1 &= \tfrac{1}{5}[-5 - 1(-\tfrac{11}{6}) - 2(\tfrac{2}{3})] = -\tfrac{9}{10} \end{aligned}\right\} \Rightarrow \quad \mathbf{x}_1 \doteq \begin{bmatrix} -0.90000 \\ -1.83333 \\ 0.66667 \end{bmatrix}$$

Note the order of the components of $\mathbf{x}_1$! A second iteration gives

$$x_2 = \tfrac{1}{6}[-11 - 1(-\tfrac{9}{10}) - 2(\tfrac{2}{3})] = -\tfrac{343}{180},$$
$$x_3 = \tfrac{1}{3}[\,2 - 1(-\tfrac{9}{10}) - 0(-\tfrac{343}{180})] = \tfrac{29}{30}, \quad \text{etc.}$$

Four iterations of (6) are shown in Table 4.3-1. The $\mathbf{x}_k$ components have been *recorded* rounded to 5*d*. However, the *stored* (i.e., most accurate) *not* rounded values of x_i were used to get successive $\mathbf{x}_k$'s. This time $\mathbf{x}_k$ (with components obtained in the order x_2, x_3, x_1) is converging to $\bar{\mathbf{x}} = [-1 \quad -2 \quad 1]^T$.

TABLE 4.3-1. Four Gauss–Seidel Iterations of (6) with $\mathbf{x}_0 = \mathbf{0}$

	$k = 0$	$k = 1$	$k = 2$	$k = 3$	$k = 4$
x_1	0	−0.90000	−1.00556	−1.00315	−1.00040
x_2	0	−1.83333	−1.90556	−1.98796	−2.00009
x_3	0	0.66667	0.96667	1.00185	1.00105

■

ALGORITHM 4.3A. GAUSS–SEIDEL ITERATION (METHOD OF SUCCESSIVE DISPLACEMENTS)

PURPOSE: To solve an $n \times n$ linear system $A\mathbf{x} = \mathbf{b}$ for $\bar{\mathbf{x}}$ to *NumDig* accurate digits. First set $\mathbf{x} = [x_1 \quad \cdots \quad x_n]^T$ to $\mathbf{x}_0$. Then repeat the iterative step

$$(*) \quad x_i \leftarrow \{\text{solve } (\text{E}_i) \text{ for } x_i\}, \qquad i = 1, \ldots, n$$

until *MaxIt* iterations are performed, or sooner if for some iteration

$$|\Delta x_i| \text{ is sufficiently small for all } i$$

In (*), $(\text{E}_1), \ldots, (\text{E}_n)$ and $x_1, \ldots, x_n$ are preselected rearrangements of the given equations and variables.

```
{Initialize} GET n, A, b, x0                    {equation parameters, initial guess}
                 MaxIt, NumDig                  {termination parameters}
x ← x0;  Tol ← 10^(-NumDig)                     {termination test tolerance}
{Iterate: x = [x1 ··· xn]^T is the current xk}
DO FOR k = 1 TO Maxit
|   Cnvrgd ← TRUE                               {boolean convergence flag}
|   DO FOR i = 1 TO n                {update xi by solving (Ei) using current x}
|   |   xprev ← xi                              {needed for Δxi}
|   |   xi ← (1/aii)[bi − Σ_{j≠i} aij xj]       {aij = coefficient of xj in (Ei)}
|   |__ IF |xi − xprev| > Tol·max(1, |xi|) THEN Cnvrgd ← FALSE
|   OUTPUT(k, x1, ..., xn)                      {x = [x1 ··· xn]^T is xk}
|__ UNTIL Cnvrgd = TRUE

IF Cnvrgd = TRUE
|   THEN OUTPUT(x` approximates x̄ with 'NumDig` accurate digits.')
|__ ELSE OUTPUT(`Convergence not evident in 'Maxit` iterations.')
STOP
```

We see from (6) that *the given equations can be solved in any order, and for any variable, as long as each variable is solved for exactly once.* This flexibility is incorporated in the pseudocode for Gauss–Seidel iteration given in Algorithm 4.3A. Note that the iteration will terminate if in getting $\mathbf{x}_k$, *all n* values of $x_i - x_{\text{prev}} = \Delta x_i$ satisfy *either*

$$\underbrace{|x_i - x_{\text{prev}}| \leq 10^{-NumDig}}_{\text{Absolute Difference Test}} \quad or \quad \underbrace{|x_i - x_{\text{prev}}| \leq 10^{-NumDig} \cdot |x_i|}_{\text{Relative Difference Test}} \tag{7}$$

The Relative Difference Test for *NumDig significant digits* was sufficient for finding a nonzero *scalar* $\bar{x}$. However a nonzero *vector* solution $\bar{\mathbf{x}}$ may well have *zero components* for which *the Relative Difference Test will fail*. If it happens that $x_i \to 0$, the Absolute Difference Test in (7) will terminate the iteration when x_i appears to have *NumDig* accurate *decimal places* (see Section 2.2B).

4.3B Improving the Likelihood of Convergence

The *i*th equation of the linear system $A\mathbf{x} = \mathbf{b}$ is

$$(\text{E}_i) \; a_{i1}x_1 + \cdots + a_{im}x_m + \cdots + a_{in}x_n = b_i \tag{8a}$$

We call x_m the **strictly dominant variable** of (E_i) if

$$|a_{im}| > \sum_{j \neq m} |a_{ij}| \qquad (a_{im} \text{ is the \textbf{strictly dominant entry} of row}_i A) \tag{8b}$$

For an entry a_{im} to be strictly dominant, its magnitude must be larger than the *sum* of the magnitudes of the other entries of $\text{row}_i A$; just having the largest magnitude of the row is *not* enough. The matrix A will be called a **strictly dominant matrix** if each row has a strictly dominant entry, *each in different column*, that is, if some row rearrangement of A is strictly *diagonally* dominant (see Section 3.2H). The importance of strict dominance is given in the following result, whose (informative) proof is given at the end of this subsection.

DOMINANCE THEOREM. *If the linear system* $A\mathbf{x} = \mathbf{b}$ *has a strictly dominant coefficient matrix, and each equation is solved for its strictly dominant variable, then Gauss–Seidel iteration will converge to* $\mathbf{x}$ *for any choice of* $\mathbf{x}_0$, *no matter how the* (E_i)*'s are rearranged.*

The matrix A of Example 4.3A is strictly dominant; and in (6), each equation is solved for its dominant variable. So the convergence seen in Table 4.3-1 could have been predicted by the Dominance Theorem.

If A is strictly *diagonally* dominant, then the Dominance Theorem assures us that Gauss–Seidel iteration will converge if we

$$\boxed{\text{solve } (\text{E}_i) \text{ for } x_i = \frac{1}{a_{ii}}\left\{b_i - \sum_{j \neq i} a_{ij}x_j\right\}} \tag{9}$$

This simplest of strategies is also a desirable one to try when the coefficient matrix is symmetric because if A happens to be *positive definite* (this means that $\mathbf{x}^T A\mathbf{x} > 0$ for all nonzero **x**), Gauss–Seidel iteration using (9) will converge whether A is diagonally dominant or not [49]. See Exercise 4-17.

Most coefficient matrices are neither strictly dominant nor positive definite. Although the Dominance Theorem does not apply in these cases, it suggests that we should reason as follows when choosing a variable to solve each (E_i) for when using Gauss–Seidel iteration.

VARIABLE SELECTION STRATEGY. *Solve as many equations as possible for the variable having the largest (in magnitude) coefficient.*

Proof of the Dominance Theorem. Since performing row interchanges on a strictly dominant matrix results in another strictly dominant matrix, there is no loss of generality in assuming that A is strictly *diagonally* dominant and (9) is used. Let $\bar{\mathbf{x}} = [\bar{x}_1 \cdots \bar{x}_n]^T$ be the exact solution of the system $A\mathbf{x} = \mathbf{b}$. Then

$$\bar{x}_i = \frac{1}{a_{ii}}\left\{b_i - \sum_{j\neq i} a_{ij}\bar{x}_j\right\} \qquad \text{for } i = 1, 2, \ldots, n$$

The error of the jth component of the current approximation $\mathbf{x} = [x_1 \cdots x_n]^T$ will be denoted by

$$\epsilon_j = \bar{x}_j - x_j \qquad \text{for } j = 1, 2, \ldots, n$$

Consequently, the error of $x_i^{(\text{new})} = \{b_i - \Sigma_{j\neq 1} a_{ij}x_j\}/a_{ii}$ satisfies

$$\epsilon_i^{(\text{new})} = \bar{x}_i - x_i^{(\text{new})} = \frac{-1}{a_{ii}}\left\{\sum_{j\neq i} a_{ij}(\bar{x}_j - x_j)\right\} = \frac{-1}{a_{ii}}\left\{\sum_{j\neq 1} a_{ij}\epsilon_j\right\}$$

So, if we let $|\epsilon_j|_{\max}$ denote the largest $|\epsilon_j|$ for $j \neq i$, then

$$|\epsilon_i^{(\text{new})}| = \frac{1}{|a_{ii}|}\left|\sum_{j\neq i} a_{ij}\epsilon_j\right| \leq \frac{\sum_{j\neq i}|a_{ij}|}{|a_{ii}|}|\epsilon_j|_{\max} \leq \delta|\epsilon_j|_{\max} \qquad \textbf{(10a)}$$

where

$$\delta = \max\left\{\frac{\sum_{j\neq 1}|a_{1j}|}{|a_{11}|}, \frac{\sum_{j\neq 2}|a_{2j}|}{|a_{22}|}, \ldots, \frac{\sum_{j\neq n}|a_{nj}|}{|a_{nn}|}\right\} \qquad \textbf{(10b)}$$

In words, (10) says that the *error of* $x_i^{(\text{new})}$ *is smaller than the error of the other components of* $\mathbf{x}^{(\text{new})}$ *by a factor of at least* δ. The convergence of $\mathbf{x}^{(\text{new})}$ to $\bar{\mathbf{x}}$ will therefore be assured if $\delta < 1$, that is, if

$$|a_{ii}| > \sum_{j\neq 1}|a_{ij}| \qquad \text{for } i = 1, 2, \ldots, n \qquad \textbf{(10c)}$$

This is precisely the condition for strict diagonal dominance.

4.3C Accelerating Convergence of Gauss–Seidel Iteration

If the ith equation of $A\mathbf{x} = \mathbf{b}$ is solved for x_j, the resulting Gauss-Seidel equation is

$$x_j^{(\text{new})} = \frac{1}{a_{ij}}\left\{b_i - \sum_{k \neq j} a_{ik}x_k\right\} = x_j + \frac{1}{a_{ij}}\left\{b_i - \sum_{k=1}^{n} a_{ik}x_k\right\}$$

It will be useful to put this in the "increment form"

$$x_j^{(\text{new})} = x_j + dx_j, \qquad \text{where } dx_j = \frac{1}{a_{ij}}\{b_i - (\text{row}_i A)\mathbf{x}\} \tag{11}$$

Note that the increment dx_j can be viewed as

$$dx_j = \frac{1}{a_{ij}}\{i\text{th component of } (\mathbf{b} - A\mathbf{x})\} = \frac{1}{a_{ij}}\{i\textbf{th residual} \text{ of } \mathbf{x}\} \tag{12}$$

Clearly, dx_j indicates how well $\mathbf{x}$ satisfies (E_i); moreover, the condition

$$\rho_j = \frac{dx_j}{\text{preceding } dx_j} \approx \text{constant} \tag{13}$$

for successive iterations indicates that the convergence of x_j to $\bar{x}_j$ (or the divergence from $\bar{x}_j$) is *linear* (see Section 2.1A).

EXAMPLE 4.3C Examine whether (13) holds for the system $A\mathbf{x} = \mathbf{b}$, where

$$[A\,\vdots\,\mathbf{b}] = \begin{bmatrix} -1 & 1 & -4 & : & 0 \\ 2 & 2 & 0 & : & 1 \\ 3 & 3 & 2 & : & \frac{1}{2} \end{bmatrix} \quad \left(\bar{\mathbf{x}} = \begin{bmatrix} 1.25 \\ -0.75 \\ -0.5 \end{bmatrix}\right) \tag{14}$$

Solution: No rearrangement of equations makes A either symmetric or diagonally dominant. In view of the Variable Selection Strategy, we should solve (E_1) for x_3; (E_2) and (E_3) can be solved for either x_1 or x_2. Three (among many) possibilities are

(a) Solve (E_3) for x_2, then (E_2) for x_1, then (E_1) for x_3.
(b) Solve (E_3) for x_1, then (E_2) for x_2, then (E_1) for x_3.
(c) Solve (E_1) for x_3, then (E_2) for x_1, then (E_3) for x_2.

Table 4.3-2 shows $\mathbf{x}^{(\text{new})}$ (to 5s) and its ρ_j's (to 3s) for several iterations of (a), (b), and (c), all starting with $\mathbf{x}_0 = [0 \quad 0 \quad 0]^T$.

In Table 4.3-2, the Aitken improvement formula [(20b) of Section 2.1F] was used when all ρ_j's were constant to about 2s for two successive k's. The "impr" entries indicate the result when this was done. In Table 4.3-2(a) for example, all three ρ_j's satisfied $\rho_j \doteq 0.667$ for $k = 3$ and $k = 4$. Thus the values for $k = 2$, 3, and 4 gives

$$(x_i)_{\text{improved}} = x_{i,4} - \frac{(x_{i,4} - x_{i,3})^2}{x_{i,4} - 2x_{i,3} + x_{i,2}}, \qquad i = 1, 2, 3 \tag{15}$$

TABLE 4.3-2 Using Gauss–Seidel Iteration to Solve a 3 × 3 System

(a) Solve (E_3) for x_2, then (E_2) for x_1, then (E_1) for x_3; $\mathbf{x}_0 = [0 \quad 0 \quad 0]^T$.

k	x_1	ρ_1	x_2	ρ_2	x_3	ρ_3
1	0.33333		0.16667		−4.16667	
2	0.63889	0.917	−0.13889	−1.83	−0.19444	3.67
3	0.84259	0.667	−0.34259	0.667	−0.29630	0.667
4	0.97840	0.667	−0.47840	0.667	−0.36420	0.667
4_{impr}	1.2500		−0.75000		−0.50000	

(b) Solve (E_3) for x_1, then (E_2) for x_2, then (E_1) for x_3; $\mathbf{x}_0 = [0 \quad 0 \quad 0]^T$.

k	x_1	ρ_1	x_2	ρ_2	x_3	ρ_3
1	0.16667		0.33333		0.041667	
2	−0.19444	−2.17	0.69444	1.08	0.22222	4.33
3	−0.67593	1.33	1.11759	1.33	0.46296	1.33
4	−1.3179	1.33	1.18179	1.33	0.78395	1.33
4_{impr}	1.2500		−0.75000		−0.50000	

(c) Solve (E_1) for x_3, then (E_2) for x_1, then (E_3) for x_2; $\mathbf{x}_0 = [0 \quad 0 \quad 0]^T$.

k	x_1	ρ_1	x_2	ρ_2	x_3	ρ_3
1	0.50000		−0.33333		0.00000	
2	0.83333	0.667	−0.52778	0.583	−0.20833	
3	1.0278	0.583	−0.63426	0.548	−0.34028	0.633
4	1.1343	0.548	−0.69059	0.529	−0.41551	0.570
5	1.1906	0.529	−0.71798	0.518	−0.45621	0.541
6	1.2198	0.518	−0.73472	0.512	−0.47759	0.525
6_{impr}	1.2512		−0.75038		−0.50125	

Specifically,

$$(x_1)_{\text{improved}} = 0.97840 - \frac{(0.97840 - 0.84259)^2}{0.97840 - 2(0.84259) + 0.63889} \doteq 1.25 = \bar{x}_1$$

$$(x_2)_{\text{improved}} = -0.47840 - \frac{(-0.47840 + 0.34259)^2}{-0.47840 - 2(-0.34259) - 0.13889} \doteq -0.75 = \bar{x}_2$$

$$(x_3)_{\text{improved}} = -0.36420 - \frac{(-0.36420 + 0.29630)^2}{-0.36420 - 2(-0.29630) - 0.19444} \doteq -0.5 = \bar{x}_3$$

The ρ_j values were so nearly constant that a single improvement gave $5s$ accuracy in all three components!

For $k = 3$ and 4 in Table 4.3-2(b), all ρ_j values were (to $3s$) $1.33 > 1$, indicating linear *divergence*. Nevertheless, one application of the Aitken formula gave to $5s$ the desired values of $\bar{x}_1$, $\bar{x}_2$, and $\bar{x}_3$ from which the components of $\mathbf{x}_k$ were diverging!

In Table 4.3-2(c), the ρ_j's did not become constant as well as they did in (a) and (b). As a result, the Aitken formula was not used until $k = 6$, and the improved components of $\mathbf{x}_6$ were only accurate to two or three significant digits. ■

An examination of the system (14) reveals no reason why the ρ_j's of Table 4.3-2 should have been more constant in (a) and (b) than in (c). Indeed, a satisfactory, general a priori strategy for determining which (E_i) should be solved for which x_j (and in which order) to ensure rapid, liner convergence has not been found. In the absence of such a strategy, it is desirable to have a computer program for Gauss–Seidel iteration, which is interactive, so that during execution the user may examine the results of a few iterations and, based on these results, modify the x_j's solved for in the (E_i)'s and/or the order in which the (E_i)'s are to be solved (see Problem C4-8). A strategy called **relaxation** consists of replacing (11) by

$$x_i^{(\text{new})} = x_i + \lambda\, dx_i, \qquad \text{where } dx_i = \frac{1}{a_{ii}}\{b_i - (\text{row}_i A)\mathbf{x}\} \tag{16}$$

Generally, either $0 < \lambda < 1$ (**underrelaxation**) or $1 < \lambda < 2$ (**overrelaxation**) is used, with $\lambda = 1$ corresponding to Gauss–Seidel iteration. It has been found that in certain situations relaxation can improve the likelihood or accelerate the rate of the convergence of Gauss–Seidel iteration, although the convergence will remain linear. Unfortunately, one effect of relaxation is to make the ρ_i's less constant so that the advantage of more rapid linear convergence is offset by the loss of the ability to use the Aitken formula effectively. The reader wishing to know more about relaxation is referred to [39].

4.4 Solving Nonlinear Systems of Equations

An equation that contains expressions such as

$$x^3, \quad y^{-2}, \quad xy, \quad \frac{\sqrt{y}}{x}, \quad (2y - z)^2, \quad \sin x, \quad e^{yz}, \quad z\sqrt{x + y}$$

is called **nonlinear** in $x, y, z, \ldots$ because it *cannot* be written as

$$ax + by + cz + \cdots = \text{constant} \qquad (\text{a } \textit{linear equation} \text{ in } x, y, z, \ldots)$$

A system of n equations in n unknowns $x_1, x_2, \ldots, x_n$ is called **nonlinear** if one or more of the equations is nonlinear. By bringing all nonzero terms to the left side of all equations, any nonlinear $n \times n$ system can be put in the general form

$$\left.\begin{aligned} (\mathrm{E}_1)\ f_1(x_1, x_2, \ldots, x_n) &= 0 \\ (\mathrm{E}_2)\ f_2(x_1, x_2, \ldots, x_n) &= 0 \\ \vdots \qquad\qquad &\ \ \vdots \\ (\mathrm{E}_n)\ f_n(x_1, x_2, \ldots, x_n) &= 0 \end{aligned}\right\} \text{ or simply } \left\{\begin{aligned} f_1(\mathbf{x}) &= 0 \\ f_2(\mathbf{x}) &= 0 \\ &\ \vdots \\ f_n(\mathbf{x}) &= 0 \end{aligned}\right., \text{ where } \mathbf{x} = \begin{bmatrix} x_1 \\ x_2 \\ \vdots \\ x_n \end{bmatrix} \tag{1}$$

The system (1) will be abbreviated as simply $\mathbf{f}(\mathbf{x}) = \mathbf{0}$. A vector $\overline{\mathbf{x}} = [\overline{x}_1 \ \ \overline{x}_2 \ \ \cdots \ \ \overline{x}_n]^T$ that satisfies $\mathbf{f}(\overline{\mathbf{x}}) = \mathbf{0}$, [i.e., whose coordinates satisfy $(\mathrm{E}_1), \ldots, (\mathrm{E}_n)$] will be called a **root** of the nonlinear system (1). The Alternative Theorem of Section 3.4F asserted that *linear*

$n \times n$ systems can have only one root or infinitely many roots. By contrast, as illustrated next, *nonlinear $n \times n$ systems often have finitely many roots.*

4.4A Introduction to 2 × 2 Nonlinear Systems

Rather than use subscripts, a general 2×2 nonlinear system will be denoted by the simplified notation

$$\begin{aligned} (E_1)\ f(x, y) &= 0, \text{ or simply } f(\mathbf{x}) = 0 \\ (E_2)\ g(x, y) &= 0, \text{ or simply } g(\mathbf{x}) = 0 \end{aligned} \qquad \text{where } \mathbf{x} = \begin{bmatrix} x \\ y \end{bmatrix} \tag{2}$$

We shall identify the vector $\mathbf{x}$ with the point (x, y) on the xy-plane in the usual way. As a result, roots $\bar{\mathbf{x}} = [\bar{x} \quad \bar{y}]^T$ of (2) can be interpreted geometrically as described in the following example.

EXAMPLE 4.4A Use graphs to estimate all roots of the 2×2 system

$$\begin{aligned} (E_1)\ ye^x - 2 &= 0 \qquad [\text{here } f(\mathbf{x}) = f(x, y) = ye^x - 2] \\ (E_2)\ x^2 + y - 4 &= 0 \qquad [\text{here } g(\mathbf{x}) = g(x, y) = x^2 + y - 4] \end{aligned} \tag{3}$$

Can you find these roots using previously obtained numerical methods?

SOLUTION: Since both (E_1) and (E_2) can be solved for y, we see that

$$\begin{aligned} \mathbf{x} \text{ satisfies } (E_1) &\Leftrightarrow \mathbf{x} \text{ lies on the exponential curve } y = 2e^{-x} \\ \mathbf{x} \text{ satisfies } (E_2) &\Leftrightarrow \mathbf{x} \text{ lies on the parabolic curve } y = 4 - x^2 \end{aligned} \tag{4a}$$

So roots $\bar{\mathbf{x}} = [\bar{x} \quad \bar{y}]^T$ correspond to points of intersection of these two curves in the xy-plane. We see from Figure 4.4-1 that there are two roots

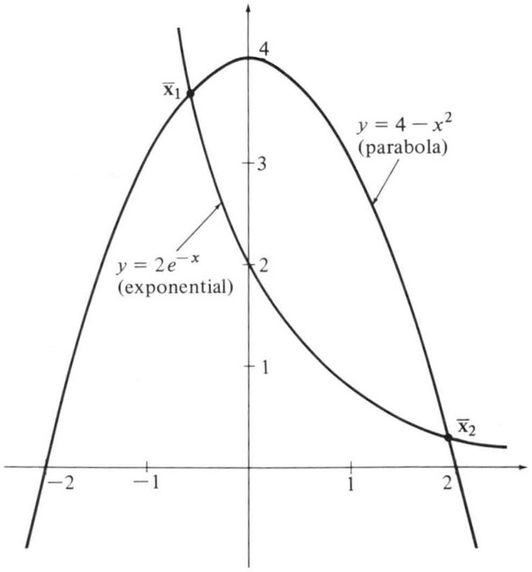

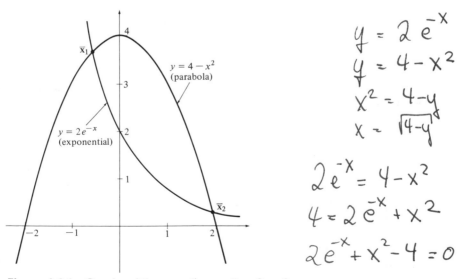

Figure 4.4-1 Graphs of the equations $ye^x - 2 = 0$, $x^2 + y - 4 = 0$.

$$\bar{\mathbf{x}}_1 \approx [-0.6 \quad 3.7]^T \qquad \text{and} \qquad \bar{\mathbf{x}}_2 \approx [1.9 \quad 0.4]^T \tag{4b}$$

Eliminating y in (4a) shows that the x-coordinates of $\bar{\mathbf{x}}_1$ and $\bar{\mathbf{x}}_2$ satisfy

$$x^2 + 2e^{-x} - 4 = 0 \tag{5}$$

This is of the form $f(x) = 0$, and so can be solved for $\bar{x}_1 \approx -0.6$ and $x_2 \approx 1.9$ using any of the iterative root-finding methods of Chapter 2 (e.g., NR or SEC). The y-coordinates of $\bar{\mathbf{x}}_1$ and $\bar{\mathbf{x}}_2$ can then be found by substituting the calculated $\bar{x}_1$ and $\bar{x}_2$ in (4a) (see Problem 4-22). ∎

It is generally good practice to reduce n by eliminating variables as we did in (5) whenever this can be done easily. However, this cannot always be done. For example, *neither x nor y can be eliminated* from the nonlinear system

$$(E_1)\ xe^y - x^5 + y + 3 = 0 \qquad (E_2)\ x + y + \tan x - \sin y = 0 \tag{6}$$

Try it. The attempt will help you appreciate the need for methods that can solve nonlinear systems in the *general* form

$$(E_1)\ f(x, y) = 0 \qquad (E_2)\ g(x, y) = 0 \tag{7}$$

4.4B The Newton–Raphson Method for 2×2 Systems (NR₂)

The graphs of the equations

$$z = f(x, y) \qquad \text{and} \qquad z = g(x, y) \tag{8}$$

are surfaces Σ_f and Σ_g whose intersection is a curve C in xyz-space, as illustrated in Figure 4.4-2(a). Setting $f(x,y)$ and $g(x,y)$ to zero corresponds to intersecting Σ_f and Σ_g with the xy-plane ($z = 0$). So *a root $\bar{\mathbf{x}}$ of (7) corresponds to a point at which the curve C meets the xy-plane.*

Suppose that $\mathbf{x}_k = (x_k, y_k)$ is a current approximation of $\bar{\mathbf{x}} = (\bar{x}, \bar{y})$. If Π_f denotes the tangent plane to Σ_f at the point $P_f(\mathbf{x}_k, f(\mathbf{x}_k))$, and Π_g denotes the tangent plane to Σ_g at the point $P_g(\mathbf{x}_k, g(\mathbf{x}_k))$, then Π_f and Π_g will intersect in a line L, as shown in Figure 4.4-2(b). If $\mathbf{x}_k$ is sufficiently close to $\bar{\mathbf{x}}$, then L will lie close to the curve C, and so the point $\mathbf{x}_{k+1}$ where L meets the xy-plane should be closer to the desired root $\bar{\mathbf{x}}$ than $\mathbf{x}_k$. (This argument should seem familiar; see Figure 2.3-3.) A point $\mathbf{x} = (x, y)$ near $\mathbf{x}_k$ can be written as

$$\mathbf{x} = (x, y), \qquad \text{where } x = x_k + dx \quad \text{and} \quad y = y_k + dy \tag{9}$$

See page 65

The z-coordinate of the points (x, y, z) on the planes Π_f and Π_g can be found using the tangent plane equation obtained in calculus.

ON Π_f: $\quad z \overset{?}{=} f(x_k, y_k) + \dfrac{\partial}{\partial x} f(x_k, y_k)\, dx + \dfrac{\partial}{\partial y} f(x_k, y_k)\, dy \tag{10a}$

$$Z = f(x_k + dx,\ y_k + dy) = f(x_k, y_k) + \cdots$$

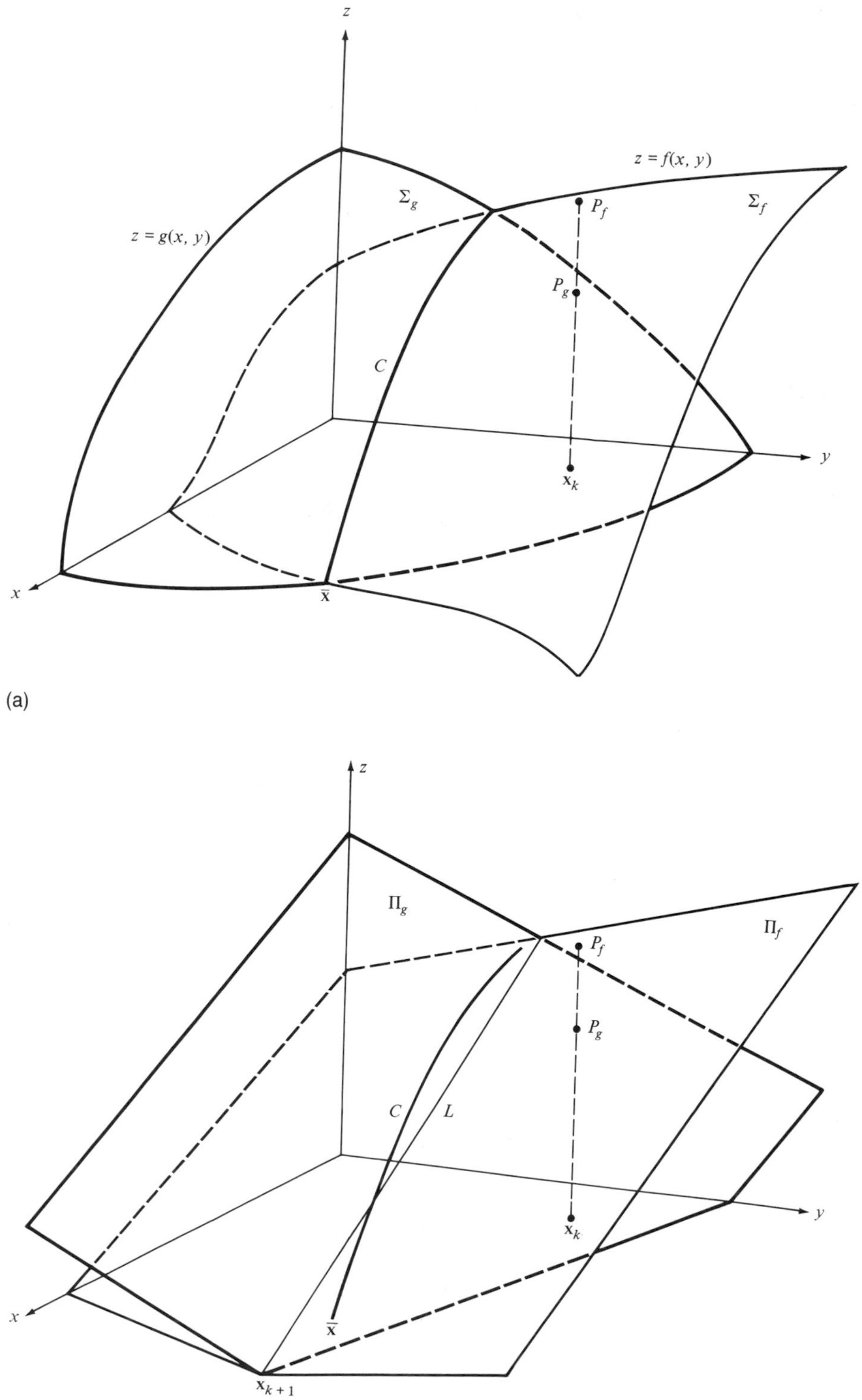

Figure 4.4-2 *(a) Σ_f and Σ_g. (b) Their tangent planes Π_f and Π_g at $\mathbf{x}_k$.*

On Π_g: $z = g(x_k, y_k) + \frac{\partial}{\partial x} g(x_k, y_k)\,dx + \frac{\partial}{\partial y} g(x_k, y_k)\,dy$ **(10b)**

Setting z to zero in (10) yields a *linear* system in the variables dx and dy. Its matrix form is $J\mathbf{dx} = -\mathbf{f}(\mathbf{x}_k)$, that is,

$$\begin{bmatrix} \frac{\partial}{\partial x} f(\mathbf{x}_k) & \frac{\partial}{\partial y} f(\mathbf{x}_k) \\ \frac{\partial}{\partial x} g(\mathbf{x}_k) & \frac{\partial}{\partial y} g(\mathbf{x}_k) \end{bmatrix} \begin{bmatrix} dx \\ dy \end{bmatrix} = -\begin{bmatrix} f(\mathbf{x}_k) \\ g(\mathbf{x}_k) \end{bmatrix}, \quad \text{where } \mathbf{x}_k = (x_k, y_k) \tag{11}$$

$$J = \mathbf{f}'(\mathbf{x}_k) \qquad \mathbf{dx} \qquad -\mathbf{f}(\mathbf{x}_k)$$

The solution of (11), which we denote by $\mathbf{dx}_k$, is what must be added to $\mathbf{x}_k$ to get the point $\mathbf{x}_{k+1}$ where L meets the xy-plane in Figure 4.4- 2(b). The 2×2 linear system $J\mathbf{dx} = -\mathbf{f}(\mathbf{x}_k)$ in (11) is best solved as $-J^{-1}\mathbf{f}(\mathbf{x})$, using the formula for J^{-1} [see (15) of Section 3.1E]. Thus

$$\mathbf{x}_{k+1} = \mathbf{x}_k + \mathbf{dx}_k, \qquad \text{where } \mathbf{dx}_k = -\mathbf{f}'(\mathbf{x}_k)^{-1}\mathbf{f}(\mathbf{x}_k) \tag{12}$$

$\Leftrightarrow [-f'(x_k)]\,dx_k = f(x_k)$

The astute reader will recognize this as the NR formula (6b) of Section 2.3C. The iteration (12) will be called the **Newton–Raphson method** for solving the 2×2 nonlinear system (7), and abbreviated as $\mathbf{NR_2}$. The 2×2 matrix

$$\mathbf{f}'(\mathbf{x}) = \begin{bmatrix} \frac{\partial}{\partial x} f(x, y) & \frac{\partial}{\partial y} f(x, y) \\ \frac{\partial}{\partial x} g(x, y) & \frac{\partial}{\partial y} g(x, y) \end{bmatrix} \tag{13}$$

evaluated at $\mathbf{x}_k$ to get J in (11) is called the **Jacobian matrix** at $\mathbf{x} = (x, y)$ for the nonlinear system (E_1) $f(x, y) = 0$, (E_2) $g(x, y) = 0$.

EXAMPLE 4.4B Use NR_2 to solve to 4s the 2×2 nonlinear system

$$(\mathrm{E}_1)\ ye^x = 2 \qquad (\mathrm{E}_2)\ x^2 + y = 4 \tag{14a}$$

Solution: Since NR_2 can only be used to solve $\mathbf{f}(\mathbf{x}) = \mathbf{0}$, we must first convert (14a) to an equivalent system of this form, say

$$(\mathrm{E}_1)\ ye^x - 2 = 0 \qquad (\mathrm{E}_2)\ x^2 + y - 4 = 0 \tag{14b}$$

We saw in Example 4.4A that this system has two roots, $\bar{\mathbf{x}}_1 \approx [-0.6 \quad 3.7]^T$ and $\bar{\mathbf{x}}_2 \approx [1.9 \quad 0.4]^T$. When graphically obtained estimates such as these are available, they should be used as initial guesses. For the iteration, we use (14b) [*not* (14a)] to get

$$\mathbf{x} = \begin{bmatrix} x \\ y \end{bmatrix}, \quad \mathbf{f}(\mathbf{x}) = \begin{bmatrix} ye^x - 2 \\ x^2 + y - 4 \end{bmatrix}, \quad \text{and} \quad \mathbf{f}'(\mathbf{x}) = \begin{bmatrix} ye^x & e^x \\ 2x & 1 \end{bmatrix} \tag{15a}$$

[see (11) and (13)]. We can now use (12) to get $\overline{\mathbf{x}}_1$ and $\overline{\mathbf{x}}_2$.

For $\overline{\mathbf{x}}_1$: We start with the initial guess $\mathbf{x}_0 = [-0.6 \quad 3.7]^T$. From (15a),

$$\mathbf{f}(\mathbf{x}_0) = \begin{bmatrix} 3.7e^{-0.6} - 2 \\ (-0.6)^2 + 3.7 - 4 \end{bmatrix} \quad \text{and} \quad \mathbf{f}'(\mathbf{x}_0) = \begin{bmatrix} 3.7e^{-0.6} & e^{-0.6} \\ 2(-0.6) & 1 \end{bmatrix} \tag{15b}$$

Evaluating to 6s (one guard digit) and substituting in (11) gives

$$\begin{bmatrix} 2.03060 & 0.548812 \\ -1.2 & 1 \end{bmatrix} \begin{bmatrix} dx \\ dy \end{bmatrix} = -\begin{bmatrix} 0.0306031 \\ 0.06 \end{bmatrix} \tag{15c}$$

This is $\mathbf{f}'(\mathbf{x}_0)\,\mathbf{dx} = -\mathbf{f}(\mathbf{x}_0)$. Solving for $\mathbf{dx}_0$ as $-\mathbf{f}'(\mathbf{x}_0)^{-1}\mathbf{f}(\mathbf{x}_0)$ gives

$$\mathbf{dx}_0 = \frac{-1}{2.68917} \begin{bmatrix} 1 & -0.548812 \\ 1.2 & 2.03060 \end{bmatrix} \begin{bmatrix} 0.0306031 \\ 0.06 \end{bmatrix} \doteq \begin{bmatrix} 0.000865 \\ -0.05896 \end{bmatrix} \tag{15d}$$

$$\mathbf{x}_1 = \mathbf{x}_0 + \mathbf{dx}_0 = \begin{bmatrix} -0.6 \\ 3.7 \end{bmatrix} + \begin{bmatrix} 0.000865 \\ -0.05896 \end{bmatrix} = \begin{bmatrix} -0.599135 \\ 3.64104 \end{bmatrix} \tag{15e}$$

The entries of $\mathbf{dx}_0$ indicate that $\mathbf{x}_0$ is accurate to about 3d. By assuming the convergence to be quadratic, $\mathbf{x}_1$ should certainly be accurate to at least the desired 4s. Another iteration (which will not be shown) confirms this.

For $\overline{\mathbf{x}}_2$: If we start with the initial guess $\mathbf{x}_0 = [1.9 \quad 0.4]^T$, a similar computation gives

$$\mathbf{dx}_0 = -\mathbf{f}'(\mathbf{x}_0)^{-1}\mathbf{f}(\mathbf{x}_0) \doteq \begin{bmatrix} 0.02672 \\ -0.111553 \end{bmatrix}, \qquad \mathbf{x}_1 = \mathbf{x}_0 + \mathbf{dx}_0 = \begin{bmatrix} 1.92672 \\ 0.288447 \end{bmatrix}$$

$$\mathbf{dx}_1 = -\mathbf{f}'(\mathbf{x}_1)^{-1}\mathbf{f}(\mathbf{x}_1) \doteq \begin{bmatrix} -0.00098 \\ 0.003086 \end{bmatrix}, \qquad \mathbf{x}_2 = \mathbf{x}_1 + \mathbf{dx}_1 = \begin{bmatrix} 1.92574 \\ 0.291533 \end{bmatrix}$$

Here again the $\mathbf{dx}_k$'s demonstrate that if $\overline{\mathbf{x}}$ is a simple root and $\mathbf{x}_0$ is accurate to about 1d, then the number of accurate decimal places will approximately double with each iteration of NR_2 (quadratic convergence), and two iterations should give more than 4s accuracy. The convergence will be slower if $\overline{\mathbf{x}}$ is a multiple root [see (iii) of Problem M4-9].

4.4C The Newton–Raphson Method for $n \times n$ Systems (NR_n)

Let $\overline{\mathbf{x}} = [\overline{x}_1 \quad \overline{x}_2 \quad \cdots \quad \overline{x}_n]^T$ be a desired root of the $n \times n$ nonlinear system $\mathbf{f}(\mathbf{x}) = \mathbf{0}$, whose ith equation is

$$f_i(\mathbf{x}) = f_i(x_1, \ldots, x_n) = 0, \qquad i = 1, \ldots, n \tag{16}$$

and suppose that $\mathbf{x}_k$ is a current approximation of $\bar{\mathbf{x}}$. Our strategy for getting an improved approximation $\mathbf{x}_{k+1}$ is to solve a *linear* system that approximates the system (16) for $\mathbf{x}$ near $\mathbf{x}_k$ [see Figure 4.3-2(b)]. Specifically, if $\mathbf{x} = \mathbf{x}_k + \mathbf{dx}$, where $\mathbf{dx} = [dx_1 \quad dx_2 \quad \cdots \quad dx_n]^T$, we shall approximate the exact equation $f_i(\mathbf{x}_k + \mathbf{dx}) = 0$ in (16) using the total differential [Appendix II.5C]:

$$f_i(\mathbf{x}_k) + \frac{\partial f_i(\mathbf{x}_k)}{\partial x_1} dx_1 + \frac{\partial f_i(\mathbf{x}_k)}{\partial x_2} dx_2 + \cdots + \frac{\partial f_i(\mathbf{x}_k)}{\partial x_n} dx_n = 0, \qquad i = 1, \ldots, n \qquad \textbf{(17a)}$$

This system is linear in $dx_1, dx_2, \ldots, dx_n$; its matrix form is

$$\underbrace{\begin{bmatrix} \frac{\partial f_1(\mathbf{x}_k)}{\partial x_1} & \frac{\partial f_1(\mathbf{x}_k)}{\partial x_2} & \cdots & \frac{\partial f_1(\mathbf{x}_k)}{\partial x_n} \\ \frac{\partial f_2(\mathbf{x}_k)}{\partial x_1} & \frac{\partial f_2(\mathbf{x}_k)}{\partial x_2} & \cdots & \frac{\partial f_2(\mathbf{x}_k)}{\partial x_n} \\ \vdots & \vdots & & \vdots \\ \frac{\partial f_n(\mathbf{x}_k)}{\partial x_1} & \frac{\partial f_n(\mathbf{x}_k)}{\partial x_2} & \cdots & \frac{\partial f_n(\mathbf{x}_k)}{\partial x_n} \end{bmatrix}}_{J = \mathbf{f}'(\mathbf{x}_k)} \underbrace{\begin{bmatrix} dx_1 \\ dx_2 \\ \vdots \\ dx_n \end{bmatrix}}_{\mathbf{dx}} = \underbrace{- \begin{bmatrix} f_1(\mathbf{x}_k) \\ f_2(\mathbf{x}_k) \\ \vdots \\ f_n(\mathbf{x}_k) \end{bmatrix}}_{-\mathbf{f}(\mathbf{x}_k)} \qquad \textbf{(17b)}$$

[see (11)]. We can therefore get $\mathbf{x}_{k+1}$ from $\mathbf{x}_k$ as

$$\mathbf{x}_{k+1} = \mathbf{x}_k + \mathbf{dx}_k, \qquad \text{where } \mathbf{dx}_k \text{ is the solution of } \mathbf{f}'(\mathbf{x}_k)\mathbf{dx} = -\mathbf{f}(\mathbf{x}_k) \qquad \textbf{(17c)}$$

[see (12)]. The matrix $J = \mathbf{f}'(\mathbf{x}_k)$ in (17b) is the **Jacobian matrix** for the nonlinear system $\mathbf{f}(\mathbf{x}) = \mathbf{0}$ at $\mathbf{x}_k$. Notice that $\text{row}_i J$ contains all partials of $f_i(\mathbf{x})$ (ith equation), whereas $\text{col}_j J$ contains all partials with respect to x_j (jth variable). Thus $J = \mathbf{f}'(\mathbf{x}_k) = (\partial f_i(\mathbf{x}_k)/\partial x_j)_{n \times n}$.

The algorithm based on (17) is called the **Newton–Raphson method for $n \times n$ systems**, abbreviated as **NR$_n$**. Pseudocode for it is shown in Algorithm 4.4C, where the **termination test** finds $\bar{x}_i$ to *NumDig* significant digits if $|x_i| \geq 1$, and to *NumDig* decimal places if $|x_i| < 1$, as described in (7) of Section 4.3A.

ALGORITHM 4.4C. NEWTON–RAPHSON METHOD FOR $n \times n$ NONLINEAR SYSTEMS (NR$_n$)

PURPOSE: To find roots $\bar{\mathbf{x}}$ of an $n \times n$ nonlinear system written as

$$f_i(x_1, x_2, \ldots x_n) = 0, \qquad i = 1, \ldots, n$$

to *NumDig* accurate digits within *MaxIt* iterations, using the arrays

$\mathbf{x}$ {stores the current approximation of $\bar{\mathbf{x}}$}
$\mathbf{f}$ {stores $\mathbf{f}(\mathbf{x}) = [f_1(\mathbf{x}) \quad f_2(\mathbf{x}) \quad \cdots \quad f_n(\mathbf{x})]^T$}
$\mathbf{dx}$ {stores the increment $[dx_1 \quad dx_2 \quad \cdots \quad dx_n]^T$}
J {stores $\mathbf{f}'(\mathbf{x})$, the Jacobian matrix at $\mathbf{x}$}

GET n, $\mathbf{x}_0$, {system size, initial guess}
    MaxIt, *NumDig* {termination parameters}
$Tol \leftarrow 10^{-NumDig}$ {termination test tolerance}
$\mathbf{x} \leftarrow \mathbf{x}_0$
DO FOR $k = 1$ TO *Maxit*
    $\mathbf{f} \leftarrow \mathbf{f}(\mathbf{x})$; $J \leftarrow \mathbf{f}'(\mathbf{x})$ {Evaluate $\mathbf{f}(\mathbf{x})$ and $\mathbf{f}'(\mathbf{x})$ at the current $\mathbf{x}$}
    $\mathbf{dx} \leftarrow$ (solution of the linear system $J\mathbf{dx} = -\mathbf{f}$) {See *Note* below.}
    $\mathbf{x} \leftarrow \mathbf{x} + \mathbf{dx}$ {$\mathbf{x} = [x_1 \quad \cdots \quad x_n]^T$ is now $\mathbf{x}_k$}
    OUTPUT($k, x_1, \ldots, x_n$)
    UNTIL $|dx_i| \leq Tol \cdot \max(1, |x_i|)$, $i = 1, \ldots, n$ {**termination test**}
IF **termination test** was satisfied
    THEN OUTPUT($\mathbf{x}$' approximates $\bar{\mathbf{x}}$ with '*NumDig*' accurate digits.')
    ELSE OUTPUT('Convergence not evident in '*Maxit*' iterations.')
STOP
{**NOTE:** If $n > 2$, the linear system $J\mathbf{dx} = -\mathbf{f}$ should be solved by a direct (i.e., elimination or factorization) method, *not* as $-J^{-1}\mathbf{f}$.}

EXAMPLE 4.4C Use NR_3 to solve 6*s* the 3×3 nonlinear system

$$\begin{aligned} &(E_1) \quad w(\xi^2 - \eta^2) + \eta^2 = \tfrac{1}{3} \\ &(E_2) \quad w(\xi^4 - \eta^4) + \eta^4 = \tfrac{1}{5} \\ &(E_3) \quad w(\xi^6 - \eta^6) + \eta^6 = \tfrac{1}{7} \end{aligned} \qquad \textbf{(18a)}$$

given that

$$\tfrac{1}{2} < w < 1 \text{ and } 0 < \xi < \eta < 1 \qquad \textbf{(18b)}$$

SOLUTION: To use NR_3, we must first put (18a) in the form $\mathbf{f}(\mathbf{x}) = \mathbf{0}$. We do this by introducing the vectors

$$\mathbf{x} = \begin{bmatrix} w \\ \xi \\ \eta \end{bmatrix} \quad \text{and} \quad \mathbf{f}(\mathbf{x}) = \begin{bmatrix} w(\xi^2 - \eta^2) + \eta^2 - \frac{1}{3} \\ w(\xi^4 - \eta^4) + \eta^4 - \frac{1}{5} \\ w(\psi^6 - \eta^6) + \eta^6 - \frac{1}{7} \end{bmatrix} \qquad \textbf{(19a)}$$

The Jacobian matrix for the system $\mathbf{f}(\mathbf{x}) = \mathbf{0}$ is the 3×3 matrix

$$\mathbf{f}'(\mathbf{x}) = \begin{bmatrix} \frac{\partial f_1(\mathbf{x})}{\partial w} & \frac{\partial f_1(\mathbf{x})}{\partial \xi} & \frac{\partial f_1(\mathbf{x})}{\partial \eta} \\ \frac{\partial f_2(\mathbf{x})}{\partial w} & \frac{\partial f_2(\mathbf{x})}{\partial \xi} & \frac{\partial f_2(\mathbf{x})}{\partial \eta} \\ \frac{\partial f_3(\mathbf{x})}{\partial w} & \frac{\partial f_3(\mathbf{x})}{\partial \xi} & \frac{\partial f_3(\mathbf{x})}{\partial \eta} \end{bmatrix} = \begin{bmatrix} (\xi^2 - \eta^2) & 2\xi w & 2(1 - w)\eta \\ (\xi^4 - \eta^4) & 4\xi^3 w & 4(1 - w)\eta^3 \\ (\xi^6 - \eta^6) & 6\xi^5 w & 6(1 - w)\eta^5 \end{bmatrix} \qquad \textbf{(19b)}$$

Note that the columns of $\mathbf{f}'(\mathbf{x})$ reflect the order of the variables in $\mathbf{x}$. In view of (18b), we take $\mathbf{x}_0 = [w_0 \quad \xi_0 \quad \eta_0]^T = [\frac{3}{5} \quad \frac{1}{3} \quad \frac{2}{3}]^T$ as our initial guess. The linear system $\mathbf{f}'(\mathbf{x}_0)\,\mathbf{dx} = -\mathbf{f}(\mathbf{x}_0)$ is then

$$\begin{bmatrix} -7/12 & 2/5 & 2/3 \\ -203/27 & 4/45 & 25/27 \\ -1729/5184 & 2/35 & 625/648 \end{bmatrix} \begin{bmatrix} d\omega \\ d\xi \\ d\eta \end{bmatrix} = -\begin{bmatrix} 1/90 \\ 1/3240 \\ -6593/816{,}480 \end{bmatrix} \tag{20a}$$

This can be solved by decomposing $A = \mathbf{f}'(\mathbf{x}_0)$ and then completing the forback matrix $[\hat{L}\backslash\hat{U}\colon\hat{\mathbf{b}}\colon\bar{\mathbf{c}}\colon\mathbf{dx}_0]$, with $\mathbf{b} = -\mathbf{f}(\mathbf{x}_0)$, either by hand or on a computer. The solution (shown rounded to $6s$) is

$$\mathbf{dx}_0 \doteq \begin{bmatrix} 0.0592037 \\ 0.0107620 \\ 0.0286794 \end{bmatrix}, \text{ so, } \mathbf{x}_1 = \mathbf{x}_0 + \mathbf{dx}_0 = \begin{bmatrix} \frac{3}{5} \\ \frac{1}{3} \\ \frac{2}{3} \end{bmatrix} + \begin{bmatrix} 0.0592037 \\ 0.0107620 \\ 0.0286794 \end{bmatrix} = \begin{bmatrix} 0.659204 \\ 0.344095 \\ 0.862013 \end{bmatrix} \tag{20b}$$

Similarly, forming and solving $\mathbf{f}'(\mathbf{x}_1)\mathbf{dx} = -\mathbf{f}(\mathbf{x}_1)$, then $\mathbf{f}'(\mathbf{x}_2)\mathbf{dx} = -\mathbf{f}(\mathbf{x}_2)$ gives

$$\mathbf{x}_2 = \mathbf{x}_1 + \begin{bmatrix} 7.09233\text{E}-3 \\ 4.07963\text{E}-3 \\ 9.03952\text{E}-4 \end{bmatrix} = \begin{bmatrix} 0.652111 \\ 0.340016 \\ 0.861109 \end{bmatrix} \tag{20c}$$

$$\mathbf{x}_3 = \mathbf{x}_2 + \begin{bmatrix} -3.37504\text{E}-5 \\ -3.46391\text{E}-5 \\ -2.75192\text{E}-5 \end{bmatrix} = \begin{bmatrix} 0.652145 \\ 0.339981 \\ 0.861136 \end{bmatrix} \tag{20d}$$

Since $\mathbf{x}_2$ and $\mathbf{x}_3$ agree to about $4s$ and the convergence is quadratic, $\mathbf{x}_3$ should (and does) have the desired $6s$ accuracy. This solution will be used to derive the four-point Gauss quadrature formula in Example 7.4A(b). ■

4.4D Implementing NR$_n$ on a Computer

From a practical standpoint, NR$_n$ has three shortcomings.

1. ***It is important to have a good initial guess***. We saw in Figure 2.4-5 that if the NR method for solving $f(x) = 0$ ($n = 1$) is used when $f'(x_k)$ is near zero, then x_{k+1} can be far from x_k. The likelihood of this happening increases as n^2 for NR$_n$, because *overshoot can occur if any one of the Jacobian matrix entries $\partial f_i(\mathbf{x}_k)/\partial x_j$ is nearly zero*. (This can be seen when $n = 2$ by examining the tangent planes in Figure 4.4- 1.) It is often the case that the context in which an actual problem arises provides some insight into the location of the desired root(s). However, even with a good initial guess, it is important to guard against overshoot. A simple remedy [see (10) of Section 2.4C] is to put a bound, say *DxMax*, on the magnitude of the absolute *and* relative error of the components of $\mathbf{dx}$:

DO FOR $i = 1$ TO n

  $Tol \leftarrow DxMax \cdot \min(1, |x_i|)$ **(21)**

  IF $|dx_i| > Tol$ THEN $dx_i \leftarrow \text{sign}(dx_i) \cdot Tol$

This strategy ensures that dx_i is "small" when $\mathbf{x}_k$ is near $\mathbf{x}_0$ but does not affect dx_i once $\mathbf{x}_k$ is near $\bar{\mathbf{x}}$. It can be incorporated in the same loop that does the **termination test** in Algorithm 4.4C.

2. ***It is a lot of work to find and program the n^2 entries of* $\mathbf{f}'(\mathbf{x})$**. This shortcoming is a serious one because it renders computers essentially useless for large n, which is precisely when you need them! (Consider the likelihood of finding *and* programming the $10^2 = 100$ entries of $\mathbf{f}'(\mathbf{x})$ without a mistake when $n = 10$ if you should be ambitious enough to try.) The natural remedy is to approximate the partial derivatives numerically. For example, it follows directly from the definition of $\partial f_i(\mathbf{x})/\partial x_j$ (the partial derivative of f_i with respect to x_j at the point $\mathbf{x} = [x_1 \ \cdots \ x_n]^T$) that

$$\frac{f_i(x_1, \ldots, x_{j-1}, x_j + h_{ij}, x_{j+1}, \ldots, x_n) - f_i(x_1, \ldots, x_n)}{h_{ij}} \tag{22a}$$

can be used to approximate $J(i, j) = \partial f_i(\mathbf{x})/\partial x_j$ at the current $\mathbf{x}$. Moreover, if the stepsize h_{ij} is taken to be that suggested by Steffensen, namely

$$h_{ij} = f_i(\mathbf{x}) \quad \text{for all } j \qquad [\text{provided that } f_i(\mathbf{x}) \neq 0] \tag{22b}$$

then (in the absence of roundoff error) quadratic convergence will result using the numerical partial derivatives (22) whenever it will occur using $J(i, j) = \partial f_i(\mathbf{x})/\partial x_j$ [12]. However, this result must be implemented with care, because *serious roundoff error can result from taking h_{ij} too small*, as we saw in Section 1.1D. Effective methods for evaluating derivatives numerically are described in Section 7.2G.

3. **$\mathbf{f}'(\mathbf{x})$ *must be evaluated and decomposed in each iteration when $n > 2$*.** A partial remedy for this shortcoming can be implemented when all partial derivatives $\partial f_i(\mathbf{x})/\partial x_j$ are continuous near $\bar{\mathbf{x}}$ so that

$$\mathbf{f}'(\mathbf{x}_{k+1}) \approx \mathbf{f}'(\mathbf{x}_k) \qquad \text{when } \mathbf{x}_{k+1} \approx \mathbf{x}_k \tag{23}$$

This suggests that we can *save the LU-decomposition of* $\mathbf{f}'(\mathbf{x}_k)$ obtained in getting $\mathbf{dx}_k$ and *reuse it for several successive iterations*. Reusing $<\hat{L}\backslash\hat{U}, \boldsymbol{\rho}_{m,p}>$ this way does not appreciably slow the convergence if $\mathbf{f}'(\mathbf{x})$ is not ill conditioned. Since each reuse eliminates about $2n^3/3$ arithmetic operations, this strategy can reduce the time needed to get a specified accuracy when n is large.

A computer implementation of NR_n that incorporates the modifications just described and that uses an accurate linear system solver as described in Section 4.2B will provide solutions to just about any $n \times n$ nonlinear system that you are likely to encounter, except possibly those arising in the solution of differential equations. Alternative methods in this special (but important) situation are described in the next section and in Section 8.5D.

4.4E Other Methods for Solving Nonlinear Systems

Even as modified in the preceding section, NR_n will still require between $O(h^2)$ and $O(h^3)$ arithmetic operations per iteration in addition to the necessary evaluations of $\mathbf{f}$ and $\mathbf{f}'$; and it will be especially inefficient when the Jacobian is either singular or nearly singular at $\bar{\mathbf{x}}$. If it happens that $\mathbf{f}'(\bar{\mathbf{x}})$ is strictly dominant (see Section 4.3B), then **Gauss–Seidel iteration** can be used effectively with only *one evaluation of* $\mathbf{f}(\mathbf{x})$ (in slightly altered form) *per iteration*, as illustrated in the following example.

EXAMPLE 4.4E Solve (to 4s) the nonlinear 3 × 3 system

$$\begin{array}{ll}(E_1) & \sqrt{x} + \ln y + \tfrac{1}{2}\cos z = 0 \\ (E_2) & y\ln(4x) - 2z - e^z + 1 = 0 \\ (E_3) & 4x + y\sin z - 1 = 0\end{array} \qquad \text{for } \mathbf{x} = \begin{bmatrix} x \\ y \\ z \end{bmatrix} \text{ near } \begin{bmatrix} \frac{1}{3} \\ \frac{1}{3} \\ \pi/6 \end{bmatrix} \tag{24}$$

SOLUTION: If we view (24) as $\mathbf{f}(\mathbf{x}) = \mathbf{0}$ and take $\mathbf{x}_0 = [\frac{1}{3} \;\; \frac{1}{3} \;\; \pi/6]^T$, then

$$\mathbf{f}'(\mathbf{x}) = \begin{bmatrix} \dfrac{1}{2\sqrt{x}} & \dfrac{1}{y} & -\dfrac{1}{2}\sin z \\ \dfrac{y}{x} & \ln(4x) & -(2 + e^z) \\ 4 & \sin z & y\cos z \end{bmatrix}; \qquad \text{so } \mathbf{f}'(\mathbf{x}_0) \doteq \begin{bmatrix} 0.866 & 3 & -0.250 \\ 0.333 & 0.288 & -3.69 \\ 4 & 0 & 0.333 \end{bmatrix} \tag{25}$$

Since $\mathbf{f}'(\mathbf{x}_0)$ is strictly dominant, the Dominance Theorem of Section 4.3B suggests that if $\mathbf{f}'(\mathbf{x})$ is sufficiently close to $\mathbf{f}'(\mathbf{x}_0)$, then Gauss–Seidel iteration will converge if we repeatedly:

$$\text{Solve } (E_3) \text{ for } x = \tfrac{1}{4}(1 - y\sin z) \tag{26a}$$

$$\text{then } (E_1) \text{ for } y = e^{-[(\cos z)/2 + \sqrt{x}]} \tag{26b}$$

$$\text{then } (E_2) \text{ for } z = \tfrac{1}{2}[1 + y\ln(4x) - e^z] \tag{26c}$$

Notice that the variable solved for can also appear on the right-hand side [see z in (26c)]. Starting with $\mathbf{x}_0 = [\frac{1}{3} \;\; \frac{1}{3} \;\; \pi/6]^T$, the iteration (26) gives (to 6s) $\mathbf{x}_1, \ldots, \mathbf{x}_4$ shown in Table 4.4-1. It is evident that the iteration is converging to $\bar{\mathbf{x}} = [\frac{1}{4} \;\; 1/e \;\; 0]^T$.

TABLE 4.4-1 Gauss–Seidel Iteration for a Nonlinear System

	$\mathbf{x}_1$	$\mathbf{x}_2$	$\mathbf{x}_3$	$\mathbf{x}_4$	$\bar{\mathbf{x}}$
x_1	0.208333	0.288244	0.233121	0.260795	0.250000
y_2	0.410882	0.367535	0.377449	0.365168	0.367879
z_3	−0.381502	0.184741	−0.114646	0.061877	0.000000

■

Gauss–Seidel iteration is used in the Liebmann Process to solve nonlinear two-point boundary value problems [Section 8.5D].

In 1956 Broyden [9] described a method for solving nonlinear systems that generalizes the Secant Method. As with the SEC Method of Section 2.3D, the convergence of this **quasi-Newton method**, although faster than linear, is a bit slower than quadratic. However, **Broyden's method** uses only $O(n^2)$ arithmetic operations in addition to the evaluation of $\mathbf{f}(\mathbf{x})$ per iteration. A family of quadratically convergent methods that require substantially less arithmetic per iteration than NR_n was later described by Brown [8]. However, these **Newton-like methods** are considerably more difficult to implement than Broyden's method. The reader wishing to see quantitative comparisons of Newton-like and quasi-Newton methods is referred to [36]. It is often the case that the components of realistic solutions $\bar{\mathbf{x}}$ must satisfy inequalities of the form $x_j \geqslant a_j$ or $x_j \leqslant b_j$. Wegstein described an adaptation of the

Secant Method that can be used to solve $\mathbf{f}(\mathbf{x}) = \mathbf{0}$ in the presence of such constraints (see [42]).

PROBLEMS

Section 4.1

4-1 Solve the given linear system using the LU-decomposition algorithm with Scaled Partial Pivoting. Enter the values of the forback matrix $[\hat{L}\backslash\hat{U}:\hat{\mathbf{b}}:\bar{\mathbf{c}}:\bar{\mathbf{x}}_{\text{calc}}]$ rounded to $3s$ as in Example 4.1B. Then reuse your LU-decomposition to perform one iterative improvement on $\bar{\mathbf{x}}_{\text{calc}}$, again rounding to $3s$ as in Example 4.1C.
(a) $(\mathrm{E}_1)\ 1.3x - 0.536y = -5.23 \quad (\mathrm{E}_2)\ -0.24x + y = 9.76\ (\bar{\mathbf{x}} = [1 \quad 10]^T)$
(b) $(\mathrm{E}_1)\ 0.34p + 2.32v = -4.64 \quad (\mathrm{E}_2)\ 0.78p + 2.68v = 5.36\ (\bar{\mathbf{x}} = [0 \quad -2])$

4-2 Same as Problem 4-1 but entering forback matrix values *chopped*. Compare $\bar{\mathbf{x}}_{\text{calc}}$ and $\bar{\mathbf{x}}_{\text{improved}}$ to those obtained in Problem 4-1.

4-3 Same as Problem 4-1 but using $2s$. Find $C_p(A)$ and use it to predict the most significant digit of $\bar{\mathbf{x}}_{\text{calc}}$ that might be inaccurate. Was it (roughly) correct?

4-4 Same as Problem 4-3 but using cond A obtained with $\|\mathbf{x}\| = \|\mathbf{x}\|_1$.

4-5 This problem shows that pivoting can be important even if A is well conditioned. Consider the 2×2 linear system

$$(\mathrm{E}_1)\ \epsilon x + \left(\frac{1}{\epsilon}\right)y = \frac{1}{\epsilon} \qquad (\mathrm{E}_2)\ x + y = 1, \quad \text{where } 0 < \epsilon < 1$$

(a) Show that the exact solution is $\bar{\mathbf{x}} = [0 \quad 1]^T$.
(b) Calculate $C_C(A)$ as $s_1 s_2/|\det A|$ [see (9) of Section 4.1B]. Deduce that A is ill conditioned for $\epsilon \approx 1$ but well conditioned for $\epsilon \approx 0$.
(c) Use the LU-Decomposition Algorithm 4.2C first with BP and then with SPP to solve the system with $\epsilon = 0.01$. Store forback matrix entries rounded to $4s$. What conclusion(s) can you draw from the result?

4-6 For the coefficient matrix in Problem 4-5, find cond A as a formula in terms of ϵ **(a)** if $\|\mathbf{x}\| = \|\mathbf{x}\|_1$ **(b)** if $\|\mathbf{x}\| = \|\mathbf{x}\|_\infty$. Does your formula give the same results as obtained in Problem 4-5(b)?

4-7 Use (7) of Section 3.1C to obtain a formula for finding $\text{row}_n\hat{A}$ from a given LU-decomposition of A *without finding all of* $\hat{A}$. Then, for parts (a)–(f) of your choice in Problem 3-25, find $C_p(A)$, using your formula to determine the nth row scale factor $\hat{s}_n$ efficiently.

4-8 Suppose that $A_{n\times n}\mathbf{x} = \mathbf{b}$ is solved for $\bar{\mathbf{x}}_{\text{calc}}$ on a $7s$ device and that C is a condition number as in (7b) and (11) of Section 4.1B. How many accurate significant digits of $\bar{\mathbf{x}}_{\text{calc}}$ should you expect? Explain. (Include the possibility of accumulated roundoff error for "large" n.)
(a) $n = 4, C = 90$ **(b)** $n = 90, C = 4$ **(c)** $n = 100, C = 750$

4-9 For the matrices A whose inverse was found in Problem 3-32: Use the Matrix Norm Representation Theorem of Section 4.1D to find $\|A\|$, $\|A^{-1}\|$, and cond A when $\|\mathbf{x}\|_1$ is used as $\|\mathbf{x}\|$.

4-10 Same as Problem 4-9 but with $\|\mathbf{x}\|_\infty$ used as the norm of $\mathbf{x}$.

Section 4.2

4-11 For the matrix A in (12) of Practical Consideration 4.2D:
(a) Verify that the Crout decomposition formulas (7) yield the $L\backslash U_1$ shown (see $L\backslash U$ obtained in Problem M3-6).
(b) For the $L\backslash D\backslash U$ shown, form a unit L_1 from L, a unit U_1 from U, and a diagonal matrix D from $\backslash D\backslash$, and show that $LDU = A$.

4-12 How would you modify the forback calculations to use $L\backslash D\backslash U$ obtained using the formulas (6) of Section 4.2B?

4-13 **(a)** Use the Crout Decomposition Algorithm 4.2D to solve the stored system (13) of Example 4.1B. Store entries rounded to 3*s* as in the example. Compare the accuracy to that obtained in Example 4.1B. Explain any differences.

(b) Perform one iterative improvement on your answer to part (a), reusing your Crout decomposition (as in Example 4.1C) to obtain $\bar{\mathbf{e}}_{calc}$. Compare the accuracy to that obtained in Example 4.1C. Explain any differences.

4-14 Find $N(\pm)$, $N(*)$, and $N(/)$ (as in Table 3.2-1 of Section 3.2G) for solving an $n \times n$ linear system $A\mathbf{x} = \mathbf{b}$ by the Choleski Factorization Algorithm 4.2E. Find the entries for $n = 2, 3, 5, 20, 100$, and 1000. Your answers should be about half as large as those of Problem 3-18(d) for large n. Are they?

4-15 For $[A:\mathbf{b}]$ given in parts (a)–(d): Use the Choleski factorization to determine if the symmetric matrix A is positive definite. If it is, use Algorithm 4.2E to solve $A\mathbf{x} = \mathbf{b}$. All systems have $\bar{\mathbf{x}} = [1 \quad -1 \quad 1]^T$ as their solution.

(a) $\begin{bmatrix} 5 & 3 & -2 & : & 0 \\ 3 & 2 & -2 & : & -1 \\ -2 & -2 & 5 & : & 5 \end{bmatrix}$ **(b)** $\begin{bmatrix} 2 & 3 & -1 & : & -2 \\ 3 & 4 & -2 & : & -3 \\ -1 & -2 & 1 & : & 2 \end{bmatrix}$

(c) $\begin{bmatrix} 18 & 0 & -8 & : & 10 \\ 0 & 8 & 12 & : & 4 \\ -8 & 12 & 22 & : & 2 \end{bmatrix}$ **(d)** $\begin{bmatrix} 3 & 0 & 1 & : & 4 \\ 0 & -3 & 0 & : & 3 \\ 1 & 0 & 3 & : & 4 \end{bmatrix}$

Section 4.3

4-16 Solve the given equations in such a way that Gauss–Seidel iteration is sure to converge. Then do four iterations starting with $\mathbf{x}_0 = \mathbf{0}$.

(a)
$$\begin{aligned} 2x - 6y - z &= -3 \\ -8x + 3y + z &= -4 \\ x + y - 3z &= 5 \end{aligned}$$

(b)
$$\begin{aligned} r - s \qquad + 3u \qquad &= 1 \\ 5r \qquad + t \qquad - v &= 3 \\ 4s + 3t - u \qquad &= -1 \\ 3t + u - 5v &= -4 \\ r + s \qquad\qquad - 4v &= -3 \end{aligned}$$

4-17 For the linear systems described in (a)–(d) of Problem 4-15: Solve (E_1) for x_1, (E_2) for x_2, and (E_3) for x_3. Then use Gauss–Seidel iteration starting with $\mathbf{x}_0 = [\frac{1}{2} \quad -\frac{1}{2} \quad \frac{1}{2}]^T$, to get $\mathbf{x}_1$, $\mathbf{x}_2$, $\mathbf{x}_3$. Should you expect the iteration to converge? Explain.

Section 4.4

4-18 Find $\mathbf{f}'(\mathbf{x})$. Then use NR_2 to find $\mathbf{x}_1$ and $\mathbf{x}_2$ for the given $\mathbf{x}_0$.

(a) (E_1) $x^2 + y^2 - 4 = 0$ (E_2) $y - (x - 1)^2 = 0$; $\mathbf{x}_0 = [2 \quad 1]^T$

(b) (E_1) $xy - 1 = 0$ (E_2) $y + x^2 - 3 = 0$; $\mathbf{x}_0 = [1 \quad 1]^T$

4-19 Do (i) and (ii) for the given 2×2 nonlinear system:

(i) Use the graphs of the equations (perhaps in an equivalent form that is easier to sketch) to obtain estimates of *all* roots.

(ii) Use NR_2 to find all roots to 4*s*. Use the estimates obtained graphically in (i) as initial guesses.

(a) $x = 2 \ln y$, $y = xy - 1$ **(b)** $v^2 = 9 - u$, $v = \ln u$ **(c)** $2 \sin r - s = 1$, $r^2 - s^2 = 4$

4-20 Rewrite the given 3×3 linear system as $\mathbf{f}(\mathbf{x}) = \mathbf{0}$, and find $\mathbf{f}'(\mathbf{x})$. Then use NR_3 to find $\mathbf{x}_1$ for the given $\mathbf{x}_0$. You should be able to solve $\mathbf{f}'(\mathbf{x}_0)\, \mathbf{dx} = -\mathbf{f}(\mathbf{x}_0)$ for $\mathbf{dx}_0$ almost by inspection.

(a) $x^2 - y^2 + z^3 = -5$, $4x + y^3 + z = 7$, $xz + 5y = 1$, $\mathbf{x}_0 = \begin{bmatrix} 1 \\ 0 \\ -1 \end{bmatrix}$ **(b)** $x^2 - 3\sin y = z^2$, $2xy - z = 1$, $e^{x \pm y} + z^2 = 0$, $\mathbf{x}_0 = \begin{bmatrix} 0 \\ 0 \\ 0 \end{bmatrix}$

4-21 Use Gauss–Seidel iteration as in Example 4.4E to solve Problem 4-20(a) for $\bar{\mathbf{x}}$ near the given $\mathbf{x}_0$ to $3s$.

4-22 Use a numerical method of Section 2.3 to solve the equation in (5) of Section 4.4A.

MISCELLANEOUS PROBLEMS

M4-1 Use any available means to solve $A_{\text{stored}}\mathbf{x} = \mathbf{b}_{\text{stored}}$ in (13) of Example 4.1B to at least $5s$. Does $\bar{\mathbf{x}}_{\text{calc}} = [.126 \quad 9.37 \quad 4.90]^T$ obtained in (16) approximate this solution to about $3s$?

M4-2 If iterative improvement finds $\bar{\mathbf{e}}$ exactly, then $\bar{\mathbf{x}}_{\text{improved}}$ will be exactly $\bar{\mathbf{x}}$. Why? Verify this for the given erroneous $\bar{\mathbf{x}}_{\text{calc}}$.
(a) $(E_1)\ x + y = -1 \quad (E_2)\ -2x + 3y = 7; \quad \bar{\mathbf{x}}_{\text{calc}} = [-1 \quad 5]^T$
(b) $(E_1)\ 2x + 2y = 4 \quad (E_2)\ 4x + 3y = 5; \quad \bar{\mathbf{x}}_{\text{calc}} = [-2 \quad 1]^T$

M4-3 Modify the Iterative Improvement Algorithm 4.1C so as to improve $(A^{-1})_{\text{calc}}$. Use $(A^{-1})_{\text{calc}}$ itself (rather than the LU-decomposition) to get the E_{calc} to add to $(A^{-1})_{\text{calc}}$ to get $(A^{-1})_{\text{improved}}$. To test your algorithm, perform *two* iterative improvements on

$$(A^{-1})_{\text{calc}} = \begin{bmatrix} -1.9 & 1.1 \\ 1.6 & -0.6 \end{bmatrix}, \qquad \text{where } A = \begin{bmatrix} 1 & 2 \\ 3 & 4 \end{bmatrix} \quad \text{for which } A^{-1} = \begin{bmatrix} -2 & 1 \\ \frac{3}{2} & -\frac{1}{2} \end{bmatrix}$$

M4-4 Show that $\|\mathbf{x}\|$ satisfies (24a)–(24c) of Section 4.1D.
(a) $\|\mathbf{x}\| = \|\mathbf{x}\|_1$ **(b)** $\|\mathbf{x}\| = \|\mathbf{x}\|_\infty$ **(c)** $\|\mathbf{x}\| = \|\mathbf{x}\|_2\ (= \sqrt{\mathbf{x}^T\mathbf{x}})$
For $\|\mathbf{x}\|_2$, use $\|\mathbf{x}^T\mathbf{y}\| \leq \|\mathbf{x}\|_2\|\mathbf{y}\|_2$ (**Cauchy–Schwarz inequality**) to prove (24d).

M4-5 Prove the following results for $\|A\| = \max\{\|A\mathbf{x}\|/\|\mathbf{x}\|\}_{\mathbf{x}\neq\mathbf{0}}$. Assume that the norm on n-space satisfies (24a)–(24c) of Section 4.1D.
(a) $\|A\|$ satisfies (24a)–(24d) **(b)** $\|A\| = \max\{\|A\mathbf{x}\|/\|\mathbf{x}\|\}_{\|\mathbf{x}\|=1}$

M4-6 For $\|A\|$ defined in (25a) of Section 4.1D: Prove that if $\|\mathbf{x}\| = \|\mathbf{x}\|_1$ then $\|A\| = \|A\|_1$, where $\|A\|_1 = \max\{\|\text{col}_j A\|_1\}_{1\leq j\leq n}$. Do so by justifying the steps of the following outline.

Step 1: Show that for all $\mathbf{x}$, $\|A\| \leq \|A\|_1$ because

$$\|A\mathbf{x}\|_1 = \Sigma_i\Sigma_j|a_{ij}x_j \leq \Sigma_i\Sigma_j|a_{ij}||x_j| = \Sigma_j(\Sigma_i|a_{ij}|)|x_j| \leq \|A\|_1\|\mathbf{x}\|_1$$

Step 2: Let $\text{col}_k A$ be the column of A having the largest absolute sum. Show that equality holds in Step 1 when $\mathbf{x}$ has $x_j = (\pm)1$, where $(\pm)$ is the sign of a_{jk}. Deduce that $\|A\| \geq \|A\|_1$, hence $\|A\| = \|A\|_1$.

M4-7 Prove that if $\|\mathbf{x}\| = \|\mathbf{x}\|_\infty$ then $\|A\| = \|A\|_\infty$, where $\|A\|_\infty = \max\{\|\text{row}_i A\|_1\}_{1\leq i\leq n}$. Use two steps, as was done in Problem M4-6.

M4-8 In what sense does every positive definite matrix have a "square root"?

M4-9 Do (i)–(iii) below for the given nonlinear 2×2 system.

(a) $x^2 + y^2 = a^2$, $xy = 1$ **(b)** $x^2 + y^2 = 1$, $2y^2 - x = a$

(i) Let $N(a)$ be the number of roots $\bar{\mathbf{x}}$ for the given a. Sketch the graphs of the equations and determine $N(a)$ for any real number a.

(ii) Solve (E_2) for y and substitute in (E_1) to eliminate y. Then use the quadratic formula to determine all positive values of a for which $N(a) = 0$, 2, and 4.

(iii) Use NR_2 with each given $\mathbf{x}_0$ to find $\mathbf{x}_1$, $\mathbf{x}_2$, $\mathbf{x}_3$, $\mathbf{x}_4$ to at least 6s. Use the values of Δx_k and Δy_k to determine if the convergence appears to be linear or quadratic.

For (a): $\mathbf{x}_0 = [2 \quad 0]^T$ with $a = 2$; $\mathbf{x}_0 = [1 \quad 0]^T$ with $a = \sqrt{2}$.
For (b): $\mathbf{x}_0 = [1 \quad 1]^T$ with $a = 0$; $\mathbf{x}_0 = [-1 \quad \frac{1}{2}]^T$ with $a = \frac{17}{8}$.

(iv) Relate your results in (iii) to your results in (ii).

M4-10 Use NR_2 to find to 4s the root near the given initial guesses $\mathbf{x}_0$.
(a) Problem 2 at the beginning of Section 1.1; $x_0 = [0.5 \quad 0.5]^T$.
(b) The system (6) of Section 4.4A; $\mathbf{x}_0 = [1 \quad -2.5]^T$ and $[-0.5 \quad 2]^T$.

M4-11 Show that Bairstow's method of Section 2.5C is an application of NR_2.

M4-12 ***Solving Complex Linear Systems Using Real Arithmetic***

(a) Suppose that $A = A_R + iA_I$ and $\mathbf{b} = \mathbf{b}_R + i\mathbf{b}_I$. Show that the solution $\bar{\mathbf{x}} = \bar{\mathbf{x}}_R + i\bar{\mathbf{x}}_I$ of the complex linear system $A\mathbf{x} = \mathbf{b}$ can be found by solving the two *real* linear systems

$$(A_R + A_I)\mathbf{x}_R = \tfrac{1}{2}(\mathbf{b}_I + \mathbf{b}_R) \qquad \text{and} \qquad (A_R - A_I)\mathbf{x}_I = \tfrac{1}{2}(\mathbf{b}_I - \mathbf{b}_R)$$

for $\bar{\mathbf{x}}_R$ and $\bar{\mathbf{x}}_I$, respectively. Use this fact to solve $A\mathbf{x} = \mathbf{b}$, where

$$A = \begin{bmatrix} 1+i & 2 & 0 \\ 2+i & 0 & i \\ 1-i & i & 1 \end{bmatrix} \qquad \text{and} \qquad \mathbf{b} = \begin{bmatrix} 1-i \\ 1+i \\ i \end{bmatrix}$$

COMPUTER PROBLEMS

NOTE: Problems C4-1–C4-4 ask you to modify code for solving a linear system using Algorithm 4.2B. You may replace the Fortran code shown in Figures 4.2-1–4.2-3 by any available code implementing Algorithm 4.2B. For each modification you make, state its advantages and disadvantages and compare the accuracy of $\bar{\mathbf{x}}_{\text{calc}}$ obtained using it to the accuracy obtained using Algorithm 4.2B for the system $H\mathbf{x} = \mathbf{b}$, in which H is the $n \times n$ **Hilbert matrix** $(h_{ij})_{n \times n}$, where $h_{ij} = 1/(i + j - 1)$, and $\mathbf{b}$ is $[1 \quad 1/2 \quad \cdots \quad 1/n]^T$ (so that $\bar{\mathbf{x}}$ is $\text{col}_1 I_n$), for $n = 3$, 5, and 10. Be warned that H is a well-known example of a symmetric matrix that gets ill conditioned rapidly as n increases.

C4-1 **(a)** Write a `SUBROUTINE GET` for the calling program in Figure 4.2-3.
(b) Modify Figure 4.2-3 so that it performs one iterative improvement if either $C_p(A) > 50$ or $n > 10$. It should output *both* $\bar{\mathbf{x}}_{\text{calc}}$ and $\bar{\mathbf{x}}_{\text{improved}}$.

C4-2 Modify `SUBROUTINE DECOMP` of Figure 4.2-1 to do some or all of parts (a)–(c). In all cases you will also have to modify `SUBROUTINE FORBAK` of Figure 4.2-2.
(a) Add a *column* pointer array `NEWCOL` to the parameter list in order to use Scaled Full Pivoting as described in Section 3.3F.
(b) Store $L\backslash D\backslash U$ and use Partial Extended Precision as described in Practical Consideration 4.2D (see Problem 4-12).
(c) Store A^T so that the summations along the rows of A vary over the *second* argument (j) of the stored `A`(i, j).

C4-3 Convert `DECOMP` and `FORBAK` to `CHLSKI` (Choleski factorization) and `CHFWBK`, respectively, so that they solve $A\mathbf{x} = \mathbf{b}$ using the Choleski Factorization Algorithm 4.2E when A is positive definite. Test using the systems given in (a) and (c) of Problem 4-15 before solving $H\mathbf{x} = \mathbf{b}$ of the note preceding Problem C4-1.

C4-4 Modify your subprogram in Problem C4-3 so that only the lower triangular part of A is stored in a one-dimensional array `A1`, having $n(n+1)/2$ entries, with a_{ij} stored in `A1`(k_{ij}), where $k_{ij} = i(i-1)/2 + j$ for $1 \leq i \leq j \leq n$. Test as in Problem C4-3.

C4-5 Explain why line 2500 of `SUBROUTINE DECOMP` (Figure 4.2-1) was coded as `ABS(A(II),M) .GT. MAXRAT*S(II)` rather than `ABS(A(II),M)/S(II) .GT. MAXRAT`.

C4-6 Modify Figure 4.2-3 (but *not* `DECOMP` and `FORBAK`) so as to solve $A\mathbf{x} = \mathbf{b}$ as described in Problem M4-12 when the entries of A and $\mathbf{b}$ are complex numbers. Test by solving the system given in Problem M4-12 and then finding A^{-1}.

C4-7 Modify Figure 4.2-3 so as to solve k linear systems using Algorithm 3.4A. If $k = 0$, it should find $(A_{n\times n}^{-1})$ one column at a time as described in Section 3.4B.

C4-8 Write a program `GAUSDL` that does up to *MaxIt* Gauss–Seidel iterations. The program should read *MaxIt*, n, $A(n, n)$, $\mathbf{b}$, and pointer arrays **Ieqn** and **Jvar** where, for example,

$$\mathbf{Ieqn} = [3 \quad 1 \quad 2] \qquad \text{and} \qquad \mathbf{Jvar} = [2 \quad 3 \quad 1] \qquad (n = 3)$$

means solve (E_3) for x_2, then (E_1) for x_3, then (E_2) for x_1. It should call a subprogram `SEIDEL` to do the actual iteration. Test using Example 4.3C.

C4-9 Modify `GAUSDL` of Problem C4-8 so as to use Aitken's formula on the last three x_j's when two successive dx_j/(previous dx_j) ratios have a *relative* difference less than *Thresh*. Test using Example 4.3C.

C4-10 Write a subprogram `NR2(MAXIT, NUMDIG, X, CONVGD)` that performs up to *MaxIt* iterations of NR_2 to find *NumDig* accurate digits of the solution near a given $\mathbf{x}_0$ (the initialized X) of a 2×2 nonlinear system. It should call:

`EVAL(X, F, J)`, provided by the user, to evaluate `F` $= \mathbf{f}(\mathbf{x})$ and `J` $= \mathbf{f}'(\mathbf{x})$ at the current $\mathbf{x}_k$ (stored in `X`).

`SOLVE(X, F, J, DX, TOL, CONVGD)` to find `DX` $=$ `-J`$^{-1}$`F` $(= \mathbf{dx}_k)$ using the formula for $J_{2\times 2}^{-1}$ (see Problem 3-10), along with `CONVGD` = TRUE if the convergence test of Algorithm 4.4C (with $Tol = 10^{-NumDig}$) is satisfied. It should also restrict the length of `DX(1)` and `DX(2)` using the strategy given in (21) of Section 4.4D.

If convergence occurs, `NR2` should return $\overline{\mathbf{x}}_{calc}$ in the array `X` and `CONVGD` = TRUE. If not, or if a singular `J` occurs, it should return `CONVGD` = FALSE. Test `NR2` using Problem 4-10.

C4-11 Modify `NR2` of Problem C4-10 to become `NRN(N, MAXIT, NUMDIG, X, NUMUSE, CONVGD)` by modifying subprograms `EVAL` and `SOLVE` as follows:

`EVAL(N, X, F, NEWJ, J)` calls `EVALF(X, F)`, provided by the user, evaluates `F` $= \mathbf{f}(\mathbf{x})$ at $\mathbf{x}_k$ stored in `X`. Then, if `NEWJ` is TRUE, `SUBROUTINE JACOB(N, X, F, J)` should be called to evaluate `J` $= \mathbf{f}'(\mathbf{x})$ *numerically* at $\mathbf{x}_k$, using Steffensen's stepsize h_{ij} to approximate $\partial f_i(\mathbf{x}_k)/\partial x_j$ using (22) of Section 4.4C but with $|h_{ij}|$ restricted to be *at most* $0.2|x_j|$ but *at least* $\max\{Tol, 0.002|x_j|\}$.

`SOLVE(N, X, F, J, DX, TOL, CONVGD)` calls subprograms such as `DECOMP` and `FORBAK` of Section 4.2B (with `EPS` = `TOL`) to solve `J`$\mathbf{dx}$ = `-F` for `DX`.

Subprogram `NRN` should set `NEWJ` to TRUE; only if k (the iteration counter) is a multiple of `NUMUSE` (see Shortcoming 3 of Section 4.4D). Test using Example 4.4C.

5

Curve Fitting and Function Approximation

5.0

Introduction

We now turn our attention from the algebraic problem of solving equations to the geometric problem of finding a **guess function** $g(x)$ whose graph has a prescribed shape. Two situations when such an approximation is needed are described in this chapter. In the first, the shape is determined by a *finite* number of experimentally determined, hence *erroneous*, data points $P_1, P_2, \ldots, P_M$ [Figure 5.0-1(a)]. This is the **discrete curve-fitting problem**. It is considered in Sections 5.1–5.3. In the second situation, the shape is determined by part of the

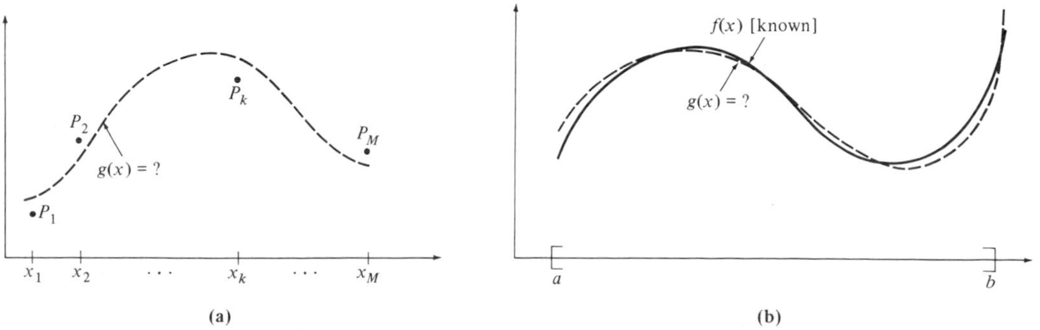

Figure 5.0-1 *The curve-fitting problem: (a) discrete; (b) continuous.*

graph of a function $f(x)$, which is *known* but *difficult to evaluate* [Figure 5.0-1(b)]. This *continuous* curve-fitting problem will be referred to as the **approximation problem**. It is considered in Section 5.4.

5.1 Curve Fitting Using the Least-Square-Error Criterion

A function $g(x)$ is said to **fit** a given set of **data points** $P_1(x_1, y_1)$, $P_2(x_2, y_2)$, ..., $P_M(x_M, y_M)$ if its graph comes near (*but not necessarily through*) the points in the xy-plane. The problem of finding a $g(x)$ to fit a given set of data arises frequently in science and engineering. This *geometric* problem is usually solved by transforming it to the *algebraic* problem of solving equations as described in this section.

5.1A The Discrete Curve-Fitting Problem

The following example illustrates a situation where there is a need for a curve that provides "good fit" to experimental data.

EXAMPLE 5.1A Calibrating a spring Let F denote the force required to stretch a spring x units from its natural length L (Figure 5.1-1). For small deflections, F will be proportional to x, that is, there will be a **spring constant**, which we will denote by κ, such that

$$F = \kappa x \qquad \text{(this is \textbf{Hooke's law})} \tag{1}$$

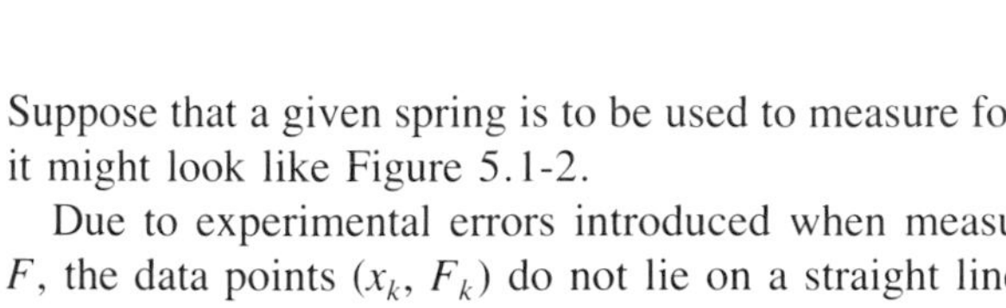

F = force needed to stretch the spring x units beyond its natural length L

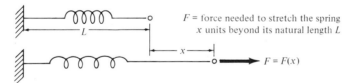

Figure 5.1-1 *Stretching a spring x units beyond its natural length.*

Suppose that a given spring is to be used to measure forces. Measurements made to calibrate it might look like Figure 5.1-2.

Due to experimental errors introduced when measuring both the length x and the force F, the data points (x_k, F_k) do not lie on a straight line. However, they lie close enough to one for $0 \leq x \leq 9$ to indicate that Hooke's law (1) holds for x's in this range. The spring constant κ can therefore be estimated *graphically* by drawing the straight line through the origin that appears to come closest to the data points for $0 \leq x \leq 9$, and then getting κ as the *slope* of this line, as shown in Figure 5.1-2. Since the x- and F-scales can be read accurately to about 3s, we can expect the graphical estimate $\kappa \approx 1.72$ to be accurate to about 3s.

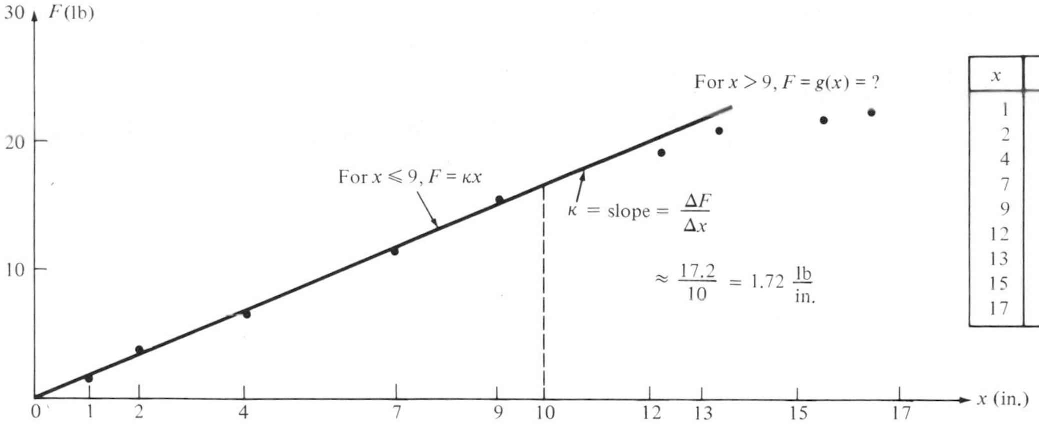

x	F
1	1.5
2	3.9
4	6.6
7	11.7
9	15.6
12	18.8
13	19.6
15	20.6
17	21.1

Figure 5.1-2 *Experimentally determined values of F for some x's.*

The calibration of the spring is not so simple for $x > 9$. Indeed, since (1) no longer applies, we must first *find* a function $g(x)$ whose graph has the general shape of the data for these x's, say

$$g(x) = \alpha + \beta \ln x \quad \text{or} \quad g(x) = \alpha x^{\beta} \quad \text{or} \quad g(x) = \frac{\alpha}{\beta + x} \tag{2a}$$

logarithmic curve power curve hyperbolic curve

and *then* determine values of the **parameters** α and β that make the graph of $g(x)$ fit the data well for $x > 9$. We can then calibrate the spring as

$$F(x) = \begin{cases} 1.72x & \text{for } 0 \leq x \leq 9 \\ g(x) & \text{for } 9 < x \leq 17 \end{cases} \tag{2b}$$

Since there is no simple geometric way to find $g(x)$, it must be found numerically. We shall do so in Example 5.2C. ■

5.1B The Least-Square-Error Criterion

Suppose we want a formula that describes how a physical quantity (which we will denote by y) varies as a function of a second quantity (which we will denote by x), and that M **data points**

$$P_1(x_1, y_1), \quad P_2(x_2, y_2), \quad \ldots, \quad P_M(x_M, y_M) \tag{3}$$

have been determined experimentally to help establish this functional relationship. Suppose further that either our understanding of the relationship or an examination of a plot of the data points leads us to select a continuous, n-parameter **guess function** $g(x)$ such that

$$g(x) \approx \text{functional dependence of } y \text{ on } x \tag{4}$$

Since the points (x_k, y_k) are experimental, they are not likely to lie on the exact (unknown) y versus x curve. However, they represent all we know about the curve; so any quantitative indicator of how well $g(x)$ fits the data must be based on the M numbers

$$\delta_k = g(x_k) - y_k = \text{the } \mathbf{deviation\ of\ } \boldsymbol{g(x)} \mathbf{\ at\ } \boldsymbol{x_k} \tag{5}$$

which measure the vertical distance from the data points $P_k(x_k, y_k)$ to the graph of g for $k = 1, \ldots, M$ (Figure 5.1-3). Three such error indicators are

The **absolute error** of g: $E_1(g) = |\delta_1| + |\delta_2| + \cdots + |\delta_k| = \sum_{k=1}^{M} |\delta_k|$ **(6a)**

The **square error** of g: $E_2(g) = \delta_1^2 + \delta_2^2 + \cdots + \delta_M^2 = \sum_{k=1}^{M} \delta_k^2$ **(6b)**

The **maximum error** of g: $E_\infty(g) = \max\{|\delta_1|, |\delta_2|, \ldots, |\delta_M|\}$ **(6c)**

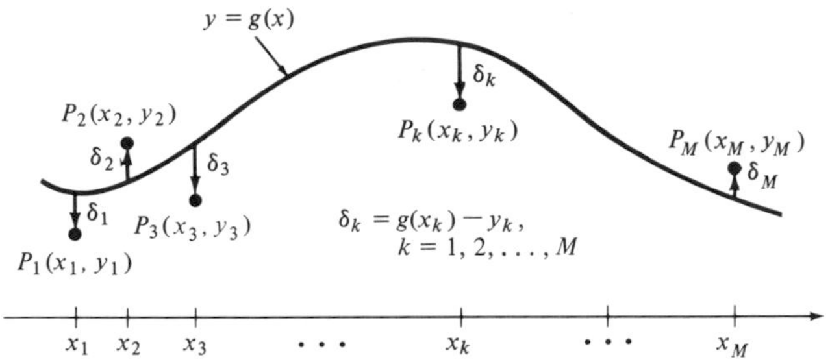

Figure 5.1-3 *M deviations, $\delta_1, \ldots, \delta_M$ at $x_1, \ldots, x_M$.*

If $E(g)$ is taken to be any of $E_1(g)$, $E_2(g)$, *or* $E_\infty(g)$, then $E(g)$ is zero when the graph of g passes *through* $P_1, \ldots, P_M$ [i.e., when $g(x_k) = y_k$ for all k], and $E(g)$ gets larger as the graph of g moves away from the P_k's. So *a small value of $E(g)$ indicates that $g(x)$ fits the data well.*

The square error $E_2(g)$ is by far the most widely used indicator of how well $g(x)$ fits a finite number of data points. There are two reasons for this. The first is pragmatic: The mathematical problem of minimizing $E(g)$ turns out to be especially easy when $E(g)$ is $E_2(g)$. The second is statistical [14]: If the error of each y_k is normally distributed (i.e., if its density function is the familiar "bell-shaped" curve) and $g(x)$ has the correct functional form, then the $g(x)$ obtained by minimizing $E_2(g)$ will approach the actual functional dependence as M (the number of data points) increases. Consequently, from this point on,

$E(g)$ will mean square error $E_2(g)$

and "best fit" will be achieved when the parameters of $g(x)$ are adjusted so as to

$$\text{minimize } E(g), \qquad \text{where } E(g) = \sum_{k=1}^{M} [g(x_k) - y_k]^2 \tag{7}$$

to derivatives w.r.t. the parameters of g, set to zero and solve that is, when $g(x)$ has the **least square error**. The $g(x)$ with the minimum $E(g)$ will be called the **least square $g(x)$** and denoted by† $\hat{g}(x)$. The following example shows how to get $\hat{g}(x)$ for a one-parameter $g(x)$.

EXAMPLE 5.1B For the data on the "linear" part of Figure 5.1-2, namely

$$P_1(1, 1.5),\quad P_2(2, 3.9),\quad P_3(4, 6.6),\quad P_4(7, 11.7),\quad P_5(9, 15.6) \tag{8}$$

(a) Find the square error $E(g)$ as a function of κ for $g(x) = \kappa x$.
(b) Find the parameter value $\hat{\kappa}$ that minimizes $E(g)$. What is $E(\hat{g})$?

SOLUTION (a): Taking $g(x) = \kappa x$, $y_k = F_k$, and $M = 5$ in (7) yields

$$E(g) = (\kappa x_1 - F_1)^2 + (\kappa x_2 - F_2)^2 + \cdots + (\kappa x_5 - F_5)^2 \tag{9}$$

$$= \kappa^2 \sum_{k=1}^{M} x_k^2 - 2\kappa \sum_{k=1}^{M} x_k F_k + \sum_{k=1}^{M} F_k^2 \tag{10}$$

For the data (8), $\Sigma_{k=1}^{M} x_k^2 = 1^2 + 2^2 + 4^2 + 7^2 + 9^2 = 151$,

$$\sum_{k=1}^{M} x_k F_k = 1(1.5) + 2(3.9) + 4(6.6) + 7(11.7) + 9(15.6) = 258$$

and $\Sigma_{k=1}^{M} F_k^2 = 441.27$. Substituting in (10) yields the desired result

$$E(g) = 151\kappa^2 - 2\kappa(258) + 441.27 \tag{11}$$

SOLUTION (b): $E(g)$ is a quadratic in κ with a positive coefficient of κ^2. So its graph is a parabola that opens upward, and the unique $\hat{\kappa}$ that minimizes $E(g)$ can be found either by completing the square or by using calculus as follows:

$$\frac{dE(g)}{d\kappa} = 2\kappa(151) - 2(258) = 0 \Leftrightarrow \kappa = \hat{\kappa} = \frac{258}{151} \doteq 1.71 \tag{12}$$

Thus for $0 \leqslant x \leqslant 9$, a desirable calibration formula for the spring is

$$F = \hat{g}(x), \qquad \text{where } \hat{g}(x) = \hat{\kappa}x \doteq 1.71x \tag{13}$$

because the graph of $\hat{g}(x)$ comes closer to the data (8) (in the square error sense) than any other straight line through the origin. This confirms that the value $\kappa = 1.72$, obtained graphically in Figure 5.1-2, was in fact accurate to about 3*s*. The least square error, $E(\hat{g}) = 0.45$, is obtained by taking $\kappa = \hat{\kappa} = 258/151$ in either (11) or (9). ■

†In this chapter, a caret (ˆ) will denote "best" in the least-square-error sense.

5.1C The General Linear Fit: $L(x) = mx + b$

To illustrate how easy a two-parameter fit can be, consider the special guess function whose graph is a straight line:

$$L(x) = mx + b \qquad (m = \text{slope},\ b = y\text{-intercept}) \tag{14}$$

Given M data points $P_1(x_1, y_1), \ldots, P_M(x_M, y_M)$, our objective is to find parameter values $\hat{m}$ and $\hat{b}$ that minimize the square error

$$E(L) = \sum_{k=1}^{M} (mx_k + b - y_k)^2 = (mx_1 + b - y_1)^2 + \cdots + (mx_M + b - y_M)^2 \tag{15}$$

We know from calculus that the desired b and m satisfy

$$\frac{\partial E(L)}{\partial b} = 0 \quad \text{and} \quad \frac{\partial E(L)}{\partial m} = 0 \tag{16}$$

Since the x_k's and y_k's are constant and $\Sigma_{k=1}^{M} b = Mb$,

$$\begin{aligned}\frac{\partial E(L)}{\partial b} &= \sum_{k=1}^{M} 2(mx_k + b - y_k)\cdot\frac{\partial}{\partial b}(mx_k + b - y_k) \\ &= \sum_{k=1}^{M} 2(mx_k + b - y_k)\cdot 1 = 2\left(Mb + m\cdot\sum_{k=1}^{M} x_k - \sum_{k=1}^{M} y_k\right)\end{aligned} \tag{17a}$$

(recall that $\partial/\partial b$ is performed with m held fixed), and similarly

$$\frac{\partial E(L)}{\partial m} = \sum_{k=1}^{M} 2(mx_k + b - y_k)x_k = 2\left(m\sum_{k=1}^{M} x_k^2 + b\sum_{k=1}^{M} x_k - \sum_{k=1}^{M} x_k y_k\right) \tag{17b}$$

Equating the partial derivatives (17) to zero yields the *linear* system†

$$Mb + (\Sigma\, x_k)m = \Sigma\, y_k \quad \text{and} \quad (\Sigma\, x_k)b + (\Sigma\, x_k^2)m = \Sigma\, x_k y_k \tag{18a}$$

The matrix form of this 2×2 linear system is

NORMAL EQUATIONS FOR $L(x)$:

$$\begin{bmatrix} M & \Sigma\, x_k \\ \Sigma\, x_k & \Sigma\, x_k^2 \end{bmatrix}\begin{bmatrix} b \\ m \end{bmatrix} = \begin{bmatrix} \Sigma\, y_k \\ \Sigma\, x_k y_k \end{bmatrix} \tag{18b}$$

Note the order (b then m) in the vector of unknowns $[b \quad m]^T$.

Once values are substituted for M, x_k, and y_k in the system (18), it can be solved easily for the **least square parameters** $\hat{m}$ and $\hat{b}$. Putting $\hat{m}$ and $\hat{b}$ in $L(x)$ then yields $\hat{L}(x)$, the **least square line** or **regression line** for the data $P_1, \ldots, P_M$. Thus

†Throughout this chapter, $\Sigma_{k=1}^{M}$, the sum over all data points, will often be abbreviated as Σ_k, or simply Σ as in (18).

$$\hat{L}(x) = \hat{m}x + \hat{b} \text{ satisfies } E(\hat{L}) \leq E(\mathrm{L}) \quad \text{for any } L(x) = mx + b \tag{19}$$

(see Problem M5-2). The least square error, $E(\hat{L})$, can then be found by putting $\hat{m}$ and $\hat{b}$ in the special formula

$$E(\hat{L}) = \Sigma\, y_k^2 - [\hat{b}(\Sigma\, y_k) + \hat{m}(\Sigma\, x_k y_k)] \tag{20}$$

[see Problem M5-8(a)]. This formula generally requires less calculation than (15) because the summations $\Sigma\, y_k$ and $\Sigma\, x_k y_k$ were obtained previously as the components of the target vector of the normal equations (18b). However, it should *only* be used for the least square line $\hat{L}(x)$.

EXAMPLE 5.1C Consider the five data points

$$P_1(1, 5.12), \quad P_2(3, 3), \quad P_3(6, 2.48), \quad P_4(9, 2.34), \quad P_5(15, 2.18) \tag{21}$$

Find $\hat{L}(x) = \hat{m}x + \hat{b}$ for these data, then use its square error $E(\hat{L})$ to assess how well the least square line $y = \hat{L}(x)$ fits the data.

SOLUTION: For the data (21), $M = 5$ and $\Sigma\, y_k^2 = 51.5928$. Also,

$$\Sigma\, x_k = 34, \quad \Sigma\, x_k^2 = 352, \quad \Sigma\, y_k = 15.12, \quad \text{and} \quad \Sigma\, x_k y_k = 82.76 \tag{22a}$$

Hence the matrix form (18b) of the normal equations is

$$\begin{bmatrix} 5 & 34 \\ 34 & 352 \end{bmatrix} \begin{bmatrix} b \\ m \end{bmatrix} = \begin{bmatrix} 15.12 \\ 82.76 \end{bmatrix} \tag{22b}$$

which can easily be solved for the least square parameters using the formula for $A_{2\times 2}^{-1}$ [see (16) of Section 3.1E] as follows:

$$\begin{bmatrix} \hat{b} \\ \hat{m} \end{bmatrix} = \frac{1}{604} \begin{bmatrix} 352 & -34 \\ -34 & 5 \end{bmatrix} \begin{bmatrix} 15.12 \\ 82.76 \end{bmatrix} = \begin{bmatrix} 4.15289 \\ -0.166027 \end{bmatrix} \tag{22c}$$

So $\hat{L}(x) \doteq -0.1660x + 4.153$; and by (20), its square error is

$$E(\hat{L}) = 51.5928 - [(4.15298)(15.12) + (-0.166027)(82.76)] \doteq 2.54 \tag{23}$$

Since $E(\hat{L})$ is rather large (for five points with y_k values between 2 and 6) we can see *without a sketch* that *a straight line does not fit the data (21) particularly well*. For a better fit, we need a guess function with a "curved" graph as described in Section 5.2. ■

Practical Consideration 5.1C (Fitting straight lines on a digital device). Hand calculators with built-in linear regression routines accumulate the sums required for (18b) during data entry and then solve (18b) for $\hat{b}$ and $\hat{m}$ using the formula for $A_{2\times 2}^{-1}$ (see Problem 3-10). Figure

```
00100        SUBROUTINE LINFIT (M, X, Y,  YCEPT, SLOPE, ROFL)
00200        DIMENSION  X(M), Y(M)
00300        DOUBLE PRECISION SUMX, SUMY, SUMXY, SUMX2, SUMY2, DET, RNUMER
00400   C - - - - - - - - - - - - - - - - - - - - - - - - - - - - - C
00500   C  INPUT:  M and the data arrays X = [x1...xM] and Y = [y1...yM]  C
00600   C  RETURNED VALUES:  SLOPE = m^,  YCEPT = b^,  and  ROFL = R(L^)  C
00700   C                    for the least square line L^(x) = m^x + b^  C
00800   C - - - - - - - - - - - - - - - - VERSION 2: 9/9/85  - - C
01000        SUMX = 0.D0
01100        SUMY = 0.D0
01200        SUMXY = 0.D0
01300        SUMX2 = 0.D0
01400        SUMY2 = 0.D0
01500        DO 10 K=1,M
01600           SUMX = SUMX + DBLE(X(K))
01700           SUMY = SUMY + DBLE(Y(K))
01800           SUMXY = SUMXY + DBLE(X(K))*Y(K)
01900           SUMX2 = SUMX2 + DBLE(X(K))**2
02000           SUMY2 = SUMY2 + DBLE(Y(K))**2
02100    10  CONTINUE
02200   C   **Solve the normal equations for SLOPE and YCEPT, and get ROFL
02300        DET =  M*SUMX2 - SUMX**2
02400        SLOPE = SNGL( (M*SUMXY - SUMX*SUMY)/DET )
02500        YCEPT = SNGL( (SUMX2*SUMY - SUMX*SUMXY)/DET )
02600        RNUMER = YCEPT*SUMY + SLOPE*SUMXY - SUMY*SUMY/M
02700        ROFL = SNGL( RNUMER/(SUMY2 - SUMY*SUMY/M) )
02800        RETURN
02900        END
```

Figure 5.1-4 *Fortran* SUBROUTINE *for fitting straight lines.*

5.1-4 shows a Fortran SUBROUTINE LINFIT that gets $\hat{m}$ (SLOPE) and $\hat{b}$ (YCEPT) this way, and returns them along with $R(\hat{L})$ (DETIND), an indicator of good fit obtained using (25) obtained in Section 5.1D. ∎

5.1D Using the Determination Index $R(g)$ to Assess Good Fit

The square error $E(g) = \sum_{k=1}^{M} [g(x_k) - y_k]^2$ has two serious deficiencies as an indicator of good fit. First, it can be large simply because M (the number of data points) is large, even if the fit is fairly good. More important, it depends on the scale used for the y values. For example, if y has units of length, then $E(g)$ can be reduced by a factor of 10,000 simply by measuring y in meters rather than centimeters!

Statisticians have shown that in the important case when $g(x)$ is a weighted sum of n functions $\phi_1(x), \ldots, \phi_n(x)$, that is, when

$$g(x) = \gamma_1\phi_1(x) + \cdots + \gamma_n\phi_n(x) \qquad [g(x) \text{ is a \textbf{linear model}}] \qquad \textbf{(24a)}$$

the deficiencies of $E(g)$ can be avoided by using the **index of determination** of $g(x)$, defined as the ratio

$$R(g) = \frac{\sum_{k=1}^{M} [g(x_k) - \hat{y}]^2}{\sum_{k=1}^{M} (y_k - \hat{y})^2}, \qquad \text{where } \hat{y} = \frac{1}{M}\left(\sum_{k=1}^{M} y_k\right) \tag{24b}$$

to measure how well $g(x)$ fits the data $P_1, \ldots, P_M$. The denominator of $R(g)$ is the **variance** of the y_k's; it is a measure of their dispersion about their **mean** $\hat{y}$. The numerator indicates how much of this dispersion is accounted for when y_k is replaced by $g(x_k)$. It can be shown [14] that *the least square* $g(x)$, that is, $\hat{g}(x)$, satisfies

$$0 \leq R(\hat{g}) \leq 1, \qquad \text{with good fit corresponding to } R(\hat{g}) \approx 1 \tag{24c}$$

Moreover, for the special case of the least square line $\hat{L}(x) = \hat{m}x + \hat{b}$ [weighted sum of $\phi_1(x) = 1$ and $\phi_2(x) = x$] the index of determination can be obtained as

$$R(\hat{L}) = \frac{\hat{b}\left(\sum_{k=1}^{M} y_k\right) + \hat{m}\left(\sum_{k=1}^{M} x_k y_k\right) - M\hat{y}^2}{\left(\sum_{k=1}^{M} y_k^2\right) - M\hat{y}^2}, \qquad \text{where } \hat{y} = \frac{1}{M}\left(\sum_{k=1}^{M} y_k\right) \tag{25}$$

For example, the determination index of $\hat{L}(x)$ obtained in Example 5.1C is

$$R(\hat{L}) = \frac{(4.152918)15.12 + (-0.166027)82.76 - 5(\frac{1}{5}\cdot 15.12)^2}{51.5928 - 5(\frac{1}{5}\cdot 15.12)^2} \doteq 0.567 \tag{26}$$

Since $R(\hat{L})$ is not near 1, we can conclude from (24c) that $\hat{L}(x)$ does *not* fit the data well. Notice that we did not have to worry about the y-scale [as we did in using (23)] to reach this conclusion.

Practical Consideration 5.1D (How many data points?). A value of $R(\hat{g})$ that is *not* close to one always indicates *poor* fit. However, $R(\hat{g}) \approx 1$ indicates *good* fit *only* if there are sufficiently many *excess* data points to "smooth" the errors in data. Statisticians will accept the fit of $L(x) = mx + b$ to M data points with P% confidence if

$$T > t_{\alpha/2}, \qquad \text{where } T = \sqrt{\frac{R(\hat{L})(M-2)}{1 - R(\hat{L})}} \qquad \text{and} \qquad \alpha = 1 - \frac{P}{100} \tag{27}$$

In (27), $(M - 2)$ is the number of "excess" data points, which statisticians refer to as the *degrees of freedom*; t is the *Student t variable*, which itself depends on M and is tabulated in any statistics book [11]; and α is the *significance level*. For 95% confidence ($\alpha = 0.05$), the values of $t_{0.025}$ for several values of M are

M	3	4	5	6	7	8	9	10
$t_{0.025}$	12.706	4.303	3.182	2.776	2.571	2.447	2.365	2.306

(28)

In Example 5.1C, where $M = 5$ and $R(\hat{L})$ turned out to be 0.567,

$$T = \sqrt{\frac{0.567(5 - 2)}{1 - 0.567}} \doteq 1.982 \quad \text{and} \quad t_{0.025} = 3.182 \tag{29}$$

Since $T < t_{0.025}$, we conclude from (27) that we cannot be 95% confident that $\hat{L}(x)$ "explains" the data. In this case, this conclusion could have been reached more easily by simply plotting the data. The test in (27) is really useful when $R(\hat{L})$ is closer to 1. ■

5.2 Fitting Monotone, Convex Data

Let us restrict our attention to data $P_1, \ldots, P_M$ that can be fit well by a guess function $g(x)$ whose graph is

Monotone, that is, either strictly increasing or strictly decreasing
Convex, that is, either concave up or concave down

on the entire **fitting interval** $[x_1, x_M]$. The graph of a $g(x)$ that is suitable for fitting monotone, convex data has *no turning points* and *no inflection points* for $x_1 \leqslant x \leqslant x_M$. Two parameters, say α and β, are generally sufficient to achieve good fit in this special case. Figure 5.2-1 presents six commonly used **two-parameter guess functions** $g(x)$ and shows how the intercepts and asymptotes of their graphs are related to α and β (see also Problem 5-8).

5.2A The Normal Equations for a Two-Parameter Guess Function

Given M data points $P_1(x_1, y_1), P_2(x_2, y_2), \ldots, P_M(x_M, y_M)$, and a guess function $g(x)$ having two parameters α and β, we wish to

$$\text{minimize } E(g), \qquad \text{where } E(g) = \sum_{k=1}^{M} [g(x_k) - y_k]^2 \tag{1}$$

This requires solving the **normal equations**

$$\frac{\partial E(g)}{\partial \alpha} = 0 \quad \text{and} \quad \frac{\partial E(g)}{\partial \beta} = 0 \tag{2}$$

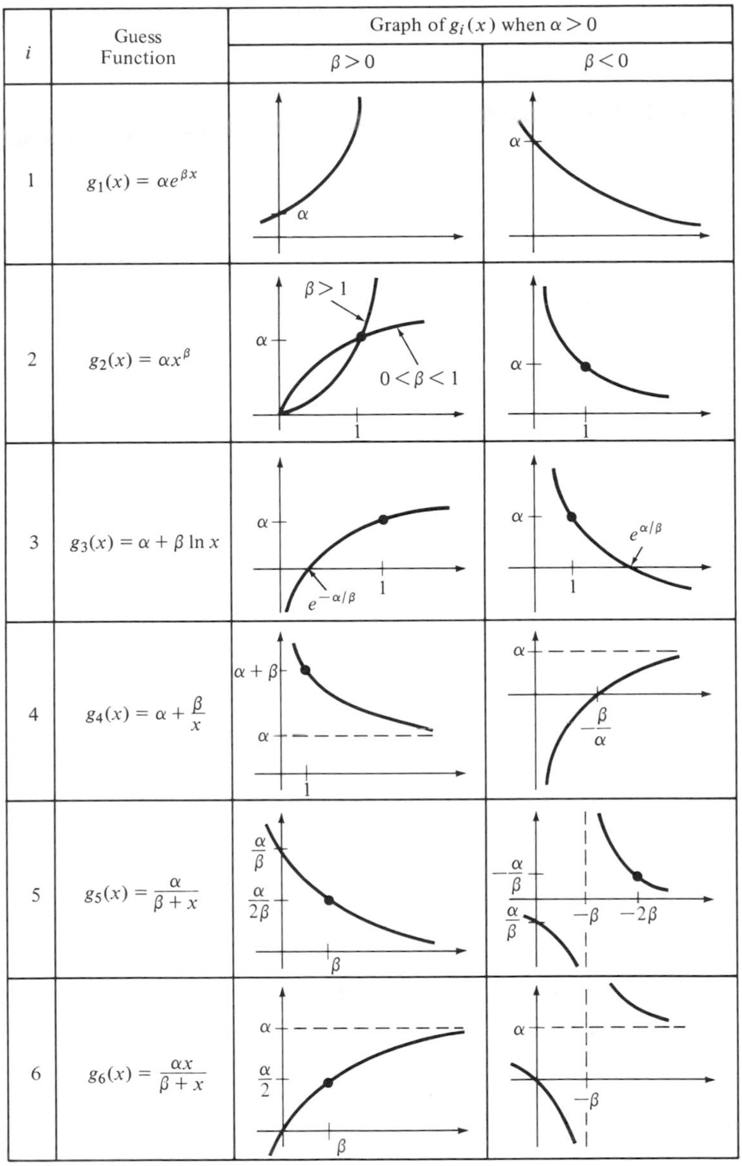

Figure 5.2-1 *Six guess functions that can fit monotone, convex data.*

for the **least square parameters** $\hat{\alpha}$ and $\hat{\beta}$ the $g(x)$ with these parameters will be denoted by $\hat{g}(x)$ and called the **least square $g(x)$** for $P_1, \ldots, P_M$ because it satisfies $E(\hat{g}) \leq E(g)$ for any such $g(x)$. Unfortunately, the normal equations (2) can be *nonlinear*, and when they are they can be rather messy to solve.

EXAMPLE 5.2A Consider the **exponential guess function** $g(x) = \alpha e^{\beta x}$.

(a) Find the normal equations for this $g(x)$. Are they linear?

(b) Find the least square exponential for the data

$$P_1(1, 5.12), \quad P_2(3, 3), \quad P_3(6, 2.48), \quad P_4(9, 2.34), \quad P_5(15, 2.18) \tag{3}$$

and compare the fit to that of $\hat{L}(x)$ obtained in Example 5.1C.

SOLUTION (a): The square error of $g(x) = \alpha e^{\beta x}$ is

$$E(g) = \sum_{k=1}^{M} (\alpha e^{\beta x_k} - y_1)^2 = (\alpha e^{\beta x_1} - y_1)^2 + \cdots + (\alpha e^{\beta x_M} - y_M)^2 \tag{4}$$

To minimize this $E(g)$, we must solve the normal equations

$$0 = \frac{\partial E(g)}{\partial \alpha} = \sum_{k=1}^{M} 2(\alpha e^{\beta x_k} - y_k)^1 \frac{\partial}{\partial \alpha}(\alpha e^{\beta x_k} - y_k) = 2\sum_{k=1}^{M} (\alpha e^{\beta x_k})e^{\beta x_k}$$

$$0 = \frac{\partial E(g)}{\partial \beta} = \sum_{k=1}^{M} 2(\alpha e^{\beta x_k} - y_k)^1 \frac{\partial}{\partial \beta}(\alpha e^{\beta x_k} - y_k) = 2\alpha\sum_{k=1}^{M} (\alpha e^{\beta x_k} - y_k)x_k e^{\beta x_k}$$

or, upon canceling 2 and 2α in the first and second equations, respectively,

$$\alpha(e^{2\beta x_1} + e^{2\beta x_2} + \cdots + e^{2\beta x_M}) - (y_1 e^{\beta x_1} + y_2 e^{\beta x_2} + \cdots + y_M e^{\beta x_M}) = 0 \tag{5a}$$

$$\alpha(x_1 e^{2\beta x_1} + \cdots + x_M e^{2\beta x_M}) - (x_1 y_1 e^{\beta x_1} + \cdots + x_M y_M e^{\beta x_M}) = 0 \tag{5b}$$

SOLUTION (b): The 2×2 system $\mathbf{f}(\mathbf{x}) = \mathbf{0}$ in (5) is *not* linear (why?), hence *cannot* be solved in one step as $A^{-1}\mathbf{b}$. It can, however, be solved *iteratively* either using NR_2 of Section 4.4B or by first eliminating α and then using the Secant Method to get β. In either case, the evaluation of $\mathbf{f}(\mathbf{x})$ is so messy that a computer is a virtual necessity if human error is to be avoided. Initial guesses for α and β can be obtained using the following geometric interpretations of α and β:

α = the y-intercept of the curve $y = \alpha e^{\beta x}$ **(6a)**

β = the **time constant** that controls the rate of growth or decay of the curve $y = \alpha e^{\beta x}$ **(6b)**

plot of the data (3) (see Figure 5.2-2) reveals that the data are monotone and convex and that the y-intercept of an exponential $g(x)$ is about $4\frac{1}{2}$. To get β, note that

$$g(1) \approx 2g(6) \qquad \text{so that } \alpha e^{\beta} \approx 2\alpha e^{6\beta} \quad \text{or} \quad e^{-5\beta} \approx 2 \tag{6c}$$

Solving yields $\beta \approx -(\ln 2)/5 \approx -\frac{1}{10}$. An iterative algorithm, starting with the initial guesses $\alpha \approx 4\frac{1}{2}$, $\beta \approx -\frac{1}{10}$, can now be used as described in Problem C5-1 to get the least square parameters $\hat{\alpha} = 4.677$ and $\hat{\beta} = -0.07472$ (to 4s). So the desired **least square exponential** for the data (3) is

$$\hat{g}(x) \doteq 4.677e^{-0.07472x}; \qquad \text{and} \qquad E(\hat{g}) = \sum_{k=1}^{5} [\hat{\alpha} e^{\beta_k} - y_k]^2 \doteq 1.84 \tag{7}$$

Notice that $E(\hat{g}) = 1.84$ is not much smaller than $E(\hat{L}) = 2.54$ obtained in (23) of Example 5.1C. Figure 5.2-2 confirms that the least square exponential does not fit the data much better than the least square line.

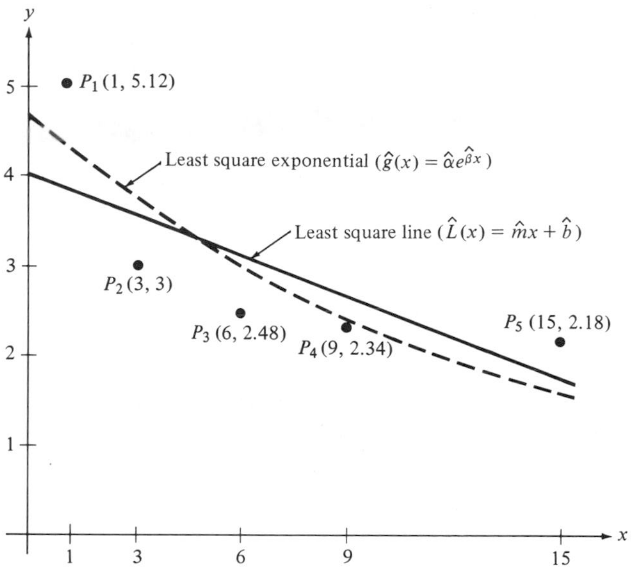

Figure 5.2-2 *Least square line and exponential for P_1, . . ., P_5.* ■

5.2B Linearizing Monotone, Convex Data

For each of the six two-parameter $g(x)$'s shown in Figure 5.2-1, the *given* variables (x, y) can be transformed into *new* variables (X, Y) in such a way that

$$y = g(x) \quad \Leftrightarrow \quad Y = mX + b, \qquad \text{where } X = X(x, y) \text{ and } Y = Y(x, y) \tag{8}$$

Geometrically, points $P(x, y)$ that lie on the graph of g in xy-space become points $Q(X, Y)$ on a *straight line* in XY-space (Figure 5.2-3). For example,

$$y = g_5(x) = \frac{\alpha}{x + \beta} \quad \Leftrightarrow \quad \beta y + xy = \alpha \quad \Leftrightarrow \quad y = -\frac{1}{\beta}xy + \frac{\alpha}{\beta} \tag{9a}$$

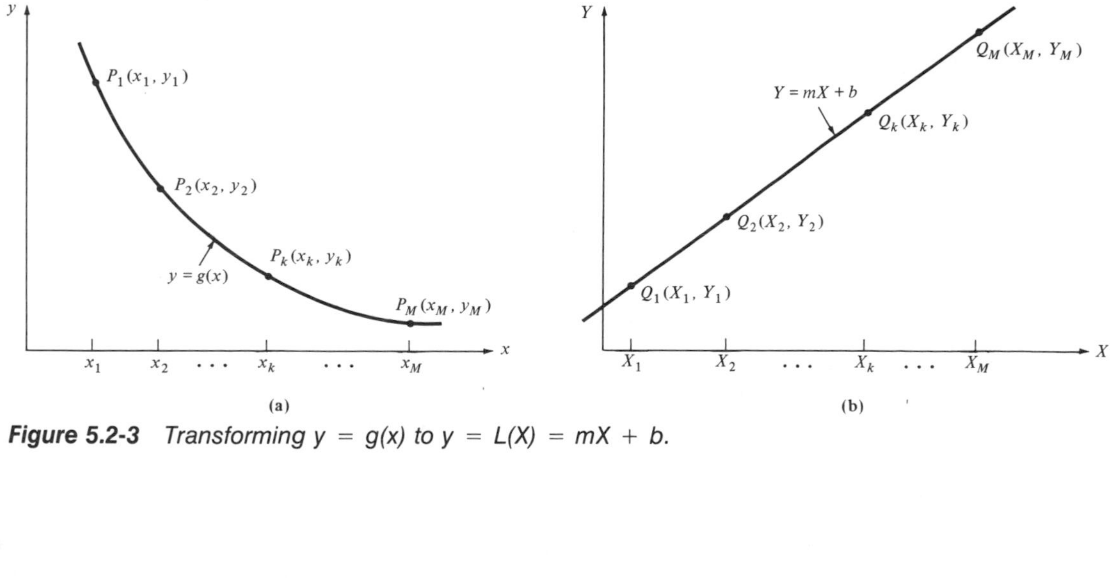

Figure 5.2-3 *Transforming $y = g(x)$ to $y = L(X) = mX + b$.*

This last expression is of the form $Y = mX + b$, with

$$Y = y, \quad m = -\frac{1}{\beta}, \quad X = xy, \quad \text{and} \quad b = \frac{\alpha}{\beta} \tag{9b}$$

Similarly, taking logarithms of both sides of $y = g_1(x) = \alpha e^{\beta x}$ yields

$$y = \alpha e^{\beta x} \quad \Leftrightarrow \quad \ln y = \ln \alpha + \beta x \quad \Leftrightarrow \quad \underbrace{\ln y}_{Y} = \underbrace{\beta}_{m} \cdot \underbrace{x}_{X} + \underbrace{\ln \alpha}_{b} \tag{10}$$

Converting $y = g(x)$ to $Y = mX + b$ as in (8) will be referred to as **linearizing** $g(x)$. Table 5.2-1 summarizes how this can be done for $g_1(x)$ through $g_6(x)$ of Figure 5.2-1. The last two columns are obtained by solving the "$m =$" and "$b =$" equations for α and β when $m = \hat{m}$ and $b = \hat{b}$. For example, we can see from (9b) that

$$\text{For } g_5(x), \quad \beta = \frac{-1}{\hat{m}} \qquad \text{hence } \alpha = \beta\hat{b} = -\frac{\hat{b}}{\hat{m}} \tag{11}$$

Figure 5.2-3 suggests that we can avoid nonlinear systems when fitting $g_1(x), \ldots, g_6(x)$ of Table 5.2-1 by reasoning as follows:

> If the transformation that linearizes $g(x)$ maps the given data points $P_k(x_k, y_k)$ into the points $Q_k(X_k, Y_k)$ and these Q_k's lie close to a line in XY-space, then the curve $y = g(x)$ that corresponds to this line should fit the P_k's well in xy-space. **(12)**

This reasoning is the basis for the **Linearization Algorithm** 5.2B.

TABLE 5.2-1 Linearizing the Six Curves in Figure 5.2-1

			Transformation Relations					
i	$y = g_i(x)$	Linearized Form $Y = L(X) = a + bX$	$X =$	$Y =$	$b =$	$m =$	$\alpha =$	$\beta =$
1	$y = \alpha e^{\beta x}$	$\ln y = \ln \alpha + \beta x$	x	$\ln y$	$\ln \alpha$	β	$e^{\hat{b}}$	$\hat{m}$
2	$y = \alpha x^{\beta}$	$\ln y = \ln \alpha + \beta(\ln x)$	$\ln x$	$\ln y$	$\ln \alpha$	β	$e^{\hat{b}}$	$\hat{m}$
3	$y = \alpha + \beta \ln x$	$y = \alpha + \beta(\ln x)$	$\ln x$	y	α	β	$\hat{b}$	$\hat{m}$
4	$y = \alpha + \dfrac{\beta}{x}$	$y = \alpha + \beta\left(\dfrac{1}{x}\right)$	$\dfrac{1}{x}$	y	α	β	$\hat{b}$	$\hat{m}$
5	$y = \dfrac{\alpha}{\beta + x}$	$y = \dfrac{\alpha}{\beta} + \dfrac{-1}{\beta}(xy)$	xy	y	$\dfrac{\alpha}{\beta}$	$-\dfrac{1}{\beta}$	$\dfrac{-\hat{b}}{\hat{m}}$	$\dfrac{-1}{\hat{m}}$
6	$y = \dfrac{\alpha x}{\beta + x}$	$y = \alpha + (-\beta)\left(\dfrac{y}{x}\right)$	$\dfrac{y}{x}$	y	α	$-\beta$	$\hat{b}$	$-\hat{m}$

ALGORITHM 5.2B. LINEARIZATION ALGORITHM

PURPOSE: To fit a two-parameter $g(x)$ to monotone, convex data $P_1(x_1, y_1), \ldots, P_M(x_M, y_M)$ when $g(x)$ is one of $g_1(x), \ldots, g_6(x)$ in Table 5.2-1.

GET M, x_k's, y_k's, and i {index of $g_i(x)$}

{***linearize***} Form the **linearized data** $Q_k(X_k, Y_k)$ from the given data $P_k(x_k, y_k)$ using the "$X=$" and "$Y=$" columns of Table 5.2-1.

{***get*** $\hat{m}$, $\hat{b}$} Set up and solve the normal equations

$$\begin{bmatrix} M & \Sigma X_k \\ \Sigma X_k & \Sigma X_k^2 \end{bmatrix}\begin{bmatrix} b \\ m \end{bmatrix} = \begin{bmatrix} \Sigma Y_k \\ \Sigma X_kY_k \end{bmatrix}$$

to get $\hat{m}$ and $\hat{b}$ for $Y = \hat{L}(X) = \hat{m}X + \hat{b}$, the least square line for the *linearized* data $Q_k(X_k, Y_k)$ [*not* $P_k(x_k, y_k)$].

{***get*** α, β} Get α and β for the $g(x)$ corresponding to $\hat{L}(X) = \hat{m}X + \hat{b}$ using the "$\alpha =$" and "$\beta =$" columns of Table 5.2-1.

OUTPUT ('α = 'α' and β = 'β' make $g(x)$ fit the data well')
STOP

NOTE: The α and β obtained in the ***get*** α, β step will generally *not* be the least square parameters $\hat{\alpha}$ and $\hat{\beta}$. However, the $g(x)$ with these parameters will provide good fit to the *given* data $P_1(x_1, y_1), \ldots, P_M(x_M, y_M)$ if $\hat{L}(X)$ fits the *linearized* data $Q_1(X_1, Y_1), \ldots, Q_M(X_M, Y_M)$ well.

EXAMPLE 5.2B Use the Linearization Algorithm to fit

$$\textbf{(a)}\ g_1(x) = \alpha e^{\beta x} \qquad \text{and} \qquad \textbf{(b)}\ g_4(x) = \alpha + \frac{\beta}{x}$$

to the five data points of Example 5.2A, namely

$$P_1(1, 5.12),\quad P_2(3, 3),\quad P_3(6, 2.48),\quad P_4(9, 2.34),\quad P_5(15, 2.18) \tag{13}$$

Compare the fit of these curves to that of $\hat{g}_1(x)$ and $\hat{g}_4(x)$.

SOLUTION (a) (LINEARIZED EXPONENTIAL FIT)

LINEARIZE. For $g_1(x) = \alpha e^{\beta x}$, $X_k = x_k$ and $Y_k = \ln y_k$. So the Q_k's (to 5s) are

$$Q_1(1, 1.6332),\quad Q_2(3, 1.0986),\quad Q_3(6, 0.90826),\quad Q_4(9, 0.85015),\quad Q_5(15, 0.77932) \tag{14}$$

GET $\hat{m}$, $\hat{b}$. The matrix form of the normal equations of the least square line for the *linearized* data (14) is

$$\begin{bmatrix} 5 & 34 \\ 34 & 352 \end{bmatrix}\begin{bmatrix} b \\ m \end{bmatrix} = \begin{bmatrix} 5.2695 \\ 29.720 \end{bmatrix}, \qquad \text{from which} \quad \begin{bmatrix} \hat{b} \\ \hat{m} \end{bmatrix} \doteq \begin{bmatrix} 1.3980 \\ -0.050601 \end{bmatrix} \tag{15a}$$

GET α, β. From Table 5.2-1, $\alpha = e^{\hat{b}} \doteq e^{1.3980} \doteq 4.047$ and $\beta = \hat{m} \doteq -0.05060$. So

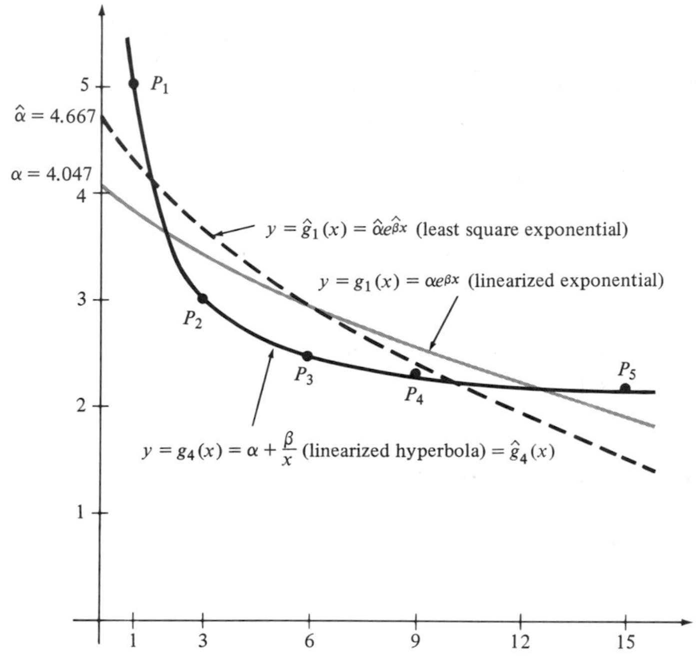

Figure 5.2-4 *Graphs of g_1, $\hat{g}_1$, and g_4 $(= \hat{g}_4)$.*

$$g_1(x) \doteq 4.047e^{-0.05060x}, \qquad \text{hence } E(g_1) = \sum_{k=1}^{5} [g_1(x_k) - \hat{y}_k]^2 \doteq 2.24 \tag{15b}$$

The parameters of this linearized $g_1(x)$ are fairly close to those of the least square exponential $\hat{g}_1(x) \doteq 4.677e^{-0.07472x}$ obtained in Example 5.2A. Figure 5.2-4 shows that $g_1(x)$ and $\hat{g}_1(x)$ appear to fit the data about as well to the eye, although $E(g_1) \doteq 2.24$ is somewhat larger than $E(\hat{g}_1) \doteq 1.84$.

SOLUTION (b) (LINEARIZED HYPERBOLIC FIT)

LINEARIZE. For $g_4(x) = \alpha + \beta/x$, $X_k = 1/x_k$ and $Y_k = y_k$. So the Q_k's are

$$Q_1(\tfrac{1}{1}, 5.12), \quad Q_2(\tfrac{1}{3}, 3), \quad Q_3(\tfrac{1}{6}, 2.48), \quad Q_4(\tfrac{1}{9}, 2.34), \quad Q_5(\tfrac{1}{15}, 2.18) \tag{16}$$

GET $\hat{m}$, $\hat{b}$. To get the least square line for (16), we solve

$$\begin{bmatrix} 5 & \frac{151}{90} \\ \frac{151}{90} & \frac{9631}{8100} \end{bmatrix} \begin{bmatrix} b \\ m \end{bmatrix} = \begin{bmatrix} 5.12 \\ 6.9387 \end{bmatrix} \quad \text{to get} \quad \begin{bmatrix} \hat{b} \\ \hat{m} \end{bmatrix} = \begin{bmatrix} 1.9681 \\ 3.1468 \end{bmatrix} \tag{17a}$$

GET α, β. For this particular guess function, $\beta = \hat{m}$ and $\alpha = \hat{b}$. So to 4*s*,

$$g_4(x) \doteq 1.968 + \frac{3.147}{x}\ ; \text{hence } E(g_4) = \sum_{k=1}^{5} \left[1.968 + \frac{3.147}{x_k} - y_k\right]^2 \doteq 0.00097 \tag{17b}$$

This small $E(g_4)$ suggests that $y = g_4(x)$ fits the data well. Since $g_4(x)$ is a linear model [with $\phi_1(x) = 1$ and $\phi_2(x) = 1/x$] and there are five data points, we can confirm this statistically by calculating

$$R(g_4) = \frac{\sum_{k=1}^{5} [\hat{b} + \hat{m}/x_k - \bar{y}]^2}{\sum_{k=1}^{5} [y_k - \bar{y}]^2} \doteq 0.9998, \quad \text{which is} \approx 1 \tag{17c}$$

(see Practical Consideration 5.1D). This good fit is shown graphically in Figure 5.2-4. In fact, the y_k values of the given data (13) were obtained by perturbing $2 + 3/x_k$ for $k = 1, \ldots, 5$. Problem 5-22 shows that $g_4(x) = \alpha + \beta/x$ is one of the two $g(x)$'s of Table 5.2 − 1 for which the Linearization Algorithm 5.2B actually yields $\hat{g}(x)$. So $g_4(x)$ is actually the least square hyperbola $\hat{\alpha} + \hat{\beta}/x$ for the data (13). ■

Practical Consideration 5.2B (Fitting curves through the origin). If the point (0, 0) is known to be a data point, then $g_2(x) = \alpha x^\beta$ $(\beta > 0)$ and $g_6(x) = \alpha x/(\beta + x)$ are suitable guess functions. However, in order to use the Linearization Algorithm 5.2B for these $g(x)$'s, *the point* (0, 0) *must be removed from the data*. (Why?) This applies whether the calculations are performed by hand *or* on a computer as described next. ■

5.2C Fitting Monotone, Convex Data on a Computer

For most computers, one can get software that calls a subprogram such as LINFIT in Figure 5.1-4 to fit a straight line and then uses the Linearization Algorithm 5.2B to fit the six monotone, convex curves of Figure 5.2-1 to any provided data. An interactive Fortran 77 calling program for such a subroutine, CURFIT, is shown in Figure 5.2-5.

```
C    Program to fit monotone, convex data by calling SUBROUTINE CURFIT
     PARAMETER (MAXM = 3Ø)
     CHARACTER FILNAM*1Ø,  XVAR, YVAR !variables used as x, y
     DIMENSION X(MAXM), Y(MAXM)
     DATA IW,IR /5, 5/                             !interactive I/O
     WRITE (IW, 1)                                 !FileName prompt
     READ  (IR, 2) FILNAM                          !<CR>=TTY; if so,
     IF (FILNAM.EQ.'          ') THEN              !get data from TTY
        CALL TTYIN  (MAXM,M,X,Y,XVAR,YVAR,IR,IW)
     ELSE                                          !get data from file
        CALL FILEIN (MAXM,M,X,Y,XVAR,YVAR,FILNAM)
        WRITE (IW, 3) FILNAM, M                    !echo back FileName
     ENDIF                                         !echo back data as
     WRITE (IW, 4) XVAR, (X(I), I=1,M)             !XVAR values
     WRITE (IW, 4) YVAR, (Y(I), I=1,M)             !YVAR values
C
     CALL CURFIT (M, X, Y, IW, IR)
C
     STOP
1    FORMAT ('ØName of data file (<CR> = No file)')
2    FORMAT (A1Ø)
3    FORMAT ('ØData file: ',A1Ø,4x,'The',I4,'  data points are:')
4    FORMAT ('Ø',A1,':'1ØG13.6,2(/,3X,1ØG13.6))
     END
```

Figure 5.2-5 *Calling program for* SUBROUTINE CURFIT.

```
Data file: SPRING          The    5  data points are:

x:  9.ØØØØØ       12.ØØØØ       13.ØØØØ       15.ØØØØ       17.ØØØØ

F:  15.6ØØØ       18.8ØØØ       19.6ØØØ       2Ø.6ØØØ       21.1ØØØ

                              Linearized
      Curve Type              Det. Index         a               b
   Ø. y = a + b*x             Ø.92Ø71         1Ø.Ø79          Ø.68641
   1. y = a*exp(b*x)          Ø.89858         11.6Ø9          Ø.37459E-Ø1
   2. y = a*(x**b)            Ø.95222         5.5229          Ø.48377
   3. y = a + b*ln(x)         Ø.96724         -3.4352          8.8265
   4. y = a + b/x             Ø.99292         27.687          -1Ø7.68
   5. y = a/(b + x)           Ø.93545         -5ØØ.32         -39.6Ø4
   6. y = a*x/(b + x)         Ø.89153         35.351           1Ø.921

      Equation #4. y = a + b/x     has the best linearized fit
```

Figure 5.2-6 *Output of calling program for* `SUBROUTINE CURFIT`.

Notice that the user can provide the name of a previously prepared file containing the value of M ($\leq$ MAXM) and the data points (x_k, y_k). When this is done, the program reads this input from the file by calling `SUBROUTINE FILEIN` (see line 1100). This option is useful when M is large, in which case modifying or reusing the data in file is a lot easier than by doing so from the keyboard. Since the statements for opening and closing files are hardware-dependent, we will not offer an example of a `SUBROUTINE FILEIN`.

Figure 5.2-6 shows the output of Figure 5.2-5 for the data points on the nonlinear part of the force versus stretch length curve indicated in Figure 5.1-2. The "Linearized Det. Index" entries give $R(\hat{L})$, the index of determination of $\hat{L}(X)$, which is an easily calculated indirect indicator of how well $g(x)$ fits the given data (see `SUBROUTINE LINFIT` in Figure 5.1-4). For the data provided, $g_4(x) = \alpha + \beta/x$ appears to give best fit.

5.3

Fitting an n-Parameter Linear Model

Data that appear to have either turning points or inflection points on the fitting interval $[x_1, x_M]$ generally require guess functions with more than two parameters for adequate fit. Most such data can be fit by a linear combination of n functions $\phi_1(x), \phi_2(x), \ldots, \phi_n(x)$, that is, by

$$g(x) = \gamma_1\phi_1(x) + \gamma_2\phi_2(x) + \cdots + \gamma_n\phi_n(x) = \sum_{j=1}^{n} \gamma_j\phi_j(x) \quad (1)$$

for a suitable n and appropriate choices of $\phi_1, \phi_2, \ldots, \phi_n$. Such a guess function will be referred to as an ***n*-parameter linear model**. The importance of linear models is that their normal equations are guaranteed to be linear, hence easily solved, for *any* n and *any* choice of ϕ_i.

5.3A The Normal Equations for $g(x) = \Sigma_{j=1}^{n} \gamma_j \phi_j(x)$

Given M data points $P_1(x_1, y_1), \ldots, P_M(x_M, y_M)$ and n prescribed functions $\phi_1(x), \phi_2(x), \ldots, \phi_n(x)$, we wish to minimize the square error

$$E(g) = \sum_{k=1}^{M} [g(x_k) - y_k]^2, \qquad \text{where } g(x_k) = \sum_{j=1}^{n} \gamma_j \phi_j(x_k) \tag{2}$$

If we view $E(g)$ as a function of the n parameters $\gamma_1, \gamma_2, \ldots, \gamma_n$, then the desired **least square parameters** $\hat{\gamma}_1, \hat{\gamma}_2, \ldots, \hat{\gamma}_n$ can be obtained by solving the n **normal equations**

$$0 = \frac{\partial E(g)}{\partial \gamma_i} = \sum_{k=1}^{M} 2[g(x_k) - y_k] \frac{\partial}{\partial \gamma_i} [g(x_k) - y_k], \qquad i = 1, \ldots, n \tag{3}$$

Since $\phi_1(x_k), \ldots, \phi_n(x_k)$ and y_k are constant and $\partial \gamma_j / \partial \gamma_i = 0$ for $j \neq i$, we see from (2) that $\partial[g(x_k) - y_k]/\partial \gamma_i = \phi_i(x_k)$. So (3) gives

$$\sum_{k=1}^{M} \left[\sum_{j=1}^{n} \gamma_j \phi_j(x_k) - y_k \right] \phi_i(x_k) = 0, \qquad i = 1, \ldots, n \tag{4a}$$

or

$$\sum_{k=1}^{M} \sum_{j=1}^{n} \gamma_j \phi_j(x_k) \phi_i(x_k) = \sum_{k=1}^{M} y_k \phi_i(x_k), \qquad i = 1, \ldots, n \tag{4b}$$

Upon interchanging the order of the j and k summations, we see that the **normal equations** of $g(x)$ are

$$\sum_{j=1}^{n} \gamma_j \left(\sum_{k=1}^{M} \phi_i(x_k) \phi_j(x_k) \right) = \sum_{k=1}^{M} \phi_i(x_k) y_k, \qquad i = 1, 2, \ldots, n \tag{5a}$$

These constitute a *linear* system whose matrix form is

$$\underbrace{\begin{bmatrix} \phi_1 \cdot \phi_1 & \phi_1 \cdot \phi_2 & \cdots & \phi_1 \cdot \phi_n \\ \phi_2 \cdot \phi_1 & \phi_2 \cdot \phi_2 & \cdots & \phi_2 \cdot \phi_n \\ \vdots & \vdots & & \vdots \\ \phi_n \cdot \phi_1 & \phi_n \cdot \phi_2 & \cdots & \phi_n \cdot \phi_n \end{bmatrix}}_{A} \underbrace{\begin{bmatrix} \gamma_1 \\ \gamma_2 \\ \vdots \\ \gamma_n \end{bmatrix}}_{\boldsymbol{\gamma}} = \underbrace{\begin{bmatrix} \phi_1 \cdot \mathbf{y} \\ \phi_2 \cdot \mathbf{y} \\ \vdots \\ \phi_n \cdot \mathbf{y} \end{bmatrix}}_{\mathbf{b}} \tag{5b}$$

where we have used the following abbreviations for a_{ij} and b_j:

$$\boldsymbol{\phi}_i \cdot \boldsymbol{\phi}_j = \sum_{k=1}^{M} \phi_i(x_k)\phi_j(x_k) \qquad \text{and} \qquad \boldsymbol{\phi}_i \cdot \mathbf{y} = \sum_{k=1}^{M} \phi_i(x_k)y_k \tag{5c}$$

If $A = (\boldsymbol{\phi}_i \cdot \boldsymbol{\phi}_j)_{n\times n}$ is nonsingular, then the normal equations (5) can be solved† for the unique **least square parameter vector** $\hat{\boldsymbol{\gamma}}$, whose components $\hat{\gamma}_1, \hat{\gamma}_2, \ldots, \hat{\gamma}_n$ are the parameters of the **least square** $\boldsymbol{g(x)}$

$$\hat{g}(x) = \hat{\gamma}_1\phi_1(x) + \hat{\gamma}_2\phi_2(x) + \cdots + \hat{\gamma}_n\phi_n(x) = \sum_{j=1}^{n} \hat{\gamma}_j\phi_j(x) \tag{6a}$$

whose name comes from the fact that its square error satisfies

$$E(\hat{g}) = \sum_{k=1}^{M} [\hat{g}(x_k) - y_k]^2 < E(g) \qquad \text{for any other } g(x) = \sum_{j=1}^{n} \gamma_j\phi_j(x) \tag{6b}$$

Once (5) is solved for $\hat{\boldsymbol{\gamma}}$, the index of determination of $\hat{g}(x)$ in (6b) can be obtained easily (see [14]) using the special formula

$$R(\hat{g}) = \frac{\hat{\boldsymbol{\gamma}}^T\mathbf{b} - M\hat{y}^2}{\sum_{k=1}^{M} y_k^2 - M\hat{y}^2}, \qquad \text{where } \hat{\boldsymbol{\gamma}}^T\mathbf{b} = \sum_{i=1}^{n} \gamma_i b_i \quad \text{and} \quad \hat{y} = \frac{1}{M}\left(\sum_{k=1}^{M} y_k\right) \tag{7}$$

In (7), $\hat{y}$ is the **mean** of the y_k's and **b** is the target vector in (5b). As noted earlier in Practical Consideration 5.1D, $R(\hat{g}) \approx 1$ should *not* be taken as an indicator of good fit *unless* there are sufficiently many data points to "smooth" the errors in the data. As a "rule of thumb," we will only take $R(\hat{g}) \approx$ as indicative of good fit if there are at least three excess data points, that is, if

$$\text{(number of data points)} \geqslant 3 + \text{(number of parameters)} \tag{8}$$

with the understanding that six to ten excess data points would generally be needed if a thorough statistical analysis of the fit were to be made. Requirement (8) makes formula (7) easier to use than the definition of $R(g)$ given in (24b) of Section 5.1D.

The following example describes a situation in which a curve fit is used as a means to an end, rather than an end in itself.

EXAMPLE 5.3A The concentration C of a desired compound t seconds after two expensive reaction components are added to a liquid solution is known to satisfy

$$C = C(t) = C_{ss} + ae^{-0.47t} + be^{-0.06t} \tag{9a}$$

†When $n = 2$, the normal equations (5) can be solved using the formula $\hat{\boldsymbol{\gamma}} = A^{-1}\mathbf{b}$; for $n > 2$, they are best be solved using a direct (i.e., factorization or elimination) method of Chapter 3.

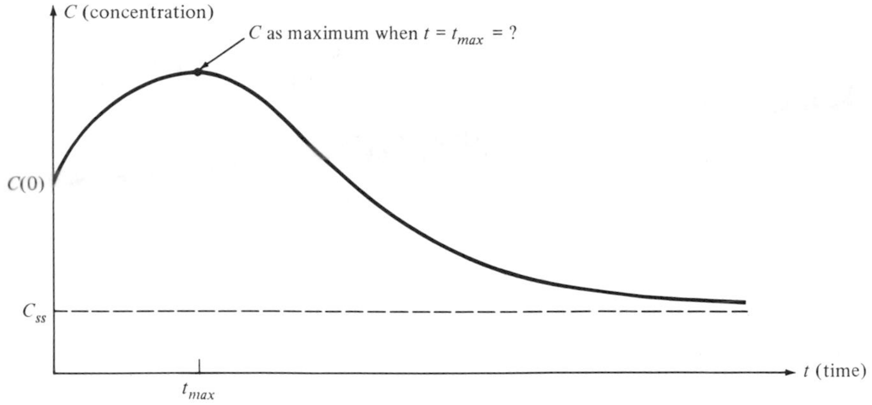

Figure 5.3-1 *Concentration of a compound after t seconds of reaction.*

(see Figure 5.3-1). To maximize the yield of the desired compound, the reaction should be stopped when C is a maximum, that is, after $t_{\max}$ seconds, where $t_{\max}$ can be found by solving

$$C'(t) = -0.47ae^{-0.47t} - 0.06be^{-0.06t} = 0 \tag{9b}$$

for

$$t = t_{\max} = \frac{\ln[(-0.47a)/(0.06b)]}{0.47 - 0.06} = \frac{1}{0.41}\ln\left(\frac{6b}{47a}\right) \tag{9c}$$

It turns out that a and b depend on the purity of the reaction components and must therefore be determined *experimentally* by obtaining samples of C at selected t_k's each time a new batch of reaction components is used. For rapid reactions, some of the desired readings [e.g., $C(0)$ at $t = 0$, just after the reaction begins] may have to be skipped. On the other hand, the measurement of the "steady-state" concentration C_{ss} (after the entire reaction is complete) requires a delay that may be unacceptable when the reaction is slow. Suppose that the only available readings $P_k(t_k, C_k)$ are

$$P_1(3, 4.1), \quad P_2(9, 4.3), \quad P_3(12, 3.9), \quad P_4(18, 3.4), \quad P_5(24, 3.1), \quad P_6(30, 2.7) \tag{10}$$

Then $t_{\max}$ lies between $t_1 = 3$ and $t_3 = 12$. Get a more accurate approximation.

SOLUTION: In view of (9a), the natural guess function is

$$g(t) = c + ae^{-0.47t} + be^{-0.06t} \tag{11a}$$

This $g(t)$ is a linear combination of the three functions

$$\phi_1(t) = 1, \quad \phi_2(t) = e^{-0.47t}, \quad \text{and} \quad \phi_3(t) = e^{-0.06t} \tag{11b}$$

So its normal equations are linear; in matrix form (5), they are

$$\begin{bmatrix} \Sigma\, 1\cdot 1 & \Sigma\, e^{-0.47t_k} & \Sigma\, e^{-0.06t_k} \\ \Sigma\, 1\cdot e^{-0.47t_k} & \Sigma\, e^{-0.94t_k} & \Sigma\, e^{-0.53t_k} \\ \Sigma\, 1\cdot e^{-0.06t_k} & \Sigma\, e^{-0.53t_k} & \Sigma\, e^{-0.12t_k} \end{bmatrix} \begin{bmatrix} c \\ a \\ b \end{bmatrix} = \begin{bmatrix} \Sigma\, 1\cdot y_k \\ \Sigma\, e^{-0.47t_k} \\ \Sigma\, e^{-0.06t_k} \end{bmatrix} \quad \textbf{(12)}$$

For the six data points (10), this becomes (to 6s)

$$\begin{bmatrix} 6 & 0.262474 & 2.64659 \\ 0.262474 & 0.0598304 & 0.214210 \\ 2.64659 & 0.214210 & 1.47298 \end{bmatrix} \begin{bmatrix} c \\ a \\ b \end{bmatrix} = \begin{bmatrix} 21.5 \\ 1.07818 \\ 10.1642 \end{bmatrix} \quad \textbf{(13a)}$$

Solving, say using LU-factorization, gives the least square parameters

$$\hat{c} \doteq 2.1187, \quad \hat{a} \doteq -4.9029, \quad \hat{b} \doteq 3.8066 \quad (\text{to } 5s) \quad \textbf{(13b)}$$

We can now take $\hat{a}$ for a and $\hat{b}$ for b in (9c) to get $t_{\max} \approx 5.638$.

NOTE: The C_k values in (10) were obtained by rounding to 2s the values at t_k of

$$C(t) = 2 - 5e^{-0.47t} + 4e^{-0.06t} \qquad \text{for which } t_{\max} = 5.565 \quad \textbf{(14)}$$

So the calculated $t_{\max}$, namely 5.638, is accurate to 2s. Since the C_k's were accurate to only 2s, this is as much accuracy as we should expect. ■

Practical Consideration 5.3A (Testing the fit empirically). An alternative to using $R(\hat{g})$ to confirm good fit is to reason as follows: *If $\hat{g}(x)$ fits the data well, then the calculated least square parameters will not change much if a data point is added to or deleted from the data.* For example, if we were able to sample the initial concentration to get $C(0) = 1.0$ in Example 5.3A, then we can add (0, 1.0) to the data (10) and repeat the procedure in (13). The resulting least square parameters would be

$$\hat{c} \doteq 2.1165, \quad \hat{a} \doteq -4.9300, \quad \hat{b} \doteq 3.8141, \qquad \text{hence } t_{\max} \approx 5.646 \quad \textbf{(15)}$$

A comparison of (13b) and (15) would then suggest (correctly) that $\hat{c}$, $\hat{a}$, $\hat{b}$, and $t_{\max}$ are all accurate to 2s. If no other data point is available, a data point could have been *deleted* (Problem 5-26). ■

5.3B Fitting Polynomials; the Polynomial Wiggle Problem

Taking $\phi_1(x) = 1$, $\phi_2(x) = x$, $\phi_3(x) = x^2, \ldots, \phi_n(x) = x^{n-1}$ in the general linear model (1) yields the $(n-1)$st degree **polynomial guess function**

$$g_n(x) = \gamma_1 + \gamma_2 x + \gamma_3 x^2 + \cdots + \gamma_n x^{n-1} = \sum_{j=1}^{n} \gamma_j x^{j-1} \quad \textbf{(16a)}$$

Since $\phi_i(x_k)\phi_j(x_k) = x_k^{i-1}x_k^{j-1} = x_k^{i+j-2}$, the matrix form (5) of the normal equations for $g_n(x)$ is

$$\underbrace{\begin{bmatrix} M & \Sigma\, x_k & \Sigma\, x_k^2 & \cdots & \Sigma\, x_k^{n-1} \\ \Sigma\, x_k & \Sigma\, x_k^2 & \Sigma\, x_k^3 & \cdots & \Sigma\, x_k^{n} \\ \Sigma\, x_k^2 & \Sigma\, x_k^3 & \Sigma\, x_k^4 & \cdots & \Sigma\, x_k^{n+1} \\ \vdots & \vdots & \vdots & & \vdots \\ \Sigma\, x_k^{n-1} & \Sigma\, x_k^{n} & \Sigma\, x_k^{n+1} & \cdots & \Sigma\, x_k^{2n-2} \end{bmatrix}}_{A_{n\times n}} \underbrace{\begin{bmatrix} \gamma_1 \\ \gamma_2 \\ \gamma_3 \\ \vdots \\ \gamma_n \end{bmatrix}}_{\gamma} = \underbrace{\begin{bmatrix} \Sigma\, y_k \\ \Sigma\, x_k y_k \\ \Sigma\, x_k^2 y_k \\ \vdots \\ \Sigma\, x_k^{n-1} y_k \end{bmatrix}}_{\mathbf{b}} \qquad \textbf{(16b)}$$

WARNING: It is reasonable to expect the fit of a polynomial $g_n(x)$ to improve as its degree (i.e., $n - 1$) increases. The following example shows how *wrong* this expectation can be. It also shows the need for "excess" data points.

EXAMPLE 5.3B Assume the following data to be reasonably accurate.

$$P_1(1, 6),\quad P_2(2, 1),\quad P_3(4, 2),\quad P_4(5, 3),\quad P_5(10, 4),\quad P_6(16, 5) \qquad \textbf{(17)}$$

(a) Find the least square polynomials of degree 1, ..., 5, that is, $\hat{g}_2(x), \ldots, \hat{g}_6(x)$.

(b) Find their indices of determination $R(\hat{g}_2), \ldots, R(\hat{g}_6)$. Do these numbers indicate how well $\hat{g}_2(x), \ldots, \hat{g}_6(x)$ fit the data?

SOLUTION (a): For $g_2(x) = \gamma_1 + \gamma_2 x$, the normal equations (16b) are

$$\begin{bmatrix} M & \Sigma\, x_k \\ \Sigma\, x_k & \Sigma\, x_k^2 \end{bmatrix}\begin{bmatrix} \gamma_2 \\ \gamma_1 \end{bmatrix} = \begin{bmatrix} \Sigma\, y_k \\ \Sigma\, x_k y_k \end{bmatrix}, \quad \text{that is,} \quad \begin{bmatrix} 6 & 38 \\ 38 & 402 \end{bmatrix}\begin{bmatrix} \gamma_1 \\ \gamma_2 \end{bmatrix} = \begin{bmatrix} 21 \\ 151 \end{bmatrix} \qquad \textbf{(18)}$$

The astute reader will recognize these as the normal equations of the straight-line guess function $L(x)$, with $\gamma_1 = b$ and $\gamma_2 = m$ [see (18b) of Section 5.1C]. The normal equations for $g_3(x), \ldots, g_6(x)$ are obtained similarly. For example, for the cubic guess function

$$g_4(x) = \gamma_1 + \gamma_2 x + \gamma_3 x^2 + \gamma_4 x^3 \qquad (n = 4) \qquad \textbf{(19a)}$$

and the data (17), the matrix form of the normal equations (16b) is

$$\begin{bmatrix} 6 & 38 & 402 & 5{,}294 \\ 38 & 402 & 5{,}294 & 76{,}434 \\ 402 & 5{,}294 & 76{,}434 & 1{,}152{,}758 \\ 5{,}294 & 76{,}434 & 1{,}152{,}758 & 17{,}797{,}002 \end{bmatrix}\begin{bmatrix} \gamma_1 \\ \gamma_2 \\ \gamma_3 \\ \gamma_4 \end{bmatrix} = \begin{bmatrix} 21 \\ 151 \\ 1{,}797 \\ 24{,}997 \end{bmatrix} \qquad \textbf{(19b)}$$

Solving (16b) for the least square parameters for $n = 2, \ldots, 6$ yields the coefficients of $\hat{g}_2(x), \ldots, \hat{g}_6(x)$ shown in Figure 5.3-2. Unfortunately, *none* of their graphs really fits the data well!

SOLUTION (b): Since $g_n(x)$ in (16a) is a linear model, the desired determination indices will be obtained using formula (7), namely

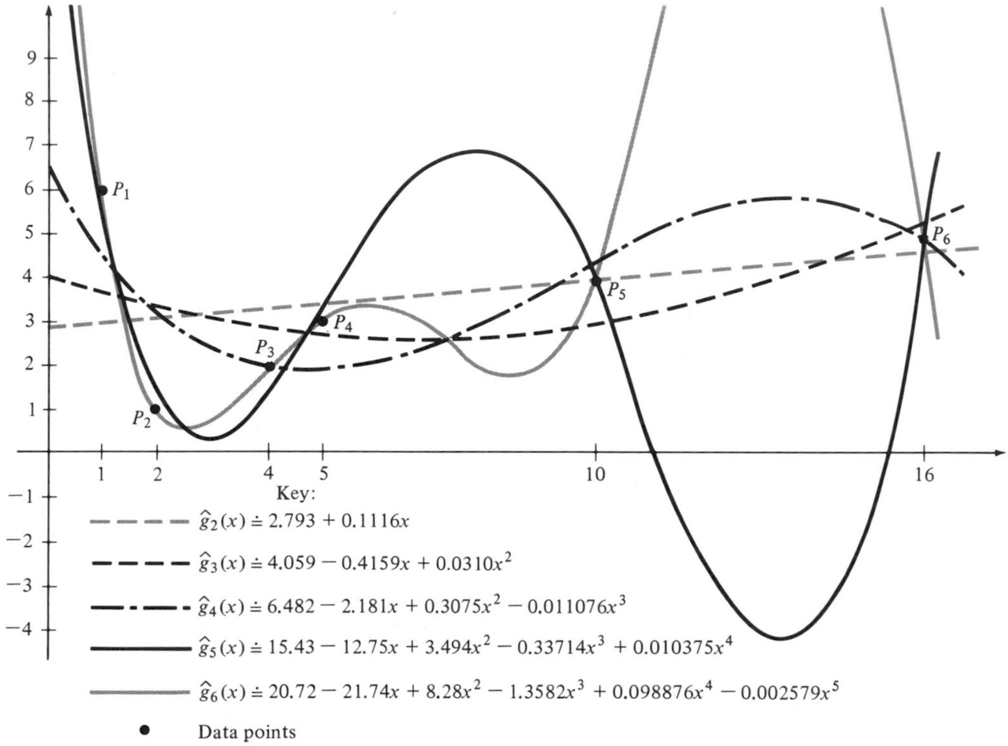

Figure 5.3-2 *Least square polynomials of degree 1, 2, 3, 4, 5.*

$$R(\hat{g}_n) = \frac{\hat{\boldsymbol{\gamma}}^T\mathbf{b} - M\hat{y}^2}{\sum_{k=1}^{M} y_k^2 - M\hat{y}^2}, \qquad \text{where } \hat{\boldsymbol{\gamma}}^T\mathbf{b} = \sum_{i=1}^{n} \hat{\gamma}_i b_i \tag{20}$$

The y-values of the data given in (17) satisfy

$$\hat{y} = \frac{1}{6}(6 + 1 + 2 + 3 + 4 + 5) = 3.5 \qquad \text{and} \qquad \sum_{k=1}^{M} y_k^2 = 91 \tag{21a}$$

For the (four-parameter) cubic guess function $g_4(x)$, the normal equations (19b) have target vector $\mathbf{b} = [21 \quad 151 \quad 1797 \quad 24{,}997]^T$ and solution

$$\hat{\boldsymbol{\gamma}} = [\underbrace{6.482161}_{\hat{\gamma}_1} \quad \underbrace{-2.181341}_{\hat{\gamma}_2} \quad \underbrace{0.3075475}_{\hat{\gamma}_3} \quad \underbrace{-0.01107595}_{\hat{\gamma}_4}]^T \quad (\text{to } 7s) \tag{21b}$$

We can therefore use formula (20) to get (to $4s$)

$$R(\hat{g}_4) = \frac{(21\hat{\gamma}_1 + 151\hat{\gamma}_2 + 1797\hat{\gamma}_3 + 24997\hat{\gamma}_4) - 6(3.5)^2}{91 - 6(3.5)^2} \doteq 0.5166 \tag{22}$$

The remaining entries of the following table are obtained similarly.

$\hat{g}_n(x)$	$\hat{g}_2(x)$	$\hat{g}_3(x)$	$\hat{g}_4(x)$	$\hat{g}_5(x)$	$\hat{g}_6(x)$
$R(\hat{g}_n)$	0.1148	0.2624	0.5166	0.9630	1.000

(23)

In view of (8), good fit can be deduced from $R(\hat{g}_n) \approx 1$ provided that $n \leq 6 - 3 = 3$. Since $R(\hat{g}_n)$ is not near 1 until $n = 5$, we can conclude from the three small values of $R(\hat{g}_n)$ in (23) that $\hat{g}_2(x)$, $\hat{g}_3(x)$, and $\hat{g}_4(x)$ give *poor* fit; but we do not have enough data to conclude that $\hat{g}_5(x)$ or $\hat{g}_6(x)$ give *good* fit. (In fact, we know from Figure 5.3-2 that they do not!) ■

It can be seen from (23) and Figure 5.3-2 that *in the absence of sufficiently many data points, $R(\hat{g}) \approx 1$ does not necessarily indicate that $\hat{g}(x)$ fits the data well on* $[x_1, x_M]$. In fact, $R(\hat{g}_6) = 1$ despite the fact that the graph of $\hat{g}_6(x)$ bears no resemblance to the curve that gave rise to the data (Figure 5.3-3)! Also, if the graph of $\hat{g}_5(x)$ were used to approximate such a curve in a realistic situation, then the highly erroneous *negative* values of $\hat{g}_5(x)$ for $11 < x < 15$ (see Figure 5.3-2) could result in serious damage and perhaps even loss of life!

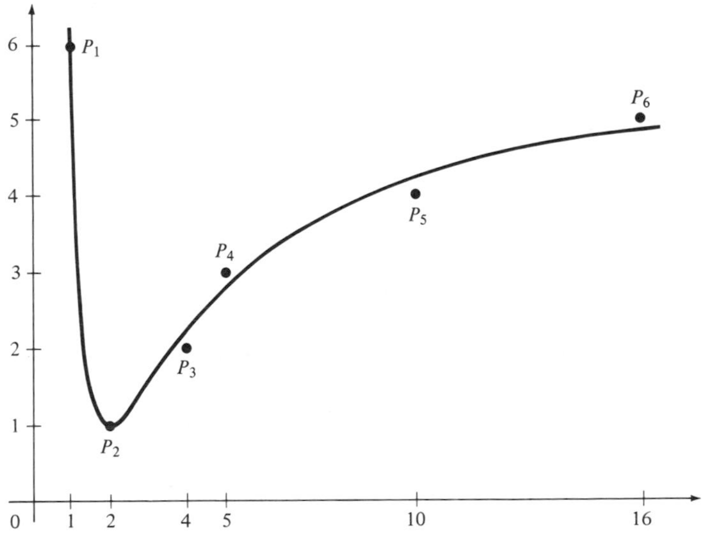

Figure 5.3-3 *Curve that gave rise to $P_1, \ldots, P_6$.*

What happened in the preceding example was this: *Each additional parameter increased the degree of $\hat{g}_n(x)$ by one and so allowed its graph to have one more inflection point.* This in turn enabled it to "wiggle" one more time in order to come closer to the data points $P_1, \ldots, P_M$. More generally, for any given set of M data points,

$$\hat{g}_n(x_k) \text{ will approach } y_k \textit{ for all } k \text{ as } n \text{ approaches } M \tag{24}$$

Hence $R(\hat{g}_n) = \Sigma_k[\hat{g}_n(x_k) - \bar{y}]^2/\Sigma_k(y_k - \bar{y})^2$ *will approach 1 as the number of parameters approaches the number of data points*. In the limiting case when $n = M$, the M normal equations in effect impose the M constraints $\hat{g}_M(x_k) = y_k$, $k = 1, \ldots, M$, so that $R(\hat{g}_M) = 1$ will result *even if the curve* $y = \hat{g}_M(x)$ *looks nothing like the actual y versus x curve!* Thus, although the least square criterion [minimizing $E(g_n)$] ensures good fit for x *near* an x_k, the values of $\hat{g}_n(x)$ may be very inaccurate for x's *between* the x_k's. This phenomenon is important enough to be worthy of a name:

The Polynomial Wiggle Problem. If the actual y versus x function is not polynomial-like, then an attempt to make the graph of a polynomial go near (or through) the data $P_k(x_k, y_k)$ will result in oscillations between successive P_k's. Such oscillations get larger as the degree of the polynomial increases. **(25)**

So *the remedy for poor polynomial fit is a more suitable guess function rather than a polynomial of higher degree*. In fact, if you think about it, the curve in Figure 5.3-3 is not "polynomial-like" because a polynomial curve does not have a vertical asymptote, and its slope does not approach zero as $|x|$ gets large. So we had no business even trying to fit a polynomial to the data! A more suitable guess function will be described in Section 5.3C.

In view of the Polynomial Wiggle Problem, *polynomial curve fitting should be done using the lowest-degree polynomial that appears to give good fit*. The following criterion ensures just this.

Degree Selection Criterion. Use the $(n - 1)$st-degree polynomial $\hat{g}_n(x) = \Sigma_{j=1}^{n} \hat{\gamma}_j x^{j-1}$ to fit M data points $P_1, \ldots, P_M$ if

$$n \le M - 3 \text{ and } R(\hat{g}_n) \approx 1, \quad \text{but} \quad R(\hat{g}_{n-1}) \text{ is } \textit{not} \approx 1 \qquad \textbf{(26)}$$

If no such n exists (as in Example 5.3B), then a more suitable guess function should be sought.

Practical Consideration 5.3B (Ill conditioning). The pivot condition numbers (using SPP) of the coefficient matrices $A_{n \times n}$ used to get $\hat{g}_n(x)$ in Example 5.3B are

n	2	3	4	5	6
$C_p(A_{n\times n})$	2.49	33.5	794	1.21E5	$\approx$ E+8!!

(27)

More generally, *the coefficient matrix A in (16b) tends to become ill conditioned rapidly as n increases*. This means that when n is large, $g_n(x)$'s with parameters quite different from the least square parameters can nevertheless have square errors about as small as $E(\hat{g}_n)$! *A* becomes ill conditioned especially rapidly with increasing n when

$$|x_1| \text{ is comparable to or larger than } |x_M - x_1| \qquad \textbf{(28)}$$

In this case rescaling the x_k's so as to lie between -2 and 2, as described in the **Polynomial**

Scaling Algorithm 5.3B, will make the normal equations less ill conditioned (see Problems 5-30 and C5-3). Alternatively (or in addition), orthogonal polynomials can be used as described in Algorithm 5.3D. In any case, *polynomial curve fitting should be performed in extended precision, and with $n < 7$, if possible*. If a high-degree polynomial appears to be needed, the fitting interval $[x_1, x_M]$ can be subdivided into two or more subintervals, with a different polynomial (of degree less than six) used over each subinterval (Problem 5-29). ■

ALGORITHM 5.3B. POLYNOMIAL SCALING ALGORITHM

PURPOSE: To calculate accurately the parameters of the least square $(n - 1)$st degree polynomial $\hat{g}_n(x)$ for the M data points $P_1(x_1, x_1), \ldots, P_M(x_M, x_M)$.

GET M, **x**, **y**, {arrays of x_k, y_k values}, n

{***Scale x:*** Map the x_k's proportionally into X_k's in the interval $[-2, 2]$}
Slope $\leftarrow 4/(x_M - x_1)$; DO FOR $k = 1$ TO M; $X_k \leftarrow -2 + \mathit{Slope}\cdot(x_k - x_1)$

{***Fit (X, y)***} Find the parameters of the least square $(n - 1)$st degree polynomial $\hat{G}_n(x)$ for the *transformed data* $Q_k(X_k, y_k)$, (**NOTE:** The y_k values are unchanged.)

{***Un-scale X***} Replace X in $\hat{G}_n(X)$ by $-2 + \mathit{Slope}\cdot(x - x_1)$ to get

$$\hat{g}_n(x) = \hat{G}_n(-2 + \mathit{Slope}\cdot(x - x_1))$$

{***Collect like powers of x***} Write $\hat{g}_n(x)$ as $\hat{\gamma}_1 + \hat{\gamma}_2 x + \hat{\gamma}_3 x^2 + \cdots + \hat{\gamma}_n x^{n-1}$

OUTPUT ('The $\hat{g}_n(x)$ coefficients (constant first) are: '$\hat{\gamma}_1, \hat{\gamma}_2, \ldots, \hat{\gamma}_n$)
STOP

5.3C The Importance of Sketching the Data

Statisticians call data points that are seriously in error **outliers** and recommend that they be removed from the data set. If the functional form of $g(x)$ is known, then it is generally easy to tell if an anomalous-looking data point is actually an outlier. If the functional form of $g(x)$ is *not* known, a repeated reading should be obtained (if possible) and/or the instrumentation examined to try to determine whether the point should be included as part of the data. For example, the data point $P_1(1, 6)$ in Figure 5.3-3 necessitates an abrupt change in the derivative of $g(x)$ for $x \approx 2$. Indeed, if this point were removed, the data would appear to be much more "well behaved" (in fact, possibly monotone and convex). But should it be removed?

Example 5.3B began with "Assume the following data to be reasonably accurate" to indicate that P_1 is *not* an outlier. Therefore, in order for $g(x)$ to fit the data well, its graph should have a *vertical asymptote near* $x = 0$ and should approach a *horizontal asymptote as x approaches* ∞ (see Figure 5.3-3). Nonconstant polynomial curves have *neither* of these properties, hence cannot possibly fit these data well! Indeed, had we sketched the data beforehand, we would not have wasted our time even trying to fit a polynomial! The general point to be made here is this:

When the functional form of the y versus x curve is not known, the data should be plotted to help determine $g(x)$'s that are suitable for good fit. **(29)**

A suitable guess function for data such as that in Figure 5.3-3 is a polynomial in $1/x$, say

$$G_n(x) = \gamma_1 + \frac{\gamma_2}{x} + \frac{\gamma_3}{x^2} + \cdots + \frac{\gamma_n}{x^{n-1}} = \sum_{j=1}^{n} \frac{\gamma_j}{x^{j-1}} \tag{30}$$

(where $n \geqslant 1$), because the graph of $G_n(x)$ has the y-axis ($x = 0$) as a vertical asymptote and the line $y = \gamma_1$ as a horizontal asymptote. The **Polynomial Transformation Algorithm** 5.3C uses reasoning similar to the Linearization Algorithm 5.2B to fit $G_n(x)$ to M data points $P_1(x_1, y_1), \ldots, P_M(x_M, y_M)$. This three-step algorithm is especially useful when software for fitting polynomials is available.

ALGORITHM 5.3C. POLYNOMIAL TRANSFORMATION

PURPOSE: To fit $G_n(x) = \sum_{j=1}^{n} \frac{\gamma_j}{x^{j-1}}$ to M data points $P_1(x_1, y_1), \ldots, P_M(x_M, y_M)$.

Step 1. Transform each $P_k(x_k, y_k)$ to $Q(X_k, y_k)$, where $X_k = 1/x_k$.

Step 2. Find the least square *polynomial* $\hat{g}_n(X)$ for $Q_1(X_1, y_1), \ldots, Q_M(X_M, y_M)$.

Step 3. The desired $\hat{G}_n(x)$ is $\hat{g}_n(1/x)$ {replace X by $1/x$ in $\hat{g}_n(X)$}. In fact, since the y_k's are not transformed, $R(\hat{G}_n) = R(\hat{g}_n)$.

EXAMPLE 5.3C For the nonpolynomial data of Figure 5.3-3, namely

$$P_1(1, 6), \quad P_2(2, 1), \quad P_3(4, 2), \quad P_4(5, 3), \quad P_5(10, 4), \quad P_6(16, 5) \tag{31}$$

(a) Use Algorithm 5.3C to get $\hat{G}_3(x) = \hat{\gamma}_1 + \hat{\gamma}_2/x + \hat{\gamma}_3/x^2$.
(b) Show that $\hat{G}_3(x)$ is the best $G_n(x)$ of the form $\Sigma_{j=1}^{n} \gamma_j/x^{j-1}$.

SOLUTION (a): The three steps of the algorithm proceed as follows:

STEP 1. For the data (31), the transformed points $Q_k(X_k, y_k)$ are

$$Q_1(\tfrac{1}{1}, 6), \quad Q_2(\tfrac{1}{2}, 1), \quad Q_3(\tfrac{1}{4}, 2), \quad Q_4(\tfrac{1}{5}, 3), \quad Q_5(\tfrac{1}{10}, 4), \quad Q_6(\tfrac{1}{16}, 5) \tag{32}$$

STEP 2. The coefficient matrix of the normal equations (19b) needed to fit the second-degree polynomial $g_3(X) = \gamma_1 + \gamma_2 X + \gamma_3 X^2$ to the data (32) is

$$A = \begin{bmatrix} M & \Sigma X_k & \Sigma X_k^2 \\ \Sigma X_k & \Sigma X_k^2 & \Sigma X_k^3 \\ \Sigma X_k^2 & \Sigma X_k^3 & \Sigma X_k^4 \end{bmatrix} = \begin{bmatrix} 6 & 2.1125 & 1.3664062 \\ 2.1125 & 1.3664062 & 1.1149869 \\ 1.3664062 & 1.1149869 & 1.0681215 \end{bmatrix} \tag{33a}$$

and target vector

$$\mathbf{b} = [\Sigma\, y_k \quad \Sigma\, X_k y_k \quad \Sigma\, X_k^2]^T = [21 \quad 8.3125 \quad 6.5545313]^T \tag{33b}$$

Solving $A\boldsymbol{\gamma} = \mathbf{b}$ yields the coefficients of

$$\hat{g}_3(X) = \underbrace{6.041638}_{\hat{\gamma}_3}X^2 - \underbrace{20.379906}_{\hat{\gamma}_2}X + \underbrace{20.347333}_{\hat{\gamma}_1} \tag{33c}$$

We can now use (33a), (33b), $\bar{y} = \frac{21}{6} = 3.5$, and $\Sigma_k y_k^2 = 91$ to get

$$R(\hat{g}_3) = \frac{\hat{\boldsymbol{\gamma}}^T\mathbf{b} - 6(3.5)^2}{91 - 6(3.5)^2} \doteq 0.9905 \tag{33d}$$

STEP 3. The desired $\hat{G}_3(x)$ and its determination index are thus

$$\hat{G}_3(x) = \hat{g}_3\left(\frac{1}{x}\right) = 6.041638 - \frac{20.379906}{x} + \frac{20.347333}{x^2}; \qquad R(\hat{G}_3) \doteq 0.9905 \tag{33e}$$

SOLUTION (b): Applying the steps (33) with $n = 2$ gives

$$\hat{G}_2(x) = \hat{g}_2\left(\frac{1}{x}\right) = 2.980468 - \frac{1.475595}{x}; \qquad R(\hat{G}_2) \doteq 0.07747 \tag{34}$$

Since $R(\hat{g}_2) << R(\hat{g}_3) \approx 1$ and $n = 3 \leq M - 3$, we can conclude from the Degree Selection Criterion (26) that the most suitable $\hat{g}_n(X)$ for the *transformed* points $Q_k(X_k, y_k)$ is $\hat{g}_3(X)$, hence the most suitable $\hat{G}_n(x)$ for the *given* points $P_k(x_k, y_k)$ is $\hat{G}_3(x)$. ■

5.3D Orthogonality and Weighted Least Squares

The reader of this subsection is assumed to be familiar with the geometric aspects of the standard **inner product** on M-space, namely

$$\mathbf{x}^T\mathbf{y} = \sum_{k=1}^{M} x_k y_k, \text{ for } \mathbf{x} = [x_1 \quad \cdots \quad x_M]^T \text{ and } \mathbf{y} = [y_1 \quad \cdots \quad y_M]^T \tag{35}$$

which is closely tied to the **euclidean norm** on M-space as follows:

$$\|\mathbf{x}\|^2 = \sum_{k=1}^{M} x_k^2 = \mathbf{x}^T\mathbf{x} \qquad \text{for } \mathbf{x} = [x_1 \quad \cdots \quad x_M]^T \tag{36}$$

We assume $\mathbf{x}$ and $\mathbf{y}$ to be obtained from data points $P_1(x_1, y_1), \ldots, P_M(x_M, y_M)$. The x_k values can be used to associate with any function $f(x)$ the **sample vector**

$$\mathbf{f} = [f(x_1) \quad f(x_2) \quad \cdots \quad f(x_M)]^T \qquad \text{(in } M\text{-space)} \tag{37}$$

Suppose that we want to fit an n-parameter linear model

$$g_n(x) = \gamma_1\phi_1(x) + \gamma_2\phi_2(x) + \cdots + \gamma_n\phi_n(x) = \sum_{i=1}^{n} \gamma_i\phi_i(x) \tag{38a}$$

to these data. It will be useful to introduce the $M \times n$ matrix F whose ith column is the sample vector associated with $\phi_i(x)$, that is,

$$F = [\boldsymbol{\phi}_1 : \boldsymbol{\phi}_2 : \cdots : \boldsymbol{\phi}_n] = \begin{bmatrix} \phi_1(x_1) & \phi_2(x_1) & \cdots & \phi_n(x_1) \\ \phi_1(x_2) & \phi_2(x_2) & \cdots & \phi_n(x_2) \\ \vdots & \vdots & & \vdots \\ \phi_1(x_M) & \phi_2(x_M) & \cdots & \phi_n(x_M) \end{bmatrix} \tag{38b}$$

If $\boldsymbol{\gamma}$ denotes the n-vector of parameters of $g_n(x)$, then by Problem 3-5(b),

$$F\boldsymbol{\gamma} = \gamma_1\boldsymbol{\phi}_1 + \gamma_2\boldsymbol{\phi}_2 + \cdots + \gamma_n\boldsymbol{\phi}_n \qquad \text{(in the span of } \boldsymbol{\phi}_1, \boldsymbol{\phi}_2, \ldots, \boldsymbol{\phi}_n) \tag{38c}$$

and so the kth component of the product $F\boldsymbol{\gamma}$ is $\Sigma_{i=1}^{n} \phi_i(x_k)\gamma_i = g_n(x_k)$. In view of (36), the square error of $g_n(x)$ can be written as

$$E(g_n) = \sum_{k=1}^{M} [g_n(x_k) - y_k]^2 = \|F\boldsymbol{\gamma} - \mathbf{y}\|^2 \tag{38d}$$

Since $\|F\boldsymbol{\gamma} - \mathbf{y}\|$ is the euclidean distance in M-space from $F\boldsymbol{\gamma}$ to $\mathbf{y}$, we see from (38d) and (38c) that the least square parameter vector $\hat{\boldsymbol{\gamma}}$ [which minimizes $E(g_n)$] is the n-vector $\boldsymbol{\gamma}$ for which the product $F\boldsymbol{\gamma}$ is as *close as possible to* $\mathbf{y}$ *in M-space* (Figure 5.3-4). Put another

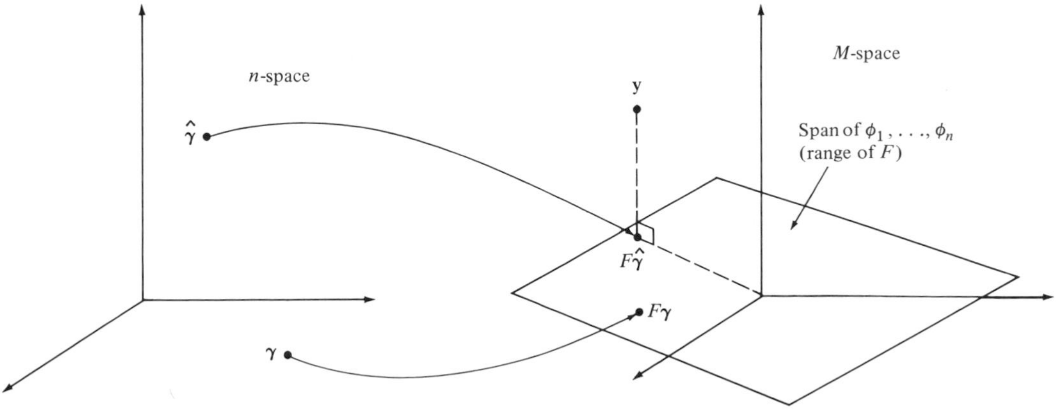

Figure 5.3-4 *$F\hat{\gamma}$ is the projection of* $\mathbf{y}$ *on the column space of F.*

way, $F\hat{\gamma}$ must be the unique *projection* of $\mathbf{y}$ on the subspace spanned by $\boldsymbol{\phi}_1, \boldsymbol{\phi}_2, \ldots, \boldsymbol{\phi}_n$. So $\mathbf{y} - F\hat{\gamma}$ must be orthogonal to $\boldsymbol{\phi}_i$, that is, $\boldsymbol{\phi}_i^T(\mathbf{y} - F\hat{\gamma}) = 0$, for all i. Since $\boldsymbol{\phi}_i = \text{col}_i F$, we must have

$$(\text{col}_i F)^T F\hat{\gamma} = (\text{col}_i F)^T \mathbf{y}, \quad \text{or} \quad (\text{row}_i F^T)F\hat{\gamma} = (\text{row}_i F^T)\mathbf{y},\ i = 1, \ldots, n \qquad \textbf{(38e)}$$

But $(\text{row}_i A)B = \text{row}_i(AB)$ [see (7) of Section 3.1C]; so (38e) says that $\hat{\gamma}$ must be a solution of the linear system

$$F^T F\boldsymbol{\gamma} = F^T\mathbf{y} \qquad \left[\text{normal equations for } g_n(x) = \sum_{i=1}^{n} \gamma_i\phi_i(x)\right] \qquad \textbf{(39)}$$

The preceding discussion is more than an alternative derivation of the normal equations (5) because it shows that $\hat{\gamma}$ actually *minimizes* $E(g_n)$. Also, since n cannot realistically be larger than M (why?), the coefficient matrix $A = F^TF$ in (39) will be nonsingular if and only if its n columns are linearly independent (see [45]). Consequently,

$F^TF\boldsymbol{\gamma} = F^T\mathbf{y}$ will have a *unique* solution $\hat{\gamma}$ if and only if $\boldsymbol{\phi}_1, \boldsymbol{\phi}_2, \ldots, \boldsymbol{\phi}_n$ are linearly dependent vectors in M-space.	**(40)**

Our second application of orthogonality deals with the possible ill conditioning of F^TF when $g_n(x)$ is a polynomial. Suppose that $p_0(x), p_1(x), \ldots, p_{n-1}(x)$ are polynomials of degree $0, 1, \ldots, n - 1$, respectively, that are **x_k-orthogonal**, by which we mean

$$\mathbf{p}_i^T\mathbf{p}_j = \sum_{k=1}^{M} p_i(x_k)p_j(x_k) = 0 \quad \text{if } i \neq j; \qquad \text{but } \mathbf{p}_i^T\mathbf{p}_i \neq 0 \text{ for all } i \qquad \textbf{(41a)}$$

Then the $(n - 1)$st degree polynomial guess function can be written as

$$g_n(x) = a_0p_0(x) + \cdots + a_{n-1}p_{n-1}(x) = \sum_{i=0}^{n-1} a_ip_i(x) \qquad \textbf{(41b)}$$

This is a linear model for which $F = [\mathbf{p}_0\vdots\cdots\vdots\mathbf{p}_{n-1}]$. Since $\mathbf{p}_0, \ldots, \mathbf{p}_{n-1}$ are mutually orthogonal [see (41a)], the normal equations (39) become the *diagonal* linear system

$$(\mathbf{p}_i^T\mathbf{p}_i)a_i = \mathbf{p}_i^T\mathbf{y}, \qquad i = 0, 1, \ldots, n - 1 \qquad \textbf{(41c)}$$

whose ith equation can be solved *by inspection* for $\hat{a}_i = \mathbf{p}_i^T\mathbf{y}/\mathbf{p}_i^T\mathbf{p}_i$, the ***i*th Fourier coefficient** of $g_n(x) = \Sigma_{i=0}^{n-1} a_ip_i(x)$. Summarizing, we obtain

If $p_0(x), p_1(x), \ldots, p_{n-1}(x)$ are x_k-orthogonal, then the least square $(n - 1)$st degree polynomial guess function $\hat{g}_n(x)$ can be represented as

$$\hat{g}_n(x) = \sum_{i=0}^{n-1} \hat{a}_i p_i(x), \qquad \text{where } \hat{a}_i = \frac{\mathbf{p}_i^T \mathbf{y}}{\mathbf{p}_i^T \mathbf{p}_i}, \quad i = 0, \ldots, n - 1 \qquad \textbf{(42)}$$

We shall call (42) the **Fourier representation**† of $\hat{g}_n(x)$.

Once $\hat{g}_{n-1}(x)$ and $p_{n-1}(x)$ are known, we can get $\hat{g}_n(x)$ in (42) by simply adding $\hat{a}_{n-1}p_{n-1}(x)$ to $\hat{g}_{n-1}(x)$, *without having to solve a (possibly ill-conditioned) linear system!* A set of x_k-orthogonal polynomials $p_0(x), p_1(x), \ldots, p_{n-1}(x)$ needed to incorporate this important strategy can be constructed *recursively* as follows (see Problem M5-12):

INITIALIZE: $p_0(x) = 1; \quad p_1(x) = x - \hat{x}, \quad \text{where } \hat{x} = \dfrac{1}{M}\left(\sum_{k=1}^{M} x_k\right)$ **(43a)**

RECURRENCE: $p_i(x) = xp_{i-1}(x) - \alpha_i p_{i-1}(x) - \beta_{i-1}p_{i-2}(x),$ **(43b)**

where

$$\alpha_i = \frac{(\mathbf{x}\mathbf{p}_{i-1})^T \mathbf{p}_{i-1}}{\mathbf{p}_{i-1}^T \mathbf{p}_{i-1}} \qquad \text{and} \qquad \beta_{i-1} = \frac{(\mathbf{x}\mathbf{p}_{i-1})^T \mathbf{p}_{i-2}}{\mathbf{p}_{i-2}^T \mathbf{p}_{i-2}} \qquad \textbf{(43c)}$$

In (43c), $(\mathbf{x}\mathbf{p}_{i-1})$ denotes $[x_1 p_{i-1}(x_1) \quad \cdots \quad x_M p_{i-1}(x_M)]^T$, the sample vector for the polynomial $xp_{i-1}(x)$. Formula (43b) is known as the **three-term recurrence relation** for x_k-orthogonal polynomials.

The **Orthogonal Polynomial Fitting Algorithm** 5.3D utilizes (43) by storing the x_k-sample vectors $\mathbf{p}_i$ rather than the $p_i(x)$'s themselves. The recursion is terminated when either $R(\hat{g}_n) \geqslant Rfit$ (typically ≈ 0.98) or $i = MaxDegree$ (typically ≈ 6).

ALGORITHM 5.3D. ORTHOGONAL POLYNOMIAL FITTING

PURPOSE: To find the Fourier representation of a least square polynomial that gives good fit to M given data points $P_1, \ldots, P_M$ (if such a polynomial exists).

†It is important to realize that $\hat{g}_n(x) = \sum_{i=0}^{n-1} \hat{a}_i p_i(x)$ in (42) is the same polynomial as $\hat{g}_n(x) = \sum_{i=1}^{n} \hat{\gamma}_i x^{i-1}$ obtained by solving (39) with $\phi_i(x) = x^{i-1}$ for $i = 1, \ldots, n$. The uniqueness of $\hat{g}_n(x)$ is immediate from (40), the uniqueness of the projection of $\mathbf{y}$ (Figure 5.3-4), and the easily proved fact that $\boldsymbol{\phi}_1, \ldots, \boldsymbol{\phi}_n$ will span the same n-dimensional subspace as $\mathbf{p}_0, \ldots, \mathbf{p}_{n-1}$ when (41a) holds. If x_k-orthogonal $\mathbf{p}_i$'s cannot be found, then F in (39) will be singular. In this case, the **pseudoinverse** of F can be used to get the (unique) solution $\hat{\boldsymbol{\gamma}}$ *having the smallest euclidean norm in n-space*. A readable discussion of pseudoinverses can be found in [45].

```
GET M, x {array of x_k values}, y {array of y_k values}, MaxDegree, Rfit
Sigy ← Σ_{k=1}^{M} y_k;   Sigy2 ← Σ_{k=1}^{M} y_k^2
x̂ ← (Σ_{k=1}^{M} x_k)/M;   ŷ ← Sigy/M                 {averages of x_k, y_k values}
p_0 ← 1; b_1 ← Sigy;   â_0 ← ŷ                 {â_0 is p_0^T y/p_0^T p_0 = ŷ; ĝ_1(x) = ŷ}
p_1 ← x − x̂1;   b_2 ← p_1^T y                  {b = [b_1  b_2]^T is the current F^T y}
â_1 ← b_2/p_1^T p_1                              {ĝ_2(x) = L̂(x) = ŷ + â_1(x − x̂)}
Rnumer ← â_0 b_1 + â_1 b_2 − M(Sigy)^2;   Rdenom ← Sigy2 − M(Sigy)^2
R ← Rnumer/Rdenom                                      {this is R(ĝ_2) = R(L̂)}
IF R < Rfit THEN                                  {try a nonlinear polynomial fit}
|  DO FOR i = 2 TO MaxDegree UNTIL R ≥ Rfit          {i = degree = n − 1}
|  |  α_i ← (xp_{i-1})^T p_{i-1}/p_{i-1}^T p_{i-1};   β_{i-1} ← (xp_{i-1})^T p_{i-2}/p_{i-2}^T p_{i-2}
|  |  p_i ← (xp_{i-1}) − α_i p_{i-1} − β_{i-1} p_{i-2}      {three-term recurrence relation}
|  |  SqLen ← p_i^T p_i;   IF SqLen = 0 THEN STOP                   {algorithm fails}
|  |  b_{i+1} ← p_i^T y;   â_i ← b_{i+1}/p_i^T p_i                {ith Fourier coefficient}
|  |  Rnumer ← Rnumer + â_i b_{i+1};   R ← Rnumer/Rdenom            {= R(ĝ_{i+1})}
|__|__ OUTPUT (i, α_i, β_{i-1}, â_i, R)           {parameters for p_i(x) and ĝ_{i+1}(x)}
IF R < Rfit THEN OUTPUT ('Polynomial fit does not seem to be suitable')
|__ ELSE OUTPUT ('Degree 'i' fit has R(ĝ) = 'R)
```

EXAMPLE 5.3D It is easy to see that the hypothetical historical data

$$(1600, 0), \quad (1700, 1), \quad (1800, 4), \quad (1900, 9) \qquad (x = \text{year}) \tag{44}$$

lie on the parabola $y = (x/100 - 16)^2 = 256 - 0.32x + 0.0001x^2$.

(a) Find $A = F^TF$ of the normal equations (39) for fitting a parabola $g_3(x) = \gamma_1 + \gamma_2 x + \gamma_3 x^2$. How accurately can (39) be solved on a 7s device?

(b) Find $\hat{g}_3(x)$ using the Orthogonal Polynomial Algorithm 5.3D.

Solution (a): For $\phi_1(x) = 1$, $\phi_2(x) = x$, $\phi_3(x) = x^2$, and the data (44),

$$F = \begin{bmatrix} 1 & 1600 & 1600^2 \\ 1 & 1700 & 1700^2 \\ 1 & 1800 & 1800^2 \\ 1 & 1900 & 1900^2 \end{bmatrix}; \qquad \text{so } A = F^TF = \begin{bmatrix} 4 & 7000 & 1.23\text{E}7 \\ 7000 & 1.23\text{E}7 & 2.17\text{E}10 \\ 1.23\text{E}7 & 2.17\text{E}10 & 3.84\text{E}13 \end{bmatrix} \tag{45}$$

For this $A_{3\times 3}$, $C_p(A)$ is about 0.3E7. So the components of $\hat{\boldsymbol{\gamma}}$ obtained on a 7s device may have erroneous leading digits!

Solution (b): We first get $\mathbf{p}_i$, $\hat{a}_i$, and $R(\hat{g}_i)$ recursively as follows:

$\boldsymbol{i = 0}$: $\mathbf{p}_0 = \mathbf{1} = [1\ 1\ 1\ 1]^T$; so $\mathbf{p}_0^T\mathbf{y} = 1\cdot 0 + 1\cdot 1 + 1\cdot 4 + 1\cdot 9 = 14$

$$\mathbf{p}_0^T\mathbf{p}_0 = 1 + 1 + 1 + 1 = 4\ (= M), \text{ and } \hat{a}_0 = \frac{\mathbf{p}_0{}^T\mathbf{y}}{\mathbf{p}_0^T\mathbf{p}_0} = \frac{7}{2}\ (= \hat{y})$$

$\boldsymbol{i = 1}$: $\hat{x} = 1750$; so $\mathbf{p}_1 = \mathbf{x} - \hat{x}\mathbf{1} = [-150\ \ -50\ \ 50\ \ 150]^T = 50[-3\ \ -1\ \ 1\ \ 3]^T$

$$\mathbf{p}_1^T\mathbf{y} = 50(-3\cdot 0 - 1\cdot 1 + 1\cdot 4 + 3\cdot 9) = 1500; \quad \mathbf{p}_1^T\mathbf{p}_1 = 50^2(20) = 50{,}000$$

$$\hat{a}_1 = \frac{\mathbf{p}_1^T\mathbf{y}}{\mathbf{p}_1^T\mathbf{p}_1} = \frac{3}{100}$$

$$Rnumer = [\hat{a}_0 \quad \hat{a}_1][\mathbf{p}_0^T\mathbf{y} \quad \mathbf{p}_1^T\mathbf{y}]^T - M\hat{y}^2 = \tfrac{7}{2}14 + \tfrac{3}{100}1500 - 4(\tfrac{49}{4}) = 45$$

$$Rdenom = \mathbf{y}^T\mathbf{y} - M\hat{y}^2 = (0^2 + 1^2 + 4^2 + 9^2) - 4(\tfrac{7}{2})^2 = 49$$

$$R(\hat{g}_2) = \frac{Rnumer}{Rdenom} = \frac{45}{49} \doteq 0.91$$

***i* = 2:**

$$(\mathbf{x}\mathbf{p}_1)^T\mathbf{p}_0 = 50[-3\cdot 1x_1 - 1\cdot 1x_2 + 1\cdot 1x_3 + 3\cdot 1x_4] = 50{,}000$$

$$(\mathbf{x}\mathbf{p}_1)^T\mathbf{p}_1 = 50^2[(-3)^2x_1 + (-1)^2x_2 + 1^2x_3 + 3^2x_4] = 87{,}500{,}000$$

$$\alpha_2 = \frac{(\mathbf{x}\mathbf{p}_1)^T\mathbf{p}_1}{\mathbf{p}_1^T\mathbf{p}_1} = \frac{87{,}500{,}000}{50{,}000} = 1750$$

$$\beta_1 = \frac{(\mathbf{x}\mathbf{p}_1)^T\mathbf{p}_0}{\mathbf{p}_0^T\mathbf{p}_0} = \frac{50{,}000}{4} = 12{,}500$$

$$\begin{aligned}\mathbf{p}_2 &= (\mathbf{x}\mathbf{p}_1) - \alpha_2\mathbf{p}_1 - \beta_1\mathbf{p}_0\\ &= 50[-3x_1 \quad -x_2 \quad x_3 \quad 3x_4]^T - 1750\cdot 50[-3 \quad -1 \quad 1 \quad 3]^T\\ &\quad -12{,}500[1 \quad 1 \quad 1 \quad 1]^T\\ &= [10{,}000 \quad -10{,}000 \quad -10{,}000 \quad 10{,}000]^T\\ &= 10^4[1 \quad -1 \quad -1 \quad 1]^T\end{aligned}$$

$$\hat{a}_2 = \frac{\mathbf{p}_2^T\mathbf{y}}{\mathbf{p}_2^T\mathbf{p}_2} = \frac{10^4(0 - 1 - 4 + 9)}{10^8(1 + 1 + 1 + 1)} = 10^{-4}$$

$$Rnumer \leftarrow Rnumer + \hat{a}_2(\mathbf{p}_2^T\mathbf{y}) = 45 + 10^{-4}(4\cdot 10^4) = 49; \quad \text{so } R(\hat{g}_3) = 1$$

We can conclude from $R(\hat{g}_3) = 1$ that the parabolic curve $y = \hat{g}_3(x)$ goes *through* (not just near) $P_1, \ldots, P_4$. And since $P_1, \ldots, P_4$ lie on the parabola $y = (x/100 - 16)^2$, it is reasonable to expect that $\hat{g}_3(x) = (x/100 - 16)^2$. To see that this is true, we shall build $\hat{g}_3(x)$ recursively using (41), (42), and our preceding work:

$$\hat{g}_1(x) = \hat{a}_0p_0(x) = \frac{7}{2} \quad \text{(least square constant)} \tag{46a}$$

$$\hat{g}_2(x) = \hat{g}_1(x) + \hat{a}_1p_1(x) = \frac{7}{2} + \frac{3}{100}(x - 1750) \; [= \hat{L}(x)] \tag{46b}$$

$$\begin{aligned}\hat{g}_3(x) &= \hat{g}_2(x) + \hat{a}_2p_2(x) = \hat{g}_2(x) + 10^{-4}[(x - 1750)p_1(x) - 12{,}500]\\ &= \frac{7}{2} + \frac{3}{100}(x - 1750) + \frac{(x - 1750)^2 - 12{,}500}{10{,}000} = \left(\frac{x}{100} - 16\right)^2\end{aligned} \tag{46c}$$

The fact that $\hat{g}_3(x) = (x/100 - 16)^2$ also follows from Section 6.1C. ■

Practical Consideration 5.3D (The need for scaling). The preceding example shows that the Orthogonal Polynomial Algorithm 5.3D can generate $\mathbf{p}_i$'s for which $\mathbf{p}_i^T\mathbf{p}_i = \|\mathbf{p}_i\|^2$ grows rapidly as i increases. There is thus a possible risk of overflow in calculating the $\hat{a}_i$'s. One way to avoid this is to scale x to X in $[-2, 2]$ initially, as described in Algorithm 5.3B. Alternatively (or in addition), one can **normalize** $\mathbf{p}_i$ by scaling it so that its length, $\|\mathbf{p}_i\|$, is near 1. These remedies are illustrated in Problems 5-30 and 5-31. ■

When we minimize $E(g) = \Sigma_{k=1}^{M} \delta_k^2$, we are giving each deviation $\delta_k = [g(x_k) - y_k]$ equal importance. The ability to "weight" certain data points more than others is sometimes needed and can be effected easily by minimizing the **weighted square error**

$$E_w(g) = \sum_{k=1}^{M} w_k[g(x_k) - y_k]^2, \qquad \text{where } w_k \geq 0 \text{ is the } \textbf{weight} \text{ for } P_k(x_k, y_k) \tag{47}$$

The resulting $\hat{g}(x)$ is called the **weighted least square solution** for the data $P_k(x_k, y_k)$.

When $g(x)$ is a linear model, say $g(x) = g_n(x) = \Sigma_{i=1}^{n} \gamma_i \phi_i(x)$, then minimizing $E_w(g_n)$ amounts to solving the **weighted normal equations**

$$(DF)^T(DF)\boldsymbol{\gamma} = (DF)^T D\mathbf{y}, \qquad \text{where } D = \text{diag}(\sqrt{w_1}, \sqrt{w_2}, \ldots, \sqrt{w_M}) \tag{48a}$$

Since left-multiplying by D has the effect of scaling the ith *row* by $\sqrt{w_i}$,

$$DF = (\sqrt{w_i}\boldsymbol{\phi}_i^T\boldsymbol{\phi}_j)_{M\times n} \qquad \text{and} \qquad D\mathbf{y} = [\sqrt{w_1}y_1 \quad \cdots \quad \sqrt{w_M}y_M]^T \tag{48b}$$

Problem 5-34 examines this generalization further.

5.4 Approximating a Known f(x) by a Simpler g(x) on an Interval I

We now turn our attention to the **approximation problem**, which can be stated as follows:

> Given a function $f(x)$, find a function $g(x)$ that is easy to evaluate and approximates $f(x)$ with a specified accuracy for all x's in a specified interval $I = [a, b]$.

"Easy to evaluate" generally means in terms of the three arithmetic operations, $+$, $-$, and $*$ [i.e., $g(x)$ = polynomial in x], and perhaps also / [i.e., $g(x)$ = a rational function of x]; it can also mean that all multiplications and divisions used to get $g(x)$ are by powers of the arithmetic base of the device, because these require only shifting the decimal (or binary or octal or hexadecimal) point. A least square method that can be used for any selected $g(x)$ is given in Section 5.4B. Special methods for $g(x)$'s that are either polynomial or rational functions are given in Sections 5.4C–5.4E.

5.4A The Approximation Problem

Algorithms for the "built-in" functions on a digital device generally use polynomial and rational approximations together with the functions' algebraic properties to calculate their values to the device accuracy for all storable values in their domain. Less accurate approximations are sometimes needed in *real-time* situations, such as tracking an enemy missile (an accurate estimate of where it will be in one second is of no value if it takes two seconds to obtain!) or controlling a chemical or nuclear reaction or an electronic circuit when the controlled variables change rapidly. A less critical real- time situation is described in the following example.

EXAMPLE 5.4A The *N*-body problem Suppose that $B_1, B_2, \ldots, B_N$ are celestial bodies (e.g., planets, stars, moons, etc.) whose gravitational fields influence each other's motion. Let us assume that all N bodies move in the same plane, so that the position of B_i can be described by coordinates (x_i, y_i). If the position and velocity of $B_1, \ldots, B_N$ are all known at an **initial time**, arbitrarily taken as $t_0 = 0$, then their trajectories can be simulated by selecting a time increment Δt sufficiently small to enable the accurate determination of *all* N positions (x_i, y_i) at the times

$$t_1 = \Delta t, \quad t_2 = 2\Delta t, \quad t_3 = 3\Delta t, \quad \ldots, \quad t_k = k\Delta t, \quad \ldots \qquad \textbf{(1)}$$

In order to determinate the positions (x_i, y_i) of the N bodies at time $t_{k+1} = t_k + \Delta t$, it is necessary to evaluate *all* $N(N - 1)/2$ interbody distances

$$r_{ij} = \sqrt{(x_i - x_j)^2 + (y_i - y_j)^2} = \text{distance}(B_i, B_j), \qquad i \neq j \qquad \textbf{(2)}$$

at time t_k. It turns out that the calculation of the square roots in (2) accounts for most of the "number crunching" time spent getting the N positions at times $t_0, t_1, \ldots$. If N is large enough to simulate a large solar system or a small galaxy and the calculation of the positions is performed during execution on a microcomputer, then the display of resulting motion will be too slow to be useful as a simulation. To speed things up, we could let

$$D = \max\{|x_i - x_j|, |y_i - y_j|\} \quad \text{and} \quad d = \min\{|x_i - x_j|, |y_i - y_j|\} \qquad \textbf{(3a)}$$

and write r_{ij} in (2) in the "normalized form,"

$$r_{ij} = \sqrt{D^2 + d^2} = D\sqrt{1 + \left(\frac{d}{D}\right)^2}, \qquad \text{where } 1 \leqslant 1 + \left(\frac{d}{D}\right)^2 \leqslant 2 \qquad \textbf{(3b)}$$

We can therefore reduce the execution time if we can find a function $g(x)$ that is easier (i.e., faster) to evaluate than $\sqrt{x}$ and satisfies

$$g(x) \approx \sqrt{x} \text{ accurately for all } x \text{ in the closed interval } [1, 2] \qquad \textbf{(4)}$$

The remainder of this section will be devoted to methods for finding a suitable $g(x)$ in a situation such as this. ■

5.4B Continuous Least Square Approximation of $f(x)$ on $I = [a, b]$

Suppose $g(x)$ is an n-parameter function that we wish to use to approximate $f(x)$ on the interval $I = [a, b]$. By analogy with the *discrete* least square discussion of Section 5.1B, we can define

$$\delta(x) = g(x) - f(x) = \text{the } \textbf{deviation of } \boldsymbol{g} \textbf{ at } \boldsymbol{x} \qquad (5)$$

and then "sum" $\delta(x)^2$ over the continuum of x's in I by integrating:

$$E(g) = \int_a^b \delta(x)^2\,dx = \int_a^b [g(x) - f(x)]^2\,dx = \text{the } \textbf{square error of } \boldsymbol{g} \qquad (6)$$

As in Section 5.1B, $E(g)$ will be zero if $g(x) = f(x)$ over I, and will increase as the total area between the graphs of g and f over I grows. To minimize $E(g)$, we view it as a function of the n parameters of $g(x)$, say $\gamma_1, \gamma_2, \ldots, \gamma_n$, and then solve the

$$\textbf{Normal Equations for } \boldsymbol{g}\textbf{:} \quad \frac{\partial E(g)}{\partial \gamma_i} = 0, \qquad \text{for } i = 1, 2, \ldots, n \qquad (7)$$

for the **least square parameters** $\hat{\gamma}_1, \hat{\gamma}_2, \ldots, \hat{\gamma}_n$ which, when substituted in $g(x)$, give the **least square $g(x)$**, denoted by $\hat{g}(x)$.

In general, the system (7) will be nonlinear and messy to solve. However, if $g(x)$ is a **linear model**, that is, if

$$g(x) = \gamma_1\phi_1(x) + \gamma_2\phi_2(x) + \cdots + \gamma_n\phi_n(x) \qquad \textbf{(8a)}$$

then an argument almost identical to that of Section 5.3A shows that the normal equations (7) will be a *linear* system whose matrix form is

$$\begin{bmatrix} \int \phi_1\phi_1 & \int \phi_1\phi_2 & \cdots & \int \phi_1\phi_n \\ \int \phi_2\phi_1 & \int \phi_2\phi_2 & \cdots & \int \phi_2\phi_n \\ \vdots & \vdots & & \vdots \\ \int \phi_n\phi_1 & \int \phi_n\phi_2 & \cdots & \int \phi_n\phi_n \end{bmatrix} \begin{bmatrix} \gamma_1 \\ \gamma_2 \\ \vdots \\ \gamma_n \end{bmatrix} = \begin{bmatrix} \int \phi_1 f \\ \int \phi_2 f \\ \vdots \\ \int \phi_n f \end{bmatrix} \qquad \textbf{(8b)}$$

where we have used the abbreviations

$$\int \phi_i\phi_j = \int_a^b \phi_i(x)\phi_j(x)\,dx \qquad \text{and} \qquad \int \phi_i f = \int_a^b \phi_i(x)f(x)\,dx \qquad \textbf{(8c)}$$

Once $\hat{g}(x)$ has been found, it is important to know the **maximum absolute error of $\hat{g}(x)$ on I**, that is,

$$\epsilon_{max} = \max\{|\hat{\delta}(x)|: a \leq x \leq b\}, \qquad \text{where } \hat{\delta}(x) = \hat{g}(x) - f(x) \tag{9a}$$

The max/min theory of calculus can be used as follows to accomplish this: First find all *critical points*† c of $\hat{\delta}(x)$ itself [rather than $|\hat{\delta}(x)|$]; then

$$\epsilon_{max} = \max\{|\hat{\delta}(c)|: c \text{ is a critical point of } \hat{\delta}(x) \text{ on } I\} \tag{9b}$$

EXAMPLE 5.4B Find to 4s the parameters of $\hat{g}(x)$ when $f(x) = \sqrt{x}$ is approximated on $I = [1, 2]$ by the quadratic $g(x) = \alpha + \beta x + \gamma x^2$. Then find the maximum deviation ϵ_{max}.

SOLUTION: This $g(x)$ is of the form (8a) with $\phi_1(x) = 1$, $\phi_2(x) = x$, and $\phi_3(x) = x^2$. So the normal equations (8b) are

$$\begin{bmatrix} \int_1^2 1\cdot 1\,dx & \int_1^2 1\cdot x\,dx & \int_1^2 1\cdot x^2\,dx \\ \int_1^2 x\cdot 1\,dx & \int_1^2 x\cdot x\,dx & \int_1^2 x\cdot x^2\,dx \\ \int_1^2 x^2\cdot 2\,dx & \int_1^2 x^2\cdot x\,dx & \int_1^2 x^2\cdot x^2\,dx \end{bmatrix} \begin{bmatrix} \alpha \\ \beta \\ \gamma \end{bmatrix} = \begin{bmatrix} \int_1^2 1\cdot\sqrt{x}\,dx \\ \int_1^2 x\cdot\sqrt{x}\,dx \\ \int_1^2 x^2\cdot\sqrt{x}\,dx \end{bmatrix} \tag{10a}$$

The formula $\int_1^2 x^k\,dx = (2^{k+1} - 1)/(k + 1)$ can now be used to convert (10a) to

$$\begin{bmatrix} 1 & \frac{3}{2} & \frac{7}{3} \\ \frac{3}{2} & \frac{7}{3} & \frac{15}{4} \\ \frac{7}{3} & \frac{15}{4} & \frac{31}{5} \end{bmatrix} \begin{bmatrix} \alpha \\ \beta \\ \gamma \end{bmatrix} = \begin{bmatrix} \frac{2}{3}(2\sqrt{2} - 1) \\ \frac{2}{5}(4\sqrt{2} - 1) \\ \frac{2}{7}(8\sqrt{2} - 1) \end{bmatrix}; \quad \text{so} \quad \begin{bmatrix} \hat{\alpha} \\ \hat{\beta} \\ \hat{\gamma} \end{bmatrix} = \begin{bmatrix} 0.449254 \\ 0.622285 \\ -0.070170 \end{bmatrix} \tag{10b}$$

(using any method of Chapter 3). The deviation of $\hat{g}(x)$ is thus

$$\hat{\delta}(x) = \hat{g}(x) - f(x) = \hat{\alpha} + \hat{\beta}x + \hat{\gamma}x^2 - \sqrt{x} \tag{10c}$$

To get ϵ_{max} for $\hat{g}(x)$, we must first solve $\hat{\delta}'(x) = 0$, that is,

$$\hat{\beta} + 2\hat{\gamma}x - \frac{1}{2\sqrt{x}} = 0 \tag{11}$$

This is not easy to do algebraically. However, a numerical method such as NR or SEC can be used to find two roots, $\bar{x}_1 \doteq 1.25942$ and $\bar{x}_2 \doteq 1.70803$, in the interior of the approximation interval $I = [1, 2]$. So by (9b),

$$\begin{aligned} \epsilon_{max} &= \max\,\{|\hat{\delta}(1)|, |\hat{\delta}(1.25942)|, |\hat{\delta}(1.70803)|, |\hat{\delta}(2)|\} \\ &\doteq \max\{|1.369\text{E}{-3}|, |-0.5660\text{E}{-3}|, |0.5076\text{E}{-3}|, |-1.0696\text{E}{-3}|\} \\ &= 1.369\text{E}{-3} \end{aligned} \tag{12}$$

We can conclude that $\hat{g}(x)$ approximates $\sqrt{x}$ with at most a small error in the third decimal place for $1 \leq x \leq 2$. For other approximations, see Problems 5-40 and M5-17. ■

†Recall that the **critical points** of $\hat{\delta}(x)$ on I are those points c in I at which (i) $\hat{\delta}'(c) = 0$, *or* (ii) $\hat{\delta}'(c)$ does not exist, *or* (iii) c is an endpoint of I.

5.4C Taylor Polynomial Approximation

In what follows, a will denote a fixed **base point** about which a given function f has a power series expansion with a positive (possibly infinite) radius of convergence R. This means that for any point x within R units of a, the function value at x is given *exactly* by the **Taylor series** expansion

$$f(x) = f(a) + \frac{f'(a)}{1!}(x - a) + \frac{f''(a)}{2!}(x - a)^2 + \cdots = \sum_{k=0}^{\infty} \frac{f^{(k)}(a)}{k!}(x - a)^k \quad \textbf{(13)}$$

This series is usually referred to as a **Maclaurin series** when $a = 0$. A brief review of Taylor series is given in Appendix II.2B.

Truncating the series (1) after the $(x - a)^n$ term gives the ***n*th Taylor approximation**

$$f(x) \approx P_n(x) = f(a) + \frac{f'(a)}{1!}(x - a) + \frac{f''(a)}{2!}(x - a)^2 + \cdots + \frac{f^{(n)}(a)}{n!}(x - a)^n \quad \textbf{(14)}$$

We shall refer to $P_n(x)$ as the ***n*th Taylor polynomial** based at a if $a \neq 0$, and the ***n*th Maclaurin polynomial** if $a = 0$. The error of the nth Taylor approximation is referred to as the ***n*th remainder** at x and denoted by $R_n(x)$. Thus

$$f(x) = P_n(x) + R_n(x), \qquad \text{so that } R_n(x) = f(x) - P_n(x) \quad \textbf{(15a)}$$

The **Lagrange form** of the nth remainder is

$$R_n(x) = \frac{f^{(n+1)}(\xi)}{(n + 1)!}(x - a)^{n+1}, \qquad \text{where } \xi \text{ lies between } a \text{ and } x \quad \textbf{(15b)}$$

$P_n(x)$ and $R_n(x)$ are illustrated graphically in Figure 5.4-1. Some of the more frequently occurring Taylor polynomials and their remainders are summarized in Table A.2-1.

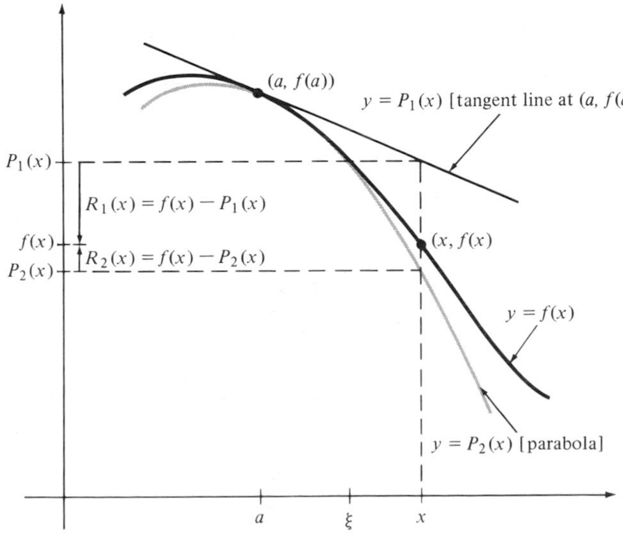

Figure 5.4-1 *$f(x) = P_n(x) + R_n(x)$ for $n = 1$ and 2.*

5.4D Rational Function Approximation for $x \approx 0$

Rational functions that approximate $f(x)$ for $x \approx 0$† are obtained in two steps as follows:

Step 1. Find an N big enough so that the Nth Maclaurin approximation

$$f(x) \approx P_N(x) = c_0 + c_1 x + c_2 x^2 + \cdots + c_N x^N \tag{16a}$$

is a bit more accurate than the desired accuracy over a prescribed interval I such as $[-R, R]$, $[0, R]$ or $[-R, 0]$.

Step 2. Find a rational function $r(x)$ such that the approximation

$$P_N(x) \approx r(x) = \frac{a_0 + a_1 x + \cdots + a_n x^n}{1 + b_1 x + \cdots + b_m x^m} \tag{16b}$$

is accurate enough on I to ensure that

$$f(x) - r(x) = \text{error of approximating } f(x) \text{ by } r(x) \tag{16c}$$

is sufficiently small for all x in I.

The denominator of $r(x)$ should have a nonzero constant term to avoid roundoff problems for $x \approx 0$. Since we can divide the numerator and denominator of $r(x)$ by this nonzero b_0, there is no loss of generality in assuming b_0 to be 1 as we did in (16b). The degrees m and n in $r(x)$ of Step 2 are generally chosen so as to achieve the desired accuracy in a way that either requires the *fewest operations* or can be evaluated in the *least amount of time*. Whatever the criterion, the following strategy, known as **Padé Approximation**, usually proves to be efficient because it ensures that

$$P_N(x) - r(x) \text{ behaves like } Cx^{N+1}, \quad \text{hence is small for } x \approx 0 \tag{17}$$

PADÉ APPROXIMATION. Choose m and n such that

$$m + n = N \qquad \textit{and} \qquad \textit{either } m = n \quad \text{or} \quad n = m + 1 \tag{18a}$$

Then choose the $N + 1$ unknowns $a_0, \ldots, a_n, b_1, \ldots, b_m$ so that the constant term and the coefficients of $x, x^2, \ldots, x^N$ in the numerator of the error

$$P_N(x) - r(x) = \frac{(c_0 + \cdots + c_N x^N)(1 + b_1 x + \cdots + b_m x^m) - (a_0 + \cdots + a_n x^n)}{1 + b_1 x + \cdots + b_m x^m} \tag{18b}$$

are all zero.

†If an approximation of $f(x)$ for $x \approx a \neq 0$ is desired, the change of variable $u = x - a$ converts the problem to that of approximating $f(a + u)$ for $u \approx 0$.

EXAMPLE 5.4D Use Padé approximation to approximate e^x to $4d$ on $I = [-\frac{1}{2}, \frac{1}{2}]$.

SOLUTION:

STEP 1. We must find an N such that the approximation

$$e^x \approx P_N(x) = 1 + x + \frac{x^2}{2!} + \frac{x^3}{3!} + \cdots + \frac{x^N}{N!} \tag{19}$$

is accurate to a bit more than $4d$ on I. We know from the Lagrange form of the Nth remainder that for any fixed x in I, the error of the approximation (19) is

$$e^x - P_N(x) = \frac{e^\xi}{(N+1)!} x^{N+1}, \qquad \text{where } \xi \text{ lies between 0 and } x \tag{20a}$$

Since e^x is increasing on I, e^ξ can be no larger than $e^{1/2}$. So

$$|e^x - P_N(x)| = \frac{e^\xi}{(N+1)!}|x|^{N+1} \leqslant \frac{e^{1/2}}{(N+1)!}\left(\frac{1}{2}\right)^{N+1} \qquad \text{for any } x \text{ in } I \tag{20b}$$

We can thus ensure $4d$ accuracy on I by taking N large enough so that

$$\frac{e^{1/2}}{2^{N+1}(N+1)!} < 0.5\cdot 10^{-4}$$

Trial and error shows that $N = 5$ is suitable In fact,

$$|e^x - P_5(x)| \leqslant \frac{e^{1/2}}{6!2^6} \doteq 0.36\cdot 10^{-4} \qquad \text{for all } x \text{ in } I = [-\tfrac{1}{2}, \tfrac{1}{2}] \tag{21}$$

STEP 2. For $N = 5$, (18a) instructs us to take $n = 3$ and $m = 2$. So (18b) becomes

$$P_5(x) - r(x) = \frac{\left(1 + \frac{x}{1} + \frac{x^2}{2} + \frac{x^3}{6} + \frac{x^4}{24} + \frac{x^5}{120}\right)(1 + b_1x + b_2x^2) - (a_0 + a_1x + a_2x^2 + a_3x^3)}{1 + b_1x + b_2x^2}$$

The coefficients we must equate to zero in the numerator to get the $N + 1$ $(=6)$ unknowns $b_1, b_2, a_0, a_1, a_2, a_3$ are

$$\begin{array}{llll}
\text{constant:} & 1 + 0\,b_1 + 0b_2 - a_0 = 0 & (\text{hence } a_0 = 1) & \\
\text{coefficient of } x^1: & \frac{1}{1} + 1\,b_1 + 0b_2 - a_1 = 0 & (\text{hence } a_1 = 1 + b_1) & \\
\text{coefficient of } x^2: & \frac{1}{2} + \frac{1}{1}\,b_1 + 1b_2 - a_2 = 0 & (\text{hence } a_2 = \frac{1}{2} + b_1 + b_2) & \textbf{(22a)} \\
\text{coefficient of } x^3: & \frac{1}{6} + \frac{1}{2}\,b_1 + \frac{1}{1}b_2 - a_3 = 0 & (\text{hence } a_3 = \frac{1}{6} + \frac{1}{2}b_1 + b_2) & \\
\text{coefficient of } x^4: & \frac{1}{24} + \frac{1}{6}\,b_1 + \frac{1}{2}b_2 = 0 & & \\
\text{coefficient of } x^5: & \frac{1}{120} + \frac{1}{24}b_1 + \frac{1}{6}b_2 = 0 & &
\end{array}$$

The last two equations are linear in b_1 and b_2. (In general, the last m will be linear in b_1, $\ldots, b_m$.) Solving them for b_1 and b_2, we get

$$\begin{bmatrix} b_1 \\ b_2 \end{bmatrix} = \begin{bmatrix} \frac{1}{6} & \frac{1}{2} \\ \frac{1}{24} & \frac{1}{6} \end{bmatrix}^{-1} \begin{bmatrix} -\frac{1}{24} \\ -\frac{1}{120} \end{bmatrix} = \frac{1}{\frac{1}{36} - \frac{1}{48}} \begin{bmatrix} \frac{1}{6} & -\frac{1}{2} \\ -\frac{1}{24} & \frac{1}{6} \end{bmatrix} \begin{bmatrix} -\frac{1}{24} \\ -\frac{1}{120} \end{bmatrix} = \begin{bmatrix} -\frac{2}{5} \\ \frac{1}{20} \end{bmatrix} \tag{22b}$$

From (22a), $a_1 = \frac{3}{5}$, $a_2 = \frac{3}{20}$, and $a_3 = \frac{1}{60}$, so

$$r(x) = \frac{1 + \frac{3}{5}x + \frac{3}{20}x^2 + \frac{1}{60}x^3}{1 - \frac{2}{5}x + \frac{1}{20}x^2} = \frac{[(\frac{1}{60}x + \frac{3}{20})x + \frac{3}{5}]x + 1}{(\frac{1}{20}x - \frac{2}{5})x + 1} \tag{23a}$$

The nested form on the right requires five multiply/divides† and five add/subtracts. If we divide numerator and denominator by $b_2 = \frac{1}{20}$ in (23a), we get

$$r(x) = \frac{\frac{1}{3}x^3 + 3x^2 + 12x + 20}{x^2 - 8x + 20} = \frac{[(\frac{1}{3}x + 3)x + 12]x + 20}{(x - 8)x + 20} \tag{23b}$$

which also requires five add/subtracts, but only four multiply/divides. This is one less than the nested form of $P_5(x)$, that is,

$$P_5(x) = \left(\left(\left(\left(\frac{x}{120} + \frac{1}{24}\right)x + \frac{1}{6}\right)x + \frac{1}{2}\right)x + 1\right)x + 1 \tag{24}$$

We can further improve the ease of calculating $r(x)$ by putting (23b) in **continued fraction** form as follows:

$$\begin{aligned} r(x) &= \frac{1}{3}\left\{\frac{x^3 + 9x^2 + 36x + 60}{x^2 - 8x + 20}\right\} = \frac{1}{3}\left\{(x + 17) + \frac{152x - 280}{x^2 - 8x + 20}\right\} \\ &= \frac{1}{3}\left\{(x + 17) + \cfrac{152}{\cfrac{x^2 - 8x + 20}{x - \frac{280}{152}}}\right\} = \frac{1}{3}\left\{(x + 17) + \cfrac{152}{(x - \frac{117}{19}) + \cfrac{\frac{3125}{361}}{x - \frac{280}{152}}}\right\} \\ &\doteq 0.33333\left\{(x + 17) + \cfrac{152}{(x - 6.1579) + \cfrac{8.6565}{x - 1.8421}}\right\} \end{aligned} \tag{25}$$

This requires only three multiply/divides! On a device that divides about as quickly as it multiplies, (25) will evaluate $r(x)$ more quickly than (23) or (24). ■

The accuracy of $r(x)$ and $P_5(x)$ is compared in Figure 5.4-2. In this example, the error of the approximation $P_5(x) \approx r(x)$ cancels much of the error of the Maclaurin approximation $e^x \approx P_5(x)$. As a result, $r(x)$ approximates e^x to $5d$ on $[-\frac{1}{2}, \frac{1}{2}]$! In general, *one can expect Padé approximation to yield a rational function r(x) that approximates f(x) with about the same accuracy as $P_N(x)$ but with fewer multiply/divides.*

Figure 5.4-2 illustrates a shortcoming of Taylor polynomial approximation. Although the Nth Taylor polynomial provides a good approximation of $f(x)$ near the base point (here it

†The divisions in the rational coefficients (e.g., $\frac{3}{5}$, $\frac{3}{20}$, $\frac{1}{60}$, etc.) are not counted because they are stored as 0.6, 0.15, 0.01666 . . . , and so on.

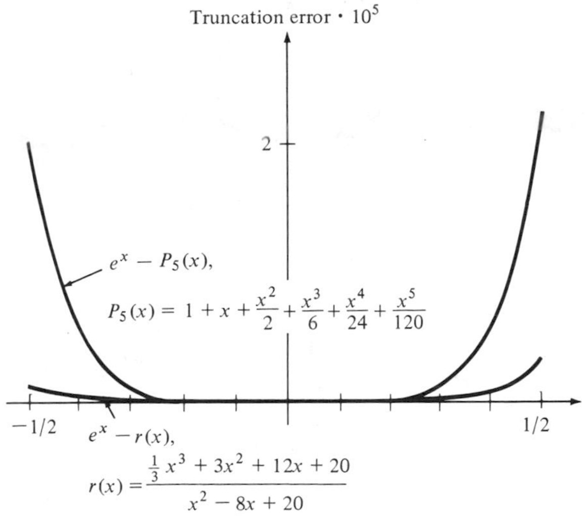

Figure 5.4-2 *Errors of $P_5(x)$ and $r(x)$ on $I = [-\frac{1}{2}, \frac{1}{2}]$.*

is 0), it is *not* well suited to approximating $f(x)$ *uniformly* on a whole interval I. A procedure for obtaining a more uniform *polynomial* approximation from $P_n(x)$ is given next.

5.4E Chebyshev Economization

We wish to get polynomials that approximate $f(x)$ with about the same maximum error all along a given interval I. The problem of **uniform approximation** on $[-1, 1]$ was studied in detail by P. L. Chebyshev, who introduced the ***n*th Chebyshev polynomial**†

$$T_n(\xi) = \cos n\theta, \qquad \text{where } \theta = \cos^{-1}\xi \text{ for } -1 \leq \xi \leq 1 \tag{26a}$$

for $n = 0, 1, 2, \ldots$. Since $|\cos u| \leq 1$ for all u, it is clear that for any n,

$$|T_n(\xi)| \leq 1 \qquad \text{for all } \xi \text{ in } [-1, 1] \tag{26b}$$

In fact, $T_n(\xi)$ oscillates between $+1$ and -1 exactly n times as ξ goes from -1 to 1. This **"equiripple property"** is illustrated in Figure 5.4-3. What is *not* clear is that $T_n(\xi)$ is a polynomial in ξ on $[-1, 1]$. If we write ξ as $\cos\theta$ in definition (26a), we get

$$T_0(\xi) = \cos 0 \equiv 1, \qquad T_1(\xi) = \cos(\cos^{-1}\xi) = \xi \tag{27a}$$

$$\begin{aligned} \xi T_n(\xi) &= \cos\theta \cdot \cos n\theta = \tfrac{1}{2}[\cos(n-1)\theta + \cos(n+1)\theta] \\ &= \tfrac{1}{2}[T_{n-1}(\xi) + T_{n+1}(\xi)] \end{aligned} \tag{27b}$$

The **three-term recurrence relation** (27) is best expressed as

$$T_0(\xi) \equiv 1, \qquad T_1(\xi) = \xi, \qquad T_{n+1}(\xi) = 2\xi T_n(\xi) - T_{n-1}(\xi) \tag{28}$$

†The letter T comes from "Tsebychev," which is the French transliteration of the man's (Russian) name.

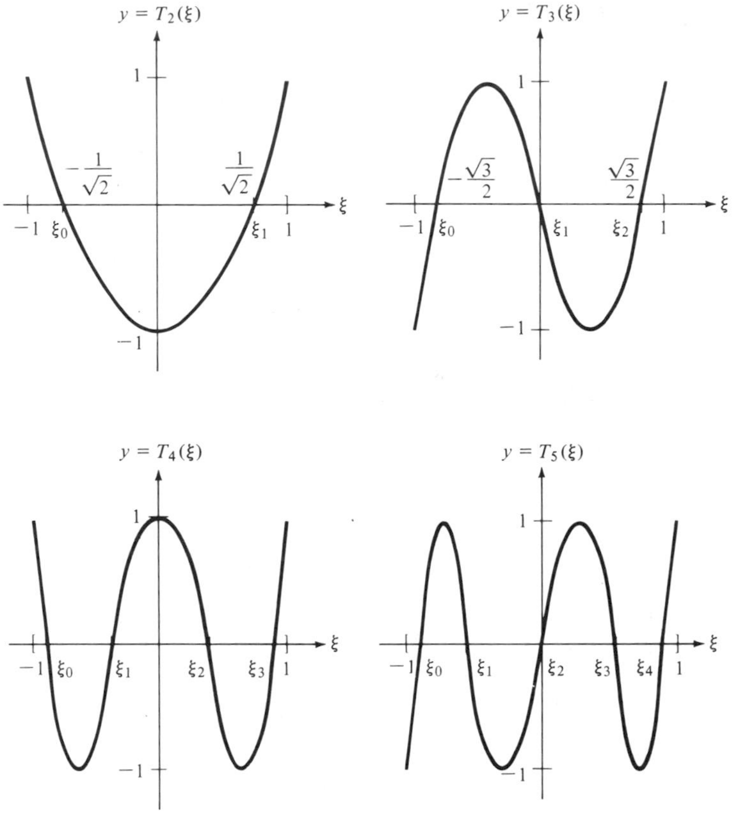

Figure 5.4-3 *Graphs of T_2, T_3, T_4, T_5.*

This generates $T_0(\xi) - T_7(\xi)$ shown in Table 5.4-1. Notice that for $n \geqslant 1$, the leading coefficient of $T_n(\xi)$ is 2^{n-1}. Solving $T_0(\xi)$, $T_1(\xi)$, $T_2(\xi)$, ... for 1, ξ, ξ^2, ... gives the expansions shown in Table 5.4-2.

Given any Nth-degree polynomial $p_N(x)$ and a prescribed interval $[a, b]$, Chebyshev polynomials can be used to obtain a lower-degree polynomial $p_{\text{econ}}(x)$, which approximates $p(x)$ uniformly on $[a, b]$. The idea is to change variables from x in $[a, b]$ to the ''normalized'' variable ξ in $[-1, 1]$, where the $T_n(\xi)$'s have the equiripple property. This procedure, called **Chebyshev economization**, is described in Algorithm 5.4E and illustrated in Figure 5.4-4.

TABLE 5.4-1 The Polynomial Form of $T_0(\xi)$, ..., $T_7(\xi)$

$T_0(\xi) = 1$
$T_1(\xi) = \xi$
$T_2(\xi) = 2\xi^2 - 1$
$T_3(\xi) = 4\xi^3 - 3\xi$
$T_4(\xi) = 8\xi^4 - 8\xi^2 + 1$
$T_5(\xi) = 16\xi^5 - 20\xi^3 + 5\xi$
$T_6(\xi) = 32\xi^6 - 48\xi^4 + 18\xi^2 - 1$
$T_7(\xi) = 64\xi^7 - 112\xi^5 + 56\xi^3 - 7\xi$

TABLE 5.4-2 The Chebyshev Expansion of 1, ξ, ..., ξ^7

$1 = T_0$
$\xi = T_1$
$\xi^2 = \frac{1}{2}(T_0 + T_2)$
$\xi^3 = \frac{1}{4}(3T_1 + T_3)$
$\xi^4 = \frac{1}{8}(3T_0 + 4T_2 + T_4)$
$\xi^5 = \frac{1}{16}(10T_1 + 5T_3 + T_5)$
$\xi^6 = \frac{1}{32}(10T_0 + 15T_2 + 6T_4 + T_6)$
$\xi^7 = \frac{1}{64}(35T_1 + 21T_3 + 7T_5 + T_7)$

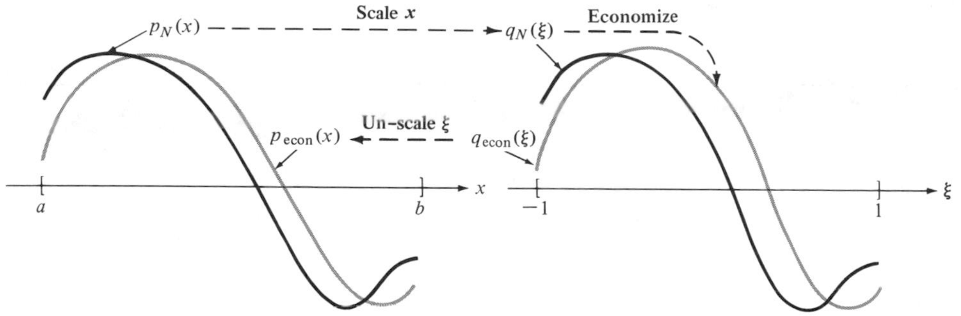

Figure 5.4-4 $p_N(x)$, $q_N(\xi)$, $q_{\text{econ}}(\xi)$, $p_{\text{econ}}(x)$.

ALGORITHM 5.4E. CHEBYSHEV ECONOMIZATION

PURPOSE: To approximate a given Nth-degree polynomial

$$p_N(x) = c_0 + c_1x + c_2x^2 + \cdots + c_Nx^N, \qquad c_N \neq 0$$

uniformly on $I = [a, b]$ by a polynomial $p_{\text{econ}}(x)$ of lower degree.

GET $a, b, N, c_0, \ldots, c_N$

{**Scale x**} Change from the x-interval $I = [a, b]$ to the normalized ξ-interval $[-1, 1]$ {where $|T_n(\xi)| \leq 1$} by forming

$$q_N(\xi) = p_N\left(a + \frac{b-a}{2}(\xi + 1)\right) \qquad \text{for } -1 \leq \xi \leq 1 \tag{29}$$

{**Expand**} Use Table 5.4-2 to get the **Chebyshev expansion**

$$q_N(\xi) = d_0 + d_1T_1(\xi) + \cdots + d_{N-1}T_{N-1}(\xi) + d_NT_N(\xi) \tag{30}$$

{**Economize**} Truncate this summation after the $d_kT_k(\xi)$ term, getting

$$q_{\text{econ}}(\xi) = d_0 + d_1T_1(\xi) + \cdots + d_kT_k(\xi), \qquad \text{where } k < N \tag{31}$$

{**Un-expand**} Use Table 5.4-1 to express this as a polynomial in ξ:

$$q_{\text{econ}}(\xi) = e_0 + e_1\xi + e_2\xi^2 + \cdots + e_k\xi^k \qquad \text{for } -1 \leq \xi \leq 1 \tag{32}$$

{This lower-degree polynomial approximates $q_N(\xi)$ uniformly on $[-1, 1]$.}

{**Un-scale ξ**} Convert back to $I = [a, b]$ by forming

$$p_{\text{econ}}(x) = q_{\text{econ}}\left(-1 + 2\frac{x-a}{b-a}\right) = \gamma_0 + \gamma_1x + \cdots + \gamma_kx^k \qquad \text{for } a \leq x \leq b \tag{33}$$

OUTPUT ('The coefficients of $p_{\text{econ}}(x)$, constant term first, are: '$\gamma_0, \ldots, \gamma_k$.)
STOP

Since $|T_j(\xi)| \leq 1$ for $j = k + 1, \ldots, N$, (30)–(33) give the following uniform bound for the error of approximating $p_N(x)$ by $p_{\text{econ}}(x)$ on $[a, b]$:

$$|p_N(x) - p_{\text{econ}}(x)| \leq |d_{k+1}| + \cdots + |d_N| \qquad \text{for all } x \text{ in } [a, b] \tag{34}$$

If $p_N(x)$ is an Nth Taylor polynomial approximation of $f(x)$, then $p_{\text{econ}}(x)$ provides a more uniform approximation of $f(x)$ than $p_N(x)$ on $[a, b]$ while requiring fewer arithmetic operations.

EXAMPLE 5.4E The fifth Maclaurin polynomial approximation of e^x is

$$e^x \approx p_5(x) = 1 + x + \frac{x^2}{2} + \frac{x^3}{6} + \frac{x^4}{24} + \frac{x^5}{120}$$

(a) Find the fourth-degree $p_{\text{econ}}(x)$ for $p_5(x)$ on $I = [-\frac{1}{2}, \frac{1}{2}]$.
(b) Compare the accuracy of $p_{\text{econ}}(x)$ to that of $p_5(x)$ on $[-\frac{1}{2}, \frac{1}{2}]$.

SOLUTION (a): We follow steps of the Chebyshev Economization Algorithm.

SCALE x. Since $[a, b] = [-\frac{1}{2}, \frac{1}{2}]$, $x = a + \frac{1}{2}(\xi + 1)/(b - a) = \frac{1}{2}\xi$. By (29),

$$q_5(\xi) = p_5\left(\frac{\xi}{2}\right) = 1 + \frac{\xi}{2} + \frac{\xi^2}{8} + \frac{\xi^3}{48} + \frac{\xi^4}{384} + \frac{\xi^5}{3840} \qquad \text{for } -1 \leq \xi \leq 1 \tag{35}$$

EXPAND. Using Table 5.4-2, we get the Chebyshev expansion

$$\begin{aligned} q_5(\xi) &= T_0 + \frac{T_1}{2} + \frac{T_0 + T_2}{8 \cdot 2} + \frac{3T_1 + T_3}{48 \cdot 4} + \frac{3T_0 + 4T_2 + T_4}{384 \cdot 8} \\ &\quad + \frac{10T_1 + 5T_3 + T_5}{3840 \cdot 16} \\ &\doteq 1.063477T_0 + 0.515788T_1 + 0.063802T_2 + 0.005290T_3 \\ &\quad + 0.000326T_4 + 0.000052T_5 \end{aligned} \tag{36}$$

ECONOMIZE AND UN-EXPAND. Dropping only the $T_5(\xi)$ term and using Table 5.4-1, we get

$$\begin{aligned} q_{\text{econ}}(\xi) &= 1.063477 + 0.515788\xi + 0.063802(2\xi^2 - 1) + 0.005290(4\xi^3 - 3\xi) \\ &\quad + 0.000326(8\xi^4 - 8\xi^2 + 1) \\ &= 1 + 0.4999186\xi + 0.125\xi^2 + 0.211589\xi^3 + 0.00260417\xi^4 \end{aligned} \tag{37}$$

UN-SCALE ξ. Finally, replacing ξ by $-1 + 2(x - a)/(b - a) = 2x$, we obtain

$$p_{\text{econ}}(x) = q_{\text{econ}}(2x) = 1 + 0.999837x + 0.5x^2 + 0.169271x^3 + 0.041667x^4 \tag{38}$$

SOLUTION (b): We know from (21) that $|e^x - p_5(x)| < 0.36\text{E}-4$ and from (34) and (36) that

$$|p_5(x) - p_{\text{econ}}(x)| \leq |d_5| \doteq 0.000052$$

We are thus assured that for $-\frac{1}{2} \leq x \leq \frac{1}{2}$,

$$|e^x - p_{econ}(x)| \leq |e^x - p_5(x)| + |p_5(x) - p_{econ}(x)| \leq 0.36\text{E}-4 + 0.52\text{E}-4$$

In fact, the two errors partly cancel (rather than add to) each other so that $p_{econ}(x)$ actually approximates e^x to within 0.5E−4 (i.e., to $4d$) on $[-\frac{1}{2}, \frac{1}{2}]$ (see Figure 5.4-5). This fourth-degree economized approximation is as efficient to evaulate as the Padé approximation (25) and is preferable for devices that perform division more slowly than multiplication. However, in this case its maximum error (near $\frac{1}{2}$) on $[-\frac{1}{2}, \frac{1}{2}]$ is larger.

NOTE: If $[a, b]$ happens to be $[-1, 1]$, then

$$x = \xi, \qquad \text{hence } q_N(\xi) = p_N(\xi) \qquad \text{and} \qquad p_{econ}(x) = q_{econ}(x)$$

So the **scale x** and **un-scale ξ** steps are not necessary (Problem 5-42).

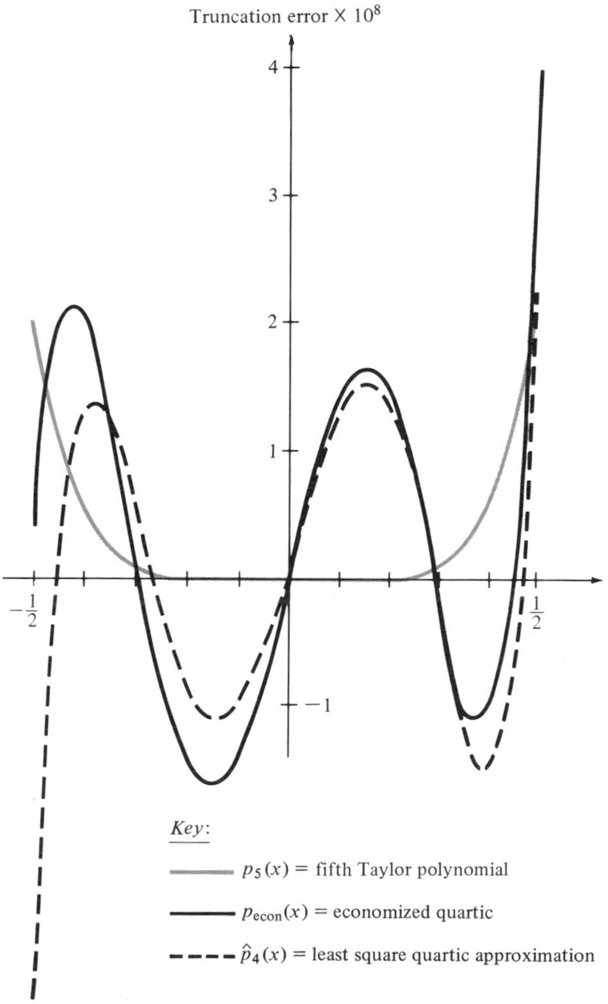

***Figure 5.4-5** Truncation error of several approximations of e^x on $[-\frac{1}{2}, \frac{1}{2}]$.*

■

Our illustrative function $f(x) = e^x$ has a Taylor series that converges rapidly as $n \to \infty$. As a result, the improved efficiency resulting from Chebyshev economization was not too dramatic. Substantial decreases in the degree needed for a prescribed accuracy have been obtained (and used internally in calculators and computers) for functions such as $\ln x$ and $\tan^{-1} x$, whose Taylor series converge slowly as $n \to \infty$. And approximations involving *both* rational functions *and* Chebyshev economization have been used successfully. The reader interested in further details is referred to [40].

PROBLEMS

Section 5.1

5-1 Consider the one-parameter guess function $g(x) = 1 + \alpha x^2$.
(a) Find the normal equation for fitting $g(x)$ to $P_1(x_1, y_1), \ldots, P_M(x_M, y_M)$. *If possible*, solve it for $\hat{\alpha}$ to get a general formula for $\hat{\alpha}$ in terms of x_k, y_k, and M.
(b) Find $\hat{\alpha}$, $\hat{g}(x)$ and $E(\hat{g})$ for $P_1(1, 0)$ and $P_2(2, 5)$.
(c) Find $\hat{\alpha}$ to 5s for $P_1(1, 0)$, $P_2(2, 5)$, and $P_3(3, 9)$. If a numerical method is needed, get an initial guess of $\hat{\alpha}$ graphically [see (6a) of Section 5.2A].

5-2 Same as Problem 5-1 but for $g(x) = e^{\alpha x}$.

5-3 Same as Problem 5-1 but for $g(x) = \alpha e^{x/2}$.

5-4 Find $\hat{c}$ that minimizes the square error of fitting the constant guess function $g(x) = c$ to $P_1(x_1, y_1), \ldots, P_M(x_M, y_M)$. Interpret $\hat{c}$ and $E(\hat{g})$ as statistical measures.

5-5 **(a)** Show that for any function $\phi(x)$, the normal equation for fitting $g(x) = \alpha\phi(x)$ to $P_1(x_1, y_1), \ldots, P_M(x_M, y_M)$ can be solved for $\hat{\alpha}$ in terms of x_k, y_k, and M.
(b) Use the formula of part (a) to fit $g(x) = \alpha \sin(\pi x/3)$ to $P_1(0, \frac{1}{2})$, $P_2(\frac{1}{2}, 1)$, and $P_3(3, 0)$ in the least square sense.

5-6 For the given data, find the least square line $\hat{L}(x) = \hat{m}x + \hat{b}$, its square error $E(\hat{L})$, and its determination index $R(\hat{L})$. Does $R(\hat{L})$ indicate good fit? Explain.
(a) $P_1(-1, 0)$, $P_2(1, 1)$, $P_3(2, 4)$
(b) $P_1(0, 1)$, $P_2(1, 2)$, $P_3(2, 4)$
(c) $P_1(1, 4)$, $P_2(2, 1)$, $P_3(3, 0)$, $P_4(4, -1)$
(d) $P_1(1, -8)$, $P_2(1.3, -3)$, $P_3(1.4, -1)$, $P_4(1.5, 1)$, $P_5(1.8, 4)$.

5-7 Einstein suggested that the threshold voltage v_0 (in volts) for the photoelectric effect varies linearly with frequency f (in Hz = 1/second) according to the equation

$$ev_0 = hf - \phi, \qquad \text{for } f > f_t = \text{threshold frequency} = \phi/h$$

where $e = 1.60219\text{E}{-19}$ coulomb (the charge of a single electron), h is Planck's constant, and ϕ is the *work function* for the particular metal being used. Fit a straight line to the following (f, v_0) data:

f (Hz $\cdot 10^{13}$)	56	70	79	83	102	120
v_0 (volts)	0.05	1.0	1.4	1.74	2.43	3.0

Then use your values of $\hat{m}$ and $\hat{b}$ to get estimates of h and f_t.

Section 5.2

5-8 ***(Supplement to Figure 5.2-1)*** Figure 5.2-1 does not show some useful possibilities for hyperbolic fit. Sketch the indicated graph with α *negative*. As in Figure 5.2-1, show all horizontal and vertical asymptotes (dashed) and all x- and y- intercepts, and describe how they are related to α and β.
(a) $g_4(x)$ for $\beta > 0$ **(b)** $g_5(x)$ for $\beta < 0$ **(c)** $g_6(x)$ for $\beta < 0$

5-9 Verify graphically that the data in parts (a)–(d) of Problem 5-6 appear to be monotone and convex. Which $g(x)$'s of Figure 5.2-1 are most likely to fit the data well, perhaps with $\alpha < 0$? (See Problem 5-8.) Are there enough data points to deduce good fit from $R(\hat{L}) \approx 1$?

5-10 In parts (a)–(d), use the Linearization Algorithm 5.2B to fit $g(x)$ to the data in the corresponding part of Problem 5-6.
(a) $g(x) = \dfrac{\alpha}{\beta + x}$ **(b)** $g(x) = \alpha e^{\beta x}$ **(c)** $g(x) = \alpha + \dfrac{\beta}{x}$ **(d)** $g(x) = \alpha + \beta \ln x$

5-11 Find the normal equations for fitting the given two-parameter guess function to M data points. If the system is linear, put it in matrix form.
(a) $g(t) = \alpha\sqrt{t} + \beta/t$ **(b)** $g(x) = \alpha x/(\beta + x)$ **(c)** $g(v) = \alpha \sin \beta v$
(d) $g(t) = \alpha e^{-t} + \beta e^{-4t}$ **(e)** $g(u) = \alpha + \beta/u^2$ **(f)** $g(x) = \alpha x^{\beta}$

5-12 Why was the Secant Method and not the Newton–Raphson method suggested for getting β in Solution (b) of Example 5.2A?

5-13 Let $g(x) = \alpha e^{\beta x}$ and $h(x) = \alpha x^{\beta}$, and consider the data

$$P_1(1, 2.3),\quad P_2(2, 6.1),\quad P_3(3, 10.7),\quad P_4(4, 16.0),\quad P_5(5, 21.9)$$

(a) Plot $P_1, \ldots, P_5$ and determine from Figure 5.2-1 which of $g(x)$ or $h(x)$ appears better suited to fit the data.
(b) Use the Linearization Algorithm 5.2B to fit $g(x)$ and $h(x)$ to the data and find $R(\hat{L})$ for the *linearized* fit of *both* g and h. Does this indirect indicator agree with your answer to part (a)? Confirm by finding $E(g)$ and $E(h)$ to see which is smaller.

5-14 Repeat Problem 5-13 for $g(x) = \alpha/(\beta + x)$, $h(x) = \alpha x/(\beta + x)$, and the data

$$P_1(0.1, 0.04),\quad P_2(1, 0.51),\quad P_3(3, 2.2),\quad P_4(4, 3.8),\quad P_5(6, 13.2)$$

NOTE: See Problem 5-8 before doing part (a) of Problem 5-14 or 5-15.

5-15 Repeat Problem 5-13 for $g(x) = \alpha + \beta \ln x$, $h(x) = \alpha + \beta/x$, and the data

$$P_1(2, 1.7),\quad P_2(4, 3.2),\quad P_3(7, 4.4),\quad P_4(12, 5.6),\quad P_5(20, 6.8)$$

5-16 **(a)** For which i's of Figure 5.2-1 does the curve $y = g_i(x)$ become a straight line when plotted on semilog paper? on log-log paper? Explain.
(b) What is the justification for Practical Consideration 5.2B?

Problems 5-17–5-20 illustrate how a transformation of x and/or y can sometimes enable you to get good fit by solving a 2×2 linear system and avoid having to solve nonlinear system or a larger linear system.

5-17 ***(Transforming x)*** The density of particles (particles/area $= N/A$) emerging from a scattering foil at an angle θ (degrees) from the perpendicular is known to satisfy

$$\text{density} = K/[\sin(\theta/2)]^n, \qquad \text{where } n \text{ is a positive integer}$$

Use the Linearization Algorithm 5.2B with $x = 1/\sin(\theta/2)$ to determine n and K from the following data:

θ (degrees)	5	15	30	40	50	75	105	150
Density (N/A)	8E+5	20,000	800	500	107	26	8	5

5-18 ***(Transforming y)*** When a V-volt charging circuit is used to charge a capacitor, the voltage v (volts) across the capacitor after t seconds satisfies $v = V - (V - v_0)e^{-t/\tau}$, where v_0 is the initial voltage (at $t = 0$) and τ is the circuit's *time constant*. A 10-volt charging circuit results in the following readings:

t (sec)	.5	1	2	3	4	5	7	9
v (volts)	6.36	6.84	7.26	8.22	8.66	8.99	9.43	9.63

Use the Linearization Algorithm 5.2B to fit $g(t) = \alpha e^{\beta t}$ to the modified data (t_k, y_k), where y_k is a suitable translation of v_k. Estimate v_0, τ, and the value of v when $t = 6$.

5-19 ***(Translating y)*** A Geiger counter measures a background reading of 232 counts in 10 minutes. The number of counts per 10-minute interval after a sample of indium (^{116}In) is introduced is

t (min)	0 – 10	10 – 20	20 – 30	30 – 40	40 – 50	50 – 60
Counts	20,511	16,174	13,904	12,514	10,775	9596

In general, emitted radiation $R = R_0 e^{-\lambda t}$, where λ is the *decay constant* of the radioactive material. Use the Linearization Algorithm 5.2B to fit $\alpha e^{\beta t}$ to the data (t_k, y_k), where $y_k =$ Counts $- 232$. (Why?) Deduce the *half life* [$= (\ln 2)/\lambda$] of indium from your result.

5-20 ***(Translating both x and y)*** For the data points $P(x_k, F_k)$ on the *nonlinear* part of Figure 5.1-2 (i.e., for $x_k \geqslant 9$), do the following:

(a) Form $T(u_k, v_k)$, where $u_k = x_k - 9$ and $v_k = 22 - F_k$.

(b) For $g(u)$ given in (i)–(iii): Fit the curve $v = g(u)$ to $T(u_k, v_k)$ of part (a). This has the effect of fitting $F = 22 - g(x - 9)$ to $P(x_k, y_k)$. Find $R(\hat{L})$ for the *linearized* fit and compare it to those shown in Figure 5.2-7.

(i) $g(u) = \alpha u^\beta$ (Use the Linearization Algorithm 5.2B.)

(ii) $g(u) = \alpha/(\beta + u)$ (Use the Linearization Algorithm 5.2B.)

(iii) $g(u) = \alpha + \beta u^2$ [Use $\hat{g}(u)$; the normal equations are linear.]

5-21 Suppose that the given data $P_k(x_k, y_k)$ are translated to $T_k(u_k, v_k)$, where $u_k = x_k - c$ and $v_k = y_k - d$ for specified values of c and d.

(a) How are the parameters of the least square line for the T_k's related to those for the P_k's? Is it worthwhile to try such a translation when fitting a straight line? Explain.

(b) When $d = 0$, find those $g_i(x)$'s of Table 5.2-1 for which using a CURFIT program to fit the T_k's might give more accuracy than using it to fit the P_k's. Justify your answer.

(c) Same as part (b) but for $d \neq 0$.

Section 5.3

5-22 Which $g_i(x)$'s of Table 5.2-1 are linear models? Show that for these $g_i(x)$'s, the α and β obtained using the Linearization Algorithm 5.2B are actually the least square parameters $\hat{\alpha}$ and $\hat{\beta}$.

5-23 Form the normal equations (5) of Section 5.3A for the given two-parameter linear model $g(x)$, then find $\hat{\alpha}$, $\hat{\beta}$, and $R(\hat{g})$ for the given data.

(a) $g(x) = \alpha x + \beta x^2$; $P_1(-1, 1)$, $P_2(1, 1)$, $P_3(2, 8)$

(b) $g(t) = \alpha e^{-t} + \beta e^{-t/5}$; $P_1(1, -0.3)$, $P_2(4, 0.4)$, $P_3(6, 0.3)$

(c) $g(u) = \alpha + \beta/u^2$; $P_1(1, 5.7)$, $P_2(2, 1.2)$, $P_3(3, 0.3)$

5-24 How many data points are needed to deduce good fit from $R(\hat{g}) \approx 1$ when n, the number of parameters of a linear model g, is 1? 2? 3? 4? 5?

5-25 For the data given in parts (a)–(d) of Problem 5-6: Find the least square quadratic $\hat{g}_3(x) = \hat{\alpha} + \hat{\beta}x + \hat{\gamma}x^2$ and its determination index $R(\hat{g}_3)$, and discuss the suitability of a quadratic fit. Could $R(\hat{g}_3)$ have been found without any calculation? Explain.

5-26 Fit $g(t) = c + ae^{-0.47t} + be^{-0.06t}$ to the data of Example 5.3B with P_4 removed. Does your answer indicate that $\hat{c}$, $\hat{a}$, and $\hat{b}$ of (13b) are accurate to $2s$? Explain.

5-27 When $g(x) = \gamma_1\phi_1(x) + \gamma_2\phi_2(x) + \gamma_3\phi_3(x)$ is fit to $P_1(0, 6)$, $P_2(1, 7)$, $P_3(3, 8)$, $P_4(6, 9)$, it turns out that $\Sigma_k \phi_1(x_k)y_k = 30$, $\Sigma_k \phi_2(x_k)y_k = -50$, $\Sigma_k \phi_3(x_k)y_k = 114$, and $\hat{\boldsymbol{\gamma}} = [5 \quad 3 \quad 2]^T$. What is $R(\hat{g})$?

5-28 Use the Degree Selection Criterion (26) of Section 5.3B to find the polynomial that is best suited to fit the given data.

(a)

x	1	2	3	4	5	6	7	8
y	−5	−12.4	−15.7	−15.1	−10.5	−1.9	10.7	27.4

(b)

x	0	1.5	2.5	4.0	6.5	8.1	9.3	11.3	13.0	15.5	17.5	19.0
y	1.2	3.5	4.5	5.3	4.5	2.3	0.7	−2.0	−3.9	−4.2	−2.6	−0.5

SUGGESTION: If you have no program to solve $A\boldsymbol{\gamma} = \mathbf{b}$, use Algorithm 5.3D.

5-29 The data $P_1, \ldots, P_6$ shown in Figure 5.3-3 (Section 5.3B) appear to be (roughly) quadratic on $[x_1, x_2]$ and $[x_4, x_6]$, and cubic on $[x_2, x_4]$. This exercise shows how to "piece together" a guess function based on this.

(a) Find $\hat{q}_L(x)$, the least square quadratic for $P_1, \ldots, P_4$ (on the left), and $\hat{q}_R(x)$, the least square quadratic for for $P_2, \ldots, P_6$ (on the right).

(b) Let $L_L(x) = (x - x_4)/(x_2 - x_4)$ and $L_R(x) = (x - x_2)/(x_4 - x_2)$. A $g(x)$ that "connects" $\hat{q}_L(x)$ and $\hat{q}_R(x)$ is

$$g(x) = \begin{cases} \hat{q}_L(x) \text{ (quadratic)} & \text{for } 1 = x_1 \leqslant x \leqslant x_2 = 2 \\ \hat{q}_L(x)L_3(x) + \hat{q}_R(x)L_4(x) \text{ (cubic)} & \text{for } 2 = x_2 < x < x_4 = 5 \\ \hat{q}_R(x) \text{ (quadratic)} & \text{for } 5 = x_4 \leqslant x \leqslant x_6 = 16 \end{cases}$$

Show that $g(x)$ is continuous, sketch its graph on [1, 16], and compare the fit visually to that of $\hat{g}_2(x), \ldots, \hat{g}_6(x)$ of Figure 5.3-2.

5-30 Consider the hypothetical historical data in Example 5.3D.

(a) Form the coefficient matrix $A_{2\times 2}$ for fitting $g_2(x) = L(x)$ and find $C_p(A_{2\times 2})$. Do $C_p(A_{2\times 2})$ and $C_p(A_{3\times 3})$ of Example 5.3D(a) demonstrate how quickly $A_{n\times n}$ can become ill conditioned as n increases?

(b) Use the Polynomial Scaling Algorithm 5.3B to fit $g_2(x)$. Compare $C_p(A_{2\times 2})$ [for fitting (X_k, y_k)] to $C_p(A_{2\times 2})$ in part (a).

(c) Repeat part (b) for for the quadratic $g_3(x)$ and $C_p(A_{3\times 3})$.

5-31 The population of a certain city was 12,500 in 1600, 14,000 in 1700, 16,000 in 1800, and 19,000 in 1900. Use the least square quadratic to predict the population in the year 2000. Two possible ways to do this without serious roundoff error are suggested in parts (a) and (b).

(a) Use the Polynomial Scaling Algorithm 5.3B.

SUGGESTION: If you did Problem 5-30(c), you can reuse some of your work.

(b) Reuse $p_0(x)$, $p_1(x)$, and $p_2(x)$ of Example 5.3D(b).

5-32 Based on a sketch of the data and your experience with the shape of the graphs of algebraic functions, suggest a linear model that might give good fit.

(a)

x	-1	0	1	2	3	4	5	6
y	-17.5	2.5	4	3.3	2.3	4	10.5	24

(b)

x	0.5	1	1.5	2	2.5	3	3.5	4	5	7	10
y	10.5	2	0.83	0.75	0.98	1.3	1.7	2.2	3.1	5.1	8

(c)

x	0.4	0.8	1.2	2.0	2.8	3.6	4.8	6.0	8.0
y	-66.1	-3.77	1.44	2.38	2.20	2.00	1.79	1.64	1.49

5-33 Deduce from Algorithm 5.3D that the least square line for $P_1, \ldots, P_M$, namely $\hat{L}(x) = \hat{m}x + \hat{b}$, must go through the point $(\hat{x}, \hat{y})$, where $\hat{x}$ and $\hat{y}$ are the means of the x_k's and y_k's, respectively.

5-34 Same as Problem 5-6 but with the data weighted as follows:

$$w_1 = 4, \quad w_2 = 9, \quad w_3 = 4, \quad w_4 = 1, \quad w_5 = 2$$

Section 5.4

5-35 For $g(x)$ in parts (a)–(c): Find the least square $g(x)$ for $f(x) = x^2$ on the interval $I = [0, 1]$ and plot the graphs of $\hat{g}$ and f for $0 \leqslant x \leqslant 1$. Then find the largest absolute error on I, $\epsilon_{\max}$, analytically as in Example 5.4B. The parameters are m and b.
(a) $g(x) = b$ (constant) **(b)** $g(x) = mx$ **(c)** $g(x) = L(x) = mx + b$

5-36 Same as Problem 5-35 but for $f(x) = \sqrt{x}$ on $I = [1, 2]$.

5-37 Same as Problem 5-35 but for $f(x) = e^x$ on $I = [0, 2]$.

5-38 ***(A One-Parameter Fit with a Nonlinear Normal Equation)***
(a) Find the normal equation for finding the least square exponential $g(x) = e^{\beta x}$ to approximate $f(x) = x^2$ on $I = [0, 2]$.
(b) Use a numerical method to solve for $\hat{\beta}$ and then to find $\epsilon_{\max}$.

5-39 Find the least square $g(x)$ for $f(x)$ on the interval I. Then find $\epsilon_{\max}$, the largest absolute error on I, as in Example 5.4B.
(a) $f(x) = 1/(1 - x)$ on $I = [0, \frac{1}{2}]$; $g(x) = x + \beta x^2$
(b) $f(x) = 1/(1 - x)$ on $I = [0, \frac{1}{2}]$; $g(x) = \alpha x + \beta x^2$
(c) $f(x) = e^{-x}$ on $I = [0, 1]$; $g(x) = L(x) = mx + b$
(d) $f(x) = e^x$ on $I = [1, 2]$; $g(x) = L(x) = mx + b$

5-40 Approximate $f(x) = \sqrt{x}$ on $I = [1, 2]$ as indicated. Then find $\epsilon_{\max}$ on I and compare it to $\epsilon_{\max}$ of Example 5.4B.
(a) Use the least square line $\hat{L}(x) = \hat{m}x + \hat{b}$
(b) Use $\hat{L}_1(x)$, the least square linear approximations of $\sqrt{x}$ on $I_1 = [1, 1.5]$ for $1 \leqslant x < 1.5$, and $\hat{L}_2(x)$, the least square linear approximations of $\sqrt{x}$ on $I_2 = [1.5, 2]$ for $1.5 \leqslant x \leqslant 2$.
(c) Use $P_2(x)$ the second Taylor polynomial based at $a = 1.44 = (1.2)^2$.
(d) Use $P_2(x)$ of part (c) economized to a first-degree polynomial.
(e) Use $P_3(x)$ the third Taylor polynomial based at $a = 1.44 = (1.2)^2$.

(f) Use $P_3(x)$ of part (e) economized to a second-degree polynomial.
(g) Use Padé approximation starting with $P_3(x)$ of part (e).

5-41 Use Padé approximation to find a rational function $r(x)$ that approximates $p_N(x) =$ for $x \approx 0$.

(a) $p_4(x) = 1 - x + \frac{x^2}{2} - \frac{x^3}{6} + \frac{x^4}{24}$ $\quad (\approx e^{-x})$

(b) $p_5(x) = x - \frac{x^3}{6} + \frac{x^5}{120}$ $\quad (\approx \sin x)$

(c) $p_4(x) = 1 + \frac{x}{4} - \frac{x^2}{32} + \frac{x^3}{128} - \frac{5x^4}{2048}$ $\quad \left(\approx \sqrt{1 + \frac{x}{2}}\right)$

(d) $p_4(x) = 1 - \frac{x^2}{2} + \frac{x^4}{24}$ $\quad (\approx \cos x)$

5-42 For parts (a)–(d) of Problem 5-41: Use Chebyshev economization to approximate $p(x)$ on $[-1, 1]$ by a polynomial $p_{\text{econ}}(x)$ obtained by removing the leading (i.e., x^N) term. Sketch the graphs of p_N and p_{econ} on $[-1, 1]$.

5-43 For parts (a) and (c) of Problem 5-41: Use Chebyshev economization to approximate $p(x)$ on $[-\frac{1}{2}, \frac{1}{2}]$ by a polynomial $p_{\text{econ}}(x)$ obtained by removing the *two* leading (i.e., x^N and x^{N-1}) terms. Sketch the graphs of p_N and p_{econ} on $[-\frac{1}{2}, \frac{1}{2}]$.

5-44 For parts (a)–(d) of Problem 5-41: Use the Lagrange form of the remainder to get a reasonable bound on the error of approximating the indicated $f(x)$ by $p_N(x)$ on $[-\frac{1}{2}, \frac{1}{2}]$ and on $[-1, 1]$.

5-45 Consider $p_2(x) = 1 + x + x^2/2$, the second-degree Maclaurin approximation of e^x.
(a) Use Chebyshev economization to get the linear $p_{\text{econ}}(x)$ on $[-1, 1]$.
(b) Sketch $p_2(x)$ and $p_{\text{econ}}(x)$ of part (a) along with the first-degree Maclaurin approximation $p_1(x) = 1 + x$ on the same axes for $-1 \leq x \leq 1$. Which straight line approximates e^x more uniformly on $[-1, 1]$?

MISCELLANEOUS PROBLEMS

M5-1 **(a)** Fit the linear model $g(x) = \alpha x + \beta x^2$ to the *entire* set of data given in Figure 5.1-2; then find $R(\hat{g})$.
(b) Discuss the advantages and disadvantages of this $\hat{g}(x)$ as compared to one of the form given in (2b) of Section 5.1A [see Figure 5.2-6 and (iii) of Problem 5-20].

M5-2 [***Calculus Proof That*** $E(\hat{L}) \leq E(L)$ ***for Any*** $L(x) = mx + b$] The solution $[\hat{b} \quad \hat{m}]^T$ of (18) of Section 5.1C is the unique critical point of $E(L)$, viewed as a function of m and b. Use the Second Derivative Test for functions of two variables to show that this vector actually minimizes $E(L)$.

M5-3 Use the trigonometric identity for $\cos(a + b)$ to convert the problem of fitting $g(x) = A\cos(\omega x + \theta)$ to an equivalent but simpler problem of fitting a two-parameter *linear* model. Explain how to get $\hat{A}$ and $\hat{\theta}$ from the solution of the equivalent problem.

M5-4 The data of Problem 5-28(b) appear to be periodic with period 20.
(a) Use Problem M5-3 to fit $g(x) = A\cos(\pi x/10 + \theta)$ to the data $(x_k, y_k - \hat{y})$, where $\hat{y} = (\Sigma\, y_k)/M$. Find $\hat{A}$, $\hat{\theta}$, $E(\hat{g})$, and $R(\hat{g})$.
(b) Estimate graphically the x value at which the data in Problem 5-28(b) appears to first cross the line $y = \hat{y}$; call it c. Then fit the two-parameter linear model $g(u) = \alpha \sin(\pi u/10) + \beta \sin(2\pi u/10)$ to the data $(u_k, y_k - \hat{y})$, where $u_k = x_k - c$. Find $R(\hat{g})$.

M5-5 The general hyperbolic guess function $g(x) = (\alpha + \beta x)/(1 + \gamma x)$ is useful for fitting monotone, convex data that appear to have a singularity (at $x = -1/\gamma$). However, it is not a linear model.

(a) Explain why good fit to the data $P_1, \ldots, P_M$ can be obtained by minimizing the "linearized square error"

$$EL(g) = \sum_{k=1}^{M} [y_k(1 + \gamma x_k) - (\alpha + \beta x_k)]^2$$

(b) Show that minimizing $EL(g)$ leads to a 3×3 linear system, and express it in matrix form.

(c) Use the result of part (b) to fit $g(x)$ to the following data:

$$P_1(0, -1), \quad P_2(1, -.5), \quad P_3(3, 1.2), \quad P_4(4, 2.8), \quad P_5(6, 12.2)$$

Then find $R(g)$. What can you conclude about the fit?

M5-6 Prove that the square error of any least square linear model $\hat{g}(x) = \Sigma_i \hat{\gamma}_i \phi_i(x)$ can be obtained as

$$E(\hat{g}) = \sum_k y_k^2 - \sum_i \hat{\gamma}_i b_i = \mathbf{y}\mathbf{y}^T - \hat{\boldsymbol{\gamma}}^T\mathbf{b}$$

where $\mathbf{y}$, $\hat{\boldsymbol{\gamma}}$, and $\mathbf{b}$ have the same meaning as in formula (20) of Section 5.3B.

OUTLINE: Apply (4a) and the definition of b_i given in (5c) of Section 5.3A after showing that

$$\begin{aligned} E(\hat{g}) &= \sum_k [\hat{g}(x_k) - y_k]\left[\sum_i \hat{\gamma}_i \phi_i(x) - y_k\right] \\ &= \sum_i \hat{\gamma}_i \left\{[\hat{g}(x_k) - y_k]\right\} - \sum_i \hat{\gamma}_i \left[\sum_k \phi_i(x_k) y_k\right] + \sum_k y_k^2 \end{aligned}$$

M5-7 Use the formula of Problem M5-6 to find $E(\hat{g})$ for Problem 5-27.

M5-8 **(a)** Deduce formula (20) of Section 5.1C from Problem M5-6.
(b) Deduce (25) of Section 5.1D from (20) of Section 5.3B.
(c) Which, if any, of the values of $E(g_1)$ or $E(g_4)$ obtained in Example 5.2B can be found using the result of Problem M5-6? Verify.

M5-9 A useful approach to solving an overdetermined system $A_{m \times n}\mathbf{x} = \mathbf{b}$ when $m < n$ is to solve $A\mathbf{x} = \mathbf{p}$, where $\mathbf{p}$ is the projection of $\mathbf{b}$ on the column space of A. Show that the desired solution satisfies the $n \times n$ linear system $A^TA\mathbf{x} = A^T\mathbf{b}$.

M5-10 Try to fit the linear model $g_2(x) = \alpha(x - 1)^2 + \beta \sin(\pi x/2)$ to the data $P_1(0, 0)$, $P_2(1, 1)$, $P_3(2, 0)$. What caused the problem?

M5-11 The discussion of (38) of Section 5.3D shows that there is a $\hat{\boldsymbol{\gamma}}$ that minimizes $E(g_n)$ even if $A = F^TF$ is singular. Explain.

M5-12 **[*Derivation of the Three-Term Recurrence Relation (43) of Section 5.3D*]**
(a) Show that $p_0(x) = 1$ and $p_1(x) = x - \hat{x}$ are x_k-orthogonal.
(b) (*Inductive step*) Assume that $p_0(x), \ldots, p_{i-1}(x)$ are x_k-orthogonal and define $p_i(x) = xp_{i-1}(x) - \alpha_i p_{i-1}(x) - \beta_{i-1}p_{i-2}(x)$, where $i > 1$. Clearly, $\mathbf{p}_i^T\mathbf{p}_j = 0$ for $j < i-2$. Show that α_i and β_i in (43c) result from setting $\mathbf{p}_i^T\mathbf{p}_{i-1} = 0$ and $\mathbf{p}_i^T\mathbf{p}_{i-2} = 0$.

M5-13 Devise an algorithm to evaluate $\hat{g}_n(x)$ at a given x using only the stored values of a_i, b_i, and $\hat{a}_i$, and the three-term recurrence relation (43) of Section 5.3D.

M5-14 Find $\hat{g}_2(x)$, $\hat{g}_3(x)$, and $\hat{g}_4(x)$ of Example 5.4B using the Orthogonal Polynomial Algorithm 5.3D. Combine like powers of x to get the the least square coefficients given in Figure 5.3-2.

M5-15 The Orthogonal Polynomial Algorithm 5.3D can be modified so that the constructed $\mathbf{p}_i$'s are *orthonormal* by replacing $\mathbf{p}_i$ by $\mathbf{p}_i/\|\mathbf{p}_i\|$ as soon as it is obtained. Note that as a result, $\hat{a}_i$ is simply $\mathbf{p}_i^T\mathbf{y}$. (Why?) Redo Example 5.4D using *orthonormal* $\mathbf{p}_i$'s. Does it remedy the problem described in Practical Consideration 5.4D?

M5-16 Adapt the Linearization Algorithm 5.2B to the problem of fitting $g_1(x), \ldots, g_6(x)$ of Table 5.2-1 to a given $f(x)$ on a given closed interval $I = [a, b]$. Is it as helpful for this *continuous* approximation problem? Why?

M5-17 Use your algorithm of Problem M5-16 to fit $g(x) = \alpha x/(\beta + x)$ to $f(x) = \sqrt{x}$ on interval $I = [1, 2]$, and find $\epsilon_{\max}$, the largest absolute error on I. Compare $\epsilon_{\max}$ to that of Example 5.4B.

M5-18 ***(Estimating a Derivative by Smoothing ''Noisy'' Data)*** Suppose that you wanted to find the sensitivity of the dial (i.e., $d\omega/d\theta$) in Figure 1.1-1(a) when $\theta = 164°$, but *only* knew the values of ω when $\theta = 50°, 100°, 150°, 200°$, and $250°$. Find a two-parameter curve $g(\theta)$ that fits this monotone, convex data well, then use $g'(164°)$ to estimate the desired rate of change. Compare to the slope shown at $\theta = 164°$ in Figure 1.1-1(b).

M5-19 ***(Estimating an Integral by Smoothing ''Noisy'' Data)*** The work in inch-pounds required to stretch the spring described in Figure 5.1-2 from 9 in. to 17 in. is the area under the F versus x curve for $9 \leqslant x \leqslant 17$. Use $\int_9^{17} g(x)\,dx$, where $g(x)$ is a two-parameter guess function that fits the data well for $9 \leqslant x \leqslant 17$, to approximate this work.

M5-20 The three-term recurrence relation for x_k-orthogonal polynomials [(43) of Section 5.3D] can be adapted to give **$w(x)$-orthogonal** polynomials on $I = [a, b]$ if we

$$\text{take the inner product } \mathbf{p}_i^T\mathbf{p}_j \quad \text{to be} \quad \int_a^b w(x)p_i(x)p_j(x)\,dx$$

where $w(x)$ is a **weight function** that is continuous on I and nonnegative on its interior. Thus,

$$p_0(x) = 1; \qquad p_1(x) = x - \hat{x}, \quad \text{where } \hat{x} = \frac{\int_a^b xw(x)\,dx}{\int_a^b w(x)\,dx}$$

and so on. Find $p_1(x), \ldots, p_{n-1}(x)$ for the given $w(x)$, I, and n.
(a) *Legendre polynomials*: $w(x) = 1$; $I = [-1, 1]$; $n = 4$
(b) *Chebyshev polynomials*: $w(x) = 1/\sqrt{1 - x^2}$; $I = [-1, 1]$; $n = 3$
(c) *Laguerre polynomials*: $w(x) = e^{-x}$; $I = [0, \infty)$; $n = 3$

M5-21 Once $w(x)$-orthogonal polynomials have been found as in Problem M5-20, the Fourier coefficients $\hat{a}_i$ of the least square ''weighted'' approximation of $f(x)$ on $I = [a, b]$ can be found as in (42) of Section 5.3D, that is, as

$$\hat{a}_i = \frac{\int_a^b w(x)p_i(x)f(x)\,dx}{\int_a^b w(x)p_i(x)p_i(x)\,dx}$$

In parts (a)–(c), find the weighted least square $(n - 1)$st degree polynomial for $f(x)$ on I using the $w(x)$, n, and I given in the corresponding part of Problem M5-20.
(a) $f(x) = \cos \pi x$ **(b)** $f(x) = x^3$ **(c)** $f(x) = x^3$

M5-22 The resistance R (in ohms) of the heating element of a toaster satisfies $R = R_0 + bi^2$, where i is the current (in amperes) through it. Determine R_0 and b from the following (i, R) data:

(1, 0.141), (2, 0.234), (4, 1.055), (6, 2.273), (9, 5.015), (12, 8.854)

M5-23 The normal equations for $g(x) = \alpha + \beta\phi(x)$ are

$$\begin{bmatrix} 4 & 1 \\ 1 & 2 \end{bmatrix} \begin{bmatrix} \alpha \\ \beta \end{bmatrix} = \begin{bmatrix} 8 \\ -5 \end{bmatrix}.$$

(a) Find *if possible*, M (number of data points), $\hat{x}$, $\hat{y}$ (average x_k, y_k), $\hat{\alpha}$, $\hat{\beta}$ (least square parameters), $E(\hat{g})$, $R(\hat{g})$.

(b) Find all values that could *not* be found in part (a), but which can be found if you also knew that $\Sigma_k\, y_k^2 = 56$.

COMPUTER PROBLEMS

C5-1 ***(Two Solutions of the Nonlinear 2 × 2 System (5) of Example 5.2A)***

(a) Eliminate α and "cross-multiply" to rewrite the resulting equation as $f(\beta) = 0$ where $f(\beta)$ uses no divisions. Use a computer program with $\beta_0 = -0.1$ [see (6c)] to solve $f(\beta) = 0$ for $\hat{\beta}$, then get $\hat{\alpha}$ from (5a).

(b) Use a computer program for NR_2 or any other suitable method with $\alpha_0 = 4\frac{1}{2}$ and $\beta_0 = -0.1$ [see (6c)] to solve for $\hat{\alpha}$ and $\hat{\beta}$.

(c) Discuss the relatative advantages and disadvantages of the solutions outlined in parts (a) and (b).

C5-2 Find the least square $g(x) = \hat{y} + A\cos(\omega x + \theta)$ for the data in Problem 5-28(b), treating A, θ, *and* ω as parameters. Find $E(\hat{g})$ and compare this solution to that of Problem M5-4(a).

C5-3 Let A denote the 4×4 coefficient matrix of the normal equations for fitting a cubic $g_4(x)$ to knots $P_1, \ldots, P_5$, for which

$$x_1 = 20, \quad x_2 = 22, \quad x_3 = 25, \quad x_4 = 28, \quad x_5 = 30$$

Use any available software to find a condition number of A. Repeat for the A resulting from the Polyomial Scaling Algorithm 5.3B, and discuss the results.

C5-4 **(a)** Write a subprogram `LINALG(M, XDAT, YDAT, OK, A, B, ROFL, YCALC)`, that finds the least square line $\hat{g}_0(x) = \hat{\alpha} + \hat{\beta}x$ and then uses the Linearization Algorithm 5.2B to fit $g_i(x)$ of Table 5.2-1 for the data $(x_1, y_1), \ldots, (x_M, y_M)$ ($M \leq 100$), stored in the arrays `XDAT` and `YDAT`. For for $i = 0, 1, \ldots, 6$, it returns

`OK`(i) A boolean flag that is `TRUE` unless the linearizing transformation for $g_i(x)$ results in dividing by zero or taking the logarithm of a nonpositive number for some (x_k, y_k). If `TRUE`, then
`A`(i) and `B`(i) are the parameters α_i and β_i for $g_i(x)$
`RLHAT`(i) $= R(\hat{L})$ for the *linearized* data (X_i, Y_i), and
`YCALC`(i, k) $= g_i(x_k)$ for $k = 1, \ldots, M$.

Subprogram `LINALG` should call the following subprograms for $i = 1, \ldots, 6$:

`LINRIZ(I, M, XDAT, YDAT, BIGX, BIGY, OK)` Returns `OK`(i); if `TRUE`, `BIGX`, `BIGY` store the *linearized* data (X_k, Y_k) for $g_i(x)$, and the following subprograms are called:

`LINFIT(M, BIGX, BIGY, MHAT, BHAT, RLHAT)` Returns `MHAT` ($= \hat{m}$) and `BHAT` ($= \hat{b}$) and `RLHAT` [$= R(\hat{L})$] for the least square line for `BIGX`, `BIGY`

FITG(I, M, XDAT, YDAT, MHAT, BHAT, A, B, YCALC) Returns α_i, β_i, and row$_i$YCALC for $g_i(x)$

NOTE: For $i = 0$, OK(i) = TRUE, A(i) = BHAT, and B(i) = MHAT.

(b) Write a calling program based on the discussion of Section 5.2C. If OK(i) is FALSE, it should output `Cannot linearize the data`. Test it using first the data of Problem 5-14 and then with (0, 0) added.

(c) Modify your calling program in part (b) so that it can print details of YCALC(i, k) (if desired) for specified values of i.

C5-5 **(a)** Write a subprogram LINMOD(M, XDAT, YDAT, N, GAMHAT, GHAT, COND, OK), that fits an n-parameter linear model $g_n(x) = \sum_{i=1}^{n} \gamma_i\phi_i(x)$ to the data $(x_1, y_1), \ldots, (x_M, y_M)$ ($M \leq 100$), stored in the arrays XDAT and YDAT. The boolean flag OK should be TRUE *unless* the coefficient matrix of the normal equations is nearly singular; if TRUE, LINMOD returns a condition number COND, the least square parameter vector GAMHAT ($= \hat{\gamma}$), and RGHAT [$= R(\hat{g}_n)$]. The user must provide a FUNCTION program PHI(I, X) that returns $\phi_i(x)$ for any $i = 1, \ldots, n$. Subprogram LINMOD should call the following subprograms:

GETAB(M, XDAT, YDAT, N, A, B, SIGY2) Calls PHI to form $F = (\phi_i(x_j))_{M \times n}$, and then uses F to get A ($= A$) and B ($= \mathbf{b}$) of the normal equations as in (39) of Section 5.3D along with SIGY2 ($= \Sigma_k y_k^2$). Summations from $k = 1$ to M should be accumulated in extended precision if available.

SOLVE(N, A, B, GAMHAT, COND, OK) Solves $A\gamma = \mathbf{b}$ for GAMHAT ($= \hat{\gamma}$) provided OK = TRUE; otherwise, it should output a `Nearly singular coefficient matrix` warning and stop.

FUNCTION DETIND(M, SIGY2, N, B, GAMHAT) Uses (7) of Section 5.3A to get DETIND = $R(\hat{g}_n)$.

(b) Write a program that gets M, XDAT, and YDAT, calls LINMOD and, if OK, outputs GAMHAT and RGHAT. Test using Examples 5.3A and 5.3D.

(c) Modify your calling program in part (b) so that it can print details as in Problem C5-4(e) (if desired) for specified values of n.

C5-6 Modify your program in Problem C5-5 so as to do polynomial curve fitting. PHI(I, X) should return x^{i-1}. Also, the calling program should allow the user to specify a starting degree NSTART and stopping degree NSTOP. Test using Example 5.3B.

C5-7 Write a scaling subprogram SCALE(M, XDAT, C, D, BIGX, SLOPE, CEPT) that uses the given values of $x_k < \cdots < x_M$ (stored in the array XDAT), c, and d to find the slope (SLOPE) and X-intercept (CEPT) of the line through (x_1, c) and (x_M, d), and returns X_k = SLOPEx_k + CEPT, $k = 1, \ldots, M$, in the array BIGX.

C5-8 **(a)** Devise an algorithm for getting the coefficients $c_0, c_1, \ldots, c_n$ when

$$p(x) = a_0 + a_1(b + mx) + a_2(b + mx)^2 + \cdots + a_n(b + mx)^n$$

is expressed in the "simplified" form $p(x) = c_0 + c_1x + c_2x^2 \cdots + c_nx^n$.

(b) Implement your algorithm in part (a) in a subprogram COLECT(N, A, M, B, C) that forms C = $[c_0 \quad c_1 \quad \cdots \quad c_n]$ from n, m, b, and A = $[a_0 \quad a_1 \quad \cdots \quad a_n]$.

C5-9 Modify your program in Problem C5-6 so as to use the Polynomial Scaling Algorithm 5.3B. It should call subprogram SCALE of Problem C5-7 for the ***Scale x*** step, and COLECT of Problem C5-8 for the ***Collect like powers of x*** step. Test using Problem 5-31.

C5-10 (a) Write a subprogram ORTPOL(M, XDAT, YDAT, MAXDEG, ALPHA, BETA, AHAT, RGHAT, RFIT, OK) that uses the Orthogonal Polynomial Algorithm 5.3D to find α_i, β_i, $\hat{a}_i$, and $R(\hat{g}_{i+1})$ for $i \leqslant$ MAXDEG until $R(\hat{g}_{i+1}) >$ RFIT, and store them in the arrays ALPHA, BETA, AHAT, and RGHAT, respectively. OK should be TRUE unless for some i, $\mathbf{p}_i^T\mathbf{p}_i = 0$; if TRUE, the $\mathbf{p}_i$'s should be normalized as in Problem M5-15.

(b) Write a program that gets M, XDAT, YDAT, and N, calls ORTPOL, and (if OK) outputs the returned values of α_i, β_i, $\hat{a}_i$, and $R(\hat{g}_{i+1})$. Test using Example 5.3D.

(c) Modify your calling program in part (b) so that it can print details as in Problem C5-4(c) (if desired) for a specified $\hat{g}_n(x)$. Use a FUNCTION subprogram GHAT(X, N, ALPHA, BETA, AHAT) that uses the result of Problem M5-13 to evaluate $\hat{g}_n(x)$. Test using $\hat{g}_2(x)$ of Example 5.3D.

C5-11 (a) Write a subprogram PADE(NP1, C, A, B) that uses the NP1 ($= N + 1$) coefficients (stored in C $= [c_0 \;\; c_1 \;\; \cdots \;\; c_N]$) of a given Nth degree polynomial $p_N(x)$ to form the coefficients of the numerator (stored in A $= [a_0 \;\; a_1 \;\; \cdots \;\; a_n]$) and denominator (stored in B $= [b_0 \;\; b_1 \;\; \cdots \;\; b_m]$) of the rational function $r(x)$ that approximates $p_N(x)$ for $x \approx 0$ using Padé approximation as described in Section 5.4D. Test using Example 5.4D.

(b) Let x_{mid} denote the midpoint, $(a + b)/2$, of an interval $[a, b]$ on which an Nth degree polynomial $p(x) = \Sigma_{i=0}^{N} a_i x^i$ is to be approximated uniformly by a rational function. Since

$$p(x) = p(x_{\text{mid}} + u) = p_N(u), \text{ where } x \approx x_{\text{mid}} \text{ corresponds to } u \approx 0$$

the calling program in part (a) can be modified as follows to accomplish this. First call COLECT of Problem C5-8 to get the coefficients c_i of $p_N(u)$ from the coefficients a_i of $p(x)$. Then use PADE of part (a) to get the coefficients A and B of the numerator and denominator, respectively, of $r(u)$ that uniformly approximates $p_N(u)$ for $u \approx 0$. Finally, use COLECT again to convert A and B to the coefficients of powers of x of $r(x - x_{\text{mid}})$, the desired rational approximation $p(x)$. Test using Problem 5-40(g).

6

Interpolation

6.0

Introduction

The **curve-fitting** methods of Chapter 5 yielded functions $g(x)$ whose graph passes *near* (but *not necessarily through*) a set of *approximate* **data points** (x_k, y_k). The **interpolation** methods of this chapter yield functions $g(x)$ whose graph passes *through* (*not just close to*) selected *accurate* **tabulated points**

$$P_i(x_i,y_i), \qquad \text{where } y_i \doteq f(x_1) \text{ rounded to the accuracy of the table}$$

Geometrically, interpolation is the game of "follow the dots"; the analytic challenge is to find functional descriptions for curves that do this. Once such a function is found, its value at z can be used to approximate $f(z)$ when z is *not* a tabulated x_i.

In the last two decades, hand-held calculators that give accurate values of trigonometric, exponential, and logarithmic functions at the touch of a button have eliminated much of the need for interpolation. Nevertheless, statisticians, scientists, and engineers still encounter situations when only certain tabulated values (from a reference book or computer printout) are available, and accurate estimates of nontabulated values are needed. Another reason to understand interpolation is that it is the basis for most numerical methods for evaluating derivatives and definite integrals and for solving differential equations.

The most natural curve to try to pass through $n + 1$ tabulated P_i's is a polynomial having $n + 1$ coefficients, that is, of degree n. This is called **polynomial interpolation**. We shall see in Section 6.1C that there is only one such polynomial, and we shall give two methods for finding it. The first, due to **Lagrange** (Section 6.1), enables us to write out an interpolating polynomial of any degree *by inspection* of the interpolated P_i's. The second, due to **Newton** (Section 6.2), is based on **difference tables** which indicate the polynomial degree that appears to be best suited for estimating $f(z)$ for a particular z. The error of polynomial interpolation is analyzed in Section 6.1E and used to locate the $n + 1$ x_i's so as to minimize the largest possible error of nth degree polynomial interpolation over a given interval (**Chebyshev interpolation**) in Section 6.1F.

The Polynomial Wiggle Problem of Section 5.3B makes polynomial interpolation undesirable when the number of interpolated P_k's is large. In this situation, a **piecewise cubic spline** can be used as described in Section 6.3.

6.1

The Unique Interpolating Polynomial for $p_k, \ldots, p_{k+n}$

A table of $f(x)$ values will be viewed as a set of **tabulated points**

$$P_i(x_i, y_i), \qquad \text{where } y_i \doteq f(x_i) \tag{1a}$$

We shall refer to the point $P_i(x_i, y_i)$ as the ***i*th knot**, and x_i as the ***i*th node**. The nodes may or may not be equally spaced and are assigned subscripts by the table's user for his or her own convenience. However, we do assume that the x_i's are in their *natural order*, that is,

$$\cdots < x_{-2} < x_{-1} < x_0 < x_1 < x_2 < x_3 < \cdots \tag{1b}$$

Our goal is to find polynomials that **interpolate** one or more of the knots, that is, whose graph goes through one or more of the P_i's, *but not necessarily all of them*.

6.1A Notation for Interpolating Polynomials

To avoid confusion, we shall use lowercase p to denote polynomials, and uppercase P to denote the knots they interpolate. In particular, for any integer $n > 0$, and any "starting index" k,

$$p_{k,k+n}(x) = \text{a polynomial of degree} \leq n \text{ that interpolates the } n+1 \textit{ consecutive} \text{ knots } P_k, P_{k+1}, \ldots, P_{k+n} \tag{2a}$$

as shown in Figure 6.1-1†. This *geometric* definition is equivalent to the $n + 1$ *algebraic* conditions

$$p_{k,k+n}(x_i) = y_i \qquad \text{for } x_k \leq x_i \leq x_{k+n} \tag{2b}$$

†It is intuitively plausible that such a polynomial can always be found. This will be confirmed in Section 6.1B.

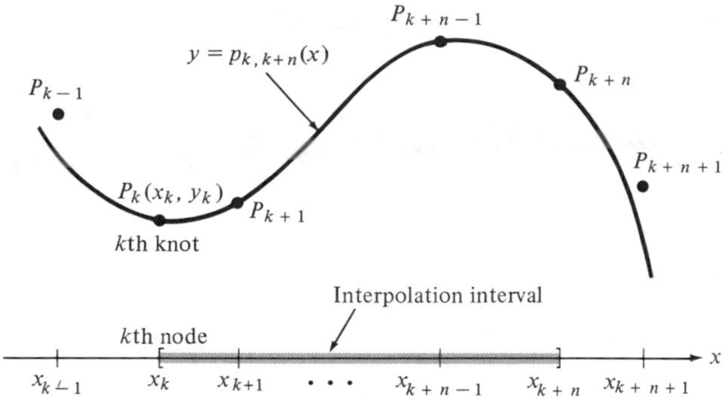

Figure 6.1-1 *Graph and interpolation interval of $p_{k,k+n}(x)$.*

However, $p_{k,k+n}(x_i)$ *need not equal y_i for any* ***other*** *tabulated x_i*, as indicated in Figure 6.1-1. The x_i's in (2b) will be called the **interpolated nodes**, and the interval $[x_k, x_{k+n}]$ the **interpolation interval**, for $p_{k,k+n}(x)$.

When $n = 0$, $p_{k,k}(x)$ is the polynomial of degree zero whose graph (a horizontal line) passes through the single knot $P_k(x_k, y_k)$; that is,

$$p_{k,k}(x) = y_k \qquad \text{(constant interpolation)} \qquad \textbf{(3a)}$$

And when $n = 1$, $p_{k,k+1}(x)$ is a polynomial of degree < 1 whose graph (a straight line) passes through $P_k(x_k, y_k)$ and $P_{k+1}(x_{k+1}, y_{k+1})$. See Figure 6.1-2. From the **point-slope form** of this unique line,

$$p_{k,k+1}(x) = y_k + \frac{y_{k+1} - y_k}{x_{k+1} - x_k}(x - x_k) \qquad \text{(linear interpolation)} \qquad \textbf{(3b)}$$

But how can we find a *quadratic* $p_{k,k+2}(x)$ whose graph passes through three consecutive knots, or a *cubic* $p_{k,k+3}(x)$ whose graph passes through four?

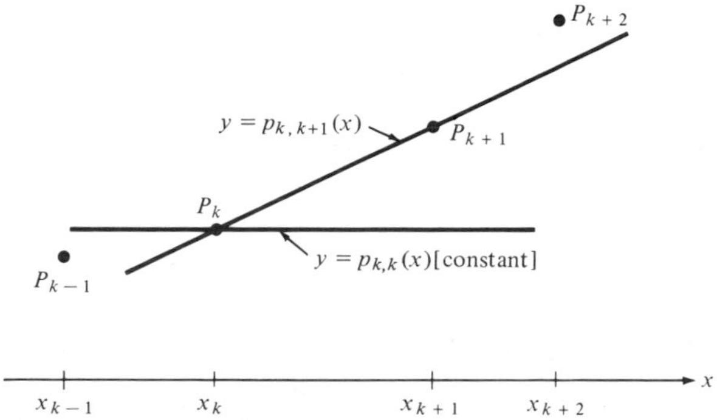

Figure 6.1-2 *Graphs of $p_{k,k}(x)$ and $p_{k,k+1}(x)$.*

EXAMPLE 6.1A Find **(a)** $p_{3,4}(x)$ and **(b)** $p_{3,5}(x)$ for the four knots

$$P_2(0, 5), \quad P_3(1, 8), \quad P_4(2, 1), \quad \text{and} \quad P_5(4, 5) \tag{4}$$

SOLUTION (a): To interpolate P_3 and P_4, we can use (3b) with $k = 3$:

$$p_{3,4}(x) = y_3 + \left(\frac{y_4 - y_3}{x_4 - x_3}\right)(x - x_3) = 8 + \frac{1 - 8}{2 - 1}(x - 1) = 15 - 7x$$

Clearly, the first degree curve $y = 15 - 7x$ passes through P_3 and P_4.

SOLUTION (b): Since the three knots P_3, P_4 and P_5 are not collinear (see Figure 6.1-3), they cannot be interpolated by a first-degree polynomial; so we try a quadratic, say

$$p_{3,5}(x) = Ax_2 + Bx + C, \qquad \text{where } A, B, C \text{ are to be determined} \tag{5a}$$

Interpolating P_3, P_4, and P_5 imposes three conditions on A, B, and C:

$$\begin{aligned} p_{3,5}(x_2) &= y_3, && \text{that is,} && A \cdot 1^2 + B \cdot 1 + C = 8 \\ p_{3,5}(x_3) &= y_4, && \text{that is,} && A \cdot 2^2 + B \cdot 2 + C = 1 \\ p_{3,5}(x_4) &= y_5, && \text{that is,} && A \cdot 4^2 + B \cdot 4 + C = 5 \end{aligned} \tag{5b}$$

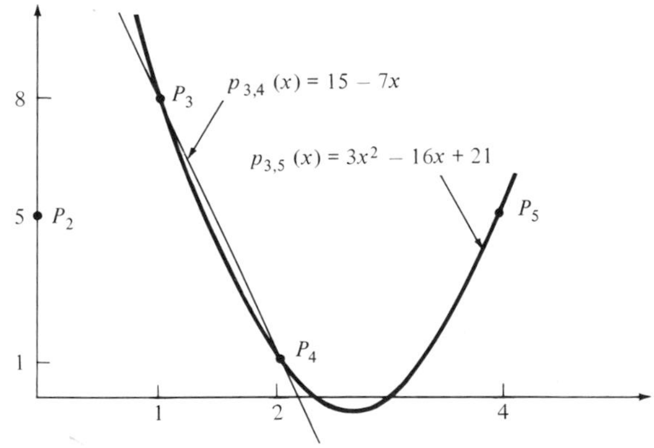

Figure 6.1-3 *Graphs of $p_{3,4}(x)$ and $p_{3,5}(x)$. [P_2 is not interpolated.]*

[see (2b)]. This *linear* system can be solved by any method of Chapter 3. Its unique exact solution, as is easily verified, is

$$A = 3, \quad B = -16, \quad \text{and} \quad C = 21, \quad \text{so that } p_{3,5}(x) = 3x^2 - 16x + 21 \tag{5c}$$

In view of (5b), this $p_{3,5}(x)$ certainly interpolates P_3, P_4, and P_5. ■

Practical Consideration 6.1A. The "brute force" method used in (5) is called the **method of undetermined coefficients**. It can always be used to convert (2b) into $n + 1$ *linear* equations in the $n + 1$ unknown coefficients of an nth degree $p_{k,k+n}(x)$ [see (5a)]. However, this method does not provide a general formula for $p_{k,k+n}(x)$ and is not convenient for hand computation for $n > 2$. Even if a computer is available to solve the linear system, the accuracy of the method is jeopardized by the fact that the system's coefficient matrix rapidly becomes ill conditioned as n increases (see Problem C6-1). Consequently, *the method of undetermined coefficients is not a computationally desirable numerical method for polynomial interpolation.* A better method is described next. ■

6.1B Existence of $p_{k,k+n}(x)$: Lagrange's Form

Suppose that we want to interpolate the four consecutive knots

$$P_2(x_2, y_2), \quad P_3(x_3, y_3), \quad P_4(x_4, y_4), \quad \text{and} \quad P_5(x_5, y_5) \tag{6}$$

It is easy to find *cubics* $L_2(x)$, $L_3(x)$, $L_4(x)$ and $L_5(x)$ that "select" x_2, x_3, x_4, and x_5, respectively, in the following sense:

$$L_2(x_2) = 1, \quad L_3(x_2) = 0, \quad L_4(x_2) = 0, \quad L_5(x_2) = 0 \tag{7a}$$
$$L_2(x_3) = 0, \quad L_3(x_3) = 1, \quad L_4(x_3) = 0, \quad L_5(x_3) = 0 \tag{7b}$$
$$L_2(x_4) = 0, \quad L_3(x_4) = 0, \quad L_4(x_4) = 1, \quad L_5(x_4) = 0 \tag{7c}$$
$$L_2(x_5) = 0, \quad L_3(x_5) = 0, \quad L_4(x_5) = 0, \quad L_5(x_5) = 1 \tag{7d}$$

For example, a cubic that "selects" x_4 from among x_2, x_3, x_4, and x_5 is

$$L_4(x) = \frac{(x - x_2)(x - x_3)(x - x_5)}{(x_4 - x_2)(x_4 - x_3)(x_4 - x_5)} \tag{8}$$

because $L_4(x_4) = 1$ (numerator equals denominator when $x = x_4$), whereas $L_4(x_2) = L_4(x_3) = L_4(x_5) = 0$ (numerator equals zero when $x = x_2$, x_3, or x_5), as shown in Figure 6.1-4. $L_2(x)$, $L_3(x)$, and $L_5(x)$ in (7) are found (by inspection!) similarly.

The y_i-weighted sum of these four cubic $L_i(x)$'s is the polynomial

$$p(x) = y_2L_2(x) + y_3L_3(x) + y_4L_4(x) + y_5L_5(x) \tag{9a}$$

of degree *at most* 3; it interpolates $P_3(x_3, y_3)$ because, by (7b),

$$p(x_3) = y_2 \cdot 0 + y_3 \cdot 1 + y_4 \cdot 0 + y_5 \cdot 0 = y_3 \tag{9b}$$

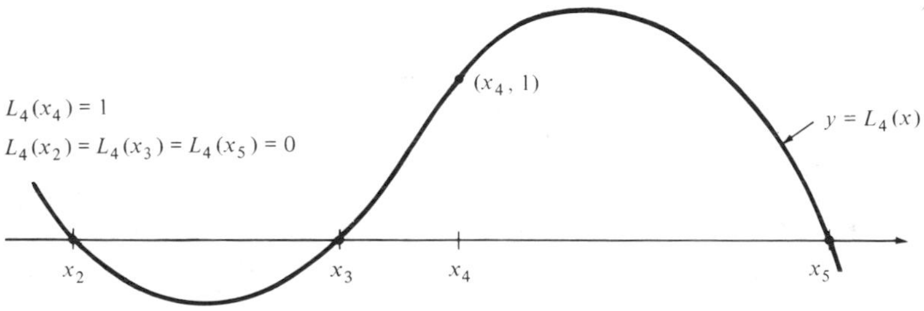

Figure 6.1-4 *$L_4(x)$ selects x_4 from among x_2, x_3, x_4, and x_5.*

Similarly, (7a), 7(c), and (7d) respectively, ensure that $p(x)$ also interpolates P_2, P_4, and P_5. So *$p(x)$ can be used as $p_{2,5}(x)$!*

More generally, a polynomial that interpolates *any* $n + 1$ consecutive knots

$$P_k(x_k, y_k), \quad P_{k+1}(x_{k+1}, y_{k+1}), \quad \ldots, \quad P_{k+n}(x_{k+n}, y_{k+n}) \tag{10}$$

can be written *by inspection* as the y_i-weighted sum

$$p_{k,k+n}(x) = y_k L_k(x) + y_{k+1} L_{k+1}(x) + \cdots + y_{k+n} L_{k+n} \tag{11a}$$

where $L_{k+i}(x)$ is the nth-degree polynomial formed as the product

$$L_{k+i}(x) = \prod_{\substack{j=0 \\ j\neq i}}^{n} \frac{(x - x_{k+j})}{(x_{k+i} - x_{k+j})} \qquad \text{for } i = 0, 1, \ldots, n \tag{11b}$$

$L_{k+i}(x)$ "selects" x_{k+i} from among $x_k, x_{k+1}, \ldots, x_{k+n}$ in the following sense:

SELECTING PROPERTY: $\begin{cases} L_{k+i}(x_{k+i}) = 1 \ [L_{k+i}(x) \text{ "selects" } x_{k+i}], \text{ but} \\ L_{k+i}(x_j) = 0 \text{ for } \textit{all other } x_j\text{'s in } [x_k, x_{k+n}] \end{cases}$ **(12)**

So *there is at least one polynomial of degree $\leq n$ that interpolates $n + 1$ given consecutive knots*. We shall show in Section 6.1C that there is *only* one. Anticipating this, we shall refer to $p_{k,k+n}(x)$ in (11) as **Lagrange's form** of ***the*** (unique) **interpolating polynomial** for $P_k, \ldots, P_{k+n}$, and we shall call the nth-degree polynomials $L_k(x), L_{k+1}(x), \ldots, L_{k+n}(x)$ in (11) and (12) the **Lagrange polynomials** for $P_k, P_{k+1}, \ldots, P_{k+n}$, in honor of the eighteenth-century French mathematician Joseph-Louis Lagrange, who was clever enough to think of them. Notice from (11b) that $L_k(x), \ldots, L_{k+n}(x)$ are formed using only the *interpolated nodes* $x_k, \ldots, x_{k+n}$. The tabulated y_{k+i} values only enter into (11a).

Taking $n = 1$ in (11) gives Lagrange's form of $p_{k,k+1}(x)$, namely

$$p_{k,k+1}(x) = y_k \frac{(x - x_{k+1})}{(x_k - x_{k+1})} + y_{k+1} \frac{(x - x_k)}{(x_{k+1} - x_k)} \tag{13}$$

whose straight-line graph interpolates $P_k(x_k, y_k)$ and $P_{k+1}(x_{k+1}, y_{k+1})$.

EXAMPLE 6.1B Find Lagrange's form of **(a)** $p_{3,4}(x)$, **(b)** $p_{3,5}(x)$, and **(c)** $p_{2,5}(x)$ for the four knots

$$P_2(0, 5), \quad P_3(1, 8), \quad P_4(2, 1), \quad \text{and} \quad P_5(4,5) \tag{14}$$

Solution (a): To interpolate P_3 and P_4, we can use (13) with $k = 3$:

$$p_{3,4}(x) = y_3\frac{(x - x_4)}{(x_3 - x_4)} + y_4\frac{(x - x_3)}{(x_4 - x_3)} = 8\frac{(x - 2)}{(1 - 2)} + 1\frac{(x - 1)}{(2 - 1)} \tag{15}$$

This simplifies to $p_{3,4}(x) = 15 - 7x$ [see Example 6.1A(a)].

Solution (b): To interpolate P_3, P_4, and P_5 ($n = 5 - 3 = 2$), we need three *quadratic* Lagrange polynomials. The first, which selects $x_3 = 1$ from among x_3, x_4, and x_5, is

$$L_3(x) = \frac{(x - x_4)(x - x_5)}{(x_3 - x_4)(x_3 - x_5)} = \frac{(x - 2)(x - 4)}{(1 - 2)(1 - 4)} \tag{16a}$$

The other two, which select $x_4 = 2$ and $x_5 = 4$, respectively, are

$$L_4(x) = \frac{(x - x_3)(x - x_5)}{(x_4 - x_3)(x_4 - x_5)} \quad \text{and} \quad L_5(x) = \frac{(x - x_3)(x - x_4)}{(x_5 - x_3)(x_5 - x_4)} \tag{16b}$$

So Lagrange's form, $p_{3,5}(x) = y_3L_3(x) + y_4L_4(x) + y_5L_5(x)$, is

$$p_{3,5}(x) = 8\frac{(x - 2)(x - 4)}{(-1)(-3)} + 1\frac{(x - 1)(x - 4)}{(2 - 1)(2 - 4)} + 5\frac{(x - 1)(x - 2)}{(4 - 1)(4 - 2)} \tag{17}$$

This simplifies to $p_{3,5}(x) = 3x^2 - 16x + 21$ [see Example 6.1A(b)].

Solution (c): To interpolate P_2, P_3, P_4, and P_5 ($n = 5 - 2 = 3$), we need four *cubic* Lagrange polynomials. By inspection of (14), (11) is

$$p_{2,5}(x) = 5\,\frac{(x - 1)(x - 2)(x - 4)}{(-1)(-3)(-5)} + 8\,\frac{(x - 0)(x - 2)(x - 4)}{(1)(-1)(-3)} + 1\,\frac{(x - 0)(x - 1)(x - 4)}{(2)(1)(-2)} + 5\,\frac{(x - 0)(x - 1)(x - 2)}{(4)(3)(2)} \tag{18a}$$

$$= 2x^3 - 11x^2 + 12x + 5 \tag{18b}$$

REMINDER: *The degree of $L_{k+i}(x)$ in (11) depends on the number of knots interpolated.* Thus $L_4(x)$ (weighted by $y_4 = 1$) is *linear* in (15), *quadratic* in (17), and *cubic* in (18a). *Keep this in mind*! ■

Practical Consideration 6.1B (Efficient evaluation of Lagrange's form). In Example 6.1B, like powers of x were collected to show that all forms of $p_{k,k+n}(x)$ are equal. However, it is so easy to evaluate the $L_i(x)$'s *in the factored form (11b)* at a particular z that we recommend leaving it that way! For a hand calculation, this allows you to *cancel before multiplying*. For example, putting $z = -1$ in (18a) gives

$$p_{2,5}(-1) = 5\,\frac{-2\cdot -3\cdot -5}{-1\cdot -2\cdot -4} + 8\,\frac{-1\cdot -3\cdot -5}{1\cdot -1\cdot -3} + 1\,\frac{-1\cdot -2\cdot -5}{2\cdot 1\cdot -2} + 5\,\frac{-1\cdot -2\cdot -3}{4\cdot 3\cdot 2} \tag{19a}$$

```
      FUNCTION PLAGR (Z, K, N, X, Y, M0, M)
      DIMENSION X(M0:M), Y(M0:M)
C - - - - - - - - - - - - - - - - - - - - - - - - - - - - - - - - - - - C
C  This subprogram evaluates p[k,k+n](x) at z, where p[k,k+n](x) is  C
C  Lagrange's form, y[k]*L[k](x) +  ...  + y[k+n]*L[k+n](x), of the  C
C  unique interpolating polynomial for (xk, yk), ..., (xk+n, yk+n).  C
C  Arrays X and Y store the M-M0+1 knots (xi, yi), i = M0, ..., M.   C
C - - - - - - - - - - - - - - - - - - - - - VERSION 2: 9/9/85  - - - - C
      PLAGR = 0.0
      DO 20 J=K,K+N
         TERMJ = Y(J)   !TERMJ accumulates yj*L[j](z)
         DO 10 I=K,K+N
            IF (I.NE.J) TERMJ = TERMJ*(Z - X(I))/(X(J) - X(I))
   10    CONTINUE
         PLAGR = PLAGR + TERMJ
   20 CONTINUE
      RETURN
      END
```

Figure 6.1-5 *Evaluating Lagrange's form of $p_{k,k+n}(x)$ at $x = z$.*

$$= \frac{75}{4} - 40 + \frac{5}{2} - \frac{5}{4} = \frac{-80}{4} = -20 \qquad \textbf{(19b)}$$

with less effort than is required to *get* and *then* evaluate (18b). On a computer, the evaluation of Lagrange's form of $p_{k,k+n}(z)$ can be programmed efficiently using a *nested loop* (see Figure 6.1-5). ■

6.1C Uniqueness of $p_{k,k+n}(x)$

For any nonzero polynomial $p(x)$, we know from algebra that

> $p(x)$ has $n + 1$ *distinct* roots, say $r_0, r_1, \ldots, r_n$, if and only if $p(x)$ has the product $\Pi_{i=0}^{n} (x - r_i)$ as a factor. **(20a)**

The degree of $\Pi_{i=0}^{n} (x - r_i)$ is $n + 1$; so it follows from (20a) that

> *if a polynomial $p(x)$ of degree $\leqslant n$ is known to have $n + 1$ distinct roots, then $p(x)$ must be identically zero.* **(20b)**

With this in mind, suppose that both $p_{k,k+n}(x)$ and $\tilde{p}_{k,k+n}(x)$ are polynomials of degree $\leqslant n$ and that both interpolate $p_k, p_{k+1}, \ldots, p_{k+n}$. Then

$$p(x) = p_{k,k+n}(x) - \tilde{p}_{k,k+n}(x) \text{ is a polynomial of degree} < n$$

and $p(x) = 0$ at the $n + 1$ interpolated nodes $x_k, x_{k+1}, \ldots, x_{k+n}$. By (20b), $p(x)$ must be the zero polynomial, so $p_{k,k+n}(x)$ and $\tilde{p}_{k,k+n}(x)$ must be identical. This proves the following important result.

UNIQUENESS OF $p_{k,k+n}(x)$ Once we have a polynomial of degree $\leq n$ that interpolates $P_k, P_{k+1}, \ldots, P_{k+n}$ (no matter how obtained), we have the *only possible one*! **(21)**

EXAMPLE 6.1C Find $p_{1,4}(x)$ for the five knots

$$P_1(-1, -20), \quad P_2(0, 5), \quad P_3(1, 8), \quad P_4(2, 1), \quad P_5(4, 5) \tag{22}$$

SOLUTION: By inspection of P_1, P_2, P_3, and P_4, Lagrange's form of $p_{0,3}(x)$ is

$$\begin{aligned} p_{1,4}(x) = & -20\frac{(x-0)(x-1)(x-2)}{(-1)(-2)(-3)} + 5\frac{(x+1)(x-1)(x-2)}{(1)(-1)(-2)} \\ & + 8\frac{(x+1)(x-0)(x-2)}{(2)(1)(-1)} + 1\frac{(x+1)(x-0)(x-1)}{(3)(2)(1)} \end{aligned} \tag{23}$$

When like terms are combined, this becomes $2x^3 - 11x^2 + 12x + 5$. Had we recalled from (18) and (19) that $P_1, \ldots, P_5$ all lie on the cubic curve $y = 2x^3 - 11x^2 + 12x + 5$, we could have avoided the preceding computation by arguing as follows: Since $2x^3 - 11x^2 + 12x + 5$ itself is of degree < 3 and interpolates the four knots $P_1, \ldots, P_4$, it must be the *unique* interpolating polynomial for them, that is, $p_{1,4}(x) = 2x^3 - 11x^2 + 12x + 5$. The same argument can be used to show that $p_{1,5}(x) = 2x^3 - 11x^2 + 12x + 5$, hence is only of degree 3 even though it interpolates five knots. ■

If we did not know beforehand that all five knots in (22) lie on a cubic curve, we would not have realized it unless we first formed $p_{1,5}(x)$ and then determined that its leading coefficient was zero. The difference tables to be described in Section 6.2 will enable us to determine the degree of the curve $y = p_{1,5}(x)$ possible without actually finding $p_{1,5}(x)$ and can then be used to get $p_{1,5}(x)$, or any other $p_{k,k+n}(x)$, even more easily than using Lagrange's form.

6.1D Lagrange's Form for *h*-Spaced Points

When the tabulated nodes x_i are h-spaced, interpolation can be simplified by introducing the "normalized" variable

$$s = \frac{x - x_k}{h} = \text{the number of } h\text{-units from a particular } x_k \text{ to } x \tag{24}$$

(See Figure 6.1-6, where $3 < s < 4$.) Then, since $x_{k+j} - x_k = hj$,

$$x - x_{k+j} = (x - x_k) - (x_{k+j} - x_k) = hs - hj = h(s - j) \tag{25}$$

So $(x - x_{k+j})/(x_{k+i} - x_{k+j}) = [h(s - j)]/[h(i - j)]$, and (11b) becomes

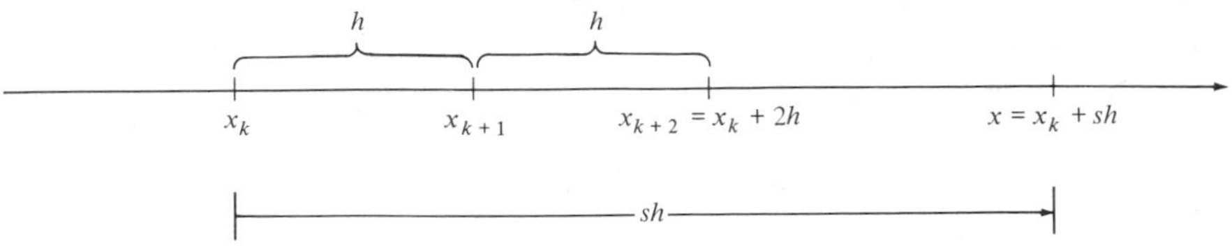

Figure 6.1-6 *Viewing x as $x_k + sh$, where h = spacing between nodes.*

$$L_{k+i}(x) = \prod_{\substack{j=0 \\ j \neq i}}^{n} \frac{(s-j)}{(i-j)}, \qquad \text{where } s = \frac{x - x_k}{h}, \quad \text{for } i = 0, 1, \ldots, n \qquad \textbf{(26)}$$

When $n = 1$, 2, and 3, Lagrange's form of $p_{k,k+n}(x)$ [see (11)] becomes

$$p_{k,k+1}(x) = \frac{(s-1)}{(0-1)} y_k + \frac{(s-0)}{(1-0)} y_{k+1} = (1-s)y_k + sy_{k+1} \qquad \textbf{(27a)}$$

$$p_{k,k+2}(x) = \frac{(s-1)(s-2)}{(-1)(-2)} y_k + \frac{(s-0)(s-2)}{(1)(-1)} y_{k+1} + \frac{(s-0)(s-1)}{(2)(1)} y_{k+2} \qquad \textbf{(27b)}$$

$$p_{k,k+3}(x) = \frac{(s-1)(s-2)(s-3)}{(-1)(-2)(-3)} y_k + \frac{s(s-2)(s-3)}{(1)(-1)(-2)} y_{k+1} + \frac{s(s-1)(s-3)}{(2)(1)(-1)} y_{k+2} + \frac{s(s-1)(s-2)}{(3)(2)(1)} y_{k+3} \qquad \textbf{(27c)}$$

where $s = (x - x_k)/h$. When these formulas are used for interpolation at $x = z$, they are best left in the factored form shown, as the following example shows.

EXAMPLE 6.1D Find **(a)** $p_{2,3}(2)$, **(b)** $p_{3,5}(0)$, and **(c)** $p_{2,5}(x)$ for the four knots

$$P_2(-1, -1), \quad P_3(1, 1), \quad P_4(3, 27), \quad \text{and} \quad P_5(5, 125) \qquad \textbf{(28)}$$

SOLUTION (a): Here $s = (2 - x_2)/2 = (2 - -1)/2 = \frac{3}{2}$. So by (27a)

$$p_{2,3}(2) = (1-s)y_2 + sy_3 = (-\tfrac{1}{2})(-1) + (\tfrac{3}{2})(1) = 2 \qquad \textbf{(29)}$$

SOLUTION (b): Here $s = (0 - x_3)/2 = -\frac{1}{2}$. So by (27b),

$$p_{3,5}(0) = \frac{(-3/2)(-5/2)}{(-1)(-2)} 1 + \frac{(-1/2)(-5/2)}{(1)(-1)} 27 + \frac{(-1/2)(-3/2)}{(2)(1)} 125 = 15 \qquad \textbf{(30)}$$

Solution (c): $p_{k,k+3}(z)$ will be z^3 for any z, by (27c) or any other formula. Why? See Problem 6-1. ■

6.1E Finding Best nth Interpolants $\hat{p}_n(z)$ for $f(z)$

Suppose that we wish to know $f(z)$, but z is not a tabulated x_i. A natural strategy in this situation is to use $p_{k,k+n}(z)$ to approximate $f(z)$. The question is

How do we choose n and k to make $p_{k,k+n}(z)$ best approximate $f(z)$?

In view of the Polynomial Wiggle Problem (Section 5.3B) the "obvious" answer, "Use all the knots," will generally *not* give the best accuracy!† So the question is not as simple as it may seem.

Once a degree n has been selected, there are many different $n + 1$ knot estimates $p_{k,k+n}(z)$, depending on k; of these, the one that is most likely to approximate $f(z)$ accurately will be denoted by $\hat{p}_n(z)$ and called the **best nth degree interpolant** for $f(z)$. For $n = 0$, the best 0th degree (constant) interpolant of $f(z)$ is‡

$$\hat{p}_0(z) = p_{k,k}(z) = y_k, \text{ where } x_k \text{ is the node } \textit{closest} \text{ to } z \tag{31a}$$

For $n = 1$, the best first-degree (straight-line) interpolant of $f(z)$ is

$$\hat{p}_1(z) = p_{k,k+1}(z), \qquad \text{where } z \text{ lies between } x_k \text{ and } x_{k+1} \tag{31b}$$

The use of $\hat{p}_1(z)$ to approximate $f(z)$ is called **linear interpolation**. This is perhaps your only previous experience with interpolation (except to play "follow the dots" as a child!) For $n > 1$, we use the following intuitively plausible criterion:

$$\hat{p}_n(z) \text{ is the } p_{k,k+n}(z) \text{ that uses the } n+1 \text{ nodes which put } z \text{ closest to the center of the interval } [x_k, x_{k+n}]. \tag{31c}$$

Thus, for a given z, we can get the successive best interpolants

$$\hat{p}_0(z), \quad \hat{p}_1(z), \quad \hat{p}_2(z), \quad \hat{p}_3(z), \quad \ldots \tag{32}$$

by adding one interpolated node at a time in such a way that for any n, $[x_k, x_{k+n}]$ *is the* $(n + 1)$*-node interval that "best centers"* z.

†This is an important difference between polynomial interpolation and curve fitting. When fitting a polynomial, increasing the number of P_i's increases the number of "excess" data points and so improves the statistical reliability of the result; when interpolating, it increases the degree and so may make the result worse.

‡If z is the midpoint of $[x_k, x_{k+1}]$, either y_k or y_{k+1} can be used as $\hat{p}_0(z)$. This is discussed further in Section 6.2G.

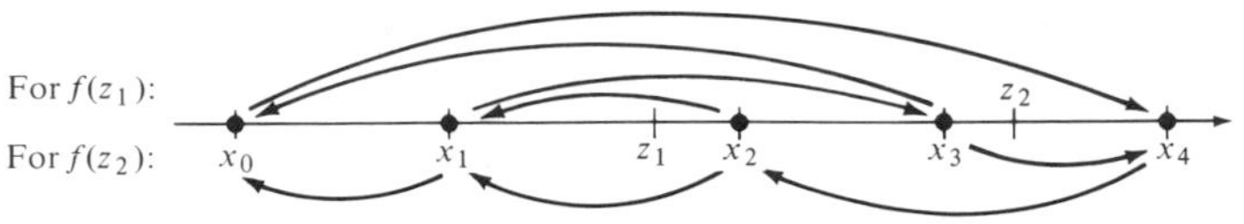

***Figure 6.1-7** Adding nodes to get best interpolants for z_1 and z_2.*

The arrows in Figure 6.1-7 indicate the order in which nodes x_i should be added to "best center" $z = z_1$ between x_1 and x_2 and $z = z_2$ between x_3 and x_4. We can see from Figure 6.1-7 that *when the spacing between nodes is reasonably uniform, the added nodes that "best center" z will come from alternate sides of z,* as long as new nodes are available to do so.

The word **extrapolation** refers to the use of $p_{k,k+n}(z)$ to approximate $f(z)$ when z lies *outside* its interpolation interval $[x_k, x_{k+n}]$. Extrapolation is necessary when z lies either to the left or right of all tabulated x_i's. In this case, $\hat{p}_n(z)$ uses the $n + 1$ tabulated nodes nearest z.

EXAMPLE 6.1E Using only the five integer-valued knots

$$P_0(0, 0), \quad P_1(1, 1), \quad P_2(4, 2), \quad P_3(9, 3), \quad P_4(16, 4) \qquad \textbf{(33)}$$

on the curve $y = f(x) = \sqrt{x}$, find $\hat{p}_0(z)$ (constant), $\hat{p}_1(z)$ (line), $\hat{p}_2(z)$ (quadratic), $\hat{p}_3(z)$ (cubic), and $\hat{p}_4(z)$ (quartic).

(a) $z = 8$ [this is interpolation; here $f(z) = \sqrt{8} \doteq 2.8284$]

(b) $z = 21$ [this is extrapolation; here $f(z) = \sqrt{21} \doteq 4.5826$]

Which of the best interpolants $\hat{p}_n(z)$ approximates $f(z)$ most accurately?

Solution (a): Since $x_2 = 4 < 8 < 9 = x_3$ and $z = 8$ lies closest to x_3,

$$\hat{p}_0(8) = y_3 = 3 \qquad \text{and} \qquad \hat{p}_1(8) = p_{2,3}(8) = \tfrac{14}{5} = 2.8 \qquad \textbf{(34)}$$

by (31a) and (31b). The calculation of $p_{2,3}(8)$ and the other values $p_{k,k+n}(z)$ used for $\hat{p}_n(z)$ is left as Problem 6-13 (or see Example 6.2E). When using (31c) for $n > 1$, *it is helpful to sketch the nodes x_i along with z on a number line*, as we have done in Figure 6.1-8.

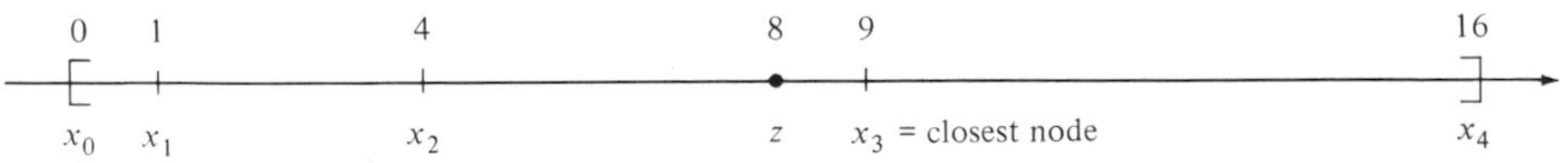

***Figure 6.1-8** Five nodes and z used in part (a) of Example 6.1E.*

There are two interpolation intervals of the form $[x_k, x_{k+2}]$ that contain $z = 8$, namely

$$[x_1, x_3] = [1, 9]\ (k = 1) \qquad \text{and} \qquad [x_2, x_4] = [4, 16]\ (k = 2)$$

Of these, [4, 16] best centers $z = 8$; however, x_4 is *farther* from z than x_1. To verify the "best centering" strategy (31c), we can find $p_{1,3}(x)$ and $p_{2,4}(x)$ and then evaluate them at $z = 8$ to get

$$p_{1,3}(8) = \frac{43}{15} \doteq 2.867 \qquad \text{and} \qquad p_{2,4}(8) = \frac{296}{105} \doteq 2.819 \tag{35}$$

Since $\sqrt{8} \doteq 2.8284$, the quadratic estimate $p_{1,3}(8)$ is less accurate than the best *linear* interpolant $\hat{p}_1(8) = 2.8$! However, $p_{2,4}(8)$, which is $\hat{p}_2(8)$ as prescribed by (31c), is more accurate than $\hat{p}_1(8)$.

Similarly, since $[x_1, x_4] = [1, 16]$ centers $z = 8$ better than $[x_0, x_3] = [0, 9]$, the best cubic interpolant of $\sqrt{8} \doteq 2.8284$ is

$$\hat{p}_3(8) = p_{1,4}(8) = \frac{128}{45} \doteq 2.844 \tag{36}$$

Although it may come as a surprise, this best cubic interpolant is *less* accurate than the best quadratic interpolant $p_{2,4}(8)$ in (35)!

The only possible five-knot estimate of $\sqrt{8} \doteq 2.8284$ is

$$\hat{p}_4(8) = p_{0,4}(8) = \frac{118}{45} = 2.622 \tag{37}$$

This is the *worst* of the five $\hat{p}_n(8)$ estimates, even though the interval [0, 16] "centers" $z = 8$ as well as an interval can!

The five values of $\hat{p}_n(8)$ are recorded for reference in Table 6.1-1, along with their errors $\hat{\epsilon}_n(8)$, where

$$\hat{\epsilon}_n(z) = f(z) - \hat{p}_n(z), \quad \text{the error of } \hat{p}_n(z) \tag{38}$$

TABLE 6.1-1 Interpolating $\sqrt{8} \doteq 2.8284$ and Extrapolating $\sqrt{21} \doteq 4.5826$

n	$\hat{p}_n(8)$	$\hat{\epsilon}_n(8)$	$\hat{p}_n(21)$	$\hat{\epsilon}_n(21)$
0	$\hat{p}_0(8) = p_{3,3}(8) = 3$	-0.172	$\hat{p}_0(21) = p_{4,4}(21) = 4$	0.583
1	$\hat{p}_1(8) = p_{2,3}(8) = 2.8$	0.028	$\hat{p}_1(21) = p_{3,4}(21) \doteq 4.714$	-0.132
2	$\hat{p}_2(8) = p_{2,4}(8) \doteq 2.819$	0.009	$\hat{p}_2(21) = p_{2,4}(21) \doteq 4.429$	0.154
3	$\hat{p}_3(8) = p_{1,4}(8) \doteq 2.844$	-0.016	$\hat{p}_3(21) = p_{1,4}(21) \doteq 5.238$	-0.656
4	$\hat{p}_4(8) = p_{1,4}(8) \doteq 2.622$	0.206	$\hat{p}_4(21) = p_{0,4}(21) \doteq -15.0$	19.583

Solution (b): Since $z = 21$ lies to the right of $x_4 = 16$, $\hat{p}_n(21)$ uses the rightmost $n + 1$ knots as shown in Table 6.1-1. Since $z = 21$ is much farther from the nearest interpolated node than $z = 8$ is, it is no surprise that the $\hat{\epsilon}_n(21)$ values are larger than those of $\hat{\epsilon}_n(8)$. What is surprising is that the best *linear* interpolant $\hat{p}_1(21) \doteq 4.714$ approximates $\sqrt{21}$ better than the best higher-degree interpolants, and that $|\hat{\epsilon}_n(21)|$ grows so rapidly for $n > 2$. ■

Example 6.1E illustrates two important characteristics of polynomial interpolation.

1. *Adding knots (i.e., increasing n) can actually result in less accuracy!*
2. *Extrapolation is generally less accurate and less reliable than interpolation.*

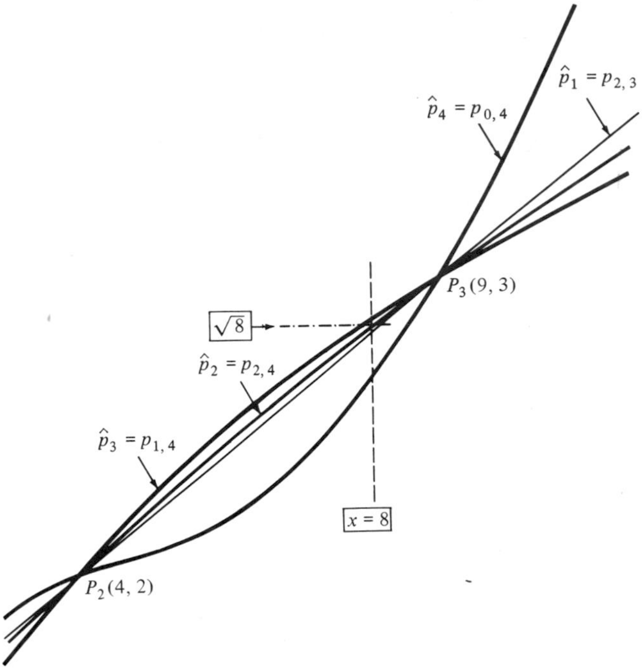

(a)

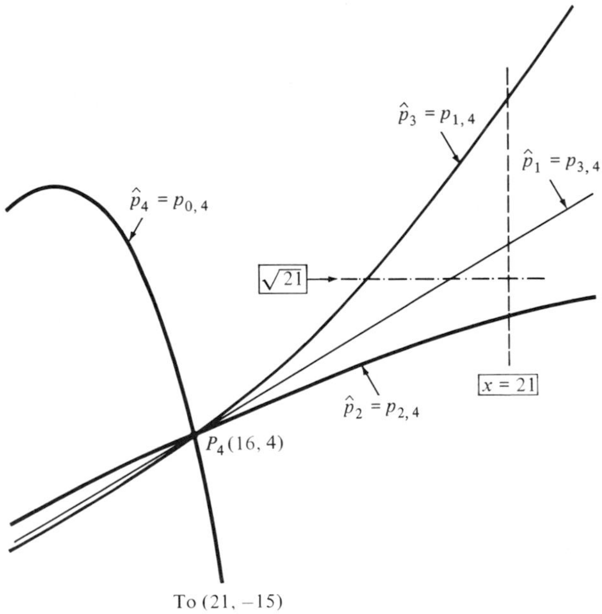

(b)

Figure 6.1-9 *Best interpolating polynomials: (a) for $\sqrt{8}$; (b) for $\sqrt{21}$.*

The villain responsible for both of these is *polynomial wiggle*. This can be seen from Figure 6.1-9, where the graphs of the $p_{k,k+n}(x)$'s used to get $\hat{p}_n(8)$ and $\hat{p}_n(21)$ are sketched for $n = 1, 2, 3,$ and 4.

To better appreciate what happened in Example 6.1E, consider the following formula (which will be derived in Section 6.1G) for the truncation error of approximating $f(z)$ by $p_{k,k+n}(z)$:

$$f(z) - p_{k,k+n}(z) = \frac{f^{(n+1)}(\xi)}{(n+1)!}(z - x_k)(z - x_{k+1}) \cdots (z - x_{k+n}) \qquad \textbf{(39)}$$

where ξ lies somewhere in the smallest closed interval containing $x_k, \ldots, x_{k+n}$ and z. It can be seen from (39) that for a given z and n, the absolute error of $p_{k,k+n}(z)$ is most likely to be *small* when the $n + 1$ distances $|z - x_k|, |z - x_{k+1}|, \ldots |z - x_{k+n}|$ are as small as possible, that is, when $x_k, \ldots, x_{k+n}$ best center z [see (31c)]. The absolute error of $p_{k,k+n}(z)$ is likely to be *large* when z is near the midpoint of two widely spaced x_i's, or even worse, when z lies to the left of x_k or to the right of x_{k+n} (extrapolation). This is why *extrapolation should be avoided if at all possible; if you must use it, expect less than the tabulated accuracy*. Note from (39) that when the x_i's are nearly equispaced, the largest absolute error of $p_{k,k+n}$ (i.e., the largest oscillation of its graph) on the interpolation interval $[x_k, x_{k+n}]$ is likely to occur near the endpoints of $[x_k, x_{k+n}]$, especially for large n.

Practical Consideration 6.1E (Achievable accuracy of polynomial interpolation). The achievable accuracy of polynomial interpolation at z depends on the location and spacing of the tabulated x_i's nearest z and the behavior of $f(x)$ in the vicinity of z. When the x_i's are sparsely spaced near z, there is generally nothing to do but expect less accuracy. This was the case in Example 6.1E; it is further illustrated in Figure 6.1-10(a). When the tabulated $f(x_i)$ values have inherent roundoff (as is usually the case), the best possible accuracy you can realistically expect is that of the tabulated $f(x_i)$ values. In any case you should *round your answer to the accuracy you think you have*. For example, if we did not know the exact values of $\sqrt{8}$ and $\sqrt{21}$, then the best we could (and should) say from Table 6.1-1 is that $\sqrt{8} \approx 2.82$ and $\sqrt{21} \approx 4.5$, with no guarantees in the last digit. If you write the answer to more places, you will only confuse yourself when you look at it at some later time! We explore this further in Section 6.2G.

In Figure 6.1-10(b), the behavior of $f(x)$ changes suddenly across x_5. Since no polynomial can behave like this, the "transition point" (x_5) should be treated as an endpoint of two separate

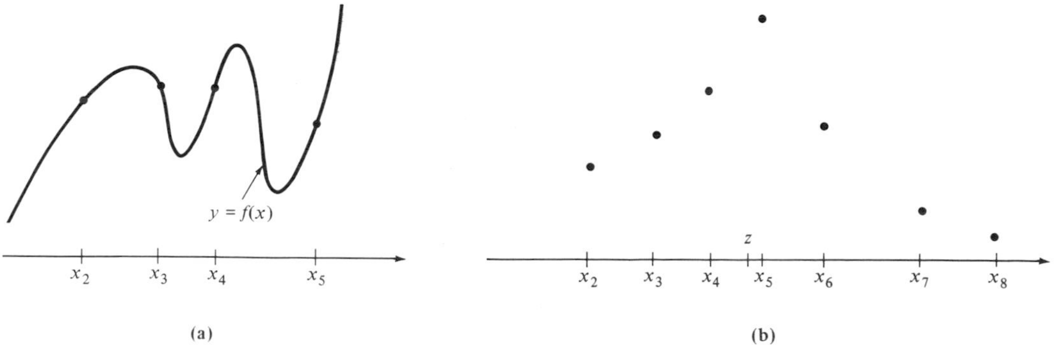

Figure 6.1-10 *(a) Sparse spacing. (b) f(x) changes across x_5.*

tables. *Never attempt to use polynomial interpolation across such a point*; doing so generally results in *less* accuracy! In Figure 6.1-10(b), for example, both $p_{4,5}(z)$ and $p_{3,5}(z)$ would be more accurate than $p_{4,6}(z)$, even though $[x_4, x_6]$ "best centers" z (see Problem M6-4). ■

6.1F Uniform Approximation Using Chebyshev Nodes

The velocity or pressure of a compressible fluid (gas or liquid) flowing through a length of pipe can be monitored by putting measuring devices (sensors) at fixed nodes along the pipe, as illustrated in Figure 6.1-11. The number of sensors used is limited by their physical size and the amount of experimental error each one introduces; this number is generally determined in advance. Once we have decided how many to use, the question is: Where should the sensors be (permanently) installed so that the readings at these nodes can be used to interpolate the values elsewhere *with uniform error* along the pipe or mathematically:

> Given an interval $[a, b]$ and a prescribed degree n, where should we locate $n + 1$ nodes $x_0, \ldots, x_n$ so as to make the *largest* interpolation error
>
> $$E_{n+1}(z) = f(z) - p_{0,n}(z), \qquad \text{for } a \leq z \leq b \tag{40a}$$
>
> *as small as possible* for any sufficiently smooth function $f(x)$?

The key to solving this problem is the error formula (39), which implies that

$$|E_{n+1}(z)| \leq \frac{\|f^{(n+1)}\|_{[a,\,b]}}{(n+1)!}\,|\Pi_{n+1}(z)| \tag{40b}$$

where $\Pi_{n+1}(x)$ denotes the $(n + 1)$st-degree polynomial

$$\Pi_{n+1}(x) = (x - x_0)(x - x_1)\cdots(x - x_n) \tag{40c}$$

and $\|f^{(n+1)}\|_{[a,\,b]}$ is largest $|f^{(n+1)}(z)|$ can be on $[a, b]$, that is,

$$\|f^{(n+1)}\|_{[a,\,b]} = \max\{|f^{(n+1)}(z)| : a \leq z \leq b\} \tag{40d}$$

Since $[a, b]$ and n are given, we have no control over either $\|f^{(n+1)}\|_{[a,\,b]}$ or $(n + 1)!$. So the desired nodes, which we shall denote by $\hat{x}_0, \ldots, \hat{x}_n$, should be chosen so as to

$$\text{minimize } \|\Pi_{n+1}\|_{[a,\,b]} = \max\{|\Pi_{n+1}(z)| : a \leq z \leq b\} \tag{41a}$$

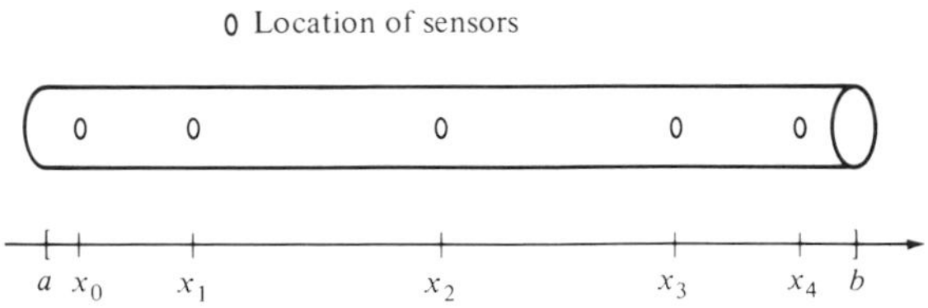

Figure 6.1-11 *Locating sensors along a length of pipe.*

The particular $\Pi_{n+1}(x)$ having these nodes as its roots will be denoted by

$$\hat{\Pi}_{n+1}(x) = (x - \hat{x}_k)(x - \hat{x}_{k+1}) \cdots (x - \hat{x}_{k+n}) \qquad \textbf{(41b)}$$

and called the **$(n + 1)$st minimax polynomial** for $[a, b]$. Thus, by definition,

$$\|\hat{\Pi}_{n+1}\|_{[a,\,b]} \le \|\Pi_{n+1}\|_{[a,\,b]} \qquad \text{for any } \Pi_{n+1}(x) = (x - x_0) \cdots (x - x_n) \qquad \textbf{(41c)}$$

The Russian mathematician P. L. Chebyshev showed that when $[a, b]$ is $[-1, 1]$, the desired minimax polynomials are related to the Chebyshev polynomials $T_n(\xi)$ described in Section 5.4E. Specifically, he proved that the $(n + 1)$st minimax polynomial for $[-1, 1]$ is

$$\hat{\Pi}_{n+1}(\xi) = (\tfrac{1}{2})^n T_{n+1}(\xi), \qquad \text{where } T_{n+1}(\xi) = \cos[(n+1)\cos^{-1}\xi] \quad \text{for } -1 \le \xi \le 1 \qquad \textbf{(41d)}$$

for $n = 0, 1, \ldots$. The $n + 1$ roots of $T_{n+1}(\xi)$ are called the **Chebyshev nodes** and denoted by $\hat{\xi}_0, \ldots, \hat{\xi}_n$. It can be seen from Figure 5.4-3 that the Cheybshev nodes are located symmetrically in $[-1, 1]$ (i.e., $\hat{\xi}_{n-i} = \hat{\xi}_i$) and are most densely spaced near the endpoints of $[-1, 1]$. It follows from (41c) that

$$T_{n+1}(\hat{\xi}_i) = 0 \quad \Leftrightarrow \quad (n+1)\cos^{-1}\hat{\xi}_i \text{ is an odd multiple of } \frac{\pi}{2}$$

So the Chebyshev nodes for a given n can be obtained using the formula

$$\hat{\xi}_i = \cos\left[\frac{2(n-i)+1}{n+1}\,\frac{\pi}{2}\right], \qquad i = 0, 1, \ldots, n \qquad \textbf{(42a)}$$

The $n + 1$ desired nodes $\hat{x}_0, \ldots, \hat{x}_n$ in $[a, b]$ can then be obtained by mapping the Chebyshev nodes *proportionally* into $[a, b]$ using the straight line through points $(-1, a)$ and $(1, b)$ (Figure 6.1-12), that is, as

$$\hat{x}_i = a + \frac{b-a}{2}(\hat{\xi}_i + 1), \qquad i = 0, 1, \ldots, n \qquad \textbf{(42b)}$$

These $\hat{x}_i$'s will be called the ***n* + 1 Chebyshev nodes on [*a*, *b*]**. Their use as interpolating nodes will be referred to as nth degree **Chebyshev interpolation** on $[a, b]$.

EXAMPLE 6.1F Suppose that the pipe shown in Figure 6.1-11 is 60 meters long. Where should the five sensors be located to make it likely that the largest error of interpolation along the pipe is as small as possible?

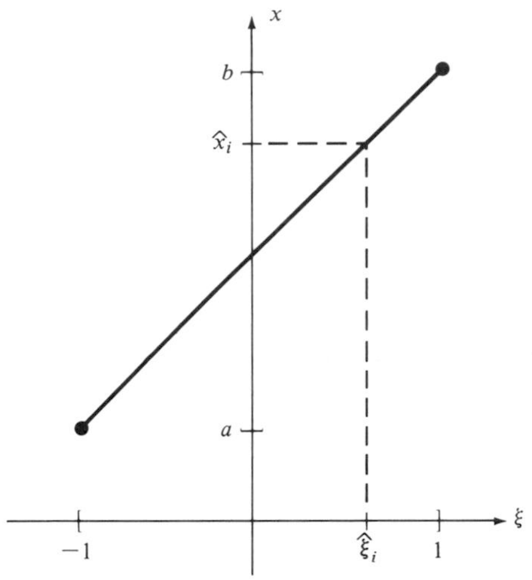

Figure 6.1-12 *Mapping $\hat{\xi}_i$ proportionally into $\hat{x}_i$.*

SOLUTION: Here $n + 1$ is 5, so $n = 4$. As a convenient scale, take $a = 0$ and $b = 60$. Then, by (42), the five sensors should be located at

$$\hat{x}_i = 30(1 + \hat{\xi}_i), \qquad \text{where } \hat{\xi}_i = \cos\left[\frac{(9 - 2i)\pi}{10}\right] \tag{43a}$$

for $i = 0, 1, \ldots, 4$. The leftmost three Chebyshev nodes on [0, 60] are thus

$$\begin{aligned} \hat{x}_0 &= 30(1 + \cos[\tfrac{9}{10}\pi]) \doteq 30(1 - 0.9511) \doteq 1.47 \text{ m} \\ \hat{x}_1 &= 30(1 + \cos[\tfrac{7}{10}\pi]) \doteq 30(1 - 0.5878) \doteq 12.37 \text{ m} \\ \hat{x}_2 &= 30(1 + \cos[\tfrac{5}{10}\pi]) = 30(1 + 0) = 30 \text{ m} \end{aligned} \tag{43b}$$

By symmetry, $\hat{x}_3 \doteq 30(1 + 0.5878) \doteq 47.63$ m and $\hat{x}_4 \doteq 30(1 + 0.9511) \doteq 58.53$ m. ■

6.1G Derivation of the Formula for $E_{k,k+n}(z)$

We wish to prove that

$$f(z) - p_{k,k+n}(z) = \frac{f^{(n+1)}(\xi)}{(n + 1)!}\Pi(z) \qquad \text{for some } \xi \text{ in } I \tag{44a}$$

where I is the smallest closed interval containing $x_k, \ldots, x_{k+n}$ and z, and

$$\Pi(x) = (x - x_k)(x - x_{k+1}) \cdots (x - x_{k+n}) \tag{44b}$$

provided that $f^{(n+1)}$ exists on the interior of I.

If z is one of the nodes $x_k, \ldots, x_{k+n}$, then $f(z) = p_{k,k+n}(z)$ and $\Pi(z) = 0$; so (44) holds trivially. Otherwise, let

$$\varphi(x) = f(x) - p_{k,k+n}(x) - \frac{\Pi(x)}{\Pi(z)}[f(z) - p_{k,k+n}(z)] \quad \textbf{(45)}$$

It is easy to see that $x_k, x_{k+1}, \ldots,, x_{k+n}$ and z are $n + 2$ distinct zeros of $\varphi(x)$, all lying in I. So, by the generalized Rolle's theorem (see Appendix II.2D),

$$\varphi^{(n+1)} \text{ has at least one zero in } I; \qquad \text{call it } \xi$$

Since degree $p_{k,k+n}(x) \leq n$ and $\Pi(x) = x^{n+1} + \cdots$,

$$\frac{d^{n+1}}{dx^{n+1}} p_{k,k+n}(x) \equiv 0 \qquad \text{and} \qquad \frac{d^{n+1}}{dx^{n+1}} \Pi(x) \equiv (n + 1)! \quad \text{(constant)}$$

But z is being held fixed, so it follows easily from (45) that

$$0 = \varphi^{(n+1)}(\xi) = f^{(n+1)}(\xi) - 0 - \frac{(n+1)!}{\Pi(z)}[f(z) - p_{k,k+n}(z)]$$

from which (44) follows.

6.2 Difference Tables of Tabulated Functions

The Polynomial Wiggle Problem [(25) of Section 5.3B] warns us that if $f(x)$ is not polynomial-like, then an attempt to pass the graph of a high-degree polynomial through several knots $P_i(x_i, f(x_i))$ can result in oscillations between successive P_i's. It is therefore necessary to find a polynomial degree that is sufficiently high to follow the shape of the graph of f, but not so high that large-amplitude oscillations occur between nodes. We now show how **difference tables** can be used to do this and also to find $p_{k,k+n}(x)$ efficiently.

6.2A Using Forward Differences $\Delta^n y_k$ When x_i's Are Equispaced

As in Section 6.1, we assume that we have a table of $f(x)$ values which we view as a set of tabulated points (i.e., knots)

$$P_i(x_i, y_i), \qquad \text{where } y_i \text{ is the tabulated value of } f(x_i) \quad \textbf{(1a)}$$

and the nodes x_i's are subscripted in their *natural order*, that is,

$$\cdots < x_{-2} < x_{-1} < x_0 < x_1 < x_2 < \cdots \quad \textbf{(1b)}$$

The **forward difference operator**, $\boldsymbol{\Delta}$, can then be defined by

$$\Delta u_k = u_{k+1} - u_k \qquad \text{for any subscripted quantity } u \quad \textbf{(2)}$$

If we apply (2) to the tabulated y-values, we get

$$\Delta y_k = y_{k+1} - y_k = \text{the } \textbf{(first) forward difference} \text{ of } y \text{ at } P_k \qquad \textbf{(3a)}$$

Applying (2) to Δy_k gives the **second forward difference** of y at P_k:

$$\Delta^2 y_k = \Delta(\Delta y_k) = \Delta y_{k+1} - \Delta y_k = y_{k+2} - 2y_{k+1} + y_k \qquad \textbf{(3b)}$$

In general, for $n = 0, 1, \ldots$ and any k, the ***n*th forward difference** of y at P_k is defined inductively by taking $\Delta^0 y_k = y_k$, then

$$\Delta^n y_k = \Delta(\Delta^{n-1} y_k) = \Delta^{n-1} y_{k+1} - \Delta^{n-1} y_k \qquad \text{for } n > 0 \qquad \textbf{(4a)}$$

Note that with this definition, $\Delta^1 y_k$ is simply Δy_k in (3a). It is easy to see that for any n and k, $\Delta^n y_k$ depends on the $n + 1$ values

$$y_k = f(x_k), \quad y_{k+1} = f(x_{k+1}), \quad \ldots, \quad y_{k+n} = f(x_{k+n}) \qquad \textbf{(4b)}$$

Needless to say, $\Delta^n y_k$ cannot be calculated unless these y_i values are all tabulated. Thus, given $P_0, \ldots, P_9$, we can calculate $\Delta^5 y_4$ but not $\Delta^5 y_5$.

If desired, $\Delta^n y_k$ can be expressed as a linear combination of $y_k, y_{k+1}, \ldots, y_{k+n}$ using the **binomial coefficients** $\binom{n}{i}$ in the nth row of **Pascal's triangle** (Figure 6.2-1). Recall that

$$\binom{n}{i} = \frac{n(n-1)(n-2)\cdots(n-i+1)}{i!} \quad \text{satisfies} \quad \binom{n}{i} = \binom{n-1}{i-1} + \binom{n-1}{i} \qquad \textbf{(5)}$$

that is, the sum of any two consecutive entries in the $(n - 1)$st row is the value of the entry below and between them in the nth row. Thus, from (4a) and the $n = 3$ and $n = 4$ rows of Figure 6.2-1(b),

$$\Delta^3 y_k = \Delta^2 y_{k+1} - \Delta^2 y_k = y_{k+3} - 3y_{k+2} + 3y_{k+1} - y_k \qquad \textbf{(6a)}$$

$$\Delta^4 y_k = \Delta^3 y_{k+1} - \Delta^3 y_k = y_{k+4} - 4y_{k+3} + 6y_{k+2} - 4y_{k+1} + y_k \qquad \textbf{(6b)}$$

Note the alternating signs. Continuing inductively yields $\Delta^n y_k = \Sigma_{i=0}^{n} \binom{n}{i} y_{k+n-i}$, that is,

$$\Delta^n y_k = y_{k+n} - \frac{n}{1!} y_{k+n-1} + \frac{n(n-1)}{2!} y_{k+n-2} - \cdots + (-1)^n y_k \qquad \textbf{(6c)}$$

	(a)	(b)
$n = 1$:	$\binom{1}{0}\ \binom{1}{1}$	1 1
$n = 2$:	$\binom{2}{0}\ \binom{2}{1}\ \binom{2}{2}$	1 2 1
$n = 3$:	$\binom{3}{0}\ \binom{3}{1}\ \binom{3}{2}\ \binom{3}{3}$	1 3 3 1
$n = 4$:	$\binom{4}{0}\ \binom{4}{1}\ \binom{4}{2}\ \binom{4}{3}\ \binom{4}{4}$	1 4 6 4 1
$n = 5$:	$\binom{5}{0}\ \binom{5}{1}\ \binom{5}{2}\ \binom{5}{3}\ \binom{5}{4}\ \binom{5}{5}$	1 5 10 10 5 1

Figure 6.2-1 *Pascal's triangle: (a) binomial coefficients; (b) their values.*

Formula (6c) is *not* the best way to find the values of the finite differences $\Delta^n y_k$. These are more easily obtained directly from definition (4a) on a **Forward Difference Table**, as shown in Figure 6.2-2(a) and illustrated for the six points

$$(x,\ 4x^3 - 6x^2 + 1), \qquad x = -1,\ -\tfrac{1}{2},\ 0,\ \tfrac{1}{2},\ 1,\ \tfrac{3}{2} \tag{7}$$

in Figure 6.2-2(b). Forward Difference Tables are filled out *one column at a time* (first y, then Δy, then $\Delta^2 y$, ...), with the subtraction in (4a) performed *visually*, using the previous column's entries as follows:

$$\begin{matrix} \Delta^{n-1}y_k & & \\ & \searrow\ominus & \\ & & \Delta^n y_k = \Delta^{n-1}y_{k+1} - \Delta^{n-1}y_k \\ & \nearrow\oplus & \\ \Delta^{n-1}y_{k+1} & & \end{matrix} \tag{8}$$

x	y	Δy	$\Delta^2 y$	$\Delta^3 y$
⋮	⋮			
x_{-2}	y_{-2}			
		Δy_{-2}		
x_{-1}	y_{-1}		$\Delta^2 y_{-2}$	
		Δy_{-1}		$\Delta^3 y_{-2}$
x_0	y_0		$\Delta^2 y_{-1}$	
		Δy_0		$\Delta^3 y_{-1}$
x_1	y_1		$\Delta^2 y_0$	
		Δy_1		$\Delta^3 y_0$
x_2	y_2		$\Delta^2 y_1$	
		Δy_2		
x_3	y_3			
⋮	⋮			

	x	y	Δy	$\Delta^2 y$	$\Delta^3 y$
x_{-3} =	-1.0	-9			
			8		
x_{-2} =	-0.5	-1		-6	
			2		3
x_{-1} =	0.0	1		-3	
			-1		3
x_0 =	0.5	0		0	
			-1		3
x_1 =	0.1	-1		3	
			2		
x_2 =	1.5	1			

Figure 6.2-2 *(a) Forward difference table; (b) values for a cubic f(x).*

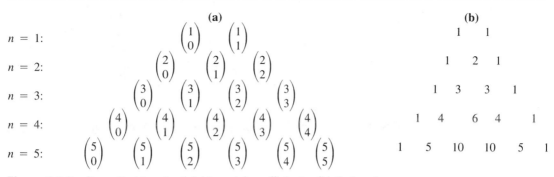

For a fixed k, all forward differences $\Delta^n y_k$ lie *diagonally down and to the right of the y_k entry of the table* [see Figure 6.2-2(a)]. In Figure 6.2-2(b), we arbitrarily took P_0 to be the point $(\frac{1}{2}, 0)$ so that $y_{-3} = -9$, $\Delta y_{-2} = \Delta y_1 = 2$, $\Delta^2 y_0 = 3$, and so on. Choosing a different tabulated point as P_0 changes the *subscripts*, k, of the difference table entries, but *not their numerical values*.

Practical Consideration 6.2A (Checking a hand calculation). If calculated correctly, difference table entries will satisfy

$$\text{sum of } \Delta^n y_i\text{'s} = (\text{bottom } \Delta^{n-1}y \text{ entry}) - (\text{top } \Delta^{n-1}y \text{ entry}) \qquad \textbf{(9)}$$

for any $n > 0$. In Figure 6.2-2(b), for example, $\Sigma\, \Delta^2 y_i = -6 - 3 + 0 + 3 = -6$, which equals $\Delta y_{-3} - \Delta y_0 = 2 - 8$. When filling out a difference table by hand, (9) can be used to check for mistakes. ■

If we did not know from (7) that $P_{-3}, \ldots, P_2$ used in Figure 6.2-2(b) all lie on a cubic curve, we could have deduced it from the constant $\Delta^3 y$ column using the following important result. In fact, (10b) could have been used with $h = \frac{1}{2}$ and $n = 3$ to deduce that the coefficient of x^3 in this cubic is $\Delta^3/[3!(\frac{1}{2})^3] = 3/\frac{6}{8} = 4$.

THEOREM 6.2A. Suppose that the x_i's are h-spaced. If certain consecutive P_i's lie on the graph of an nth degree polynomial of the form

$$p(x) = a_n x^n + (\text{polynomial of degree at most } n - 1) \qquad \textbf{(10a)}$$

then the $\Delta^n y$ column of the forward difference table for these P_i's will be constant; in fact, all $\Delta^n y_i$ entries will satisfy

$$\Delta^n y_i = a_n n! h^n = \text{constant} = \Delta^n \qquad \text{so that } a_n = \frac{\Delta^n}{n!h^n} \qquad \textbf{(10b)}$$

Conversely, if several successive entries of the $\Delta^n y$ column of the forward difference table have the same value, Δ^n, then the P_i's used to get these entries all lie on the graph of an nth degree polynomial whose leading coefficient, a_n, satisfies (10b).

Proof: That (10a) implies (10b) follows from the easily proved

$$\textbf{Linearity of the } \Delta \textbf{ Operator: } \Delta(\alpha u_k + \beta v_k) = \alpha \Delta u_k + \beta \Delta v_k \qquad \textbf{(11)}$$

(Problem M6-6) and the fact that for *any* exponent $m = 1, 2, \ldots,$

$$\Delta(x_k^m) = (x_k + h)^m - x_k^m = m x_k^{m-1} h + (\text{lower-degree terms in } x_k) \qquad \textbf{(12)}$$

[because $(x_k + h)^m = x_k^m + m x_k^{m-1} h + (\text{lower-degree terms in } x_k)$]. So if $y_k = p(x_k)$ as in (10a), then applying (11) and (12) n times gives

$$\Delta y_k = a_n n x_k^{n-1} h + (\text{lower-degree terms in } x_k) \qquad \textbf{(13a)}$$

$$\Delta^2 y_k = \Delta(\Delta y_k) = a_n n(n-1)x_k^{n-2}h^2 + (\text{lower-degree terms in } x_k) \qquad \textbf{(13b)}$$

and so on [using $\Delta(\text{constant}) = 0$], until (10b) is obtained. The converse, that the P_i's generating a constant $\Delta^n y$ column all lie on the graph of an nth degree polynomial, is proved in the footnote of Section 6.2C.

The following example shows that *difference tables do **not** give the degree of the tabulated knots when the x_i's are **not** equispaced.*

EXAMPLE 6.2A It is easy to verify that all six knots

$$P_1(-1, -20), \quad P_2(0, 5), \quad P_3(1, 8), \quad P_4(2, 1), \quad P_5(4, 5), \quad P_6(5, 40) \qquad \textbf{(14)}$$

lie on the cubic curve $y = 2x^3 - 11x^2 + 12x + 5$, and that all successive nodes *except* x_4 and x_5 are one unit apart. However, the $\Delta^3 y$ column of their difference table (Figure 6.2-3) is *not* constant.

x	y	Δy	$\Delta^2 y$	$\Delta^3 y$	$\Delta^4 y$
$x_1 = -1$	-20				
		25			
$x_2 = 0$	5		-22		
		3		12	
$x_3 = 1$	8		-10		$9 \neq 0$
		-7		21	
$x_4 = 2$	1		11		$-1 \neq 0$
		4		20	
$x_5 = 4$	5		31	↑	
		35		not constant!	
$x_6 = 5$	40				

Figure 6.2-3 *Forward difference table (x_i's are not equispaced).*

To get polynomial degree information when the x_i's are *not* equally spaced requires the *divided difference* tables. These are described next.

6.2B Using Divided Differences $\boldsymbol{\Delta}^n y_k$

Suppose that we have a table of values $y_i = f(x_i)$. The x_i's need *not* be equispaced, but we still assume that

$$\cdots < x_{-3} < x_{-2} < x_{-1} < x_0 < x_1 < x_2 < \cdots \qquad \textbf{(15)}$$

The **first divided difference** (abbreviated **1st DD**) at P_k is the number

$$\boldsymbol{\Delta} y_k = \frac{\Delta y_k}{\Delta x_k} = \frac{y_{k+1} - y_k}{x_{k+1} - x_k} \qquad \textbf{(16)}$$

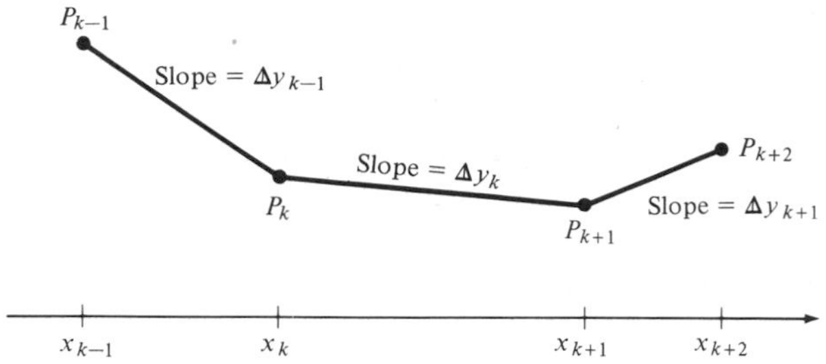

Figure 6.2-4 *Geometric interpretation of* Δy_{k-1}, Δy_k, *and* Δy_{k+1}.

Since Δy_k is the slope of the line segment $\overline{P_kP_{k+1}}$ (see Figure 6.2-4),

$$\Delta y_k > 0 \text{ suggests that } p_{k,k+1}(x) \text{ is } \textit{increasing on } [x_k, x_{k+1}] \qquad \textbf{(17a)}$$

$$\Delta y_k < 0 \text{ suggests that } p_{k,k+1}(x) \text{ is } \textit{decreasing on } [x_k, x_{k+1}] \qquad \textbf{(17b)}$$

The **second divided difference** (abbreviated **2nd DD**) at P_k is

$$\Delta^2 y_k = \frac{\Delta y_{k+1} - \Delta y_k}{x_{k+2} - x_k} \qquad \textbf{(18)}$$

Since $\Delta^2 y_k$ has units of (change in slope)/(change in x), the number $\Delta^2 y_k$ should give information about the second derivative of f over the interval $[x_k, x_{k+2}]$. In fact, by analogy with (17),

$$\Delta^2 y_k > 0 \text{ suggests that } p_{k,k+2}(x) \text{ is } \textit{concave up} \text{ on } [x_k, x_{k+1}] \qquad \textbf{(19a)}$$

$$\Delta^2 y_k < 0 \text{ suggests that } p_{k,k+2}(x) \text{ is } \textit{concave down} \text{ on } [x_k, x_{k+1}] \qquad \textbf{(19b)}$$

For example, to get $\Delta^2 y_3$ for the three knots

$$P_3(1, 8), \quad P_4(2, 1), \quad \text{and} \quad P_5(4, 5) \qquad \textbf{(20)}$$

we first get the two forward slopes

$$\Delta y_3 = \frac{y_4 - y_3}{x_4 - x_3} = \frac{1 - 8}{2 - 1} = -7 \quad \text{and} \quad \Delta y_4 = \frac{y_5 - y_4}{x_5 - x_4} = \frac{5 - 1}{4 - 2} = 2 \qquad \textbf{(21a)}$$

and we then use them in (18) to get

$$\Delta^2 y_3 = \frac{\Delta y_4 - \Delta y_3}{x_5 - x_3} = \frac{2 - (-7)}{4 - 1} = \frac{9}{3} = 3 \qquad \textbf{(21b)}$$

This positive 2nd DD suggests that the quadratic $p_{3,5}(x)$ is concave up on $[x_3, x_5] = [1, 4]$. In fact, 3 is its leading coefficient (see Figure 6.1-3).

The **nth divided difference** (abbreviated **nth DD**) at P_k is defined by taking $\Delta^0 y_k = y_k$ ($n = 0$), then recursively for $n > 0$† as

$$\Delta^n y_k = \frac{\Delta^{n-1} y_{k+1} - \Delta^{n-1} y_k}{x_{k+n} - x_k} \tag{22a}$$

With this definition, $\Delta^1 y_k$ is Δy_k of (16). Note that the number $\Delta^n y_k$ depends only on the $n + 1$ consecutive knots‡

$$P_k(x_k, y_k), \quad P_{k+1}(x_{k+1}, y_{k+1}), \quad \ldots, \quad P_{k+n}(x_{k+n}, y_{k+n}) \tag{22b}$$

Since the calculation of any nth DD requires subtracting two successive $(n - 1)$st DD's, it is convenient to form a **divided difference** (or simply **DD**) **table** as shown in the following example.

EXAMPLE 6.2B Make a DD table for the six knots of Example 6.2A, namely

$$P_1(-1, -20), \quad P_2(0, 5), \quad P_3(1, 8), \quad P_4(2, 1), \quad P_5(4, 5), \quad P_6(5, 40) \tag{23}$$

SOLUTION: The DD table, up to the $\Delta^4 y$ column, is shown in Figure 6.2-5. It was obtained *one column at a time* (first Δy, then $\Delta^2 y$, and so on). The four entries needed to calculate a desired DD $\Delta^n y_k$ can be located *by inspection* of the table by *retracing the "diagonals" terminating at the $\Delta^n y_k$ location back to the $\Delta^{n-1} y$ and x columns*, as shown by the arrows

x	$y = f(x)$	Δy = 1st DD	$\Delta^2 y$ = 2nd DD	$\Delta^3 y$ = 3rd DD	$\Delta^4 y$ = 4th DD
$x_1 = (-1)$ (circled) →	-20				
		↘ $\Delta y_1 = \frac{5 + 20}{0 - (-1)} = 25$			
$x_2 = 0$	5		↘ $\Delta^2 y_1 = \frac{3 - 25}{1 - (-1)} = -11$		
		$\Delta y_2 = \frac{8 - 5}{1 - 0} = 3$		↘ $\Delta^3 y_1 = \frac{-5 + 11}{2 - (-1)} = (2)$ (circled)	
$x_3 = 1$	8		$\Delta^2 y_2 = \frac{-7 - 3}{2 - 0} = -5$		↘ $\Delta^4 y_1 = \frac{2 - 2}{4 - (-1)} = 0$
		$\Delta y_3 = \frac{1 - 8}{2 - 1} = -7$		$\Delta^3 y_2 = \frac{3 + 5}{4 - 0} = (2)$ (circled) ↗	
$x_4 = 2$	1		↗ $\Delta^2 y_3 = \frac{2 + 7}{4 - 1} = 3$	↗	$\Delta^4 y_2 = \frac{2 - 2}{5 - 0} = 0$
		↗ $\Delta y_4 = \frac{5 - 1}{4 - 2} = 2$		$\Delta^3 y_3 = \frac{11 - 3}{5 - 1} = 2$	
$x_5 = (4)$ (circled) →	5		$\Delta^2 y_4 = \frac{35 - 2}{5 - 2} = 11$		
		$\Delta y_5 = \frac{40 - 5}{5 - 4} = 35$			
$x_6 = 5$	40				

$$\Delta^4 y_1 = \frac{\Delta^3 y_1 - \Delta^3 y_2}{x_{4+1} - x_1} \quad \left(\text{circled entries}\right)$$

Figure 6.2-5 *Divided difference table (x_i's are not equispaced).*

†The nth DD $\Delta^n y_k$ is often denoted by $f[x_k, x_{k+1}, \ldots, x_{k+n}]$. In this alternative notation, Δy_3 becomes $f[x_3, x_4]$, $\Delta^2 y_1$ becomes $f[x_1, x_2, x_3]$, and so on. We shall use the $\Delta^n y_k$ notation, because it better depicts the relation between $\Delta^n y_k$ and $\Delta^n y_k$ [see (25)], and the "|" and the "Δ" that make up the symbol Δ better represent "divided difference" [see (16) and (22)].

‡The number $\Delta^n y_k$ in (22a) will result even if the subscripts of the $n + 1$ nodes in (22b) are rearranged (see Problem M6-9). However, we shall continue to assume the natural order (15) as a convenience.

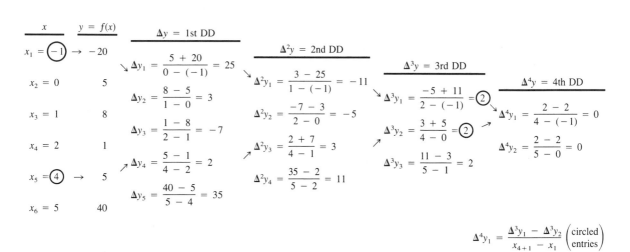

terminating at $\Delta^4 y_1$ in Figure 6.2-5. So you do not actually have to substitute in formula (22a) when filling out a DD table by hand. For illustrative purposes, intermediate calculations were shown on the DD table itself. From now on, we will do this on the side, putting only the numerical value of $\Delta^n y_k$ in the appropriate place in the table (i.e., n places diagonally down from y_k). ■

We noted in Example 6.2A that the knots in (23) lie on the curve $y = 2x^3 - 11x^2 + 12x + 5$. Even if this were not known beforehand, the following theorem would enable us to deduce from the constant $\Delta^3 y$ column that the points P_i all lie on a cubic curve with leading coefficient $a_3 = \Delta^3 = 2$. The proof of this important result is given in Section 6.2I and the footnote of Section 6.2C.

THEOREM 6.2B. If several consecutive knots P_i all lie on the graph of an nth degree polynomial, say

$$p(x) = a_n x^n + (\text{a polynomial of degree at most } n - 1) \qquad \textbf{(24a)}$$

then *any* nth divided difference obtained from them satisfies

$$\Delta^n y_k = a_n = \text{leading coefficient of the polynomial} \qquad \textbf{(24b)}$$

Conversely, if several successive entries of the $\Delta^n y$ column of a divided difference table all equal Δ^n, then the P_i's used to get these entries all lie on the graph of an nth-degree polynomial whose leading coefficient is Δ^n.

The relationship between the divided difference $\Delta^n y_k$ and the forward difference $\Delta^n y_k$ can be obtained by equating a_n in (9b) and (24b):

$$\text{If the } x_i\text{'s are } h\text{-spaced,} \qquad \text{then } \Delta^n y_k = \frac{\Delta^n y_k}{n!h^n} \qquad \textbf{(25)}$$

If we view the single entry $\Delta^n y_k$ as a "constant column," we can conclude from Theorem 6.1B that *$\Delta^n y_k$ is the leading coefficient of the interpolating polynomial for P_k, P_{k+1}, ..., P_{k+n}*, that is,

$$p_{k,k+n}(x) = \Delta^n y_k x^n + \text{terms of degree less than } n \qquad \textbf{(26)}$$

Differentiating n times gives the following illuminating result.

$$\frac{d}{dx} p_{k,k+n}(x) = \text{constant} = n!\Delta^n y_k = \frac{\Delta^n y_k}{h^n} \qquad \textbf{(27)}$$

Hence given a table of values of $f(x)$, either the number $n!\Delta^n y_k$ obtained from a divided difference table or (if the x_i's are h-spaced) the number $\Delta^n y_k/h^n$ obtained from a forward difference table can be used as a (generally crude) approximation of $f^{(n)}(x)$ over the interval $[x_k, x_{k+n}]$.

Pseudocode for forming a divided difference table in the upper-triangular part of a square matrix DD is shown in Algorithm 6.2B.

ALGORITHM 6.2B. FORMING A DIVIDED DIFFERENCE TABLE

PURPOSE: To form the divided difference table for m given knots $P_1(x_1, y_1), \ldots, P_m(x_m, y_m)$ by storing $\Delta^n y_k$ in the k,n-th entry of the matrix DD.

```
GET m, x, y   {x = [x_1  x_2  ···  x_m] and y = [y_1  y_2  ···  y_m]}
DO FOR k = 1 TO m                               {Put Δ^0 values in col_0 DD}
└  DD(k, 0) ← y_k
DO FOR n = 1 TO m − 1                           {Put Δ^n values in col_n DD}
│  DO FOR k = 1 TO m − n
└  └  DD(k,n) ← [DD(k, n) − DD(k, n − 1)]/(x_{k+n} − x_k)
OUTPUT ('The Δ^0, Δ^1, ..., Δ^{m−1} values are col_0 DD, ..., col_{m−1} DD.')
STOP
```

6.2C Formulas for Forward and Backward Interpolation

We saw in (26) that

$$p_{k,k+n}(x) = \Delta^n y_k x^n + \text{terms of degree less than } n \tag{28}$$

This observation makes it especially easy to get $p_{k,k+n}(x)$ once

$$p_{k,k+n-1}(x) = \text{interpolating polynomial for } P_k, P_{k+1}, \ldots, P_{k+n-1} \tag{29}$$

has been obtained. Indeed, since $p_{k,k+n-1}(x)$ is of degree $\leqslant n - 1$, we see from (28) that the difference, $\delta_n(x)$, between $p_{k,k+n-1}(x)$ and $p_{k,k+n}(x)$ satisfies

$$\delta_n(x) = p_{k,k+n}(x) - p_{k,k+n-1}(x) = \Delta^n y_k x^n + \text{lower-degree terms} \tag{30}$$

Moreover, for any "previously interpolated" x_i (i.e., for $k \leqslant i \leqslant k + n - 1$),

$$\delta_n(x_i) = p_{k,k+n}(x_i) - p_{k,k+n-1}(x_i) = y_i - y_i = 0 \tag{31}$$

In words, the n previously interpolated x_i's are distinct roots of $\delta_n(x)$. It therefore follows from (20) of Section 6.1C and (30) that

$$\delta_n(x) = \Delta^n y_k(x - x_k)(x - x_{k+1}) \cdots (x - x_{k+n-1}) \tag{32}$$

We can therefore rewrite (30) as the **forward interpolation formula**

$$p_{k,k+n}(x) = p_{k,k+n-1}(x) + \Delta^n y_k(x - x_k)(x - x_{k+1})\cdots(x - x_{k+n-1}) \quad \textbf{(33)}$$

This important formula enables us to "build" $p_{k,k+n}(x)$ **recursively**, that is, by adding interpolated nodes *one at a time* in the "forward order" $x_k, x_{k+1}, \ldots, x_{k+n}$, as follows:

$$p_{k,k}(x) = y_k \quad \text{(constant interpolating polynomial for } P_k) \quad \textbf{(34a)}$$

$$p_{k,k+1}(x) = y_k + \Delta y_k(x - x_k) \quad \text{(point-slope form!)} \quad \textbf{(34b)}$$

$$p_{k,k+2}(x) = y_k + \Delta y_k(x - x_k) + \Delta^2 y_k(x - x_k)(x - x_{k+1}) \quad \textbf{(34c)}$$

and so on, until $p_{k,k+n}(x)$ is obtained†. Similarly, $p_{k,k+n}(x)$ can be obtained from $p_{k+1,k+n}(x)$ using the **backward interpolation formula**

$$p_{k,k+n}(x) = p_{k+1,k+n}(x) + \Delta^n y_k(x - x_{k+n})\cdots(x - x_{k+2})(x - x_{k+1}) \quad \textbf{(35)}$$

This formula enables us to "build" $p_{k,k+n}(x)$ by adding interpolated nodes in the "backward order" $x_{k+n}, x_{k+n-1}, \ldots, x_{k+1}, x_k$ to get

$$p_{k,k+n}(x) = y_{k+n} + \Delta y_{k+n-1}(x - x_{k+n}) + \Delta^2 y_{k+n-2}(x - x_{k+n})(x - x_{k+n-1}) + \cdots \quad \textbf{(36)}$$

The recursive formulas (33) and (35) are illustrated in Figure 6.2-6. The leading coefficients used to "build" $p_{k,k+n}(x)$ by *forward interpolation* (33) lie *at and diagonally down from* y_k on divided difference table, whereas those used to "build" $p_{k,k+n}(x)$ by *backward interpolation* (35) lie *at and diagonally up from* y_{k+n}. The following example shows how easily this can be done.

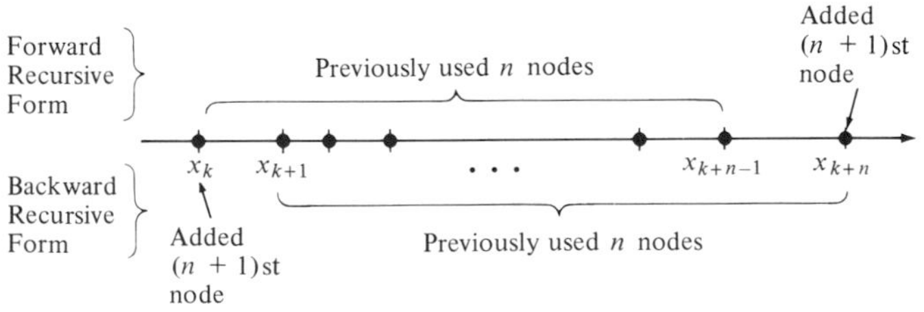

Figure 6.2-6 *Getting $p_{k,k+n}(x)$ recursively using (33) or (35).*

†It now follows immediately that any knots $P_i(x_i, y_i)$ that generate a constant, nonzero Δ^n (or Δ^n) column of a difference table must all lie on a polynomial curve of degree n.

EXAMPLE 6.2C Figure 6.2-7 shows the DD table for the six knots

$$P_1(-1, -20), \quad P_2(0, 5), \quad P_3(1, 8), \quad P_4(2, 1), \quad P_5(4, 5), \quad P_6(5, 40) \tag{37}$$

on the cubic curve $y = 2x^3 - 11x^2 + 12x + 5$. Use it to:

(a) Find $p_{1,4}(x)$ and $p_{2,5}(x)$ using forward interpolation.
(b) Find $p_{1,4}(x)$ using backward interpolation.
(c) Find $p_{1,5}(x)$ using any previously obtained $p_{k,k+3}(x)$.
(d) Find $p_{3,6}(x)$ by adding nodes in the order x_5, x_4, x_6, x_3.

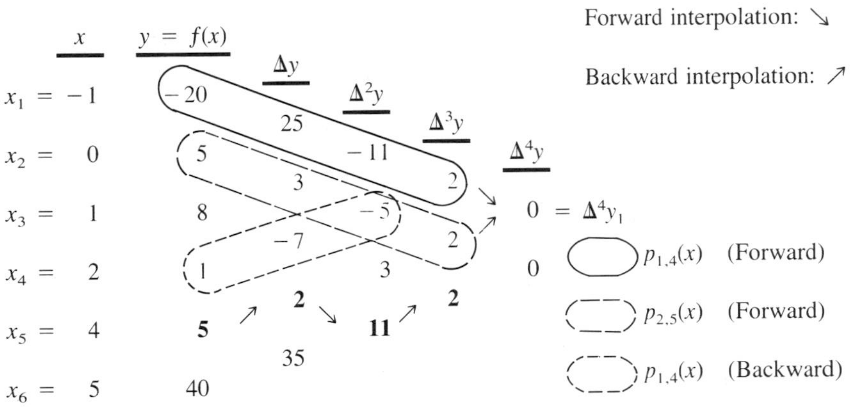

Figure 6.2-7 *Divided difference table for Example 6.2C.*

SOLUTION (a): Adding x_1, then x_2, then x_3, then x_4 [see (34)] gives

$$p_{1,1}(x) = y_1 = -20 \qquad \text{(constant interpolating polynomial for } P_1\text{)} \tag{38a}$$
$$p_{1,2}(x) = y_1 + \Delta y_1(x - x_1) = -20 + 25(x + 1) = 25x + 5 \tag{38b}$$
$$p_{1,3}(x) = (25x + 5) + (-11)(x + 1)(x - 0) = -11x^2 - 14x + 5 \tag{38c}$$
$$p_{1,4}(x) = (-11x^2 - 14x + 5) + 2(x + 1)(x - 0)(x - 1) \tag{38d}$$

The coefficients used in (38) lie at and diagonally down from $y_1 = -20$ in Figure 6.2-7. Similarly, adding x_2, then x_3, then x_4, then x_5 gives

$$p_{2,5}(x) = 5 + 3(x - 0) + (-5)x(x - 1) + 2x(x - 1)(x - 2) \tag{39}$$

These coefficients lie at and diagonally down from $y_2 = 5$ in Figure 6.2-7.

SOLUTION (b): To get $p_{1,4}(x)$ using backward interpolation, we use (35) to add nodes in the "backward order" x_4, x_3, x_2, x_1 as follows:

$$\begin{aligned} p_{1,4}(x) &= y_4 + \Delta y_3(x - x_4) + \Delta^2 y_2(x - x_4)(x - x_3) \\ &\quad + \Delta^3 y_1(x - x_4)(x - x_3)(x - x_2) \end{aligned} \tag{40a}$$
$$\begin{aligned} &= 1 + (-7)(x - 2) + (-5)(x - 2)(x - 1) \\ &\quad + 2(x - 2)(x - 1)x \end{aligned} \tag{40b}$$

These coefficients lie at and diagonally up from $y_4 = 1$ in Figure 6.2-7.

Solution (c): Using $p_{1,4}(x)$ and *forward interpolation* (33) gives

$$p_{1,5}(x) = p_{1,4}(x) + \Delta^4 y_1 (x + 1)x(x - 1)(x - 2) \quad \textbf{(41a)}$$

Alternatively, using $p_{2,5}(x)$ and *backward interpolation* (35) gives

$$p_{1,5}(x) = p_{2,5}(x) + \Delta^4 y_1 (x - 4)(x - 2)(x - 1)x \quad \textbf{(41b)}$$

The "↘" and "↗" arrows that terminate at $\Delta^4 y_1$ in Figure 6.2-7 represent (41a) and (41b), respectively.

Solution (d): Starting with $x_5 = 4$, then using (35), then (33), then (35) to add $x_4 = 2$, $x_6 = 5$, and $x_3 = 1$, respectively, gives

$$p_{3,6}(x) = 5 + 2(x - 4) + 11(x - 4)(x - 2) + 2(x - 4)(x - 2)(x - 5) \quad \textbf{(42)}$$

The four coefficients used are shown in **boldface** in Figure 6.2-7. Since the knots in (36) are known to lie on the cubic curve $y = 2x^3 - 11x^2 + 12x + 5$, uniqueness (see Example 6.1C) guarantees that $p_{1,4}(x)$, $p_{2,5}(x)$, $p_{3,6}(x)$, and $p_{1,5}(x)$ *all* simplify to $2x^3 - 11x^2 + 12x + 5$. You can confirm this directly from (38d)–(42) if you desire. ■

Practical Consideration 6.2C [Nested evaluation of $p_{k,k+n}(z)$]. The result of the recursive formulas (33) and (35) can be written in **nested form** as was done in Section 1.4B. For example, (42) and (39) can be rewritten, respectively, as

$$p_{3,6}(x) = 5 + (x - 4)\{2 + (x - 2)[11 + 2(x - 5)]\} \quad \textbf{(43a)}$$

$$p_{2,5}(x) = 5 + (x - 0)\{3 + (x - 1)[-5 + 2(x - 2)]\} \quad \textbf{(43b)}$$

where the most deeply nested factors, $(x - 5)$ in (43a) and $(x - 2)$ in (43b), are the last ones added in (42) and (39). At $z = -1$, (43b) gives

$$p_{2,5}(-1) = 5 + (-1)\{3 + (-2)[-5 + 2(-3)]\} = -20 \quad \textbf{(44)}$$

This calculation should be compared to that of Practical Consideration 6.1B. ■

It will be useful to combine (33) and (35) into a single **general recursive formula**

$$p_{k,k+n}(x) = p_{\text{prev}}(x) + \delta_n(x), \qquad \text{where } \delta_n(x) = \Delta^n y_k \Pi(x - x_{\text{prev}}) \quad \textbf{(45)}$$

In (45), x_{prev} in the product $\Pi(x - x_{\text{prev}})$ varies over the n interpolated nodes of $p_{\text{prev}}(x)$, which can be either $p_{k,k+n-1}(x)$ or $p_{k+1,k+n}(x)$.

6.2D Recursive Formulas for h-Spaced x_i's: Newton's Form of $p_{k,k+n}(x)$

Suppose that the nodes x_i are all h-spaced. We saw in (24) and (25) Section 6.1D that in this case the introduction of the "normalized" variable

$$s = \frac{x - x_k}{h} = \text{the number of } h\text{-units from } x_k \text{ to } x \quad \textbf{(46a)}$$

enables us to write

$$(x - x_{k+j}) = h(s - j) \quad \text{and} \quad (x_{k+i} - x_{k+j}) = h(i - j) \qquad \textbf{(46b)}$$

Consequently, the binomial coefficient notation (5) of Section 6.2A can be extended to any real number, s, to get the abbreviation

$$\frac{(x - x_k)(x - x_{k+1})\cdots(x - x_{k+n-1})}{n!h^n} = \frac{s(s-1)\cdots(s-n+1)}{n!} = \binom{s}{n} \qquad \textbf{(47)}$$

This, and the fact that $\Delta^n y_k = \Delta^n y_k/n!h^n$ [(27) of Section 6.2B], allows us to write the forward interpolation formula (33) as simply

$$p_{k,k+n}(x) = p_{k,k+n-1}(x) + \binom{s}{n}\Delta^n y_k, \qquad \text{where } s = \frac{x - x_k}{h} \qquad \textbf{(48a)}$$

Starting with $p_{k,k}(x) = y_k$, then using (48a) with $n = 1, 2, \ldots,$ gives

$$p_{k,k+n}(x) = y_k + \frac{s}{1!}\Delta y_k + \frac{s(s-1)}{2!}\Delta^2 y_k + \cdots + \binom{s}{n}\Delta^n y_k \qquad \textbf{(48b)}$$

This is the celebrated **forward Newton** (or **Newton–Gregory**) **form** of the unique interpolating polynomial for $P_k, \ldots, P_{k+n}$. A backward formula can be obtained similarly (Problem M6-10) and combined with it as in (45) to get a single **recursive formula for h-spaced nodes:**

$$p_{k,k+n}(x) = p_{\text{prev}}(x) + \delta_n(x), \qquad \text{where } \delta_n(x) = \Delta^n y_k \frac{\Pi(s - i_{\text{prev}})}{n!} \qquad \textbf{(49)}$$

In (49), $p_{\text{prev}}(x)$ is either $p_{k,k+n-1}(x)$ or $p_{k+1,k+n}(x)$, i_{prev} varies over the i of the *previously interpolated* nodes x_{k+i}, and $s = (x - x_k)/h$. This formula makes it especially easy to evaluate $p_{k,k+n}(z)$.

EXAMPLE 6.2D Figure 6.2-8 shows the forward difference table for

$$P_2(-1, -1), \quad P_3(1, 1), \quad P_4(3, 27), \quad \text{and} \quad P_5(5, 125) \qquad \textbf{(50)}$$

(a) Evaluate $p_{3,5}(0)$ using forward and backward interpolation.
(b) Evaluate $p_{2,5}(4)$ by adding nodes in the order x_3, x_4, x_2, x_5.

SOLUTION (a): Here $k = 3$ and $z = 0$, hence $s = (0 - x_3)/2 = -\frac{1}{2}$. So by (48b), which amounts to adding x_{3+0}, then x_{3+1}, then x_{3+2} (forward),

$$p_{3,5}(0) = y_3 + \frac{(-\frac{1}{2})}{1}\Delta y_3 + \frac{(-\frac{1}{2})(-\frac{3}{2})}{1\cdot 2}\Delta^2 y_3 \qquad \textbf{(51a)}$$

	x	y	Δy	$\Delta^2 y$	$\Delta^3 y$
	$x_2 = -1$	-1			
$z = 0 \longrightarrow$			2		
$(s = -\frac{1}{2})$	$x_3 = 1$	1		24	
			26		48
	$x_4 = 3$	27		72	
$z = 4 \longrightarrow$			98		
$(s = \frac{5}{2})$	$x_5 = 5$	125			

Figure 6.2-8 *Forward difference table for Example 6.2D (h = 2).*

$$= 1 - \tfrac{1}{2}\cdot 26 + \tfrac{3}{8}\cdot 72 = 15 \tag{51b}$$

Alternatively, using (49) to add x_{3+2}, then x_{3+1}, then x_{3+0} (backward) gives

$$p_{3,5}(0) = 125 + \frac{(-\frac{1}{2} - 2)}{1} 98 + \frac{(-\frac{1}{2} - 2)(-\frac{1}{2} - 1)}{1\cdot 2} 72 = 15 \tag{52}$$

SOLUTION (b): Here $k = 2$ and $z = 4$, hence $s = (4 - x_2)/2 = \frac{5}{2}$. Adding x_{2+1}, then x_{2+2}, then x_{2+0}, then x_{2+3} (so that $i_{\text{prev}} = 1, 2, 0, 3$) successively gives

$$\begin{aligned} p_{2,5}(4) = 1 &+ (\tfrac{5}{2} - 1)\cdot 26 + \tfrac{1}{2}(\tfrac{5}{2} - 1)(\tfrac{5}{2} - 2)\cdot 24 \\ &+ \tfrac{1}{6}(\tfrac{5}{2} - 1)(\tfrac{5}{2} - 2)(\tfrac{5}{2} - 0)\cdot 48 \end{aligned} \tag{53a}$$

Since the knots (50) lie on the cubic curve $y = x^3$, $p_{2,5}(4) = 4^3 = 64$. You can verify this quickly by adding the terms in (53a) to get

$$p_{2,2}(4) = 1, \quad p_{2,3}(4) = 1 + 39 = 40, \quad p_{1,3}(4) = 49, \quad p_{1,4}(4) = 64 \tag{53b}$$

with ease not possible using Lagrange's form, at least by hand. ■

6.2E Efficient Evaluation of Best Interpolants $\hat{p}_n(z)$

The general recursive interpolation formulas of the preceding sections make it easy to find the best nth-degree interpolants

$$\hat{p}_0(z), \quad \hat{p}_1(z), \quad \hat{p}_2(z), \quad \hat{p}_3(z), \quad \ldots \tag{54}$$

which approximate $f(z)$ for a particular z [see Section 6.1E]. In fact, once we have found the best $(n - 1)$st-degree interpolant of $f(z)$, we can use formula (45) to get the best nth-degree interpolant as

$$\hat{p}_n(z) = \hat{p}_{n-1}(z) + \hat{\delta}_n(z), \qquad \text{where } \hat{\delta}_n(z) = \Delta^n y_k \Pi(z - x_{\text{prev}}) \tag{55}$$

In (55), x_k is the leftmost interpolated node and x_{prev} varies over the n interpolated used to get the *previous* best interpolant, that is, $\hat{p}_{n-1}(z)$. It is assumed, of course, that each added node "best centers" z, as described in (31c) of Section 6.1E.

EXAMPLE 6.2E Consider the five integer-valued knots

$$P_0(0,\ 0),\quad P_1(1,\ 1),\quad P_2(4,\ 2),\quad P_3(9,\ 3),\quad P_4(16,\ 4) \tag{56}$$

on the curve $y = f(x) = \sqrt{x}$. Find $\hat{p}_0(z)$, $\hat{p}_1(z)$, $\hat{p}_2(z)$, $\hat{p}_3(z)$, and $\hat{p}_4(z)$ for approximating $\sqrt{z}$ **(a)** when $z = 8$ and **(b)** when $z = 21$.

DISCUSSION: The divided difference table for the knots (56) is shown in exact arithmetic (i.e., without roundoff) in Figure 6.2-9. Note that the entries down the Δ, Δ^2, and Δ^3 columns approach zero and so do *not* remain approximately constant. This (correctly) suggests that no polynomial of degree $\leqslant 3$ approximates $f(x) = \sqrt{x}$ well over the *entire* tabulated interval [0, 16]. In view of this, the "wiggle" seen in Figure 6.1-9 of Example 6.1E could have been expected.

	x	$y = \sqrt{x}$	Δy	$\Delta^2 y$	$\Delta^3 y$	$\Delta^4 y$
	$x_0 = 0$	0				
			1			
	$x_1 = 1$	1		$-\frac{1}{6}$		
			$\frac{1}{3}$		$\frac{1}{60}$	
	$x_2 = 4$	2		$-\frac{1}{60}$		$\boldsymbol{-\frac{1}{1008}}$
			$\boldsymbol{\frac{1}{5}}$		$\boldsymbol{\frac{1}{1260}}$ ↗	
$z = 8 \longrightarrow$	$x_3 = 9$	**3** ↗	↘	$\boldsymbol{-\frac{1}{210}}$ ↗		
			$\boldsymbol{\frac{1}{7}}$			
$z = 21 \longrightarrow$	$x_4 = 16$	4				

Figure 6.2-9 *Divided difference table for five knots on* $y = \sqrt{x}$.

SOLUTION **(a):** Since $x_2 = 4 < 8 < 9 = x_3$, with z being closest to x_3, the nodes needed to get $\hat{p}_0(8), \ldots, \hat{p}_4(8)$ successively are

$$x_3 = 9,\quad x_2 = 4,\quad x_4 = 16,\quad x_1 = 1,\quad x_0 = 0\quad \text{(in that order)} \tag{57a}$$

So the products needed for $\hat{\delta}_n(x)$ are the n leftmost factors of

$$(8 - 9)(8 - 4)(8 - 16)(8 - 1) = (-1)(4)(-8)(7) \tag{57b}$$

The successive leading coefficients needed to add the nodes (57a) are shown in **boldface** in Figure 6.2-9; the "↘" and "↗" arrows indicate forward and backward interpolation, respectively. Note how the leading coefficients "zigzag" on alternate sides of the level of the "$z = 8$" arrow until the edge of the table is reached. Starting with $\hat{p}_0(8) = y_3 = 3$ (because $x_3 = 9$ is closest to $z = 8$), then using (57) and the recursive formula

$$\hat{p}_n(8) = \hat{p}_{n-1}(8) + \hat{\delta}_n(8), \qquad \text{where } \hat{\delta}_n(8) = \Pi(8 - x_{\text{prev}}) \tag{58}$$

for $n = 1, 2, 3, 4$ yields

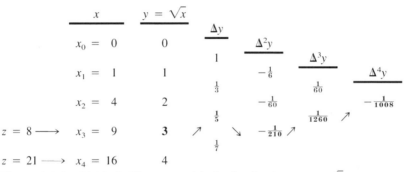

$$\overbrace{}^{\hat{\delta}_1(8) = -\frac{1}{5}} \quad \overbrace{}^{\hat{\delta}_2(8) = \frac{2}{105}} \quad \overbrace{}^{\hat{\delta}_3(8) = \frac{8}{315}} \quad \overbrace{}^{\hat{\delta}_4(8) = -\frac{2}{9}}$$

$$3 + \frac{1}{5}(-1) + \frac{-1}{210}(-1)(4) + \frac{1}{1260}(-1)(4)(-8) + \frac{-1}{1008}(-1)(4)(-8)(7) \qquad \textbf{(59)}$$

$$\hat{p}_1(8) = \tfrac{14}{5} \quad \hat{p}_2(8) = \tfrac{296}{105} \quad \hat{p}_3(8) = \tfrac{128}{45} \quad \hat{p}_4(8) = \tfrac{118}{45}$$

Solution (b): Since $z = 21$ lies to the right of the tabulated nodes, we must *extrapolate* by adding nodes in the *backward* order. So the products needed for $\hat{\delta}_n(21)$ will come from the leftmost n factors of

$$(21 - 16)(21 - 9)(21 - 4)(21 - 1) = (5)(12)(17)(20) \qquad \textbf{(60a)}$$

and the coefficients needed lie *diagonally up* from $y_4 = 4$ in Figure 6.2-9. A calculation comparable to (59), starting with $\hat{p}_0(21) = y_4 = 4$, and using (49), gives

$$\hat{\delta}_1(21) = \tfrac{5}{7} \quad \hat{\delta}_2(21) = -\tfrac{2}{7} \quad \hat{\delta}_3(21) = \tfrac{17}{21} \quad \hat{\delta}_4(21) = -\tfrac{425}{21}$$

$$4 + \frac{1}{7}(5) + \frac{-1}{210}(5)(12) + \frac{1}{1260}(60)(17) + \frac{-1}{1008}(1020)(20) \qquad \textbf{(60b)}$$

$$\hat{p}_1(21) = \tfrac{33}{7} \quad \hat{p}_2(21) = \tfrac{31}{7} \quad \hat{p}_3(21) = \tfrac{110}{21} \quad \hat{p}_4(21) = -15$$

The $\hat{p}_n(z)$ values in (59) and (60) are discussed in Example 6.1E. ■

6.2F Error Propagation on Difference Tables

Situations requiring interpolation typically involve knots of the form

$$P_i(x_i, y_i), \qquad \text{where } y_i \doteq f(x_i) \text{ correctly rounded to all places shown}$$

so that *the tabulated y_i values have inherent error*. Figure 6.2-10 shows how a small error ϵ in y_0 propagates to the (otherwise correct) $\Delta^n y_i$ entries of a forward difference table. Note that the errors in any column alternate sign, and the magnitude of these errors grows as the errors propagate to the right. When several y_i values have inherent error, the net effect on $\Delta^n y_k$ is the sum of the propagated errors of $y_k, y_{k+1}, \ldots, y_{k+n}$ used to calculate it.

$\vdots$	$\vdots$		$\vdots$		$\vdots$
x_{-2}	y_{-2}	$\vdots$	$\Delta^2 y_{-3}$	$\vdots$	$\Delta^4 y_{-4} + \epsilon$
		Δy_{-2}		$\Delta^3 y_{-3} + \epsilon$	
x_{-1}	y_{-1}		$\Delta^2 y_{-2} + \epsilon$		$\Delta^4 y_{-3} - 4\epsilon$
		$\Delta y_{-1} + \epsilon$		$\Delta^3 y_{-2} - 3\epsilon$	
x_0	$y_0 + \epsilon$		$\Delta^2 y_{-1} - 2\epsilon$		$\Delta^4 y_{-2} + 6\epsilon$
		$\Delta y_0 - \epsilon$		$\Delta^3 y_{-1} + 3\epsilon$	
x_1	y_1		$\Delta^2 y_0 + \epsilon$		$\Delta^4 y_{-1} - 4\epsilon$
		Δy_1		$\Delta^3 y_0 - \epsilon$	
x_2	y_2	$\vdots$	$\Delta^2 y_1$	$\vdots$	$\Delta^4 y_0 + \epsilon$
$\vdots$	$\vdots$		$\vdots$		$\vdots$

Figure 6.2-10 *Error propagation on a forward difference table.*

To appreciate the significance of Figure 6.2-10, suppose that the x_i's are equispaced and that $f(x)$ is actually a cubic (i.e., third-degree) polynomial. By Theorem 6.2A, the $\Delta^3 y$ column should be constant, and so the $\Delta^4 y$ column should be zero. Instead, the $\Delta^3 y$ entries with subscripts -3, -2, -1, and 0 in Figure 6.2-10 will be *nearly* constant, and the $\Delta^4 y$ entries with subscripts -4, -3, -2, -1, and 0 will be *small* and will have *alternating sign*. A similar analysis shows that an error in y_0 affects a divided difference table the same way.

Since nth differences give the leading coefficients of nth-degree interpolating polynomials, they should not be used if they are likely to be dominated by roundoff. We therefore offer the following strategy when using tables of rounded y_i values.

Roundoff Error Strategy. If the first m columns of a forward or divided difference table have entries that appear to vary smoothly, but the $(m + 1)$st column entries vary erratically, then a polynomial of degree *at most* m should be used to approximate $f(z)$. And if the mth column is approximately constant, then a polynomial of degree *exactly* m should be used.

EXAMPLE 6.2F What polynomial degree (if any) appears to be most suitable for accurately interpolating $f(z)$ using the points $P_i(x_i, y_i)$ that generated the divided difference table of Figure 6.2-11?

x	$y = f(x)$	Δ^1	Δ^2	Δ^3	Δ^4
0.0	0.5523				
		-1.8290			
0.1	0.3694		0.97167		
		-1.5375		-0.016667	
0.3	0.0619		0.96500		0.077382
		-1.2480		0.037500	
0.4	-0.0629		0.98750		-0.067858
		-0.85300		-0.010000	
0.7	-0.3188		0.98250		0.38532
		-0.46000		0.25972	
0.8	-0.3648		1.1383		0.069444
		-0.11850		0.32222	
1.0	-0.3885		1.3317		-0.037302
		0.54733		0.29238	
1.3	-0.2243		1.5363		0.041429
		1.3155		0.34210	
1.5	0.0389		1.8784		
		2.6304			
2.0	1.354				

Figure 6.2-11 *$f(x) \approx$ quadradic over [0, 0.8] and cubic over [0.7, 2].*

Solution: Since no entire column is nearly constant, there is no single polynomial degree that is best suited for interpolation over the *entire* range of x_i's. However, the top four entries of the $\Delta^2 y$ column are all nearly 0.975, and the $\Delta^3 y$ entries they generate vary erratically. This suggests that $p_{k,k+2}(z)$ (degree $<$ 2) should approximate $f(z)$ accurately *when all the interpolated x_i's lie in the interval* [0, 0.8] (the smallest interval containing the x_i's used to get the first four $\Delta^2 y$ column entries). Similarly, although the last three $\Delta^3 y$ entries are not quite as constant as the first four $\Delta^2 y$ ones, the erratic behavior at the bottom of the $\Delta^4 y$ column indicates that $\hat{p}_{k,k+n}(z)$ (degree $n \leq 3$) should approximate $f(z)$ well *when all the interpolated x_i's lie in the interval* [0.7, 2]. ■

Practical Consideration 6.2F (Recording divided difference entries). When filling out a divided difference table by hand, use *stored* (not rounded) calculator values to calculate the table's entries. The *recorded* entries may be rounded, but *to at least one "guard digit" more than the accuracy of the y_i's* (as in Figure 6.2-11) to avoid introducing harmful errors when the DD's are used as coefficients of $\hat{p}_n(z)$. ■

6.2G Interpolation in the Presence of Error

Suppose that we have rounded, slowly varying tabulated values $y_i \doteq f(x_i)$ at equally spaced nodes x_i, and we wish to get accurate approximations of $f(z)$, where z is not tabulated. Since we know that $\hat{p}_n(z)$ can become less accurate (due to "wiggle" or roundoff) when n is increased, we need a computable estimate of its error. The estimate

$$\Delta x_k = x_{k+1} - x_k \approx \epsilon_k = \text{the error of } x_k \ (\textit{not } x_{k+1}) \tag{61}$$

which was so useful in Chapter 2, suggests that we use the increment

$$\hat{\delta}_n(z) = \hat{p}_n(z) - \hat{p}_{n-1}(z) \tag{62a}$$

as an estimate the error of $\hat{p}_{n-1}(z)$ [*not* $\hat{p}_n(z)$], that is,

$$\hat{\delta}_n(z) \approx \hat{\epsilon}_{n-1}(z), \qquad \text{where } \hat{\epsilon}_{n-1}(z) = f(z) - \hat{p}_{n-1}(z) = \text{error of } \hat{p}_{n-1}(z) \tag{62b}$$

Our experience with (61) suggests that $\hat{\delta}_n(z)$ will be a reliable estimator of $\hat{\epsilon}_{n-1}(z)$ *as long as $\hat{p}_n(z)$ is more accurate than $\hat{p}_{n-1}(z)$.*

Also, if the nodes are h-spaced, we can use (49) and (55) to write the increment $\hat{\delta}_n(z)$ as

$$\hat{\delta}_n(z) = \frac{\Delta^n y_k}{n!} \Pi(s - i_{\text{prev}}), \qquad \text{where } s = \frac{z - x_k}{h} \tag{63}$$

and i_{prev} varies over the i's of the *previously used* nodes x_{k+i}.

EXAMPLE 6.2G The **normalized cumulative distribution function**

$$\Phi(x) = \frac{1}{\sqrt{2\pi}} \int_{-\infty}^{x} e^{-t^2/2}\, dt \tag{64}$$

plays an important role in statistics. Since $e^{-t^2/2}$ has no antiderivative expressible in terms of elementary functions, any needed values of $\Phi(x)$ must be obtained from tables. Two forward difference tables for $x_i = 0(0.2)1.2$†, are shown "superimposed" in Figure 6.2-12. One (in regular type) uses $\Phi(x_i)$ values rounded to $5d$; the other (in **boldface**) uses $\Phi(x_i)$ values rounded to $3d$. Asterisks (*) indicate $3d$ entries which, due to propagated roundoff, are not accurate to $3d$.

(a) What do the tables tell you about the maximum possible degree for *reliable* polynomial interpolation?

(b) Estimate $\Phi(0.32)$ using both the $3d$ and $5d$ tables. Compare the accuracy obtained.

(c) Repeat part (b) for the extrapolation of $\Phi(2)$.

index	x	$y = \Phi(x)$	Δy	$\Delta^2 y$	$\Delta^3 y$	$\Delta^4 y$	$\Delta^5 y$
0	0	0.50000					
		0.500	0.07926				
			0.079				
				−0.00310			
1	0.2	0.57926		**−0.003**			
		0.579	0.07616		−0.00273		
			0.076		**−0.002***		
$z = 0.32 \longrightarrow$							
				−0.00583		0.00062	
2	0.4	0.65542		**−0.005***		**−0.002***	
		0.655	0.07033		−0.00211		0.00024
			0.071*		**−0.004***		**0.006***
				−0.00794		0.00086	
3	0.6	0.72575		**−0.009***		**0.004***	
		0.726	0.06239		−0.00125		0.00003
			0.062		**−0.000***		**0.004***
				−0.00919		0.00083	
4	0.8	0.78814		**−0.009**		**0.000***	
		0.788	0.05320		−0.00042		
			0.053		**−0.000***		
				−0.00961			
5	1.0	0.84134		**−0.009***			
		0.841	0.04359				
			0.044				
6	1.2	0.88493					
		0.885					
$z = 2 \longrightarrow$							

Figure 6.2-12 *Forward differences of $\Phi(x)$ for 0(0.2)1.2.*

SOLUTION (a): The rather large change in the $\Delta^5 y$ column of the $5d$ table indicates the presence of significant roundoff error; so $\hat{p}_0(z), \hat{p}_1(z), \ldots, \hat{p}_4(z)$ can be found without serious roundoff error. On the other hand, the erratic behavior in the $\Delta^4 y$ column of the $3d$ table indicates the presence of some roundoff error in the $\Delta^3 y$ column. So only $\hat{p}_0(z)$, $\hat{p}_1(z)$, and $\hat{p}_2(z)$ can be found without serious roundoff error, although $\hat{p}_3(z)$ is worth looking at. We will use $|\hat{\delta}_n(z)|$ to estimate the absolute error of $\hat{p}_{n-1}(z)$ for $n \leqslant 4$ for the $5d$ table, and for $n \leqslant 3$ for the $3d$ table [see (62b)].

†The notation $a(h)b$ is often used to abbreviate $a, a + h, a + 2h, \ldots, a + nh = b$.

SOLUTION (b): The circled entries of Figure 6.2-12 give the locations of the leading coefficients $\hat{p}_0(0.32)$, $\hat{p}_1(0.32)$, ..., $\hat{p}_5(0.32)$. If we use $k = 1$ in (63), then the added nodes are x_{1+i}, where (in the order added)

$$i = 1, 0, 2, -1, 3, 4; \quad \text{and} \quad \text{for } z = 0.32, \quad s = \frac{0.32 - 0.2}{0.2} = 0.6 \quad \text{(65a)}$$

The resulting $\hat{\delta}_n(0.32)$ terms, obtained by (63), will involve the leftmost n factors of the fifth-degree product

$$\frac{(s - 1)s(s - 2)(s + 1)(s - 3)}{1 \cdot 2 \cdot 3 \cdot 4 \cdot 5} = \frac{-0.4}{1} \cdot \frac{0.6}{2} \cdot \frac{-1.4}{3} \cdot \frac{-1.6}{4} \cdot \frac{-2.4}{5} \quad \text{(65b)}$$

and the circled $3d$ entries of Figure 6.2-12. The calculation is

$$\overbrace{\hat{\delta}_1(0.32)} \qquad \overbrace{\hat{\delta}_2(0.32)} \qquad \overbrace{\hat{\delta}_3(0.32)}$$

$$\Phi(0.32) \approx \underbrace{0.655}_{\hat{p}_0(0.32)} + \underbrace{0.076\left(\frac{-0.4}{1}\right)}_{\hat{p}_1(0.32)} + \underbrace{-0.005(-0.4)\left(\frac{0.6}{2}\right)}_{\hat{p}_2(0.32)} + \underbrace{-0.002(-0.12)\left(\frac{-1.4}{3}\right)}_{\hat{p}_3(0.32)} \quad \text{(65c)}$$

We stopped at $\hat{p}_3(0.32)$ because we know from part (a) that $\hat{p}_4(0.32)$ and $\hat{p}_5(0.32)$ are not likely to be more accurate. However, for reference (and confirmation), their values are entered in Table 6.2-1(a). The shrinking $\hat{\delta}_n(0.32)$ values suggest that $\Phi(0.32) \approx \hat{p}_2(0.32)$ $\doteq 0.625$. The entries of Table 6.2-1(b) are obtained similarly using the $5d$ entries of Figure 6.2-12; in this case, the shrinking $\hat{\delta}_n(0.32)$ values suggest that $\Phi(0.32) \approx \hat{p}_4(0.32) \doteq$ 0.62552.

The exact value of $\Phi(0.32)$ is 0.625516 ($6d$); so the $5d$ estimate, 0.62552, is accurate to the tabulated $5d$ accuracy. The error estimate, $\delta_5 = -0.0000026$, correctly predicted this accuracy. However, the $3d$ estimate, 0.625, has a slight error in the last digit. Since the tabulated $\Phi(x)$ values have inherent roundoff, this is as good as we can realistically expect. Note that *in both cases, the most accurate value of $\hat{p}_n(0.32)$ was **not** the one that used the most tabulated points.*

TABLE 6.2-1   Interpolating $\Phi(0.32)$ Using (a) $3d$ Values and (b) $5d$ Values

(a)		(b)	
$\hat{\delta}_n(0.32)$	$\hat{p}_n(0.32)$	$\hat{\delta}_n(0.32)$	$\hat{p}_n(0.32)$
	$\hat{p}_0 = 0.655$		$\hat{p}_0 = 0.65542$
$\hat{\delta}_1 = -0.03040$	$\hat{p}_1 = 0.6246$	$\hat{\delta}_1 = -0.0304640$	$\hat{p}_1 = 0.624956$
$\hat{\delta}_2 = 0.00060$	$\hat{p}_2 = 0.6252$	$\hat{\delta}_2 = 0.0006996$	$\hat{p}_2 = 0.625656$
$\hat{\delta}_3 = -0.00011$	$\hat{p}_3 = 0.6251$	$\hat{\delta}_3 = -0.0001529$	$\hat{p}_3 = 0.625503$
$\hat{\delta}_4 = -0.00004$	$\hat{p}_4 = 0.6250$	$\hat{\delta}_4 = 0.0000139$	$\hat{p}_4 = 0.625517$
$\hat{\delta}_5 = -0.00006$	$\hat{p}_5 = 0.6250$	$\hat{\delta}_5 = -0.0000026$	$\hat{p}_5 = 0.625514$

SOLUTION (c): The node nearest $z = 2$ is $x_6 = 1.2$; the nodes added for the required *backward* extrapolation have subscripts $6 + i$ ($k = 6$), where

$$i = 0,\ -1,\ -2,\ -3,\ -4; \qquad \text{and} \quad \text{for } z = 2, \quad s = \frac{2 - 1.2}{0.2} = 4 \tag{66a}$$

The resulting $\hat{\delta}_n(2)$ terms will involve the leftmost n factors of

$$\frac{s(s+1)(s+2)(s+3)(s+4)}{1\cdot 2\cdot 3\cdot 4\cdot 5} = \frac{4}{1}\cdot\frac{5}{2}\cdot\frac{6}{3}\cdot\frac{7}{4}\cdot\frac{8}{5} \tag{66b}$$

Table 6.2-2(b) can now be obtained using (63) and the $5d$ entries of Figure 6.2-12:

$$\underbrace{0.88493}_{\hat{p}_0(2)} + \overbrace{0.05359\left(\frac{4}{1}\right)}^{\hat{\delta}_1(2)} \underbrace{\vphantom{)}}_{\hat{p}_1(2)} + \overbrace{-0.00961\left(4\cdot\frac{5}{2}\right)}^{\hat{\delta}_2(2)} \underbrace{\vphantom{)}}_{\hat{p}_2(2)} + \overbrace{-0.00042\left(10\cdot\frac{6}{3}\right)}^{\hat{\delta}_3(2)} \underbrace{\vphantom{)}}_{\hat{p}_3(2)} + \overbrace{0.00083\left(20\cdot\frac{7}{4}\right)}^{\hat{\delta}_4(2)} \underbrace{\vphantom{)}}_{\hat{p}_4(2)} \tag{66c}$$

and so on. Although the $\hat{\delta}_n(2)$ values do not shrink uniformly, the small value $\hat{\delta}_5(2) = 0.00168$ suggests that $\Phi(2) \approx \hat{p}_4(0.32) \doteq 0.984^-$. Since $\hat{\delta}_5(z) \approx \epsilon_4(z)$, it would be risky (and in fact wrong) to infer any more accuracy. The entries of Table 6.2-2(a) are obtained similarly using the $3d$ entries of Figure 6.2-12, and give the estimate $\hat{\Phi}(2) \approx \hat{p}_2(2) \doteq 0.971$. The exact value of $\hat{\Phi}(2)$ is 0.977250 (to $6d$); so the $5d$ estimate, 0.984^-, is no better than the $3d$ estimate, 0.971, and not nearly as good as the $5d$ estimate of $\Phi(0.32)$. This is simply the price one pays for having to extrapolate. ■

TABLE 6.2-2 Interpolating $\hat{\Phi}(2)$ Using (a) $3d$ Values and (b) $5d$ Values

(a)		*(b)*	
$\hat{\delta}_n(2)$	$\hat{p}_n(2)$	$\hat{\delta}_n(2)$	$\hat{p}_n(2)$
	$\hat{p}_0 = 0.885$		$\hat{p}_0 = 0.88493$
$\hat{\delta}_1 = 0.176$	$\hat{p}_1 = 1.061$	$\hat{\delta}_1 = 0.17436$	$\hat{p}_1 = 1.05929$
$\hat{\delta}_2 = -0.090$	$\hat{p}_2 = 0.971$	$\hat{\delta}_2 = -0.09610$	$\hat{p}_2 = 0.96319$
$\hat{\delta}_3 = 0.000$	$\hat{p}_3 = 0.971$	$\hat{\delta}_3 = -0.00840$	$\hat{p}_3 = 0.95479$
$\hat{\delta}_4 = 0.000$	$\hat{p}_4 = 0.971$	$\hat{\delta}_4 = 0.02905$	$\hat{p}_4 = 0.98384$
$\hat{\delta}_5 = -0.224$	$\hat{p}_5 = 0.747$	$\hat{\delta}_5 = -0.00168$	$\hat{p}_5 = 0.98216$

Practical Consideration 6.2G (Bessel's formula for z near a midpoint). Suppose that the x_i's are equispaced and that z lies near the midpoint of $[x_k, x_{k+1}]$ as indicated in Figure 6.2-13. If we introduce the "normalized" variable $s = (z - x_k)/h$ and form $\hat{p}_n(z)$ by adding nearest nodes starting with x_k we would get

$(s = 0)\ x_k$	y_k		$\Delta^2 y_{k-1}$		$\Delta^4 y_{k-2}$		$\Delta^6 y_{k-3}$
$z \longrightarrow$		Δy_k		$\Delta^3 y_{k-1}$		$\Delta^5 y_{k-2}$	
$(s = 1)\ x_{k+1}$	y_{k+1}		$\Delta^2 y_k$		$\Delta^4 y_{k-1}$		$\Delta^6 y_{k-2}$

Figure 6.2-13 *Forward differences used in Bessel's formula.*

$$y_k + s\Delta y_k + \frac{s(s-1)}{2!}\Delta^2 y_{k-1} + \frac{s(s-1)(s+1)}{3!}\Delta^3 y_{k-1} + \frac{s(s-1)(s+1)(s-2)}{4!}\Delta^4 y_{k-2} + \cdots \tag{67a}$$

Alternatively, we could add nearest nodes starting with x_{k+1} to get

$$y_{k+1} + (s-1)\Delta y_k + \frac{(s-1)s}{2!}\Delta^2 y_k + \frac{(s-1)s(s-2)}{3!}\Delta^3 y_{k-1} + \frac{(s-1)s(s-2)(s+1)}{4!}\Delta^4 y_{k-1} + \cdots \tag{67b}$$

When n is *odd*, say $n = 2m - 1$, $\hat{p}_n(z)$ will be $p_{k+1-m,k+m}(z)$ whether (67a) or (67b) is used. However, when n is *even*, say $n = 2m$, $\hat{p}_n(z)$ will be $p_{k-m,k+m}(z)$ if (67a) is used, and $p_{k+1-m,k+1+m}(z)$ if (67b) is used. Since their $n + 1$ point interpolating intervals $[x_{k-m}, x_{k+m}]$ and $[x_{k+1-m}, x_{k+1+m}]$ "center" z about equally well, their *average* should approximate $f(z)$ no worse than either $\hat{p}_n(z)$. Upon combining like terms, the average, $[p_{k-m,k+m}(z) + p_{k+1-m,k+1+m}(z)]/2$, can be written as

$$\frac{y_{k+1}+y_k}{2} + (s-\tfrac{1}{2})\Delta y_k + \frac{s(s-1)}{2!}\left\{\frac{\Delta^2 y_k + \Delta^2 y_{k-1}}{2} + (s-\tfrac{1}{2})\frac{\Delta^3 y_{k-1}}{3}\right\} + \frac{s(s-1)(s+1)(s-2)}{4!}\left\{\frac{\Delta^4 y_{k-1} + \Delta^4 y_{k-2}}{2} + (s-\tfrac{1}{2})\frac{\Delta^5 y_{k-2}}{5}\right\} + \cdots \tag{68}$$

This is **Bessel's formula**. The forward differences used in it are those shown in Figure 6.2-13; and the i's of the products $\Pi(s - i)$ vary over those of the *previously used* nodes x_{k+i}.

In Example 6.2G(b), roundoff error made it necessary to use $\hat{p}_2(0.32) \doteq 0.655$ to approximate $\Phi(0.32)$. Bessel's formula gives

$$\Phi(0.32) \approx \frac{0.597 + 0.655}{2} + (0.6 - 0.5)(0.076) + \frac{(0.6)(-0.4)}{2}\cdot\frac{-0.005 + -0.003}{2} \tag{69}$$

This also becomes 0.655 when rounded to $3d$. In this case, the roundoff error in $\Delta^2 y_1 = \mathbf{-0.005}$ could not be offset by the more accurate formula. In general, however, (69) is more likely to provide the tabulated accuracy than either (67a) or (67b) when n is even and z lies nearly midway between x_k and x_{k+1}. ■

Summary: Using Polynomial Interpolation to Approximate $f(z)$

1. *Make a difference table*. If the x_i's are equispaced, use *finite differences*. Otherwise, use *divided differences*, carrying one or two guard digits to avoid introducing additional roundoff error.
2. *Case 1: A nearly constant column, say the* mth, *is evident*. Use $\hat{p}_n(z)$ to approximate $f(z)$. If there is an $(m + 1)$st column, use $\hat{\delta}_{m+1}(z)$ as a rough estimate of its error.

   Case 2: No nearly constant column is evident. Accumulate the best interpolants $\hat{p}_0(z)$, $\hat{p}_1(z)$, $\hat{p}_2(z)$, ... and note their differences $\hat{\delta}_1(z)$, $\hat{\delta}_2(z)$, ... as we did in this and the preceding section; or use Bessel's formula *if appropriate*, that is, if n is even *and* the

x_i's are uniformly spaced *and* z lies near the midpoint x_k and x_{k+1}. Stop when either $|\hat{\delta}_n(z)|$ stops decreasing (wiggle) or the new leading coefficient would come from a column with erratic changes (roundoff).

3. *Round your approximation* to the accuracy you think you have, and no more than the accuracy of the tabulated $f(x)$ values used.

It should be noted that Bessel's formula, along with those of Gauss and Stirling, were developed for equispaced points to give approximations of $f(z)$ which are at least as accurate as those obtained by polynomial interpolation. The improved accuracy can be useful in actuarial work (where interpolation is referred to as **collocation**); however, it is not viewed as significant enough to warrant further consideration here. The interested reader is referred to [26].

6.2H Inverse Interpolation

Suppose that we have a table of values (x_i, y_i), where $y_i \doteq f(x_i)$, and we need the value of $x = z$ for which $f(z) = c$. Mathematically, we want to find $z = f^{-1}(c)$. Two methods of doing this are possible:

Method 1: Determine a polynomial $p(x)$ that appears to approximate $f(x)$ well near where z should be; then use a method of Chapter 2 to

$$\text{solve } p(x) - c = 0 \quad \text{for} \quad \bar{x} \approx \text{desired } z = f^{-1}(c) \tag{70}$$

Method 2: Use the knots (y_i, x_i) on the *inverse* curve $x = f^{-1}(y)$ to estimate $z = f^{-1}(c)$ using interpolation as summarized in Section 6.2G.

EXAMPLE 6.2H From the tabulated values of $f(x) = e^x$ shown in Figure 6.2-14, find an accurate estimate of the z for which $f(z) = 2$.

SOLUTION:

METHOD 1. From the tabulated values, we can see that $z \approx 0.7$. Also, since the $\Delta^5 y$ column is nearly constant, $\hat{p}_5(z)$ should approximate e^x well over [0, 1.2]. Since $n = 5$ is odd,

x	$f(x) = e^x$	Δy	$\Delta^2 y$	$\Delta^3 y$	$\Delta^4 y$	$\Delta^5 y$
$x_0 = 0$	1.					
		0.22140				
$x_1 = 0.2$	1.22140		0.04902			
		0.27042		0.01086		
$x_2 = 0.4$	1.49182		0.05988		0.00238	
		0.33030		0.01324		0.00058
$x_3 = 0.6$	1.82212		0.07312		0.00296	
$z = ?\longrightarrow$		0.40342		0.01620		0.00062
$x_4 = 0.8$	2.22554		0.08932		0.00358	
		0.49274		0.01978		
$x_5 = 1.0$	2.71828		0.10910			
		0.60184				
$x_6 = 1.2$	3.32012					

Figure 6.2-14 *5s forward difference table for e^x.*

Bessel's formula will give the same result as $\hat{p}_5(z) = p_{1,6}(z)$. We shall get $p_{1,6}(x)$ using (67a) with $k = 3$, that is, starting with x_3. In *nested form*,

$$p_{1,6}(x) = 1.82212 + \frac{s}{1}\Bigg(0.40342 + \frac{s-1}{2}\Bigg(0.07312 + \frac{s+1}{3}\Bigg(0.01620 + \frac{s-2}{4}\Bigg(0.00296 + \frac{s+2}{5}\Big(0.00062\Big)\Bigg)\Bigg)\Bigg)\Bigg) \tag{71}$$

where $s = (x - x_3)/h = (x - 0.6)/0.2$. Since this nested form is easy to evaluate but hard to differentiate, the Secant Method (*not* NR) can be used to

$$\text{solve } p_{1,6}(x) - 2 = 0 \quad \text{for} \quad \bar{s} = 0.4657321 = \frac{\bar{x} - 0.6}{0.2} \tag{72}$$

(This requires only three iterations, starting with $x_{-1} = 0.4$ and $x_0 = 0.5$.) So $z \approx \bar{x} = 0.2\bar{s} + 0.6 = 0.6931464$.

y	$f^{-1}(y)$	Δx	$\Delta^2 x$	$\Delta^3 x$	$\Delta^4 x$	$\Delta^5 x$
$y_0 = 1.$	0					
		0.90334				
$y_1 = 1.2214$	0.2		−0.33295			
		0.73959		0.13350		
$y_2 = 1.4918$	0.4		−0.22320		−0.049107	
		0.60551		0.073317		0.015702
$y_3 = 1.8221$	0.6		−0.14958		−0.022127	
$c = 2 \longrightarrow$		0.49576		0.040195		0.0058191
$y_4 = 2.2255$	0.8		−0.10028		−0.0099144	
		0.40589		0.022069		
$y_5 = 2.7183$	1.0		−0.067222			
		0.33231			$\Delta^6 y_0 =$	−0.0042595
$y_6 = 3.3201$	1.2					

Figure 6.2-15 *DD table for $f^{-1}(y)$ = ln y (entries rounded to 5s).*

METHOD 2. Since the y_i values of Figure 6.2-14 are *not* equispaced, we need a *divided* difference table for f^{-1}. This is shown in Figure 6.2-15, where no column is nearly constant and there is no evidence of serious propagated roundoff error. We must therefore add nodes as shown in Table 6.2-3 and examine the values of $|\hat{\delta}_n(2)|$ to determine the $\hat{p}_n(2)$ that appears to be most accurate. The results are tabulated on Table 6.2-3. From $\hat{\delta}_5(2)$ and $\hat{\delta}_6(2)$, it appears that $z \doteq 0.69311^+$.

TABLE 6.2-3 Interpolating $f^{-1}(2)$ from a DD Table for (y_i, x_i)

New Node	$\hat{\delta}_n(2) = \hat{p}_n(2) - \hat{p}_{n-1}(2)$	$\hat{p}_n(2)$
y_3		$\hat{p}_0(2) = x_3 = 0.6$
y_4	$\hat{\delta}_1(2) = 0.0881860$	$\hat{p}_1(2) = 0.688186$
y_2	$\hat{\delta}_2(2) = 0.0060010$	$\hat{p}_2(2) = 0.694187$
y_5	$\hat{\delta}_3(2) = -0.0008195$	$\hat{p}_3(2) = 0.693367$
y_1	$\hat{\delta}_4(2) = -0.0003240$	$\hat{p}_4(2) = 0.693043$
y_6	$\hat{\delta}_5(2) = 0.0000663$	$\hat{p}_5(2) = 0.693110$
y_0	$\hat{\delta}_6(2) = 0.0000641$	$\hat{p}_6(2) = 0.693174$

The exact z for this example is ln 2 = 0.693147 (6d), so Method 1 found z more accurately than Method 2. ■

Practical Consideration 6.2H (Selecting the best method). If a difference table for the given (x_i, y_i) knots has a nearly constant column (as in Example 6.2H), then Method 1 is recommended. If not, make a difference table for (y_i, x_i) and use Method 2, because it gives a better indication of accuracy attained than Method 1. ■

6.2I Proof That $\Delta^n y_k$ Is the Leading Coefficient of $p_{k,k+n}(x)$

Let

$$p(x) = \frac{1}{x_{k+n} - x_k}\{(x - x_k)p_{k+1,k+n}(x) - (x - x_{k+n})\,p_{k,k+n-1}(x)\} \qquad \textbf{(73)}$$

Since

$$p_{k,k+n-1}(x_i) = y_i \qquad \text{for } i = k, \ldots, k + n - 1$$

and

$$p_{k+1,k+n}(x_i) = y_i \qquad \text{for } i = k + 1, \ldots, \quad k + n$$

we see from (73) that $p(x)$ interpolates $P_k, \ldots, P_{k+n}$:

$$p(x_k) = \frac{1}{x_{k+n} - x_k}\{0 \cdot p_{k+1,k+n}(x_k) - (x_k - x_{k+n})y_k\} = y_k$$

$$p(x_i) = \frac{1}{x_{k+n} - x_k}\{(x_i - x_k)y_i - (x_i - x_{k+n})y_i\} = y_i, \qquad i = k + 1, \ldots, k + n - 1$$

$$p(x_{k+n}) = \frac{1}{x_{k+n} - x_k}\{(x_{k+n} - (x_i - x_k)y_{k+n} - 0 \cdot p_{k,k+n-1}(x_{k+n})\} = y_{k+n}$$

Moreover, since $p_{k+1,k+n}(x)$ and $p_{k,k+n-1}(x)$ both have degree $\leqslant n - 1$, it follows from (73) that degree $p(x) \leqslant n$. So, by the uniqueness of the interpolating polynomial (Section 6.1C), $p(x)$ must be $p_{k,k+n}(x)$, that is,

$$p_{k,k+n}(x) = \frac{1}{x_{k+n} - x_k}\{(x - x_k)p_{k+1,k+n}(x) - (x - x_{k+n})p_{k,k+n-1}(x)\} \qquad \textbf{(74)}$$

This result, due to Neville, allows us prove by induction on n that

$$\text{the coefficient of } x^n \text{ in } p_{k,k+n}(x) \text{ is } \Delta^n y_k \qquad \textbf{(75)}$$

We saw in (3b) of Section 6.1 that (75) is true when $n = 1$. Let us assume it to be true for interpolating polynomials of degree $\leqslant n - 1$. Then the coefficient of x^{n-1} in $p_{k+1,k+n}(x)$ and $p_{k,k+n-1}(x)$ are $\Delta^{n-1}y_{k+1}$ and $\Delta^{n-1}y_k$, respectively. In view of (74), the coefficient of x^n in $p_{k,k+n}(x)$ is $(\Delta^{n-1}y_{k+1} - \Delta^{n-1}y_k)/(x_{k+n} - x_k)$. This is the definition of $\Delta^n y_k$ and so completes the proof of (75).

6.3

Interpolation Using Piecewise Cubic Splines

As before, suppose that we are given $n + 1$ knots on the graph of f, that is,

$$P_k(x_k, y_k), \qquad \text{where } y_k \doteq f(x_k), \qquad k = 0, 1, \ldots, n \tag{1}$$

The nodes x_k need not be equispaced, but we do assume that

$$x_0 < x_1 < \cdots < x_k < x_{k+1} < \cdots < x_n \tag{2}$$

We saw in Figure 5.3-1 that unless the knots lie on the graph of a polynomial, $p_{0,n}(x)$ tends to "wiggle" between knots, especially near the endpoints of the interpolating interval $[x_0, x_n]$ if n is large.

A strategy that has been found to be effective for large n is to "piece together" the graphs of lower-degree polynomials $q_k(x)$, where $q_k(x)$ interpolates the two successive knots P_k and P_{k+1} for $k = 0, 1, \ldots, n - 1$ as shown in Figure 6.3-1. [The fact that the "pieces" $q_k(x)$ and $q_{k+1}(x)$ are "tied together" at P_{k+1} accounts for the name "knots."] The purpose of this section is to develop computational procedures for this strategy called **piecewise polynomial interpolation**.

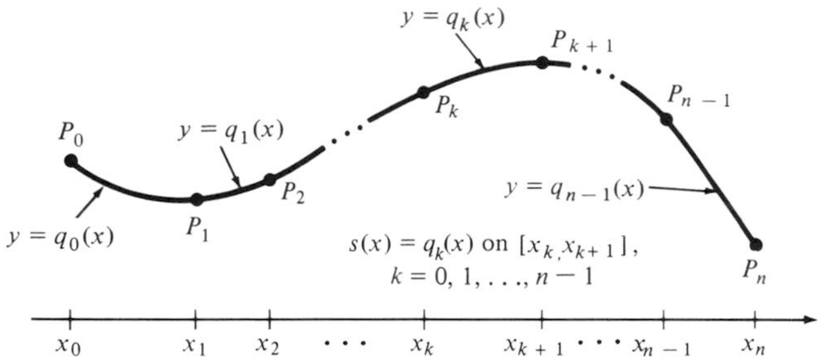

Figure 6.3-1 *Piecewise polynomial interpolation of P_0, . . ., P_n.*

6.3A Piecewise Linear Interpolation

The simplest piecewise polynomial strategy is to simply connect consecutive knots with straight lines as shown in Figure 6.3-2. This strategy, called **piecewise linear interpolation,** can be described analytically using Lagrange's form of $p_{k,k+1}(x)$ to define the interpolating function $s(x)$ as follows:

$$s(x) = q_k(x) = y_k\left(\frac{x - x_{k+1}}{x_k - x_{k+1}}\right) + y_{k+1}\left(\frac{x - x_k}{x_{k+1} - x_k}\right) \qquad \text{for } x_k \leq x \leq x_{k+1} \tag{3}$$

The resulting curve, shown in Figure 6.3-2, is continuous but is not well suited for interpolating smooth (i.e., differentiable) functions.

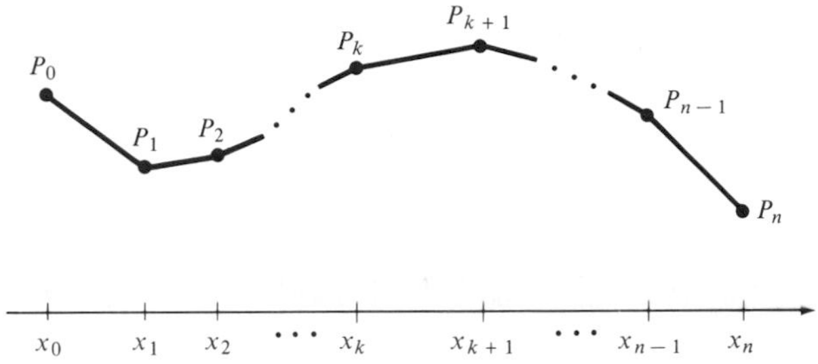

Figure 6.3-2 Piecewise linear interpolation.

6.3B Piecewise Cubic Splines

If we are to allow for the possibility that $f(x)$ has an inflection point between P_k and P_{k+1}, then $q_k(x)$ must be at least third degree (i.e., cubic). Experience has shown that for most smooth functions $f(x)$, cubic $q_k(x)$'s provide adequately accurate interpolation.

Terminology. The function $s(x)$ is called **piecewise cubic** on $[x_0, x_n]$ if there exist cubics $q_0(x), \ldots, q_{n-1}(x)$ such that

$$s(x) = q_k(x) \text{ on } [x_k, x_{k+1}] \qquad \text{for } k = 0, 1, \ldots, n-1 \tag{4}$$

In order for $s(x)$ to interpolate $P_0, \ldots, P_n$, the $q_k(x)$'s must satisfy

$$\textbf{S0: } q_k(x_k) = y_k \quad \text{and} \quad q_k(x_{k+1}) = y_{k+1} \qquad \text{for } k = 0, 1, \ldots, n-1 \tag{5a}$$

We shall call $s(x)$ in (4) a **cubic spline** (or simply a **spline**) if the "pieces" $q_k(x)$ have the *same slope* and *same concavity* at the knots where they are joined, that is, if

$$\textbf{S1: } q'_{k-1}(x_k) = q'_k(x_k)\ [= s'(x_k)] \qquad \text{for } k = 1, 2, \ldots, n-1 \tag{5b}$$

$$\textbf{S2: } q''_{k-1}(x_k) = q''_k(x_k)\ [= s''(x_k)] \qquad \text{for } k = 1, 2, \ldots, n-1 \tag{5c}$$

The $2n$ conditions in S0, together with the $n-1$ conditions in each of S1 and S2, ensure that $s(x)$ and both its first and second derivatives are continuous on $[x_0, x_n]$. Consequently, its graph, $y = s(x)$, is *smooth* and has a continuously turning tangent. Any $s(x)$ satisfying S0–S2 will be called an **interpolating spline** for $P_0, \ldots, P_n$.

6.3C Equations for Finding $q_0(x), \ldots, q_{n-1}(x)$

If $s(x)$ is piecewise cubic on $[x_0, x_n]$, then its second derivative $s''(x)$ is piecewise linear on $[x_0, x_n]$; in particular, by S2, $q''_k(x)$ is linear and interpolates $(x_k, s''(x_k))$ and $(x_{k+1}, s''(x_{k+1}))$ on $[x_k, x_{k+1}]$. So

$$q''_k(x) = s''(x_k)\left(\frac{x - x_{k+1}}{x_k - x_{k+1}}\right) + s''(x_{k+1})\left(\frac{x - x_k}{x_{k+1} - x_k}\right), \qquad k = 0, 1, \ldots, n-1 \tag{6}$$

by (3). If we denote the increment from x_k to x_{k+1} by

$$h_k = x_{k+1} - x_k \qquad \text{for } k = 0, 1, \ldots, n-1 \tag{7}$$

and we denote the second derivative of s at x_k by

$$\sigma_k = s''(x_k) \qquad \text{for } k = 0, 1, \ldots, n \tag{8}$$

then (6) can be rewritten as

$$q_k''(x) = \frac{\sigma_k}{h_k}(x_{k+1} - x) + \frac{\sigma_{k+1}}{h_k}(x - x_k), \qquad k = 0, 1, \ldots, n-1 \tag{9}$$

where the h_k's and σ_k's are constants, with the σ_k's to be determined. Integrating (9) twice the respect to x gives for $k = 0, 1, \ldots, n-1$,

$$q_k(x) = \frac{\sigma_k}{h_k}\frac{(x_{k+1} - x)^3}{6} + \frac{\sigma_{k+1}}{h_k}\frac{(x - x_k)^3}{6} + \lambda_k(x) \tag{10a}$$

where $\lambda_k(x) = C_k + D_k x$. In view of the form of the σ_k and σ_{k+1} terms of (10a), $\lambda_k(x)$ is best rewritten in the alternative form

$$\lambda_k(x) = A_k(x - x_k) + B_k(x_{k+1} - x), \qquad A_k, B_k \text{ are arbitrary constants} \tag{10b}$$

Putting (10) in S0, we get for $k = 0, 1, \ldots, n-1$,

$$y_k = \frac{\sigma_k}{6}h_k^2 + B_k h_k \qquad \text{and} \qquad y_{k+1} = \frac{\sigma_{k+1}}{6}h_k^2 + A_k h_k \tag{11}$$

Finally, solving this for A_k and B_k and substituting in (10) gives

$$\begin{aligned} q_k(x) = {} & \frac{\sigma_k}{6}\left[\frac{(x_{k+1} - x)^3}{h_k} - h_k(x_{k+1} - x)\right] \\ & + \frac{\sigma_{k+1}}{6}\left[\frac{(x - x_k)^3}{h_k} - h_k(x - x_k)\right] \\ & + y_k\left[\frac{x_{k+1} - x}{h_k}\right] + y_{k+1}\left[\frac{x - x_k}{h_k}\right], \qquad k = 0, 1, \ldots, n-1 \end{aligned} \tag{12}$$

This formula can be used to evaluate $s(x)$ [as $q_k(x)$] for $x_k \leqslant x \leqslant x_{k+1}$ *once we know the values of* σ_k and σ_{k+1}. Thus, to be able to use $s(x)$ to approximate $f(x)$ on $[x_0, x_n]$, we must find the second derivatives

$$\sigma_0, \sigma_1, \ldots, \sigma_n \qquad (n + 1 \text{ unknowns})$$

To this end we impose S1. Differentiating (12) gives

$$q_k'(x) = \frac{\sigma_k}{6}\left[\frac{-3(x_{k+1}-x)^2}{h_k} + h_k\right] + \frac{\sigma_{k+1}}{6}\left[\frac{3(x-x_k)^2}{h_k} - h_k\right] + \Delta y_k$$

where $\Delta y_k = (y_{k+1} - y_k)/h_k$. Hence, for $k = 0, 1, \ldots, n-1$,

$$q_k'(x_k) = \frac{\sigma_k}{6}[-2h_k] + \frac{\sigma_{k+1}}{6}[-h_k] + \Delta y_k \qquad \textbf{(13a)}$$

$$q_k'(x_{k+1}) = \frac{\sigma_k}{6}[h_k] + \frac{\sigma_{k+1}}{6}[2h_k] + \Delta y_k \qquad \textbf{(13b)}$$

Replacing k by $k - 1$ in (13b) to get $q_{k-1}'(x_k)$, and equating to (13a) gives

$$\begin{aligned}(\mathrm{E}_k)\ h_{k-1}\sigma_{k-1} + 2(h_{k-1} + h_k)\sigma_k + h_k\sigma_{k+1} \\ = 6[\Delta y_k - \Delta y_{k-1}], \qquad k = 1, \ldots, n-1\end{aligned} \qquad \textbf{(14)}$$

where $\Delta y_{k-1} = (y_k - y_{k-1})/h_{k-1}$. If the x_k's are equispaced, say $h_k = h$ for all k, then (14) simplifies to

$$(\mathrm{E}_k)\ \sigma_{k-1} + 4\sigma_k + \sigma_{k+1} = \frac{6}{h}[\Delta y_k - \Delta y_{k-1}], \qquad k = 1, \ldots, n-1 \qquad \textbf{(15)}$$

EXAMPLE 6.3C Find $(\mathrm{E}_1) - (\mathrm{E}_3)$ for the five logarithmic knots

$$P_0(1, \ln 1), \quad P_1(2, \ln 2), \quad P_2(3, \ln 3), \quad P_3(4, \ln 4), \quad P_4(6, \ln 6)$$

SOLUTION: Here $x_0 = 1$, $x_1 = 2$, $x_2 = 3$, $x_3 = 4$, and $x_4 = 6$, hence

$$h_0 = h_1 = h_2 = 1 \qquad \text{and} \qquad h_3 = 6 - 4 = 2$$

Since $n = 4$, the system (14) consists of the $n - 1 = 3$ equations

$$(\mathrm{E}_1)\ 1\sigma_0 + 2(1+1)\sigma_1 + 1\sigma_2 = 6\left[\frac{\ln(\frac{3}{2})}{1} - \frac{\ln(\frac{2}{1})}{1}\right] \doteq -1.72609$$

$$(\mathrm{E}_2)\ 1\sigma_1 + 2(1+1)\sigma_2 + 1\sigma_3 = 6\left[\frac{\ln(\frac{4}{3})}{1} - \frac{\ln(\frac{3}{2})}{1}\right] \doteq -0.70670 \qquad \textbf{(16)}$$

$$(\mathrm{E}_3)\ 1\sigma_2 + 2(1+2)\sigma_3 + 2\sigma_4 = 6\left[\frac{\ln(\frac{6}{4})}{2} - \frac{\ln(\frac{4}{3})}{1}\right] \doteq -0.50970$$

where we have used the identify $\ln a - \ln b = \ln(a/b)$. ■

6.3D Endpoint Strategies

As seen in the preceding example, the system (14) is *linear* in the $n + 1$ unknowns $\sigma_0, \ldots, \sigma_n$. However, since it has only $n - 1$ equations, it is underdetermined, and hence has infinitely many solutions (Section 3.3E). Table 6.3-1 summarizes four strategies for eliminating σ_0 from (E_1) and σ_n from (E_{n-1}), yielding an $(n - 1) \times (n - 1)$ *tridiagonal system* (see Section 3.2H) in the variables $\sigma_1, \sigma_2, \ldots, \sigma_{n-1}$.

TABLE 6.3-1 Equations for Implementing Four Endpoint Strategies

Strategy	*Endpoint Condition*	*Equations for Eliminating σ_0 or σ_n*
I	Specify the value of $s''(x)$ at the endpoint	$\sigma_0 = s''(x_0)$ $\sigma_n = s''(x_n)$
II	Assume that $s''(x)$ is *constant* near the endpoint	$\sigma_0 = \sigma_1$ $\sigma_n = \sigma_{n-1}$
III	Assume that $s''(x)$ is *linear* near the endpoint	$\sigma_0 = \frac{1}{h_1}\{(h_0 + h_1)\sigma_1 - h_0\sigma_2\}$ $\sigma_n = \frac{1}{h_{n-2}}\{-h_{n-1}\sigma_{n-2} + (h_{n-2} + h_{n-1})\sigma_{n-1}\}$
IV	Specify the value of $s'(x)$ at the endpoint	$\sigma_0 = \frac{3}{h_0}[\Delta y_0 - s'(x_0)] - \frac{1}{2}\sigma_1$ $\sigma_n = \frac{3}{h_{n-1}}[s'(x_n) - \Delta y_{n-1}] - \frac{1}{2}\sigma_{n-1}$

It is important to note that *it is not necessary to use the same strategy at both endpoints* as long as some strategy is used at each of x_0 and x_n.

In geometric terms, *strategy I imposes a desired concavity of $s(x)$ at x_0 or x_n; strategy II imposes quadratic behavior on $s(x)$ over the "double interval" $[x_0, x_2]$ or $[x_{n-2}, x_n]$; strategy III imposes the behavior of a single cubic on $s(x)$ over $[x_0, x_2]$* or $[x_{n-2}, x_n]$; and *strategy*

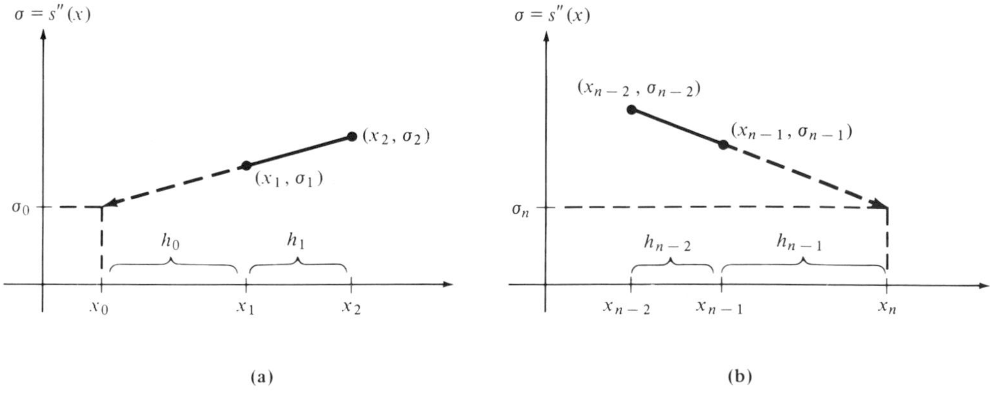

Figure 6.3-3 *Strategy III: (a) at x_0; (b) at x_n.*

IV imposes a desired slope of s(x) at x_0 or x_n. The strategy III equations are obtained by taking $k = 1$ and $k = n - 2$ in (9) and using the identities

$$x_2 - x_0 = h_0 + h_1 \qquad \text{and} \qquad x_n - x_{n-2} = h_{n-2} + h_{n-1} \tag{17}$$

(see Figure 6.3-3). The strategy IV equations are obtained by taking $k = 0$ in (13a) and $k = n - 1$ in (13b) (see Problem 6-34).

If we use Table 6.3-1 to eliminate σ_0 from

$$(\mathrm{E}_1)\ h_0\sigma_0 + 2(h_0 + h_1)\sigma_1 + h_1\sigma_2 = 6[\Delta y_1 - \Delta y_0]$$

[the first equation in (14)], then upon combining terms we get

$$\begin{aligned}
&(\mathrm{E}_1)_{\mathrm{I}}\ \ 2(h_0 + h_1)\sigma_1 + h_1\sigma_2 &&= 6[\Delta y_1 - \Delta y_0] - h_0 s''(x_0)\\
&(\mathrm{E}_1)_{\mathrm{II}}\ \ (3h_0 + 2h_1)\sigma_1 + h_1\sigma_2 &&= 6[\Delta y_1 - \Delta y_0]\\
&(\mathrm{E}_1)_{\mathrm{III}}\ \ (h_0 + 2h_1)\sigma_1 + (h_1 - h_0)\sigma_2 &&= \frac{6h_1}{h_0 + h_1}[\Delta y_1 - \Delta y_0]\\
&(\mathrm{E}_1)_{\mathrm{IV}}\ \ \left(\frac{3}{2}h_0 + 2h_1\right)\sigma_1 + h_1\sigma_2 &&= 3[2\Delta y_1 - 3\Delta y_0 + s'(x_0)]
\end{aligned} \tag{18a}$$

where the subscript on (E_1) corresponds to the endpoint strategy used to eliminate σ_0. Similarly, if we use Table 6.3-1 to eliminate σ_n from

$$(\mathrm{E}_{n-1})\ h_{n-2}\sigma_{n-2} + 2(h_{n-2} + h_{n-1})\sigma_{n-1} + h_{n-1}\sigma_n = 6[\Delta y_{n-1} - ,\Delta y_{n-2}]$$

[the last equation in (14)], then upon combining terms we get

$$\begin{aligned}
&(\mathrm{E}_{n-1})_{\mathrm{I}}\ \ h_{n-2}\sigma_{n-2} + 2(h_{n-2} + h_{n-1})\sigma_{n-1}\\
&\qquad\qquad = 6[\Delta y_{n-1} - \Delta y_{n-2}] - h_{n-1}s''(x_n)\\
&(\mathrm{E}_{n-1})_{\mathrm{II}}\ \ h_{n-2}\sigma_{n-2} + (2h_{n-2} + 3h_{n-1})\sigma_{n-1} = 6[\Delta y_{n-1} - \Delta y_{n-2}]\\
&(\mathrm{E}_{n-1})_{\mathrm{III}}\ \ (h_{n-2} - h_{n-1})\sigma_{n-2} + (2h_{n-2} + h_{n-1})\sigma_{n-1}\\
&\qquad\qquad = \frac{6h_{n-2}}{h_{n-2} + h_{n-1}}[\Delta y_{n-1} - \Delta y_{n-2}]\\
&(\mathrm{E}_{n-1})_{\mathrm{IV}}\ \ h_{n-2}\sigma_{n-2} + \left(2h_{n-2} + \frac{3}{2}h_{n-1}\right)\sigma_{n-1}\\
&\qquad\qquad = 3[3\Delta y_{n-1} - 2\Delta y_{n-2} - s'(x_n)]
\end{aligned} \tag{18b}$$

Note that strategies I–IV affect *only* (E_1) and (E_{n-1}). *The "interior" equations* (E_2)–(E_{n-2}) *are always given by (14).*

EXAMPLE 6.3D Illustrate strategies I–IV for the five logarithmic knots

$$P_0(1, \ln 1),\quad P_1(2, \ln 2),\quad P_2(3, \ln 3),\quad P_3(4 \ln 4),\quad P_4(6, \ln 6)$$

SOLUTION [SEE (16)]: At $x_0 = 1$, since $h_1 = h_2 = 1$, (18a) gives

$$\left.\begin{aligned}
&(E_1)_{I} \quad 2(1+1)\sigma_1 + 1\sigma_2 &&= -1.72609 - 1s''(x_0)\\
&(E_1)_{II} \quad (3\cdot 1 + 2\cdot 1)\sigma_1 + 1\sigma_2 &&= -1.72609\\
&(E_1)_{III} \quad (1 + 2\cdot 1)\sigma_1 + (1-1)\sigma_2 &&= \frac{6\cdot 1}{1+1}[-1.72609]\\
&(E_1)_{IV} \quad \left(\frac{3}{2}\cdot 1 + 2\cdot 1\right)\sigma_1 + 1\sigma_2 &&= 3[2(0.40547) - 3(0.69315) + 1s'(x_0)]
\end{aligned}\right\} \qquad \textbf{(19a)}$$

At $x_n = x_4 = 6$, since $h_{n-2} = h_2 = 1$ and $h_{n-1} = h_3 = 2$, (18b) gives

$$\left.\begin{aligned}
&(E_3)_{I} \quad 1\sigma_2 + 2(1+2)\sigma_3 &&= -0.50970 - 2s''(x_4)\\
&(E_3)_{II} \quad 1\sigma_2 + (2\cdot 1 + 3\cdot 2)\sigma_3 &&= -0.50970\\
&(E_3)_{III} \quad (1-2)\sigma_2 + (2\cdot 1 + 2)\sigma_3 &&= \frac{6\cdot 1}{1+2}[-0.50970]\\
&(E_3)_{IV} \quad 1\sigma_2 + \left(2\cdot 1 + \frac{3}{2}\cdot 2\right)\sigma_3 &&= 3[2(0.20273) - 3(0.28768) - s'(x_4)]
\end{aligned}\right\} \qquad \textbf{(19b)}$$

Whatever strategies are used at x_0 and x_4, the second (and only remaining) equation is given by (16), that is,

$$(E_2)\ 1\sigma_1 + 2(1+1)\sigma_2 + 1\sigma_3 = 6[\Delta y_2 - \Delta y_1] = -0.70670 \qquad \textbf{(20)}$$

With four choices for (E_1) in (19a) and another four for (E_{n-1}) in (19b), there are 16 possible endpoint strategy combinations, with further flexibility possible in specifying values of s'' of s' if strategy I or IV is used. So strategies I–IV offer considerable latitude in describing the endpoint behavior of $s(x)$. ■

6.3E Finding $\sigma_0, \ldots, \sigma_n$

No matter which of the strategies (18) are used to eliminate σ_0 from (E_1) and σ_n from (E_{n-1}), the resulting equations can be written as

$$(E_k)\ l_k\sigma_{k-1} + d_k\sigma_k + u_k\sigma_{k+1} = b_k, \qquad k = 1, 2, \ldots, n-1 \qquad \textbf{(21)}$$

where $a_1 = 0$ [σ_0 is removed from (E_1)] and $u_{n-1} = 0$ [σ_n is removed from (E_{n-1})]. In matrix form, (21) is the $(n-1)\times(n-1)$ *tridiagonal* system

$$T\boldsymbol{\sigma} = \mathbf{b}, \qquad \text{where } \boldsymbol{\sigma} = [\sigma_1 \quad \sigma_2 \quad \cdots \quad \sigma_{n-1}]^T \text{ and } T = \text{trid}(\mathbf{l}, \mathbf{d}, \mathbf{u}) \qquad \textbf{(22)}$$

Since $h_0, \ldots, h_{n-1}$ are positive, it follows from (14) and (18) that *the coefficient matrix T is diagonally dominant*; so (21) can always be solved uniquely for $\sigma_1, \ldots, \sigma_{n-1}$ using Gaussian Elimination Algorithm 3.2H; Table 6.3-1 can then be used to get σ_0 and σ_n.

EXAMPLE 6.3E Find $\sigma_0, \ldots, \sigma_4$ for the spline that fits

$$P_0(1, \ln 1), \quad P_1(2, \ln 2), \quad P_2(3, \ln 3), \quad P_3(4, \ln 4), \quad P_4(6, \ln 6) \tag{23}$$

using strategy IV with $s'(1) = 1$ at $x_0 = 1$ and strategy III at $x_n = x_4 = 6$. Then use $s(3.7)$ to estimate $f(3.7) = \ln(3.7) \doteq 1.3083$.

Solution: From (19) and (20), the system $T\boldsymbol{\sigma} = \mathbf{b}$ that we must solve is

$$\begin{aligned}
&(E_1)_{IV} \ \tfrac{7}{2}\sigma_1 + 1\sigma_2 &&= 3[-1.26851 + 1] = -0.80553\\
&(E_2) \ \ 1\sigma_1 + 4\sigma_2 + 1\sigma_3 &&= -0.70670\\
&(E_3)_{III} \ (-1)\sigma_2 + \ \ 4\sigma_3 &&= -0.16990
\end{aligned}$$

Using Gaussian elimination for tridiagonal systems (Section 3.2H), we get (to $5d$)

$$[T{:}\mathbf{b}] \rightarrow \begin{bmatrix} \frac{7}{2} & 1 & 0 & : & -0.80553 \\ & \frac{26}{7} & 1 & : & -0.47655 \\ & & \frac{111}{26} & : & -0.29820 \end{bmatrix} \Rightarrow \begin{bmatrix} \sigma_1 \\ \sigma_2 \\ \sigma_3 \end{bmatrix} \begin{matrix} = \\ = \\ = \end{matrix} \begin{bmatrix} -0.19887 \\ -0.10950 \\ -0.06985 \end{bmatrix} \tag{24a}$$

To get σ_0 and σ_4, we use Table 6.3-1:

$$\begin{aligned}
\sigma_0 &= 3\left[\frac{0.69315 - 1}{1}\right] - \frac{1}{2}(-0.19887) = -0.82111\\
\sigma_4 &= \frac{-2}{2}(-0.10950) + \frac{1+2}{1}(-0.06985) = 0.00945
\end{aligned} \tag{24b}$$

Having found $\sigma_0, \ldots, \sigma_4$, we can now interpolate $\ln z$ on $[x_0, x_n] = [1, 6]$. In particular, since $z = 3.7$ satisfies

$$x_2 = 3 < z = 3.7 < 4 = x_3$$

we can use formula (12) with $k = 2$ to approximate $\ln(3.7)$ as

$$\begin{aligned}
s(z) = q_2(z) &= \frac{\sigma_2}{6}\left[\frac{(x_3 - z)^3}{h_2} - h_2(x_3 - z)\right] + \frac{\sigma_3}{6}\left[\frac{(z - x_2)^3}{h_2} - h_2(z - x_2)\right]\\
&\qquad + y_2\left[\frac{x_3 - z}{h_2}\right] + y_3\left[\frac{z - x_2}{h_2}\right]\\
&= \frac{-0.10950}{6}\left[\frac{(0.3)^3}{1} - 1(0.3)\right] + \frac{-0.06985}{6}\left[\frac{(0.7)^3}{1} - 1(0.7)\right]\\
&\qquad + \ln 3\left[\frac{0.3}{1}\right] + \ln 4\left[\frac{0.7}{1}\right]\\
&= 1.3091 \quad (\text{Error} = 1.3083 - 1.3091 = -0.0008)
\end{aligned} \tag{25}$$

Notice that although the exact value $f'(1) = 1/1$ was used for $s'(1)$ in $(E_1)_{IV}$, the values of $\sigma_k = s''(x_k)$ agree with the exact values $f''(x_k) = -1/x_k^2$ to at most $2s$; and σ_4 actually has the wrong sign [$f''(6) = -1/36$]. Nevertheless, for $z = 3.7$ (near the middle of the interpolating interval [1, 6]), $s(z)$ approximated $f(z) = \ln z$ to almost 4s. ■

Practical Consideration 6.3E (Selecting endpoint strategies). Strategy I is frequently used with $\sigma_0 = \sigma_n = 0$. These so-called *free boundary conditions* result in a $y = s(x)$ curve having the shape that would be assumed by a flexible rod (such as draftsman's spline) if it were bent around pegs at the knots but allowed to maintain its natural (straight-line) shape outside $[x_0, x_n]$. This $s(x)$ is therefore referred to as a **natural spline**. A free boundary condition should only be used if it is known that the curve $y = f(x)$ approaches a straight line (i.e., flattens) or has an inflection point as x approaches the endpoint.

There are situations (e.g., beam problems in civil engineering) where $s'(x_0)$ or $s'(x_n)$ are known. When these *clamped boundary conditions* are given, strategy IV should be used.

At endpoints where neither $s'(x)$ or $s''(x)$ is known, strategy II or III should be used. If the $y = f(x)$ curve appears to have an inflection point near the endpoint, strategy III (cubic behavior) is recommended; otherwise, use strategy II (quadratic behavior). This would have avoided the incorrect sign of σ_4 (at the right endpoint) in Example 6.3E. ■

6.3F Summary: Algorithm for Piecewise Cubic Spline Interpolation

Once endpoint strategies have been selected at x_0 and x_n, proceed as in Algorithm 6.3F. As with polynomial interpolation, one should try to avoid approximating $f(z)$ for z outside $[x_0, x_n]$. If this must be done when $s(x)$ is available, one should use

$$f(z) \approx q_0(z) \text{ for } z < x_0 \qquad \text{and} \qquad f(z) \approx q_{n-1}(z) \text{ for } z > x_n \tag{26}$$

with the understanding that $s(x)$ may not behave like $f(x)$ outside $[x_0, x_n]$.

If several values of $s(z)$ are needed [e.g., to make a table of approximate $f(z)$ values], or if derivatives or integrals of $s(x)$ are to be used to approximate those of $f(x)$, then it is worthwhile to make the substitution

$$x_{k+1} - z = (x_{k+1} - x_k) - (z - x_k) = h_k - (z - x_k) \tag{27}$$

in $q_k(z)$ and then collect powers of $(z - x_k)$ to get

$$q_k(z) = y_k + c_{1,k}(z - x_k) + c_{2,k}(z - x_k)^2 + c_{3,k}(z - x_k)^3 \tag{28a}$$

where

$$c_{1,k} = \Delta y_k - \frac{h_k}{6}(\sigma_{k+1} + 2\sigma_k), \quad c_{2,k} = \frac{\sigma_k}{2}, \quad \text{and} \quad c_{3,k} = \frac{\sigma_{k+1} - \sigma_k}{6h_k} \tag{28b}$$

For example, the $k = 2$ values of Example 6.3E, namely

$$h_2 = 1, \quad \Delta y_2 = \frac{\ln(\frac{4}{3})}{1} = 0.28768, \quad \sigma_2 = -0.10950, \quad \sigma_3 = -0.06985$$

[see (24a)] could be used in (28b) to obtain

ALGORITHM 6.3F. INTERPOLATION USING A PIECEWISE CUBIC SPLINE

PURPOSE: To evaluate $s(z)$, where s is the interpolating cubic spline for $n + 1$ given knots $P_0(x_0, y_0), \ldots, P_n(x_n, y_n)$ that satisfies specified endpoint conditions (at x_0 and x_n), and z is a specified point between x_0 and x_n.

GET n, **x**, **y**, $\{\mathbf{x} = [x_0 \quad x_1 \quad \cdots \quad x_n]$ and $\mathbf{y} = [y_0 \quad y_1 \quad \cdots \quad y_n]\}$
z, {point at which interpolated valued is desired}
parameters {if needed} for endpoint conditions

DO FOR $k = 0$ TO $n - 1$
$h_k \leftarrow x_{k+1} - x_k$

{**Form [T:b]**} Use (14) to form (E_2) ..., (E_{n-2}), (18a) to form (E_1), and (18b) to form (E_{n-1}) for the tridiagonal system $T\boldsymbol{\sigma} = \mathbf{b}$ in (22).
OUTPUT ([T:**b**])

{**Get σ's**} Use Gaussian Elimination (Algorithm 3.2H) to solve $T\boldsymbol{\sigma} = \mathbf{b}$ for $\sigma_1, \sigma_2, \ldots, \sigma_{n-1}$; then use Table 6.3-1 to get σ_0 and σ_n.
OUTPUT ($\sigma_0, \sigma_1, \ldots, \sigma_n$)

{**Interpolate**} Find k such that $x_k \leqslant z < x_{k+1}$; then get $s(z)$ as

$$SofZ \leftarrow \frac{\sigma_k}{6}\left[\frac{(x_{k+1} - z)^3}{h_k} - h_k(x_{k+1} - z)\right] + \frac{\sigma_{k+1}}{6}\left[\frac{(z - x_k)^3}{h_k} - h_k(z - x_k)\right] + y_k\left(\frac{x_{k+1} - z}{h_k}\right) + y_{k+1}\left(\frac{z - x_k}{h_k}\right)$$

OUTPUT ('The interpolated value $s(z)$ is' *SofZ*)
STOP

$$c_{1,2} = 0.28768 = \frac{1}{6}[-0.06985 + 2(-0.10950)] \doteq 0.33582$$

$$c_{2,2} = \frac{1}{2}(-0.10950) = -0.05475 \tag{29}$$

$$c_{3,2} = \frac{-0.06985 - (-0.10950)}{6 \cdot 1} = 0.00661$$

from which (28a) (in nested form) with $z - x_k = 3.7 - 3.0 = 0.7$ gives

$$q_2(3.7) = \{[c_{3,2}(0.7) + c_{2,2}](0.7) + c_{1,2}\}(0.7) + \ln 3 = 1.3091 \tag{30}$$

in agreement with the value obtained in (25).

6.3G Polynomial Versus Cubic Spline Interpolation

For hand calculation of an approximation of $f(z)$, polynomial interpolation as described in Section 6.1E [i.e., finding best mth interpolants $\hat{p}_0(z)$, $\hat{p}_1(z)$, ... until the $\hat{\hat{\delta}}_m(z)$'s stop decreasing] is the preferred method. A computer program that does this is a bit easier to use than one for cubic spline interpolation because there is no need to select endpoint strategies.

If a single smooth curve is to interpolate a large number of knots (say $n > 6$), cubic splines are preferable because they are not as prone to "wiggle" between knots. Consequently, of the two approximations

$$f'(z) \approx \hat{p}_m'(z) \qquad \text{and} \qquad f'(z) \approx s'(z) \tag{31}$$

the latter is generally more accurate and, if (28) is used, easy to obtain. Similarly, once the $c_{i,k}$'s in (28) are found, the approximation

$$\int_a^b f(x)\,dx \approx \int_a^b s(x)\,dx \qquad \text{for } x_0 \leqslant a < b \leqslant x_n \tag{32}$$

can be used to obtain accurate results easily (see Problem 6-33).

However, since $s''(x)$ is only piecewise linear, it cannot be expected to be very accurate no matter what endpoint strategies are used. If a second- or higher-order derivative is needed, the best general strategy is to *fit* (rather than interpolate) a suitable curve as in Section 5.2 or Section 5.3 and then use its derivative to approximate that of f.

For a more extensive discussion of spline approximations, see [13].

PROBLEMS

Section 6.1

6-1 For the four cubic knots $P_0(0, 0)$, $P_1(1, 1)$, $P_2(2, 8)$, and $P_3(3, 27)$:

(a) Find $p_{1,3}(x)$ using the method of undetermined coefficients.

(b) Find Lagrange's form of $p_{1,3}(x)$ and of $p_{0,2}(x)$.

(c) Find $p_{0,3}(x)$ by inspection of $P_0, \ldots, P_3$. Explain your reasoning (see Section 6.1C).

6-2 Same as Problem 6-1 but for the quadratic knots $P_0(0, 0)$, $P_1(1, 1)$, $P_2(2, 4)$, and $P_3(3, 9)$.

6-3 Find the point-slope form (3b) and Lagrange's form (13) of the interpolating polynomial for the two given knots and show that they are equal.

(a) $P_2(-2, 3)$ and $P_3(1, 0)$ **(b)** $P_0(3, 1)$ and $P_1(5, 6)$ **(c)** $P_1(2, 2)$ and $P_2(4, 8)$

6-4 Consider the knots $P_0(-2, -15)$, $P_1(-1, -2)$, $P_2(0, 1)$, $P_3(2, 4)$, $P_4(3, 10)$. Find Lagrange's form (unsimplified) of the indicated $p_{k,k+n}(x)$, then evaluate $p_{k,k+n}(1)$ as described in Practical Consideration 6.1B.

(a) $p_{2,3}(x)$ **(b)** $p_{1,3}(x)$ **(c)** $p_{0,3}(x)$ **(d)** $p_{0,4}(x)$

6-5 Give the value of *only* those of $p_{0,2}(0)$, $p_{0,2}(2)$, $p_{2,3}(2)$, $p_{2,3}(3)$, $p_{2,3}(0)$, $p_{1,4}(4)$ that can evaluated *by inspection* of $P_0, \ldots, P_4$ of Problem 6-4 [see 7(2)].

6-6 Let $L_2(x)$ and $L_3(x)$ denote the second and third Lagrange polynomials used to get the given $p_{k,k+n}(x)$. Give the value of *only* those of $L_2(0)$, $L_2(2)$, $L_2(3)$, $L_3(0)$, $L_3(2)$, $L_3(3)$, $L_3(4)$ that can be found *by inspection* of $P_0, \ldots, P_4$ of Problem 6-4 using the Selecting Property (12).
(a) $p_{2,3}(x)$ **(b)** $p_{0,3}(x)$ **(c)** $p_{2,4}(x)$ **(d)** $p_{0,4}(x)$

6-7 Do Problem 6-4 for the knots $P_0(-3, 1)$, $P_1(0, 7)$, $P_2(2, 9)$, $P_3(3, 7)$, $P_4(5, 1)$.

6-8 Same as Problem 6-5 but for $P_0, \ldots, P_4$ of Problem 6-7.

6-9 Same as Problem 6-6 but for $P_0, \ldots, P_4$ of Problem 6-7.

6-10 For $P_2, \ldots, P_5$ of Example 6.1D, use (27c) to show that $p_{2,5}(0) = 0$ and $p_{2,5}(2) = 8$.

6-11 For the knots $P_1(-1, 21)$, $P_2(-\frac{1}{2}, 7)$, $P_3(0, 1)$, $P_4(\frac{1}{2}, -3)$, $P_5(1, -11)$, use (27) to find **(a)** $p_{1,3}(1)$; **(b)** $p_{2,4}(\frac{1}{4})$; **(c)** $p_{1,4}(-2)$; **(d)** $p_{2,5}(2)$.

6-12 For $x_{-1} = -5$, $x_0 = -2$, $x_1 = 0$, $x_2 = 1$, $x_3 = 3$, and $x_4 = 6$, use the "best centering strategy" (31c) to find the indices k and $k + n$ of the $p_{k,k+n}(z)$ that should be used to get $\hat{p}_n(z)$.
(a) $\hat{p}_0(-3)$ **(b)** $\hat{p}_3(-1)$ **(c)** $\hat{p}_2(2)$ **(d)** $\hat{p}_3(4)$ **(e)** $\hat{p}_3(-6)$

6-13 Use Lagrange's form to get the values of **(a)** $\hat{p}_n(8)$ and **(b)** $\hat{p}_n(21)$ given in Table 6.1-1 of Example 6.1E.

6-14 From the five logarithmic knots $P_0(\frac{1}{4}, \ln \frac{1}{4})$, $P_1(1, 0)$, $P_2(4, \ln 4)$, $P_3(9, \ln 9)$, $P_4(16, \ln 16)$, make a table showing $\hat{p}_0(z), \hat{p}_1(z), \ldots, \hat{p}_4(z)$ as in Example 6.1E for **(a)** $z = 7$; **(b)** $z = 13$. Is $\hat{p}_4(z)$ the most accurate approximation of $\ln z$? If not, explain.

6-15 Same as 6-14 but for $P_0(\frac{1}{4}, 2)$, $P_1(1, 1)$, $P_2(4, \frac{1}{2})$, $P_3(9, \frac{1}{3})$, $P_4(16, \frac{1}{4})$ on the curve $y = x^{-1/2}$.

6-16 **(a)** Find the five Chebyshev nodes $\hat{x}_0, \hat{x}_1, \ldots, \hat{x}_4$, for $[\frac{1}{4}, 16]$.
(b) Use the five knots $(\hat{x}_0, f(\hat{x}_0)), \ldots, (\hat{x}_4, f(\hat{x}_4))$ with $f(x) = \sqrt{x}$ to get the best interpolants $\hat{p}_0(z), \hat{p}_1(z), \ldots$ for $z = 8$. Compare to Example 6.1E(a).
(c) Same as part (b) but using $f(x) = \ln x$ and $z = 7$. Compare to Problem 6-14(a).

Section 6.2

6-17 Determine the numerical values of the missing entries.

x	y	Δy	$\Delta^2 y$	$\Delta^3 y$
$x_0 = -3$	y_0			
		Δy_0		
$x_1 = -2$	y_1		$\Delta^2 y_0$	
		Δy_1		
$x_2 = -1$	y_2		17	
		1		$\Delta^3 y_1$
$x_3 = 0$	y_3		$\Delta^2 y_2$	
		49		$\Delta^3 y_2$
$x_4 = 3$	y_4		67	
		Δy_4		
$x_5 = 4$	463			

6-18 Determine the leading coefficients of $p_{1,3}(x)$, $p_{2,5}(x)$, and $p_{0,4}(x)$ *by inspection* of the following Divided Difference table.

x	y	Δy	$\Delta^2 y$	$\Delta^3 y$	$\Delta^4 y$	$\Delta^5 y$
$x_0 = -3$	-71					
		92				
$x_1 = -2$	21		-34			
		-10		9		
$x_2 = 0$	1		2		-2	
		-4		-1		1
$x_3 = 1$	-3		-2		4	
		-8		19		
$x_4 = 2$	-11		55			
		102				
$x_5 = 3$	91					

6-19 For the DD table of Problem 6-18, build $p_{0,5}(x)$ by adding nodes in the indicated order; put your answer in nested form, and use it to evaluate $p_{0,5}(-1)$.
(a) $x_0, x_1, x_2, x_3, x_4, x_5$ (forward) **(b)** $x_2, x_1, x_3, x_4, x_5, x_0$
(c) $x_5, x_4, x_3, x_2, x_1, x_0$ (backward) **(d)** $x_3, x_2, x_1, x_4, x_0, x_5$

6-20 By inspection of the DD table given in Problem 6-18, find $p''_{2,4}(x)$, $p'''_{1,4}(x)$, $p'''_{2,5}(x)$, $p^{(v)}_{0,5}(x)$ [see (27) of Section 6.2B.]

6-21 It is evident from the table of Problem 6-23(c) that $f(x) = 10$ for an $\bar{x}$ between $x = 5$ and $x = 10$. Refer to Section 6.2H.
(a) Why should Method 1, rather than Method 2, be used to find $\bar{x}$ accurately? Use Method 1 to find $\bar{x}$ to 5*s*.
(b) Use Method 2 to approximate $\bar{x}$. How accurate do you think it is?

6-22 Refer to the Forward Difference table shown in Figure 6.2-2 of Section 6.2A.
(a) Find the leading coefficient of $p_{1,2}(x)$, $p_{-2,0}(x)$, $p_{0,2}(x)$, and $p_{-1,2}(x)$.
(b) Find $p_{0,2}(0)$ and $p_{-1,2}(-1)$ using Newton's forward interpolation formula.
(c) Find $p_{0,2}(0)$ and $p_{-1,2}(-1)$ using Newton's backward interpolation formula.
(d) Find $p_{-2,1}(\frac{1}{4})$ by adding nodes in the order x_{-1}, x_0, x_{-2}, x_1.
(e) Express $p_{0,2}(x)$ and $p_{-1,2}(x)$ as polynomials in s. Note that s is different in each case!

6-23 Make a Forward Difference table if appropriate, or a Divided Difference table otherwise, to determine the degree of the polynomial on which the knots $(x, f(x))$ lie.

(a) x	$f(x)$	**(b)** x	$f(x)$	**(c)** x	$f(x)$	**(d)** x	$f(x)$
-6	-60	0	288	0	2	-3	245
-4	-9	4	195	5	7	-2	46
-3	0	8	160	10	128	-1	0
-1	0	12	144	15	189	0	2
0	-3	16	96	20	226	2	30
2	0	20	13	25	487	3	146
3	12	24	0	30	1432		

6-24 For parts (a)–(d) of Problem 6-23: Find $p''_{2,4}(x)$, $p'''_{1,4}(x)$, $p'''_{2,5}(x)$, $p^{(v)}_{0,5}(x)$. Assume that the first tabulated knot is $P_0(x_0, y_0)$.

6-25 Same as Problem 6-24 but for a DD table for the knots in Problem 6-23(b), with $(-4, -9)$ taken as P_0.

6-26 On the Forward Difference table for the knots in Problem 6-23(b), take (4, 195) as P_0 and then use (49) of Section 6.2D to form $p_{0,5}(x)$ (in terms of s) as indicated in parts (a)–(d) of Problem 6-19.

6-27 Same as Problem 6-26 but for the knots in Problem 6-23(c), with (0, 2) taken as P_0.

Section 6.3

6-28 Four cubic splines are to passed through the knots $P_0(-2, -8)$, $P_1(0, 0)$, $P_2(1, 1)$, and $P_3(2, 8)$ on the curve $y = x^3$. The endpoint conditions for these cubic splines are:
(I) Strategy I with $s''(-2) = -12$, $s''(2) = 12$ [exactly $d^2(x^3)/dx^2$]
(II) Strategy II at both endpoints
(III) Strategy III at both endpoints
(IV) Strategy IV with $s'(-2) = s'(2) = 12$ [exactly $d(x^3)/dx$]
(a) Which of (I)–(IV) will result in $s(x) = x^3$ exactly on $[-2, 2]$? Explain.
(b) Set up the equations in σ_0 and σ_1 imposed by each of (I)–(IV).
(c) Solve the four systems in part (b) and get $\sigma_0, \ldots, \sigma_3$ for each of (I)–(IV). For which of (I)–(IV) is σ_k exactly $d^2(x^3)/dx^2$ at x_k for $k = 0, 1, 2, 3$?

6-29 **(a)** Find the cubic spline with the "natural" endpoint conditions $s''(0) = s''(16) = 0$ for $P_0, \ldots, P_4$ of Example 6.1E.
(b) Evaluate $s(x)$ obtained in part (a) at $x = 8$ and $x = 21$. Compare to Example 6.1E.

6-30 Same as Problem 6-29 but using the (exact) endpoint conditions $s'(0) = 20$, $s'(16) = \frac{1}{8}$.

6-31 Same as Problem 6-29 but using strategy III at $x_0 = 0$ and strategy II at $x_4 = 16$.

6-32 Use (28) of Section 6.3F to express $q_k(x)$ as $y_k + c_{1,k}(x - x_k) + c_{2,k}(x - x_k)^2 + c_{3,k}(x - x_k)^3$.
(a) $q_1(x)$ $(0 \leq x \leq 1)$, for $s(x)$ in (I) of Problem 6-28.
(b) $q_0(x)$ $(-2 \leq x \leq 0)$, for $s(x)$ in (III) of Problem 6-28.
(c) $q_2(x)$ $(4 \leq x \leq 9)$, for $s(x)$ in Problem 6-29(a).
(d) $q_3(x)$ $(9 \leq x \leq 16)$, for $s(x)$ in Problem 6-31(a).

6-33 Suppose that $s(x) = q_k(x)$ for $x_k \leq x \leq x_{k+1}$ for $k = 0, 1, \ldots, n-1$, where

$$q_k(x) = y_k + c_{1,k}(x - x_k) + c_{2,k}(x - x_k)^2 + c_{3,k}(x - x_k)^3$$

Derive formulas involving x_k y_k, $c_{i,k}$, z, a, and b for evaluating
(a) $s'(z)$ $[\approx f'(z)]$ **(b)** $\int_a^b s(x)\,dx$ $[\approx \int_a^b f(x)\,dx]$

6-34 Use Figure 6.3-3 to derive the strategy III equations of Table 6.3-1.

MISCELLANEOUS PROBLEMS

M6-1 Without writing out expressions for $L_{k+j}(x)$, show that the Lagrange polynomials for any $n+1$ distinct nodes $x_k, x_{k+1}, \ldots, x_{k+n}$ must satisfy
(a) $L_k(x) + \cdots + L_{k+n}(x) = 1$ and **(b)** $x_kL_k(x) + \cdots + x_{k+n}L_{k+n}(x) = x$ for all x.
NOTE: Use the uniqueness of Section 6.1C with $y_{k+j} = 1$ (degree 0) in part (a), and with $y_{k+j} = x_{k+j}$ (degree 1) in part (b).

M6-2 Verify Problem M6-1 for **(a)** $L_3(x)$ and $L_4(x)$ of Example 6.1B(a) **(b)** $L_3(x)$, $L_4(x)$, and $L_5(x)$ of Example 6.1B(b), **(c)** $L_2(x)$, $L_3(x)$, $L_4(x)$, and $L_5(x)$ of Example 6.1B(c).

M6-3 In (a)–(c), find *only* those values that can be found *by inspection* of the four knots $P_1(0, 5)$, $P_2(1, 2)$, $P_3(4, 0)$, $P_4(6, -5)$.
(a) $p_{1,3}(2)$, $p_{1,3}(0)$, $p_{2,3}(1)$, $p_{3,4}(1)$, $p_{1,3}(3)$, $p_{1,3}(4)$

(b) $L_1(1)$, $L_1(2)$, $L_2(1)$, $L_3(3)$, $L_3(4)$, $L_4(3)$, $L_4(4)$, if $L_i(x)$ is the ith Lagrange polynomial used to find $p_{2,4}(x)$.

(c) Same as part (b) but for the $L_i(x)$'s used to find $p_{1,3}(x)$.

M6-4 In the following table, $f(x)$ is described by *different* differentiable functions on either side of $x = 0$ but has a "cusp," hence is *not* differentiable at $x = 0$, as a sketch of the tabulated points $(x, f(x))$ will show.

x_i	-6	-5	-4	-3	-2	-1	0	1	2	3	4	5	6
$f(x)$	932	487	226	89	28	7	2	7	10	5	-14	-53	-118

(a) You should *only* approximate $f(z)$ as $p_{k,k+n}(z)$, where x_k *and* x_{k+n} are either both $\geqslant 0$ or both $\leqslant 0$. Why?

(b) If the first node $(-6, 932)$ is taken to be P_1 and $z = 0.6$, which $p_{k,k+n}(z)$ should be used as $\hat{p}_n(z)$ for $n = 0, 1, \ldots, 5$? Evaluate $\hat{p}_0(z)$, $\hat{p}_1(z)$, $\hat{p}_2(z)$, $\ldots$ by any method, stopping when you think you have an accurate approximation of $f(z)$.

(c) Same as (b) but for $z = -1.2$.

M6-5 A strategy for getting an approximation of $f(z)$ from $\hat{p}_0(z)$, $\hat{p}_1(z)$, $\ldots$, is as follows: If $|\hat{\delta}_m(z)| = |\hat{p}_m(z) - \hat{p}_{m-1}(z)|$ is not significantly smaller than $|\hat{\delta}_{m-1}(z)|$, then use either $\hat{p}_{m-1}(z)$ or $\hat{p}_m(z)$ or their average. Discuss how well this strategy works in (a) Example 6.1E; (b) Problem 6-14; (c) Problem 6-15.

M6-6 Use definition (2) of Section 6.2A to prove that $\Delta(au_k + bv_k) = a\Delta u_k + b\Delta v_k$.

M6-7 What can you conclude about the knots $P_2, \ldots, P_5$ if $\Delta^2 y_3 = 0$? if $\Delta^3 y_2 = 0$?

M6-8 The nth divided difference at x_k is often written as $f[x_k, \ldots, x_{k+n}]$ rather than $\Delta^n y_k$.

(a) Write in the alternative notation: $\Delta^2 y_3$, $\Delta^2 y_1$, $\Delta^1 y_2$, $f[x_2, x_3]$, $f[x_1, x_2, x_3]$, $f[x_3, \ldots, x_6]$.

(b) Write the definition (22a) of Section 6.2B in $f[x_k, \ldots, x_{k+n}]$ notation.

(c) Write the forward *and* backward difference formulas for $p_{1,4}(x)$ and $p_{2,5}(x)$ using the $f[x_k, \ldots, x_{k+n}]$ notation.

M6-9 Deduce from Theorem 6.2B that the value of $\Delta^n y_k$ does not depend upon the order in which subscripts are assigned to the generating knots $P_k, \ldots, P_{k+n}$ are arranged. Verify by showing that the constant Δ^3 still results when $P_1, \ldots,$ of Figure 6.2-5 are entered in the order P_3, P_6, P_1, P_4, P_2, P_5.

M6-10 Derive the Newton backward form of $p_{k,k+n}(x)$ from (48a) of Section 6.2D.

M6-11 The **gamma function** $\Gamma(x)$ is a smooth function of x for $x > 0$ that satisfies $\Gamma(n) = (n - 1)!$ when n is a positive integer. Use this fact to approximate $\Gamma(\frac{3}{2})$ and $\Gamma(\frac{5}{2})$ as accurately as you can.

M6-12 Your boss knows what the gross revenue (in millions of dollars) when the following number of items is sold:

items sold	100,000	200,000	300,000	400,000	500,000
gross revenue	11.2	15.9	17.1	16.3	15.0

You will lose your job unless you provide an accurate estimate of how many items to sell (not necessarily a multiple of 100,000) in order that your boss get the maximum possible gross revenue. How many items would you recommend selling? Justify.

M6-13 The following table shows the volume V of a gas in a cylinder as the pressure P on a piston is increased.

P (lbf/in²)	60	80	100	120	140	160	180
V (in³)	80.0	69.2	60.0	52.0	45.0	38.6	32.5

(a) Approximate V as accurately as you can when $P = 95$ lbf/in.².
(b) Can V be approximated reliably by a polynomial in P of degree 4 or less over the interval $60 \leqslant P \leqslant 180$? Justify.
(c) Find as accurately as you can the value of P for which $V = 50$.

M6-14 Thermodynamics often requires the use of tables relating pressure P (psi) and specific volume v (ft³/lbm) at a particular temperature. These table entries are usually spaced closely enough so that linear interpolation provides sufficient accuracy. Confirm this by finding $\hat{p}_1(z)$ and $\hat{p}_2(z)$ and examining whether they agree to about the tabulated accuracy.
(a) Find P when $v = 3.959$ for superheated water vapor at 600°F using the tabulated (P, v) entries (120.0, 5.169), (140.0, 4.412), (160.0, 3.848), (180.0, 3.409).
(b) Find P when $v = 0.558$ for saturated freon-12 vapor at 200°F using the tabulated (P, v) entries (80.0, 0.6910), (90.0, 0.6094), (100.0, 0.5441), (125.0, 0.4264).

COMPUTER PROBLEMS

C6-1 Show that if $p_{1,n}(x) = C_1 + C_2x + \cdots + C_nx^{n-1}$, then the coefficient matrix of the linear system $p_{1,n}(i) = y_i$ $(i = 1, \ldots, n)$ is $A_n = (i^{j-1})_{n\times n}$. Then use any available computer program to get a condition number for A_n for $n = 2, 3, 4$, and 5. Is A_n becoming ill conditioned as n increases?

C6-2 Write a computer program that reads in a set of tabulated knots and then calls a subprogram such as `FUNCTION PLAGR` of Figure 6.1-5 to make a table as in Table 6.1-1 (Example 6.1E) for one or more values of z.

C6-3 **(a)** Write a subprogram `DIFF(KODE, MAXNP, NP, X, Y, TABLE)` that forms the kth differences for the `NP` knots whose coordinates are stored in the arrays `X` and `Y` and stores them in the entries of the kth row of the array `TABLE(MAXNP, NP)` for $k = 0, 1, \ldots,$ `NP` $-$ `1`. It should be a Forward or Divided Difference table according as `KODE` is 0 or 1.
(b) Write a computer program that calls `DIFF` and then prints the table for x_k between x_0 and x_n specified by the user.
WARNING: Unless you are a gifted programmer, be content to print the table as a right triangle rather than an isoceles triangle as you would by hand. Test using Problems 6-17 and 6-23(b).
(c) Modify your calling program in (b) so as to use `TABLE` along with the recursive formulas (33), (35), and (49) of Section 6.3 to make a table as in Table 6.1-1 for one or more values of z. Test using Example 6.1E

C6-4 **(a)** Write a subprogram `SPLINE(N, X, Y, STRATØ, ENDØ, STRATN, ENDN, SIGMA)` where the `INTEGER` variables `STRATØ` and `STRATN` (= 1, 2, 3, or 4) and, if necessary, the values `ENDØ` [$= s'(x_0)$ or $s''(x_0)$] and `ENDN` [$= s'(x_n)$ or $s''(x_n)$] impose the desired endpoint conditions. The subprogram should form and solve the appropriate tridiagonal system to get $\sigma_0, \ldots, \sigma_n$ (in the array `SIGMA`) as described in the **Form [*T*:b]** and **Get σ's** steps of Algorithm 6.3F. Test using Problem 6-28.
(b) Write a `FUNCTION` program `INTRPS(Z, N, X, Y, SIGMA)` that uses `SIGMA` returned by subprogram `SPLINE` of part (a) to calculate $s(z)$ as *SofZ* of the **Interpolate** step of Algorithm 6.3F for a specified value of z. Test using Problem 6-28(I) for which $s(z)$ is exactly z^3.

C6-5 **(a)** Write a program `SPLINC(N, X, Y, STRATØ, ENDØ, STRATN, ENDN, C)` that calls `SPLINE` of Problem C6-4(a) and then forms and returns the $3 \times n$ matrix `C` as in (28) of Section 6.3F.

(b) Same as in Problem C6-4(b) but using `C` returned by `SPLINC` rather than `SIGMA` returned by `SPLINE`.

(c) Write a program that uses `C` returned by `SPLINC` of part (a) to find $s'(z)$ or $\int_a^b s(x)\,dx$ for any z, a, and b in or near the interval $[x_0, x_n]$ (see Problem 6-33). Test using Problem 6-28(I) for which $s(x)$ is exactly x^3.

C6-6 Use any available cubic spline program to interpolate the knots given in Example 5.3B. Get enough values to plot $s(x)$ on $[1, 16]$. Does $s(x)$ look like the generating curve shown in Figure 5.3-2?

C6-7 When the origin of an xy-plane is drawn through the rear tip of a cross section of an airplane wing, its bottom contour is a smooth curve through the following points, where x is the horizontal distance toward the front of the wing, and x and y are in feet.

x	0	3	5	7	9	11	12	13	14	15
y	0.0	1.8	2.5	2.9	2.8	2.3	1.8	1.2	1.0	1.8

(a) Fit a cubic spline with endpoint conditions $s''(0) = 0$, $s'(15) = 20$ through these points.

(b) Use Problem 6-33 to estimate (i) the slope of the bottom contour at the rear wing tip; (ii) the area between the bottom contour and the x-axis; (iii) the lowest point of the contour for x between 13 and 15.

7

Numerical Methods for Differentiation and Integration

7.0

Introduction

We now turn our attention to numerical procedures for approximating the two fundamental quantities of calculus, namely *the derivative of f at z,* defined as the limit

$$f'(z) = \lim_{h \to 0} \frac{\Delta f(z)}{h}, \qquad \text{where } \frac{\Delta f(z)}{h} = \frac{f(z + h) - f(z)}{h} \tag{1a}$$

that is, $\Delta f(z)/h$ is the **difference quotient of f at z**, and the *definite integral of f over* $[a, b]$, given by the limit

$$\int_a^b f(x)\,dx = \lim_{h \to 0} R[h], \qquad \text{where } R[h] = \sum_{k=1}^{n} f(x_{k-1})h \tag{1b}$$

that is, $R[h]$ is the **Riemann sum** using the left endpoint of n subintervals of equal length $h = (b - a)/n$ (see Appendix II.4A).

In calculus, one learns various rules of differentiation that allow $f'(z)$ to be calculated exactly as long as $f(x)$ is expressible in terms of the elementary functions (i.e., algebraic, trigonometric, and exponential functions and their inverses). One also learns to use the

Fundamental Theorem of Calculus, which asserts that any function $f(x)$ that is continuous on $[a, b]$ can be integrated from a to b as

$$\int_a^b f(x)\,dx \approx F(b) - F(a) = F(x)]_a^b, \quad \text{where } F(x) = \int f(x)\,dx \tag{2}$$

that is, $F(x)$ is any antiderivative of $f(x)$. Unfortunately, these differentiation and integration techniques are of no use when $f(x)$ is known only at certain tabulated values (as in Chapter 6) and when there is no known formula for $f(x)$ (as in Example 1.1A). Think about it! Moreover, there are perfectly reasonable looking functions $f(x)$ such as

$$\sqrt{\cos^2 x + 3\sin^2 x}, \quad e^{\alpha x^2}, \quad \text{or} \quad \frac{\sin x}{x} \tag{3}$$

whose antiderivative *cannot* be found on a table of integrals! How, then, do you integrate such a function from a to b?

A general approach to approximating derivatives and *proper* definite integrals as in (1b) is given in Section 7.1. This is followed by more detailed discussions of methods for approximating derivatives (Section 7.2), proper integrals (1b) (Sections 7.3 and 7.4), improper integrals (Section 7.4), and multiple integrals (Section 7.5). A powerful general formula due to *Richardson* is obtained in Section 7.2 for the purpose of improving the accuracy of approximations of derivatives. As it turns out, this important formula also provides the basis for *Romberg integration* in Section 7.3 and is a valuable tool for getting improved approximations of solutions of differential equations in Chapter 8.

7.1 Approximating $f^{(k)}(z)$ and $\int_a^b f(x)\,dx$ as $\sum_{i=0}^n w_i f(x_i)$

This chapter is concerned with finding approximations of

$$f^{(k)}(z) = \text{the } k\text{th derivative of } f(x) \text{ at a particular } z \tag{1a}$$

and

$$\int_a^b f(x)\,dx = \text{the definite integral of } f \text{ from } a \text{ to } b \tag{1b}$$

when there is no formula for $f(x)$ as in Example 1.1A, or when $f(x)$ is known only at certain tabulated values. The approximations we seek are *weighted sums* of $n + 1$ **sampled values** $f(x_i)$, that is, approximation formulas of the following form:

$$f^{(k)}(z) \approx w_0 f(x_0) + w_1 f(x_1) + \cdots + w_n f(x_n) = \sum_{i=0}^n w_i f(x_i) \tag{2a}$$

$$\int_a^b f(x)\,dx \approx w_0 f(x_0) + w_1 f(x_1) + \cdots + w_n f(x_n) = \sum_{i=0}^n w_i f(x_i) \tag{2b}$$

We assume that the sampled $f(x)$'s are either known or can be calculated as needed. The **sampled nodes** x_i need *not* be equispaced, although they often are. Formula (2a) is based on the intuitive idea that sampled *values of the function at or near z* can be used to get approximations of derivatives of *f at z*. Similarly, (2b) is based on the idea that sampled values of the integrand function *on or near* the interval $[a, b]$ can be used to approximate its integral *over a* to *b*. If the weighted sum $\Sigma_{i=0}^n w_i f(x_i)$ in (2a) or (2b) is abbreviated as $\Sigma(f)$, then for any scalars α and β,

$$\Sigma\,(\alpha f + \beta g) = \alpha\,\Sigma\,(f) + \beta\,\Sigma\,(g) \tag{2c}$$

We shall refer to this as the **linearity** of the operator Σ.

Once $n + 1$ sampled nodes x_i have been selected in (2a) or (2b), the challenge is to find $n + 1$ **sample weights** w_i, which make the approximation accurate in some sense. The most general accuracy criterion is the following: An approximation formula is said to have ***n*th-degree exactness** if it is guaranteed to be exact (i.e., equality) when $f(x)$ is a *polynomial* of degree n or less. As a rule (*but not always*), the higher the degree of exactness, the more accurate the formula.

7.1A Formulas Involving One or Two Sampled Nodes

The first derivative $f'(z)$ can be interpreted geometrically as the *slope of the tangent line to the graph of f at the point* $(z, f(z))$. It can therefore be seen from Figure 7.1-1 that approximations of $f'(z)$ can be obtained by selecting a small (*but nonzero*) **stepsize** h, and then using one of the following two-point formulas:

$$f'(z) \approx \frac{\Delta f(z)}{h}, \qquad \text{where } \frac{\Delta f(z)}{h} = \frac{f(z+h) - f(z)}{h} \tag{3}$$

$$f'(z) \approx \frac{\delta f(z)}{2h}, \qquad \text{where } \frac{\delta f(z)}{2h} = \frac{f(z+h) - f(z-h)}{2h} \tag{4}$$

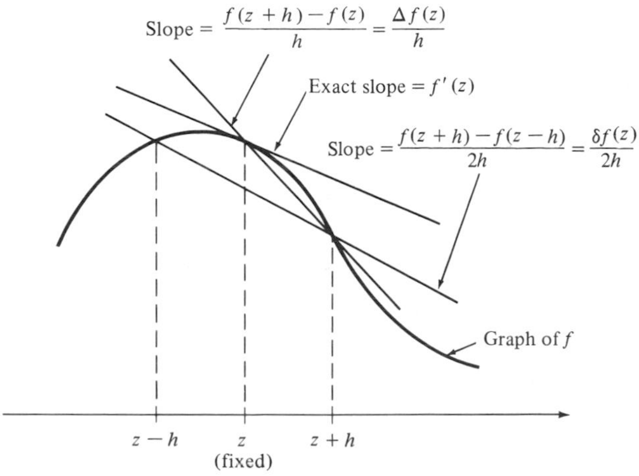

Figure 7.1-1 Geometric interpretation of $\Delta f(z)/h$ and $\delta f(z)/2h$.

The fact that $f'(z)$ is *defined* as the limit of $\Delta f(z)/h$ as $h \to 0$, makes $\Delta f(z)/h$ a reasonable formula for approximating $f'(z)$. Geometrically, $\Delta f(z)/h$ is the slope of a "forward secant" when $h \to 0$, and the slope of a "backward secant" when $h < 0$, whereas $\delta f(z)/2h$ is the slope of a "symmetric secant" whether h is positive or negative.

It is evident from Figure 7.1-1 that $\Delta f(z)/h$ *cannot* be expected to equal $f'(z)$ *unless* the graph of f is a straight line. So *$\Delta f(z)/h$ has first-degree exactness*. One can also see that $\delta f(z)/2h$ will generally approximate the exact tangent slope $f'(z)$ more accurately than $\Delta f(z)/h$. In fact, *$\delta f(z)/2h$ has quadratic exactness* [see Example 7.1A(a)].

In much the same way, approximations of the number $\int_a^b f(x)\,dx$ can be obtained from its geometric interpretation as the *net signed area between the graph of f and the interval* [*a*, *b*] *on the x-axis*, where area above [*a*, *b*] is positive, and area below it is negative. Three simple approximation formulas suggested by this are

LEFT ENDPOINT RULE: $\int_a^b f(x)\,dx \approx L]_a^b = (b - a)f(a)$ **(5)**

MIDPOINT RULE: $$\int_a^b f(x)\,dx \approx M]_a^b = (b - a)f\left(\frac{a + b}{2}\right) \qquad (6)$$

TRAPEZOIDAL RULE: $$\int_a^b f(x)\,dx \approx T]_a^b = \frac{b - a}{2}[f(a) + f(b)] \qquad (7)$$

These approximations are illustrated (for $f > 0$ on $[a, b]$) in Figure 7.1-2. Geometrically, $L]_a^b$ and $M]_a^b$ are the areas of approximating *rectangular* regions, whereas $T]_a^b$ is the area of an approximating *trapezoidal* region.

It can be seen from Figure 7.1-2 that $L]_a^b$ is guaranteed to be exact only if the graph of f is a *horizontal* line (exactness degree 0), whereas *both* $M]_a^b$ and $T]_a^b$ will be exact when the graph is *any* straight line (exactness degree 1). Formulas (3)–(7) thus illustrate the following general rule:

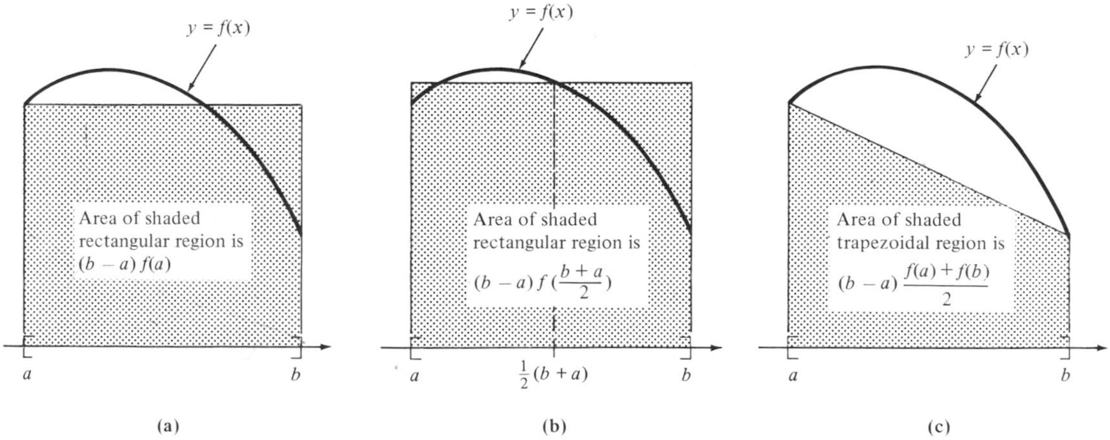

Figure 7.1-2 *(a) Left-endpoint rule. (b) Midpoint rule. (c) Trapezoidal rule.*

An approximation formula that samples $n + 1$ values of $f(x)$ can be expected to have exactness degree n, although it may actually be higher. **(8)**

EXAMPLE 7.1A Approximate as indicated, and compare the accuracy of the approximations.

(a) $d(x^2)/dx]_{x=z}$ using $\Delta f(z)/h$ and $\delta f(z)/2h$, for any z and h.
(b) $d(e^x)/dx]_{x=1}$ using $\Delta f(1)/h$ and $\delta f(1)/2h$ with $h = 0.01$.
(c) $\int_0^1 e^x\,dx$ using $L]_0^1$, $M]_0^1$ and $T]_0^1$.
(d) $\int_a^b x^2\,dx$ using $M]_a^b$ and $T]_a^b$, for any a and b.

SOLUTION (a): Here $f(x) = x^2$ and $f'(z) = 2z$ (exactly). By (3) and (4),

$$\frac{\Delta f(z)}{h} = \frac{(z + h)^2 - z^2}{h} = \frac{2zh + h^2}{h} = 2z + h \qquad (\text{error} = -h) \tag{9a}$$

$$\frac{\delta f(z)}{2h} = \frac{(z + h)^2 - (z - h)^2}{2h} = \frac{4zh}{2h} = 2z \qquad (\text{error} = 0) \tag{9b}$$

Thus, if $f(x) = x^2$, then regardless of z, $\Delta f(z)/h$ has an error proportional to h, whereas $\delta f(z)/2h$ is exactly $f'(z)$. Since $\delta f(z)/2h$ is exact when $f(x) = mx + b$, it follows from the linearity (2c) that it is exact for *any* quadratic $f(x)$.

SOLUTION (b): Here $f(x) = e^x$ and $f'(1) = e^1 \doteq 2.71828$; with $h = 0.01$,

$$\frac{\Delta f(1)}{h} = \frac{e^{1.01} - e^1}{0.01} \doteq 3.73192 \quad \text{and} \quad \frac{\delta f(1)}{2h} = \frac{e^{1.01} - e^{0.99}}{0.02} \doteq 2.71833 \tag{10}$$

The quadratic exact formula $\delta f(1)/2h$ is accurate to $5s$, whereas the first-degree exact formula $\Delta f(1)/h$ is accurate to less than $3s$. This indicates that a quadratic approximates $f(x) = e^x$ much better on $[-0.01, 0.01]$ than a straight line does on $[0, 0.01]$.

SOLUTION (c): Here $[a, b]$ is $[0, 1]$ and $\int_0^1 e^x\,dx = e^1 - e^0 \doteq 1.71828$.

$$\int_0^1 e^x\,dx \approx L]_0^1 = (1 - 0)e^0 = 1.00000 \qquad (\text{error} = 0.71833) \tag{11a}$$

$$\int_0^1 e^x\,dx \approx M]_0^1 = (1 - 0)e^{(1+0)/2} = 1.64872 \qquad (\text{error} = 0.06956) \tag{11b}$$

$$\int_0^1 e^x\,dx \approx T]_0^1 = \frac{1 - 0}{2}[e^0 + e^1] = 1.85914 \qquad (\text{error} = -0.14086) \tag{11c}$$

None of these approximations is very accurate, although $M]_a^b$ is best.

SOLUTION (d): Here $[a, b]$ is $[0, 1]$ and $\int_a^b x^2\,dx = (b^3 - a^3)/3$, whereas

$$M]_a^b = (b - a)\left(\frac{b + a}{2}\right)^2 \qquad \left[\text{error} = \frac{(b - a)^3}{12}\right] \tag{12a}$$

$$T]_a^b = \frac{b - a}{2}(b^2 + a^2) \qquad \left[\text{error} - \frac{(b - a)^3}{-6}\right] \tag{12b}$$

The errors in (12a) and (12b) were obtained by subtracting $M]_a^b$ and $T]_a^b$, respectively, from $(b^3 - a^3)/3$. Thus, for $f(x) = x^2$ and *any* interval $[a, b]$, the magnitude of the error of $T]_a^b$ will be twice as large as that of $M]_a^b$. ■

Practical Consideration 7.1A ($M]_a^b$ or $T]_a^b$?). We can only expect $M]_a^b$ and $T]_a^b$ to be accurate when the graph of f looks like a straight line over $[a, b]$. In this case, the rectangle in Figure 7.1- 2(b) generally overestimates the desired area over part of $[a, b]$, and underestimates it over the rest. This helpful cancellation generally does *not* occur for the trapezoid in Figure 7.1-2(c). Consequently, $M]_a^b$, which samples the *one midpoint* of $[a, b]$, is generally more accurate than $T]_a^b$, which samples its *two endpoints*, as we saw in (11) and (12). So, when first-degree exactness is wanted, $M]_a^b$ is preferable to $T]_a^b$ whenever the midpoint of $[a, b]$ can be sampled. ■

None of the approximations (11) is very accurate because e^x does not behave like a straight line on the interval $[0, 1]$. To approximate $\int_a^b f(x)\,dx$ with better than first-degree exactness, it is necessary to sample more than two values of $f(x)$. A general strategy for doing this for derivatives as well as integrals is described next.

7.1B Using Lagrange Polynomials to Get Polynomial Exactness

We wish to find $n + 1$ sample weights w_i, which ensure at least nth-degree exactness of an $(n + 1)$-point approximating formula of the form

$$w_0 f(x_0) + \cdots + w_n f(x_n), \qquad \text{that is, } \sum_{i=0}^{n} w_i f(x_i) \tag{13}$$

The simplest and perhaps most natural strategy for doing this is to approximate $f(x)$ by Lagrange's form of the unique interpolating polynomial for the $n + 1$ knots $P_0(x_0, f(x_0))$, $\ldots$, $P_n(x_n, f(x_n))$ on its graph, that is, to use the approximation (see Section 6.1B)

$$f(x) \approx p_{0,n}(x) = f(x_0)L_0(x) + \cdots + f(x_n)L_n(x) \tag{14a}$$

where $L_i(x)$ is the ith *Lagrange polynomial* for the $n + 1$ sampled nodes given by

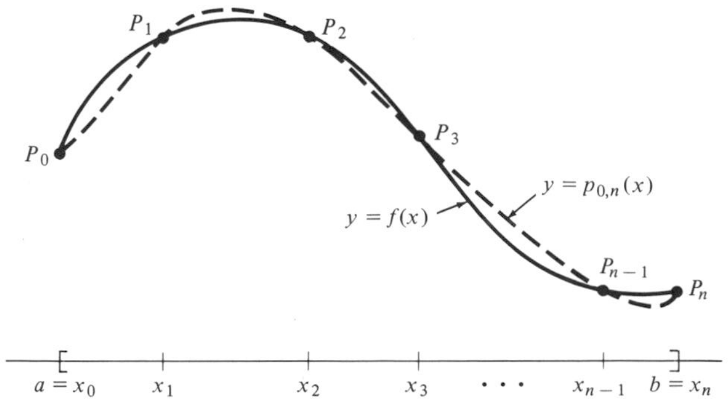

Figure 7.1-3 *Approximating $f(x)$ by $p_{0,n}(x)$.*

$$L_i(x) = \prod_{\substack{j=0 \\ j\neq i}}^{n} \frac{(x - x_j)}{(x_i - x_j)} \qquad \text{for } i = 0, 1, \ldots, n \tag{14b}$$

This strategy is illustrated graphically in Figure 7.1-3. By either differentiating (14a) k times or integrating it from a to b, we get the result we seek.

POLYNOMIAL EXACTNESS THEOREM. For any given $n + 1$ sampled nodes x_i, the following $(n + 1)$-point approximation formulas have exactness degree at least n:

$$f^{(k)}(z) \approx p_{0,n}^{(k)}(z) = \underbrace{L_0^{(k)}(z)}_{w_0} \cdot f(x_0) + \cdots + \underbrace{L_n^{(k)}(z)}_{w_n} \cdot f(x_n) \tag{15a}$$

$$\int_a^b f(x)\,dx \approx \int_a^b p_{0,n}(x)\,dx = \underbrace{\int_a^b L_0(x)\,dx}_{w_0} \cdot f(x_0) + \cdots + \underbrace{\int_a^b L_n(x)\,dx}_{w_n} \cdot f(x_n) \tag{15b}$$

The guaranteed exactness when $f(x)$ is a polynomial of degree $\leqslant n$ follows immediately from the fact that $p_{0,n}(x)$ must equal $f(x)$ in this case. (This is the uniqueness of Section 6.1C.)

The Polynomial Exactness Theorem says that nth degree exactness of (13) will be assured if we take the ith weight w_i to be the nth derivative of $L_i(x)$ at z when approximating $f^{(k)}(z)$, and the integral of $L_i(x)$ from a to b when approximating $\int_a^b f(x)\,dx$. Formulas (15) are very general and can be used whether or not the x_i's are equispaced, and whether or not z in (15a) and a and b in (15b) are sampled nodes x_i. They are also easy to use, as the following example shows.

EXAMPLE 7.1B

(a) Find sample weights w_0, w_1, and w_2 which ensure quadratic exactness when $f'(0)$, $f''(1)$, and $\int_0^1 f(x)\,dx$ are approximated by

$$w_0 f(-1) + w_1 f(0) + w_2 f(3) \qquad (\text{i.e., } x_0 = -1,\ x_1 = 0,\ x_2 = 3) \tag{16}$$

(b) Use your formulas with $f(x) = e^x$, and discuss the accuracy achieved.

SOLUTION (a): The Lagrange polynomials for the three sampled nodes in (16) are

$$L_0(x) = \frac{x(x-3)}{(-1)(-4)}, \qquad L_1(x) = \frac{(x+1)(x-3)}{(1)(-3)}, \qquad L_2(x) = \frac{(x+1)x}{(4)(3)} \tag{17a}$$

In order to use (15), we combine like powers of x to get

$$L_0(x) = \frac{x^2 - 3x}{4}, \qquad L_1(x) = \frac{x^2 - 2x - 3}{-3}, \qquad L_2(x) = \frac{x^2 + x}{12} \tag{17b}$$

By (15a), first with $k = 1$ and $z = 0$, then with $k = 2$ and $z = 1$:

$$f'(0) \approx p'_{0,2}(0) = L'_0(0)f(-1) + L'_1(0)f(0) + L'_2(0)f(3)$$
$$= \frac{-3}{4} f(-1) + \frac{2}{3} f(0) + \frac{1}{12} f(3) \quad \textbf{(18a)}$$

$$f''(1) \approx p''_{0,2}(1) = L''_0(1)f(-1) + L''_1(1)f(0) + L''_2(1)f(3)$$
$$= \frac{1}{2} f(-1) - \frac{2}{3} f(0) + \frac{1}{6} f(3) \quad \textbf{(18b)}$$

Similarly, (15b) with $a = 0$ and $b = 2$ gives $\int_0^1 f(x)\,dx \approx \int_0^1 p_{0,2}(x)\,dx$, that is,

$$\int_0^1 f(x)\,dx \approx \int_0^1 L_0(x)\,dx \cdot f(-1) + \int_0^1 L_1(x)\,dx \cdot f(0) + \int_0^1 L_2(x)\,dx \cdot f(3)$$
$$= \frac{-7}{24} f(-1) + \frac{11}{9} f(0) + \frac{5}{72} f(3) \quad \textbf{(18c)}$$

Solution (b): Taking $f(x) = e^x$ in (18a), (18b), and (18c), respectively, gives

$$f'(0) \approx \frac{-3}{4} e^{-1} + \frac{2}{3} e^0 + \frac{1}{12} e^3 \doteq 2.065 \qquad \text{(exact is } e^0 = 1) \quad \textbf{(19a)}$$
$$f''(1) \approx \frac{1}{2} e^{-1} - \frac{2}{3} e^0 + \frac{1}{6} e^3 \doteq 2.865 \qquad \text{(exact is } e \doteq 2.718) \quad \textbf{(19b)}$$
$$\int_0^1 e^x\,dx \approx \frac{-7}{24} e^{-1} + \frac{11}{9} e^0 + \frac{5}{72} e^3 \doteq 2.510 \qquad \text{(exact is } e - 1 \doteq 1.718) \quad \textbf{(19c)}$$

The reason for the poor accuracy in (19) is simply that the interpolating quadratic for $P_0(-1, e^{-1})$, $P_1(0, 1)$, and $P_2(3, e^3)$ does not approximate $f(x) = e^x$ well over the interpolation interval $[-1, 3]$ (see Figure 7.1-4). ■

It is easy to verify that the formulas (3)–(7) of Section 7.1A could have been obtained using (15) (Problem 7-4). Notice that the three-point approximation (19c) is *less* accurate than the one- and two-point estimates in (11). This illustrates the fact that *increasing the number of sampled nodes does not necessarily increase the accuracy of an approximation (15)!* Indeed, if sampled nodes must be added, they are most likely to improve the accuracy when they are close to z in (15a), or close to $[a, b]$ in (15b).

Practical Consideration 7.1B (Checking a hand calculation). Both (15a) and (15b) must be exact when $f(x) = 1$ (constant, degree 0). Consequently, for any k in (15a) and any a and b in (15b):

> In (15a), $w_0 + \cdots + w_n$ must equal $0 \left(= \frac{d^k 1}{dx^k}\right)$. **(20a)**
>
> In (15b), $w_0 + \cdots + w_n$ must equal $b - a \left(= \int_a^b 1\,dx\right)$. **(20b)**

For example, the sum $w_0 + w_1 + w_2$ is 0 in (18a) and (18b), and 1 $(= b - a)$ in (18c).

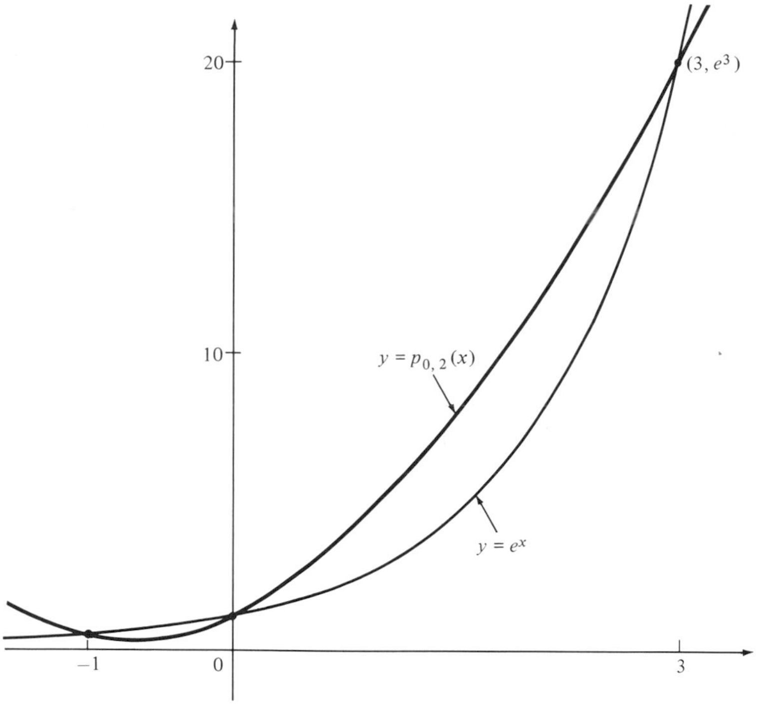

Figure 7.1-4 *$p_{0,2}(x)$ approximates e^x poorly on [−1, 3].*

When using (15a) or (15b), (20a) or (20b) can be used either as a check or as a simplified means of getting the last w_i. ■

7.1C Truncation Error Formulas

Example 7.1B demonstrates the need for error estimates for the approximations (15). The truncation error of approximating $f(x)$ by $p_{0,n}(x)$ is

$$f(x) - \sum_{i=0}^{n} w_i f(x_i) = \frac{f^{(n+1)}(\xi)}{(n+1)!} \prod_{i=0}^{n} (x - x_i) \tag{21}$$

(see Section 6.1G). So the errors of (15a) and (15b) are, respectively,

$$f^{(k)}(z) - \sum_{i=0}^{n} w_i f(x_i) = \frac{d^k}{dx^k} \left\{ \frac{f^{(n+1)}(\xi)}{(n+1)!} \prod_{i=0}^{n} (x - x_i) \right\} \Bigg]_{x=z} \tag{22a}$$

$$\int_a^b f(x)\,dx - \sum_{i=0}^{n} w_i f(x_i) = \int_a^b \frac{f^{(n+1)}(\xi)}{(n+1)!} \prod_{i=0}^{n} (x - x_i)\,dx \tag{22b}$$

where ξ lies somewhere in the smallest closed interval containing the sampled nodes x_i and either z [in (22a)] or a and b [in (22b)]. Unfortunately, formulas (22) are of limited practical value because ξ in (21) is a (generally unknown) function of x. The following theorem describes conditions under which they reduce to more usable formulas.

THEOREM 7.1C [FORMULAS FOR THE TRUNCATION ERROR OF $\Sigma_{i=0}^{n} w_i f(x_i)$].

(a) If $\Sigma_{i=0}^{n} w_i f(x_i)$ of (15a) is used to approximate $f'(z)$ when z is one of the sampled nodes, say x_j, then the truncation error of the approximation satisfies

$$f'(x_j) - \sum_{i=0}^{n} w_i f(x_i) = C\frac{f^{(n+1)}(\xi)}{(n+1)!}, \qquad \text{where } C = \frac{d}{dx}\left\{\prod_{i=0}^{n}(x - x_i)\right\}\Bigg]_{x=x_j} \tag{23a}$$

for at least one ξ in the smallest interval containing all x_i's.

(b) If the product $\Pi_{i=0}^{n} (x - x_i)$ does not change sign on $[a, b]$, then the truncation error of $\Sigma_{i=0}^{n} w_i f(x_i)$ of (15b) satisfies

$$\int_a^b f(x)\,dx - \sum_{i=0}^{n} w_i f(x_i) = C\frac{f^{(n+1)}(\hat{\xi})}{(n+1)!}, \qquad \text{where } C = \int_a^b \prod_{i=0}^{n}(x - x_i)\,dx \tag{23b}$$

for at least one $\hat{\xi}$ in the smallest interval containing a, b, and all x_i's.

(c) The constant C in either (23a) and (23b) can be obtained as the exact error of the approximation when $f(x) = x^{n+1}$, that is, as

$$\underbrace{(n+1)x_j^n - \sum_{i=0}^{n} w_i x_i^{n+1}}_{C \text{ in (23a)}} \quad \text{or} \quad \underbrace{\frac{b^{n+2} - a^{n+2}}{n+2} - \sum_{i=0}^{n} w_i x_i^{n+1}}_{C \text{ in (23b)}} \tag{23c}$$

Formula (23a) is obtained by applying the product rule for derivatives to (22a) and noting that $\Pi_{i=0}^{n} (x_j - x_i) = 0$; formula (23b) is a direct application of the Integral Mean Value Theorem [see Appendix II.4C]; and (23c) is true because when $f(x) = x^{n+1}$, $f^{(n+1)}(x) = (n+1)!$ (constant), hence $f^{(n+1)}(\xi)/(n+1)! = 1$ *regardless of the value of* ξ in (23a) or (23b). The details are left as an exercise (Problem M7-4).

Theorem 7.1C can provide analytic insight into the accuracy of a formula of the form (22). For example, the approximation $\Delta f(z)/h$ samples $x_0 = z$ and $x_1 = z + h$ $(n = 1)$ and [see (12a)] has an error of $-h$ $(= C)$ when $f(x) = x^2$. So, by (23a) and (23c),

$$f'(z) - \frac{\Delta f(z)}{h} = -\frac{f''(\xi)}{2}h \qquad \text{for some } \xi \text{ between } z \text{ and } z + h \tag{24}$$

Thus the error shrinks approximately in proportion to h, with constant of proportionality $-f''(\xi)/2$. So, for a given h, the formula works best when $|f''(x)| \approx 0$ (i.e., when the graph is nearly a straight line) near z. Similarly, the approximation $\int_a^b f(x)\,dx \approx T]_a^b$ samples

$x_0 = a$ and $x_1 = b$ ($n = 1$), and [see (12b)] has an error of $-(b - a)^3/6$ when $f(x) = x^2$. Since $\Pi_{i=0}^{1} = (x - a)(x - b)$ stays nonnegative on $[a, b]$, we can conclude from (23b) and (23c) that

$$\int_a^b f(x)\,dx - T]_a^b = \frac{(b - a)^3}{-12} f''(\xi) \qquad \text{for some } \xi \text{ between } a \text{ and } b \tag{25}$$

Theorem 7.1C can also be used to obtain truncation error bounds. Consider, for example, the three-point ($n = 2$) approximations in (19), for which

$$\prod_{i=0}^{2} (x - x_i) = (x + 1)x(x - 3) = x^3 - 2x^2 - 3x \quad \text{and} \quad f(x) = e^x \tag{26a}$$

hence $f^{(3)}(x) = e^x$. Since $z = 0$ is x_1, we can apply (23a) to the approximation of $f'(0) = e^0$ in (19a) to get

$$f'(0) - 2.065 = \frac{e^\xi}{3!}\cdot\frac{d}{dx}(x^3 - 2x^2 - 3x)\Big]_{x=0} = \frac{e^\xi}{3!}(3x^2 - 4x - 3)\Big]_{x=0} = \frac{-e^\xi}{2} \tag{26b}$$

where $-1 < \xi < 3$. Unfortunately, the best we can say from (26b) is that the error of (19a) is between $-e^{-1}/2 \doteq -0.184$ and $-e^3/2 \doteq -10.04$, which is a rather imprecise estimate. Similarly, since $\Pi_{i=0}^{2} (x - x_i) \le 0$ on $[0, 1]$, we can apply (23b) to the approximation of $\int_0^1 e^x\,dx$ in (19c) to get

$$\int_0^1 e^x\,dx - 2.510 = \frac{e^{\hat{\xi}}}{3!}\int_0^1 x^3 - 2x^2 - 3x\,dx = \frac{-13}{12} e^{\hat{\xi}} \tag{26c}$$

Since $\hat{\xi}$ also lies between -1 and 3, (26c) is even less precise than (26b)!

We see from (26) that the truncation error estimates (23) do not necessarily give precise error bounds. In the following sections, we shall show how Taylor series can be used to get more accurate approximations of $f^{(k)}(z)$ and $\int_a^b f(x)\,dx$ as well as more precise estimates of their truncation errors.

7.2 Numerical Differentiation and Richardson's Formula

The approach to be taken in this section is motivated by the two fundamental definitions of calculus, namely

$$f'(z) = \lim_{h \to 0} \frac{\Delta f(z)}{h}, \qquad \text{where } \frac{\Delta f(z)}{h} = \frac{f(z + h) - f(z)}{h} \tag{1a}$$

$$\int_a^b f(x)\,dx = \lim_{h \to 0} R[h], \qquad \text{where } R[h] = h \cdot \sum_{i=0}^{n-1} f(x_i) \tag{1b}$$

(see Appendix II.2A and II.4A). Both $f'(z)$ and $\int_a^b f(x)\,dx$ are special cases of a numerical quantity Q that is either defined or can be expressed as

$$Q = \lim_{h \to 0} F[h], \qquad \text{where } F[h] \text{ is a formula for approximating } Q \tag{2a}$$

In the absence of roundoff error, the value of such a Q can be found to any desired accuracy using the approximation

$$Q \approx F[h] \text{ for a sufficiently small } (\textit{but nonzero})\ h \tag{2b}$$

The error of this approximation is called the **truncation error** (or **discretization error**) of $F[h]$ and denoted by $\tau[h]$. Thus

$$\tau[h] = Q - F[h] = \text{the truncation error of } F[h] \tag{2c}$$

We will consistently use square brackets for approximation formulas that depend on h (e.g., $R[h]$, $F[h]$) to distinguish them from the functions $f(x)$ for which they are used. Also, notation such as $F[0.2]$ will be used as a convenient way to denote the value of the formula when $h = 0.2$.

7.2A Convergence of $\Delta f(z)/h$ and $\delta f(z)/2h$

We saw in Section 7.1A that $Q = f'(z)$ can be approximated by either $F[h] = \Delta f(z)/h$ or $F[h] = \delta f(z)/2h$, where

$$\frac{\Delta f(z)}{h} = \frac{f(z+h) - f(z)}{h} \quad \text{and} \quad \frac{\delta f(z)}{2h} = \frac{f(z+h) - f(z-h)}{2h} \tag{3}$$

We also saw that for a given small (but nonzero) h, $\delta f(z)/2h$ is generally more accurate than $\Delta f(z)/h$. The following example will enable us draw some quantitative conclusions about how quickly their accuracy *improves* (in the absence of roundoff error) as h shrinks to zero.

EXAMPLE 7.2A Let $f(x) = e^x$ and take $z = 1$. Use both $\Delta f(z)/h$ and $\delta f(z)/2h$ with stepsizes $h = \pm 0.2, \pm 0.02, \pm 0.01$, and ± 0.002 to approximate $f'(z) = e^1 = 2.7182818$ ($8s$), and discuss the accuracy obtained.

SOLUTION: When $h = 0.2$, the values of $F[h]$ and $\tau[h]$, evaluated in extended precision and rounded to $8s$, are

$$\frac{\Delta f(z)}{h} = \frac{e^{1.2} - e^1}{0.2} \doteq 3.0091755 \quad \text{and} \quad \tau[h] = f'(z) - \frac{\Delta f(z)}{h} = -0.2908936 \tag{4a}$$

$$\frac{\delta f(z)}{2h} = \frac{e^{1.2} - e^{0.8}}{0.4} \doteq 2.7364400 \quad \text{and} \quad \tau[h] = f'(z) - \frac{\delta f(z)}{2h} = -0.0181582 \tag{4b}$$

The other entries of Table 7.2-1 are obtained similarly. Underlined digits are those that would be in error when rounded. Notice that the value of $\delta f(z)/2h$ is the same for h and

TABLE 7.2-1 Approximating $f'(z) = e$ by $\Delta f(z)/h$ and $\delta f(z)/2h$

h	$F[h]$ $\Delta f(z)/h$	$\tau[h]$ $e - \Delta f(z)/h$	$F[h]$ $\delta f(z)/2h$	$\tau[h]$ $e - \delta f(z)/2h$
0.2	3.0091755	−0.2908936	2.7364400	−0.0181582
0.02	2.7456468	−0.0273649	2.7184631	−0.0001813
0.01	2.7319187	−0.0136368	2.7183271	−0.0000453
0.002	2.7210019	−0.0027201	2.7182836	−0.0000018
−0.002	2.7155654	−0.0027165	2.7182836	−0.0000018
−0.02	2.7047356	−0.0135462	2.7183271	−0.0000453
−0.01	2.6912793	−0.0270025	2.7184631	−0.0001813
−0.2	2.4637045	−0.2545773	2.7364400	−0.0181582

$-h$. As expected, both $\Delta f(z)/h$ and $\delta f(z)/2h$ become more accurate as $|h|$ gets smaller, and $\delta f(z)/2h$ is more accurate than $\Delta f(z)/h$ for all values of h. Indeed, the most accurate value of $\Delta f(z)/h$ (when $h = -0.002$) is accurate to only 3s, as compared to the almost 7s accuracy of $\delta f(z)/2h = 2.7182836$.

■

7.2B The Order of an Approximation; "Big O" Notation

Let us examine *quantitatively* the rates at which the errors in Table 7.2-1 get smaller as $h \to 0$. From the $h = 0.2$ and $h = 0.02$ entries, we find that

$$\text{For } F[h] = \frac{\Delta f(z)}{h}: \quad \frac{\tau[0.2]}{\tau[0.02]} \approx \frac{-0.2908936}{-0.0273649} \approx 10 = \left(\frac{0.2}{0.02}\right)^1 \tag{5a}$$

$$\text{For } F[h] = \frac{\delta f(z)}{2h}: \quad \frac{\tau[0.2]}{\tau[0.02]} \approx \frac{-0.0181582}{-0.0001813} \approx 100 = \left(\frac{0.2}{0.02}\right)^2 \tag{5b}$$

In fact, as you can verify, the following holds for *all* of the table's "stepsize ratios" $r = 0.2/0.02 = 10$, $r = 0.02/0.01 = 2$, or $r = 0.01/0.002 = 5$:

$$\text{For } F[h] = \frac{\Delta f(z)}{h}: \quad \frac{\tau[rh]}{\tau[h]} \approx r \quad \text{so that } \tau[rh] \approx r^1\tau[h] \tag{6a}$$

$$\text{For } F[h] = \frac{\delta f(z)}{2h}: \quad \frac{\tau[rh]}{\tau[h]} \approx r^2 \quad \text{so that } \tau[rh] \approx r^2\tau[h] \tag{6b}$$

In words, reducing h by a factor of r reduces the error of $\Delta f(z)/h$ by a factor of about r; but it reduces the error of $\delta f(z)/2h$ by a factor of about r^2. So $\delta f(z)/2h$ *is not only more* *accurate than $\Delta f(z)/h$ for a given h, but its accuracy improves more rapidly as h approaches zero.*

Notation (Rate of Convergence of $F[h]$ to Q) Suppose $F[h] \to Q$ as $h \to 0$, and $\tau[h] = Q - F[h]$ is the truncation error of the approximation $Q \approx F[h]$. We call $\tau[h]$ an ***n*th-order truncation error** and write

$$\tau[h] = O(h^n) \qquad \text{if for any } r,\ \tau[rh] \approx r^n \tau[h] \text{ for } h \approx 0 \tag{7}$$

that is, if for small h the error of $F[rh]$ is about r^n times that of $F[h]$. In this case, we shall say that $F[h]$ is an ***n*th-order** [abbreviated $\boldsymbol{O(h^n)}$] **approximation** of Q. In view of (5), $\Delta f(z)/h$ will henceforth be referred to as the **$O(h)$ forward/backward difference approximation of $f'(z)$**, and $\delta f(z)/2h$ as the **$O(h^2)$ central difference approximation of $f'(z)$**.

The effect of multiplying, adding, and scaling truncation errors of known order is summarized in the following theorem (Problem M7-5):

THEOREM 7.2B. If $\tau_n[h] = O(h^n)$ and $\tau_m[h] = O(h^m)$, then

$$\tau_n[h]\tau_m[h] = O(h^{n+m}); \text{ in particular } h^n\tau_m[h] = O(h^{n+m}) \tag{8a}$$

$$\text{If } ab \neq 0, \text{ then } a\tau_n[h] + b\tau_m[h] = O(h^{\text{smaller of } m \text{ and } n}) \tag{8b}$$

$$\tau_n[rh] = O(h^n) \text{ (the order of } \tau_n[h]) \quad \text{for any fixed } r \tag{8c}$$

The simplest analytic tool for determining the order of $\tau[h]$ is the following:

$$\text{If } \tau[h] \approx Ch^n \ (C = \text{constant}) \text{ for small } h, \qquad \text{then } \tau[h] = O(h^n) \tag{9}$$

Indeed, if $\tau[h] \approx Ch^n$, then $\tau[rh] \approx C(rh)^n \approx r^n \cdot Ch^n \approx r^n \cdot \tau[h]$, that is, (7) holds. The C in (9) will be called the **convergence constant** of $\tau[h]$. It can be seen from Figure 7.2-1 that *the larger the order n, the faster the formula* $F[h]$ *approaches* Q *as* $h \to 0$. This makes higher-order formulas more desirable than lower-order ones. And if two approximations are of the same order, the one with the smaller convergence constant should be more accurate for a given (small) h.

When $F[h]$ approximates Q where Q is either a derivative or integral of $f(x)$, the order of $\tau[h]$ is usually found by assuming that f has a Taylor series representation

$$f(x + h) = f(x) + hf'(x) + \frac{h^2}{2!}f''(x) + \frac{h^3}{3!}f'''(x) + \frac{h^4}{4!}f^{(\text{iv})}(x) + \cdots \tag{10a}$$

which converges for values of h sufficiently close to zero, and then using this series to express Q as

$$Q = F[h] + (Ch^n + Dh^m + Eh^p + \cdots), \qquad \text{where } n < m < p < \cdots \tag{10b}$$

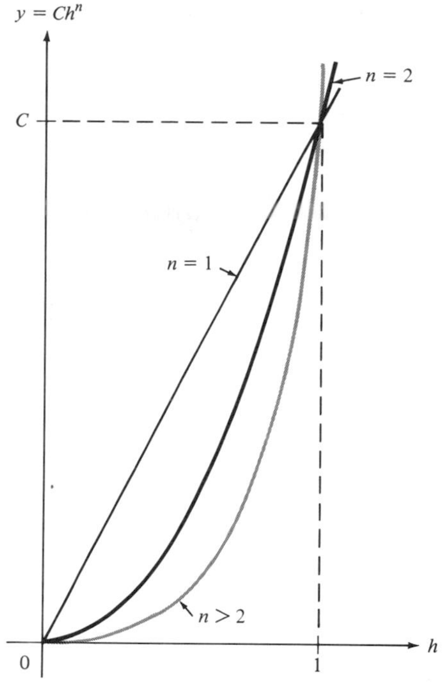

Figure 7.2-1 *Ch^n versus h for C > 0.*

Since $Q = F[h] + \tau[h]$, the Maclaurin series in parentheses must satisfy

$$\tau[h] = Ch^n + Dh^m + Eh^p + \cdots \qquad \text{so that } \tau[h] \approx Ch^n \text{ for } h \approx 0 \tag{10c}$$

(see Appendix II.3). Finally, combining (9) with (10c) gives:

> *If the leading term of the Maclaurin series for $\tau[h]$ is Ch^n, then $F[h]$ is an* n*th-order approximation of Q.* **(11)**

Although the ideas discussed in (7)–(11) may seem somewhat theoretical, they provide the basis for the powerful *practical* procedures that follow. So, it is advisable to *review them carefully before proceeding.*

7.2C Truncation Errors of $\Delta f(z)/h$, $\delta f(z)/2h$, and $\delta^2 f(z)/h^2$

We saw *empirically* in (5a) that $\Delta f(z)/h$ is an $O(h)$ approximation of $f'(z)$. To show this *analytically*, we simply solve the Taylor series

$$f(z + h) = f(z) + hf'(z) + \frac{h^2}{2!} f''(z) + \frac{h^3}{3!} f'''(z) + \frac{h^4}{4!} f^{(iv)}(z) + \cdots \tag{12a}$$

for $f'(z)$ by subtracting $f(z)$ from both sides, dividing by h, and then transposing all terms except $f'(z)$ to one side to get

$$\underbrace{f'(z)}_{Q} = \underbrace{\frac{f(z+h)-f(z)}{h}}_{F[h] = \frac{\Delta f(z)}{h}} \underbrace{- \frac{f''(z)}{2}h - \frac{f'''(z)}{6}h^2 - \frac{f^{(iv)}(z)}{24}h^3 + \cdots}_{\tau[h] = f'(z) - \frac{\Delta f(z)}{h}} \tag{12b}$$

Since the leading term of $\tau[h]$ is Ch^1 $(n = 1)$, we conclude from (11) that $F[h] = \Delta f(z)/h$ is an $O(h)$ approximation of $f'(z)$.

Similarly, to show analytically that $\delta f(z)/2h$ is a second-order approximation of $f'(z)$, first replace h by $-h$ in (12a) to get

$$f(z-h) = f(z) - hf'(z) + \frac{h^2}{2!}f''(z) - \frac{h^3}{3!}f'''(z) + \frac{h^4}{4!}f^{(iv)}(z) - \cdots \tag{13a}$$

then subtract (13a) from (12a), divide by $2h$, and transpose as in (12) to get

$$\underbrace{f'(z)}_{Q} = \underbrace{\frac{f(z+h)-f(z-h)}{2h}}_{F[h] = \frac{\delta f(z)}{2h}} \underbrace{- \frac{f'''(z)}{6}h^2 - \frac{f^{(v)}(z)}{120}h^4 - \frac{f^{(vii)}(z)}{5040}h^6 - \cdots}_{\tau[h] = f'(z) - \frac{\delta f(z)}{2h}} \tag{13b}$$

Since the leading term in the Maclaurin series for $\tau[h]$ is Ch^2 $(n = 2)$, we conclude from (11) that $\delta f(z)/2h$ is an $O(h^2)$ approximation of $f'(z)$. Alternatively, *adding* (13a) to (12a) and then solving for $f''(z)$ gives

$$f''(z) \approx \frac{f(z-h) - 2f(z) + f(z+h)}{h^2} + \left[\frac{-f^{(iv)}(z)}{12}h^2 - \frac{f^{(vi)}(z)_4}{360} - \cdots\right] \tag{13c}$$

Since Ch^2 is the leading term of the Maclaurin series for $\tau[h]$,

$$\frac{\delta^2 f(z)}{h^2} = \frac{f(z-h) - 2f(z) + f(z+h)}{h^2} \text{ is an } O(h^2) \text{ approximation of } f''(z) \tag{14}$$

We shall call $\delta^2 f(z)/h^2$ the **$O(h^2)$ central difference approximation of $f''(z)$**.

EXAMPLE 7.2C Use $\delta^2 f(z)/h^2$ with $h = 0.2$, 0.02, 0.01, and 0.002 to approximate $d^2(e^x)/dx^2]_{x=1}$. Verify empirically that $\tau[h] = O(h^2)$.

Solution: Here $f(x) = e^x$, $z = 1$, and $Q = f''(z) = e^1 = 2.7182818$ $(8s)$. Taking $h = 0.2$ in $F[h] = \delta^2 f(z)/h^2$ gives

$$F[0.2] = \frac{e^{0.8} - 2e^1 + e^{1.2}}{(0.2)^2} \doteq 2.72736 \quad \text{and} \quad \tau[0.2] = e - F[0.2] = 0.00907 \tag{15}$$

The remaining $F[h]$ and $\tau[h]$ values, obtained to 13s and entered to 8s, are given in Table 7.2-2, where it is shown that $\tau[rh] \approx r^2\tau[h]$; so $\tau[h] = O(h^2)$.

TABLE 7.2-2 Approximating $f''(z) = e$ by $\delta^2 f(z)/h^2$

	h	$F[h] = \delta^2 f(x)/h^2$	Error $= \tau[h] = e - \delta^2 f(x)/h^2$	
$r = 10$	0.2	2.7273549	−0.0090730	$\tau[0.2] \approx 10^2\,\tau[0.02]$
$r = 2$	0.02	2.7183724	−0.0000906	$\tau[0.02] \approx 2^2\,\tau[0.01]$
$r = 5$	0.01	2.7183044	−0.0000226	$\tau[0.01] \approx 5^2\,\tau[0.002]$
	0.002	2.7182827	−0.0000009	

An approximation formula $F[h]$ will be called **exact** if $F[h]$ is exactly Q, that is, if $\tau[h] = 0$. Since the lowest-order derivatives appearing in the series for the truncation errors for $\delta f(z)/2h$ and $\delta^2 f(z)/h^2$ are, respectively, $f'''(z)$ [see (13b)] and $f^{(iv)}(z)$ [see (13c)],

$$\frac{\delta f(z)}{2h} \text{ is exact for quadratics, } \quad \text{and} \quad \frac{\delta^2 f(z)}{h^2} \text{ is exact for cubics} \tag{16}$$

7.2D The Stepsize Dilemma

At this point we assume that the reader has carefully read Section 1.3C, which describes the effects of roundoff error on the calculation of $\Delta f(z)/h$. We saw there that any digital device will calculate meaningless values if h is allowed to get too small. In such a situation, the actual error is the net effect of truncation error and roundoff error, that is,

$$\underbrace{\text{actual error}}_{\epsilon[h]} = \underbrace{\text{truncation error}}_{\tau[h]} + \underbrace{\text{roundoff error}}_{\rho[h]} \tag{17}$$

where $\tau[h]$ shrinks to zero (at a rate that depends only on the *approximation formula* $F[h]$), whereas $\rho[h]$ tends to get large (at a rate that depends on both *the formula and the device*) as $h \to 0$. This is illustrated in Figure 7.2-2, which shows that $\rho[h]$ rapidly becomes the dominant term in (17) if h is allowed to get smaller a "threshold stepsize," $h_{\tau=\rho}$. We are thus faced with the challenge of finding a stepsize small enough to make $\tau[h]$ acceptably small, yet large enough to avoid having the calculated $F[h]$ contaminated by roundoff error. This difficult challenge, which we will refer to as the **Stepsize Dilemma**, must be faced whenever we try to approximate *any* derivative of f at z using a formula that samples values of f at or near z.

The most important conclusion to be drawn from the preceding discussion is that *for any given device, there will be a limit to the accuracy that can be achieved using a particular formula* $F[h]$. This maximum accuracy corresponds to the minimum $\epsilon[h]$, denoted by ϵ_{min}

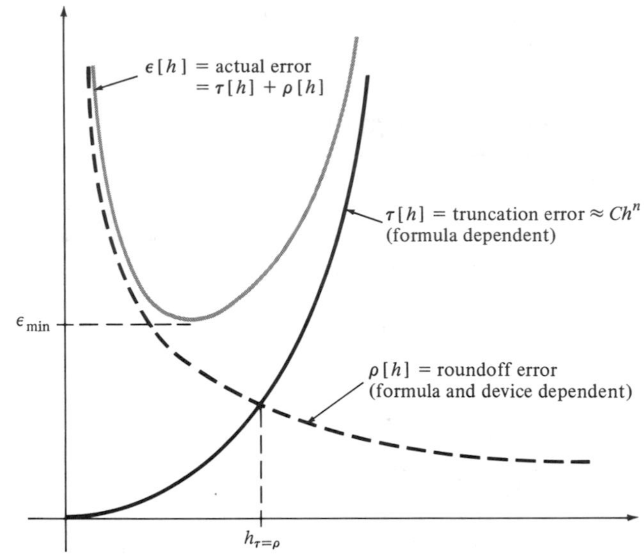

Figure 7.2-2 *Actual error = truncation error + roundoff error.*

in Figure 7.2-2. If the most accurate estimate is not acceptable, we can try one of the following:

Remedy 1: Use a higher-order formula (i.e., increase n). This will reduce ϵ_{min} by lowering the $\tau[h]$ curve in Figure 7.2-2.

Remedy 2: Use more significant digits in the calculation. This will reduce ϵ_{min} by lowering the $\rho[h]$ curve in Figure 7.2-2.

Unfortunately, *both remedies require repeating the calculation*. And Remedy 2 may require using a different digital device as well! That's the bad news. The good news is that there is an "improvement formula" similar to Aitken's formula of Section 2.1F which, if used correctly, enables you to get more accurate estimates of Q from the *previously calculated estimates*. This alternative, which should be tried before either Remedy 1 or Remedy 2, will be described in the next section. Before doing so, however, consider the following example, which illustrates the problem quantitatively.

EXAMPLE 7.2D Approximate $d(e^x)/dx]_{x=1}$ by both $\Delta f(z)/h$ and $\delta f(z)/2h$ with $h = 0.2, 0.02,$ $0.1, 0.002, 0.0005$, and 0.0002. Start with values of e^x rounded to $7s$, but do all intermediate calculations in extended precision. How accurate are the best of these approximations? The exact answer is $e^1 = 2.7182818$ $(8s)$.

Solution: The results for $\Delta f(z)/h$ and $\delta f(z)/2h$ are shown, respectively, in Tables 7.2-3 and 7.2-4. The underlined digits indicate the presence of roundoff error (see Table 7.2-1). Notice that the error estimate $\epsilon[h] \approx Ch^n$ gets better as $|h|$ decreases, until $h \approx 0.002$ in Table 7.2-3 and $h \approx 0.02$ in Table 7.2-4 (i.e., $h \approx h_{\tau=\rho}$). For smaller $|h|$, roundoff error exceeds the truncation error. Also, although we started with $7s$ accuracy, we can only approximate $d(e^x)/dx]_{x=1}$ to about $4s$ using $\Delta f(z)/h$ (with $h \approx 0.0005$), and to about $5s$ using $\delta f(z)/2h$ (with $h \approx 0.002$). And ϵ_{min} for these most accurate approximations consists mostly of roundoff error! (See Problems M7-6 and M7-7.) ■

TABLE 7.2-3 Approximating $d(e^x)/dx|_{x=1}$ by $\Delta f(z)/h$ on a 7s Device

h	$F[h]$ $\Delta f(z)/h$	Ch^n in (12b) $-f'(z)h/2 = -eh/2$	$\epsilon[h]$ $e - \Delta f(z)/h$	
0.2	3.009175	−0.271828	−0.290893	$\epsilon[h] \approx \tau[h]$
0.02	2.745650	−0.027183	−0.027368	
0.01	2.731900	−0.013591	−0.013618	
0.002	2.721000	−0.002718	−0.002718	
0.0005	2.718000	−0.000679	−0.000282	$\epsilon[h] \approx \rho[h]$
0.0002	2.719997	−0.000272	−0.001715	

TABLE 7.2-4 Approximating $d(e^x)/dx|_{x=1}$ by $\delta f(z)/2h$ on a 7s Device

h	$F[h]$ $\delta f(z)/2h$	Ch^n in (13b) $-f''(z)h^2/6 = -eh^2/6$	$\epsilon[h]$ $e - \delta f(z)/2h$	
0.2	2.736440	−0.018122	−0.018158	$\epsilon[h] \approx \tau[h]$
0.02	2.718475	−0.000181	−0.000193	
0.01	2.718350	−0.000045	−0.000068	$\epsilon[h] \approx \rho[h]$
0.002	2.718250	−0.000002	−0.000032	
0.0005	2.718000	−0.000000	−0.000282	
0.0002	2.720000	−0.000000	−0.001718	

7.2E Richardson's Improvement Formula

Consider again the general approximation of an exact quantity Q by an nth-order formula $F[h]$. Suppose we know that in fact†

$$Q = F[h] + Ch^n + O(h^m) \qquad \text{so that } \tau[h] = Ch^n + O(h^m) \tag{18a}$$

where $C \neq 0$ and $n < m$, and that we have used the formula twice to get two approximations, which we denote by

$$F[h] \text{ and } F[h_{\text{larger}}], \qquad \text{where } h_{\text{larger}} \text{ is the larger stepsize} \tag{18b}$$

Since $F[h] \to Q$ as $h \to 0$, $F[h]$ should be more accurate than $F[h_{\text{larger}}]$. In this situation, it is useful to introduce the **stepsize ratio** r, where

$$r = \frac{h_{\text{larger}}}{h} \qquad \text{so that } h_{\text{larger}} = rh \text{ where } r > 1 \tag{18c}$$

Replacing h by $rh = h_{\text{larger}}$ in (18a), and using (9c), gives

$$Q = F[rh] + r^nCh^n + O(h^m), \qquad \text{that is, } \tau[rh] = r^nCh^n + O(h^m) \tag{18d}$$

Subtracting (18d) from $r^n \cdot$(18a) eliminates the leading Ch^n terms of the truncation errors $\tau[h]$ and $\tau[rh]$, and leaves

†In (18a) and elsewhere, "$+ O(h^m)$" should be read as "plus terms of degree at least m."

$$r^nQ - Q = r^nF[h] - F[rh] + O(h^m) \quad \text{or} \quad Q = \frac{r^nF[h] - F[rh]}{r^n - 1} + O(h^m) \quad \textbf{(18e)}$$

[see (9a) and (9b)], thus proving the following important result.

RICHARDSON'S IMPROVEMENT THEOREM. If $F[h]$ is an nth-order approximation of Q, which we have used twice to get the approximations

$$F[h] \text{ and } F[h_{\text{larger}}], \qquad \text{where } \frac{h_{\text{larger}}}{h} = r > 1$$

then an improved approximation of Q can be obtained as

$$F_1[h] = \frac{r^nF[h] - F[rh]}{r^n - 1}, \qquad \text{where } rh = h_{\text{larger}} \qquad \textbf{(19a)}$$

Moreover, if the truncation of $F[h]$ actually satisfies $\tau[h] = Ch^n + Dh^m + O(h^p)$, where $n < m < p$, then $F_1[h]$ will be an $O(h^m)$ approximation of Q, so any two approximations $F_1[h]$ and $F_1[rh]$ can be used to get

$$F_2[h] = \frac{r^mF_1[h] - F_1[rh]}{r^m - 1}, \qquad \text{an } O(h^p) \text{ approximation of } Q \qquad \textbf{(19b)}$$

Formula (19a) will be referred to as **Richardson's Improvement Formula.**† It tells us to weight the more accurate approximation, $F[h]$, by $r^n/(r^n - 1)$ and the less accurate one, $F[h_{\text{larger}}]$, by $-1/(r^n - 1)$, where *n is the order of the formula being improved* and *r is the stepsize ratio of the two values being improved.* If you remember this, then (19b) is simply a special case of (19a). Note that the order of $F_1[h]$ comes from the term *after Ch^n* in the truncation error for the formula being improved.

The only thing you must know to use (19a) is the order, n, of the formula being improved. *You do not have to know m or p.* For example, if $F[h]$ is an $O(h^3)$ approximation of Q ($n = 3$) and you have values of $F[0.2]$ and $F[0.8]$, so that $r = (0.8)/(0.2) = 4$, then an improved (at least fourth-order) approximation of Q is

$$F_1[0.2] = \frac{4^3F[0.2] - F[0.8]}{4^3 - 1} = \frac{64F[0.2] - F[0.8]}{63} \qquad \textbf{(20)}$$

However, if *several* terms of the series for $\tau[h]$ are known, then the ability to "improve the improvement" as in (19b) can (*and should!*) be used to get more accuracy than that of the values of $F[h]$ used in (19a), as shown in the next section.

7.2F Using Richardson's Formula Iteratively

We saw in (12b) that the $O(h)$ forward/backward difference approximation $f'(z) \approx \Delta f(z)/h$ satisfies

†The use of (19a) is also referred to as **Richardson extrapolation** or the **deferred approach to the limit**.

$$\underbrace{f'(z)}_{Q} = \underbrace{\Delta f(z)/h}_{F[h]} + \underbrace{Ch + Dh^2 + Eh^3 + \cdots}_{\tau[h]} \tag{21}$$

So Richardson's theorem can be used with $n = 1$, $m = 2$, $p = 3$, ... to improve values of $\Delta f(z)/h$. It is convenient to get the improved values $F_1[h]$, $F_2[h]$, $F_3[h]$, ... *after calculating* $F[h]$ and entering them to the right of $F[h]$ in a **Richardson table** as shown for the $\Delta f(z)/h$ values of Table 7.2-3 in Table 7.2-5. The $F_1[h]$, $F_2[h]$, and $F_3[h]$ values were obtained as

FORMULA	TYPICAL CALCULATION (TO 8s, ONE GUARD DIGIT)	
$F_1[h] = \dfrac{r^1F[h] - F[rh]}{r^1 - 1}$	$F_1[0.02] = \dfrac{10^1(2.745650) - 3.009175}{10^1 - 1} = 2.7163694$	**(22a)**
$F_2[h] = \dfrac{r^2F_1[h] - F_1[rh]}{r^2 - 1}$	$F_2[0.01] = \dfrac{2^2(2.7181500) - 2.7163694}{2^2 - 1} = 2.7187435$	**(22b)**
$F_3[h] = \dfrac{r^3F_2[h] - F_2[rh]}{r^3 - 1}$	$F_3[0.002] = \dfrac{5^3(2.7182802) - 2.7187435}{5^3 - 1} = 2.7182765$	**(22c)**

TABLE 7.2-5 Richardson Table for Improving $F[h] = \Delta f(z)/h$; $f'(z) \doteq 2.7182818$

	h	$F[h]$, $O(h)$ $\Delta f(z)/h$	$F_1[h]$, (19a) $\tau[h] = O(h^2)$	$F_2[h]$, (19b) $\tau[h] = O(h^3)$	$F_3[h]$, (19b) $\tau[h] = O(h^4)$
	0.2	3.009175			
$r = 10$					
	0.02	2.745650	2.7163694		
$r = 2$					
	0.01	2.731900	2.7181500	2.7187435	
$r = 5$					
	0.002	2.721000	2.7182750	2.7182802	2.7182765
$r = 4$					
	0.005	2.718000	2.7170000	2.7169150	2.7168933

Similarly, we saw in (13b) that the $O(h^2)$ central difference approximation $f'(z) \approx \delta f(z)/2h$ satisfies

$$\underbrace{f'(z)}_{Q} = \underbrace{\delta f(z)/2h}_{F[h]} + \underbrace{Ch^2 + Dh^4 + Eh^6 + \cdots}_{\tau[h]} \tag{23}$$

So Richardson's theorem can be used to improve values of $\delta f(z)/2h$. The Richardson table for the $\delta f(z)/2h$ values of Table 7.2-4 (before serious roundoff error was evident) is shown in Table 7.2-6. This time $n = 2$, $m = 4$, $p = 6$, ..., so the values of $F_1[h]$, $F_2[h]$, $F_3[h]$ were calculated as

$$F_1[h] = \frac{r^2F[h] - F[rh]}{r^2 - 1}, \qquad F_2[h] = \frac{r^4F_1[h] - F_1[rh]}{r^4 - 1}, \qquad F_3[h] = \frac{r^6F_2[h] - F_2[rh]}{r^6 - 1} \tag{24}$$

with $r = 10$, 2, 5, or 4 as needed.

TABLE 7.2-6 Richardson Table for Improving $F[h] = \delta f(z)/2h$; $f'(z) \doteq 2.7182818$

h		$\delta f(z)/2h$	$F_1[h]$, $O(h^4)$	$F_2[h]$, $O(h^6)$	$F_3[h]$, $O(h^8)$
	0.2	2.736440			
$r = 10$					
	0.02	2.718475	2.7182935		
$r = 2$					
	0.01	2.718350	2.7183083	2.7183093	
$r = 5$					
	0.002	2.718250	2.7182458	2.7182457	2.718247

The most accurate entry of Table 7.2-5 is $F_2[0.002] = 2.7182802$; that of Table 7.2-6 is $F_1[0.02] = 2.7182935$. These did *not* occur when h was smallest. Also, the most accurate value of all, $F_2[0.002] = 2.7182802$, is on the table for the lower-order formula $\Delta f(z)/h$. Why? Simply because $F[0.002] = 2.721$ managed to get calculated with negligible roundoff error (in its sixth decimal place). In fact, the most accurate entries of *both* Tables 7.2-5 and 7.2-6 are in the row of the $F[h]$ for which Ch^n best approximates $\epsilon[h]$ (see Tables 7.2-3 and 7.2-4), *not* the row with the most accurate $F[h]$.

Practical Consideration 7.2F (Using a Richardson table effectively). The derivation (18) of Richardson's formula was based solely on the behavior of the truncation error $\tau[h]$, with no provision for roundoff error. Indeed, since formula (19a) "weights" $F[h]$ r^n times as much as $F[rh]$, *any error in* $F[h]$ *will simply propagate to the entries to its right in the table!* And the larger the values of r and n, the more faithfully the error of $F[h]$ gets duplicated as it propagates! Consequently,

Richardson's formula is most effective when all $F[h]$ entries used have negligible roundoff error. **(25a)**

The best way to ensure this is to *start with a fairly large initial h* and *use small stepsize ratios*, say, between 1 and 4. After a new $F[h]$ is obtained, the calculation of $F_1[h]$, $F_2[h]$, $F_3[h]$, ... (to its right) should be stopped when the **termination test**

$F_i[h]$ and $F_{i-1}[h]$ agree to about the desired accuracy **(25b)**

is satisfied. The availability of $|F_i[h] - F_{i-1}[h]|$ as an estimate of the error of $F_{i-1}[h]$ (*not* $F_i[h]$) "along the way" is an especially valuable feature of Richardson's formula. Be warned, however, that roundoff error adversely affects its reliability. For example, the Absolute Error Test for 5*d* accuracy (i.e., $|F_i[h] - F_{i-1}[h]| < 0.5\text{E}-6$), would correctly terminate the iteration in Table 7.2-5 after $F_3[0.002] = 2.7182765$. But in Table 7.2-6 it would do so after $F_2[0.01] = 2.7183093$, which is accurate to only 4*d*. ■

7.2G Approximating $f'(z), \ldots, f^{(iv)}(z)$ Accurately

The general formula (15a) of Section 7.1B can be used to get the following $O(h^2)$ formulas in which $f_j = f(z + jh)$ (Problems M7-9 and M7-10).

$O(h^2)$ FORWARD/BACKWARD DIFFERENCE FORMULAS: $\tau[h] = Ch^2 + Dh^3 + Eh^4 + \cdots$

$$f'(z) \approx \frac{1}{2h}[-3f_0 + 4f_1 - f_2] \tag{26a}$$

$$f''(z) \approx \frac{1}{h^2}[2f_0 - 5f_1 + 4f_2 - f_3] \tag{26b}$$

$$f'''(z) \approx \frac{1}{2h^3}[-5f_0 + 18f_1 - 24f_2 + 14f_3 - 3f_4] \tag{26c}$$

$$f^{(iv)}(z) \approx \frac{1}{h^4}[3f_0 - 14f_1 + 26f_2 - 24f_3 + 11f_4 - 2f_5] \tag{26d}$$

$O(h^2)$ CENTRAL DIFFERENCE FORMULAS: $\tau[h] = Ch^2 + Dh^4 + Eh^6 + \cdots$

$$f'(z) \approx \frac{1}{2h}[-f_{-1} + 0f_0 + f_1] = \frac{\delta f(z)}{2h} \tag{27a}$$

$$f''(z) \approx \frac{1}{h^2}[f_{-1} - 2f_0 + f_1] = \frac{\delta^2 f(z)}{h^2} \tag{27b}$$

$$f'''(z) \approx \frac{1}{2h^3}[-f_{-2} + 2f_{-1} + 0f_0 - 2f_1 + f_2] \tag{27c}$$

$$f^{(iv)}(z) \approx \frac{1}{h^4}[f_{-2} - 4f_{-1} + 6f_0 - 4f_1 + f_2] \tag{27d}$$

Richardson's theorem can be applied to these formulas to remedy the Stepsize Dilemma of Section 7.2D. In the absence of roundoff error, it works more effectively when $F[h]$ is a central difference formula in (27) because these are more accurate than the forward/backward difference formulas (26) for a given h, and their accuracy improves more rapidly with each application of Richardson's formula.

EXAMPLE 7.2G Suppose you did not know that $d^3(e^x)/dx^3]_{x=1}$ is $e^1 = 2.71828183$ (8*s*). How accurately can you find it on an Apple IIe that is a 9*s* device?

SOLUTION: To get the most accuracy before serious roundoff occurs, we shall (i) use (27c) as $F[h]$, (ii) start with a rather *large stepsize* which is a *binary number*, say $h = 1$, and (iii) use the *small shrinking ratio* $r = 2$. This will keep h as large as possible and will also ensure that h is calculated without roundoff error. The resulting Richardson table is shown in Table 7.2-7. A typical $F[h]$ calculation is

$$F\left[\frac{1}{4}\right] = \frac{1}{2(\frac{1}{4})^3}[-e^{1/2} + 2e^{3/4} - 2e^{5/4} + e^{3/2}] \doteq 2.7610214 \tag{28}$$

Since $n = 2$, $m = 4$, and $p = 6$, the $F_i[h]$ values were obtained using (24) with $r = 2$. The Relative Error Test for 7*s* accuracy would (correctly) stop the iteration after $F_3[\frac{1}{8}]$. Since $F[\frac{1}{8}]$ is only accurate to 3*s*, Richardson's theorem improved the accuracy substantially.

This is fortunate, because some roundoff error has crept into the seventh significant digit of $F[\frac{1}{16}]$! (How can you tell?) ■

TABLE 7.2-7 Richardson Table for $F[h]$ in (27c); $f'''(z) \doteq 2.71828183$

	h	$F[h]$, $O(h^2)$ $\delta f(z)/2h$	$F_1[h]$, (29a) $\tau[h] = O(h^4)$	$F_2[h]$, (29b) $\tau[h] = O(h^6)$	$F_3[h]$, (29b) $\tau[h] = O(h^8)$
	1	3.469773			
$r = 2$					
	1/2	2.892482	2.7000518		
$r = 2$					
	1/4	2.761024	2.7172012	2.7182445	
$r = 2$					
	1/8	2.728917	2.7182149	2.7182825	2.7182816
$r = 2$					
	1/16	2.720939	2.7182798	2.7182841	2.7182841

The roundoff error that occurs in (26) and (27) for sufficiently small h results from *subtractive cancellation* in the numerator, which gets *magnified* when divided by the small denominator. The best safeguard against this is to use f_i values and intermediate calculations that carry about 50% more significant digits than the desired accuracy of $f^{(k)}(z)$. If this can be done, Richardson's formula will generally result in the impressive improvement seen in Table 7.2-7.

Practical Consideration 7.2G (Richardson versus a higher-order formula). Formula (26a) can be derived by applying Richardson's formula with $r = 2$ to the $O(h)$ formula $F[h] = \Delta f(z)/h$ to get the $O(h^2)$ approximation

$$f'(z) \approx F_1[h] = \frac{2F[h] - F[2h]}{2 - 1} = 2\frac{f_1 - f_0}{h} - \frac{f_2 - f_0}{2h} = \frac{1}{2h}[-3f_0 + 4f_1 - f_2] \tag{29}$$

Thus, if we wanted an $O(h^2)$ approximation of $d(e^x)/dx]_{x=1}$ with $h = 0.1$, we could either use (26a) directly to get

$$f'(1) \approx \frac{1}{0.2}[-3e^1 + 4e^{1.1} - e^{1.2}] \doteq 2.708508 \tag{30a}$$

or we could use Richardson's formula as in (28) to get

$$F[0.2] = \frac{e^{1.2} - e^1}{0.2} \doteq 3.0091747$$
$$F[0.1] = \frac{e^{1.1} - e^1}{0.1} \doteq 2.8588419 \qquad F_1[0.1] = \frac{2F[0.1] - F[0.2]}{2 - 1} \doteq 2.708508 \tag{30b}$$

Both solutions (30) yield the same numerical value. But which is better? The "obvious" answer, (30a), is *wrong* because *it does not indicate how many digits of 2.708508 are accurate!* On the other hand, the reward for the slight extra effort of (30b) is the ability to reason as follows: Since $|F_1[0.1] - F[0.1]|$ is about 0.15, $F[0.1]$ should be accurate to almost 2s; so $F_1[h] = 2.708508$ is likely to be accurate to almost 3s (i.e., about 2.71). *The ability to estimate the* *accuracy of $F_1[h]$ makes the Richardson approach as in (30b) preferable to an equivalent single formula* in a realistic situation when you do not know the exact value of the number being approximated (see Problems 7-11 and 7-19)! ■

7.2H Approximating Derivatives of Tabulated Functions

Richardson's formula can be thought of as a prescription for using a *less accurate* approximation $F[rh]$ to improve a *more accurate* approximation $F[h]$, with the result being the *improved approximation* $F_1[h]$. This interpretation can be very useful when we only know (*accurate*) values of $f(x)$ at tabulated, h_0-spaced nodes and we cannot get greater accuracy by using $h < h_0$. We illustrate this with a function for which the exact derivatives are known for comparison.

EXAMPLE 7.2H Suppose that the only available values of $f(x) = e^x$ are given (correctly rounded) to $7s$ at $x = 0, 0.2, 0.4, 0.6, 0.8$, and 1 ($h_0 = 0.2$). How accurately can you estimate **(a)** $f'''(1)$ and **(b)** $f''(0.2)$?

REMARK: Since $e^{-x} = 1/e^x$, we actually know $f(x)$ to $7s$ at the following 0.2-spaced nodes:

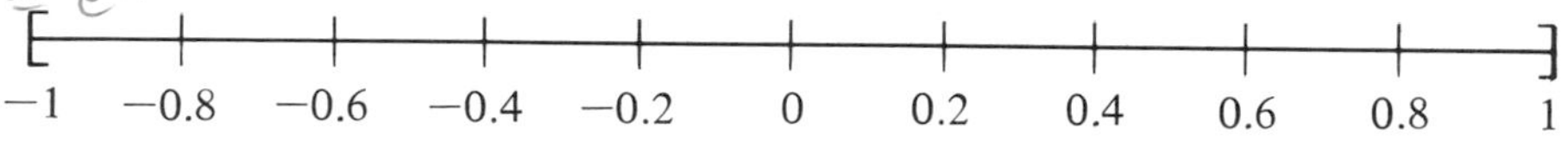

(31)

SOLUTION (a): Since $z = 1$ is right most tabulated node in (31), we must use $7s$ values of e^x in the $O(h^2)$ backward difference formula (26c) (with $h < 0$) as $F[h]$. Taking $h = -h_0 = -0.2$ gives

$$F[-h_0] = \frac{1}{2(-0.2)^3}[-5e^1 + 18e^{0.8} - 24e^{0.6} + 14e^{0.4} - 3e^{0.2}] \doteq 2.5741874 \quad \textbf{(32a)}$$

to improve this approximation, we use the next larger possible h, namely $-2h_0 = -0.4$, to get the (worse) approximation

$$F[-2h_0] = \frac{1}{2(-0.4)^3}[-5e^1 + 18e^{0.6} - 24e^{0.2} + 14e^{-0.2} - 3e^{-0.6}] \doteq 2.2745594 \quad \textbf{(32b)}$$

Richardson's formula can now be used to get the improved approximation

$$F_1[-h_0] = \frac{2^2F[-h_0] - F[-2h_0]}{2^2 - 1} \doteq 2.6740634 \quad \textbf{(32c)}$$

Since $|F_1[-h_0] - F[-h_0]|$ is about 0.1, we can conclude that $F[-h_0]$ is accurate to about $2s$. The improved estimate $F_1[-h_0]$ is only $O(h^3)$ (why?); so if we round $F_1[-h_0]$ to $3s$ to get the approximation $f'''(1) \approx 2.67$, we should expect some error in the least significant digit [see (30b)].

SOLUTION (b): Since $z = 0.2$ has several tabulated nodes to both its right *and* its left in (31), we can use the 7*s* values of e^x in the $O(h^2)$ central difference approximation $F[h] = \delta^2 f(z)/h^2$ in (27b) to get approximations in the following order [using (27b) and (24)]:

$$F[h_0] = \frac{e^0 - 2e^{0.2} + e^{0.4}}{(0.2)^2} \doteq 1.225475 \qquad \textbf{(33a)}$$

$$F[2h_0] \doteq 1.237774, \qquad \text{then } F_1[h_0] = \frac{2^2F[h_0] - F[2h_0]}{3} \doteq 1.221375 \qquad \textbf{(33b)}$$

$$F[3h_0] \doteq 1.258486, \qquad \text{then } F_1[2h_0] \doteq 1.221024 \text{ and } F_2[h_0] \doteq 1.221386 \qquad \textbf{(33c)}$$

as shown in Table 7.2-8. Since $F_2[h_0]$ and $F_1[h_0]$ agree to about 6*s*, we can conclude that $f''(0.2) = 1.22139$ to within reasonable roundoff in the last significant digit. Had this accuracy not been evident, we could have added $F[4h_0] = F[0.8]$, then $F_1[3h_0]$ (using $r = \frac{4}{3}$), then $F_2[2h_0]$ (using $r = \frac{3}{2}$), and finally $F_3[h_0]$ (using $r = \frac{2}{1} = 2$). In this way, each added entry (at the *top* of the $F[h]$ column) uses available nodes *closest to z*.

TABLE 7.2-8 Richardson Table for $F[h] = \delta^2 f(z)/h^2$; $f''(0.2) \doteq 1.22140$

h		$F[h]$, $O(h^2)$ $\delta^2 f(z)/h^2$	$F_1[h]$ $\tau[h] = O(h^4)$	$F_2[h]$ $\tau[h] = O(h^6)$
	$3h_0 = 0.6$	1.258486		
$r = \frac{3}{2}$				
	$2h_0 = 0.4$	1.237774	1.221204	
$r = 2$				
	$h_0 = 0.2$	1.225475	1.221375	1.221386

■

A comparison of the accuracy achieved in (a) and (b) of the preceding example shows why forward/backward difference approximations of $f^{(k)}$ should *only* be used when z is near the beginning or end of a table and there are not enough tabulated nodes on either side of z to use a central difference formula.

One last comment. The approximation of derivatives is very sensitive to errors in the sampled $f(x_i)$ values. So if you must get an approximation of $f^{(k)}(z)$ from *experimental* (hence erroneous) data, *do not use any method of this chapter!* Instead, find a $g(x)$ that *fits* the data well and use the derivative of $g(x)$ to approximate that of $f(x)$ (see Problem M5-18).

7.3 Composite Rules and Romberg Integration

A formula that approximates a definite integral $\int_a^b f(x)\,dx$ as a weighted sum of sampled values of the integrand is often referred to as a **quadrature formula**. It is a **closed** formula if *both* a and b are sampled nodes (e.g., the Trapezoidal Rule), and an **open** formula if *neither* is (e.g., the Midpoint Rule). In this section we examine quadrature formulas that sample $n + 1$ *equispaced* nodes

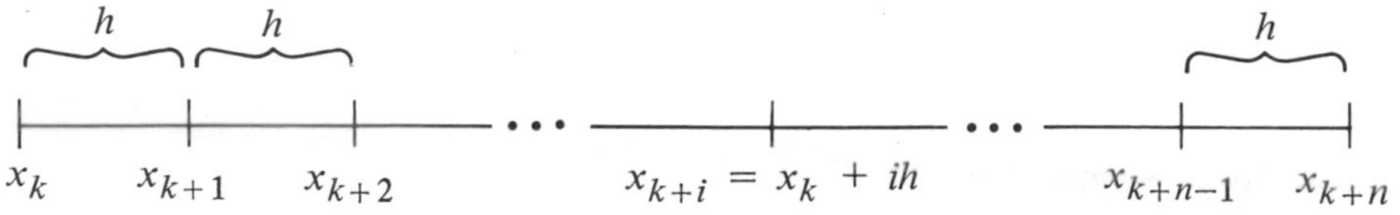

(1a)

with x_k denoting the *leftmost sampled node*. So our general formula is

$$\int_a^b f(x)\,dx \approx w_k f_k + \cdots + w_{k+n} f_{k+n}, \qquad \text{where } f_{k+i} = f(x_{k+i}) \qquad \textbf{(1b)}$$

$$w_{k+i} = \int_a^b L_{k+i}(x)\,dx, \qquad \text{where } L_{k+i}(x) = \prod_{\substack{j=0\\ j\neq i}}^{n} \frac{(x - x_{k+j})}{(x_{k+i} - x_{k+j})} \qquad \textbf{(1c)}$$

We saw in Section 7.1B that when w_{k+i} in (1b) is taken to be the integral from a to b of the $(k + i)$th Lagrange polynomial for the sampled nodes as in (1c), then (1b) becomes the approximation $\int_a^b f(x)\,dx \approx \int_a^b p_{k,k+n}(x)\,dx$, hence will be exact (i.e., equality) whenever $f(x)$ is a polynomial of degree n or less; also [see (23b) of Section 7.1C] *if the product* $\Pi_{i=0}^{n} (x - x_{k+i})$ *does not change sign on the interval* $[a, b]$, then its truncation error can be represented as

$$\tau = \int_a^b f(x)\,dx - \sum_{i=0}^{n} w_{k+i} f_{k+i} = \frac{f^{(n+1)}(\xi)}{(n+1)!} \int_a^b \prod_{i=0}^{n} (x - x_{k+i})\,dx \qquad \textbf{(1d)}$$

where ξ lies somewhere in the smallest closed interval containing a, b, and all sampled nodes x_i.

7.3A Quadrature Formulas for *h*-Spaced Nodes

Suppose that both a and b lie an integer number of h-steps from x_k, say,

$$a = x_k + jh \quad \text{and} \quad b = x_k + mh \qquad (j \text{ and } m \text{ are integers}) \qquad \textbf{(2a)}$$

Then the length of $[a, b]$ is $(m - j)h$, and a quadrature formula for $\int_a^b f(x)\,dx$ will be referred to as an **$(m - j)$-panel quadrature formula**. In this case (see Section 6.1D) the change of variable from x to

$$s = \frac{x - x_k}{h} \quad \text{so that } ds = \frac{dx}{h} \quad \text{or, equivalently,} \quad dx = h\,ds \qquad \textbf{(2b)}$$

changes $\int_{x=a}^{b} \phi(x)\,dx$ to $\int_{s=j}^{m} \phi(x_k + hs)(h\,ds)$. So (1c) becomes

$$w_{k+i} = h\int_j^m L_{k+i}(x_k + hs)\,ds, \qquad \text{where } L_{k+i}(x_k + hs) = \prod_{\substack{j=0 \\ j\neq i}}^{n} \frac{s-j}{i-j} \tag{2c}$$

[see (26) of Section 6.1D]; and since $(x - x_{k+i}) = h(s - i)$, (1d) becomes

$$\tau = C\frac{f^{(n+1)}(\xi)}{(n+1)!}h^{n+2}, \qquad \text{where } C = \int_j^m \prod_{i=0}^{n}(s-i)\,ds \tag{2d}$$

whenever $\Pi_{i=0}^{n}(s - i)$ does not change sign on the s-interval $[j, m]$. Table 7.3-1 gives $L_{k+i}(x_k + hs)$ and $\Pi_{i=0}^{n}(s - i)$ in a form ready to integrate [see (27) of Section 6.1D] for $n = 2$, 3, and 4.

TABLE 7.3-1 Getting w_{k+i} and τ When Sampling $n + 1$ h-Spaced Nodes

n	$L_{k+i}(x_k + hs)$ in (2c)	$\Pi_{i=0}^{n}(s - i)$ in (2d)
2	$L_{k+0} = \frac{1}{2}(s^2 - 3s + 2)$ $L_{k+1} = -1(s^2 - 2s)$ $L_{k+2} = \frac{1}{2}(s^2 - s)$	$s^3 - 3s^2 + 2s$
3	$L_{k+0} = -\frac{1}{6}(s^3 - 6s^2 + 11s - 6)$ $L_{k+1} = \frac{1}{2}(s^3 - 5s^2 + 6s)$ $L_{k+2} = -\frac{1}{2}(s^3 - 4s^2 + 3s)$ $L_{k+3} = \frac{1}{6}(s^3 - 3s^2 + 2s)$	$s^4 - 6s^3 + 11s^2 - 6s$
4	$L_{k+0} = \frac{1}{24}(s^4 - 10s^3 + 35s^2 - 50s + 24)$ $L_{k+1} = -\frac{1}{6}(s^4 - 9s^3 + 26s^2 - 24s)$ $L_{k+2} = \frac{1}{4}(s^4 - 8s^3 + 19s^2 - 12s)$ $L_{k+3} = -\frac{1}{6}(s^4 - 7s^3 + 14s^2 - 8s)$ $L_{k+4} = \frac{1}{24}(s^4 - 6s^3 + 11s^2 - 6s)$	$s^5 - 10s^4 + 35s^3 - 50s^2 + 24s$

For example, the four-point ($n = 3$), one-panel quadrature formula

$$\int_{x_j}^{x_{j+1}} f(x)\,dx \approx w_{j-1}f_{j-1} + w_jf_j + w_{j+1}f_{j+1} + w_{j+2}f_{j+2} \tag{3a}$$

is a cubic exact approximation over a *single* panel $[x_j, x_{j+1}]$. Here x_k (the leftmost sampled node) is x_{j-1}, so x_{j-1}, x_j, x_{j+1}, and x_{j+2} correspond to $s = 0$, 1, 2, and 3, respectively, and the sample weights in (3a), obtained using (2c) and Table 7.3-1, are

$$w_{j-1} = -\frac{h}{6}\int_1^2 (s^3 - 6s^2 + 11s - 6)\,ds = \frac{-1}{24}h \quad (= w_{j+2}) \tag{3b}$$

$$w_j = \frac{h}{2}\int_1^2 (s^3 - 5s^2 + 6s)\,ds = \frac{13}{24}h \quad (= w_{j+1}) \tag{3c}$$

The equalities in parentheses follow from the symmetry of the sampled nodes with respect to the panel $[x_j, x_{j+1}]$. As a check, note that the sum of the four weights is h (the length of the integration interval $[x_j, x_{j+1}]$), as it should by (20b) of Section 7.1B. So the desired formula is

$$\int_{x_j}^{x_{j+1}} f(x)\,dx \approx \frac{h}{24}[-f_{j-1} + 13(f_j + f_{j+1}) - f_{j+2}] \tag{3d}$$

And since $\Pi_{i=0}^{3}(s - i) = s(s-1)(s-2)(s-3)$ does not change sign over the s-interval $[1, 2]$, we can use (2d) and Table 7.3-1 to get

$$\tau[h] = C\,\frac{f^{(iv)}(\xi)}{4!}h^5, \qquad \text{where } C = \int_1^2 (s^4 - 6s^3 + 11s^2 - 6s)\,ds = \frac{11}{30} \tag{3e}$$

The work in (3) is a derivation of (5) below. The other important formulas shown in (4)–(10) can be obtained similarly. In the number lines shown next to the formula's name, the shaded portions are the intervals of integration, and the circled number over the sampled node x_i gives the *relative weight* of $f_i = f(x_i)$ as compared to the weights of the other sampled values. As before, ξ in the truncation error formulas lies in the smallest interval containing all sampled nodes *and* the interval of integration.

ONE-PANEL FORMULAS

Trapezoidal Rule (T):

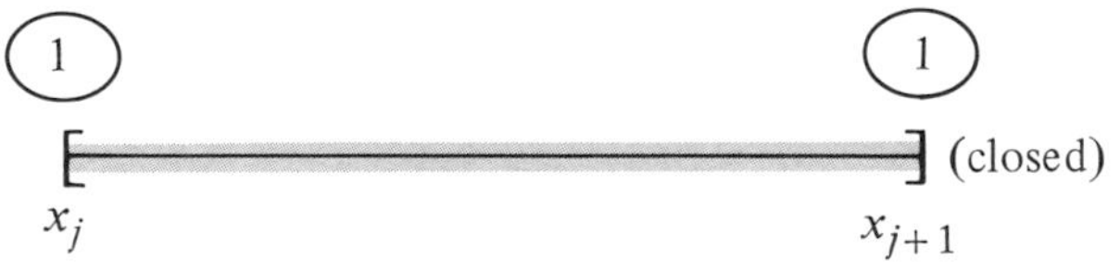

$$\int_{x_j}^{x_{j+1}} f(x)\,dx \approx T]_{x_j}^{x_{j+1}} = \frac{h}{2}[f_j + f_{j+1}] \tag{4a}$$

$$\tau_T = \frac{-1}{12}h^3 f''(\xi) = O(h^3) \quad \text{(exact for straight lines)} \tag{4b}$$

Central Cubic Rule (CC):

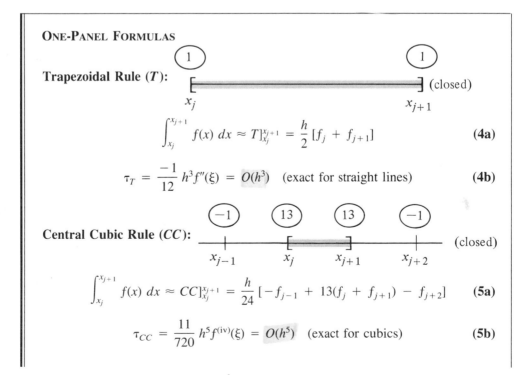

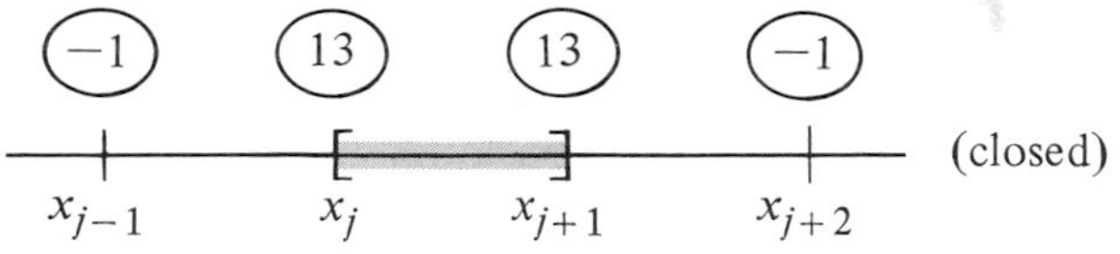

$$\int_{x_j}^{x_{j+1}} f(x)\,dx \approx CC]_{x_j}^{x_{j+1}} = \frac{h}{24}[-f_{j-1} + 13(f_j + f_{j+1}) - f_{j+2}] \tag{5a}$$

$$\tau_{CC} = \frac{11}{720}h^5 f^{(iv)}(\xi) = O(h^5) \quad \text{(exact for cubics)} \tag{5b}$$

Adams $\frac{1}{24}$-Corrector (AC):

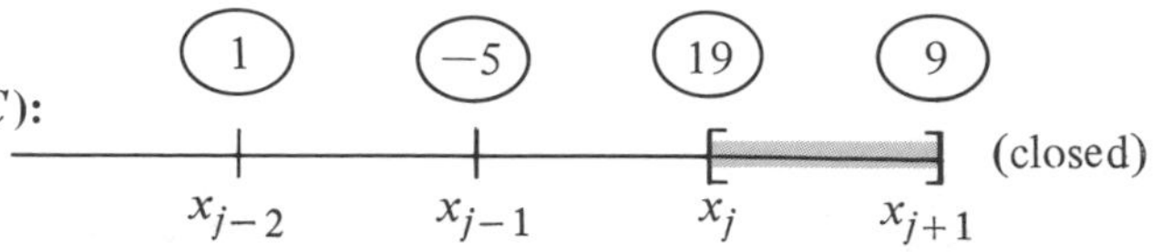

$$\int_{x_j}^{x_{j+1}} f(x)\,dx \approx AC]_{x_j}^{x_{j+1}} = \frac{h}{24}[f_{j-2} - 5f_{j-1} + 19f_j + 9f_{j+1}] \qquad \textbf{(6a)}$$

$$\tau_{AC} = \frac{-19}{720} h^5 f^{(\text{iv})}(\xi) = O(h^5) \quad \text{(exact for cubics)} \qquad \textbf{(6b)}$$

Adams $\frac{1}{24}$-Predictor (AP):

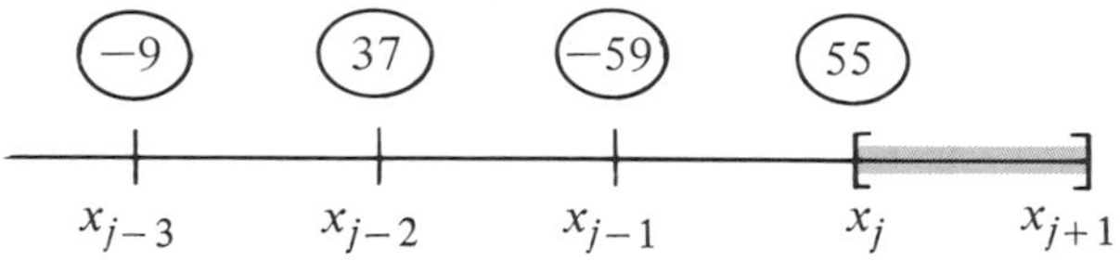

$$\int_{x_j}^{x_{j+1}} f(x)\,dx \approx AP]_{x_j}^{x_{j+1}} = \frac{h}{24}[-9f_{j-3} + 37f_{j-2} - 59f_{j-1} + 55f_j] \qquad \textbf{(7a)}$$

$$\tau_{AP} = \frac{251}{720} h^5 f^{(\text{iv})}(\xi) = O(h^5) \quad \text{(exact for cubics)} \qquad \textbf{(7b)}$$

TWO-PANEL FORMULAS

Midpoint Rule (M):

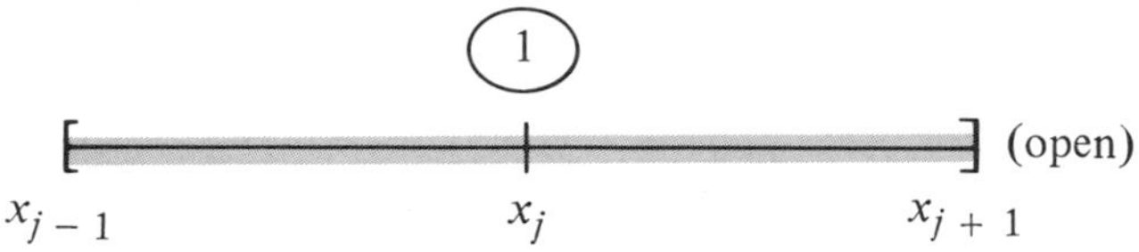

$$\int_{x_{j-1}}^{x_{j+1}} f(x)\,dx \approx M]_{x_{j-1}}^{x_{j+1}} = 2hf_j \qquad \textbf{(8a)}$$

$$\tau_M = \frac{1}{3} h^3 f''(\xi) = O(h^3) \quad \text{(exact for straight lines)} \qquad \textbf{(8b)}$$

Simpson's $\frac{1}{3}$-Rule ($S_{1/3}$):

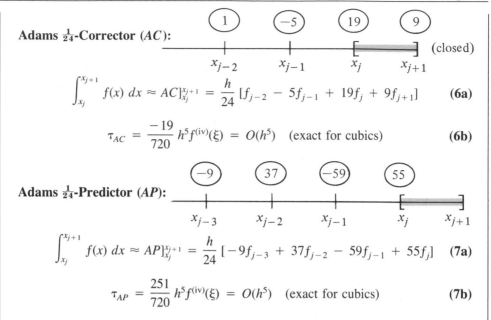

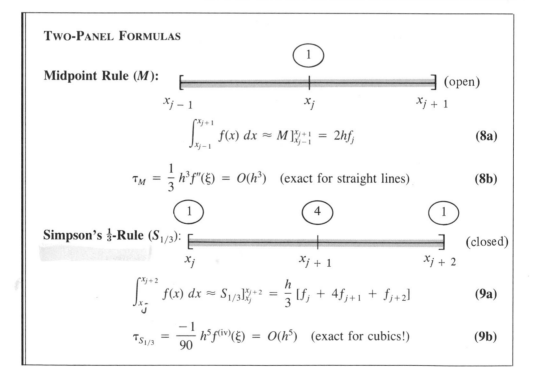

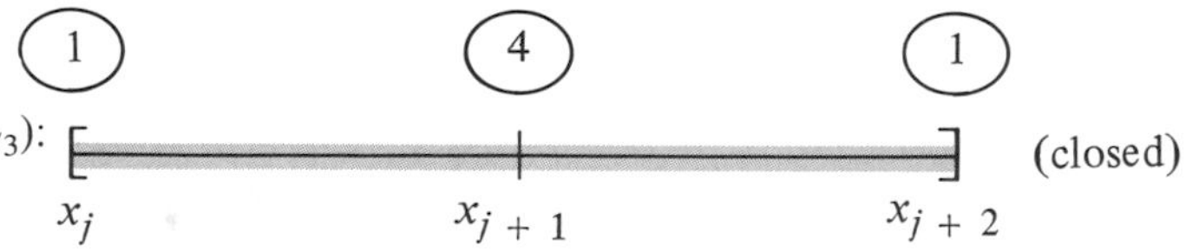

$$\int_{x_j}^{x_{j+2}} f(x)\,dx \approx S_{1/3}]_{x_j}^{x_{j+2}} = \frac{h}{3}[f_j + 4f_{j+1} + f_{j+2}] \qquad \textbf{(9a)}$$

$$\tau_{S_{1/3}} = \frac{-1}{90} h^5 f^{(\text{iv})}(\xi) = O(h^5) \quad \text{(exact for cubics!)} \qquad \textbf{(9b)}$$

THREE-PANEL FORMULA

Simpson's $\frac{3}{8}$-Rule ($S_{3/8}$):

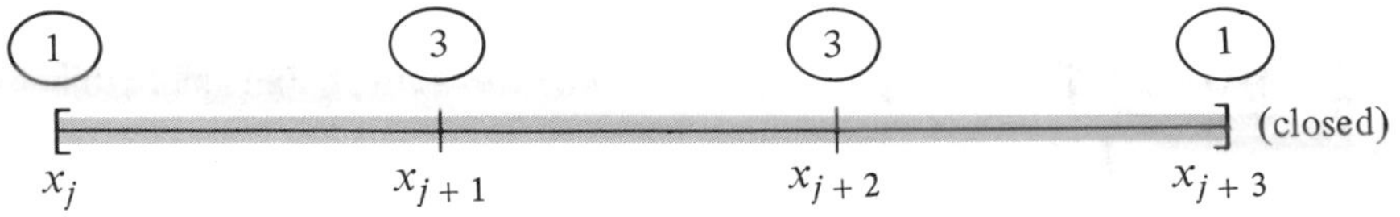

$$\int_{x_j}^{x_{j+3}} f(x)\,dx \approx S_{3/8}]_{x_j}^{x_{j+3}} = \frac{3}{8}h[f_j + 3(f_{j+1} + f_{j+2}) + f_{j+3}] \qquad \textbf{(10a)}$$

$$\tau_{S_{3/8}} = \frac{-1}{80}h^5 f^{(iv)}(\xi) = O(h^5) \quad \text{(exact for cubics)} \qquad \textbf{(10b)}$$

The truncation error of the one-panel formulas can be obtained using (2d) as in (3e); the others are considered in our discussion of Newton–Cotes formulas at the end of this section. a quadrature formula for which $\tau = Ch^{n+2}f^{(n+1)}(\xi)$ will be exact (because $\tau = 0$) when $f(x)$ is a polynomial of degree $\leqslant n$, as indicated in (4b)–(10b). Note that the Midpoint and Simpson's $\frac{1}{3}$-rules have one degree of exactness *more* than we might expect. This extra degree of exactness often occurs when the *midpoint* of the integration interval is sampled.

Among the formulas for which $\tau = Ch^5 f^{(iv)}(\xi)$, those with the smaller $|C|$ are generally more accurate for a given h. For example, among the cubic exact one-panel formulas, CC ($C = 11/720$) and AC ($C = -19/720$) should be noticably more accurate than AP ($C = 251/720$). In view of the Polynomial Wiggle Problem (Section 5.3B) and the fact that AP integrates over the *extrapolated* part of the interpolating polynomial $p_{j-3,j}(x)$, this should not be too surprising.

EXAMPLE 7.3A Use quadrature formulas to estimate the following integrals as accurately as you can, assuming that the only available values of e^x are at the 0.4-spaced sample nodes shown in Figure 7.3-1.

$$\textbf{(a)} \int_{5.2}^{6.0} e^x\,dx \qquad \textbf{(b)} \int_{4.8}^{6.0} e^x\,dx \qquad \textbf{(c)} \int_{5.6}^{6.0} e^x\,dx \qquad \textbf{(d)} \int_{6.0}^{6.8} e^x\,dx$$

Underline digits that would be incorrect after rounding. The exact values to 7s are (a) 222.1566; (b) 281.9184; (c) 133.0024; (d) 494.4185.

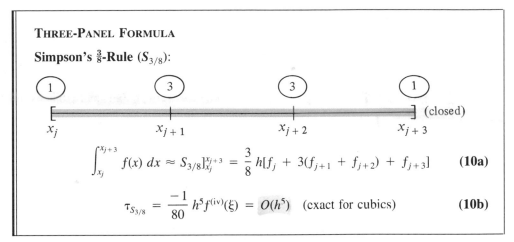

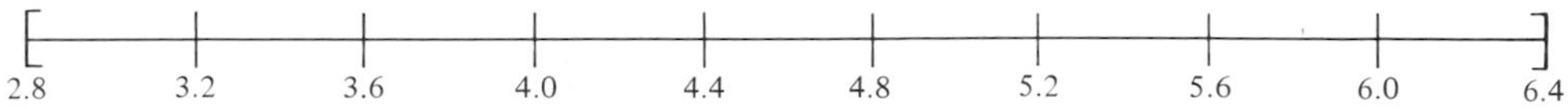

Figure 7.3-1 *Ten nodes at which the value of e^x is known.*

SOLUTION (a): Simpson's $\frac{1}{3}$-Rule (9a) with $h = 0.4$ and $x_j = 5.2$ gives

$$\int_{5.2}^{6.0} \approx S_{1/3}]_{5.2}^{6.0} = \frac{0.4}{3}[e^{5.2} + 4e^{5.6} + e^{6.0}] \doteq 222.1\underline{876} \qquad \textbf{(11a)}$$

For comparison, the Midpoint Rule (8a) with $h = 0.4$ and $x_j = 5.6$ gives

$$\int_{5.2}^{6.0} \approx M]_{5.2}^{6.0} = (6.0 - 5.2)e^{5.6} \doteq 2\underline{16.3411} \tag{11b}$$

the cubic exact Simpson's $\frac{1}{3}$-Rule yielded about two correct digits more than the linear exact Midpoint Rule.

Solution (b): Simpson's $\frac{3}{8}$-Rule (10a) with $h = 0.4$ and $x_j = 4.8$ gives

$$\int_{4.8}^{6.0} \approx S_{3/8}]_{4.8}^{6.0} = (0.4)\frac{3}{8}[e^{4.8} + 3(e^{5.2} + e^{5.6}) + e^{6.0}] \doteq 282.\underline{0053} \tag{12a}$$

For comparison, using the Trapezoidal Rule (4a) over each panel gives

$$\int_{4.8}^{6.0} \approx T]_{4.8}^{5.2} + T]_{5.2}^{5.6} + T]_{5.6}^{6.0}$$
$$= 6\underline{0.55653} + 9\underline{0.33973} + 13\underline{4.7710} \doteq 2\underline{85.6673} \tag{12b}$$

It is important to observe that both solutions used the same four sampled nodes. However, weighting them by the cubic-exact Simpson weights resulted in about two correct digits more than weighting them by the linear-exact trapezoidal ones!

Solution (c): To approximate the integral over this single h_0-panel, we can use the cubic exact formulas CC, AC, and AP with $h = 0.4$ to get

$$\int_{5.6}^{6.0} \approx CC]_{5.6}^{6.0} = \frac{0.4}{24}[e^{5.2} - 13(e^{5.6} + e^{6.0}) - e^{6.4}] \doteq 132.9\underline{500} \tag{13a}$$

$$\int_{5.6}^{6.0} \approx AC]_{5.6}^{6.0} = \frac{0.4}{24}[e^{4.8} - 5e^{5.2} + 19e^{5.6} + 9e^{6.0}] \doteq 133.0\underline{685} \tag{13b}$$

$$\int_{5.6}^{6.0} \approx AP]_{5.6}^{6.0} = \frac{0.4}{24}[-9e^{4.4} + 37e^{4.8} - 59e^{5.2} + 55e^{5.6}] \doteq 13\underline{2.3536} \tag{13c}$$

Note that CC and AC are somewhat more accurate than AP, and that all three yielded more accurate values than $T]_{5.6}^{6.0} \doteq 13\underline{4.7710}$ obtained in (12b).

Solution (d): How can we get $\int_{6.0}^{6.8}$ if the value of $f(6.8)$ is not available? (Think about it!) To extrapolate *beyond* the tabulated values, we can use the Adams Predictor (7a) with $h = 0.8$ and $x_j = 6.0$:

$$\int_{6.0}^{6.8} \approx AP]_{6.0}^{6.8} = \frac{0.8}{24}[-9e^{3.6} + 37e^{4.4} - 59e^{5.2} + 55e^{6.0}] \doteq 4\underline{72.5940} \tag{14}$$

This is less accurate than (13c) because the stepsize used is twice as large. Can you think of a way to get comparable accuracy? ■

Practical Consideration 7.3a (One-panel integration). When using h_0-spaced *tabulated* function values, the Central Cubic Rule (5a) will generally yield the most accurate cubic exact approximation of a one-panel integration. The Adams formulas should be used with $h > 0$ when one of the limits of integration is the rightmost tabulated node, and with $h < 0$ when one of the limits of integration is the leftmost tabulated node; the more accurate *AC* formula should be used whenever both endpoints can be sampled. ■

A closed quadrature formula for $\int_a^b f(x)\,dx$ which samples the $n+1$ nodes that partition $[a, b]$ into n panels each of length $h = (b-a)/n$ is an

n-PANEL NEWTON–COTES (NC_n) FORMULA:

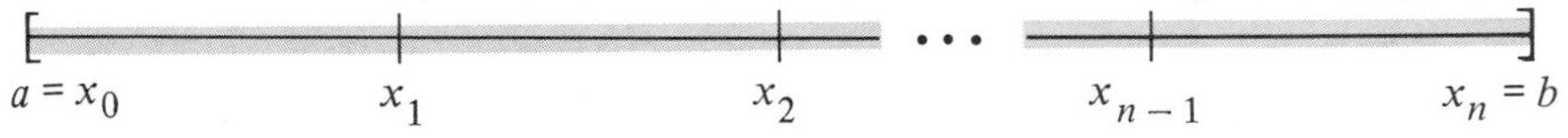

$$\int_{a=x_j}^{b=x_{j+n}} f(x)\,dx \approx NC_n]_a^b = h[w_0 f_0 + w_1 f_1 + \cdots + w_n f_n] \tag{15a}$$

where $w_0, \ldots, w_n$ are obtained using (2c). Note that $NC_n]_a^b$ is $T]_a^b$ when $n = 1$, $S_{1/3}]_a^b$ when $n = 2$, and $S_{3/8}]_a^b$ when $n = 3$. The truncation error of $NC_n]_a^b$ is

$$\tau_{NC_n} = \begin{cases} C_n f^{(n+1)}(\xi)h^{n+2} \text{ (exactness degree } n) & \text{if } n \text{ is } odd \\ C_n f^{(n+2)}(\xi)h^{n+3} \text{ (exactness degree } n+1) & \text{if } n \text{ is } even \end{cases} \tag{15b}$$

Notice the "bonus" extra degree of exactness when n is even. For example, taking $n = 4$ gives the following five-point formula:

FOUR-PANEL NEWTON–COTES FORMULA:

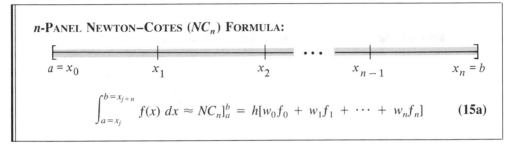

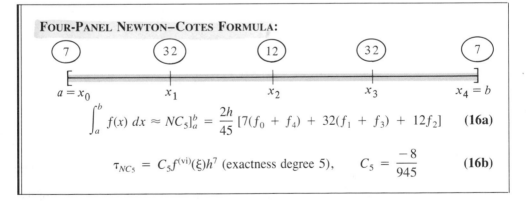

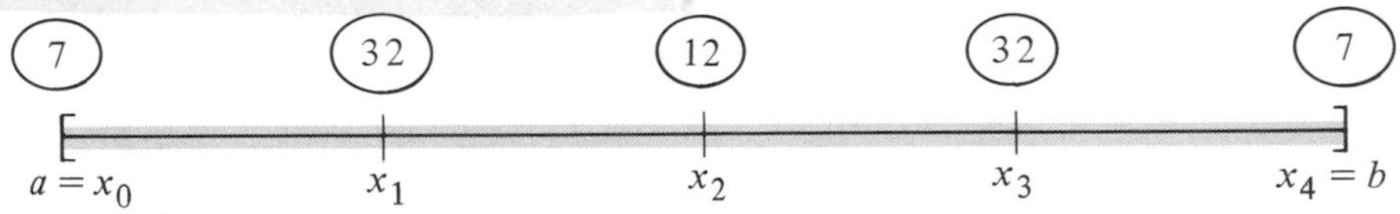

$$\int_a^b f(x)\,dx \approx NC_5]_a^b = \frac{2h}{45}[7(f_0 + f_4) + 32(f_1 + f_3) + 12f_2] \tag{16a}$$

$$\tau_{NC_5} = C_5 f^{(\text{vi})}(\xi)h^7 \text{ (exactness degree 5)}, \qquad C_5 = \frac{-8}{945} \tag{16b}$$

The most direct way to get the error expressions (15) is to use Newton's forward difference form [(48b) of Section 6.2D] of the interpolating polynomial $p_{0,m}(x)$, where $m \geq n+1$. We illustrate its use for Simpson's $\frac{1}{3}$-Rule ($n = 2$):

$$\int_a^b f(x)\,dx = \underbrace{\int_a^b p_{0,2}(x)\,dx}_{NC_2 = S_{1/3}]_a^b} + \underbrace{\Delta^3 f_0 \int_0^2 \binom{s}{3} h\,ds + \Delta^4 f_0 \int_0^2 \binom{s}{4} h\,ds + O(h^6)}_{\tau_{NC_2}} \tag{17a}$$

where $\int_0^1 \binom{s}{3}\,ds = \left(\frac{1}{3!}\right)\int_0^1 s(s-1)(s-2)\,ds = 0$, and similarly, $\int_0^1 \binom{s}{4}\,ds = -\frac{1}{90}$. So (27) of Section 6.2C can be used to get

$$\tau_{NC_2} = \frac{-h}{90}\Delta^4 f_0 + O(h^6) = \frac{-h}{90}h^4[f^{(\text{iv})}(a) + O(h)] + O(h^6) \qquad \textbf{(17b)}$$

This can be put in the "Lagrange Remainder form" (15b) [see (9c)] by a usual Intermediate Value Theorem argument (see Appendix II.1C). Truncation errors are discussed further in Problem M7-16 and Section 7.3E.

Newton–Cotes formulas are not recommended for $n > 5$ because the Polynomial Wiggle Problem of Section 5.3B will adversely affect the accuracy of the approximation $f(x) \approx p_{0,n}(x)$ for sufficiently large n *unless* $f(x)$ turns out to be a polynomial (in which case a quadrature formula is not needed!). Better weights for the $n + 1$ sampled values f_i in (15a) generally result from the *composite rules* described next.

7.3B Composite Trapezoidal and Simpson Rules: $T[h]$ and $S_{1/3}[h]$

We are now ready to describe efficient methods for approximating $\int_a^b f(x)\,dx$ to a prescribed accuracy when f is continuous on $[a, b]$. We first partition $[a, b]$ into n **panels** of equal length as follows:

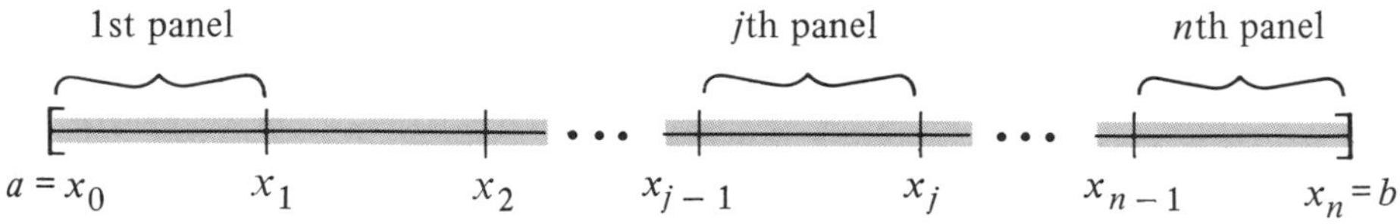

(18a)

so that the $n + 1$ equispaced sampled nodes x_j are given by the formula

$$x_j = a + jh, \qquad \text{where } h = \frac{b - a}{n} \quad \text{for } j = 0, 1, \ldots, n \qquad \textbf{(18b)}$$

It will be convenient to denote the sum of the *interior* sampled values as

$$\sum_{\text{interior}}[h] = f_1 + f_2 + \cdots + f_{n-1}, \qquad \text{where } f_i = f(x_i) \qquad \textbf{(18c)}$$

If the Trapezoidal Rule (4a) is used to approximate the integral over each of the n panels $[x_{j-1}, x_j]$, the resulting approximation is

$$\int_a^b f(x)\,dx \approx \frac{h}{2}[f(a) + f_1] + \frac{h}{2}[f_1 + f_2] + \cdots + \frac{h}{2}[f_{n-1} + f(b)] \qquad \textbf{(19)}$$

Upon combining (interior) f_i terms and factoring h, this becomes the Composite Trapezoidal Rule $T[h]$.†

†The notation $T[h]$ and $\tau_T[h]$ (with h in square brackets) will be used consistently to distinguish this *composite* Trapezoidal Rule from the one-panel rule T and its error τ_T (without brackets) in (4).

Composite Trapezoidal Rule $T[h]$:

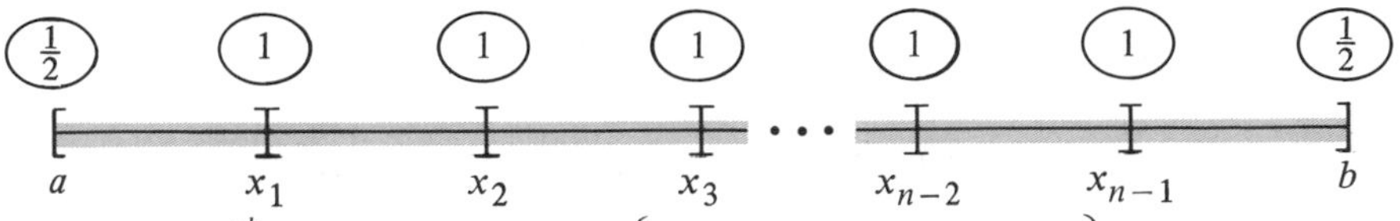

$$\int_a^b f(x)\,dx \approx T[h] = h\left\{\frac{f(a)+f(b)}{2} + \sum_{\text{interior}}[h]\right\} \tag{20a}$$

$$\tau_T[h] = \frac{-1}{12} f^{(\text{iv})}(\xi)(b-a)h^2, \qquad \text{where } a < \xi < b \tag{20b}$$

The formula for the truncation error $\tau_T[h]$ rests on the assumption that $f^{(\text{iv})}(x)$ is continuous on $[a, b]$. In this case, (4b) and the Intermediate Value Theorem (Appendix II.1C) can be used to get

$$\tau_T[h] = \frac{-1}{12} h^3 \sum_{j=1}^{n} f^{(\text{iv})}(\xi_j) = \frac{-1}{12} h^3[nf^{(\text{iv})}(\xi)] = \frac{-1}{12} f^{(\text{iv})}(\xi)(b-a)h^2 \tag{20c}$$

The last equality used the "order reducing" identity $nh = b - a$.

To get a composite Simpson's $\frac{1}{3}$-Rule, *the number of panels, n, must be an even integer.* We can then use (9a) over each of the $n/2$ "double panels"

$$[a, x_2], \quad [x_2, x_4], \quad [x_4, x_6], \quad \ldots, \quad [x_{n-2}, b] \qquad (b = x_n,\ n\ \textit{even})$$

If we then combine (interior) f_i terms and factor $h/3$, we get the

Composite Simpson's Rule $S_{1/3}[h]$:

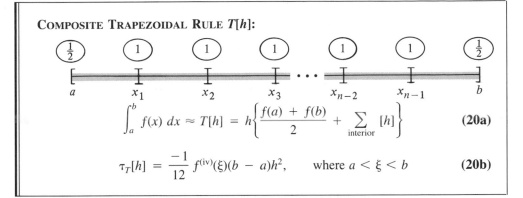

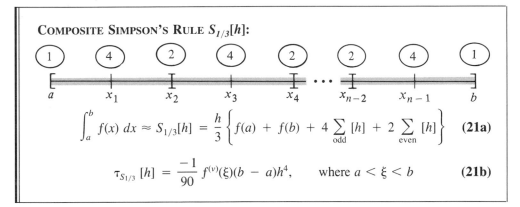

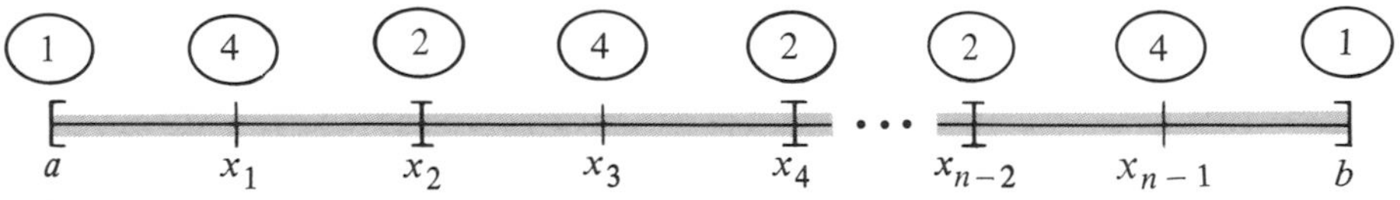

$$\int_a^b f(x)\,dx \approx S_{1/3}[h] = \frac{h}{3}\left\{f(a) + f(b) + 4\sum_{\text{odd}}[h] + 2\sum_{\text{even}}[h]\right\} \tag{21a}$$

$$\tau_{S_{1/3}}[h] = \frac{-1}{90} f^{(\text{v})}(\xi)(b-a)h^4, \qquad \text{where } a < \xi < b \tag{21b}$$

In (21a), $\Sigma_{\text{odd}}[h]$ and $\Sigma_{\text{even}}[h]$ denote the "interior" summations

$$\sum_{\text{odd}}[h] = f_1 + f_3 + \cdots + f_{n-1} \quad \text{(double-panel midpoints)} \tag{22a}$$

$$\sum_{\text{even}}[h] = f_2 + f_4 + \cdots + f_{n-2} \quad (\textit{their interior}\ \text{endpoints}) \tag{22b}$$

Note from $f^{(iv)}(\xi)$ in (20b) and $f^{(v)}(\xi)$ in (21b) that $T[h]$ has exactness degree *one* and $S_{1/3}[h]$ has exactness degree *three*, the same as $T]_a^b$ in (4) and $S_{1/3}]_a^b$ in (9). However, as a result of the use of the identity $nh = b - a$ [see (20c)], $T[h]$ *is* $O(h^2)$ and $S_{1/3}[h]$ *is* $O(h^4)$, *one order less* than $T]_a^b$ and $S_{1/3}]_a^b$.

EXAMPLE 7.3B Use **(a)** $T[0.8]$, **(b)** $T[0.4]$, and **(c)** $S_{1/3}[0.4]$ to approximate

$$\int_{2.8}^{5.2} e^x \, dx \qquad (\text{exact value} = e^{5.2} - e^{2.8} \doteq 164.82760) \qquad \textbf{(23)}$$

SOLUTION (a): **(24a)**

$$T[0.8] = 0.8\left[\frac{e^{2.8} + e^{5.2}}{2} + (e^{3.6} + e^{4.4})\right] \qquad \textbf{(24b)}$$

$$\doteq 173.52604 \quad (\tau_T[0.8] = 164.82760 - 173.52604 \doteq -8.698) \qquad \textbf{(24c)}$$

SOLUTION (b): **(25a)**

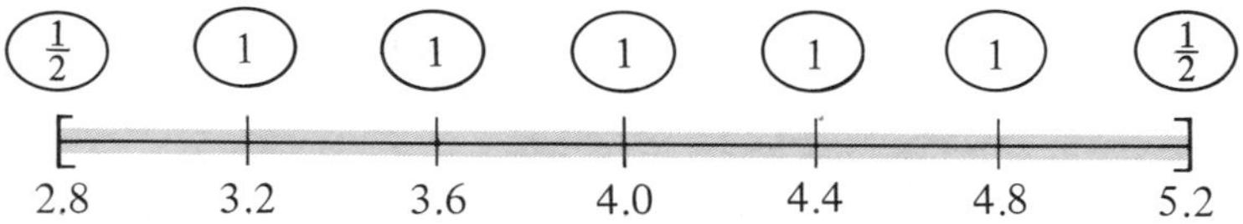

$$T[0.4] = 0.4\left[\frac{e^{2.8} + e^{5.2}}{2} + (e^{3.2} + e^{3.6} + e^{4.0} + e^{4.4} + e^{4.8})\right] \qquad \textbf{(25b)}$$

$$\doteq 167.01946 \quad (\tau_T[0.4] = 164.82760 - 167.01946 \doteq -2.192) \qquad \textbf{(25c)}$$

SOLUTION (c): **(26a)**

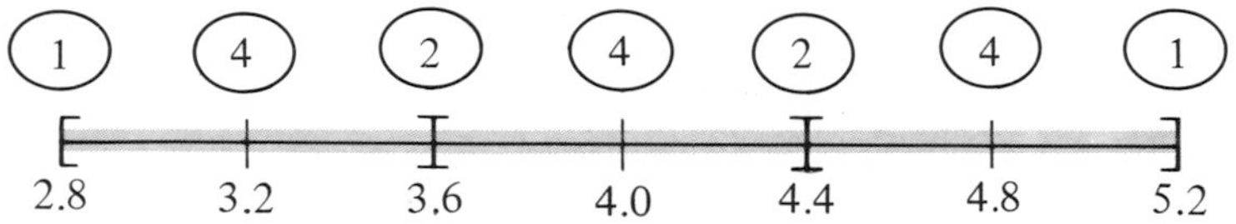

$$S_{1/3}[0.4] = \frac{0.4}{3}\,[e^{2.8} + e^{5.2} + \underbrace{4(e^{3.2} + e^{4.0} + e^{4.8})}_{\Sigma_{\text{odd}}[0.4]} + \underbrace{2(e^{3.6} + e^{4.4})}_{\Sigma_{\text{even}}[0.4]}] \qquad \textbf{(26b)}$$

$$= 164.85060 \quad (\tau_{S_{1/3}}[.4] = 164.82760 - 164.85060 = -0.023) \qquad \textbf{(26c)}$$

Thus weighting the seven sampled values $e^{2.8}, \ldots, e^{5.2}$ as in (26) produces an error about 100 times smaller than weighting them as in (25)! ■

Note from (24c) and (25c) that $\tau_T[0.8] \approx 2^2\tau_T[0.4]$, confirming that $T[h]$ is an $O(h^2)$ approximation of $\int_a^b f(x)\,dx$, as asserted in (20b). So we can use Richardson's formula with $n = 2$ to get

$$T_1[0.4] = \frac{2^2T[0.4] - T[0.8]}{2^2 - 1} = \frac{4(167.01946) - 173.52604}{3} \doteq 164.85060 \qquad \textbf{(27)}$$

It is no accident that $T_1[0.4]$ equals $S_{1/3}[0.4]$ in (26c), because in general:

> If $T[2h]$ is used in Richardson's formula to improve $T[h]$, the result is the Composite Simpson approximation $S_{1/3}[h]$. **(28)**

To see why, look at Figure 7.3-2, which shows the effect of halving the panel width from $2h$ to h (i.e., doubling n). Note that $\sum_{\text{odd}}[h]$ is the sum of the "new" sampled nodes (i.e., the midpoints of the $2h$-panels), whereas

$$\sum_{\text{even}}[h] = \sum_{\text{interior}}[2h]; \quad \text{also} \quad \sum_{\text{interior}}[h] = \sum_{\text{even}}[h] + \sum_{\text{odd}}[h] \tag{29}$$

Using this in Richardson's formula with $r = (2h)/h = 2$ and $n = 2$ gives

$$T_1[h] = \frac{2^2T[h] - T[2h]}{2^2 - 1} \tag{30a}$$

$$= \frac{1}{3}\left\{4h\left(\frac{f(a) + f(b)}{2} + \sum_{\text{odd}}[h] + \sum_{\text{even}}[h]\right) - 2h\left(\frac{f(a) + f(b)}{2} + \sum_{\text{even}}[h]\right)\right\}$$

$$= \frac{h}{3}\left\{f(a) + f(b) + 4\sum_{\text{odd}}[h] + 2\sum_{\text{even}}[h]\right\} = S_{1/3}[h] \tag{30b}$$

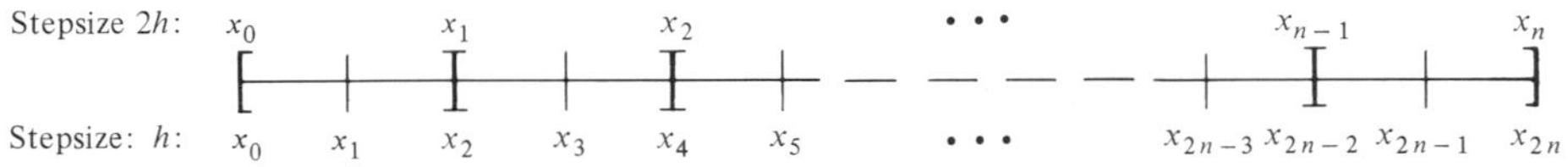

Figure 7.3-2 *Comparing subscripts for stepsizes h and 2h.*

There is another important consequence of Figure 7.3-2. In view of (29), the Composite Trapezoidal Rule (20a) can be written as

$$T[h] = h\left\{\frac{f(a) + f(b)}{2} + \sum_{\text{interior}}[2h] + \sum_{\text{odd}}[h]\right\} \tag{31a}$$

$$= \frac{1}{2}\left\{2h\left(\frac{f(a) + f(b)}{2} + \sum_{\text{interior}}[2h]\right) + 2h\sum_{\text{odd}}[h]\right\} \tag{31b}$$

Substituting $T[2h]$ for its formula in (31b) yields the

> **Recursive Form of $T[h]$:** $$T[h] = \frac{1}{2}\left(T[2h] + 2h\sum_{\text{odd}}[h]\right) \tag{32}$$

We can thus obtain $T[h]$ by averaging $T[2h]$ with the $2h$-weighted sum of the sampled values at the *midpoints* of the $2h$-panels. For example, $T[0.4]$ in Example 7.3A(b) could have been obtained using $T[0.8]$ of part (a) as follows:

$$T[0.4] = \tfrac{1}{2}\{T[0.8] + 0.8(e^{3.2} + e^{4.0} + e^{4.8})\} = 167.01946 \tag{33}$$

Practical Consideration 7.3B (Evaluating $T[h]$ and $S_{1/3}[h]$). The recursive form (32) avoids duplicating the $T[2h]$ calculation, hence is the recommended way to find $T[h]$ when $T[2h]$ is known. Also, the value of $S_{1/3}[h]$ is best calculated as $T_1[h]$ as in (27), rather than directly using (21b), because its calculation as $T_1[h]$ will indicate its accuracy (see Practical Consideration 7.2G). ■

7.3C Romberg Integration

We shall see in Section 7.3E that the truncation error of the Composite Simpson's Rule $S_{1/3}[h]$ satisfies

$$\tau_{S_{1/3}}[h] = Cf^{(iv)}(\xi_4)h^4 + Df^{(vi)}(\xi_6)h^6 + Ef^{(viii)}(\xi_8)h^8 + \cdots \qquad \textbf{(34)}$$

where each of ξ_4, ξ_6, ξ_8, ... lies somewhere between a and b. So if $f^{(iv)}$, $f^{(vi)}$, $f^{(viii)}$, ... are nearly constant on $[a, b]$, then the $O(h^2)$ Composite Trapezoidal estimates

$$T[h_0], \quad T[\tfrac{1}{2}h_0], \quad T[\tfrac{1}{4}h_0], \quad T[\tfrac{1}{8}h_0], \quad \ldots \qquad (r = 2) \qquad \textbf{(35a)}$$

can be improved using a Richardson table as shown in Table 7.3-2. When Richardson's theorem is used this way (i.e., with $F[h] = T[h]$ and $r = 2$), the procedure is called **Romberg integration** and the resulting table is called a **Romberg table**. In the $T[h]$ column of the table, all but the first $T[h]$ entry (i.e., $T[h_0]$) are most efficiently obtained from $T[2h]$ using the Recursive Form (32).

TABLE 7.3-2 Romberg Table

h	$T[h]$ $O(h^2)$	$T_1[h] = S_{1/3}[h]$ $O(h^4)$	$T_2[h]$ $O(h^6)$	$T_3[h]$ $O(h^8)$	$T_4[h]$ $O(h^{10})$
h_0	$T[h_0]$				
$\frac{1}{2}h_0$	$T[\frac{1}{2}h_0]$	$T_1[\frac{1}{2}h_0]$			
$\frac{1}{4}h_0$	$T[\frac{1}{4}h_0]$	$T_1[\frac{1}{4}h_0]$	$T_2[\frac{1}{4}h_0]$		
$\frac{1}{8}h_0$	$T[\frac{1}{8}h_0]$	$T_1[\frac{1}{8}h_0]$	$T_2[\frac{1}{8}h_0]$	$T_3[\frac{1}{8}h_0]$	
$\frac{1}{16}h_0$	$T[\frac{1}{16}h_0]$	$T_1[\frac{1}{16}h_0]$	$T_2[\frac{1}{16}h_0]$	$T_3[\frac{1}{16}h_0]$	$T_4[\frac{1}{16}h_0]$

The calculations indicated by the arrows in Table 7.3-2 are

$$\begin{array}{ccccccccccc} T[2h] & & & & T_1[2h] & & & & T_2[2h] & & \\ & \searrow{*2^2} & & & & \searrow{*2^4} & & & & \searrow{*2^6} & \\ & & & \div 3 & & & & \div 15 & & & \div 63 \\ T[h] & \rightarrow & \ominus & \longrightarrow & T_1[h] & \rightarrow & \ominus & \longrightarrow & T_2[h] & \rightarrow \ominus \longrightarrow & T_3[h] \end{array} \qquad \textbf{(35b)}$$

and so on. Thus $T_1[h]$, $T_2[h]$, ... can be obtained from the formula

$$T_i[h] = \frac{4^i T_{i-1}[h] - T_{i-1}[2h]}{4^i - 1}, \qquad i = 1, 2, \ldots \qquad \textbf{(35c)}$$

with $T_0[h]$ denoting the trapezoidal estimate $T[h]$.

Practical Consideration 7.3C (Starting and stopping Romberg integration). The initial panel length h_0 should be small enough so that the graph of f has *at most* one turning point and one inflection point over each panel but not much smaller. If this is done and $f(x)$ is continuous on $[a, b]$, then any desired accuracy (which is reasonable for the device) should be achievable using at most nine rows of the Romberg table. The algorithm should be terminated when two successive improvements are sufficiently close, that is, when

$$|T_i[h] - T_{i-1}[h]| \text{ indicates the desired accuracy} \tag{35d}$$

$T_i[h]$ can then be taken as the desired approximation of $\int_a^b f(x)\,dx$. ■

EXAMPLE 7.3C Use Romberg integration to evaluate $\int_1^{2.2} \ln x\,dx$ to $6s$.

SOLUTION: Since $f(x) = \ln x$ is increasing and concave down on $[1, 2.2]$, we could start with only one panel. However, for a bit more initial accuracy, we shall start with $n = 2$ so that $h_0 = (2.2 - 1)/2 = 0.6$. The following calculations (carrying one guard digit, i.e., $7s$) give the Romberg table shown in Table 7.3-3. We begin with $T[h_0]$:

$$\mathbf{k = 0:}\quad T[0.6] = 0.6\left\{\frac{\ln 2.2 + \ln 1}{2} + \ln 1.6\right\} = 0.5185394 \tag{36}$$

Using (20a) for $T[h]$, then Richardson's formula [or (35b)] for $T_i[h]$ gives:

$$\mathbf{k = 1:}\quad T[0.3] = \tfrac{1}{2}\{T[0.6] + 0.6(\ln 1.3 + \ln 1.9)\} = 0.5305351 \tag{37a}$$

$$T_1[0.3] = \frac{4T[0.3] - T[0.6]}{3} = \frac{4(0.5305351) - 0.5185394}{3} = 0.5345337 \tag{37b}$$

Since $T_1[0.3]$ and $T[0.3]$ do not agree to about $6s$, we form another row.

$$\mathbf{k = 2:}\quad T[0.15] = \tfrac{1}{2}\{T[0.3] + 0.3(\ln 1.15 + \ln 1.45 + \ln 1.75 + \ln 2.05)\} \tag{38a}$$
$$= 0.5335847$$

$$T_1[0.15] = \frac{4T[0.15] - T[0.3]}{3} = \frac{4(0.5335847) - 0.5305351}{3} = 0.5346013 \tag{38b}$$

$$T_2[0.15] = \frac{16T_1[0.15] - T_1[0.3]}{15} = \frac{16(0.5346013) - 0.5345337}{15} = 0.5346058 \tag{38c}$$

Since $T_2[0.15]$ and $T_1[0.15]$ agree to almost $6s$, $T_2[0.15] = 0.5346058$ should be accurate to $6s$. In fact, the exact integral (to $7s$) is 0.5346062. ■

TABLE 7.3-3 Romberg Table for $\int_1^{2.2} \ln x\,dx$; $h_0 = 0.6$ (Two Panels)

h	$T[h]$, $O(h)^2$	$T_1[h] = S_{1/3}[h]$ $\tau[h] = O(h^4)$	$T_2[h]$, $O(h^6)$
0.6	0.5185394		
0.3	0.5305351	0.5345337	
0.15	0.5335847	0.5346013	0.5346058

Note from Table 7.3-3 that $T[0.15]$ is accurate to only almost $3s$. This demonstrates improvement in accuracy that can be achieved using Romberg integration when the integrand is sufficiently smooth. In fact, in the absence of roundoff error, Romberg integration will converge to $I = \int_a^b f(x)\,dx$ even if f is not particularly smooth; all that is required is for $T[h_0/2^k]$ to approach I as $k \to \infty$ (see [40]). Another way of describing the improvement follows from (34):

$$T_i[h] \text{ will be exact for polynomials of degree} \leq 2i + 1 \tag{39}$$

Pseudocode for Romberg integration is given in Algorithm 7.3C where for $k = 0, 1, \ldots, MaxRows$, the $k + 1$ entries when $h = h_0/2^k$, namely

$$T[h], \quad T_1[h], \quad T_2[h], \quad \ldots, \quad T_k[h] \tag{40a}$$

are stored in the kth row of a matrix T using the subscripts

$$T_{k,0}, \quad T_{k,1}, \quad T_{k,2}, \quad \ldots, \quad T_{k,k} \tag{40b}$$

If the termination test (35d) is not satisfied in *MaxRows* (say 9) rows, you should check to see if f has a singularity on $[a, b]$. If one is detected, use a method of Section 7.4. If you cannot find one, repeat Algorithm 7.3C with a *larger* h_0 if roundoff error is evident, and a *smaller* h_0 otherwise.

ALGORITHM 7.3C. ROMBERG INTEGRATION

PURPOSE: To find $\int_b^a f(x)\,dx$ to *NumSig* significant digits by forming a Romberg table stored in a matrix T. The kth iteration (for which $h = h_0/2^k$) yields

$$\text{row}_k T = [T_{k,0} \quad T_{k,1} \quad \cdots \quad T_{k,k-1} \quad T_{k,k}], \qquad k = 0, 1, \ldots, MaxRows$$

GET a, b, {endpoints of the interval of integration}
 n, {initial number of panels}
 MaxRows, *NumSig* {termination parameters}
$h \leftarrow (b - a)/n$ {this is h_0}
$T_{0,0} \leftarrow h\left(\dfrac{f(a) + f(b)}{2} + \sum_{\text{interior}} [h]\right)$ {Composite Trapezoidal Rule}
$RelTol \leftarrow 10^{-NumSig}$
DO FOR $k = 1$ TO *MaxRows* UNTIL $Convgd$ = TRUE {Boolean convergence flag}
 $h \leftarrow h/2$ {h is now $h_0/2^k$}
 $T_{k,0} \leftarrow \dfrac{1}{2}\left(T_{k-1,0} + 2h\sum_{\text{odd}}[h]\right)$ {Recursive Trapezoidal Rule}
 DO FOR $i = 1$ TO k UNTIL $Convgd$ = TRUE
  $T_{k,i} \leftarrow \dfrac{4^i T_{k,i-1} - T_{k-1,i-1}}{4^i - 1}$ {this is $T_i[h]$}
  $Convgd \leftarrow (|T_{k,i} - T_{k,i-1}| \leq RelTol \cdot \max\{1, |T_{k,i}|\})$

IF *Convgd*
 THEN OUTPUT ('$\int_b^a f(x)\,dx$ = '$T_{k,i}$' to '*NumSig*' significant digits')
 ELSE OUTPUT (*MaxRows*' rows did not yield the desired accuracy')
STOP

7.3D Integrating Tabulated Functions

How can we get accurate approximations of $\int_a^b f(x)\,dx$ when $f(x)$ is known only at tabulated nodes $x = x_i$? If the nodes are not equispaced, the general formulas of Section 7.1B can be used to get a formula that ensures suitable polynomial exactness. If they are, but a or b is not tabulated, then Table 7.3-1 can be used as in Section 7.3A to do this more efficiently.

This section is concerned with evaluating $\int_a^b f(x)\,dx$ when $f(x)$ *is tabulated at h-spaced nodes*, and *both a and b are tabulated*, so that $[a, b]$ consists of n h-panels, where $n = (b - a)/h$. The general strategy in this situation is to *partition* $[a, b]$ *into several subintervals, each consisting of a finite number of panels*, and use a suitable approximation formula over each subinterval. Two general guidelines for doing this effectively are

1. The error will be mostly that of the least accurate method, so all formulas used should have the *same exactness degree*.
2. When seeking cubic exactness, *try to use the (more accurate) Simpson's $\frac{1}{3}$-Rule* over as many "double panels" as possible.

With these suggestions in mind, let k be the highest power of 2 that divides into n, so that

$$n = 2^k p, \qquad \text{where } p \text{ is an odd integer} \tag{41}$$

Table 7.3-4 recommends strategies designed to get at least cubic exactness. These strategies are illustrated in Example 7.3D.

TABLE 7.3-4 Strategies for Integrating over n h-panels; $n = 2^k p$

k	*Suggested Integration Strategy over* $[a, b]$	*Exactness*
$k = 0$ (n is odd)	Use Simpson's $\frac{3}{8}$-Rule over *either* the first *or* last three panels, and $S_{1/3}[h]$ over the rest.	Cubic
$k > 0$ (n is even)	Use Romberg integration with $h_0 = 2^k h$. You can get to $T_k[h_0/2^k] = T_k[h]$ on the table.	Degree $2k + 1$

EXAMPLE 7.3D Approximate the following integrals with at least cubic exactness accuracy, assuming that the only available values of e^x are at the 0.4-spaced nodes shown in Figure 7.3-3.

(a) $\int_{2.0}^{5.6} e^x\,dx$ (9 panels; exact value $= e^{5.6} - e^{2.0} \doteq 263.0374$)
(b) $\int_{2.0}^{6.8} e^x\,dx$ (12 panels; exact value $= e^{6.8} - e^{2.0} \doteq 890.4583$)

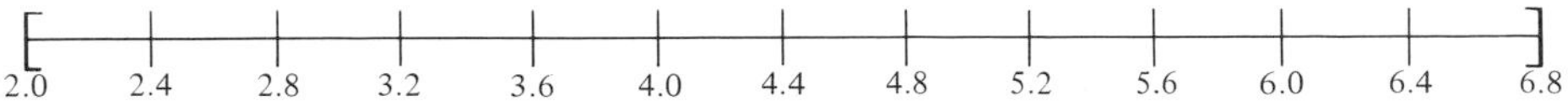

Figure 7.3-3 *Thirteen nodes at which values of e^x are known.*

SOLUTION (a): Since the interval of integration consists of an odd number of 0.4-panels, we can get cubic exactness as follows:

$$\int_{2.0}^{5.6} e^x\,dx \approx S_{1/3}[0.4]_{2.0}^{4.4} + S_{3/8}]_{4.4}^{5.6} \qquad \textbf{(42a)}$$

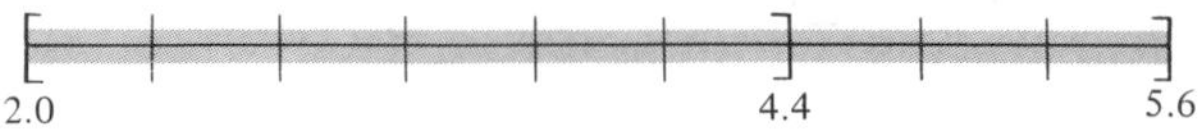

$$= \frac{0.4}{3}\,[e^{2.0} + e^{4.4} + 4(e^{2.4} + e^{3.2} + e^{4.0}) + 2(e^{2.8} + e^{3.6})]$$

$$+ \tfrac{3}{8}(0.4)[e^{4.4} + 3(e^{4.8} + e^{5.2}) + e^{5.6})]$$

$$= 74.07215 + 189.03379 = 263.1059 \qquad \textbf{(42b)}$$

Alternatively, we can use Simpson's $\frac{3}{8}$-Rule on the leftmost three panels thus:

$$\int_{2.0}^{5.6} e^x\,dx \approx S_{3/8}]_{2.0}^{3.2} + S_{1/3}[0.4]_{3.2}^{5.6} \qquad \textbf{(43a)}$$

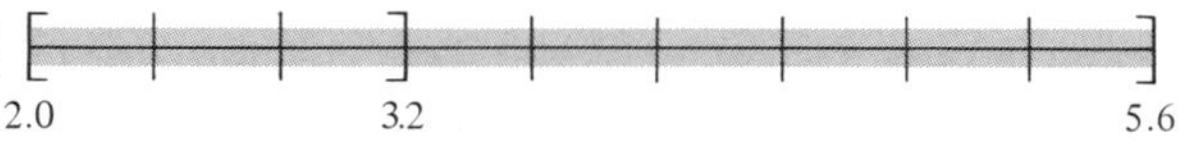

$$\doteq 17.14876 + 245.92819 = 263.0770 \qquad \textbf{(43b)}$$

Either strategy (42) or (43) could thus have been used to approximate $\int_{2.0}^{5.6} e^x\,dx$ to $4s$. However, it is advisable to get *both* because doing so enables you to use the number of digits to which they agree *with each other* (here four) as an estimate of their accuracy.

SOLUTION (b): Here $12 = 2^2 3$ is even, with $k = 2$ and $p = 3$ in (41). So we can use Romberg integration, starting with three panels of length $2^k h = 4(0.4) = 1.6$ (Table 7.3-5). Since $T_2[0.4] = 890.4654$ agrees with $T_1[0.4] = 890.5835 = S_{1/3}[0.4]$ to almost $4s$, we could have concluded *without knowing the exact answer* that $T_2[0.4]$ is accurate to about $5s$. This fifth-degree exact approximation has about one more accurate digit than the cubic exact approximations (42) and (43). ■

TABLE 7.3-5 Romberg Table (Rounded to 7s) for $\int_{2.0}^{6.8} e^x\,dx$

h	$T[h]$, $O(h^2)$	$T_1[h] = S_{1/3}[h]$ $\tau[h] = O(h^4)$	$T_2[h]$, $O(h^6)$
$h_0 = 1.6$	1072.782		
0.8	937.4504	892.3399	
0.4	902.2995	890.5825	890.4654

Problem M7-38 examines the possibility of getting better than cubic exactness when integrating over an odd number of panels.

7.3E The Truncation Error of $S_{1/3}[h]$ and $M]_a^b$

To get a series for the truncation error of Simpson's $\frac{1}{3}$-Rule, we introduce the indefinite integral function for the integrand $f(x)$, namely

$$I(x) = \int_{x_j}^{x} f(u)\,du \qquad \text{for which } I(x_j) = 0 \quad \text{and} \quad I'(x) = \frac{dI(x)}{dx} = f(x) \qquad \textbf{(44)}$$

The Taylor series for $I(x_j + h) = I(x_{j+1})$ and $I(x_j - h) = I(x_{j-1})$ are

$$I(x_{j+1}) = I(x_j) + I'(x_j)h + \frac{I''(x_j)}{2!}h^2 + \frac{I'''(x_j)}{3!}h^3 + \frac{I^{(\text{iv})}(x_j)}{4!}h^4 + \cdots \tag{45a}$$

$$I(x_{j-1}) = I(x_j) - I'(x_j)h + \frac{I''(x_j)}{2!}h^2 - \frac{I'''(x_j)}{3!}h^3 + \frac{I^{(\text{iv})}(x_j)}{4!}h^4 - \cdots \tag{45b}$$

where, by (44), $I'(x_j) = f_j$, $I''(x_j) = f'(x_j)$, and so on. Since $I(x_{j+1}) - I(x_{j-1})$ is the definite integral of $f(x)$ over $[x_{j-1}, x_{j+1}]$ (this is the Fundamental Theorem of Calculus), we can subtract (45b) from (45a) to get

$$\int_{x_{j-1}}^{x_{j+1}} f(x)\,dx = \underbrace{2hf_j}_{\text{Midpoint Rule}} + \underbrace{\frac{2}{3!}f''(x_j)h^3 + \frac{2}{5!}f^{(\text{iv})}(x_j)h^5 + \frac{2}{7!}f^{(\text{vi})}(x_j)h^7 + \cdots}_{\tau_M \,=\, \text{its truncation error}} \tag{46}$$

If we replace $f''(x_j)$ in (46) by the Taylor series obtained in (13c) of Section 7.2C, namely

$$f''(x_j) = \frac{f_{j-1} - 2f_j + f_{j+1}}{h^2} - \frac{1}{12}f^{(\text{iv})}(x_j)h^2 - \frac{1}{360}f^{(\text{vi})}(x_j)h^4 - \cdots \tag{47}$$

and then combine like powers of h, we get the series

$$\int_{x_{j-1}}^{x_{j+1}} f(x)\,dx = \underbrace{\frac{h}{3}[f_{j-1} + 4f_j + f_{j+1}]}_{\text{Simpson's } \frac{1}{3}\text{-Rule}} + \underbrace{\frac{-1}{90}f^{(\text{iv})}(x_j)h^5 + Df^{(\text{vi})}(x_j)h^7 + \cdots}_{\tau_{S1/3} \,=\, \text{its truncation error}} \tag{48}$$

So if $[a, b]$ is partitioned into n h-panels, where n is even, then (48) can be used over each of the *double panels* $[x_0, x_2], [x_2, x_4], \ldots, [x_{n-2}, x_n]$ (where $a = x_0$ and $b = x_n$); summing the $n/2$ resulting τ_S terms and applying the Intermediate Value Theorem as in (20c) then yields the series for the truncation error of $S_{1/3}[h]$ given in (34).

7.4

Methods Based on Unevenly Spaced Nodes

The integration methods of the preceding sections were based on formulas for equispaced sampled nodes. These methods are very efficient for differentiable integrands $f(x)$ that can be approximated fairly well by a single polynomial over the entire integration interval $[a, b]$. However, if the graph of f has nearly constant slope over part of the interval but changes rapidly over other parts, then methods based on uniform spacing are inefficient. This is especially true if f has an infinite discontinuity (i.e., a singularity) on $[a, b]$.

In this section we describe two methods that sample values of $f(x)$ efficiently, namely *Gauss quadrature* (Sections 7.4A and 7.4B) and *adaptive quadrature* (Section 7.4F), and we propose strategies for dealing with improper integrals (Sections 7.4D and 7.4E).

7.4A Gauss Quadrature on $[-1, 1]$

We first restrict consideration to the "normalized" interval $[-1, 1]$. For any $n \geqslant 1$, our objective is to find n sample points

$$\xi_1, \xi_2, \ldots, \xi_n \qquad \text{in } [-1, 1] \tag{1a}$$

(which are *not* selected in advance) and n corresponding weights

$$\gamma_1, \gamma_2, \ldots, \gamma_n \tag{1b}$$

such that the **n-point Gauss quadrature formula on $[-1, 1]$,** namely

$$\int_{-1}^{1} f(\xi)\, d\xi \approx \gamma_1 f(\xi_1) + \gamma_2 f(\xi_2) + \cdots + \gamma_n f(\xi_n) \tag{2}$$

is exact for polynomials of degree $\leqslant 2n - 1$. We shall call the ξ_k's in (1a) the **Gaussian sample points** of $[-1, 1]$, and the γ_k's in (1b) their **Gaussian weights.**

Making (2) exact for $f(\xi) = 1, \xi, \xi^2, \ldots, \xi^{2n-1}$, yields $2n$ equations

$$(E_k)\ \gamma_1\xi_1^{k-1} + \gamma_2\xi_2^{k-1} + \cdots + \gamma_n\xi_n^{k-1} = \int_{-1}^{1} \xi^{k-1}\, d\xi, \qquad k = 1, 2, \ldots, 2n \tag{3}$$

where the kth equation (E_k) imposes the exactness of (2) for $f(\xi) = \xi^{k-1}$. From elementary calculus

$$\int_{-1}^{1} \xi^{k-1}\, d\xi = \left.\frac{\xi^k}{k}\right]_{-1}^{1} = \begin{cases} 0 & \text{if } k \text{ is even} \\ 2/k & \text{if } k \text{ is odd} \end{cases} \tag{4}$$

The system (3) is nonlinear in the $2n$ variables $\xi_1, \ldots, \xi_n, \gamma_1, \ldots, \gamma_n$. Its solution generally requires a numerical procedure such as NR_n (Section 4.4C).

EXAMPLE 7.4A Derive the **(a)** two-point and **(b)** four-point Gauss quadrature formulas on $[-1, 1]$.

Solution (a): Taking $n = 2$ in (3) and (4) yields a nonlinear system in $\xi_1, \xi_2, \gamma_1, \gamma_2$:

$$\mathbf{f(\xi) = 1:} \qquad (\mathrm{E}_1)\ \gamma_1 1 + \gamma_2 1 = \int_{-1}^{1} 1\, d\xi = \frac{2}{1}$$

$$\begin{aligned} f(\xi) = \xi: &\quad (\mathrm{E}_2)\ \gamma_1\xi_1 + \gamma_2\xi_2 = \int_{-1}^{1} \xi\, d\xi = 0 \\ f(\xi) = \xi^2: &\quad (\mathrm{E}_3)\ \gamma_1\xi_1^2 + \gamma_2\xi_2^2 = \int_{-1}^{1} \xi^2\, d\xi = \frac{2}{3} \\ f(\xi) = \xi^3: &\quad (\mathrm{E}_4)\ \gamma_1\xi_1^3 + \gamma_2\xi_2^3 = \int_{-1}^{1} \xi^3\, d\xi = 0 \end{aligned} \tag{5}$$

This *nonlinear* system can be solved analytically if we argue as follows: Since the interval $[-1, 1]$ is *symmetric* about 0, formula (2) should give equal importance to the values of f on $[-1, 0]$ and $[0, 1]$; that is, we should have

$$\xi_1 = -\xi_2 \text{ and } \gamma_2 = \gamma_1$$

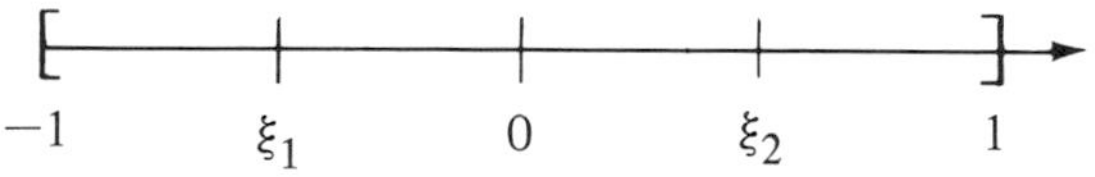

It then follows from (E_1) that γ_1 and $\gamma_2 = 1$, and from (E_3) that

$$1(\xi_1)^2 + 1(-\xi_1)^2 = \frac{2}{3} \iff \xi_1^2 = \frac{1}{3} \iff \xi_1 = -\frac{1}{\sqrt{3}},\ \xi_2 = \frac{1}{\sqrt{3}}$$

So the **two-point Gauss quadrature formula on $[-1, 1]$** is simply

$$\int_{-1}^{1} f(\xi)\, d\xi \approx f\left(-\frac{1}{\sqrt{3}}\right) + f\left(\frac{1}{\sqrt{3}}\right), \qquad \text{where } \frac{1}{\sqrt{3}} \doteq 0.5773502692 \tag{6}$$

By construction [see (5)], (6) is exact for polynomials of degree ≤ 3.

SOLUTION (b): Taking $n = 4$ in (3) and (4) and using the symmetry relations

$$\xi_3 = -\xi_2\ (=\xi), \quad \xi_4 = -\xi_1\ (=\eta), \quad \gamma_2 = \gamma_3\ (=w), \quad \gamma_1 = \gamma_4$$

along with the exactness condition for $f(\xi) = 1$, namely $2\gamma_1 + 2\gamma_2 = 2$ yields the 3×3 linear system of Example 4.4C. The components of $\bar{\mathbf{x}} = [\bar{w}\ \bar{\xi}\ \bar{\eta}]^T$ obtained there agree with the $n = 4$ values for γ_2, ξ_3, and ξ_4 given in Table 7.4-1. An alternative (preferable) method for getting the entries of Table 7.4-1 is described in Problem M7-21. ■

TABLE 7.4-1 ξ_k **and** γ_k **for** $\int_{-1}^{1} f(\xi)\, d\xi \approx \sum_{k=1}^{n} \gamma_k f(\xi_k)$, $n = 2, \ldots, 6$

n	ξ_k	γ_k	n	ξ_k	γ_k
2	$\pm 1/\sqrt{3}$	1	5	±0.9061798459 ±0.5384693101 0	0.2369268850 0.4786286705 0.5688888889
3	$\pm\sqrt{0.6}$ 0	5/9 8/9			
4	±0.8611363116 ±0.3399810436	0.3478548451 0.6521451549	6	±0.9324695142 ±0.6612093865 ±0.2386191861	0.1713244924 0.3607615730 0.4679139346

Similarly, the **three-point Gauss quadrature formula on $[-1, 1]$** is

$$\int_{-1}^{1} f(\xi)\,d\xi \approx \frac{1}{9}[5f(-\sqrt{0.6}) + 8f(0) + 5f(\sqrt{0.6})],$$
$$\text{where } \sqrt{0.6} \doteq 0.7745966692 \qquad \textbf{(7)}$$

This formula is exact for polynomials of degree $\leq 2\cdot 3 - 1 = 5$.

A table of Gauss sample points ξ_k and their weights γ_k for $n = 2, 3, 4, 5, 6$ is given in Table 7.4-1.

Notice that (i) the Gaussian sample points ξ_k are symmetrically located in the *open* interval $(-1, 1)$ so that (2) *is an open formula*, (ii) the density of the Gaussian sample points ξ_k is greatest near the endpoints of $(-1, 1)$, and (iii) the γ_k's are all positive. More extensive tables and properties of the ξ_k's and γ_k's can be found in [46].

7.4B Gauss Quadrature on $[a, b]$

To use Gauss quadrature to integrate $f(x)$ over the arbitrary closed interval $[a, b]$, we simply map the ξ-interval $[-1, 1]$ into the x-interval $[a, b]$ using the linear transformation

$$x = a + \frac{b-a}{2}(\xi + 1), \qquad dx = \frac{b-a}{2}\,d\xi \qquad \textbf{(8)}$$

(see Figure 6.1-12). Making this substitution in $\int_a^b f(x)\,dx$ gives

$$\begin{aligned}\int_{x=a}^{b} f(x)\,dx &= \int_{\xi=-1}^{1} f\left(a + \frac{b-a}{2}(\xi+1)\right)\cdot\frac{b-a}{2}\,d\xi \\ &= \frac{b-a}{2}\int_{\xi=-1}^{1} f\left(a + \frac{b-a}{2}(\xi+1)\right)d\xi\end{aligned} \qquad \textbf{(9)}$$

If we use a Gauss formula on $[-1, 1]$ to approximate this last integral, we get

$$\int_{x=a}^{b} f(x)\,dx \approx \frac{b-a}{2}\{\gamma_1 f(x_1) + \gamma_2 f(x_2) + \cdots + \gamma_n f(x_n)\} \qquad \textbf{(10a)}$$

where γ_k is the tabulated Gaussian weight associated with the tabulated Gaussian sample point ξ_k in $[-1, 1]$, and x_k is obtained from ξ_k as follows:

$$x_k = a + \frac{b-a}{2}(\xi_k + 1), \qquad k = 1, 2, \ldots, n \qquad \textbf{(10b)}$$

These x_k's will be referred to as the ***n* Gaussian sample points of $[a, b]$; *n*-point Gauss quadrature on $[a, b]$** thus consists of two steps:

Step 1: Use (10b) to get the n sample points $x_1, \ldots, x_n$.

Step 2: Put $x_1, \ldots, x_n$ in (10a) to get the desired integral.

EXAMPLE 7.4B Use Gauss quadrature to approximate

$$\textbf{(a)} \int_0^{0.5} e^x\,dx \qquad \textbf{(b)} \int_1^{2.2} \ln x\,dx$$

SOLUTION (a): From (10b), the Gaussian sample points on [0, 0.5] are

$$x_k = 0 + \frac{0.5 - 0}{2}(\xi_k + 1) = \frac{1}{4}(\xi_k + 1), \qquad k = 1, 2, \ldots, n \tag{11a}$$

Since e^x has no inflection points, it can be approximated well by a cubic on the ''thin'' interval [0, 0.5]; so we expect n-point Gauss quadrature to give good accuracy when $2n - 1 = 3$ (i.e., $n = 2$). From (11a) and Table 7.4-1,

$$x_1 = \frac{1}{4}\left(-\frac{1}{\sqrt{3}} + 1\right) \doteq 0.105662 \qquad \text{and} \qquad x_2 = \frac{1}{4}\left(\frac{1}{\sqrt{3}} + 1\right) \doteq 0.394338 \tag{11b}$$

Taking these values in the general Gauss formula (10a), we get

$$\int_0^{0.5} e^x\,dx \approx \frac{0.5 - 0}{2}\{1e^{x_1} + 1e^{x_2}\}$$
$$\doteq 0.648712 \qquad (\text{error} = 0.648721 - 0.648712 = 0.000009)$$

This open two-point Gauss formula gives almost 5s accuracy, comparable to the six-point Simpson's rule calculations in (c) of Example 7.3B!

SOLUTION (b): From (10b), the Gauss sample points on [1, 2.2] are

$$x_k = 1 + \frac{2.2 - 1}{2}(\xi_k + 1) = 1 + 0.6(\xi_k + 1) = 1.6 + 0.6\xi_k \tag{12a}$$

Again $f(x)$ has no inflection points. However, [1, 2.2] is not a ''thin'' interval; so we shall use the three-point formula. From Table 7.4-1 and (12a),

$$x_1 = 1.6 + 0.6(-\sqrt{0.6}) \doteq 1.135242, \quad x_2 = 1.6, \quad x_3 = 1.6 + 0.6(\sqrt{0.6}) \doteq 2.064758$$

Hence, by the general Gauss quadrature formula (10a),

$$\int_1^{2.2} \ln x\,dx \approx \frac{2.2 - 1}{2}\left\{\frac{5}{9}\ln x_1 + \frac{8}{9}\ln x_2 + \frac{5}{9}\ln x_3\right\}$$
$$\doteq 0.534622 \qquad (\text{error} = 0.534606 - 0.534622 = -0.000016) \tag{12b}$$

This open three-point Gauss formula is more accurate than the five-point estimate $S[0.3]$ obtained in (37b) of Example 7.3C. It should be evident that *Gauss quadrature can attain a specified accuracy using about half as many sample points as the Newton–Cotes-based methods of* Section 7.3A. This reflects the fact that the degree of exact polynomial fit of an n-point Gauss formula is about twice that of an n-point Newton–Cotes formula.

7.4C Choosing a Method for Estimating a Proper Integral

The method used to evaluate a proper integral $\int_a^b f(x)\,dx$ depends on the situation.

Situation 1: *The number of sample points n is specified, but their location can be determined by the user.* Such situations arise when sensors that read the *rate of flow* of a liquid or gas are to be *permanently* installed along a fixed length of uniform pipe or tubing as in Figure 6.1-11 and what is wanted is the *volume* through the pipe (i.e., the *integral* of the rate along the pipe). *For maximum accuracy in such a situation, Gaussian quadrature should be used*; for the tubing problem, the sensors should be put at the Gaussian sample points along the pipe, and their readings weighted as in the Gauss quadrature formula (10a) (Problem 7-44).

Situation 2: *The function values are known only at a discrete set of x's.* Such situations arise when dealing with tabulated functions or values generated by a computer printout. *In such situations, one has no choice but to use the method of Section 7.1B if the x_k's are unevenly spaced, or the composite rules of Section 7.3D if they are equispaced* (see also Problem M7-17).

Situation 3: *$f(x)$ can be evaluated wherever necessary to determine $\int_a^b f(x)\,dx$ to a specified accuracy.* In this commonly occurring situation, one should first draw a sketch if necessary to ensure that the integral is proper (i.e., that $f(x)$ has no singularities on $[a, b]$). In order to determine that the desired accuracy has been achieved, it is necessary to obtain at least one additional, more accurate approximation. It is here that the greater accuracy of the Gauss formulas is offset by the fact that, *unlike Romberg integration, one cannot reuse previous sample values*. Thus, to test the accuracy of an n-point Gauss formula on $[a, b]$, the reasonable approach of bisecting $[a, b]$ and using the formula over each half requires $2n$ *additional* evaluations, more than twice as many as the $n - 1$ "new" evaluations needed to get $T[h/2]$ for Romberg integration [see Section 7.3C]. And Romberg integration does not require transforming ξ_k into x_k. So, *in this situation, Romberg integration is generally preferable for hand calculation*, provided that the function behaves "uniformly" on $[a, b]$; if not, an adaptive method should be used (see Section 7.4F).

The accuracy of all quadrature formulas considered so far rests on the ability to approximate the integrand $f(x)$ by a polynomial on the interval of integration $[a, b]$. Since polynomials are continuous for all x, special care is needed to estimate $\int_a^b f(x)\,dx$ accurately if $f(x)$ has discontinuities on $[a, b]$. Indeed, quadrature formulas that are generally $O(h^n)$ converge much more slowly as $h \to 0$ if $f(x)$ has singularities on $[a, b]$. *It is therefore important to examine the integrand carefully to see if it has any discontinuities on $[a, b]$ before using any numerical integration method.* Similarly, special care is needed if the interval of integration is unbounded, that is $[a, \infty)$, $(-\infty, b]$, or $(-\infty, \infty)$. We now consider these cases.

7.4D Integrals with Infinite Discontinuities on $[a, b]$

If the graph of f has a vertical asymptote on $[a, b]$, then $\int_a^b f(x)\,dx$ cannot exist as a proper integral; instead, one defines it as an **improper integral** as follows:

$$\text{If } \lim_{x \to a^+} |f(x)| = \infty, \qquad \text{then } \int_a^b f(x)\,dx = \lim_{\epsilon \to 0^+} \int_{a+\epsilon}^b f(x)\,dx \tag{13a}$$

$$\text{If } \lim_{x \to b^-} |f(x)| = \infty, \qquad \text{then } \int_a^b f(x)\,dx = \lim_{\epsilon \to 0^+} \int_a^{b-\epsilon} f(x)\,dx \tag{13b}$$

An improper integral is called **convergent** if the appropriate limit as $\epsilon \to 0^+$ converges (to a real number), and **divergent** otherwise. A strategy that can always be employed when $|f(x)| \to \infty$ somewhere on $[a, b]$ is to **"isolate" the singularity** by evaluating $\int_a^b f(x)\,dx$ as

$$\int_a^b f(x)\,dx = \int_a^{a+\epsilon} f(x)\,dx + \int_{a+\epsilon}^b f(x)\,dx \qquad \text{if } \lim_{x \to a^+} |f(x)| = \infty \tag{14a}$$

or as

$$\int_a^b f(x)\,dx = \int_a^{b-\epsilon} f(x)\,dx + \int_{b-\epsilon}^b f(x)\,dx \qquad \text{if } \lim_{x \to b^-} |f(x)| = \infty \tag{14b}$$

with Gauss quadrature used over the thin ϵ-interval near the singularity, that is, $[a, a + \epsilon]$ in (14a) or $[b - \epsilon, b]$ in (14b). Any convenient method (e.g., Romberg integration) can be used for the remaining *proper* integral (see Figure 7.4-1). For convergent integrals, the integral over the ϵ-interval will approach zero as $\epsilon \to 0^+$.

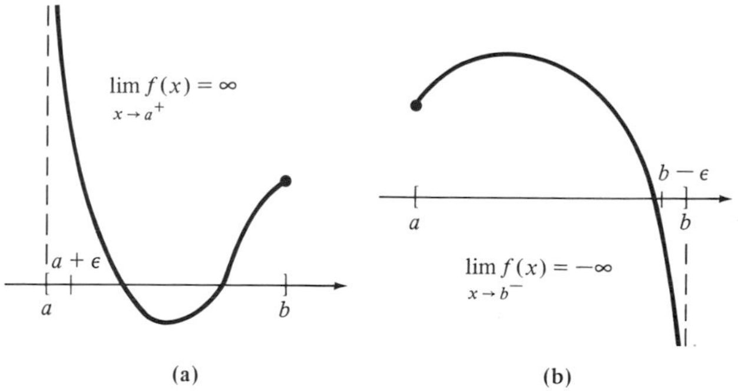

***Figure 7.4-1** Isolating a singularity: (a) at x = a; (b) at x = b.*

EXAMPLE 7.4D Find **(a)** $I_1 = \int_0^{\pi/2} \tan x\,dx$ and **(b)** $I_2 = \int_0^1 dx/\sqrt{x}$ (if they converge).

SOLUTION (a): Since $\tan x \to \infty$ as $x \to \pi/2^-$, (14b) with $\epsilon = 0.01$ gives

$$I_1 = \int_0^{\beta_1} \tan x\,dx + \int_{\beta_1}^{\pi/2} \tan x\,dx, \qquad \beta_1 = \frac{\pi}{2} - 0.01 \doteq 1.5607963$$

The (proper) first integral is 4.605187 ($6d$); five-point Gauss on $[\beta_1, \pi/2]$ gives

$$\int_{\beta_1}^{\pi/2} \tan x\, dx \approx \frac{0.01}{2}\left[\sum_{i=1}^{5} \gamma_i \tan\left(\beta_1 + \frac{0.01}{2}(\xi_i + 1)\right)\right] \doteq 4.566650 \qquad \textbf{(15)}$$

Since (15) is comparable in size to $\int_0^{\beta_1} \tan x\, dx$, we suspect that the given integral may diverge. To check this, we further "isolate" $\pi/2$:

$$\int_{\beta_1}^{\pi/2} \tan x\, dx = \underbrace{\int_{\beta_1}^{\beta_2} \tan x\, dx}_{\doteq\, 4.605153} + \underbrace{\int_{\beta_2}^{\pi 2} \tan x\, dx}_{\approx\, 4.566667}, \qquad \beta_2 = \frac{\pi}{2} - 0.0001 \doteq 1.5706963 \qquad \textbf{(16)}$$

where five-point Gauss was used for the integral on $[\beta_2, \pi/2]$. We conclude from (15) and (16) that $\int_{\pi/2-\epsilon}^{\pi/2} \tan x\, dx$ does *not* approach 0 as $\epsilon \to 0^+$, and hence that $\int_0^{\pi/2} \tan x\, dx$ diverges. In fact, by the definition (13b),

$$I_1 = \int_0^{\pi/2} \tan x\, dx = \lim_{\epsilon\to 0^+}\left\{\int_0^{\pi/2-\epsilon} \tan x\, dx\right\} = \lim_{\epsilon\to 0^+}\left\{-\ln\left|\cos\left(\frac{\pi}{2} - \epsilon\right)\right|\right\} = +\infty.$$

Solution (b): Since $1/\sqrt{x} \to \infty$ as $x \to 0^+$,

$$I_2 = \int_0^1 \frac{dx}{\sqrt{x}} = \int_0^{0.01} \frac{dx}{\sqrt{x}} + \int_{0.01}^1 \frac{dx}{\sqrt{x}}$$

The (proper) integral over $[0.01, 1]$ is 1.8; five-point Gauss on $[0, 0.01]$ gives

$$\int_0^{0.01} \frac{dx}{\sqrt{x}} \approx \frac{0.01}{2}\left\{\sum_{i=1}^{5} \gamma_i \frac{1}{\sqrt{0.005(\xi_i + 1)}}\right\} \doteq 0.184160$$

To check the apparent convergence, we further "isolate" the singularity:

$$\int_0^{0.01} \frac{dx}{\sqrt{x}} = \int_0^{0.0001} \frac{dx}{\sqrt{x}} + \int_{0.001}^{0.01} \frac{dx}{\sqrt{x}}$$
$$\doteq 0.018416 + 0.18 \text{ (using five-point Gauss on } [0, 0.0001])$$

Clearly, the integral is convergent and its value is approximately

$$I_2 \approx 1.8 + 0.18 + 0.018416 = 1.9984616$$ ■

In fact, we can use definition (13a) to see that $I_2 = \int_0^1 dx/\sqrt{x} = 2$. The poor accuracy of five-point Gauss quadrature on $[0, 0.0001]$ results from the poor fit of a polynomial of degree $\leq 2\cdot 5 - 1 = 9$ to $f(x) = 1/\sqrt{x}$ near $x = 0$.

If f has a singularity at an interior point of $[a, b]$, say as in Figure 7.4-2, then (14) can be applied over each of the subintervals. This improper integral will diverge if any one of $\int_a^c$, $\int_c^d$, or $\int_d^b$ diverges.

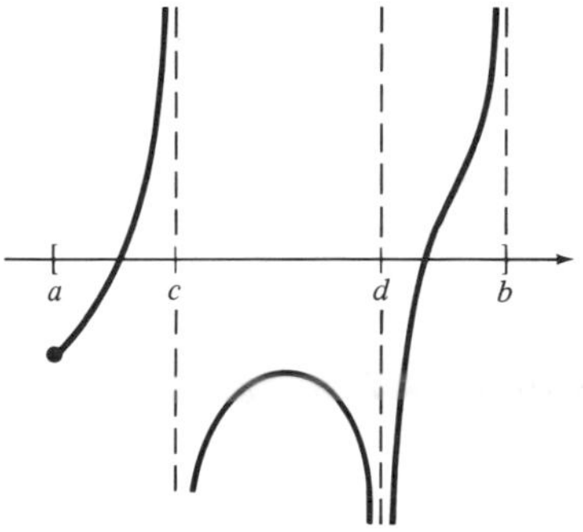

Figure 7.4-2 *Singularities at interior points of [a, b].*

7.4E Integrals over $[a, \infty)$, $(-\infty, b]$, and $(-\infty, \infty)$

Definite integrals of a continuous $f(x)$ over the unbounded intervals $[a, \infty)$ and $(-\infty, b]$ cannot be proper but are defined as the following **improper integrals**:

$$\int_a^\infty f(x)\,dx = \lim_{b\to\infty} \int_a^b f(x)\,dx \tag{17a}$$

$$\int_{-\infty}^b f(x)\,dx = \lim_{a\to-\infty} \int_a^b f(x)\,dx \tag{17b}$$

Such an improper integral is **convergent** if the appropriate limit of the proper integral exists (as a real number); otherwise, it is **divergent.**

Two general strategies for evaluating (17) are

Strategy I (Transformation): Make a substitution such as

$$x = u^{-n} \qquad \text{so that} \qquad dx = -nu^{-(n+1)}\,du \tag{18}$$

or

$$u = e^{nx} \qquad \text{so that} \qquad x = \frac{\ln u}{n},\ dx = \frac{du}{nu} \tag{19}$$

to transform (17) into a u-integral over a bounded interval; then use an appropriate method.

Strategy II (Isolation) [For $\int_a^\infty f(x)\,dx$]: Sketch $f(x)$ to determine intervals

$$[a, b_1],\quad [b_1, b_2],\quad [b_2, b_3],\quad \cdots \tag{20a}$$

over which $f(x)$ appears to behave like a polynomial. Then evaluate the integral of $f(x)$ over these intervals in sequence, stopping when it seems clear that

$$\int_a^{b_1} f(x)\,dx + \int_{b_1}^{b_2} f(x)\,dx + \cdots + \int_{b_{n-1}}^{b_n} f(x)\,dx \tag{20b}$$

either has the accuracy that is desired or is diverging. An analogous procedure applies to $\int_{-\infty}^{b} f(x)\,dx$.

Integrals over $(-\infty, \infty)$ can be separated into integrals over $(-\infty, c)$ and (c, ∞) for some convenient c. Such integrals converge if and only if both $\int_{-\infty}^{c}$ and $\int_{c}^{\infty}$ converge.

EXAMPLE 7.4E Evaluate to 5*s*: **(a)** $I_1 = \int_1^\infty x^2 e^{-x^2}\,dx$; **(b)** $I_2 = \int_0^\infty \sqrt{x}\, e^{-x}\,dx$.

SOLUTION (a): We try the substitution $u = 1/x^2$ $(x = 1/\sqrt{u},\ dx = -\frac{1}{2}u^{-3/2}\,du)$:

$$I_1 = \int_{x=1}^{\infty} x^2 e^{-x^2}\,dx = \int_{u=1}^{0} \frac{1}{u} e^{-1/u}\left(-\frac{1}{2} u^{-3/2}\,du\right) = \frac{1}{2}\int_0^1 \frac{e^{-1/u}}{u^{5/2}}\,du \tag{21a}$$

Using L'Hospital's rule, one can show that

$$f(u) = \frac{e^{-1/u}}{u^{5/2}} \quad \Rightarrow \quad f(0^+) = \lim_{u \to 0^+} \frac{e^{-1/u}}{u^{5/2}} = 0 \tag{21b}$$

hence that $f(u)$ has a *removable* discontinuity at $u = 0$. One can therefore take $f(0) = 0$ and use Romberg integration to get $I_1 \doteq 0.25364$.

Alternatively, strategy II could have been used to get

$$\begin{aligned} I_1 &= \int_1^3 x^2 e^{-x^2}\,dx + \int_3^{10} x^2 e^{-x^2}\,dx + \int_{10}^{50} x^2 e^{-x^2}\,dx + \cdots \\ &\doteq 0.253446 + 0.000195 + 0 \doteq 0.25364 \end{aligned} \tag{22}$$

with Romberg integration used over the subintervals (see also Section 7.4F).

SOLUTION (b): We try the substitution $u = e^{-x}$ $(du = -e^{-x}\,dx,\ x = -\ln u)$:

$$I_2 = \int_{x=0}^{\infty} \sqrt{x}\, e^{-x}\,dx = \int_{u=1}^{0} \sqrt{-\ln u}\,(-du) = \int_0^1 \sqrt{\ln (1/u)}\,du \tag{23a}$$

This time $f(u) = \sqrt{\ln (1/u)} \to \infty$ as $u \to 0^+$; so the transformed I_2 is improper. Isolating the singularity at 0^+ as in Section 7.3D, we get

$$I_2 = \int_{0.01}^{1} \sqrt{\ln\left(\frac{1}{u}\right)}\,du + \int_{0.0001}^{0.01} \sqrt{\ln\left(\frac{1}{u}\right)}\,du + \int_{0}^{0.0001} \sqrt{\ln\left(\frac{1}{u}\right)}\,du \tag{23b}$$

$$\doteq 0.862649 + 0.023273 + 0.000319 \doteq 0.88624\ (5s) \tag{23c}$$

Alternatively, I_2 could have been obtained to 5*s* by integrating $\sqrt{x}e^{-x}$ to 6*s* over [0, 2], [2, 10], and [10, 50] (strategy II). ■

An entertaining presentation of strategies for evaluating improper integrals is given in [1].

7.4F Adaptive Quadrature

Let us abbreviate $\int_a^b f(x)\,dx$ as simply $\int_a^b$. We saw in Section 7.3C that Romberg integration can be used to approximate $\int_a^b$ to any (reasonable) desired accuracy using equally spaced sampled nodes. This uniform spacing can result in many needless samples of $f(x)$ when the value of $\int_a^b$ is determined primarily by the values of $f(x)$ in a relatively thin subinterval of $[a, b]$ such as $[c, d]$ in Figure 7.4-3(a) or $[a, c]$ in Figure 7.4-3(b). If we took the trouble to graph these integrands over $[a, b]$ *before* integrating numerically, we could have obtained points c and d, as shown, and then integrated as

$$\int_a^b = \int_a^c + \int_c^d + \int_d^b \text{ in (a)} \qquad \text{and} \qquad \int_a^b = \int_a^c + \int_c^b \text{ in (b)} \tag{24}$$

with a large or small stepsize *as needed* used over each subinterval. A preferable strategy, which does not require a sketch, is to *have the algorithm itself adjust the stepsize* so as to use a *smaller* h where the slope of f is varying *rapidly*, and a *larger* h where it is varying *slowly*. Quadrature methods incorporating this strategy are called **adaptive** or **variable stepsize** methods.

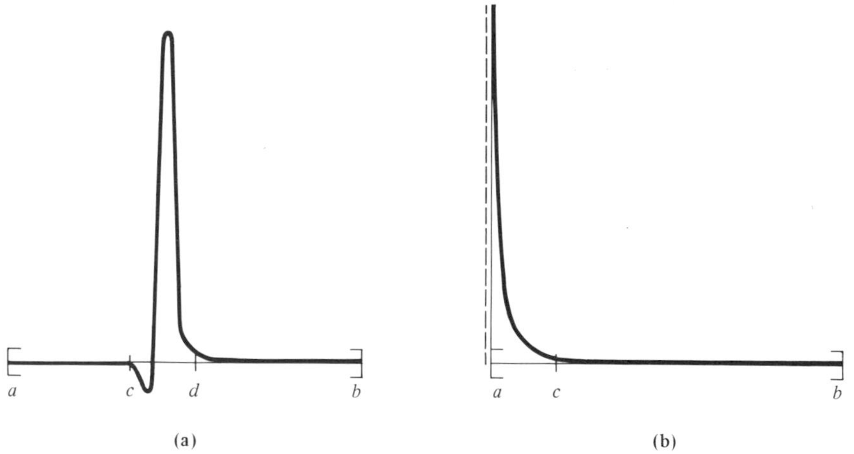

Figure 7.4-3 *(a) A "spike" on [a, b]; (b) an asymptote near [a, b].*

Adaptive methods require a quadrature formula, say QF, that can approximate the integral of f over subintervals $[\alpha, \beta]$ of $[a, b]$ with an error for which there is an *easily calculated estimate*, *ErrEst*, so that

$$QF \approx \int_\alpha^\beta \qquad \text{and} \qquad ErrEst \approx \int_\alpha^\beta - QF \tag{25}$$

We shall view QF as being applied to a sequence of "trial intervals," which are placed one above the other in a **trial stack** as follows:

If I lies directly above J in the trial stack, then I lies directly to the left of J on the real line, and the length of I is either the same as or half of the length of J **(26a)**

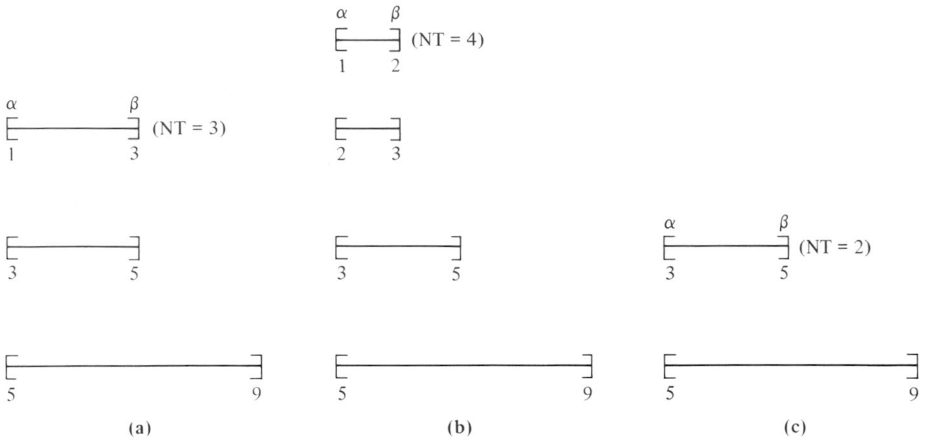

Figure 7.4-4 *(a) Trial Stack. (b) Halve* [α, β]. *(c) Remove* [α, β].

as illustrated by the three-interval stack for $[a, b] = [1, 9]$ shown in Figure 7.4-4(a). At any given time,

$$[\alpha, \beta] \text{ will denote the trial interval on top of the stack} \tag{26b}$$

Initially, $[\alpha, \beta]$ is $[a, b]$ itself. Intervals are then either added to or removed from the *top* of the trial stack as described in Algorithm 7.4F. *MinLen* is needed to prevent $[\alpha, \beta]$ from being halved [Figure 7.4-4(b)] indefinitely, as might otherwise occur if $f(x)$ has an infinite discontinuity somewhere in $[a, b]$. Indeed, if $[\alpha, \beta]$ in Figure 7.4-4(b) is too small to be halved, then $[\alpha, \beta]$ will be removed from the trial stack [Figure 7.4-4(c)] *whether QF is sufficiently accurate or not*. This ensures that the trial stack will eventually be depleted, hence that the algorithm will terminate.

The important steps of Algorithm 7.4F can be implemented as follows:

A Strategy for the Accuracy Test

Suppose that we want *Val* of Algorithm 7.4F to approximate $\int_a^b$ to within $\pm\epsilon$. Since $[a, b]$ can be viewed as being partitioned into the $[\alpha, \beta]$'s that reach the top of the trial stack, a simple proportionality argument suggests that the accumulated absolute error will be less than ϵ if each time *QF* is used,

$$QF \text{ approximates } \int_\alpha^\beta \text{ to within } \pm\frac{\beta - \alpha}{b - a}\,\epsilon \tag{27}$$

The boolean accuracy flag *NotAcc* will be set to TRUE when (27) *fails*.

Formulas for *QF* and *ErrEst* A particularly efficient way to get these formulas is to use the $O(h^4)$ Composite Simpson's $\frac{1}{3}$-Rule with only two panels, that is,

ALGORITHM 7.4F. ADAPTIVE SIMPSON QUADRATURE

PURPOSE: To approximate $\int_a^b = \int_a^b f(x)\,dx$ to a prescribed accuracy, sampling closely spaced nodes only where necessary.

GET a, b, accuracy test parameters,
    MinLen {minimum allowable length on top of stack}

Val ← 0   {*Val* accumulates the estimated value of $\int_a^b$}
Err ← 0   {*Err* accumulates the estimated error of *Val*}
Put $[a, b]$ in the trial stack   {$[\alpha, \beta]$ is now $[a, b]$}

DO UNTIL the trial stack is empty   {$[\alpha, \beta]$ is on top of the stack}
    Form *QF* and *AbsEst*   {approximations of $\int_a^b$ and $|\int_a^b - QF|$}
    NotAcc = (*AbsEst* is not sufficiently small)   {accuracy test flag}
    IF (*NotAcc* AND $|\beta - \alpha|/2 \geqslant$ *MinLen*)
        THEN   {replace $[\alpha, \beta]$ by $[\alpha, m]$ over $[m, \beta]$ in the trial stack}
            $m = (\alpha + \beta)/2$ {m is the midpoint of $[\alpha, \beta]$}
            $[\alpha, \beta] \leftarrow [\alpha, m]$ over $[m, \beta]$ {Figure 7.4-4(b)}
        ELSE   {*QF* has desired accuracy *or* $[\alpha, \beta]$ cannot be halved}
            Val ← *Val* + *QF*; *Err* ← *Err* + *AbsEst*
            Remove $[\alpha, \beta]$ from the trial stack   {Figure 7.4-4(c)}
            IF *NotAcc* THEN OUTPUT('|error| on '$[\alpha, \beta]$' ≈ '*AbsEst*)

OUTPUT(*Val*' approximates $\int_a^b$ with an absolute error of at most '*Err*)
STOP

$$S_\alpha^\beta = \frac{\beta - \alpha}{6}\,[f(\alpha) + 4f(m) + f(\beta)] \qquad h = \left(\frac{\beta - \alpha}{2}\right) \tag{28}$$

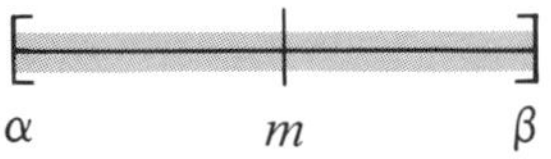

to improve the four-panel Composite Simpson's $\frac{1}{3}$-Rule approximation

$$QF = S_\alpha^m + S_m^\beta \qquad \left(h = \frac{\beta - \alpha}{4}\right) \tag{29a}$$

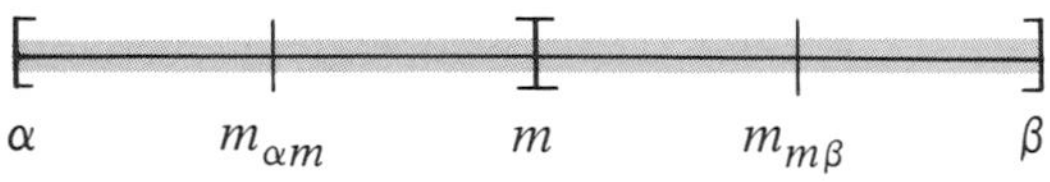

where

$$S_\alpha^m = \frac{\beta - \alpha}{12}\,[f(\alpha) + 4f(m_{\alpha m}) + f(m)] \tag{29b}$$

and

$$S_m^\beta = \frac{\beta - \alpha}{12}\,[f(m) + 4f(m_{m\beta}) + f(\beta)] \tag{29c}$$

using Richardson's formula (with $n = 4$ and $r = 2$) as follows:

$$\int_{\alpha}^{\beta} \approx QF_{\text{improved}} = \frac{2^4 QF - S_{\alpha}^{\beta}}{2^4 - 1} = \frac{16(S_{\alpha}^{m} + S_{m}^{\beta}) - S_{\alpha}^{\beta}}{15} \tag{30}$$

We can then take *ErrEst* to be ($QF_{\text{improved}} - QF$) which, when simplified, is

$$ErrEst = \frac{1}{15}(S_{\alpha}^{m} + S_{m}^{\beta} - S_{\alpha}^{\beta}) \tag{31}$$

Storing the Trial Stack Since our objective is to avoid performing unnecessary function evaluations, we will represent $[\alpha, \beta]$ in the trial stack as the "double panel" shown in (29a), that is, by storing α, m, and β *along with* the (previously obtained) values of $f(\alpha)$, $f(m)$, $f(\beta)$, *and* S_{α}^{β}. This will enable QF in (29a) and *ErrEst* in (31) to be calculated without having to reevaluate these values. We shall also store

$$Nhalve = \text{number of times } [a, b] \text{ was halved to get } [\alpha, \beta] \tag{32a}$$

and remove $[\alpha, \beta]$ from the top of the trial stack [whether or not QF satisfies the accuracy test (27)] if

$$Nhalve = MxHalv \text{ (typically 6–20)} \tag{32b}$$

where

$$MxHalv = \text{the maximum allowable number of halvings of } [a, b] \tag{32c}$$

as specified by the user. The test (32b) has the same effect as using $MinLen = |b - a|/2^{MxHalv}$ in Algorithm 7.4F, but it involves only `INTEGER` variables, hence is a bit easier to implement. Thus the trial stack will be represented in memory using eight one-dimensional arrays

`NHALV`, `A`, `M`, `B`, `FA`, `FM`, `FB`, and `SAB` **(33)**

to store *Nhalve*, α, m, β, $f(\alpha)$, $f(m)$, $f(\beta)$, and S_{α}^{β}, respectively. Notice that the maximum height of the trial stack is $Nhalve + 1$. (Why?) We shall use an `INTEGER` pointer `NT` to give the index of the current $[\alpha, \beta]$ in the arrays (33). Removing $[\alpha, \beta]$ from the trial stack is accomplished by simply replacing `NT` by `NT` − 1. An empty trial stack corresponds to `NT` = 0.

The method that results when QF in Algorithm 7.4F is taken to be the four-panel Composite Simpson's $\frac{1}{3}$-Rule as described in (27)–(33) will be called **Adaptive Simpson Quadrature**. A Fortran 77 implementation, `SUBROUTINE ADSIM`, is shown in Figure 7.4-5. It calls `SUBROUTINE SIMP13` to use Simpson's $\frac{1}{3}$-Rule to integrate over a specified subinterval and `SUBROUTINE ADDTRY` to add (or replace) an interval at a specifed height (`N`) of the trial stack. These two subprograms are shown in Figure 7.4-6.

```
ØØ1ØØ          SUBROUTINE ADSIM(F,AA,BB,EPS,MXHALV,VAL,ERR,DTAILS,IW)
ØØ2ØØ          REAL A(1Ø),M(1Ø),B(1Ø),FA(1Ø),FM(1Ø),FB(1Ø),SAB(1Ø),MM,MAM,MMB
ØØ3ØØ   C - - - - - - - - - - - - - - - - - - - - - - - - - - - - - - - - - C
ØØ4ØØ   C PURPOSE: To integrate F on [AA, BB] by Adaptive Simpson Quadrature. C
ØØ5ØØ   C          For the Nth interval [a,b] in the Trial Stack, locations   C
ØØ6ØØ   C            A(N), M(N), B(N), FA(N), FM(N), FB(N), SAB(N), NHALV(N)   C
ØØ7ØØ   C          store a, (a+b)/2, b, F(a), F(m), F(b), S1/3[a,b], and the   C
ØØ8ØØ   C          number of times [AA, BB] was halved to get [a,b].          C
ØØ9ØØ   C  INPUT: AA, BB, EPS = desired absolute error,                       C
Ø1ØØØ   C            MXHALV = maximum number times [AA, BB] can be halved     C
Ø11ØØ   C  OUTPUT: VAL = approximation obtained of the desired integral,      C
Ø12ØØ   C          ERR = approximate upper bound on the absolute error of VAL C
Ø13ØØ   C SUBROUTINES CALLED: SIMP13 to use Simpson's 1/3-rule,               C
Ø14ØØ   C                     ADDTRY to add to the Trial Stack               C
Ø15ØØ   C - - - - - - - - - - - - - - - - - - - - - VERSION 1: 9/9/85 - C
Ø16ØØ          COMMON /STACK/ A,M,B,FA,FM,FB,SAB,NHALV(1Ø)
Ø17ØØ          EXTERNAL F
Ø18ØØ          LOGICAL  DTAILS,  NOTACC  !accuracy flag for SAB(NT)
Ø19ØØ          RATIO = EPS/ABS(BB - AA)  !desired absolute error per unit x
Ø2ØØØ          VAL = Ø.Ø                 !accumulates the desired integral
Ø21ØØ          ERR = Ø.Ø                 !accumulates a bound on error of VAL
Ø22ØØ          NT = 1                    !number of intervals in Trial Stack
Ø23ØØ          FAA = F(AA)               !initialize Trial Stack with [AA, BB]:
Ø24ØØ          FBB = F(BB)
Ø25ØØ          CALL SIMP13(F,AA,BB,FAA,FBB,MM,FMM,SAABB)    !SAABB = S1/3[AA,BB]
Ø26ØØ          CALL ADDTRY(NT,Ø,AA,MM,BB,FAA,FMM,FBB,SAABB) !now SAB(1) = SAABB
Ø27ØØ   C
Ø28ØØ   C    MAIN LOOP: Test the accuracy of SAB(NT) = S1/3[a,b] until NT = Ø
Ø29ØØ      1Ø CALL SIMP13(F,A(NT),M(NT),FA(NT),FM(NT),MAM,FAM,SAM) !for [a,m]
Ø3ØØØ          CALL SIMP13(F,M(NT),B(NT),FM(NT),FB(NT),MMB,FMB,SMB) !for [m,b]
Ø31ØØ          ABSEST = ABS(SAM + SMB - SAB(NT))/15.Ø
Ø32ØØ          NOTACC = (ABSEST .GT. RATIO*ABS(B(NT) - A(NT)))
Ø33ØØ          IF (NOTACC .AND. (NHALV(NT).LT.MXHALV)) THEN     !HALVE [a,b]:
Ø34ØØ             CALL ADDTRY(NT+1,NHALV(NT)+1,A(NT),MAM,M(NT), !Put [a,m]
Ø35ØØ     &                           FA(NT),FAM,FM(NT), SAM) ! above [a,b];
Ø36ØØ             CALL ADDTRY(NT,  NHALV(NT)+1,M(NT),MMB,B(NT), !replace [m,b]
Ø37ØØ     &                           FM(NT),FMB,FB(NT), SMB) ! by [m,b].
Ø38ØØ             NT = NT + 1
Ø39ØØ          ELSE                      !REMOVE [a,b]:
Ø4ØØØ             VAL = VAL + SAM + SMB
Ø41ØØ             ERR = ERR + ABSEST
Ø42ØØ             IF (DTAILS) WRITE (IW,1) A(NT),B(NT),SAB(NT),ABSEST,NOTACC
Ø43ØØ             NT = NT - 1
Ø44ØØ          ENDIF                  !end of main loop
Ø45ØØ          IF (NT.GT.Ø) GOTO 1Ø  !ELSE Trial Stack is empty, so ...
Ø46ØØ                          RETURN
Ø47ØØ      1 FORMAT(' S1/3[',G11.6,', ',G11.6,'] =',G13.7,
Ø48ØØ     &              ' ABSEST =',G11.5,'  NOTACC =',L2)
Ø49ØØ          END
```

Figure 7.4-5 *Fortran 77* SUBROUTINE *for Adaptive Simpson Quadrature*

The sample calling program shown in Figure 7.4-7 uses Adaptive Simpson Quadrature with ϵ (EPS) $= 0.001$ and *MxHalv* $= 6$ to integrate $f(x) = 1/(x + 0.1)^2$ from 0 to 4. The name of the FUNCTION subprogram that evaluates $f(x)$ (FTEST in Figure 7.4-7) must be declared EXTERNAL in the calling program so that it will be used as the function F (rather than interpreted as a REAL variable) by SUBROUTINE ADSIM.

```
ØØ1ØØ          SUBROUTINE SIMP13(F, A, B, FA, FB, XM, FM, SIMP)
ØØ2ØØ   C - - - - - - - - - - - - - - - - - - - - - - - - - - - - - - - - - C
ØØ3ØØ   C  PURPOSE:  To use passed values of A, B, FA, and FB to form       C
ØØ4ØØ   C            XM = (A+B)/2, FM = F(XM), and SIMP13 = S1/3[A,B]        C
ØØ5ØØ   C - - - - - - - - - - - - - - - - - - - - - VERSION 1: 9/9/85 - - C
ØØ6ØØ          XM = (A + B)/2
ØØ7ØØ          FM = F(XM)
ØØ8ØØ          SIMP = (B - A)*(FA + 4.*FM + FB)/6.
ØØ9ØØ          RETURN
Ø1ØØØ          END
Ø11ØØ
Ø12ØØ          SUBROUTINE ADDTRY (N, NHALVN, AN, MN, BN, FAN, FMN, FBN, SANBN)
Ø13ØØ          REAL A(1Ø), M(1Ø), B(1Ø), FA(1Ø), FM(1Ø), FB(1Ø), SAB(1Ø), MN
Ø14ØØ   C - - - - - - - - - - - - - - - - - - - - - - - - - - - - - - - - - C
Ø15ØØ   C  PURPOSE: To add [AN, BN] to the Trial Stack at index (level) N.   C
Ø16ØØ   C - - - - - - - - - - - - - - - - - - - - - VERSION 1: 9/9/85 - C
Ø17ØØ          COMMON /STACK/ A,M,B,FA,FM,FB,SAB,NHALV(1Ø)
Ø18ØØ          NHALV(N) = NHALVN
Ø19ØØ          A(N) = AN
Ø2ØØØ          M(N) = MN
Ø21ØØ          B(N) = BN
Ø22ØØ          FA(N) = FAN
Ø23ØØ          FM(N) = FMN
Ø24ØØ          FB(N) = FBN
Ø25ØØ          SAB(N) = SANBN
Ø26ØØ          RETURN
Ø27ØØ          END
```

Figure 7.4-6 SUBROUTINEs SIMP13 *and* ADDTRY

The output of Figure 7.4-7 is shown in Figure 7.4-8(a). The calculated value, *Val*, is the sum of the *QF* values on the 10 subintervals shown in Figure 7.4-8(b). Since the graph of $f(x) = 1/(x + 0.1)^2$ looks like Figure 7.4-3(b) over $[a, b] = [0, 4]$, we see that the algorithm used smaller subintervals on the left part of $[0, 4]$ (i.e., nearest the vertical asymptote at $x = -0.1$) and larger subintervals on the right part [where $f(x)$ is nearly constant]. The output indicates that the accuracy test (27) failed on the leftmost two subintervals, namely $[0, \frac{1}{16}]$ and $[\frac{1}{16}, \frac{1}{8}]$. Nevertheless, the calculated value $Val = 9.75305$ is very close to being within $\pm\epsilon = \pm0.001$ of the exact value,

$$\int_0^4 \frac{dx}{(x + 0.1)^2} = \frac{-1}{x + 0.1}\Bigg]_0^4 \doteq 9.75610 \qquad (34)$$

```
C  Program to integrate from AA to BB by calling SUBROUTINE ADSIM
      EXTERNAL  FTEST
      REAL A(1Ø), M(1Ø), B(1Ø), FA(1Ø), FM(1Ø), FB(1Ø), SAB(1Ø)
      COMMON /STACK/ A, M, B, FA, FM, FB, SAB, NHALV(1Ø)
      COMMON /COUNT/ NEVALS
      DATA AA, BB, EPS, MXHALV, IW /Ø.Ø, 4.Ø, 1.E-Ø3, 6, 5/
      NEVALS = Ø
C
      CALL ADSIM (FTEST, AA, BB, EPS, MXHALV, VAL, ERR, .TRUE., IW)
C
      WRITE (IW, 1) VAL, ERR, NEVALS
    1 FORMAT('ØIntegral =',G14.7,'  |error| < ',G9.4,I7,' evaluations')
      STOP
      END

      FUNCTION FTEST(X)              !becomes F in SUBROUTINE ADSIM
      COMMON /COUNT/ NEVALS          !optional
      FTEST = 1.Ø / (Ø.1 + X)**2
      NEVALS = NEVALS + 1            !counts evaluations of FTEST(X)
      RETURN
      END
```

Figure 7.4-7 *Sample user-provided code for* SUBROUTINE ADSIM

```
S1/3[.ØØØØØØE+ØØ, .625ØØØE-Ø1] = 3.854889    ABSEST = .53798E-Ø3   NOTACC = T
S1/3[.625ØØØE-Ø1, .125ØØØ     ] = 1.71Ø194    ABSEST = .49161E-Ø4   NOTACC = T
S1/3[.125ØØØ    , .1875ØØ     ] = .9663282    ABSEST = .9ØØ43E-Ø5   NOTACC = F
S1/3[.1875ØØ    , .25ØØØØ     ] = .6211566    ABSEST = .24Ø8ØE-Ø5   NOTACC = F
S1/3[.25ØØØØ    , .375ØØØ     ] = .75215Ø1    ABSEST = .168ØØE-Ø4   NOTACC = F
S1/3[.375ØØØ    , .5ØØØØØ     ] = .43865Ø7    ABSEST = .33754E-Ø5   NOTACC = F
S1/3[.5ØØØØØ    , .75ØØØØ     ] = .49Ø4937    ABSEST = .18461E-Ø4   NOTACC = F
S1/3[.75ØØØØ    , 1.ØØØØØ     ] = .2674286    ABSEST = .3Ø473E-Ø5   NOTACC = F
S1/3[1.ØØØØØ    , 2.ØØØØØ     ] = .43595Ø6    ABSEST = .18572E-Ø3   NOTACC = F
S1/3[2.ØØØØØ    , 4.ØØØØØ     ] = .2341596    ABSEST = .11375E-Ø3   NOTACC = F

Integral =  9.7573Ø5      |error| < .9397E-Ø3     41 evaluations
```

(a)

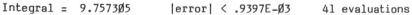

(b)

Figure 7.4-8 *(a) Output of Figure 7.4-7. (b) Intervals used as* $[\alpha, \beta]$ *(not to scale)*

Moreover, this accuracy was achieved by sampling $f(x)$ at only 41 nodes [namely the 11 endpoints shown in Figure 7.4-8(b) and three equispaced nodes interior to each of the 10 subintervals shown]. Comparable accuracy using Romberg integration starting with two panels would require 129 $\frac{1}{32}$-spaced sampled nodes!

If the accuracy test (27) is not assured over all subintervals for a given ϵ, *MxHalv* should be increased.

7.5

Multiple Integration

We know from calculus that if $f(x, y)$ is continuous over a bounded region R such as that of Figure 7.5-1(a), then the **double integral over *R*** is defined as the limit of the Riemann sums:

$$\iint_R f(x, y)\, dA = \lim_{\text{all } \Delta A_i \to 0} \left(\sum_i f(x_i, y_i)\, \Delta A_i \right) \qquad \textbf{(1)}$$

where the index i varies over all rectangles that intersect R.

If the **region of integration *R*** can be described as

$$R = \{(x, y): a \leq x \leq b, \quad F_1(x) \leq y \leq F_2(x) \qquad \text{for each } \textit{fixed } x\} \qquad \textbf{(2)}$$

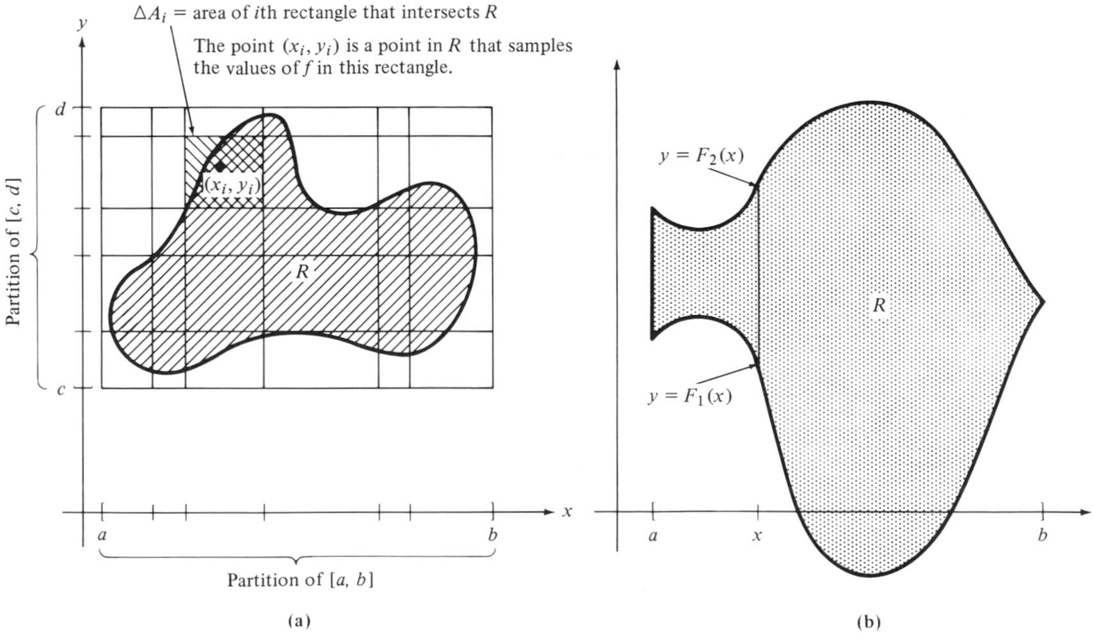

Figure 7.5-1 *(a) Double integration. (b) Iterated integration.*

[Figure 7.5-1(b)], then the double integral (1) can be evaluated more easily as the **iterated integral**

$$\iint_R f(x, y)\, dA = \int_{x=a}^{b} \left[\int_{y=F_1(x)}^{F_2(x)} f(x, y)\, dy \right] dx \tag{3a}$$

The **inner integral** with respect to y [in square brackets in (3a)] is carried out with x held fixed, so its value generally depends on this fixed x. We express this dependence explicitly by rewriting the iterated integral in (3a) as

$$\int_a^b \int_{F_1(x)}^{F_2(x)} f(x, y)\, dy\, dx = \int_a^b g(x)\, dx, \qquad \text{where } g(x) = \int_{y=F_1(x)}^{F_2(x)} f(x, y)\, dy \tag{3b}$$

This suggests the following procedure.

7.5A General Numerical Procedure for Approximating $I = \int_a^b \int_{F_1(x)}^{F_2(x)} f(x, y)\, dy\, dx$

Step 1: *Sketch the region R by inspection of*

$$\int_{x=a}^{b} \int_{y=F_1(x)}^{F_2(x)}$$

Step 2: *After assessing how much g(x) in* (3b) *may vary over* $[a, b]$, *select a quadrature formula for the x-integral:*

$$I = \int_a^b g(x)\, dx \approx w_1 g(x_1) + \cdots + w_n g(x_n) \tag{4a}$$

Step 3: *For* $k = 1, \ldots, n$, *assess how much* $f(x_k, y)$ *(viewed as a function of y) varies over* $[F_1(x_k), F_2(x_k)]$, *and use a suitable quadrature formula to approximate*

$$g(x_k) = \int_{y = F_1(x_k)}^{F_2(x_k)} f(x_k, y)\, dy \tag{4b}$$

Step 4: *Put the approximations* (4b) *in* (4a) *to get the desired approximation.*

Note 1: Different quadrature formulas may be used for different k's in (4b), depending on the behavior of $f(x_k, y)$ for $F_1(x_k) \leq y \leq F_2(x_k)$. In particular, the Midpoint Rule often suffices when $F_1(x_k) \approx F_2(x_k)$.

Note 2: The accuracy of n-point Gauss quadrature as compared to other n-point formulas makes it desirable for hand calculation of (4a) and (4b).

EXAMPLE 7.5A Estimate $I = \int_0^1 \int_1^{e^x} \left(x + \frac{1}{y}\right) dy\, dx$ to about 4s.

SOLUTION: Here $I = \int_0^1 g(x)\, dx$, where $g(x) = \int_1^{e^x} \left(x + \frac{1}{y}\right) dy$.

STEP 1. The region of integration is shown shaded in Figure 7.5-2.

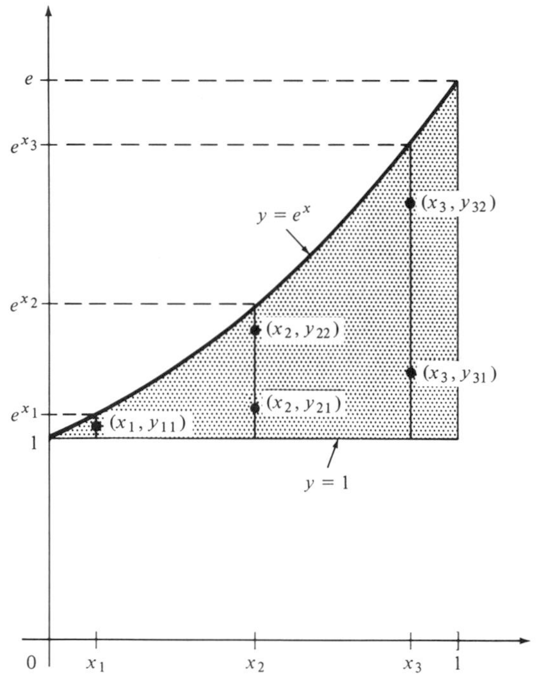

Figure 7.5-2 *$R = \{(x, y): 0 \leq x \leq 1,\ 1 \leq y \leq e^x$ for each fixed $x\}$.*

STEP 2. Since $g(x)$ involves e^x and x, which vary like low-degree polynomials for $0 \leq x \leq 1$, we try three-point Gauss quadrature for the x-integral. By (7) and (10) of Section 7.4,

$$I = \int_0^1 g(x)\, dx \approx \left(\frac{1-0}{2}\right)\left[\frac{5}{9}\, g(x_1) + \frac{8}{9}\, g(x_2) + \frac{5}{9}\, g(x_3)\right] \qquad \textbf{(5a)}$$

where x_1, x_2, and x_3 are the three Gauss sample points on [0, 1], that is,

$$x_1 = \tfrac{1}{2}(1 - \sqrt{0.6}) \doteq 0.11270, \qquad x_2 = \tfrac{1}{2}, \qquad x_3 = \tfrac{1}{2}(1 + \sqrt{0.6}) \doteq 0.88730 \qquad \textbf{(5b)}$$

STEP 3. Since $1/y$ varies slowly for $1 \leq y \leq e^{x_k}$, $k = 1, 2, 3$, we approximate (5a) by obtaining $g(x_1)$, $g(x_2)$, and $g(x_3)$, as follows:

$$g(x_1) = \int_1^{e^{x_1}} \left(x_1 + \frac{1}{y}\right) dy: \quad \text{use the Midpoint Rule}$$

$$g(x_2) = \int_1^{e^{x_2}} \left(x_2 + \frac{1}{y}\right) dy: \quad \text{use two-point Gauss quadrature}$$

$$g(x_3) = \int_1^{e^{x_3}} \left(x_3 + \frac{1}{y}\right) dy: \quad \text{use two-point Gauss quadrature}$$

The five required sample points of R are shown in Figure 7.5-2, where

$$y_{11} = \text{midpoint of } [1, e^{x_1}] = \frac{e^{x_1} + 1}{2} \doteq 1.05965$$

y_{21} and y_{22} are the two Gauss sample points of $[1, e^{x_2}]$, that is,

$$y_{21} = 1 + \frac{e^{x_2} - 1}{2}\left(1 - \frac{1}{\sqrt{3}}\right) \doteq 1.13709, \quad y_{22} = 1 + \frac{e^{x_2} - 1}{2}\left(1 + \frac{1}{\sqrt{3}}\right) \doteq 1.51163$$

and y_{31} and y_{32} are the two Gauss sample points of $[1, e^{x_3}]$, that is,

$$y_{31} = 1 + \frac{e^{x_3} - 1}{2}\left(1 - \frac{1}{\sqrt{3}}\right) \doteq 1.30189, \quad y_{32} = 1 + \frac{e^{x_3} - 1}{2}\left(1 + \frac{1}{\sqrt{3}}\right) \doteq 2.12667$$

[see (10b) of Section 7.4B]. So the Midpoint Rule on $[1, e^{x_1}]$ gives

$$g(x_1) \approx (e^{x_1} - 1)\left(x_1 + \frac{1}{y_{11}}\right) \doteq 0.126026 \qquad (\text{exact value} \doteq 0.126145)$$

and two-point Gauss quadrature on $[1, e^{x_2}]$ and $[1, e^{x_3}]$ gives

$$g(x_2) \approx \frac{e^{x_2} - 1}{2}\left[\left(x_2 + \frac{1}{y_{21}}\right) + \left(x_2 + \frac{1}{y_{22}}\right)\right] \doteq 0.824192 \qquad (\text{exact value} = 0.824361)$$

$$g(x_3) \approx \frac{e^{x_3} - 1}{2}\left[\left(x_3 + \frac{1}{y_{31}}\right) + \left(x_3 + \frac{1}{y_{32}}\right)\right] \doteq 2.15208 \qquad (\text{exact value} = 2.15486)$$

STEP 4. Putting these values in (5), we get $I \approx 0.999115$. The exact value is

$$I = \int_0^1 \left(xy + \ln y\right]_{y=1}^{e^x} dx = \int_0^1 xe^x\, dx = (x - 1)e^x\Big]_0^1 = 0e^1 - (-1)e^0 = 1.0000$$

The accuracy of Gauss quadrature made it possible to get almost 3s accuracy using the values of $f(x, y)$ at only five sample points of R. ■

In general, *the quadrature formulas used in* (4b) *should be selected so as to try to ensure that* $g(x_1), \ldots, g(x_n)$ *are all accurate to at least the decimal place accuracy expected in* (4a). In the preceding example, where about 4s was desired, $g(x_1)$ and $g(x_2)$ were sufficiently accurate (to 3^+ d), but $g(x_3)$ was not. If we used three-point Gauss quadrature (rather than two) on the longer y-interval $[1, e^{x_3}]$, we would have obtained

$$g(x_3) = \frac{e^{x_3} - 1}{2}\left[\frac{5}{9}\left(x_3 + \frac{1}{1.16001}\right) + \frac{8}{9}\left(x_3 + \frac{1}{1.71428}\right) + \frac{5}{9}\left(x_3 + \frac{1}{2.26756}\right)\right] \doteq 2.15473$$

Putting this $g(x_3)$, together with $g(x_1)$ and $g(x_2)$ obtained above, in (5) gives $I \approx 0.999851$, which is accurate to almost 4s. It can thus be seen that *the selection of quadrature formulas for* (4b) *should take into account the lengths of the intervals* $[F_1(x_k), F_2(x_k)]$ *as well as the behavior of* $f(x_k, y)$ *on these intervals and the accuracy expected in* (4a). The ability to use this information to select quadrature formulas efficiently when doing a hand computation comes with experience.

7.5B Approximating $I = \int_a^b \int_{F_1(x)}^{F_2(x)} f(x, y)\,dy\,dx$ Accurately on a Computer

If a subroutine for Romberg integration is available, it can be used as follows:

Romberg Integration for Evaluating $I = \int_a^b \int_{F_1(x)}^{F_2(x)} f(x, y)\,dy\,dx$

Step 1: *Write a* FUNCTION *subprogram that uses Romberg integration to evaluate*

$$g(x) = \int_{F_1(x)}^{F_2(x)} f(x, y)\,dy \qquad \text{for any fixed } x$$

to a bit more than the prescribed accuracy.

Step 2: *Evaluate* $I = \int_a^b g(x)\,dx$ *to the desired accuracy using Romberg integration.*

EXAMPLE 7.5B Suppose that we wanted the value of the iterated integral I of Example 7.5A to 5s. Using the Composite Trapezoidal Rule $T[h]$ with $n = 2, 4, 8$ panels to approximate $I = \int_0^1 g(x)\,dx$ [with Romberg integration used to get the $g(x_k)$'s to 6s], we get

$$n_0 = 2:\quad T\left[\frac{1}{2}\right] = \frac{1}{2}\left\{\frac{g(0) + g(1)}{2} + g\left(\frac{1}{2}\right)\right\} \doteq \frac{1}{2}\{2.18350\} = 1.09175$$

$$2n_0 = 4:\quad T\left[\frac{1}{4}\right] = \frac{1}{2}\left\{1.09175 + \frac{1}{2}\sum_{\text{odd}}\left[\frac{1}{4}\right]\right\} \doteq \frac{1}{2}\{2.04614\} = 1.02307$$

$$4n_0 = 8:\quad T\left[\frac{1}{8}\right] = \frac{1}{2}\left\{1.02307 + \frac{1}{4}\sum_{\text{odd}}\left[\frac{1}{8}\right]\right\} \doteq \frac{1}{2}\{2.01155\} = 1.00578$$

The Romberg table for I is given in Table 7.5-1. It is clear from the last row that $I \doteq 1.0000$ (5s).

Practical Consideration 7.5B (Recursive calls). If Steps 1 and 2 are implemented in a language such as Pascal or PL/I, which allows a subroutine to call itself (such calls are called *recursive*

TABLE 7.5-1 Romberg Table for $I = \int_0^1 g(x)\,dx$

h	$T[h]$	$T_1[h] = S[h]$	$T_2[h]$
$\frac{1}{2}$	1.09175		
$\frac{1}{4}$	1.02307	1.00018	
$\frac{1}{8}$	1.00578	1.00002	1.00001

calls), only one Romberg integration subprogram is needed. However, if implemented in Fortran (which does not allow recursive calls), separate Romberg integration algorithms are needed for Steps 1 and 2. ■

PROBLEMS

Section 7.1

7-1 Find w_0, w_1, and w_2 of a quadratic exact approximation formula of the form $Q \approx w_0 f(-1) + w_1 f(1) + w_2 f(2)$; then use your formula with $f(x) = x^2$, $f(x) = x^3$, and $f(x) = e^x$. See Example 7.1B.
(a) $Q = f'(0)$ **(b)** $Q = f''(1)$ **(c)** $Q = \int_0^1 f(x)\,dx$ **(d)** $Q = \int_{-1}^1 f(x)\,dx$

7-2 For the approximations of Q in (a)–(d) of Problem 7-1: Use (23c) (if applicable) to find C of the truncation error formula $\tau = Cf'''(\xi)/3!$, then get a bound on $|\tau|$ when $f(x) = e^x$.

7-3 The approximation obtained in (a)–(c) of Problem 7-1 should be more accurate than that of Example 7.1B. Why? Confirm this for $f(x) = e^x$.

7-4 Use (15a) or (15b) as appropriate to get the sample weights of
(a) $\Delta f(z)/h$; **(b)** $\delta f(z)/2h$; **(c)** $L]_a^b$ **(d)** $T]_a^b$.

In (b), assume that z is sampled; the sample weight for $f(z)$ will turn out to be 0.

7-5 For (a)–(d) of Problem 7-4: Use (23c) to find C, in terms of either z or a and b, of a truncation error of the form $Cf^{(n+1)}(\xi)$, where n is the exactness degree of the formula.

7-6 Let $F[h] = \Delta f(z)/h$. For the given $f(x)$, z, h_1, and h_2, find $F[h_1]$ and $F[h_2]$ and their truncation errors $\tau[h_1]$ and $\tau[h_2]$, where $\tau[h] = f'(z) - F[h]$. Then find an *integer n* for which $\tau[h_2] \approx (h_2/h_1)^n \tau[h_1]$.
(a) $f(x) = \sin x$ (x in radians), $z = \pi/3$ [$f'(z) = \frac{1}{2}$]; $h_1 = 0.05$, $h_2 = 0.1$
(b) $f(x) = \ln x$, $z = 1$ [$f'(z) = 1$]; $h_1 = \frac{1}{40}$, $h_2 = \frac{1}{10}$
(c) $f(x) = e^{2x}$, $z = 0$ [$f'(z) = 2$]; $h_1 = 0.02$, $h_2 = 0.05$

7-7 Same as Problem 7-6 but with $F[h] = \delta f(z)/2h$.

7-8 Which of $L]_0^3$, $M]_0^3$, or $T]_0^3$ would you expect to approximate $\int_0^3 f(x)\,dx$ most accurately? least accurately? Explain. Is your conjecture correct when $f(x) = (x - 1)^2$? Explain.

Section 7.2

7-9 For (a)–(c) of Problem 7-6: Use Richardson's formula to find $F_1[h_1]$. How much more accurate is $F_1[h_1]$ than $F[h]$?

7-10 Same as Problem 7-9 but for (a)–(c) of Problem 7-7.

7-11 Let $F[h] = \delta f(z)/2h$. Use $F[2h]$ in Richardson's formula to get an $O(h^4)$ improvement of $F[h]$ as in (40b). Use your formula for $F_1[h]$ with $h = 0.05$ to approximate $d(\sin x)/dx]_{x=\pi/3}$. Is this solution preferable to $F_1[h_2]$ obtained in Problem 7-10(a)? Explain.

7-12 Suppose that $F[h] \to Q$ as $h \to 0$, and you know that $F[0.8] = 2$, $F[0.4] = 1$, and $F[0.1] = 0.5$. Use Richardson's formula to find $F_1[0.4]$, $F_1[0.1]$, and $F_2[0.1]$ under the given assumption.

(a) $\tau[h] = 3h^2 + O(h^5)$ **(b)** $F[h] = \dfrac{\Delta f(z)}{h}$

(c) $Q = F[h] + 2h + 3h^3 + O(h^4)$ **(d)** $F[h] = \dfrac{\delta f(z)}{2h}$

7-13 Show that 7*s* accuracy would *not* have been achieved in Table 7.2-7 of Example 7.2G if we started with $h = \frac{1}{2}$ rather than $h = 1$.

7-14 Make a Richardson table for the indicated $F[h]$ values. If the most accurate value is exact, explain why it might have been expected; if not, round to the places that appear accurate *based only on the values on the table.*

	$F[h]$	$f(x)$	z	Values of h to Use for $F[h]$
(a)	$\Delta f(z)/h$	$x^2 - 2x$	0	$h = 1$ and $\frac{1}{2}$
(b)	$\delta^2 f(z)/h^2$	x^4	1	$h = 2$ and 1
(c)	$\Delta f(z)/h$	e^{-x}	0	$h = 0.5, 0.1$, and 0.05
(d)	$\Delta f(z)/h$	$x^{3/2}$	0	$h = 0.5, 0.1, 0.05$, and 0.02
(e)	$\delta f(z)/2h$	$\sqrt{x}$	$\frac{1}{4}$	$h = \frac{1}{4}, \frac{1}{8}$, and $\frac{1}{32}$
(f)	(27c)	e^{-x}	0	$h = 0.2, 0.1$, and 0.04
(g)	(26a)	e^{-x}	0	$h = -0.2, -0.1$, and -0.04
(h)	$\delta^2 f(z)/h^2$	$\ln x$	1	$h = 0.6, 0.2$, and 0.1
(i)	(26b)	$\cos x$	0	$h = 0.3, 0.1, 0.05$, and 0.02

7-15 Explain why Richardson's formula improves $F[h] = \Delta f(z)/h$ more slowly when $f(x) = x^{3/2}$ than when $f(x) = e^{-x}$. See parts (c) and (d) of Example 7-14.

7-16 The following points $(x, f(x))$ are known to lie on a cubic curve.

x	-7	-6	-5	-4	-3	-2	-1	0	1	2	3	4	5	6	7
$f(x)$	-108	-60	-28	-9	0	2	0	-3	-4	0	12	35	72	126	200

Without finding a formula for $f(x)$, find the *exact* value of
(a) $f'(-6)$ **(b)** $f''(0)$ **(c)** $f'''(5)$ **(d)** $f''(7)$

7-17 The following points $(x, f(x))$ are known to lie on a fifth-degree curve.

x	0.0	0.4	0.8	1.2	1.6	2.0	2.4	2.8	3.2	3.6	4.0	4.4	4.8
$f(x)$	-1760	-135	384	406	288	195	160	144	96	13	0	330	1504

Without finding a formula for $f(x)$, find exactly *if possible* (or round to the places you think are correct) the value of
(a) $f'(0.4)$ **(b)** $f'(3.6)$ **(c)** $f''(4.0)$ **(d)** $f''(4.8)$ **(e)** $f'''(2.4)$

7-18 On the following table, $f(x)$ is described by *different* differentiable formulas on either side of $x = 0$ but has a "cusp," hence is *not* differentiable at $x = 0$, as a sketch of the points $(x, f(x))$ will show.

x	-6	-5	-4	-3	-2	-1	0	1	2	3	4	5	6
$f(x)$	932	487	226	89	28	7	2	7	10	5	-14	-53	-118

Without trying to find formulas for $f(x)$, approximate the following derivatives. Round to the number of places that appear to be accurate; give reasons for rounding as you did.
(a) $f''(-3)$ **(b)** $f'(1)$ **(c)** $f''(0^-)$ **(d)** $f'(0^+)$ **(e)** $f'''(3)$

7-19 Let $f(x) = x^2e^{1.3x}/\cos x$ (x in radians) and $z = 0$. Use a Richardson table for an $O(h^2)$ central difference formula (27), as in Example 7.2G, to approximate to 5s (a) $f'(z)$; (b) $f''(z)$; (c) $f'''(z)$. To avoid harmful roundoff error, be sure that your initial stepsize h_0 is not too small and use at least an 8s device.

7-20 Verify that the given formulas are $O(h^4)$ by using them with $h = 0.4$ and $h = 0.2$ for $f(x) = e^x$ at $z = 0$. [The exact values $f^{(k)}(0)$ are 1.]

(a) $f'''(x) \approx \dfrac{1}{8h^3}[f_{-3} - 8f_{-2} + 13f_{-1} + 0f_0 - 13f_1 + 8f_2 - f_3]$

(b) $f^{(iv)}(x) \approx \dfrac{1}{6h^4}[-f_{-3} + 12f_{-2} - 39f_{-1} + 56f_0 - 39f_1 + 12f_2 - f_3]$

7-21 Use the 5s values of $\Phi(x)$ shown in Figure 6.2-12 to try to approximate the indicated derivative with the prescribed accuracy.
(a) $\Phi'(0)$ to 3s **(b)** $\Phi'(2.0)$ to 3s **(c)** $\Phi''(0.1)$ to 2s

7-22 Use the fact that $\Phi(-x) = 0.5 - \Phi(x)$, if necessary, to get the derivatives in parts (a)–(c) of Problem 7-21 to about 5s accuracy.

Section 7.3

7-23 Show that the given approximation is actually exact, that is, an equality.
(a) $\int_1^3 4 - x\,dx \approx M]_1^3$ **(b)** $\int_1^2 x^3\,dx \approx CC]_1^2$ **(c)** $\int_0^2 x^3\,dx \approx S_{1/3}]_0^2$
(d) $\int_2^3 x^3\,dx \approx AC]_2^3$ **(e)** $\int_3^4 x^3\,dx \approx AP]_3^4$ **(f)** $\int_0^3 x^3\,dx \approx S_{3/8}]_0^3$

7-24 Approximate the given integral as indicated, using only the cubic $f(x)$ values tabulated in Problem 7-16. Is it exact? Explain.
(a) $\int_{-1}^1 f(x)\,dx \approx M]_{-1}^1$ **(b)** $\int_1^2 f(x)\,dx \approx CC]_1^2$ **(c)** $\int_{-6}^{-4} f(x)\,dx \approx AC]_{-6}^{-4}$
(d) $\int_0^2 f(x)\,dx \approx S_{1/3}]_0^2$ **(e)** $\int_1^4 f(x)\,dx \approx S_{3/8}]_1^4$ **(f)** $\int_6^8 f(x)\,dx \approx AP]_6^8$

7-25 Repeat parts (a)–(f) of Problem 7-24 for the function of Problem 7-18. Then suggest a better approximation formula if you can think of one. Explain.

7-26 **(a)** Which of $T[0.2]$, $S_{1/3}[0.2]$, and $NC_5]_0^8$ [(16a) of Section 7.3A] would you expect to approximate $\int_0^{0.8} f(x)\,dx$ most accurately? Give reasons.
(b) Approximate $\int_0^{0.8} e^x\,dx$ by $T[0.2]$, $S_{1/3}[0.2]$, and $NC_5]_0^{0.8}$. Do their values confirm your conjecture in part (a)?

7-27 Find all sample weights of the specified formula of Section 7.3A by using (2c) (along with Table 7.3-1) exactly N times. Use (20b) of Section 7.1B and symmetry as appropriate.
(a) Simpson's $\frac{1}{3}$-Rule ($N = 1$) **(b)** Simpson's $\frac{3}{8}$-Rule ($N = 1$)
(c) Adams Corrector ($N = 3$) **(d)** Adams Predictor ($N = 3$)
(e) The four-panel Newton–Cotes formula (16a) ($N = 2$)

7-28 Show that (2d) applies, then use it along with Table 7.3-1 to get the truncation error of **(a)** $T]_a^b$; **(b)** $AC]_a^b$; **(c)** $AP]_a^b$.

7-29 Let $F[h]$ be the Composite Trapezoidal Rule $T[h]$. Approximate the given integral using $F[h_0]$ and $F[h_0/2]$. Then use Richardson's formula to get $F_1[h_0/2]$.
(a) $\int_1^4 x^4\,dx$; $h_0 = 1$ **(b)** $\int_1^3 \ln x\,dx$; $h_0 = 1$
(c) $\int_0^4 dx/(x + 0.1)^2$; $h_0 = 2$ **(d)** $\int_0^{\pi/2} \cos x\,dx$; $h_0 = \pi/2$
(e) $\int_0^4 \sqrt{x}\,dx$; $h_0 = 4$ **(f)** $\int_0^{\pi/4} \tan x\,dx$; $h_0 = \pi/4$
(g) $\int_0^2 \sqrt{4 - x^2}\,dx$; $h_0 = \frac{1}{2}$ **(h)** $\int_0^{\pi/2} \sin^2 x\,dx$; $h_0 = \pi/6$

7-30 Same as Problem 7-29 but with $F[h]$ taken to be $S_{1/3}[h]$ and $h_0 = (b - a)/2$ (two panels).

7-31 For parts (a)–(h) of Problem 7-29, use (20b) *if it applies* to obtain bounds for both $|\tau_T[h_0]|$ and $|\tau_T[h_0/2]|$.

7-32 Same as Problem 7-31 but using (21b) for the truncation errors of $S_{1/3}[h_0]$ and $S_{1/3}[h_0/2]$.

7-33 For parts (a)–(h) of Problem 7-29: Use Romberg integration starting with $T[h_0]$, up to $T_3[h_0/8]$; stop sooner if 5*s* accuracy is indicated (based on table values) before getting $T_3[h_0/8]$. If $T_3[h_0/8]$ does not appear to have 5*s* accuracy, suggest a reason why.

7-34 Use any convenient method to get the *exact* integral using only the tabulated (cubic) values given in Problem 7-16. Show your method clearly.
(a) $\int_{-3}^{3} f(x)\,dx$ **(b)** $\int_0^5 f(x)\,dx$ **(c)** $\int_0^1 f(x)\,dx$ **(d)** $\int_{-7}^{-6} f(x)\,dx$
(e) $\int_6^7 f(x)\,dx$ **(f)** $\int_7^8 f(x)\,dx$ **(g)** $\int_6^9 f(x)\,dx$ **(h)** $\int_8^9 f(x)\,dx$

7-35 Use any convenient method to get the *exact* integral *if possible* using only the tabulated (fifth-degree polynomial) values given in Problem 7-17. If exactness is not possible, round to the last digit that appears to be accurate. Show your reasoning clearly.
(a) $\int_{2.0}^{3.6} f(x)\,dx$ **(b)** $\int_{2.0}^{2.8} f(x)\,dx$ **(c)** $\int_{2.0}^{4.0} f(x)\,dx$
(d) $\int_{2.0}^{4.4} f(x)\,dx$ **(e)** $\int_{0.4}^{4.0} f(x)\,dx$ **(f)** $\int_{2.0}^{2.4} f(x)\,dx$

7-36 Evaluate as accurately as you can using only the tabulated values given in Problem 7-18. Round to the last digit that appears to be accurate. Show your reasoning clearly.
(a) $\int_{-1}^{1} f(x)\,dx$ **(b)** $\int_{-3}^{3} f(x)\,dx$ **(c)** $\int_{-1}^{5} f(x)\,dx$ **(d)** $\int_{-2}^{1} f(x)\,dx$

7-37 **(a)** Derive a Composite Simpson's $\frac{3}{8}$-Rule, $S_{3/8}[h]$, that can be used when the number of panels n is divisible by 3 (so that $[a, b]$ can be viewed as $n/3$ "triple panels").
(b) Deduce the trunction error of $S_{3/8}[h]$ from (10b) [see (20c)].
(c) Use $S_{3/8}[0.4]$ to approximate $\int_{2.0}^{5.6} e^x\,dx$. Why should you expect it to be *less accurate* than (42) and (43) of Example 7.3D(a)? Is it?

7-38 Two fifth-degree exact approximations of $\int_{2.8}^{5.2} f(x)\,dx$ are given in (a) and (b). Which should be more accurate? Explain.
(a) Use $S_{1/3}[0.8]$ to improve $S_{1/3}[0.4]$
(b) Use $S_{3/8}[0.6]$ to improve $S_{3/8}[0.4]$ (see Problem 7-37).

Perform (a) and (b) with $f(x) = e^x$ [see Example 7.3B(c)]. Did you predict the more accurate approximation?

Section 7.4

7-39 Show that making the one-point Gauss quadrature formula $\int_{-1}^{1} f(\xi)\,d\xi$ exact for $f(\xi) = 1$ and $f(\xi) = \xi$ as in Example 7.4A yields the Midpoint Rule. (This explains why $M]_a^b$ is exact for straight lines.)

7-40 What is the smallest n for which n-point Gauss quadrature is exact for seventh-degree polynomials? Verify your answer for $\int_0^2 x^7\,dx$.

7-41 Use n-point Gauss quadrature to approximate the given integral.
(a) $\int_0^3 x^3\,dx,\ n = 2$ **(b)** $\int_{-1}^{2} e^{-x^2}\cos x\,dx,\ n = 3$ **(c)** $\int_{-1}^{1} e^{-x^2}\sin x\,dx,\ n = 4$

7-42 For (a)–(h) of Problem 7-29: Approximate the integral using $(n_0 + 1)$-point Gauss quadrature, where n_0 is the number of h_0-panels.

7-43 Use (a) three-point and (b) five-point Gauss quadrature to estimate $\int_0^4 \sqrt{x}\,dx$. Why is the improvement so slight? [See Problem 7-29(e).]

7-44 Where would you locate five sensors along a 60-meter tube if you want the *integral* of the sampled quantity along the entire tube? How would you weight the sampled readings?

7-45 State why the integral is improper and describe in detail one or more strategies for determining either that it diverges or what its value is to 4*s*.

(a) $\displaystyle\int_0^{\pi} \frac{\sin x}{x^{3/2}}\,dx$ **(b)** $\displaystyle\int_0^1 \frac{e^x}{x^2}\,dx$

(c) $\displaystyle\int_0^1 \frac{\ln x\,dx}{\sqrt{1-x^2}}$ **(d)** $\displaystyle\int_0^1 \frac{e^x}{\sqrt[3]{x}}\,dx$

(e) $\displaystyle\int_0^{\infty} \frac{dx}{e^x+e^{-x}}$ **(f)** $\displaystyle\int_1^{\infty} \frac{dx}{\sqrt{x}(1+x)}$

(g) $\displaystyle\int_0^{\infty} e^{-x^2}\ln x\,dx$ **(h)** $\displaystyle\int_{-\infty}^{\infty} \left(\frac{\cos x}{e^{-x}}\right)^2 dx$

7-46 Use the Adaptive Simpson Quadrature Algorithm 7.4F to approximate the given integral to within $\pm\epsilon$. Show all intermediate steps.

(a) $\int_0^4 x^{3/2}\,dx$, $\epsilon = 0.002$ **(b)** $\int_1^4 x^{1/2}\,dx$, $\epsilon = 0.001$

Section 7.5

7-47 Approximate $\int_a^b g(x)\,dx$ as indicated, showing the sample points on a sketch of the region of integration R as in Figure 7.5-2.

(a) $\displaystyle\int_{-1}^{1}\int_0^1 (e^{3x}-y)\,dy\,dx$ — Use four-point Gauss for $\int_{-1}^{1}\cdot\,dx$; two-point Gauss for $g(x_i)$.

(b) $\displaystyle\int_{-1}^{1}\int_0^{\pi/3} e^x\tan y\,dy\,dx$ — Use two-point Gauss for $\int_{-1}^{1}\cdot\,dx$; the Midpoint Rule for $g(x_1)$, $g(x_2)$.

(c) $\displaystyle\int_2^4\int_{1/2}^{\sqrt{x}} xy\,dy\,dx$ — Use two-point Gauss for $\int_2^4\cdot\,dx$; the Midpoint Rule for $g(x_1)$, $g(x_2)$.

(d) $\displaystyle\int_0^4\int_{x^2/2}^{2x} (e^x-y)\,dy\,dx$ — Use three-point Gauss for $\int_0^4\cdot\,dx$; Simpson's $\frac{1}{3}$-Rule for $g(x_2)$; the Midpoint Rule for $g(x_1)$, $g(x_2)$.

(e) $\displaystyle\int_0^{\pi/2}\int_{\tan(x/2)}^{\sin x} (x+y)\,dy\,dx$ — Use Simpson's $\frac{3}{8}$-Rule for $\int_0^{\pi/2}\cdot\,dx$; Simpson's $\frac{3}{8}$-Rule for $g(x_1)$ and $g(x_2)$.

NOTE: Here $g(x_0) = g(0) = 0$, and $g(x_3) = g(\pi/2) = 0$. (Verify.)

(f) $\displaystyle\int_0^1\int_{x^2}^{x} e^{xy}\,dy\,dx$ {Try to get 2*s* accuracy.}

(g) $\displaystyle\int_0^2\int_{1-x}^{e^{-x}} \sqrt{x+y}\,dy\,dx$ {Try to get 2*s* accuracy.}

MISCELLANEOUS PROBLEMS

M7-1 Suppose that $\Sigma_i w_i f_i$ approximates Q with exactness degree n, where $n > 0$. What must the values of $\Sigma_i w_i$ and $\Sigma_i w_i x_i$ be if Q is **(a)** $f'(z)$; **(b)** $f^{(k)}(z)$, $k \geq 2$; **(c)** $\int_a^b f(x)\,dx$?

M7-2 Let $f(x) = \ln x$ and $z = 1$. Show that $\Delta f(z)/h$ approximates $f'(z) = 1$ more accurately than $\delta f(z)/2h$ when $h = 0.95$. Does this mean that the $O(h)$ formula is a more accurate formula than the $O(h^2)$ one? Explain.

M7-3 Do Problem 7-6 for $f(x) = \sin x$ (x in radians), $z = 0$, $h_1 = 0.002$, and $h_2 = 0.02$. Your result should indicate that $\Delta f(z)/h$ is an $O(h^2)$ [*not* $O(h)$!] approximation of $f'(z) = 0$. What happened?

M7-4 Derive the truncation error formulas (23a)–(23c) of Theorem 7.1C.

M7-5 Derive (8a), (8b), and (8c) of Theorem 7.2A from definition (7).

M7-6 ***(The Effect of Roundoff Error on $\delta f(z)/2h$ and $\Delta f(z)/h$)***

(a) In using the formula $f'(z) \approx \delta f(z)/2h = (f_1 - f_{-1})/2h$, suppose that $|f'''(x)| \leq M$ for $|x - z| \leq h$ and that f_1 and f_{-1} are in error by at most $\pm\epsilon_\rho$. Show that $\epsilon[h] = f'(z) - \delta f(z)/2h$ satisfies $|\tau[h]| \leq Mh^2/6 + \epsilon_\rho/h$. How is this related to Figure 7.2-2?

(b) Find a bound for $|\epsilon[h]|$ similar to that obtained in part (a) but for the approximation $f'(z) \approx \Delta f(z)/h = (f_1 - f_0)/h$.

M7-7 Let $f(x) = e^x$ and $z = 1$. Use the error bounds obtained in Problem M7-6 with $M = 3$ ($\geq e^1$) to estimate $h_{\tau=\rho}$ of Figure 7.2-2 when approximating $d(e^x)/dx]_{x=1}$ by **(a)** $\delta f(z)/2h$ and **(b)** $\Delta f(z)/h$. Do your results predict the emergence of roundoff error in Tables 7.2-3 and 7.2-4 of Example 7.2D?

M7-8 Show that if $x_{j+i} = x_j + ih$, then (15a) of Section 7.1B becomes

$$(*)\quad f^{(k)}(z) = \frac{1}{h^k}(\Sigma_i\, w_i f_{j+i}), \qquad \text{where } w_i = \frac{d^k}{ds^k} L_{j+i}(x_j + hs)\,\Big]_{s=(z-x_j)/h}$$

where $L_{j+i}(x_j + hs)$ is given in Table 7.3-1 for $n = 2$, 3, and 4.

M7-9 Use (*) of Problem M7-8 to get the sample weights of the $O(h^2)$ forward/backward difference formula for **(a)** $f'(z)$; **(b)** $f''(z)$; **(c)** $f'''(z)$ given in (26) or Section 7.2G.

M7-10 Same as Problem M7-9 but for the $O(h^2)$ central difference formulas (27).

M7-11 ***(Derivatives at the Midpoint of $[x_j, x_{j+1}]$)*** Let $z = (x_j + x_{j+1})/2$. Use (*) of Problem M7-8 to get sample weights of cubic exact formulas of the form $f^{(k)} \approx (w_{-1}f_{j-1} + w_0 f_j + w_1 f_{j+1} + w_2 f_{j+2})/h^k$ when $f^{(k)}(x)$ is **(a)** $f'(z)$; **(b)** $f''(z)$; **(c)** $f'''(z)$. As a check, verify exactness when $f(x) = x^3$, $z = 0$, and $h = 1$.

M7-12 Use a formula of Problem M7-11 to approximate **(a)** $f'(1)$, **(b)** $f''(1.8)$, and **(c)** $f'''(3)$ using only the $f(x)$ values tabulated in Problem 7-17.

M7-13 Suppose that you had to make a table of 7s values of $f(x) = e^{x^2}$ for h-spaced nodes in the interval $0 \leq x \leq 3$. Determine the largest "convenient" decimal value of h (e.g., 0.5, 0.2, 0.1, 0.05, ...) that is small enough to ensure that the indicated approximation will be accurate to 5d, that is, $|\tau| \leq 0.5\text{E}-5$ for all x in [0, 3].

(a) $\delta f(z)/2h$ for z any interior sampled node; $\tau = f'(z) - \delta f(z)/2h$

(b) $\delta^2(z)/h^2$ for z any interior sampled node; $\tau = f''(z) - \delta^2 f(z)/h^2$

(c) $T[h]_a^b$ for any sampled nodes a and b; $\tau = \int_a^b f(x)\,dx - T[h]_a^b$

M7-14 Let $[a, b]$ be $[x_j, x_{j+4}]$ (four h-panels). Find a formula in terms of f_j, f_{j+1}, f_{j+2}, f_{j+3}, and h that results from using $S_{1/3}[2h]$ to improve $S_{1/3}[h]$. Should it be more accurate than NC_5 in (16a) of Section 7.3B? Explain.

M7-15 Show that if the truncation error of an approximation of $\int_a^b f(x)\,dx$ is known to be of the form $Cf^{(n+1)}(\xi)/(n+1)!$, then C can be obtained as the exact value of τ when $f(x) = x^{n+1}$ and $a = 0$ regardless of the sign of $\Pi(x - x_i)$ on $[a, b]$.

M7-16 Use Problem M7-15 to get the following truncation errors:

(a) τ_M in (8c) of Section 7.1A, assuming that $\tau_M = Cf''(\xi)/2!$

(b) $\tau_{S1/3}$ in (9c) of Section 7.3A, assuming that $\tau_{S1/3} = Cf^{(iv)}(\xi)/4!$

(c) $\tau_{S3/8}$ in (10c) of Section 7.3A, assuming that $\tau_{S3/8} = Cf^{(iv)}(\xi)/4!$

M7-17 Consider the **three-point four-panel open Newton–Cotes Formula**

$$\int_{x_j}^{x_{j+4}} f(x)\,dx \approx \frac{4h}{3}[2f_{j+1} - f_{j+2} + 2f_{j+3}]; \qquad \tau = Ch^5 f^{(iv)}(\xi)/4!$$

(a) Verify that it is exact for any h when $f(x) = x^3$ and $x_j = 0$.

(b) Get all sample weights using (2c) and Table 7.3-1 *only once*.
(b) Find C of the given truncation error using Problem M7-15 with $h = 1$.
(c) Should this formula be more accurate than $S_{1/3}[h]$? than NC_5 in (16a) of Section 7.3A? Explain. Confirm your answer for $\int_0^{0.8} e^x\,dx$ [see Problem 7-26].

M7-18 Consider the **five-point two-panel formula**

$$\int_{x_j}^{x_{j+2}} f(x)\,dx \approx \frac{h}{90}[-(f_{j-1} + f_{j+3}) + 34(f_j + f_{j+2}) + 114f_{j+1}]$$

(a) Show that it is exact for any h when $f(x) = x^5$ and $x_j = 0$.
(b) Get all sample weights using (2c) and Table 7.3-1 only twice.
(c) Assuming that its truncation error is of the form $Ch^7f^{(vi)}(\xi)/6!$, set $h = 1$, then use Problem M7-15 to find C.
(d) Should this formula be more accurate than Simpson's $\frac{1}{3}$-Rule? Confirm your answer for $\int_{5.2}^{6.0} e^x\,dx$ [see Example 7.3A(a)].

M7-19 The truncation error of $S_{3/8}[h]$ of Problem 7-37 can be shown to be of the form $Ch^4 + Dh^6 + Eh^8 + \cdots$.
(a) Devise a method for getting better than cubic exactness when integrating over n panels, when n is odd and divisible by 3^k, where $k > 1$. (This method is preferable to "guideline 1" at the beginning of Section 7.3D when it applies.)
(b) Use the method of part (a) to get a more accurate solution than that obtained in Example 7.3D(a).

M7-20 Use Lagrange's form of the remainder [Appendix II.2D] in (45) of Section 7.3E to get (8c) and (9c) of Section 7.3A.

M7-21 ***(Obtaining Gaussian Sample Points on [−1, 1] as Roots of Polynomials)*** The **nth Legendre polynomials $P_n(\xi)$** are generated by the three-term recurrence relation

$$P_0(\xi) = 1, \quad P_1(\xi) = \xi, \quad P_{n+1}(\xi) = \frac{1}{n+1}[(2n+1)\xi P_n(\xi) - nP_{n-1}(\xi), \; n \geq 1$$

(a) Deduce from the three-term recurrence relation that $P_n(1) = 1$ for all n.
(b) The n Gaussian sample points on $[-1, 1]$ are the n roots of $P_n(\xi)$ [40]. Verify this for $n = 2$, 3, and 4 by first using the three-term recurrence relation to find $P_n(\xi)$ and then showing that the n Gaussian sample points ξ_k given in Table 7.4-1 satisfy $P_n(\xi_k) = 0$.

M7-22 Sketch the graph of f over $[a, b]$ and verify that f is continuous at all but at most a finite number of points of $[a, b]$, where it has a removable or jump discontinuity. Then subdivide $[a, b]$ into subintervals whose endpoints are the points of discontinuity of f, integrate $f(x)$ as indicated (n_0 = initial number of panels) over *each* subinterval, and add your answers to get $\int_a^b f(x)\,dx$ to the stated accuracy.

	$f(x)$	a	b	Method	Accuracy of I
(a)	$(e^x - 1)/x$	0	2	Romberg ($n_0 = 1$)	4s
(b)	$\lvert x^3 - 1\rvert$	0	2	Three-point Gauss	Exact
(c)	$(1 - \cos x)/x^2$	0	1	Romberg ($n_0 = 1$)	4s
(d)	e^{-x^2} (Be careful!)	0	50	Romberg ($n_0 = 2$)	4s
(e)	$(x^3 - 5x - 1)\,\mathrm{sgn}\,(x)$	−1	1	Two-point Gauss	Exact
(f)	$(x^4 - 1)/(x + 1)$	−1	0	Romberg ($n_0 = 1$)	Exact

M7-23 Get the most accurate answer you can from the (P, V) table given in Problem M6-13.
(a) The work W (in lbf·in) done by the piston is $\int_{32.5}^{80} P\,dV$. Find it.
(b) Estimate dV/dP when $P = 60$ and when $P = 120$.

M7-24 The moment of inertia I of a horizontal beam of length L varies with x, measured from the left end, in such a way that $I(L - x) = I(x)$ (symmetry about the center). The **moment area method** of structural analysis finds the sag angle θ (in radians) at either end by integrating M/EI from $x = 0$ to $x = L/2$, where M is the moment about the left endpoint, and E is the beam's modulus of elasticity. If $L = 40$ ft, and $E = 30{,}000$ Ksi, find θ as accurately as you can from the following values of (x [ft], I [lbf·in^4], M [lbf·ft]):

$$(0, 500, 0),\ (5, 505, 116),\ (10, 520, 175),\ (15, 545, 197),\ (20, 580, 200)$$

M7-25 Refer to the points tabulated in Problem C6-7. Use the methods of this chapter to obtain with at least cubic-exact accuracy **(a)** the slope at the wing tip and **(b)** the area between the lower contour and the x-axis. Compare this approach to the use of splines as in Problem C6-7.

M7-26 A landscaper measures the distance across a "pear-shaped" lot along nine parallel lines spaced 3 yards apart. The readings (in yards) obtained are: 0, 5, 7, 10, 15, 18, 19, 16, and 9. What is the minimum number of square yards of lawn sod that must be purchased to cover the lot?

COMPUTER PROBLEMS

C7-1 For (a)–(h) of Problem 7-29: Use any available computer program to evaluate the integral to 5s.

C7-2 For (a)–(h) of Problem 7-45: Use any available computer program to evaluate the improper integral to 5s or to show that it is divergent.

C7-3 This problem requires a program implementing the Adaptive Simpson Quadrature Algorithm 7.4F.

(a) Describe the effect of changing `MXHALV` in Figure 7.4-7 from 6 to 7.

(b) For what degree polynomials will Algorithm 7.4F be exact if $Val \leftarrow Val + QF$ is replaced by $Val \leftarrow Val + QF_{\text{improved}}$? [See (30) of Section 7.4F.] Modify your code to implement this strategy, and compare the accuracy to that of Algorithm 7.4F (unmodified) when the integration of Section 7.4F is solved with various values of ϵ and $MxHalv$.

C7-4 Suppose that the truncation error of the approximation $F[h] \approx Q$ is known to satisfy

$$\tau[h] = Ch^{n_0} + Dh^{n_0+\Delta n} + Eh^{n_0+2\Delta n} + Fh^{n_0+3\Delta n} + \cdots$$

(a) Write a subroutine `EXTRAP(K, ROW, FHR, R, NO, DELTAN)` that uses the $(k - 1)$st row of a Richardson table passed as the `REAL` array

$$\texttt{ROW} = [F[h]\quad F_1[h]\quad F_2[h]\quad \cdots\quad F_{k-1}[h]]$$

and the scalar $F[h/r]$ (= `FHR`) to form the kth row and return it as

$$\texttt{ROW} = [F[h/r]\quad F_1[h/r]\quad F_2[h/r]\quad \cdots\quad F_k[h/r]]$$

(b) Let `F(H, other parameters)` be a `FUNCTION` subprogram that forms $F[h]$. Use `F` and `EXTRAP` of part (a) as follows in a `FUNCTION` subprogram

```
RICH(F, MAXROW, NO, DELTAN, HO, R, NUMSIG, OK)
```

which returns the k,kth entry of a Richardson table formed as follows:
$h \leftarrow h_0$; **row**(0) $\leftarrow F[h]$

```
DO FOR k = 1 TO MaxRow UNTIL OK
|  h ← h/r;  Fhr ← F[h]                          {Fhr now stores F[h_0/r^k]}
|  Invoke EXTRAP(k, row, Fhr, r, n_0, Δn)
└─ OK ← (|row(k − 1) − row(k)| ≤ 10^(−NumSig)|row(k)|)
RICH ← row(k)                                     {row(k) stores F_k[h/r^k]}
RETURN
{IF OK THEN RICH approximates Q to NumSig significant digits}
```

(c) Write a `FUNCTION` subprogram `FWDBK(H, FUNC, Z)` that returns $F[h] = \Delta f(z)/h$, where $f(x)$ is evaluated using `FUNCTION FUNC(X)`. Test `FUNCTION RICH` in part (b) using `FWDBK`, with $f(x) = e^x$ and $z = 1$, as `F(H, other parameters)`. Take $h_0 = 0.64$, $r = 4$ and $MaxRow = 5$, and set OK to TRUE when `RICH` appears to approximate Q $(= e)$ to $6s$.

(d) ***(Romberg Integration)*** Write a `FUNCTION` subprogram `RTRAP(H, FUNC, A, B, TRAP)` that forms `RTRAP` $= T[h]$ from `TRAP` $= T[2h]$ using the Recursive Form of the Trapezoidal Rule, that is, $T[h] = \frac{1}{2}\{T[2h] + 2h\ \Sigma_{\text{odd}}[h]\}$. Use `FUNCTION RICH` with `RTRAP` taken as `F(H, other parameters)` to get the result of Example 7.3C.

C7-5 Write a `FUNCTION` subprogram `GAUSS(FUNC, A, B)` that returns the result of using five-point Gauss quadrature to integrate $f(x)$ (= `FUNC(X)`) from a (= `A`) to b (= `B`). Test using $\int_1^{2.2} \ln x\, dx$, and compare the accuracy to that of Example 7.4B.

C7-6 **(a)** Write pseudocode for incorporating the result of Problem C7-4(d) to evaluate the two-step procedure for evaluating a double integral given in Section 7.5B.

(b) Implement your pseudocode in part (a) in a `FUNCTION` subprogram `DBLINT(FCN2, A, B, F1, F2, NUMSIG, MAXROWS)`. Test it using Example 7.5B. If you are using a programming language that allows recursive subroutine calls, try to use one.

8

Numerical Methods for Ordinary Differential Equations

8.0

Introduction

The world rarely reveals itself to us as observable relationships among the relevant variables. What it does make evident are relationships that describe how both the *variables and their rates of change* (i.e., *derivatives*) affect each other (i.e., **differential equations**).

Given a differential equation involving one or more physical quantities (e.g., displacement, temperature, voltage, concentration, population, etc.), the problem of interest is to find the relationship it imposes upon the variables themselves (i.e., to find its *solution*). There are special differential equations for which explicit analytic solutions can be found; unfortunately, however, these are unlikely to arise in nontrivial engineering and scientific applications. Consequently, it is important to know how to solve differential equations *numerically*.

Our attention will be confined to systems for which all dependent variables (denoted by y's) depend on a *single* independent variable (denoted by t), that is, *ordinary differential equations*. The simple *initial value problem* (IVP)

$$\frac{dy}{dt} = f(t, y), \qquad y = y_0 \quad \text{when } t = t_0$$

is considered first in Sections 8.1–8.3. The two major classes of methods, *Runge–Kutta* and *Predictor–Corrector*, are described and compared, and practical software is presented and used.

Then, in Section 8.4, matrix notation is used to generalize these methods to coupled *systems of n first-degree differential equations* involving n dependent variables, and this in turn is used to solve nth-order IVPs of the form

$$y^{(n)} = f(t, y, \ldots, y^{(n-1)}), \qquad y^{(k)}(t_0) = y_{0k}, \quad k = 0, 1, \ldots, n - 1$$

Finally, in Section 8.5 we describe the *Shooting Method* and *Finite Difference Method* for solving the two-point *Boundary Value Problem* (BVP)

$$y'' = f(t, y, y'), \qquad y(a) = \alpha, \quad y(b) = \beta$$

Unlike the preceding chapters, a programmable device is a virtual necessity if accurate numerical solutions of ordinary differential equations are to be found without an inordinate amount of work.

8.1 The Initial Value Problem (IVP)

In this section we lay the groundwork for efficient methods for solving the general first-order **initial value problem**

$$\text{(IVP)} \qquad \frac{dy}{dt} = f(t, y) \quad \text{subject to} \quad y = y_0 \text{ when } t = t_0 \tag{1}$$

where $f(t, y)$ is continuous for (t, y) near (t_0, y_0).

8.1A Existence and Uniqueness of Solutions

A **solution** of (IVP) is a function $y(t)$ that satisfies

$$y'(t) = \frac{d}{dt}[y(t)] = f(t, y(t)) \qquad \text{and} \qquad y(t_0) = y_0 \tag{2}$$

The condition $y'(t) = f(t, y(t))$ means that the function of t obtained by replacing y by $y(t)$ in $f(t, y)$ is the derivative of $y(t)$. Geometrically (Figure 8.1-1) it means that the slope of $y(t)$ at the point (t, y) on its graph can be obtained as $f(t, y)$. We shall therefore call $f(t, y)$ the **slope function** for (IVP). The **initial condition** $y(t_0) = y_0$ forces the graph of $y(t)$ to go through the point (t_0, y_0).

The fact that the graph of a solution is ''nailed down'' at (t_0, y_0) and gets traced with slope $f(t, y)$ at each point (t, y) on it suggests that $y(t)$ exists and is unique for any (IVP). The following discussion examines the extent to which this is true.

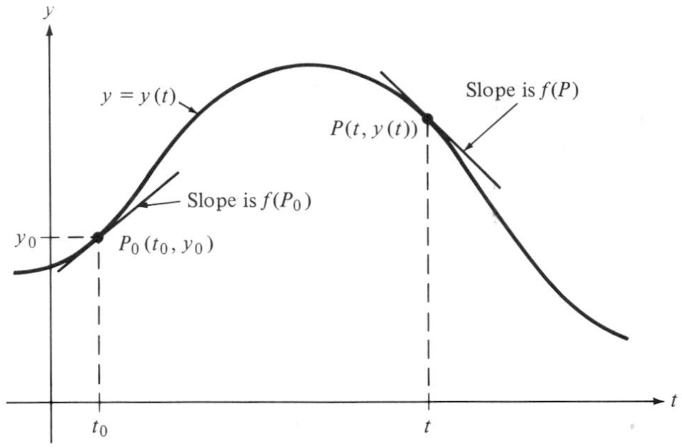

Figure 8.1-1 *Graph of the solution of (IVP).*

EXAMPLE 8.1A Discuss the existence and uniqueness of $y(t)$ for the IVP

$$\frac{dy}{dt} = -ty^2 \quad \text{subject to} \quad y = 1 \text{ when } t = 2 \quad [\text{i.e., } y(2) = 1] \tag{3}$$

SOLUTION: Separating variables and integrating shows that $y = y(t)$ must satisfy

$$-\int \frac{dy}{y^2} = \int t\,dt \quad \Leftrightarrow \quad \frac{1}{y} = \frac{1}{2}t^2 + C \quad \Leftrightarrow \quad y = \frac{2}{t^2 + 2C} \tag{4a}$$

The initial condition $y(2) = 1$ can be satisfied only if $C = -1$. So

$$y(t) = \frac{2}{t^2 + 2} \qquad \text{for all } t > \sqrt{2} \tag{4b}$$

This can be verified by showing that (2) holds for $t > \sqrt{2}$:

$$\frac{d}{dt}[y(t)] = \frac{-4t}{(t^2 - 2)^2} = -t[y(t)]^2 = f(t, y(t)) \qquad \text{and} \qquad y(2) = \frac{2}{4 - 2} = 1 \tag{5}$$

However, the IVP conditions (3) do *not* uniquely determine $y(t)$ to the left of the singularity of $y(t) = 2/(t^2 - 2)$ at $t = \sqrt{2}$. In fact, $y(t)$ in (4b) can be extended to the left of $t = \sqrt{2}$ using (4a) with *any* value of C for $t < \sqrt{2}$, as shown in Figure 8.1-2 where $C = \frac{1}{2}$ there. ■

Similarly, it is easy to verify that for *any* real value of C,

$$y(t) = \begin{cases} t^3 + 2t^2 & \text{for } t \geq 0 \\ t^3 + Ct^2 & \text{for } t < 0 \end{cases} \quad \text{satisfies} \quad \frac{dy}{dt} = 2\frac{y}{t} + t^2, \quad y(1) = 3 \tag{6}$$

for *all* t (Problem M8-1). So the IVP conditions $dy/dt = 2y/t + t^2$, $y(1) = 3$ do not

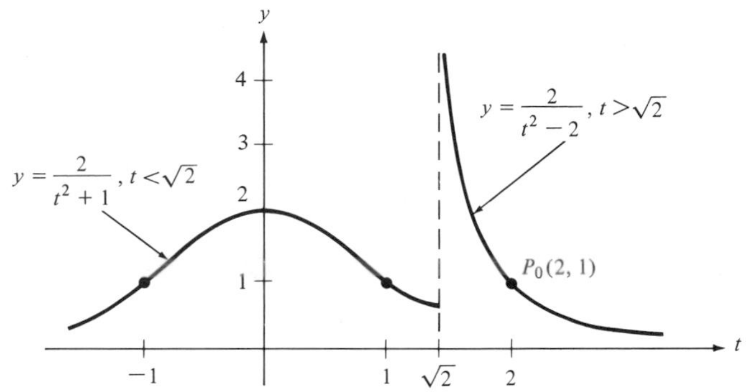

Figure 8.1-2 *Nonuniqueness of a solution across a singularity of y(t).*

uniquely determine $y(t)$ to the left of the singularity of $f(t, y) = 2y/t + t^2$ at $t = 0$. This example and Example 8.1A show that:

> A unique solution cannot be guaranteed across a singularity of *either* the *solution* $y(t)$ *or* the *slope function* $f(t, y)$. As a rule, however, $y(t)$ *exists and is unique* over any interval containing t_0 and over which *both* $y(t)$ and $f(t, y)$ are defined and continuous in t. **(7)**

More precise existence and uniqueness statements can be found in [5].

To further illustrate (7), consider the simple-looking IVP

$$y' = t^2 + y^2, \qquad y(0) = 1 \tag{8}$$

Since $f(t, y) = t^2 + y^2$ is continuous for all (t, y), $y(t)$ exists and is unique over any interval containing $t_0 = 0$ and over which $y(t)$ is continuous. And since the slope of $y(t)$ at $P(t, y)$ is the square of the distance from P to the origin, it must satisfy $y' > y^2$ for $t \neq 0$; so for $t > 0$, $y(t)$ must satisfy

$$y(t) > \hat{y}(t) = \frac{1}{1 - t} = \text{the solution of } y' = y^2, \qquad y(0) = 1 \tag{9}$$

(see Figure 8.1-3). Since $y(t) > \hat{y}(t)$ and $\hat{y}(t)$ has a singularity at $t = 1$, $y(t)$ must have a singularity at t_s somewhere between 0 and 1.

If the IVP (8) described a realistic situation, then the singularity at t_s would represent a catastrophic event (e.g., a beam or wing cracking, a chemical or nuclear explosion, an eardrum bursting, and so on). It would therefore be important to know the value of t_s with some precision, say $4s$. Unfortunately, no one has found a formula for $y(t)$ in terms of the elementary functions! In the absence of an *analytic* expression for $y(t)$, t_s would simply have to be found *numerically*. Since the slope functions $f(t, y)$ that arise in science and engineering are generally more complex than $t^2 + y^2$, the need for accurate, reliable numerical methods for solving (IVP) should be clear.

In recent years there have been significant advances in the area of **symbolic computation**; indeed, computer programs now exist that actually perform the algebraic manipulations

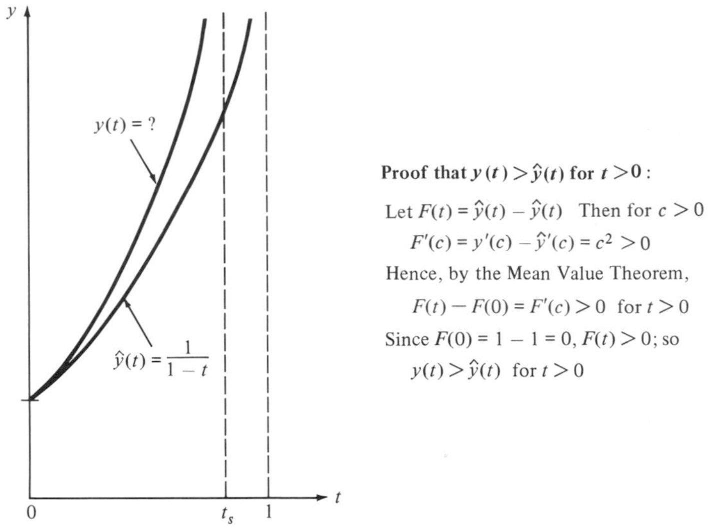
y
$y(t) = ?$
$\hat{y}(t) = \frac{1}{1-t}$
0
t_s
1
t
Proof that $y(t) > \hat{y}(t)$ for $t > 0$:
Let $F(t) = \hat{y}(t) - \hat{y}(t)$ Then for $c > 0$
$F'(c) = y'(c) - \hat{y}'(c) = c^2 > 0$
Hence, by the Mean Value Theorem,
$F(t) - F(0) = F'(c) > 0$ for $t > 0$
Since $F(0) = 1 - 1 = 0$, $F(t) > 0$; so
$y(t) > \hat{y}(t)$ for $t > 0$

Figure 8.1-3 *Graphs of y(t) and $\hat{y}(t) = 1/(1 - t)$ for $t \geq 0$.*

needed to solve *algebraic* and *transcendental* equations *as formulas* [24]. However, it is unrealistic to expect a digital device to output a formula for the solution of a *differential* equation (i.e., an equation involving *both* variables *and* their derivatives or differentials) such as

$$\text{(IVP)} \qquad y' = f(t, y), \qquad y(t_0) = y_0 \tag{10a}$$

What is reasonable to expect is a "profile" consisting of points (t_j, y_j), which lie *near* the (unknown) graph of $y(t)$ in the sense that

$$y_0 = y(t_0) \text{ (given)} \qquad \text{and} \qquad y_j \approx y(t_j) \quad \text{for } j = 1, 2, \ldots \tag{10b}$$

where either $t_0 < t_1 < t_2 < \cdots$ or $t_0 > t_1 > t_2 > \cdots$. The simplest way of generating y_1, y_2, ... is described next.

8.1B Euler's Method

Let us write t_{j+1} as $t_j + h$, where h is a small **stepsize**. If y_j is exactly $y(t_j)$, then the value of $y(t_{j+1})$ can be approximated by the y value at t_{j+1} on the tangent line at the point $P_j(t_j, y_j)$, as shown in Figure 8.1-4(a). This strategy is the basis for

> **EULER'S METHOD:** $y(t_{j+1}) \approx y_{j+1}^{(E)} = y_j + hf_j$, where $f_j = f(t_j, y_j)$ **(11a)**

We shall refer to f_j as the **Euler slope**†, and the line through (t_j, y_j) with slope f_j as the

†Unlike Chapters 6 and 7, *y_j and f_j now have different meanings*: y_j approximates the exact *value* of $y(t)$ at t_j, whereas f_j approximates the exact *slope* of $y(t)$ at t_j.

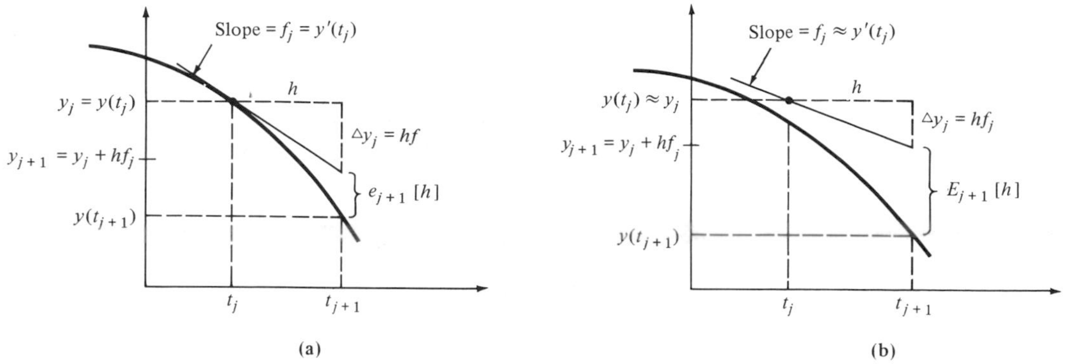

Figure 8.1-4 *Euler's method: (a) $y_j = y(t_j)$; (b) $y_j \approx y(t_j)$.*

Euler line. If y_j is exactly $y(t_j)$ as shown in Figure 8.1-4(a), then f_j is exactly $y'(t_j)$, the tangent slope at t_j. If not, the assumed continuity of $f(t, y)$ assures us that

$$f_j = f(t_j, y_j) \quad \text{will approximate } y'(t_j) \qquad \text{if } y_j \approx y(t_j) \tag{11b}$$

Figure 8.1-4(b) illustrates Euler's method in this case.

In Figure 8.1-4(a), Euler's method is used to illustrate the **per-step** (or **local**) **truncation error** at t_{j+1} defined as

$$e_{j+1}[h] = y(t_{j+1}) - y_{j+1} \quad \text{when} \quad y_j \text{ is } \textit{exactly } y(t_j) \tag{12a}$$

In Figure 8.1-4(b), Euler's method is used to illustrate the **accumulated** (or **global**) **truncation error** at t_{j+1} defined as

$$E_{j+1}[h] = y(t_{j+1}) - y_{j+1} \qquad [\text{whether } y_j = y(t_j) \text{ or not}] \tag{12b}$$

Note that $E_1[h]$ will always be $e_1[h]$ because y_0 is exactly $y(t_0)$. However, for $j > 0$, the accumulated error $E_{j+1}[h]$ generally includes *additional* error due to the fact that y_j is not exactly $y(t_j)$. Unfortunately, it is $E_{j+1}[h]$ that is actually observed at t_{j+1}.

The easiest way to implement a numerical method for (IVP) is to specify a **terminal *t***, t_N, and then proceed from t_0 to t_N in N h-steps as follows

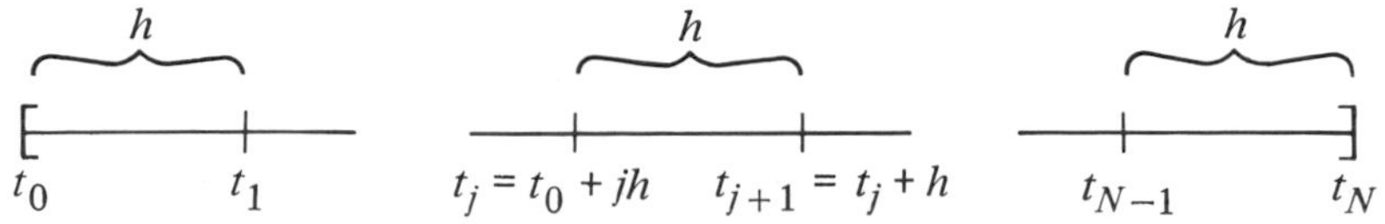

(13a)

The (fixed) stepsize h and the number of steps N are related to t_N by

$$h = \frac{t_N - t_0}{N}, \quad \text{or equivalently,} \quad N = \frac{t_N - t_0}{h} \tag{13b}$$

So specifying any *two* of h, N, and t_N determines the third. Note that h will be negative if t_N lies to the left of t_0. The following example illustrates how easy it is to use Euler's method with a fixed stepsize h.

EXAMPLE 8.1B Use Euler's method over the interval [2, 3], first with $h = 0.1$ then with $h = 0.05$, for the IVP of Example 8.1A, namely

$$y' = -ty^2, \qquad y(2) = 1 \qquad \left[y(t) = \frac{2}{t^2 - 2} \text{ for } t > \sqrt{2}\right] \tag{14}$$

Use your results to get an improved estimate of $y(3)$.

SOLUTION: Since $t_0 = 2$ and $f(t, y) = -ty^2$, the Euler's method formula (11a) is

$$y(t_{j+1}) \approx y_{j+1}^{(E)} = y_j + h(-t_j y_j^2), \qquad \text{where } t_j = 2 + jh \tag{15}$$

Using (15) with $h = 0.1$, starting with $t_0 = 2$, $y_0 = 1$ and $f_0 = -2 \cdot 1^2 = -2$, gives

$$\begin{aligned}
y(2.1) &\approx y_1^{(E)} = y_0 + hf_0 = 1 + 0.1(-2) = 0.8; \quad \text{so } f_1 = -2.1(0.8)^2 = -1.344\\
y(2.2) &\approx y_2^{(E)} = y_1 + hf_1 = 0.8 + 0.1(-1.344) = 0.6656; \quad \text{so } f_2 = -2.2(0.6656)^2\\
y(2.3) &\approx y_3^{(E)} = y_2 + hf_2 \doteq 0.6656 + 0.1(-0.97465) \doteq 0.56813, \quad \text{etc.}
\end{aligned}$$

Similarly, using (15) with $h = 0.05$ yields the following results:

$$\begin{aligned}
y(2.05) &\approx y_1^{(E)} = y_0 + hf_0 = 1 + 0.05(-2) = 0.9; \quad f_1 = -2.05(0.9)^2 = -1.6605\\
y(2.10) &\approx y_2^{(E)} = 0.9 + 0.05(-1.6605) \doteq 0.81698; \quad f_2 = -2.10(0.81698)^2 = -1.4017\\
y(2.15) &\approx y_3^{(E)} \doteq 0.81698 + 0.05(-1.4017) \doteq 0.74690, \quad \text{etc.}
\end{aligned}$$

Since two 0.05-steps are needed for each 0.1-step, the tabulated y_j's in Table 8.1-1 are $y_0, y_1, \ldots, y_{10}$ when $h = 0.1$, but $y_0, y_2, \ldots, y_{20}$ when $h = 0.05$. The exact $y(t_j)$ values were obtained as $2/(t_j^2 - 2)$.

An examination of $E[0.1]$ and $E[0.05]$ for any particular tabulated t reveals that

$$E[0.05] \approx \tfrac{1}{2}E[0.1], \qquad \text{that is, } E[(\tfrac{1}{2}h)] \approx \tfrac{1}{2}E[h] \quad (\text{with } h = 0.1) \tag{16}$$

This suggests that for any j (see Section 7.2B),

$$F[h] = y_{j+1}^{(E)} \text{ is an } O(h) \text{ approximation of } Q = y(t_{j+1}) \tag{17}$$

We will confirm (17) analytically in the next section. Assuming it for now, we can use Richardson's formula with $n = 1$ and $r = 0.1/0.05 = 2$ to get the following improved estimate of $Q = y(3)$ from the bottom row of Table 8.1-1:

TABLE 8.1-1 Using Euler's Method with $h = 0.1$ and $h = 0.05$

	Exact	Using $h = 0.1$		Using $h = 0.05$	
t_j	$y(t_j)$	$y_j[0.1]$	$E_j[0.1]$	$y_j[0.05]$	$E_j[0.05]$
$t_0 = 2.0$	1	1	0	1	0
2.1	0.8299	0.8000	0.0299	0.8170	0.0129
2.2	0.7042	0.6656	0.0386	0.6869	0.0173
2.3	0.6079	0.5681	0.0398	0.5879	0.0182
2.4	0.5319	0.4939	0.0380	0.5142	0.0177
2.5	0.4706	0.4354	0.0352	0.4539	0.0167
2.6	0.4202	0.3880	0.0322	0.4048	0.0154
2.7	0.3781	0.3488	0.0292	0.3640	0.0141
2.8	0.3425	0.3160	0.0265	0.3297	0.0128
2.9	0.3120	0.2880	0.0240	0.3003	0.0117
$t_N = 3.0$	0.2857	0.2640	0.0217	0.2751	0.0106

$$F_1[0.05] = \frac{2^1 F[0.05] - F[0.1]}{2^1 - 1} = 2(0.2751) - 0.2640 = 0.2862 \tag{18}$$

This is more accurate than $F[0.05]$ by one decimal place. ■

8.1C The Order of a Numerical Method

When y_j is exactly $y(t_j)$ [see Figure 8.1-4(a)], the Euler's method approximation $y_{j+1}^{(E)}$ is $y(t_j) + hy'(t_j)$. Since this is the first Taylor polynomial approximation of $y(t_{j+1}) = y(t_j + h)$, we can get the truncation error of y_{j+1} *by inspection* of the series

$$\underbrace{y(t_j + h)}_{y(t_{j+1})} = \underbrace{y(t_j) + hy'(t_j)}_{y_j + hf_j = y_{j+1}^{(E)}} + \underbrace{\frac{h^2}{2!} y''(t_j) + \frac{h^3}{3!} y'''(t_j) + \cdots}_{\tau[h] = e_{j+1}^{(E)}[h] = O(h^2)} \tag{19}$$

Thus, for Euler's method, the *per-step* truncation error $e_{j+1}^{(E)}[h]$ is $O(h^2)$. That the *accumulated* truncation error $E_{j+1}^{(E)}[h]$ is only $O(h)$ (one order less) follows from the following general result.

> If the per-step truncation error $e[h]$ of a method is $O(h^{n+1})$, then the accumulated truncation error $E[h]$ will be $O(h^n)$. **(20)**

To see why, let us examine the error that accumulates at t_j. Clearly,

$$t_j = t_0 + jh \Rightarrow j\ h\text{-steps are required to get from } t_0 \text{ to } t_j$$

If h is reduced to h/r, where r is a positive integer, then

$$t_j = t_0 + (rj)\frac{h}{r} \quad \Rightarrow \quad rj \ \frac{h}{r}\text{-steps are required to get from } t_0 \text{ to } t_j$$

Assume now that $e[h]$ is $O(h^{n+1})$. Then [see (7) of Section 7.2B]

$$e\left[\frac{h}{r}\right] \approx \frac{1}{r^{n+1}}\, e[h] \qquad \text{for } \textit{each } \frac{h}{r}\text{-step from } t_0 \text{ to } t_j \tag{21a}$$

Since the number of steps from t_0 to t_j is multiplied by r while each *per-step* error is divided by r^{n+1}, we expect the *accumulated* error at t_j to satisfy

$$E\left[\frac{h}{r}\right] \approx r\left(\frac{1}{r^{n+1}}\right) E[h] = \frac{1}{r^n} E[h], \qquad \text{that is, } E[h] = O(h^n) \tag{21b}$$

A numerical method is called ***n*th-order** if $E[h] = O(h^n)$, that is, if $e[h] = O(h^{n+1})$. It follows from (19) and (20) that *Euler's method is first order*, as we noted in (17).

To appreciate the need for higher-order numerical methods for solving (IVP), note that the r needed to approximate $y(t)$ with two additional digits of accuracy must satisfy

$$E\left[\frac{h}{r}\right] \approx \frac{1}{10^2} E[h] \quad \text{or, by (21b),} \quad \frac{1}{r^n} \leqslant \frac{1}{100}, \quad \text{that is,} \quad r^n \geqslant 100 \tag{22}$$

Thus r must be about 100 if $n = 1$, 10 if $n = 2$, and 5 if $n = 3$, but only 3 if $n = 4$. For example, the $E_j[0.05]$ column of Table 8.1-1 shows that twenty 0.05-steps produced about $2d$ accuracy at $t_N = 3$. Since Euler's method is only first order (i.e., $n = 1$), it would take $20 \cdot 100 = 2000$ steps (with $h = 0.0005$) to increase the accuracy to about $4d$. Aside from the time and energy that this would entail, *there is the risk that increased accumulated roundoff error may offset some or all of the reduced accumulated truncation error!* This risk would be reduced substantially if the method were fourth order, in which case only $20 \cdot 3 = 60$ steps (with $h \doteq 0.017$) would be needed to increase the accuracy to about $4d$.

We now present two easily obtained second-order methods.

8.1D The Modified Euler and Huen Methods

Most numerical methods for solving (IVP) approximate $y(t)$ at $t_{j+1} = t_j + h$ using a formula of the form

$$y(t_{j+1}) \approx y_{j+1} = y_j + h\phi, \qquad \text{where } \phi = \phi(t_j, y_j, h, f_j, \ldots) \tag{23a}$$

where "..." in (23a) means "and perhaps other sampled slope values $f(t, y)$." For Euler's method, ϕ is just the Euler slope f_j. To get more accurate formulas for ϕ, note that when $y_j = y(t_j)$,

$$h\phi = \Delta y_j = y_{j+1} - y_j \quad \text{approximates} \quad \Delta y(t) = y(t_{j+1}) - y(t_j) \tag{23b}$$

We know from the Fundamental Theorem of Calculus that when $y'(t)$ is continuous, $\Delta y(t)$ is just the integral of $y'(t)$ from t_j to t_{j+1}. Since $y'(t) = f(t, y(t))$, we can thus view (23a) as follows:

$$y(t_{j+1}) \approx y_{j+1} = y_j + h\phi, \qquad \text{where } h\phi \text{ approximates } \int_{t_j}^{t_{j+1}} f(t, y(t))\, dt \qquad \textbf{(24a)}$$

A quadrature formula for the integral in (24a) will be of the form

$$h\phi = h(\Sigma_i\, C_i m_i), \qquad \text{where } m_i = f(\tilde{t}_i, \tilde{y}_i) \approx y'(\tilde{t}_i) \qquad \textbf{(24b)}$$

with $\tilde{t}_i$ *lying in or near* the integration interval $[t_j, t_{j+1}]$. When all sampled $\tilde{t}_i$'s actually lie between t_j and t_{j+1} (inclusive), the method based on (24) will be called a **Runge–Kutta method**.

If we take $h\phi$ in (24a) to be the *Left-Endpoint Rule* [Figure 8.1-5(a)], but with $y'(t_j)$ replaced by $f_j = f(t_j, y_j)$, we get

$$y(t_{j+1}) \approx y_{j+1}^{(E)} = y_j + hf_j \qquad \text{(this is Euler's method)} \qquad \textbf{(25)}$$

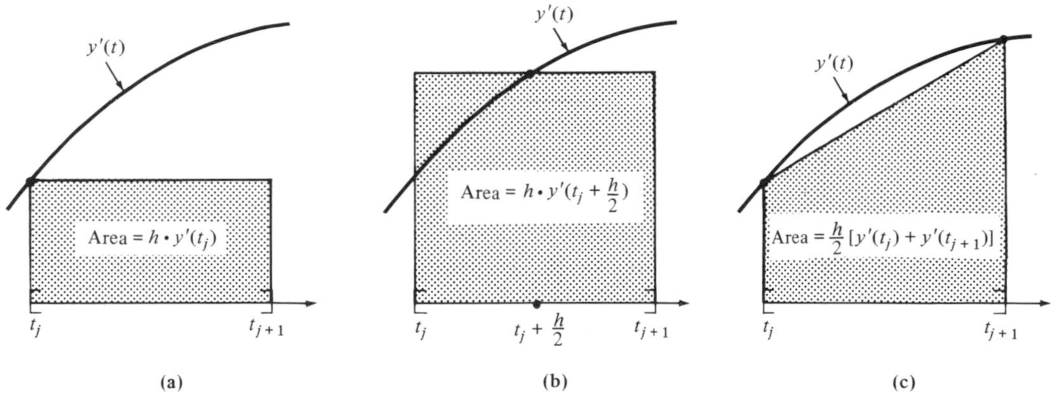

Figure 8.1-5 *Three approximations of $\int_{t_j}^{t_{j+1}} y'(t)\,dt$.*

When $y_j = y(t_j)$, f_j becomes exactly $y'(t_j)$, hence hf_j becomes exactly the Left-Endpoint Rule; so we could have deduced from its truncation error (Problem 7-5) that the per-step truncation error of $y_{j+1}^{(E)}$ is $O(h^2)$ in agreement with (19).

To ensure that the ***i*th sampled slope** m_i is a good approximation of $y'(\tilde{t}_i)$ when $\tilde{t}_i \neq t_j$, we must ensure that $\tilde{y}_i \approx y(\tilde{t}_i)$. *The challenge is to do this without knowing* $y(t)$! (Indeed, if $y(t)$ were known, then we would not need a numerical method!) If h is chosen small enough so that $y'(t)$ is nearly constant on $[t_j, t_{j+1}]$, then a convenient way to approximate $y(\tilde{t}_i)$ is to use the y-value on the Euler line as follows (Figure 8.1-6):

$$\text{If } \tilde{t}_i = t_j + \alpha h, \qquad \text{then } \tilde{y}_i = y_{j+\alpha}^{(E)} = y_j + \alpha h f_j \text{ approximates } y(\tilde{t}_i) \qquad \textbf{(26)}$$

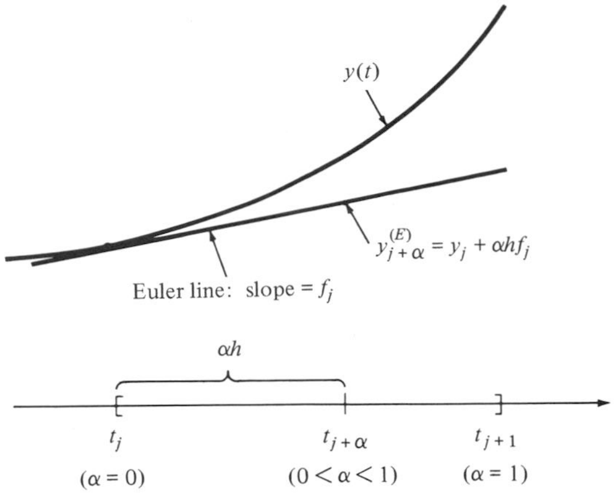

Special Cases:

α	t	$y_{j+\alpha}^{(E)}$
0	t_j	y_j
$\frac{1}{2}$	$t_{j+1/2}$	$y_j + \frac{h}{2} f_j = y_{j+1/2}^{(E)}$
1	t_{j+1}	$y_j + hf_j = y_{j+1}^{(E)}$

Note: $t_{j+1/2}$ is the midpoint of $[t_j, t_{j+1}]$.

Figure 8.1-6 *Approximating $y(t_{j+\alpha})$ by $y_{j+\alpha}^{(E)}$ on the Euler line.*

For example, if we take $h\phi$ in (24) to be the Midpoint Rule [Figure 8.1-5(b)], but with $f(t_{j+1/2}, y_{j+1/2}^{(E)})$ used to approximate $y'(t_{j+1/2})$, the result is the

MODIFIED EULER METHOD: $y(t_{j+1}) \approx y_{j+1}^{(ME)} = y_j + hf(t_j + \frac{1}{2}h,\ y_j + \frac{1}{2}hf_j)$ **(27)**

Alternatively, we can take $h\phi$ to be the Trapezoidal Rule [Figure 8.1-5(c)], but with f_j and $f(t_{j+1}, y_{j+1}^{(E)})$ used to approximate $y'(t_j)$ and $y'(t_{j+1})$, respectively. The result is

HUEN'S METHOD: $y(t_{j+1}) \approx y_{j+1}^{(H)} = y_j + \frac{1}{2}[f_j + f(t_j + h,\ y_j + hf_j)]$ **(28)**

In (27) and (28), the $h\phi$ terms do not become exactly the Midpoint and Trapezoidal rules when $y_j = y(t_j)$, because $y^{(E)}(t)$ on the Euler line is not exactly $y(t)$ [see Figure 8.1-6]. However, they are close enough so that the *per-step* truncation errors of both $e^{(ME)}[h]$ and $e^{(H)}[h]$ are of the form

$$Cy'''(t_j)h^3, \qquad \text{where } C \text{ depends on } f_y(t_j, y_j) \tag{29}$$

which is similar to (4b) and (8b) of Section 7.3A. So *the Modified Euler method and Huen's method are both second-order Runge–Kutta methods.*

EXAMPLE 8.1D Use **(a)** the Modified Euler method and **(b)** Huen's method with $h = 0.1$ to approximate $y(2.1) \doteq 0.8299$ and $y(2.2) \doteq 0.7042$ for

$$y' = -ty^2, \quad y(2) = 1 \qquad [\text{exact } y(t) = 2/(t^2 - 2)] \tag{30}$$

SOLUTION (a): Since $f(t, y) = -ty^2$, the Modified Euler formula (27) is

$$y(t_{j+1}) \approx y_{j+1}^{(ME)} = y_j - h\left(t_j + \frac{h}{2}\right)\left[y_j + \frac{h}{2}f_j\right]^2 \qquad \textbf{(31)}$$

where $f_j = -t_j y_j^2$ and $h = 0.1$. Since $f_0 = -t_0 y_0^2 = -2 \cdot 1^2 = -2$, $y(t)$ can be approximated at $t_1 = t_0 + h = 2.1$ by

$$y_1^{(ME)} = 1 - 0.1(2.05)[1 + 0.05(-2)]^2 = 0.83395$$

with truncation error $E_1[0.1] \doteq -0.0041$. Similarly, $f_1 = -t_1 y_1^2 = -2.1(0.83395)^2 \doteq -1.4605$, so $y(t_2) = y(2.2)$ can be approximated by

$$y_2^{(ME)} = 0.83395 - 0.215[0.83395 + 0.05(-1.4605)]^2 \doteq 0.70946$$

with truncation error $E_2[0.1] \doteq -0.0053$.

SOLUTION (b): Since $f(t, y) = -ty^2$, the Huen's method formula (28) is

$$y(t_{j+1}) \approx y_{j+1}^{(H)} = y_j + \frac{h}{2}[f_j - (t_j + h)(y_j + hf_j)^2] \qquad \textbf{(32)}$$

where $h = 0.1$. Since $f_0 = -2 \cdot 1^2 = -2$, $y(t_1) = y(2.1)$ can be approximated by

$$y_1^{(H)} = 1 + 0.05[-2 - 2.1(1 + 0.1(-2))^2] = 0.8328 \qquad (E_1[0.1] = -0.0029)$$

Similarly, $f_1 = -2.1(0.8328)^2 \doteq -1.4565$, hence $y(t_2) = y(2.2)$ can be approximated by

$$y_2^{(H)} = 0.8328 + 0.05[f_1 - 2.2(0.8328 + 0.1f_1)^2] \doteq 0.70804 \qquad (E_2[0.1] \doteq -0.0042)$$

These $O(h^2)$ approximations have about one more accurate significant digit than the $O(h)$ Euler's method approximations of Table 8.1-1. ■

Runge–Kutta formulas can also be obtained using Taylor series. This more general approach will be described in Section 8.2C.

8.2

Self-Starting Methods: Runge–Kutta and Taylor

In this section we continue to examine IVP solvers of the form

$$y_{j+1} = y_j + h\phi, \qquad \phi = \Sigma f(\tilde{t}_i, \tilde{y}_i) \qquad \textbf{(1)}$$

where all sampled slope values $f(\tilde{t}_i, \tilde{y}_i)$ are evaluated *after* y_j is obtained. Such methods are called **self-starting** because, as we have seen, they can be used with $j = 0$ to get y_1. Methods that are not self-starting because they require *previously* obtained sample slopes are considered in Section 8.3.

8.2A The Fourth-Order Runge–Kutta (RK4) Method

The most popular general purpose IVP solvers are ones for which $h\phi$ integrates y' with cubic exactness. Such methods will generally integrate y' accurately from t_j to t_{j+1} even if y' has a local maximum, a local minimum, *and* an inflection point on the integration interval! (Why?) Also, as we shall presently see, there are Runge–Kutta methods that sample four slopes to get $y(t)$ exactly when $f(t, y)$ is cubic (in t and y); however, it has been shown [11] that exactness for a fourth-degree $f(t, y)$ requires at least six sampled slopes. So the tradeoff between accuracy and computational effort is particularly efficient for the cubic exact, fourth-order methods. The best known of these is the following method.

FOURTH-ORDER RUNGE–KUTTA (RK4) METHOD: $[\overbrace{\qquad\qquad}^{h/2}+\overbrace{\qquad\qquad}^{h/2}]$ with points t_j, $t_{j+1/2}$, t_{j+1}

$$y_{j+1}^{(RK4)} = y_j + \frac{h}{6}[m_1 + 2(m_2 + m_3) + m_4], \qquad \text{where } m_1 = f_j = f(t_j, y_j) \qquad \textbf{(2a)}$$

and m_2, m_3, and m_4 are obtained successively as follows:

$$\begin{aligned} m_2 &= f\left(t_j + \frac{h}{2}, y_j + \frac{h}{2}m_1\right) \\ m_3 &= f\left(t_j + \frac{h}{2}, y_j + \frac{h}{2}m_2\right) \\ m_4 &= f(t_j + h, y_j + hm_3) \end{aligned} \qquad \textbf{(2b)}$$

Notice that m_1 (the Euler slope) approximates $y'(t_j)$; m_4 approximates $y'(t_{j+1})$; and both m_2 and m_3 approximate $y'(t)$ at the midpoint $t_{j+1/2}$. Notice too that if $f(t, y)$ depends only on t, say $f(t, y) = g(t)$, then (2) amounts to integrating $y'(t) = g(t)$ from t_j to t_{j+1} using Simpson's $\frac{1}{3}$-Rule. This is why (2) is sometimes referred to as the **Runge–Kutta Simpson method**.

EXAMPLE 8.2A Use RK4 with $h = 0.1$ on [2, 3] for the IVP

$$y' = -ty^2, \quad y(2) = 1 \qquad \left[\text{exact solution is } y(t) = \frac{2}{t^2 - 2}\right] \qquad \textbf{(3)}$$

SOLUTION: Starting with $t_0 = 2$, $y_0 = 1$, (2) gives

$$\begin{aligned} m_1 &= f(2.0, 1) = -(2)(1)^2 = -2 \\ m_2 &= f(2.05, 1 + 0.05(-2)) = -(2.05)(0.9)^2 = -1.6605 \\ m_3 &= f(2.05, 1 + 0.05(-1.6605)) = -(2.05)(0.916975)^2 \doteq -1.72373 \\ m_4 &= f(2.1, 1 + 0.1(-1.72373)) = -(2.1)(0.82763)^2 \doteq -1.43843 \\ y_1^{(RK4)} &= y_0 - \frac{0.1}{6}\{2 + 2(1.6605 + 1.72373) + 1.43843\} \doteq 0.829885 \end{aligned}$$

```
ØØ1ØØ    C        PROGRAM CALRK4    ! Call SUBROUTINE RK4 to solve an IVP
ØØ2ØØ             DOUBLE PRECISION Y
ØØ3ØØ             EXTERNAL FT821                           ! for Table 8.2-1
ØØ4ØØ             DATA IW, IR /5, 5/                       ! interactive I/O
ØØ5ØØ             WRITE (IW, 1)                            ! prompt for input
ØØ6ØØ             READ (IR, *) T, Y, TN, NSTEPS, NPRINT    ! input from TTY
ØØ7ØØ             IF (NPRINT.GT.Ø) WRITE (IW, 2)           ! column headings
ØØ8ØØ    C
ØØ9ØØ             CALL RK4 (FT821, T, TN, NSTEPS, NPRINT, Y, IW)
Ø1ØØØ    C
Ø11ØØ             IF (NPRINT.EQ.Ø .OR. MOD(NSTEPS,NPRINT).GT.Ø)
Ø12ØØ         &     WRITE (IW, 3) Y, T                     ! ycalc at t = tN
Ø13ØØ             STOP
Ø14ØØ    1 FORMAT ('ØINPUT tØ, yØ, tN, NSTEPS, NPRINT')
Ø15ØØ    2 FORMAT ('Ø',3X,'t',13X,'y',/,2x,9('-'),2X,14('-'))
Ø16ØØ    3 FORMAT ('Øy =',G14.7,' when   t = tN =',G1Ø.3)
Ø17ØØ             END
Ø18ØØ
Ø19ØØ             FUNCTION FT821(T,Y)
Ø2ØØØ             FT821 = -T*Y*Y          ! for Table 8.2-1
Ø21ØØ             RETURN
Ø22ØØ             END


ØØ1ØØ             SUBROUTINE RK4(F, T, TN, NSTEPS, NPRINT, Y, IW)
ØØ2ØØ             REAL M1, M2, M3, M4
ØØ3ØØ             DOUBLE PRECISION Y
ØØ4ØØ    C - - - - - - - - - - - - - - - - - - - - - - - - - - - - - - C
ØØ5ØØ    C This subroutine integrates from tØ to tN the first order IVP   C
ØØ6ØØ    C         y' = f(t, y),   y(tØ) = yØ (initial condition)         C
ØØ7ØØ    C Using NSTEPS steps of the fourth-order Runge–Kutta method. It  C
ØØ8ØØ    C prints T and Y on device IW every NPRINT steps if NPRINT > Ø.  C
ØØ9ØØ    C In the calling program, you must declare the variable used     C
Ø1ØØØ    C for Y DOUBLE PRECISION and the FUNCTION used for F EXTERNAL.    C
Ø11ØØ    C - - - - - - - - - - - - - - - - - - VERSION 2: 9/9/85 - - - - C
Ø12ØØ             IF (NPRINT.GT.Ø) WRITE (IW,1) T, Y                != tØ, yØ
Ø13ØØ             H = (TN - T)/NSTEPS
Ø14ØØ             DO 1Ø J=1, NSTEPS
Ø15ØØ                M1 = F(T, Y)
Ø16ØØ                M2 = F(T + .5*H, Y + .5*H*M1)
Ø17ØØ                M3 = F(T + .5*H, Y + .5*H*M2)
Ø18ØØ                M4 = F(T + H, Y + H*M3)
Ø19ØØ                Y = Y + H*(M1 + 2*(M2 + M3) + M4)/6            != yj
Ø2ØØØ                T = T + H                                      != tj
Ø21ØØ                IF (NPRINT.GT.Ø.AND.MOD(J,NPRINT).EQ.Ø) WRITE (IW,1) T, Y
Ø22ØØ    1Ø CONTINUE
Ø23ØØ    1 FORMAT (G1Ø.3, 3X, G14.7)
Ø24ØØ             RETURN
Ø25ØØ             END
```

Figure 8.2-1   *Fortran program for solving $y' = -ty^2$, $y(2) = 1$ by RK4.*

The other $y_j^{(RK4)}$ entries of Table 8.2-1 are obtained similarly. Note that the tabulated $y_j^{(RK4)}$'s agree with the exact values of $y(t_j)$ to about 5*s*. ■

The reasons for the popularity of RK4 are evident from the $y_j^{(RK4)}$ values tabulated in Table 8.2-1 and the Fortran program shown in Figure 8.2-1, which was used to get them. Clearly, RK4 is accurate. Moreover, `SUBROUTINE RK4`, which yielded this accuracy, is short, straightforward, and easy to use. Indeed, if $f(t, y)$ is changed, then one need only modify the `FUNCTION` subprogram used for `F` (here `FT821`), but *not* `SUBROUTINE RK4`. Any other changes can be dealt with by simply changing the input data for the calling program (here `CALRK4`).

Practical Consideration 8.2A (Partial extended precision). When using a formula of the general form $y_{j+1} = y_j + h\phi$ with a small h, the $h\phi$ term is likely to be much smaller than y_j. Consequently, there is the risk of negligible addition when adding $h\phi$ to y_j (see Section 1.3B). This risk can be avoided by storing the current y_j value in extended precision, as was done for `Y` in `SUBROUTINE RK4`. This application of the *Partial Extended Precision* strategy of Section 1.4C can significantly improve the accuracy of *any* method when N, the number of steps from t_0 to t_N, is large. ■

TABLE 8.2-1 RK4 Values for $y' = -ty^2$, $y(2) = 1$ ($h = 0.1$)

t_j	$y_j^{(RK4)}$	$y(t_j)$
2.0	1.000000	1.000000
2.1	0.829885	0.829876
2.2	0.704237	0.704225
2.3	0.607914	0.607903
2.4	0.531924	0.531915
2.5	0.470596	0.470588
2.6	0.420175	0.420168
2.7	0.378078	0.378072
2.8	0.342471	0.342466
2.9	0.312017	0.312012
3.0	0.285718	0.285714

8.2B The Fourth-Order Runge–Kutta–Fehlberg (RKF4) Method

The RK4 method does have some shortcomings. One is the need for four $f(t, y)$ evaluations at each step. This can be serious when the slope function is complicated or when only a limited time is available to integrate from t_0 to t_N. In such situations, a fourth-order *predictor–corrector method* such as APC4 of Section 8.3 should be tried.

A second, more serious shortcoming of RK4 is that the method provides no indication of the accuracy of $y_{j+1}^{(RK4)}$ at each step. If time allows, an obvious way to remedy this is to pause every few steps and repeat the integration from the current t_j to $t_{j+1} = t_j + h$ using *two* $h/2$-steps, and then use this more accurate approximation to determine if the stepsize h provided sufficient accuracy. Note, however, that each such ''simple'' check requires *eight* new $f(t, y)$ evaluations; and unless the check is performed at every step there is the possibility that you will ''miss'' the step when the error becomes unacceptably large.

An alternative approach for a fourth-order method is to sample more than four (but fewer than eight) $f(t, y)$ values per step and use this extra information to get *two* estimates of $y(t_{j+1})$. The first, to be used for y_{j+1}, should have a per-step error $e_{j+1}[h]$, which is $O(h^5)$. The second should have a *higher-order* per-step error so as to enable the calculation of

$$ErrEst = \text{a reliable, computable approximation of } e_{j+1}[h] \qquad \textbf{(4a)}$$

This estimate can then be used to recalculate the stepsize h for the next step. The idea is to take large steps when $y'(t)$ varies slowly near t_j, and small steps when it varies rapidly due either to rapid oscillations of $y(t)$ or a vertical asymptote at t_s near t_j. This can be done effectively using the variables

$$Ratio = |ErrEst/h| \quad \text{(the absolute error per unit } h) \qquad \textbf{(4b)}$$
$$Rmax = \text{the largest acceptable value of } Ratio \qquad \textbf{(4c)}$$

Rmax is generally set to $C \cdot 10^{-s}$, where s is the desired number of accurate significant digits of y_{j+1}. Two strategies that use (4) to control h are described in (5) and (6). In the pseudocode, t is the current t_j and y_{calc} is the calculated y_{j+1} whose accuracy is being tested.

HALVE/DOUBLE/LEAVE STEPSIZE STRATEGY FOR AN *n*TH-ORDER METHOD:

```
IF Ratio > Rmax                                    {insufficient accuracy}
    THEN h ← h/2                                {halve h; but do not bump t}
    ELSE t ← t + h;   y ← y_calc            {bump t; accept y_calc as y_j+1}   (5)
        IF Ratio ≤ 0.5Rmax/2^n               {more accuracy than needed}
            THEN h ← 2h                           {double h for next y_calc}
```

CONTINUOUS CHANGE STEPSIZE STRATEGY FOR AN *n*TH-ORDER METHOD:

```
IF Ratio ≤ Rmax                                      {sufficient accuracy}
    THEN t ← t + h;   y ← y_calc            {bump t; accept y_calc as y_j+1}
        Ratio ← max(SmallRatio, Ratio)               {limit shrinking of h}   (6)
    ELSE Ratio ← min(LargeRatio, Ratio)                {limit growth of h}
h ← h(Rmax/Ratio)^(1/n)                            {adjust h for next y_calc}
```

In (6), *SmallRatio* (typically 10^{-n}) and *LargeRatio* (typically 4^n) ensure that h does not change too rapidly (typically, $h/10 \leq \text{new } h \leq 4h$).

One of the more effective fourth-order methods that uses (4) is due to E. Fehlberg [17]. It will be referred to as the **fourth-order Runge–Kutta–Fehlberg** (abbreviated **RKF4**) **method**. This method uses six $f(t, y)$ evaluations per step to get a fourth-order y_{j+1} and a reliable *ErrEst*, as described in Algorithm 8.2B. The **accuracy test** shown is the Continuous Change Stepsize Strategy (6).

Figure 8.2-2 shows the result of replacing `SUBROUTINE RK4` by one that performs the `RKF4` method. Notice how h increases as $y(t)$ "levels off" near $t_N = 4$. The smaller final value of h was adjusted so that the last t_{j+1}, here t_9, is exactly $t_N = 4$.

Figure 8.2-3 illustrates the behavior of RKF4 when $y(t)$ has a singularity at some value t_s between t_0 and t_N. Notice how h decreases as the graph becomes more vertical. Even if

j	t	h	y
0	2.0000		1.0000000
1	2.1000	0.10000	0.8298735
2	2.2115	0.11149	0.6918748
3	2.3496	0.13811	0.5680785
4	2.5204	0.17081	0.4595053
5	2.7342	0.21378	0.3652406
6	3.0050	0.27078	0.2844989
7	3.3529	0.34789	0.2164084
8	3.8076	0.45478	0.1600241
9	4.0000	0.19237	0.1428565

Figure 8.2-2 *RKF4 values on [2, 4] for $y'(t) = -ty^2$, $y(2) = 1$.*

no singularity was anticipated (see Figure 8.1-3), the output would reveal that one exists near $\doteq 0.97$.

It should be clear that RKF4 is preferable to RK4 unless there is a need to use a constant value of h. (One such situation occurs when ''starting'' a fourth-order multistep method as described in Section 8.3A.) If ''nicer'' values of t_j are desired, RKF4 can be used with the Halve/Double/Leave Strategy (5).

j	t	h	y
0	0.00000		1.000000
1	0.10000	0.10000	1.111464
2	0.30103	0.20103	1.441909
3	0.43511	0.13408	1.810050
4	0.53137	0.96255E-01	2.219266
5	0.60302	0.71657E-01	2.666468
6	0.65794	0.54911E-01	3.149561
7	0.70103	0.43093E-01	3.667295
8	0.73553	0.34500E-01	4.218957
9	0.76362	0.28090E-01	4.804153
10	0.78682	0.23205E-01	5.422662
	⋮		
20	0.90230	0.55783E-02	14.79469
30	0.93415	0.19875E-02	28.03953
40	0.94747	0.95068E-03	44.76584
50	0.95440	0.53253E-03	64.93517
	⋮		
240	0.96888	0.72901E-05	1081.229
250	0.96895	0.64912E-05	1167.584
260	0.96901	0.58052E-05	1257.205
270	0.96906	0.52128E-05	1350.089

Apparent singularity near t = 0.9691

Figure 8.2-3 *RKF4 values for $y'(t) = t^2 + y^2$, $y(0) = 1$.*

ALGORITHM 8.2B. RKF4 (FOURTH-ORDER RUNGE–KUTTA–FEHLBERG METHOD)

PURPOSE: To solve to a prescribed accuracy on $[t_0, t_N]$ the IVP

$$y' = f(t, y), \qquad y = y_0 \text{ when } t = t_0$$

GET $t_0, t_N, y_0,$	{parameters of IVP}
$Rmax,$	{accuracy control parameter}
$ScaleMin, ScaleMax$	{stepsize control parameters}
$h \leftarrow (Rmax)^{1/4}$	{initial stepsize}
$Hmin \leftarrow h \cdot 10^{-4}$	{minimum allowable stepsize}
$t \leftarrow t_0; \quad y \leftarrow y_0$	{(t, y) is current (t_j, y_j)}

DO UNTIL $t = t_N$ OR $h < Hmin$
    IF $t + h > t_N$ THEN $h \leftarrow t_N - t$ {stepsize for final step}

    {Estimate e[h] for next y}
    $k_1 \leftarrow hf(t_j, y_j)$
    $k_2 \leftarrow hf(t_j + \tfrac{1}{4}h, y_j + \tfrac{1}{4}k_1)$
    $k_3 \leftarrow hf(t_j + \tfrac{3}{8}h, y_j + \tfrac{3}{32}k_1 + \tfrac{9}{32}k_2)$
    $k_4 \leftarrow hf(t_j + \tfrac{12}{13}h, y_j + \tfrac{1932}{2197}k_1 - \tfrac{7200}{2197}k_2 + \tfrac{7296}{2197}k_3)$
    $k_5 \leftarrow hf(t_j + h, y_j + \tfrac{439}{216}k_1 - 8k_2 + \tfrac{3680}{513}k_3 - \tfrac{845}{4104}k_4)$
    $k_6 \leftarrow hf(t_j + \tfrac{1}{2}h, y_j - \tfrac{8}{27}k_1 + 2k_2 - \tfrac{3544}{2565}k_3 + \tfrac{1859}{4104}k_4 - \tfrac{11}{40}k_5)$
    $ErrEst \leftarrow \tfrac{1}{360}k_1 - \tfrac{128}{4275}k_3 - \tfrac{2097}{75240}k_4 + \tfrac{1}{50}k_5 + \tfrac{2}{55}k_6$ {$\approx e[h]$}

    {Accuracy test}
    $Ratio \leftarrow |ErrEst/h|$
    IF $Ratio \leq Rmax$ THEN   {accuracy of next y is acceptable}
        $t \leftarrow t + h$ {$t = t_N$ for final step}
        $y \leftarrow y + \tfrac{25}{216}k_1 + \tfrac{1408}{2565}k_3 + \tfrac{2197}{4104}k_4 - \tfrac{1}{5}k_5$ {Now $y \approx y(t)$.}
        OUTPUT (t, h, y)

    {Set next h: $h \cdot ScaleMin \leq$ next $h \leq h \cdot ScaleMax$}
    $ScaleFactor \leftarrow 0.84 \cdot (Rmax/Ratio)^{1/4}$
    IF $ScaleFactor < ScaleMin$ THEN $ScaleFactor \leftarrow ScaleMin$
    IF $ScaleFactor > ScaleMax$ THEN $ScaleFactor \leftarrow ScaleMax$
    $h \leftarrow ScaleFactor \cdot h$

IF $t = t_N$ THEN OUTPUT ('$y = $ 'y 'approximates $y(t_N)$ to the desired accuracy')
           ELSE OUTPUT ('$h < Hmin$ occurred; apparent singularity near $t = $ 't)
           STOP

8.2C Taylor's Method

The first-order Euler's method uses the first-degree Taylor polynomial at t_j to approximate $y(t_j + h)$ (see Figure 8.1-4) . The natural extension of this is to use the nth Taylor polynomial at t_j, that is, the $O(h^{n+1})$ approximation

$$y(t_j + h) \approx y(t_j) + hy'(t_j) + \frac{h^2}{2!}y''(t_j) + \cdots + \frac{h^n}{n!}y^{(n)}(t_j) \qquad (7)$$

to get the ***n*th-order**† **Taylor's method** formula

$$y_{j+1} = y_j + h\left[f_j + \frac{h}{2}y_j'' + \frac{h^2}{3!}y_j''' + \cdots + \frac{h^{n-1}}{n!}y_j^{(n)}\right] \qquad (8)$$

In (8), y_j'', y_j''', ..., $y_j^{(n)}$ are the computable estimates of $y''(t_j)$, $y'''(t_j)$, ..., $y^{(n)}(t_j)$ *obtained by differentiating* $y' = f(t, y)$ *implicitly* with respect to t, that is, by treating $f(t, y)$ as $f(t, y(t))$ even though $y(t)$ is not known, as shown in the following example.

EXAMPLE 8.2C Use Taylor's method as specified in (a)–(c) to get estimates of $y(t)$ from $t = t_0 = 2$ to $t = t_N = 3$ for the IVP

$$y' = -ty^2, \qquad y(2) = 1 \qquad [\text{exact } y(t) \text{ is } 2/(t^2 - 2)] \qquad (9)$$

(a) second order, $h = 0.1$ **(b)** second order, $h = 0.05$ **(c)** third order, $h = 0.1$

Discussion: The necessary derivatives of y (up to y''') can be obtained implicitly using the product rule for derivatives as follows:

$$y' = f(t, y) = -ty^2 \qquad (\text{where } y \text{ is a function of } t) \qquad (10a)$$

$$y'' = \frac{dy'}{dt} = -(t\cdot 2yy' + 1\cdot y^2) = -y(2ty' + y) \qquad (10b)$$

$$y''' = \frac{dy''}{dt} = -y[2(ty'' + y') + y'] - y'(2ty' + y)$$

$$= -2[y(ty'' + 2y') + t(y')^2] \qquad (10c)$$

Solution (a): Putting (10a) and (10b) in (8) with $n = 2$ gives

$$y_{j+1} = y_j + h\left[f_j + \frac{h}{2}y_j''\right], \qquad \text{where } f_j = -t_jy_j^2,\ y_j'' = -y_j(2t_jf_j + y_j) \qquad (11)$$

Starting with $t_0 = 2$, $y_0 = 1$, and taking $h = 0.1$ in (11) gives

$$y_1 = y_0 + 0.1[f_0 + 0.05y_0''] \qquad f_0 = -2\cdot 1^2 = -2,\ y_0'' = -1[2\cdot 2(-2) + 1] = 7$$
$$= 1 + 0.1[-2 + 0.05(7)] = 0.835 \qquad [\approx y(t_1), \text{ where } t_1 = 2.1]$$

$$y_2 = y_1 + 0.1[f_1 + 0.05y_1''] \qquad f_1 = -2.1(0.835)^2,\ y_1'' = -0.835[2\cdot 2.1f_1 + 0.835]$$
$$\doteq 0.835 + 0.1[-1.4642 + 0.05(4.4376)] \doteq 0.71077 \qquad [\approx y(t_2),\ t_2 = 2.2]$$

The remaining entries of Table 8.2-2(a) are obtained similarly.

†When y_j is exactly $y(t_j)$, y_{j+1} in (8) becomes exactly the nth Taylor polynomial in (7); so $e_{j+1}[h]$ is $O(h^{n+1})$. It then follows from (20) of Section 8.1C that (8) does in fact define an nth-order method.

SOLUTION (b): Using (9) with $h = 0.05$ gives

$$y_1 = y_0 + 0.05[f_0 + 0.025y_0''] \qquad f_0 = -2,\ y_0'' = 7 \text{ (as above)}$$
$$= 1 + 0.05[-2 + 0.025(7)] = 0.90875 \qquad [\approx y(t_1),\ \text{where } t_1 = 2.05]$$

$$y_2 = y_1 + 0.05[f_1 + 0.025y_1'']; \quad f_1 = -2.1y_1^2, \quad y_1'' = -y_1[2\cdot 2.05f_1 + y_1]$$
$$\doteq 0.90875 + 0.05[-1.6929 + 0.025(5.4819)] \doteq 0.83096 \qquad [\approx y(t_2),\ t_2 = 2.1]$$

The remaining entries of Table 8.2-2(b) are obtained similarly.

SOLUTION (c): Putting (10a), (10b), and (10c) in (8) with $n = 3$ gives

$$y_{j+1} = y_j + h\left[f_j + \frac{h}{2}y_j'' + \frac{h^2}{6}y_j'''\right], \qquad y_j''' = -2[y_j(t_jy_j'' + 2f_j) + t_j(f_j)^2] \quad \textbf{(12)}$$

where f_j and y_j'' are as in (11). Taking $h = 0.1$, and starting with $t_0 = 2$, $y_0 = 1$, $f_0 = -2$, and $y_0'' = 7$ gives

$$y_1 = y_0 + 0.1\left[f_0 + 0.05y_0'' + \frac{0.01}{6}y_0'''\right] \qquad y_0''' = -2[y_0(t_0y_0'' + 2f_0) + t_0(f_0)^2]$$
$$= 1 + 0.1\left[-2 + 0.05\cdot 7 + \frac{0.01}{6}(-36)\right] = 0.829 \qquad [\approx y(t_1),\ \text{where } t_1 = 2.1]$$
$$y_2 = 0.829 + 0.1\left[(-1.4432) + 0.05(4.3377) + \frac{0.01}{6}(-19.065)\right] \doteq 0.70319$$

The remaining entries of Table 8.2-2(c) are obtained similarly. As one would expect, these y_j's are the most accurate in Table 8.2-2. ■

More accurate estimates can be obtained from the second-order ($n = 2$) Taylor's method values in (a) and (b) of Table 8.2-2 by using Richardson's formula with $r = 0.1/0.05 = 2$.

TABLE 8.2-2 Second- and Third-Order Taylor's Method Values

	Exact	(a) $n = 2, h = 0.1$		(b) $n = 2, h = 0.05$		(c) $n = 3, h = 0.01$	
t_j	$y(t_j)$	$y_j[0.1]$	$E_j[0.1]$	$y_j[0.5]$	$E_j[0.5]$	$y_j[0.5]$	$E_j[0.5]$
$t_0 = 2.0$	1	1	0	1	0	1	0
2.1	0.8299	0.8350	0.0051	0.8310	0.0011	0.8290	0.0009
2.2	0.7042	0.7108	0.0065	0.7056	0.0014	0.7032	0.0010
2.3	0.6079	0.6145	0.0066	0.6093	0.0014	0.6069	0.0010
2.4	0.5319	0.5380	0.0061	0.5332	0.0013	0.5310	0.0009
2.5	0.4706	0.4761	0.0055	0.4718	0.0012	0.4698	0.0007
2.6	0.4202	0.4250	0.0049	0.4212	0.0010	0.4195	0.0006
2.7	0.3781	0.3823	0.0043	0.3790	0.0009	0.3775	0.0005
2.8	0.3425	0.3462	0.0037	0.3433	0.0008	0.3420	0.0005
2.9	0.3120	0.3153	0.0033	0.3127	0.0007	0.3116	0.0004
$t_N = 3.0$	0.2857	0.2886	0.0029	0.2862	0.0006	0.2854	0.0003

For example, at $t = 2.2$,

$$F_1[0.05] = \frac{2^2F[0.05] - F[0.1]}{2^2 - 1} = \frac{4(0.7056) - 0.7108}{3} = 0.7039 \tag{13}$$

Since $y(2.2) = 0.7042$ (4s), this improvement of a second-order $F[h]$ is more accurate than $y_2[0.1] \doteq 0.7032$ obtained by the third-order Taylor's method.

Practical Consideration 8.2C (Shortcomings of Taylor's method). The problem with Taylor's method is that the user must first differentiate $f(t, y)$ to get formulas for $y_j'', y_j''', \ldots, y_j^{(n)}$ whether performing it by hand or on a computer. One can differentiate $y' = f(t, y)$ with respect to t (see Appendix II.5E) and get the general formulas

$$y'' = f_t + f_y f \qquad \left(\text{where } f = f(t,y) = y',\ f_t = \frac{\partial f}{\partial t}, \text{ and } f_y = \frac{\partial f}{\partial y}\right) \tag{14a}$$

$$y''' = f_{tt} + 2f_{ty}f + f_{yy}f^2 + f_y(f_t + f_y f), \quad \text{etc.} \tag{14b}$$

These formulas reduce the problem to that of finding *partial* derivatives, but the inconvenience and risk of a human error when differentiating still remain, especially for complicated $f(t, y)$. Recent advances in symbolic computation enable the computer itself to use the formulas (14) for *some* $f(t, y)$'s. At the present time, however, the most desirable methods are those that sample *only* the slope function $f(t, y)$ itself. The following section shows how Taylor's method can be used as the starting point for getting such methods. ■

8.2D General Runge–Kutta Formulas

In view of (14a), the second-order Taylor's method formula can be written as

$$y_{j+1} = y_j + h\phi_{T,2} \qquad \text{where } \phi_{T,2} = f_j + \frac{h}{2}[f_t(t_j, y_j) + f_j f_y(t_j, y_j)] \tag{15}$$

At the end of the nineteenth century, the German mathematician C. Runge observed that this expression for $\phi_{T,2}$ looks like the two-variable Taylor approximation

$$f(t_j + ph, y_j + qhf_j) \approx f_j + phf_t(t_j, y_j) + qhf_j f_y(t_j, y_j) + O(h^2) \tag{16}$$

Indeed, comparing $\phi_{T,2}$ in (15) to (16) with $p = q = \frac{1}{2}$ reveals that

$$\phi_{T,2} = f\left(t_j + \frac{h}{2}, y_j + \frac{h}{2}f_j\right) + O(h^2) \tag{17}$$

Substituting (17) in (15) yields the Modified Euler method formula [(27) of Section 8.1D]. This procedure for getting a second-order Runge–Kutta formula can be generalized by seeking "weights" a_1 and a_2 and "scale factors" p and q for which

$$\phi_{T,2} = a_1 f_j + a_2 f(t_j + ph, y_j + qhf_j) + O(h^2) \tag{18}$$

If we replace $\phi_{T,2}$ by (15) and $f(t_j + ph, y_j + qhf_j)$ by (16), and then equate coefficients of $f_t(t_j, y_j)$ and $f_y(t_j, y_j)$, we see that (18) is possible if and only if $a_1 + a_2 = 1$ and $a_2 p = a_2 q = \frac{1}{2}$. In summary, y_{j+1} given by the formula

$$y_{j+1} = y_j + h[a_1 f(t_j, y_j) + a_2 f(t_j + ph, y_j + qhf_j)] \tag{19a}$$

will differ from y_{j+1} in (15) by $h \cdot O(h^2) = O(h^3)$, provided that

$$a_1 = 1 - a_2 \quad \text{and} \quad p = q = \frac{1}{2a_2} \tag{19b}$$

for *any* nonzero weight a_2. Since (15) is itself an $O(h^3)$ approximation of $y(t_{j+1})$, the per-step error $e_{j+1}[h]$ of y_{j+1} in (19) is $O(h^3)$. So any method based on (19) will be a second-order method. And if $a_2 \geqslant \frac{1}{2}$, it will be a Runge–Kutta method because both t_j and $t_j + ph$ will lie in the interval $[t_j, t_{j+1}]$. So the Modified Euler method ($a_2 = 1$) and Huen's method ($a_2 = \frac{1}{2}$) are but two of *infinitely many* second-order Runge–Kutta methods!

Fourth-order Runge–Kutta formulas are obtained similarly. One starts with the fourth-order Taylor's formula

$$y_{j+1} = y_j + h\phi_{T,4}, \qquad \text{where } \phi_{T,4} = f_j + \frac{h}{2} y_j'' + \frac{h^2}{3!} y_j''' + \frac{h^3}{4!} y_j^{(\text{iv})} \tag{20}$$

and then uses the formulas (14) and (16) to find "weights" a_i and "scale factors" p_i and q_{ij} for which

$$\phi_{T,4} = a_1 m_1 + a_2 m_2 + a_3 m_3 + a_4 m_4 + O(h^4), \qquad \text{where } m_1 = f_j \tag{21a}$$

and m_2, m_3, and m_4 are defined recursively by

$$\begin{aligned} m_2 &= f(t_j + p_2 h, y_j + h[q_{21} m_1]) \\ m_3 &= f(t_j + p_3 h, y_j + h[q_{31} m_1 + q_{32} m_2]) \\ m_4 &= f(t_j + p_4 h, y_j + h[q_{41} m_1 + q_{42} m_2 + q_{43} m_3]) \end{aligned} \tag{21b}$$

The (messy) details can be found in [40]. We merely note that the a_i's and q_{ij}'s can be determined from p_2 and p_3, which can be chosen arbitrarily. So RK4 of Section 8.2A, which corresponds to $p_2 = p_3 = \frac{1}{2}$, is but one of infinitely fourth-order Runge–Kutta formulas.

8.2E Stability

Solutions of differential equations are subject to several types of instabilities in the sense that small changes of the input can produce disproportionately large changes in the solution. In this section we examine stability as it applies to exact solutions and numerical methods for initial value problems.

A solution $y(t)$ of the IVP $y' = f(t, y)$, $y(t_0) = y_0$ is called **stable** if it is well conditioned with respect to y_0, that is, if small relative changes in y_0 produce correspondingly small relative changes in $y(t)$ for $t \neq t_0$. Our standard for "good" stability is the

TEST EQUATION: $y' = At$, $y(0) = y_0 \neq 0$ [exact $y(t)$ is $y_0 e^{At}$]	**(22)**

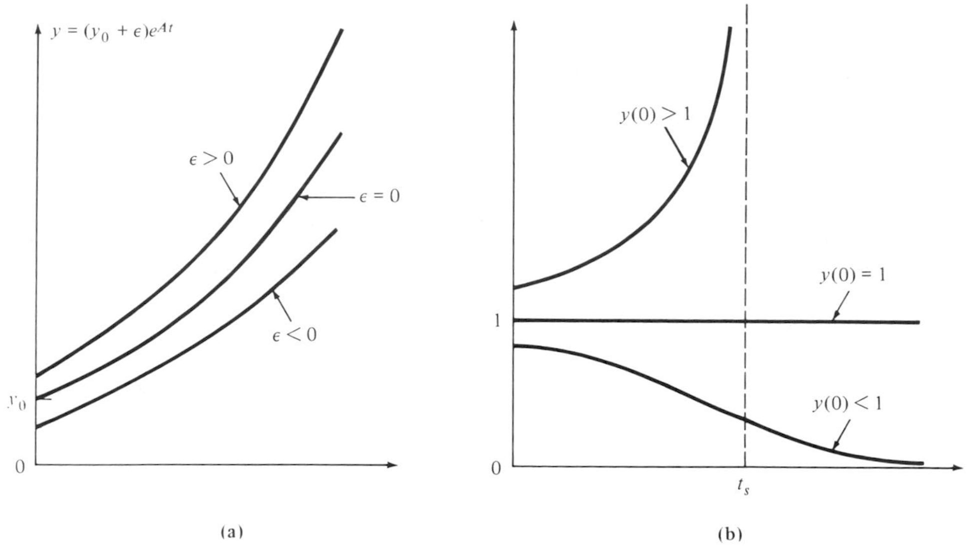

Figure 8.2-4 *(a) y(t) is stable; (b) y(t) is unstable when* $y_0 = 1$.

changing $y_0 \neq 0$ to $y_0 + \epsilon$ where $|\epsilon| < |y_0|$ changes $y(t)$ from y_0e^{At} to $(y_0 + \epsilon)e^{At}$ [Figure 8.2-4(a)]. So the *relative* change of $y(t)$ is the same as that of y_0, hence $y(t)$ is stable for any nonzero y_0.

To illustrate instability of $y(t)$, consider

$$y(t) = \frac{y_0}{y_0 + (1 - y_0)e^t}, \quad \text{the solution of } y' = y(1 - y), \quad y(0) = y_0 \tag{23}$$

It is evident from (23) that $y(t) = 1$ (constant) when $y_0 = 1$, but that $y(t)$ decreases to 0 as with increasing t if $y_0 < 1$, whereas $y(t)$ increases without bound as $t \to t_s = \ln[y_0/(y_0 - 1)]$ if $y_0 > 1$ [Figure 8.2-4(b)]. So $y(t)$ is unstable when $y_0 = 1$; it is stable, however, for all other values of y_0.

If the *exact* solution $y(t)$ is unstable, then the IVP will be difficult to solve accurately by any numerical method because *either* roundoff *or* truncation error will have the same (possibly misleading) effect as changing y_0. Unfortunately, as the following example shows, a *numerical* solution y_j can diverge from $y(t_j)$ as t_j moves from t_0 even if the exact $y(t)$ is stable.

EXAMPLE 8.2E Obtain a formula for y_j as a function of h and j if

$$\textbf{(a)}\ y_{j+1} = y_j + hf_j \quad \textbf{(b)}\ y_{j+1} = y_{j-1} + 2hf_j$$

is used with a fixed stepsize h for the Test Equation (22). How well does y_j approximate the exact solution $y(t_j) = y_0e^{Ah_j}$ as $j \to \infty$?

Solution (a): This is Euler's method. For $f(t, y) = Ay$ it becomes

$$y_{j+1} = y_j + hAy_j = (1 + hA)y_j, \qquad j = 0, 1, \ldots \tag{24}$$

so $y_1 = (1 + hA)y_0$, $y_2 = (1 + hA)y_1 = (1 + hA)^2y_0$, and so on. Hence

$$y_j = y_0(1 + hA)^j \qquad \text{for } j = 0, 1, 2, \ldots \tag{25a}$$

Since $1 + hA$ is the first Taylor approximation of e^{hA} and $t_j = hj$,

$$y_j = y_0(1 + hA)^j \text{ should approximate } y_0(e^{hA})^j = y(t_j) \tag{25b}$$

But does it? When hA is positive, $1 < 1 + hA < e^{hA}$; hence y_j will grow exponentially, although more slowly than $y(t_j)$. This is perfectly reasonable. However, when hA is negative, $y(t_j)$ shrinks to zero as $j \to \infty$; hence y_j will *not* give a reasonable approximation *unless* $|1 + hA| < 1$. Thus, when $hA < 0$, y_j will be a reasonable approximation of $y(t_j)$ *only if* $|h|$ is small enough so that $-2 < hA$.

SOLUTION (b): The formula of interest, namely

$$y_{j+1} = y_{j-1} + 2hf_j \tag{26a}$$

is known as the **Midpoint Predictor**, because $h\phi = 2hf_j$ is the Midpoint Rule approximation of the integral of $y'(t)$ from t_{j-1} to t_{j+1}. Upon taking $f_j = Ay_j$ and rearranging, it becomes

$$y_{j+1} - 2hAy_j - y_{j-1} = 0, \qquad j = 1, 2, \ldots \tag{26b}$$

This is a **homogeneous finite difference equation** whose **general solution**† is

$$y_j = c_1r^j + c_2s^j, \qquad \text{where } c_1 + c_2 = y_0 \tag{27a}$$

and r and s are the roots of its **associated characteristic equation**

$$\lambda^2 - 2hA\lambda - 1 = 0 \tag{27b}$$

The quadratic formula shows that these roots are real and given by

$$r = hA + \sqrt{1 + h^2A^2} \qquad \text{and} \qquad s = hA - \sqrt{1 + h^2A^2} \tag{27c}$$

and the $O(h^2)$ Maclaurin approximations of $\sqrt{1 + u}$ and e^u ($u = hA$) give

$$r = hA + [1 + \tfrac{1}{2}h^2A^2 + O(h^4)] = e^{hA} + O(h^3) \tag{27d}$$

and similarly, $s = -e^{-hA} + O(h^3)$. So (27a) can be written as

$$y_j = (y_0 - c_2)(e^{hA})^j + c_2(-1)^j(e^{-hA})^j + O(h^3) \tag{27e}$$

where c_2 is determined by a second initial condition, say $y_1 = y(t_1)$. Since $hj = t_j$, (27e) can be written as

†In this and the next section, reference will be made to some standard results about finite difference equations. The reader who is unfamiliar with these results is urged to accept them as true and concentrate on their consequences. A readable introduction to finite difference equations can be found in [25].

$$y_j = (y_0 - c_2)e^{At_j} - c_2(-1)^j e^{-At_j} + O(h^3) \qquad \textbf{(27f)}$$

which is an $O(h^3)$ approximation of $y(t_j) = y_0 e^{At_j}$ when $c_2 = 0$. Suppose, however, that $c_2 \neq 0$. If c_2 is small compared to y_0, it introduces only a small relative error in the leftmost term. And if hA is positive, then the middle term will shrink to zero as j increases. But when hA is negative [in which case $y(t_j) = y_0 e^{hAj} \to 0$ as $j \to \infty$], the middle term oscillates *with increasing amplitude* as j increases, and eventually becomes the dominant term in (27f). So y_j resulting from (26b) will provide an $O(h^3)$ approximation of $y(t_j)$ when $hA > 0$, but will bear no resemblance to $y(t_j)$ for large j when $hA < 0$ *regardless of how small we choose h.* ■

The $c_2(-1)^j e^{-At_j}$ term in (27f) is called a **parasitic term** of the numerical solution (27f) because it is unrelated to either the exact solution of the IVP or the per-step error of the method. In Example 8.2E(a), where there are no parasitic terms, the numerical solution y_j will behave like $y(t_j)$ for sufficiently small $|h|$. On the other hand, *the parasitic term in Example 8.2E(b) is not affected by changes in* $|h|$. It is therefore important to know if a numerical solution has parasitic terms, and if so, if they shrink to zero as j increases. The following discussion will help us do this.

Virtually all known numerical methods for solving IVP use a formula that can be written as a finite difference equation of the form

$$y_{j+1} = a_1 y_j + a_2 y_{j-1} + \cdots + a_m y_{j+1-m} + h\phi \qquad \textbf{(28a)}$$

where $h\phi$ represents a numerical integration of $y'(t)$ that will be exact when $y(t) = 1$ (constant); in this case, $h\phi$ will be zero, hence the sum $\Sigma_{i=1}^{m} a_i$ must be 1. This in turn ensures that $\lambda = 1$ must be a root of the **characteristic equation** of (28a), namely

$$\lambda^m - a_1\lambda^{m-1} - \cdots - a_{m-1}\lambda - a_m = 0 \qquad \textbf{(28b)}$$

If the m roots of (28b) are $\lambda_1, \lambda_2, \ldots, \lambda_{m-1}, \lambda_m = 1$, then the solution of (28a) will be of the form

$$y_j = c_1\lambda_1 + \cdots + c_{m-1}\lambda_{m-1} + \{\text{terms that} \to y(t_j) \text{ as } h \to 0\} \qquad \textbf{(28c)}$$

where $c_1\lambda_1, \ldots, c_{m-1}\lambda_{m-1}$ are the parasitic terms of the numerical solution (28c). It is important to realize that *these parasitic terms have nothing to do with roundoff error*. They are part of the *exact solution* of (28a), the finite difference approximation of the IVP. The method (28a) is called **unstable** if one of the roots of (28b) satisfies $|\lambda_i| > 1$, **weakly stable** if they all satisfy $|\lambda_i| \leq 1$ and those roots for which $|\lambda_i| = 1$ are simple, and **strongly stable** if roots except $\lambda_m = 1$ satisfy $|\lambda_i| < 1$. Note that if a method is unstable, then one of the parasitic terms in (28c) will dominate y_j as j increases. However, all parasitic terms of a strongly stable method will shrink to zero as $j \to \infty$, and the remaining terms will approximate $y(t_j)$ for sufficiently small h.

Since self-starting methods are all of the form $y_{j+1} = y_j + h\phi$ ($m = 1, a_1 = 1$), equation (28b) for them is simply $\lambda - 1 = 0$. Consequently,

All self-starting methods are strongly stable. **(29)**

In particular, all Runge–Kutta methods are strongly stable. This explains the results of Example 8.2E(a). On the other hand, equation (14b) for the Midpoint Predictor formula (26a) ($m = 2$, $a_1 = 0$, $a_2 = 1$), is $\lambda^2 - 1 = 0$, which has two simple roots $\lambda = \pm 1$, both of magnitude one. So this method is only weakly stable. Example 8.2E(b) thus shows that a weakly stable method can produce harmful parasitic terms even if the exact solution $y(t)$ is stable.

For the Test Problem, $f(t, y)$ is Ay, hence hA is $h\partial f(t, y)/\partial y$. This simple observation suggests how the Test Problem can be used to determine how small to take h when applying *any* numerical method to *any* IVP.

> Suppose that when a numerical method is used with a fixed stepsize for the Test Problem, $y' = Ay$, $y(0) = y_0 \neq 0$, its solution y_j approximates $y(t_j) = y_0 e^{Ahj}$ well for $\alpha < hA < \beta$. Then, when this method is used for the IVP $y = f(t, y)$, $y(t_0) = y_0$, its numerical solution y_j should approximate the exact solution well whenever the product $h\partial f(t, y)/\partial y$ lies between α and β for all (t, y) on the solution curve for t between t_0 and t_N. **(30)**

Thus we see from the results of Example 8.2E that Euler's method should be used with an h for which $h \cdot \partial f(t, y)/\partial y > -2$, and the Midpoint Predictor should be used with an h for which $h \cdot \partial f(t, y)/\partial y > 0$.

8.2F Stiffness

We have just seen that any self-starting method will trace out the shape of $y(t)$, provided that h is taken sufficiently small. Unfortunately, this assurance is of limited practical value if the IVP is **stiff**, that is, if its solution has "transient" terms that decay to zero while other terms remain essentially constant. The IVP's encountered in chemical process control, circuit theory, and vibrations are frequently stiff because their solutions have exponential terms of the form

$$c_i e^{-\lambda_i t}, \qquad \text{where } 0 \leqslant (\lambda_i)_{\min} << (\lambda_i)_{\max} \tag{31}$$

In this case, the $(\lambda_i)_{\min}$ term will remain essentially constant during the time it takes for the $(\lambda_i)_{\max}$ term to decay to essentially zero. The following concrete example will be used to demonstrate the serious computational difficulties that stiff IVP's present.

EXAMPLE 8.2F Obtain formulas for the numerical solution y_j of the IVP

$$y' = -50y + 100, \; y(0) = y_0, \qquad \text{for which } y(t) = (y_0 - 1)e^{-50t} + 2 \tag{32}$$

using

(a) Euler's method $\quad y_{j+1} = y_j + hf_j$ **(33a)**

(b) Backward Euler method $\quad y_{j+1} = y_j + hf(t_{j+1}, y_{j+1})$ **(33b)**

For what positive values of h will y_j approximate $y(t_j)$ well for $t > 0$?

NOTE: Formulas (33a) and (33b) use, respectively, the Left- and Right-Endpoint rules to integrate $y'(t)$ from t_j to t_{j+1}.

Solution (a): For $f_j = -50y_j + 100$, the Euler formula (33a) becomes

$$y_{j+1} = (1 - 50h)y_j + 100h, \qquad j = 0, 1, 2, \cdots \tag{34}$$

Let $c = 1 - 50h$ and $d = 100h$. Then the first few y_j's after y_0 are

$$y_1 = cy_0 + d, \quad y_2 = c^2y_0 + (c + 1)d, \quad y_3 = c^3y_0 + (c^2 + c + 1)d, \quad \text{etc.}$$

By induction, $y_j = c^jy_0 + (c^{j-1} + \cdots + c + 1)d$. Replacing c by $1 - 50h$, d by $100h$, and $(c^{j-1} + \cdots + c + 1)$ by $(c^j - 1)/(c - 1)$ and then simplifying gives

$$y_j = (1 - y_0)(1 - 50h)^j + 2, \qquad j = 0, 1, \ldots \tag{35}$$

as the Euler's method solution of (32). Since $y(t_j)$ approaches 2 as $j \to \infty$, y_j will be a reasonable approximation of $y(t_j)$ only if $|1 - 50h| < 1$, that is, $h < 0.04$.

Solution (b): When $f_j = -50y_j + 100$, the Backward Euler formula (33b) becomes

$$y_{j+1} = y_j + h(-50y_{j+1} + 100) \tag{36a}$$

This linear equation in y_{j+1} is easily solved for

$$y_{j+1} = \frac{y_j + 100h}{1 + 50h}, \qquad j = 0, 1, 2, \ldots \tag{36b}$$

Reasoning as in part (a), but with $c = 1/(1 + 50h)$ and $d = 100h/(1 + 50h)$, gives

$$y_j = \frac{1 - y_0}{(1 + 50h)^j} + 2, \qquad j = 0, 1, \ldots \tag{37}$$

as the Backward Euler solution of (32). Unlike (35), this y_j will approximate $y(t_j)$ well for *any* positive value of h. ■

To illustrate the computational difficulty posed by this example, let $y_0 = 0$, so that $y(t) = 2 - e^{-50t}$. The "transient" term e^{-50t} decays to zero so rapidly that by the time $t = 0.1$, $y(t)$ will be within 1% of its "steady state" value of 2. Clearly, even Euler's method can approximate this essentially constant $y(t)$ accurately using a large h for $t > 0.1$. However, we saw in Example 8.2F(a) that Euler's method must be used with $h < 0.04$ if y_j is to approach 2 with increasing j. No such restriction is necessary for the Backward Euler method of Example 8.2F(a).

More generally, if a self-starting method is used with a fixed stepsize h for a stiff IVP, then the size of h will be limited by a term in the solution that is effectively zero after a few h-steps. And the more rapidly this transient becomes insignificant, the greater the restriction it places on h! As a result, the number of steps needed to integrate from $t = t_0$ to t_N might be so large as to entail a prohibitive amount of time (even on a computer) or to cause an intolerable amount of propagated roundoff error. A variable stepsize strategy as described in Section 8.2D will be of some help but will still require smaller stepsizes long after they are really needed.

The methods that have been used to solve stiff IVP's successfully all do so by "looking ahead" by sampling y' at t_{j+1}, that is, by using an **implicit formula** of the form

$$y_{j+1} = y_j + h\phi(t_j, y_j, f_j, \ldots, \textit{and } f(t_{j+1}, y_{j+1})) \tag{38}$$

where y_{j+1} appears on both sides. The $O(h^2)$ Backward Euler formula of Example 8.2F is the simplest implicit formula and results in a first-order method which, as we have seen, can be used *with a reasonable stepsize h* for stiff IVP's. Alternatively, an $O(h^2)$ method for solving stiff IVP's is obtained by using the

$$\textsc{Trapezoidal Rule:}\quad y_{j+1} = y_j + \frac{h}{2}[f_j + f(t_{j+1}, y_{j+1})] \tag{39}$$

Practical Consideration 8.2F (Implementing an implicit formula). The Backward Euler formula (33b) was deceptively easy to use in (36) because $f(t, y)$ happened to be linear in y. When $f(t, y)$ is nonlinear in y, finding y_{j+1} might require rewriting (33b) as

$$F(y_{j+1}) = 0, \qquad \text{where } F(y_{j+1}) = y_j + hf(t_{j+1}, y_{j+1}) - y_{j+1} \tag{40}$$

and then using a root-finding method such as NR or SEC to find y_{j+1} (to a desired accuracy) as a root of F for *each j*! (See Problem C8-10.) This complication arises when using *any* implicit formula. Consequently, methods based on implicit formulas are recommended *only* for IVP's that are known to be (or suspected of being) stiff. Also, reliable computer code for an implicit method is difficult to write, especially for systems (see Problems C8-12 and C8-13). Stiff IVP's should therefore be solved using only well-tested, well-documented software such as the DGEAR subroutine of the IMSL library. Surveys of such software are given in [27] and [28]. A thorough mathematical analysis of stiff IVP's is given in [22]. ■

8.3 Multistep (Predictor–Corrector) Methods

The methods considered so far were **self-starting** in that all $f(t, y)$ evaluations needed to get y_{j+1} are performed *after* y_j is obtained. We now consider **multistep methods** that reuse previously obtained sampled slopes f_i for $i < j$ in getting y_{j+1}. Such methods are *not* self-starting because they cannot be used when $j = 0$. We now show, however, that once enough previous f_i values are available, *multistep methods can give nth-order accuracy using fewer than n evaluations of $f(t, y)$ at each step.*

8.3A Predictor–Corrector (PC) Strategies

Multistep methods usually approximate $y(t_{j+1})$ using a **predictor–corrector** (abbreviated **PC**) **strategy**. All such strategies begin with the following two steps: First get a **predicted approximation** of $y(t_{j+1})$:

$$p_{j+1} = y_i + h\phi_p(f_i, f_{i+1}, \ldots, f(t_j, y_j)), \qquad \text{where } i \leq j \tag{1a}$$

in which $h\phi_p$ is a quadrature formula that integrates $y'(t)$ from t_i to t_{j+1} *without sampling* $y'(t_{j+1})$. Then get a **corrected approximation** of $y(t_{j+1})$

$$c_{j+1} = y_j + h\phi_c(f_{i+1}, \ldots, f_j, f(t_{j+1}, p_{j+1})) \tag{1b}$$

in which $h\phi_c$ is a *closed* quadrature formula† for integrating $y'(t)$ from t_j to t_{j+1}. Notice that one new $f(t, y)$ evaluation is needed for each of (1a) and (1b). If $h\phi_p$ and $h\phi_c$ are both $(n + 1)$st-order quadrature formulas, say

$$e_{j+1}[h]_p = y(t_{j+1}) - p_{j+1} \approx C_p h^{n+1} y^{(n+2)}(\xi_p) \tag{2a}$$

$$e_{j+1}[h]_c = y(t_{j+1}) - c_{j+1} \approx C_c h^{n+1} y^{(n+2)}(\xi_c) \tag{2b}$$

and $y^{(n+2)}$ is approximately constant over $[t_i, t_{j+1}]$, then it follows from (2) that $C_c[y(t_{j+1}) - p_{j+1}] \approx C_p[y(t_{j+1}) - c_{j+1}]$, hence

$$y(t_{j+1}) \approx \frac{C_c p_{j+1} - C_p c_{j+1}}{C_c - C_p} \tag{3a}$$

Substituting this for $y(t_{j+1})$ in (2b) yields the per-step error estimate

$$e_{j+1}[h]_c \approx \delta_{j+1}, \qquad \text{where } \delta_{j+1} = C_c[p_{j+1} - c_{j+1}] \tag{3b}$$

For example, the cubic exact Adams formulas Section 7.3A yield the **fourth-order Adams Predictor–Corrector method**‡, abbreviated **APC4**.

ADAMS PREDICTOR:
$$p_{j+1} = y_j + \frac{h}{24}[-9f_{j-3} + 37f_{j-2} - 59f_{j-1} + 55f(t_j, y_j)] \tag{4a}$$

ADAMS CORRECTOR:
$$c_{j+1} = y_j + \frac{h}{24}[f_{j-2} - 5f_{j-1} + 19f_j + 9f(t_{j+1}, c_{j+1})] \tag{4b}$$

ERROR ESTIMATE:
$$\delta_{j+1} = \frac{-19}{270}[p_{j+1} - c_{j+1}] \tag{4c}$$

This method is illustrated graphically in Figure 8.3-1. The error estimate in (4c) follows from (3b) and the truncation error formulas $e_{j+1}[h]_p = 251/270h^5y^{(v)}(\xi_p)$ $(C_p = 251/270)$ and $e_{j+1}[h]_c = -19/270h^5y^{(v)}(\xi_p)$ $(C_c = -19/720)$ [see (6b) and (7b) of Section 7.3A].

†To sample $y'(t_{j+1})$, corrector formulas must be *implicit* formulas for approximating $y(t_{j+1})$ as described in Section 8.2F.

‡APC4 is also called the **Adams–Bashforth–Moulton** method.

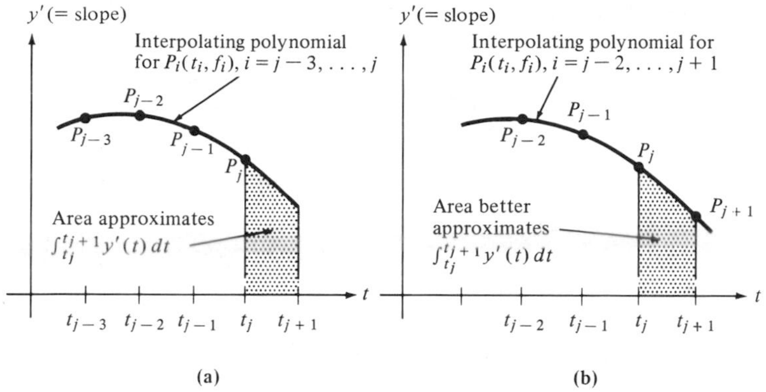

Figure 8.3-1 *(a) Adams predictor; (b) Adams corrector.*

If c_{j+1} in (3b) is used as y_{j+1}, then APC4 is a fourth-order method which, once started, requires only *two* new $f(t, y)$ evaluations per step.

If δ_{j+1} in (3b) indicates that c_{j+1} is sufficiently accurate, then c_{j+1} can be taken as y_{j+1}; if not, an "improved" corrected estimate c_{j+1} can be obtained by replacing p_{j+1} by c_{j+1} in (1b). The following example shows how this can be done for APC4.

EXAMPLE 8.3A Use APC4 with $h = 0.1$ on the interval [2, 3] for the IVP

$$y' = -ty^2,\ y(2) = 1 \qquad \left[\text{exact solution is } y(t) = \frac{2}{t^2 - 2}\right] \tag{5}$$

Try to get 4*s*. Start the method using the exact (to 7*s*) values

$$\begin{aligned}
t_0 &= 2.0, & y_0 &= y(2.0) = 1.000000, & f_0 &= -2.000000\\
t_1 &= 2.1, & y_1 &= y(2.1) \doteq 0.8298755, & f_1 &\doteq -1.446256\\
t_2 &= 2.2, & y_2 &= y(2.2) \doteq 0.7042254, & f_2 &\doteq -1.091053\\
t_3 &= 2.3, & y_3 &= y(2.3) \doteq 0.6079027 & &
\end{aligned}$$

Solution: We can now begin APC4, starting at $t = t_3$ $(j = 3)$.

$$p_4 = y_3 + \frac{h}{24}\left[-9f_0 + 37f_1 - 59f_2 + 55(-t_3p_3^2)\right] \doteq 0.5333741 \tag{6a}$$

$$c_4 = y_3 + \frac{h}{24}\left[f_1 - 5f_2 + 19f_3 + 9(-t_4p_4^2)\right] \doteq 0.5317149 \tag{6b}$$

$$\delta_4 = \frac{-19}{270}\left[p_4 - c_4\right] \doteq 0.0001144 \tag{6c}$$

Since δ_4 indicates possible inaccuracy in the fourth decimal place (which is the fourth significant digit) of y_4, we replace p_4 by c_4 in (6b) to get the following improved estimate of $y(t_4)$

$$y_4 = y_3 + \frac{h}{24}\left[f_1 - 5f_2 + 19f_3 + 9(-t_4c_4^2)\right] \doteq 0.5318739 \tag{6d}$$

The local truncation error estimate of y_4 obtained this way is

$$\delta_4 = \frac{-19}{270}[c_4 - y_4] \doteq -0.0000112 \tag{6e}$$

This suggests that y_4 in (6d) is accurate to about 5*s*. The next ($j = 4$) step is

$$p_5 = y_4 + \frac{h}{24}[-9f_1 + 37f_2 - 59f_3 + 55(-t_4y_4^2)] \doteq 0.4712642 \tag{7a}$$

$$c_5 = y_4 + \frac{h}{24}[f_2 - 5f_3 + 19f_4 + 9(-t_5p_5^2)] \doteq 0.4704654 \tag{7b}$$

$$\delta_5 = \frac{-19}{270}[p_5 - c_5] \doteq 0.0000562 \tag{7c}$$

Again, there apears to be some error in the fourth significant digit of c_5. Replacing $-t_5p_5^2$ by $-t_5c_5^2 \doteq -0.5535098$ in (7b) gives $y_5 \doteq 0.4705358$, for which $\delta_5 = -\frac{19}{270}[c_5 - y_5] \doteq -0.0000050$. This suggests that y_5 is accurate to about 5*s*. The remaining APC4 y_j entries of Table 8.3-1 are obtained similarly. ■

We can see from Table 8.3-1 that y_4 and y_5 just obtained are actually accurate to only about 4*s*. Moreover, had we "improved" y_4 by replacing c_4 by y_4 in (6d), the resulting new y_4 would have been *less* accurate than 0.531874! In fact, if the iterated "improvement" converges, it does so to a fixed point of (6d) (for which $y_4 = c_4$). Since (6d) only *approximates* $y(t_{j+1})$, there is no reason to expect this fixed point to approximate the exact $y(t_4)$ any more accurately than the y_4 shown in (6d). In fact, experience will confirm that:

> If one replacement of p_{j+1} by c_{j+1} in the corrector formula does not appear to yield the desired accuracy, you should reduce the stepsize rather than iterate the replacement. **(8)**

TABLE 8.3-1 APC4 and RK4 Values for $y' = -ty^2$, $y(2) = 1$ ($h = 0.1$)

t_j	Exact $y(t_j)$	Using APC4 ($h = 0.1$) y_j	$E_j[h]$	Using RK4 ($h = 0.1$) y_j	$E_j[h]$
$t_0 = 2.0$	1.000000	Exact	—	Exact	—
2.1	0.829876	Exact	—	0.829885	−0.000009
2.2	0.704225	Exact	—	0.704237	−0.000012
2.3	0.607903	Exact	—	0.607914	−0.000011
2.4	0.531915	0.531874	0.000041	0.531924	−0.000009
2.5	0.470588	0.470536	0.000052	0.470596	−0.000008
2.6	0.420168	0.420114	0.000054	0.420175	−0.000007
2.7	0.378072	0.378020	0.000052	0.378078	−0.000006
2.8	0.342466	0.342419	0.000047	0.342471	−0.000005
2.9	0.312012	0.311971	0.000041	0.312017	−0.000005
$t_N = 3.0$	0.285714	0.285674	0.000040	0.285718	−0.000004

In Table 8.3-1, the last APC4 step used $y_{10} = c_{10}$ and so required only two $f(t, y)$ evaluations [at (t_9, y_9) for p_{10}, and at (t_{10}, p_{10}) for c_{10}]. All other steps required a third "improvement" evaluation [at (t_{j+1}, c_{j+1}) as in (6d)]. In view of (8), APC4 requires a smaller stepsize than $h = 0.1$ to achieve 5*s* accuracy per step. This accuracy was achieved by RK4 because $f(t, y) = -ty^2$ happens to be a polynomial of degree ≤ 4 in t and y (see Section 8.2D). On the other hand, APC4 with accurate starting values will generally be more accurate than RK4 when $y(t)$ happens to be a polynomial in t of degree ≤ 4. (Why?)

The APC4 algorithm is described in Algorithm 8.3A. *MaxIt* is usually set to 2 [see (8)].

ALGORITHM 8.3A. FOURTH-ORDER ADAMS PREDICTOR–CORRECTOR (APC4) METHOD

PURPOSE: To solve to *Nsig* accurate digits on the interval $[t_0, t_N]$, the IVP

$$y' = f(t, y), \qquad y = y_0 \text{ when } t = t_0$$

using *NumSteps* h-steps. A three-dimensional array $\mathbf{F} = [F(1) \quad F(2) \quad F(3)]$ is used to store the three previously sampled slopes f_{j-1}, f_{j-2}, and f_{j-3}.

```
GET t_0, t_N, y_0,   {parameters of IVP}
    NumSteps,        {from t_0 to t_N; at the jth step, t_j = t_0 + jh}
    Nsig,            {the desired significant digit accuracy all of y_{j+1}'s}
    MaxIt            {maximum number of iterations of the corrector}

{Start}
t ← t_0;  y ← y_0;  h ← (t_N − t_0)/NumSteps;  RelTol ← 0.5·10^(−Nsig)
DO FOR j = 0 TO 2
|   F(3 − j) ← f(t, y)                                  {initialize F as [f_2, f_1, f_0]}
|   Use RK4 to get y ≈ y(t + h)                          {y is now y_{j+1}}
|__ t ← t_0 + (j+1)h;   OUTPUT (t, y)                    {(t, y) is now (t_{j+1}, y_{j+1})}
                                                         {Use APC4}
DO FOR j = 3 TO (NumSteps − 1)
|   f ← f(t, y);  t ← t_0 + (j + 1)h                     {current f_j and t_{j+1}}
|   p ← y + (h/24)[−9F(3) + 37F(2) − 59F(1) + 55f]       {predicted y_{j+1}}
|   Ok ← FALSE                                           {accuracy flag}
|   DO FOR k = 1 TO MaxIt UNTIL Ok = TRUE
|   |   c ← y + (h/24)[F(2) − 5F(1) + 19f + 9f(t, p)]    {corrected y(t + h)}
|   |   ErrEst ← (−19/270)[p − c]                        {estimate of e_{j+1}[h]_c}
|   |   IF |ErrEst| ≤ RelTol·min(1, |c|)                 {termination test}
|   |   |   THEN Ok ← TRUE;  exit loop                   {c ≈ y_{j+1} with desired accuracy}
|   |___|__ ELSE p ← c                                   {prepare for another iteration}
|   y ← c;  OUTPUT (t, y, 'Ok = 'Ok)
|__ F(3) ← F(2);  F(2) ← F(1);  F(1) ← f                 {prepare for next step}
STOP
```

In general, the order of the method used to start a PC method should be at least as large as the PC method itself. RK4 is usually used to start APC4.

Practical Consideration 8.3A (Try $y_{j+1} \approx c_{j+1} + \delta_{j+1}$?). It may have occurred to the reader that since δ_{j+1} is an estimate of $e_{j+1}[h]_c$ (the local truncation error of c_{j+1}), one can get an "improved corrector" by adding δ_{j+1} to c_{j+1}. Although this strategy is enticing, it tends to make the method unstable and therefore is *not* recommended. ■

8.3B Stepsize Control and Stability of PC Methods

All effective predictor–corrector strategies use quadrature formulas $h\phi_p$ and $h\phi_c$, which are of the same order and so yield the per-step corrector error estimate $ErrEst = \delta_{j+1}$ shown in (3). This allows the possibility of using the method with a variable stepsize h. Unfortunately, the need to restart the method each time h is changed effectively removes any possible benefits of the Continuous Change Stepsize Strategy [(5) of Section 8.2B]. However, if sufficiently many previous f_i values are stored, polynomial interpolation can be used to implement the Halve/Double/Leave Stepsize Strategy [(6) of Section 8.2B] without using another method to restart each time h is changed. For the APC4 method, one need only enlarge **F** of Algorithm 8.3A to include $F(4) = f_{j-4}$, and then vary h as follows:

TO HALVE h: Use the interpolating fourth-degree polynomial for the five nodes $(t_{j-4}, f_{j-4}), \ldots, (t_j, f_j)$ to get the bisecting values

$$f_{j-1/2} = \tfrac{1}{128}\,[-5f_{j-4} + 28f_{j-3} - 70f_{j-2} + 140f_{j-1} + 35f_j] \qquad \textbf{(9a)}$$

$$f_{j-3/2} = \tfrac{1}{64}\,[3f_{j-4} - 16f_{j-3} + 54f_{j-2} + 24f_{j-1} - f_j] \qquad \textbf{(9b)}$$

Then reinitialize **F** as $[f_{j-1/2} \;\; f_{j-1} \;\; f_{j-3/2} \;\; f_{j-2}]$ and proceed from (t_j, y_j) with stepsize $h/2$.

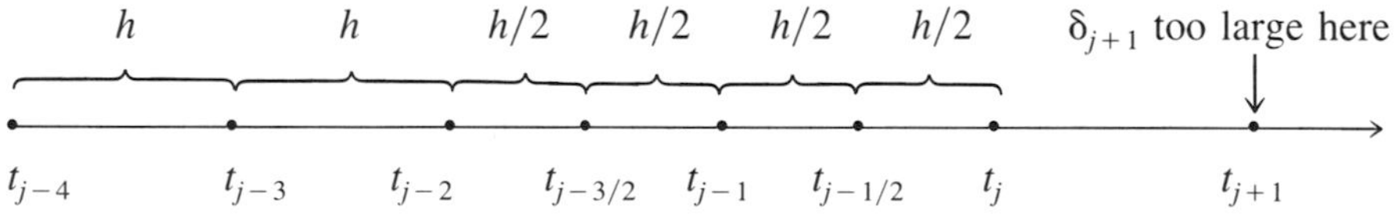

TO DOUBLE h. Take one more h-step to get y_{j+2} and f_{j+2}. Then reinitialize **F** as $[f_{j+2} \;\; f_j \;\; f_{j-2} \;\; f_{j-4}]$ and proceed from (t_{j+2}, y_{j+2}) with stepsize $2h$.

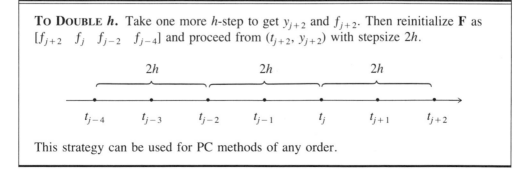

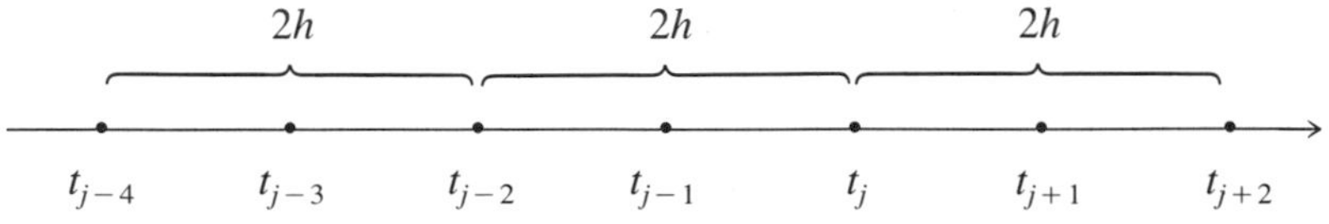

This strategy can be used for PC methods of any order.

A stability analysis of predictor–corrector methods must take into account the combined effects of both the predictor and corrector formulas, and so is more complex than that of a self-starting method. We shall examine the stability of only one class of PC methods. The **n-Step Adams–Bashforth formula** uses the interpolating polynomial for the n knots $(t_{j-n+1}, f_{j-n+1}), \ldots, (t_j, f_j)$, to get the approximation

$$y(t_{j+1}) \approx y_{j+1}^{(ABn)} = y_j + h(w_{j-n+1}f_{j-n+1} + \cdots + w_j f_j) \qquad \textbf{(10a)}$$

where the weights w_i are chosen so that $h(w_{j-n+1}f_{j-n+1} + \cdots + w_j f_j)$ approximates the integral of $y'(t)$ from t_j to t_{j+1} with exactness degree $n - 1$. Similarly, the **n-Step Adams–Moulton formula** uses the interpolating polynomial for the n knots $(t_{j-n+2}, f_{j-n+2}), \ldots, (t_{j+1}, f_{j+1})$, to get the approximation

$$y(t_{j+1}) \approx y_{j+1}^{(AMn)} = y_j + h(w_{j-n+2}f_{j-n+2} + \cdots + w_{j+1}f_{j+1}) \qquad \textbf{(10b)}$$

where the weights w_i are also chosen so that $h(w_{j-n+2}f_{j-n+2} + \cdots + w_{j+1}f_{j+1})$ approximates the integral of $y'(t)$ from t_j to t_{j+1} with exactness degree $n - 1$. The PC method that uses $y_{j+1}^{(ABn)}$ to get p_{j+1} and $y_{j+1}^{(AMn)}$ to get c_{j+1} is called the **nth-order Adams Predictor–Corrector (APCn) method**. By (28b) of Section 8.2E, the characteristic equation of *both* $y_{j+1}^{(ABn)}$ and $y_{j+1}^{(AMn)}$ is

$$\lambda^n - \lambda^{n-1} = 0 \qquad \text{or} \qquad \lambda^{n-1}(\lambda - 1) = 0 \qquad \textbf{(11)}$$

Since (11) has $\lambda = 1$ as a simple root, and $\lambda = 0$ as a root of multiplicity $n - 1$, *both $y_{j+1}^{(ABn)}$ and $y_{j+1}^{(AMn)}$ are strongly stable for any n* (see Section 8.2E). This strong stability ensures that when APCn is used as illustrated for APC4 in Algorithm 8.3A, any parasitic terms of the numerical solution will decay to zero with increasing j, and accounts for the popularity of APCn as a PC method for solving IVP. There are nth-order PC methods that have smaller per-step errors but lack this important assurance, hence are generally not recommended. In this regard, see Practical Consideration 8.3A.

8.3C Choosing a Method for Solving IVP

Unlike the methods of the preceding sections, it is unlikely that an IVP arising in a realistic situation can be solved adequately by hand. So we will limit out discussion to computer implemention and fourth-order methods. Taylor's method will not be included in our discussion because computer code for it written in Fortran, Pascal, PL/I, BASIC, APL, or Algol is generally not convenient to use (see Practical Consideration 8.2C). Of the general-purpose IVP solvers we have discussed, this leaves only RK4, RKF4, and APC4. These are summarized in Table 8.3-2. All are strongly stable, hence will track $y(t)$ well for

TABLE 8.3-2 Fourth-Order IVP Solvers Compared

Method (Section)	*Self-Starting?*	*Per-Step Error*	*Accumulated Error*	*f(t, y) Eval./Step*	*Stepsize Control*
RK4 (8.2A)	Yes	$O(h^5)$	$O(h^4)$	4	Awkward
RKF4 (8.2B)	Yes	$O(h^5)$	$O(h^4)$	6	Easy
APC4 (8.3A)	No	$O(h^5)$	$O(h^4)$	2–3	Possible

sufficiently small h, provided that $f(t, y)$ has continuous first partial derivatives on and near the solution curve $y = y(t)$.

If $f(t, y)$ is expensive to evaluate, or in a real-time situation where the amount of time available to integrate from t_0 to t_N is limited, APC4 might be preferable to RKF4 or RK4. And if a singularity of $y(t)$ is a possibility on $[t_0, t_N]$, then a method with convenient stepsize control such as RKF4 of APC4 should be tried. The foregoing remarks apply only to IVP's that are not stiff. If you know that the IVP is stiff, or an unreasonably small stepsize appears to be needed to solve it, then reliable code for an implicit method should be used (Practical Consideration 8.2F).

It should be noted that in addition to Runge–Kutta and predictor–corrector methods, there are **extrapolation methods** that use Richardson's formula, *as needed* to get a prescribed accuracy at each step. Of these the methods, those of Gragg [21] and Bullirsch and Stoer [10] are the best known. Gragg's method is based on the (nontrivial) observation that after an adjustment involving f_j and $f(t_{j+1}, y_{j+1})$, the accumulated error of the modified Euler method becomes $Ch^2 + Dh^4 + Eh^6 + \cdots$. The Bullirsch–Stoer method uses the error when a rational function (rather than a polynomial) interpolation for $y'(t)$ is integrated. The algorithms for extrapolation methods are somewhat more complex than those for Runge–Kutta and predictor–corrector methods, and often require more $f(t, y)$ evaluations per step; however, they can usually get a desired accuracy using a fixed stepsize. Extrapolation methods are discussed more thoroughly in [31].

8.4 First-Order Systems and *n*th-Order IVP's

For most physical systems, several dependent variables are needed to describe its state at any time t. For example, the **state variables** for a thermodynamic system might include temperature, pressure, volume, and entropy; the state variables of a mechanical system might include the x, y, and z displacements of certain points; the state variables of an electrical circuit might include the voltages and currents that determine the energy delivered to certain critical components, and so on. In this section we develop numerical methods for solving differential equations that describe such systems when all state variables, and perhaps their derivates, are known at an **initial time** t_0. This is a generalization of the initial value problem considered in the preceding sections.

8.4A Notation and Terminology

For example, suppose that the amount of three interreacting chemicals, say q, r, and s, at any given time t is governed by the equations

$$\begin{aligned} \frac{dq}{dt} &= r^2 - 3tq, & q(1) &= 0 \\ \frac{dr}{dt} &= 2q + s, & r(1) &= 3 \\ \frac{ds}{dt} &= rs + 3e^{-t}, & s(1) &= -2 \end{aligned} \tag{1a}$$

The state of this system at any instant of time can be described using the three state variables q, r, and s, each of which varies with t. The equations (1a) constitute a **coupled system** of differential equations, because the *rate* at which any one of the state variables changes depends on t and the values of *all* state variables at that time (not just the variable itself). The $3e^{-t}$ term in the ds/dt equation indicates that aside from the amount of s formed by the reaction, an additional amount is being added at a rate $3e^{-t}$. A *solution* of (1a) consists of *three* functions $q(t)$, $r(t)$, and $s(t)$ that satisfy the conditions stated in (1a).

The state of the system (1a) is best described using a single **state variable vector** $\mathbf{y} = [q \quad r \quad s]^T$. Its solution can then be viewed as the **solution vector** $\mathbf{y}(t) = [q(t) \quad r(t) \quad s(t)]^T$. In fact, the system can be put in the **vector form**

$$\underbrace{\begin{bmatrix} q' \\ r' \\ s' \end{bmatrix}}_{\mathbf{y}' = d\mathbf{y}/dt} = \underbrace{\begin{bmatrix} r^2 - 3tq \\ 2q + s \\ rs + 3e^{-t} \end{bmatrix}}_{\mathbf{f}(t,\, \mathbf{y})}, \quad \underbrace{\begin{bmatrix} q(1) \\ r(1) \\ s(1) \end{bmatrix}}_{\mathbf{y}(1)} = \underbrace{\begin{bmatrix} 0 \\ 3 \\ -2 \end{bmatrix}}_{\mathbf{y}_0}, \qquad \text{where } \mathbf{y} = \begin{bmatrix} q \\ r \\ s \end{bmatrix} \tag{1b}$$

Our "neutral" notation for the state variables of a system of differential equations will be $y_1, y_2, \ldots, y_n$. The order in which the subscripts are assigned is up to the user; once assigned, however, they determine the meaning of the components of the state variable vector $\mathbf{y} = [y_1 \quad y_2 \quad \cdots \quad y_n]^T$. In (1), for example, we made the "natural" assignment $q = y_1$, $r = y_2$, $s = y_3$. Reversing the first and third equations corresponds to the alternative assignment $s = y_1$, $r = y_2$, $q = y_3$; in this case, the state $\mathbf{y} = [4 \quad 1 \quad -3]^T$ occurs when $q = -3$, $r = 1$, and $s = 4$.

We are now ready to describe numerical methods for solving the general **first-order initial value problem**, abbreviated IVP_n, given by

$$(\text{IVP})_n \quad \begin{aligned} \frac{dy_1}{dt} &= f_1(t, y_1, y_2, \ldots, y_n), & y_1(t_0) &= y_{1,0} \\ \frac{dy_2}{dt} &= f_2(t, y_1, y_2, \ldots, y_n), & y_2(t_0) &= y_{2,0} \\ &\vdots & &\vdots \\ \frac{dy_n}{dt} &= f_n(t, y_1, y_2, \ldots, y_n), & y_n(t_0) &= y_{n,0} \end{aligned} \tag{2a}$$

Since $f_i(t, y_1, y_2, \ldots, y_n)$ determines the rate of change of the ith state variable y_i, we shall call it the ***i*th slope function**. Thus, for the $(\text{IVP})_3$ given in (1), $f_3(t, q, r, s) - rs + 3e^{-t}$. The vector form of (2a) is

$$\underbrace{\begin{bmatrix} y_1' \\ y_2' \\ \vdots \\ y_n' \end{bmatrix}}_{\mathbf{y}' = d\mathbf{y}/dt} = \underbrace{\begin{bmatrix} f_1(t, \mathbf{y}) \\ f_2(t, \mathbf{y}) \\ \vdots \\ f_n(t, \mathbf{y}) \end{bmatrix}}_{\mathbf{f}(t,\, \mathbf{y})}, \quad \underbrace{\begin{bmatrix} y_1(t_0) \\ y_2(t_0) \\ \vdots \\ y_n(t_0) \end{bmatrix}}_{\mathbf{y}(t_0)} = \underbrace{\begin{bmatrix} y_{1,0} \\ y_{2,0} \\ \vdots \\ y_{n,0} \end{bmatrix}}_{\mathbf{y}_0 \,=\, \text{initial state}}, \qquad \text{where } \mathbf{y} = \begin{bmatrix} y_1 \\ y_2 \\ \vdots \\ y_n \end{bmatrix} \tag{2b}$$

or simply

$$(\text{IVP})_n \quad \mathbf{y}' = \mathbf{f}(t, \mathbf{y}), \qquad \mathbf{y}(t_0) = \mathbf{y}_0 \tag{2c}$$

A solution of (2) is a vector function $\mathbf{y}(t) = [y_1(t) \quad \cdots \quad y_n(t)]^T$ whose derivative vector $\mathbf{y}'(t) = [y_1'(t) \quad \cdots \quad y_n'(t)]^T$ satisfies

$$\mathbf{y}'(t) = \mathbf{f}(t, \mathbf{y}(t)), \quad \text{and} \quad \mathbf{y}(t_0) = \mathbf{y}_0 \tag{3}$$

As in Section 8.1B, we assume solutions of $(\text{IVP})_n$ to be uniquely determined on any t-interval containing t_0 and over which all $y_i(t)$'s and all $f_i(t, y)$'s remain continuous in t.

8.4B Numerical Methods for Solving $(\text{IVP})_n$

A **numerical solution** of $(\text{IVP})_n$ on the interval $[t_0, t_N]$ is a sequence of vectors $\mathbf{y}_0, \mathbf{y}_1, \mathbf{y}_2, \ldots, \mathbf{y}_N$, where $\mathbf{y}_j$ approximates the state at $t = t_j$, that is,

$$\mathbf{y}_j = \begin{bmatrix} y_{1,j} \\ y_{2,j} \\ \vdots \\ y_{n,j} \end{bmatrix} \quad \text{approximates} \quad \mathbf{y}(t_j) = \begin{bmatrix} y_1(t_j) \\ y_2(t_j) \\ \vdots \\ y_n(t_j) \end{bmatrix} \quad \text{for } j = 0, 1, \ldots, N \tag{4}$$

Note that (IVP) of the preceding sections becomes $(\text{IVP})_n$ in (2c) upon formally replacing the scalars y, y', $f(t, y)$, and y_0 by the vectors $\mathbf{y}$, $\mathbf{y}'$, $\mathbf{f}(t, \mathbf{y})$, and $\mathbf{y}_0$. *Performing these same replacements in a method for solving IVP will result in a method for solving IVP_n.* For example, the second-order Modified Euler method for solving $(\text{IVP})_n$ is

$$\mathbf{y}_{j+1}^{(ME)} = \mathbf{y}_j + h\mathbf{f}\left(t_j + \frac{h}{2}, \mathbf{y}_j + \frac{h}{2}\mathbf{f}_j\right), \qquad \text{where } \mathbf{f}_j = \mathbf{f}(t_j, \mathbf{y}_j) \tag{5}$$

Similarly, the fourth-order Runge–Kutta method for solving $(\text{IVP})_n$, denoted by $\mathbf{RK4}_n$, is

$$\mathbf{y}_{j+1}^{(RK4)} = \mathbf{y}_j + \frac{h}{6}\{\mathbf{m}_1 + 2(\mathbf{m}_2 + \mathbf{m}_3) + \mathbf{m}_4\}, \qquad \text{where } \mathbf{m}_1 = \mathbf{f}_j = \mathbf{f}(t_j, \mathbf{y}_j) \tag{6a}$$

and the remaining "sampled slope vectors" $\mathbf{m}_2$, $\mathbf{m}_3$, and $\mathbf{m}_4$ are obtained recursively as

$$\begin{aligned} \mathbf{m}_2 &= \mathbf{f}(t_j + \tfrac{1}{2}h, \mathbf{y}_j + \tfrac{1}{2}h\mathbf{m}_1) \\ \mathbf{m}_3 &= \mathbf{f}(t_j + \tfrac{1}{2}h, \mathbf{y}_j + \tfrac{1}{2}h\mathbf{m}_2) \\ \mathbf{m}_4 &= \mathbf{f}(t_j + h, \mathbf{y}_j + h\mathbf{m}_3) \end{aligned} \tag{6b}$$

EXAMPLE 8.4B Use one step of **(a)** the Modified Euler method and **(b)** RK4$_n$ to estimate x and y when $t = 0.2$ if x and y satisfy

$$\frac{dx}{dt} = y, \qquad x(0) = 1, \qquad [f_1(t, x, y) = x, \qquad x_0 = \quad 1]$$
$$\frac{dy}{dt} = x + t, \quad y(0) = -1, \quad [f_2(t, x, y) = y + t, \quad y_0 = -1] \tag{7a}$$

SOLUTION (a): Taking $\mathbf{y} = [x \quad y]^T$, that is, $y_1 = x$, $y_2 = y$, gives the vector form

$$\underbrace{\begin{bmatrix} x' \\ y' \end{bmatrix}}_{\mathbf{y}'} = \underbrace{\begin{bmatrix} y \\ x + t \end{bmatrix}}_{\mathbf{f}(t,\,\mathbf{y})}, \quad \underbrace{\begin{bmatrix} x(0) \\ y(0) \end{bmatrix}}_{\mathbf{y}(0)} = \underbrace{\begin{bmatrix} 1 \\ -1 \end{bmatrix}}_{\mathbf{y}_0} \tag{7b}$$

Since $t_0 = 0$, $t = 0.2$ will be t_1 if $h = 0.2$. To approximate $\mathbf{y}(t_1)$ as

$$\mathbf{y}_1^{(ME)} = \mathbf{y}_0 + h\mathbf{f}(t_0 + \tfrac{1}{2}h, \mathbf{y}_0 + \tfrac{1}{2}h\mathbf{f}_0)$$

we must first get the components of $\mathbf{y}_0 + \frac{1}{2}h\mathbf{f}_0$ as follows:

$$\mathbf{f}_0 = \mathbf{f}(t_0, \mathbf{y}_0) = \mathbf{f}\left(0, \begin{bmatrix} 1 \\ -1 \end{bmatrix}\right) = \begin{bmatrix} -1 \\ 1 + 0 \end{bmatrix} = \begin{bmatrix} -1 \\ 1 \end{bmatrix}$$

hence

$$\mathbf{y}_0 + \frac{h}{2}\mathbf{f}_0 = \begin{bmatrix} 1 \\ -1 \end{bmatrix} + \frac{0.2}{2}\begin{bmatrix} -1 \\ 1 \end{bmatrix} = \begin{bmatrix} 1 - 0.1 \\ -1 + 0.1 \end{bmatrix} = \begin{bmatrix} 0.9 \\ -0.9 \end{bmatrix}$$

So

$$\mathbf{y}_1^{(ME)} = \begin{bmatrix} 1 \\ -1 \end{bmatrix} + 0.2\mathbf{f}\left(0.1, \begin{bmatrix} 0.9 \\ -0.9 \end{bmatrix}\right) = \begin{bmatrix} 1 \\ -1 \end{bmatrix} + 0.2\begin{bmatrix} -0.9 \\ 0.9 + 0.1 \end{bmatrix} = \begin{bmatrix} 0.82 \\ -0.88 \end{bmatrix}$$

SOLUTION (b): From part (a), $\mathbf{m}_1 = \mathbf{f}_0 = [-1 \quad 1]^T$ and $\mathbf{y}_0 + \frac{1}{2}h\mathbf{m}_1 = [0.9 \quad -0.9]^T$. So (6b) gives

$$\mathbf{m}_2 = \mathbf{f}\left(t_0 + \frac{h}{2}, \mathbf{y}_0 + \frac{h}{2}\mathbf{m}_1\right) = \begin{bmatrix} -0.9 \\ 0.9 + 0.1 \end{bmatrix} = \begin{bmatrix} -0.9 \\ 1.0 \end{bmatrix}; \quad \mathbf{y}_0 + \frac{h}{2}\mathbf{m}_2 = \begin{bmatrix} 0.91 \\ -0.9 \end{bmatrix}$$

$$\mathbf{m}_3 = \mathbf{f}\left(t_0 + \frac{h}{2}, \mathbf{y}_0 + \frac{h}{2}\mathbf{m}_2\right) = \begin{bmatrix} -0.9 \\ 0.91 + 0.1 \end{bmatrix} = \begin{bmatrix} -0.9 \\ 1.01 \end{bmatrix}; \quad \mathbf{y}_0 + h\mathbf{m}_3 = \begin{bmatrix} 0.82 \\ -0.798 \end{bmatrix}$$

$$\mathbf{m}_4 = \mathbf{f}(t_0 + h, \mathbf{y}_0 + h\mathbf{m}_3) = \begin{bmatrix} -0.798 \\ 0.82 + 0.2 \end{bmatrix} = \begin{bmatrix} -0.798 \\ 1.02 \end{bmatrix}$$

Hence, by (6a), the desired $RK4_n$ approximation of $\mathbf{y}(0.2)$ is

$$\mathbf{y}_1 = \begin{bmatrix} 1 \\ -1 \end{bmatrix} + \frac{0.2}{6}\left\{\begin{bmatrix} -1 \\ 1 \end{bmatrix} + 2\left(\begin{bmatrix} -0.9 \\ 1.0 \end{bmatrix} + \begin{bmatrix} -0.9 \\ 1.01 \end{bmatrix}\right) + \begin{bmatrix} -0.798 \\ 1.02 \end{bmatrix}\right\} \doteq \begin{bmatrix} 0.820067 \\ -0.798667 \end{bmatrix}$$

It is easy to verify that the exact solution of (7) is

$$\mathbf{y}(t) = \begin{bmatrix} x(t) \\ y(t) \end{bmatrix} = \begin{bmatrix} -t + \frac{1}{2}(e^t + e^{-t}) \\ -1 + \frac{1}{2}(e^t - e^{-t}) \end{bmatrix}; \quad \text{so } \mathbf{y}(0.2) = \begin{bmatrix} x(0.2) \\ y(0.2) \end{bmatrix} = \begin{bmatrix} 0.820067 \\ -0.798665 \end{bmatrix} \qquad \textbf{(8)}$$

Thus, for this simple $\mathbf{f}(t, \mathbf{y})$ (for which both f_1 and f_2 are linear in x and y), $RK4_n$ yields $5s$ accuracy with a rather large stepsize, $h = 0.2$.

8.4C Solving an *n*th-Order IVP; Degree Reduction

Initial value problems often involve second-order and higher differential equations. For example, suppose s represents displacement of a particle of mass m along a line, and that the displacement s_0 and velocity v_0 are known at $t = 0$. Then, since a = acceleration is s'', Newton's second law of motion, $F = ma$, can be written as the second-order IVP

$$s'' = \frac{1}{m} F(t, s, s'), \qquad s(0) = s_0, \ s'(0) = v_0 \qquad \textbf{(9)}$$

where $F(t, s, s')$ indicates that the force F can vary with time and the particle's displacement (e.g., if attached to a spring) and velocity (e.g., if there were friction).

Our ''neutral'' notation for an ***n*th-order initial value problem** is

$$y^{(n)} = f(t, y, y', y'', \ldots, y^{(n-1)}) \qquad \textbf{(10a)}$$

$$y(t_0) = y_0, \quad y'(t_0) = y_0', \quad \ldots, \quad y^{(n-1)}(t_0) = y_0^{(n-1)} \qquad \textbf{(10b)}$$

We are assuming in (10a) that the differential equation can be (and, in fact, has been) solved for its highest derivative $y^{(n)}$, and in (10b) that all n **initial conditions** are at the *same*† t_0. Any nth-order IVP (10) can be solved by introducing the state variable vector:

$$\mathbf{y} = \begin{bmatrix} y \\ y' \\ y'' \\ \vdots \\ y^{(n-2)} \\ y^{(n-1)} \end{bmatrix} \begin{matrix} \leftarrow y_1 \text{ is } y; & \text{so } y_1' & = y' = y_2 \\ \leftarrow y_2 \text{ is } y'; & \text{so } y_2' & = y'' = y_3 \\ \leftarrow y_3 \text{ is } y''; & \text{so } y_3' & = y''' = y_4 \\ & & \\ \leftarrow y_{n-1} \text{ is } y^{(n-2)}; & \text{so } y_{n-1}' & = y^{(n-1)} = y_n \\ \leftarrow y_n \text{ is } y^{(n-1)}; & \text{so } y_n' & = y^{(n)} = f(t, y_1, y_2, \ldots, y_n) \end{matrix} \qquad \textbf{(11a)}$$

†When the n conditions on $y, y', \ldots, y^{(n-1)}$ involve more than one t, the problem becomes a **boundary value problem**. These are considered in Section 8.5.

This simple strategy, known as **degree reduction**, transforms any nth-order IVP (10) to the first-order coupled system

$$\begin{aligned} y_1' &= y_2 \\ y_2' &= y_3 \\ y_3' &= y_4 \\ &\vdots \\ y_{n-1}' &= y_n \\ y_n' &= f(t, y_1, y_2, \ldots, y_n) \end{aligned} \qquad \begin{aligned} y_1(t_0) &= y_0 \\ y_2(t_0) &= y_0' \\ y_3(t_0) &= y_0'' \\ &\vdots \\ y_{n-1}(t_0) &= y_0^{(n-2)} \\ y_n(t_0) &= y_0^{(n-1)} \end{aligned} \tag{11b}$$

which can then be solved by *any* $(\mathrm{IVP})_n$ method of Section 8.4B, with the output looking like Table 8.4-1. Since y is y_1, the desired approximations of $y(t_j)$ are in the $y_{1,j}$ column. As an added bonus, numerical solutions for the *derivatives* $y'(t_j), \ldots, y^{(n-1)}(t_j)$ are obtained in the $y_{2,j}, \ldots, y_{n,j}$ columns! Thus, if (9) were solved using degree reduction, the $y_{1,j}$ and $y_{2,j}$ columns of solution of the equivalent $(\mathrm{IVP})_2$ would give displacement *and velocity* at $t = t_j$.

TABLE 8.4-1 Output of an $(\mathrm{IVP})_n$ Solver

t_j	$y_{1,j}$	$y_{2,j}$	$\cdots$	$y_{n,j}$	
t_0	$y_{1,0}$	$y_{2,0}$	$\cdots$	$y_{n,0}$	$\leftarrow$ components of $\mathbf{y}_0 \approx \mathbf{y}(t_0)$
t_1	$y_{1,1}$	$y_{2,1}$	$\cdots$	$y_{n,1}$	$\leftarrow$ components of $\mathbf{y}_1 \approx \mathbf{y}(t_1)$
t_2	$y_{1,2}$	$y_{2,2}$	$\cdots$	$y_{n,2}$	$\leftarrow$ components of $\mathbf{y}_2 \approx \mathbf{y}(t_2)$
$\vdots$	$\vdots$	$\vdots$		$\vdots$	
t_N	$y_{1,N}$	$y_{2,N}$	$\cdots$	$y_{n,N}$	$\leftarrow$ components of $\mathbf{y}_N \approx \mathbf{y}(t_N)$

Note that the $(\mathrm{IVP})_2$ of Example 8.4B results when the degree reduction strategy is applied with $\mathbf{y} = [x \quad y]^T$, where $y = x'$, to the second-order IVP

$$x'' = x + t, \qquad x(0) = 1, \quad x'(0) = -1 \tag{12}$$

However, before the strategy can be applied to the third-order IVP

$$ty''' + t^2y' + yy'' = 2te^{-3t}, \qquad y(1) = -2, \quad y'(1) = 1, \quad y''(1) = 0 \tag{13a}$$

the differential equation must be solved for the highest derivative

$$y''' = 2e^{-3t} - ty' - \frac{1}{t}yy'' \tag{13b}$$

Taking $y_1 = y$, $y_2 = y'$, $y_3 = y''$ then gives the desired $(\mathrm{IVP})_3$

$$\begin{aligned} y_1' &= y_2, & y_1(1) &= -2 \\ y_2' &= y_3, & y_2(1) &= 1 \\ y_3' &= 2e^{-3t} - ty_2 - \frac{1}{t}y_1y_3, & y_3(1) &= 0 \end{aligned} \tag{13c}$$

EXAMPLE 8.4C Use the fourth-order Adams predictor–corrector method, starting with exact values at $t = 2, 2.1, 2.2,$ and 2.3, to solve the second-order IVP

$$y'' = \frac{-y'}{y^2}, \qquad y(2) = 2, \quad y'(2) = \frac{1}{2} \tag{14}$$

whose exact solution, as is easily verified, is

$$y(t) = \sqrt{2t}, \qquad \text{so that } y'(t) = 1/\sqrt{2t} = 1/y(t) \tag{15}$$

For reference, we note that the use of APC4 to solve the first order system

$$(\text{IVP})_n \qquad \mathbf{y}' = \mathbf{f}(t, \mathbf{y}), \quad \mathbf{y}(t_0) = \mathbf{y}_0$$

requires the following $\mathbf{APC4}_n$ equations [see (3) of Section 8.3A]:

$$\mathbf{p}_{j+1} = \mathbf{y}_j + \frac{h}{24}[-9\mathbf{f}_{j-3} + 37\mathbf{f}_{j-2} - 59\mathbf{f}_{j-1} + 55\mathbf{f}(t_j, \mathbf{y}_j)] \tag{16a}$$

$$\mathbf{c}_{j+1} = \mathbf{y}_j + \frac{h}{24}[\mathbf{f}_{j-2} - 5\mathbf{f}_{j-1} + 19\mathbf{f}_j + 9\mathbf{f}(t_{j+1}, \mathbf{p}_{j+1})] \tag{16b}$$

$$\boldsymbol{\delta}_{j+1} = \frac{-19}{270}[\mathbf{p}_{j+1} - \mathbf{c}_{j+1}] \tag{16c}$$

SOLUTION: Putting $y_1 = y$, $y_2 = y'$ transforms (14) to the following $(\text{IVP})_2$:

$$\underbrace{\begin{bmatrix} y_1' \\ y_2' \end{bmatrix}}_{\mathbf{y}'} = \underbrace{\begin{bmatrix} y_2 \\ -\dfrac{y_2}{y_1^2} \end{bmatrix}}_{\mathbf{f}(t,\,\mathbf{y})}, \qquad \underbrace{\begin{bmatrix} y_1(2) \\ y_2(2) \end{bmatrix}}_{\mathbf{y}(2)} = \underbrace{\begin{bmatrix} 2 \\ \dfrac{1}{2} \end{bmatrix}}_{\mathbf{y}_0} \tag{17}$$

Using (15), we get $\mathbf{y}_j$ as $\mathbf{y}(t_j) = \mathbf{y}(2 + 0.1j)$ to $7s$, $j = 0, 1, 2, 3$:

$$\mathbf{y}_0 = \begin{bmatrix} 2 \\ 0.5 \end{bmatrix}, \quad \mathbf{y}_1 = \begin{bmatrix} 2.049390 \\ 0.4879500 \end{bmatrix}, \quad \mathbf{y}_2 = \begin{bmatrix} 2.097618 \\ 0.4767313 \end{bmatrix}, \quad \mathbf{y}_3 = \begin{bmatrix} 2.144761 \\ 0.4662524 \end{bmatrix} \tag{18a}$$

Since $\mathbf{f}_j = [y_{2j} \quad -y_{2j}/y_{1j}^2]^T$, where $\mathbf{y}_j = [y_{1j} \quad y_{2j}]^T$,

$$\mathbf{f}_0 = \begin{bmatrix} 0.5 \\ -0.125 \end{bmatrix}, \quad \mathbf{f}_1 = \begin{bmatrix} 0.4879500 \\ -0.1161786 \end{bmatrix}, \quad \mathbf{f}_2 = \begin{bmatrix} 0.4767313 \\ -0.1083480 \end{bmatrix} \tag{18b}$$

With $h = 0.1$ and $t_3 = 2.3$, we get $\mathbf{y}_4$ as follows:

$$\mathbf{p}_4 = \mathbf{y}_3 + \frac{h}{24}\{-9\mathbf{f}_0 + 37\mathbf{f}_1 - 59\mathbf{f}_2 + 55\mathbf{f}(t_3, \mathbf{y}_3)\} \doteq \begin{bmatrix} 2.190890 \\ 0.4564364 \end{bmatrix}$$

So

$$\mathbf{f}(t_4, \mathbf{p}_4) = \begin{bmatrix} 0.4564364 \\ \dfrac{-0.4564364}{(2.190890)^2} \end{bmatrix} \doteq \begin{bmatrix} 0.4564364 \\ -0.09509094 \end{bmatrix}$$

hence

$$\mathbf{y}_4 = \mathbf{c}_4 = \mathbf{y}_3 + \frac{h}{24}\{\mathbf{f}_1 - 5\mathbf{f}_2 + 19\mathbf{f}_3 + 9\mathbf{f}(t_4, \mathbf{p}_4)\} \doteq \begin{bmatrix} 2.190890 \\ 0.4564354 \end{bmatrix}$$

$$\boldsymbol{\delta}_4 = -\frac{19}{270}\{\mathbf{c}_4 - \mathbf{p}_4\} = -\frac{19}{270}\left\{\begin{bmatrix} 2.190890 \\ 0.4564354 \end{bmatrix} - \begin{bmatrix} 2.190890 \\ 0.4564364 \end{bmatrix}\right\} \doteq \begin{bmatrix} 0 \\ 7.0\ \mathrm{E}{-8} \end{bmatrix}$$

It appears that this $\mathbf{y}_4$, without further iteration, is accurate to about 7*s*. In fact, by (15), the exact components of $\mathbf{y}(t_4)$ are

$$y(t_4) = \sqrt{2(2.4)} \doteq 2.1908902 \qquad \text{and} \qquad y'(t_4) = \frac{1}{y(t_4)} \doteq 0.45643546 \tag{19}$$

■

The excellent accuracy attained in the preceding example should not be too surprising because $\mathbf{y}(t) = [\sqrt{2t} \quad 1/\sqrt{2t}]^T$ varies slowly on $[t_0, t_4] = [2.0, 2.4]$. More generally, given a suitable h, APC4$_n$ can usually be expected to solve (IVP)$_n$ about as accurately as RK4$_n$; and it will do so with between $2n$ and $3n$ scalar function evaluations [of $f_i(t, \mathbf{y})$] per step, compared to $4n$ for RK4$_n$.

8.4D Solving (IVP)$_n$ on a Computer

Most computer installations have in their libraries at least one program for solving

$$(\mathrm{IVP})_n \quad \mathbf{y}' = \mathbf{f}(t, \mathbf{y}), \qquad \mathbf{y}(t_0) = \mathbf{y}_0$$

The program DIFSUB in the IMSL package is a rather sophisticated, variable stepsize and variable order method that has proven to be especially reliable. We shall illustrate computer code for a general purpose (IVP)$_n$ solver using the less sophisticated, but still reliable, Fortran `SUBROUTINE RKF4` shown in Figure 8.4-1

`SUBROUTINE RKF4` is based on Algorithm 8.2B. The **estimate next** $\boldsymbol{e[h]}$ steps are performed by calling `SUBROUTINE SUMK` [Figure 8.4-2(a)] to put $\mathbf{k}_1, \ldots, \mathbf{k}_6$ in columns 1, ..., 6 of the `DOUBLE PRECISION` $n \times 6$ matrix `K`. `RATIO` is then set to the largest (in magnitude) component of `ERREST` (lines 2500–2900) and used in line 3000 to determine whether the array `Y` calculated in lines 3200–3300 will be a sufficiently accurate approxi-

```
ØØ1ØØ          SUBROUTINE RKF4 (EVALF, RMAX, SCAMIN, SCAMAX, NPRINT, IW)
ØØ2ØØ          DOUBLE PRECISION Y(1Ø), K(1Ø, 6)
ØØ3ØØ          COMMON N, NSTEPS, T, TN, H, Y
ØØ4ØØ          EXTERNAL EVALF
ØØ5ØØ   C - - - - - - - - - - - - - - - - - - - - - - - - - - - - - - - C
ØØ6ØØ   C PURPOSE: To integrate n (n <= 1Ø) coupled first-order IVP's, i.e.,   C
ØØ7ØØ   C          (IVP)n  Y' = F(t,Y),  Y(tØ) = YØ  (tØ,YØ are initial t,Y)   C
ØØ8ØØ   C from tØ to tN (which can be < tØ) with a relative error tolerance   C
ØØ9ØØ   C RMAX, using the fourth-order Runge-Kutta-Fehlberg (RKF4) algorithm. C
Ø1ØØØ   C If NPRINT > Ø, then j, tj, and Yj are printed every NSTEPS steps.   C
Ø11ØØ   C The returned T will be tN, or ts (singularity of a yi) if evident.  C
Ø12ØØ   C SUBROUTINES CALLED:  SUMK to put the n-vector Kj in colj(K), j=1,6   C
Ø13ØØ   C                      EVALF (user-provided) to evaluate F(t,Y)       C
Ø14ØØ   C In EVALF, declare the arrays Y(n) and F(n) DOUBLE PRECISION.        C
Ø15ØØ   C In calling program, declare EVALF EXTERNAL, Y(n) DOUBLE PRECISION.  C
Ø16ØØ   C - - - - - - - - - - - - - - - - - - - - - - - VERSION 2: 9/9/85 - - C
Ø17ØØ          H = SIGN(RMAX**.25, TN-T)     !initial h
Ø18ØØ          HMIN = Ø.5E-4*ABS(H)          !minimum allowable |h|
Ø19ØØ          NSTEPS = Ø                    !counts steps taken
Ø2ØØØ          IF (NPRINT.GT.Ø) WRITE(IW,1) NSTEPS,T,H,(Y(I),I=1,N)  !tØ,hØ,YØ
Ø21ØØ   C
Ø22ØØ   C      MAIN LOOP: Step by adjusted H until T = TN or |H| < HMIN
Ø23ØØ      1Ø IF (H*(T+H-TN) .GE. Ø.) H = TN-T  !adjust final stepsize
Ø24ØØ          CALL SUMK (EVALF, K)             !get K1=coll(K) ,.., K6=col6(K)
Ø25ØØ          RATIO = Ø.Ø                      !becomes max |ErrEst(i)/h|
Ø26ØØ          DO 2Ø I=1,N
Ø27ØØ             ERREST = K(I,1)/36Ø-128*K(I,3)/4275-2197*K(I,4)/7524Ø+
Ø28ØØ         &                              K(I,5)/5Ø+2*K(I,6)/55
Ø29ØØ      2Ø   RATIO = AMAX1(RATIO, ABS(ERREST/H))
Ø3ØØØ          IF (RATIO .LE. RMAX) THEN   !use Y to approximate Y(t+h)
Ø31ØØ          T = T + H                   !step to next T; now get Y:
Ø32ØØ          DO 3Ø I=1,N
Ø33ØØ      3Ø   Y(I) = Y(I)+25*K(I,1)/216+14Ø8*K(I,3)/2565+
Ø34ØØ         &                      2197*K(I,4)/41Ø4-K(I,5)/5
Ø35ØØ          NSTEPS = NSTEPS + 1
Ø36ØØ          IF (NPRINT.GT.Ø .AND. (MOD(NSTEPS,NPRINT).EQ.Ø .OR. T.EQ.TN))
Ø37ØØ         &    WRITE(IW,1) NSTEPS, T, H, (Y(I), I=1,N)
Ø38ØØ          ENDIF
Ø39ØØ          IF (T .EQ. TN) RETURN                !adjusted final step; ELSE
Ø4ØØØ          SCALE = Ø.84*(RMAX/RATIO)**.25       !adjust h even if RATIO>RMAX
Ø41ØØ          IF (SCALE.LT.SCAMIN) SCALE = SCAMIN  !now SCALE >= SCAMIN
Ø42ØØ          IF (SCALE.GT.SCAMAX) SCALE = SCAMAX  !and SCALE <= SCAMAX
Ø43ØØ          H = SCALE*H                          !adjusted h
Ø44ØØ          IF (ABS(H) .GE. HMIN) GOTO 1Ø        !ELSE exit loop, due to
Ø45ØØ          WRITE(IW,2) T                        !apparent singularity
Ø46ØØ          RETURN
Ø47ØØ        1 FORMAT(I4, 2G13.5, 8G14.7, /, 24X, 2G14.7)
Ø48ØØ        2 FORMAT('Ø  Apparent singularity near t =',G11.4)
Ø49ØØ          END
```

Figure 8.4-1 SUBROUTINE RKF4.

```
      SUBROUTINE SUMK (EVALF, K)
      REAL P(6)
      DOUBLE PRECISION Y(1Ø), Q(1Ø), K(1Ø, 6), SUM(1Ø), F(1Ø)
      COMMON N, NSTEPS, T, TN, H, Y
C - - - - - - - - - - - - - - - - - - - - - - - - - - - - - - - - - - C
C  PURPOSE: To evaluate col[j]K = h*F(t+pj*h, SUM), j=1,...,6, where  C
C             SUM(i) = Y(i) + h*(qi1*col[1]K + ... + qi,j-1*col[j-1]K) C
C  using pj and qik of the O(h4) Runge-Kutta-Fehlberg (RKF4) method.  C
C - - - - - - - - - - - - - - - - - - - - - - - - - VERSION 2: 9/9/85 - C
      DATA (P(J),J=1,6) /Ø., .25, .375, .923Ø7692, 1., .5/   !p4=12/13
      DO 3Ø J=1,6         !get col[j]K = Kj
         DO 1Ø I=1,N
            IF (J.EQ.1) SUM(I) = Y(I)
            IF (J.EQ.2) SUM(I) = Y(I)+K(I,1)/4
            IF (J.EQ.3) SUM(I) = Y(I)+(3*K(I,1)+9*K(I,2))/32
            IF (J.EQ.4) SUM(I) = Y(I)+(1932*K(I,1)-72ØØ*K(I,2)+
     &                                           7296*K(I,3))/2197
            IF (J.EQ.5) SUM(I) = Y(I)+439*K(I,1)/216-8*K(I,2)+
     &                                  368Ø*K(I,3)/513-845*K(I,4)/41Ø4
            IF (J.EQ.6) SUM(I) = Y(I)-8*K(I,1)/27+2*K(I,2)-3544*
     &                             K(I,3)/2565+1859*K(I,4)/41Ø4-11*K(I,5)/4Ø
   1Ø    CONTINUE
         CALL EVALF(T+P(J)*H, SUM, F)   !F now stores Kj/h
         DO 2Ø I=1,N
   2Ø       K(I,J) = H*F(I)             !colj(K) now stores Kj
   3Ø CONTINUE                          !K now stores K1, ..., K6
      RETURN
      END
```

(a)

```
C  Simple program to integrate (IVP)n using SUBROUTINE RKF4
      PARAMETER (SCAMIN=Ø.1, SCAMAX=4.Ø, RMAX=Ø.ØØØ1, IW=5)
      DOUBLE PRECISION Y(1Ø)
      COMMON N, J, T, TN, H, Y
      EXTERNAL FF822                !for Figure 8.2-2
      DATA N, T, TN, NPRINT, Y(1) /1, 2, 4, 1, 1./
      WRITE (IW,1) T, TN, NPRINT  !T,Y are initial t,Y (i.e., tØ,YØ)
      CALL RKF4 (FF822, RMAX, SCAMIN, SCAMAX, NPRINT, IW)
      IF (NPRINT.EQ.Ø) WRITE (IW,2) J,T,H,(Y(I),I=1,N) !last j,t,h,Y
      STOP
    1 FORMAT('ØIntegrating from',G14.6,' to',G14.6,' printing every',
     &       I4,'  steps'//'   j',5X,'t',12X,'h',7X,'------Y------>')
    2 FORMAT(I4, 2G13.5, 8G14.7, /, 24X, 2G14.7)
      END

      SUBROUTINE FF822 (T, Y, F)    !for Figure 8.2-2
      DOUBLE PRECISION Y(1), F(1)
      F(1) = -T*Y(1)*Y(1)
      RETURN
      END
```

(b)

Figure 8.4-2*(a)* SUBROUTINE SUMK; *(b) calling program for* SUBROUTINE RKF4.

mation of $\mathbf{y}(t + h)$ [**accuracy test** steps], and in line 4000 to set the size of the next h [**set h** steps]. Note that the arrays `Y` and `F` are stored in `DOUBLE PRECISION` to guard against the accumulated effects of negligible addition (see Practical Consideration 8.2A).

The user of `SUBROUTINE RKF4` must provide a `SUBROUTINE EVALF(T, Y, F)` to form the array `F` $= \mathbf{f}(t, \mathbf{y})$ from `T` $(= t)$ and `Y` $(= \mathbf{y})$. `SUBROUTINE RKF4` was used to obtain the values shown in Figures 8.2-2 [see Figure 8.4-2(b)] and 8.2-3. An application when $n = 2$ is given in Section 8.5A.

8.4E Linearity

A first-order system is **linear** if for $i = 1, \ldots, n$ the ith equation is of the form

$$y_i' = a_{i1}(t)y_1 + a_{i2}(t)y_2 + \cdots + a_{in}(t)y_n + F_i(t) \tag{20a}$$

In matrix notation, a linear system is of the form

$$\mathbf{y}' = A(t)\mathbf{y} + \mathbf{F}(t), \qquad \text{where } A(t) = (a_{ij}(t))_{n\times n} \text{ and } \mathbf{F}(t) = [F_1(t) \cdots F_n(t)]^T \tag{20b}$$

Such a system is **homogeneous** if $\mathbf{F}(t) \equiv \mathbf{0}$, and **nonhomogeneous** otherwise. For example, in Example 8.4B, we considered a nonhomogeneous linear system whose matrix form is

$$\underbrace{\begin{bmatrix} x' \\ y' \end{bmatrix}}_{\mathbf{y}'} = \underbrace{\begin{bmatrix} 0 & 1 \\ 1 & 0 \end{bmatrix}}_{A(t)} \underbrace{\begin{bmatrix} x \\ y \end{bmatrix}}_{\mathbf{y}} + \underbrace{\begin{bmatrix} 0 \\ t \end{bmatrix}}_{\mathbf{F}(t)} \qquad [A(t) \equiv \text{constant}] \tag{21}$$

An nth-order differential equation is **linear** if it is of the form

$$y^{(n)} = \phi_1(t)y + \phi_2(t)y' + \cdots + \phi_n(t)y^{(n-1)} + F(t) \tag{22}$$

It is **homogeneous** if $F(t) \equiv 0$ as well. The second-order IVP of Example 8.4C is not linear. (Why?) It is easy to see that the procedure in (10) of Section 8.4C yields a linear (homogeneous) nth-order system when applied to a linear (homogeneous) nth-order differential equation (Problem M8-9).

If the entries of $A(t)$ and $\mathbf{F}(t)$ are continuous, then one can introduce the **antiderivative matrix**

$$M(t) = \int A(t)\, dt = \left(\int a_{ij}(t)\, dt\right)_{n\times n} \tag{23}$$

(where $\int dt$ is performed entrywise, ignoring constants of integration), and the $n \times n$ **exponential matrices** $e^{M(t)}$ and $e^{-M(t)}$, where

$$e^{\pm M(t)} = I \pm M(t) + \frac{1}{2!}M(t)^2 \pm \frac{1}{3!}M(t)^3 + \cdots = \sum_{n=0}^{\infty} \frac{(\pm 1)^n}{n!} M(t)^n \tag{24}$$

Both $e^{M(t)}$ and $e^{-M(t)}$ are defined and continuous whenever $M(t)$ is, in which case $e^{-M(t)}$ can be used as an integrating factor to show that the solution of the nonhomogeneous linear $(\text{IVP})_n$

$$\mathbf{y}' = A(t)\mathbf{y} + \mathbf{F}(t), \qquad \mathbf{y}(t_0) = \mathbf{y}_0 \tag{25}$$

is given by the matrix formula

$$\mathbf{y}(t) = e^{M(t)}\, e^{-M(t_0)}\mathbf{y}_0 + e^{M(t)} \int_{t_0}^{t} e^{-M(\tau)}\mathbf{F}(\tau)\, d\tau \tag{26}$$

Formulas (24)–(26) are pursued in more detail in Problems M8-10–M8-12. Although generally awkward to use computationally, they are of considerable theoretical importance. For example, it follows from formula (26) that *the linear* $(\text{IVP})_n$ (22) *has a unique solution over every interval on which all entries of* $\mathbf{F}(t)$ *and* $A(t)$ *are continuous.*

Formula (26) is but one indication of the fact that when trying to solve an IVP *analytically*, it is important to know whether it is linear or nonlinear. On the other hand, the *numerical procedures* described in this section apply equally well whether the IVP is linear or nonlinear! The *boundary* value problems described next, on the other hand, are easier to solve numerically when they are linear.

8.5

Boundary Value Problems

Consider the problem of solving the nth-order differential equation

$$y^{(n)} = f(t, y, y', y'', \ldots, y^{(n-1)}) \tag{1a}$$

subject to n constraints on y and/or its derivatives of the form

$$y^{(k_1)}(t_1) = \beta_1, \quad y^{(k_2)}(t_2) = \beta_2, \quad \ldots, \quad y^{(k_n)}(t_n) = \beta_n \tag{1b}$$

If all t_j's are the same, say t_0, then (1) is an nth-order IVP. However, if the **boundary conditions** (1b) involve m distinct t_i's, where $m > 1$, then (1) is an nth-order, m-point **boundary value problem** (abbreviated **BVP**). These arise naturally in the study of beam deflections, heat flow, and various dynamic systems.

An analysis of the general nth-order BVP is beyond the scope of this book. We shall focus our attention on the second-order, two-point BVP

$$(\text{BVP})_2 \quad y'' = f(t, y, y'), \qquad y(a) = \alpha, \qquad y(b) = \beta \tag{2}$$

In $(\text{BVP})_2$, the boundary conditions "nail down" the solution at (a, α) and (b, β).

8.5A The Shooting Method

We wish to use our ability to solve a second-order IVP as a tool for solving $(\text{BVP})_2$. Toward this end, consider the associated IVP

$$(\text{IVP})_x \quad y'' = f(t, y, y'), \qquad y(a) = \alpha, \qquad y'(a) = x \tag{3}$$

If we denote the solution of $(\text{IVP})_x$ by $y_x(t)$, then

$$x = \text{the initial slope (at } t = a\text{) of } y_x(t) \tag{4}$$

Our objective is to find that value $\bar{x}$ for which

$$y_{\bar{x}}(b) = \beta = \text{desired boundary value at } t = b \tag{5}$$

If the solution curve $y = y_{\bar{x}}(t)$ is thought of as a "trajectory," then the condition $y_{\bar{x}}(b) = \beta$ corresponds to hitting the "target value" β at $t = b$. When this occurs, $y_{\bar{x}}(t)$ satisfies $(\text{BVP})_2$, and hence is the desired $y(t)$ (see Figure 8.5-1). This is why the strategy of searching for an $\bar{x}$ for which $y_{\bar{x}}(t)$ satisfies (5) is called the **Shooting Method** for solving $(\text{BVP})_2$.

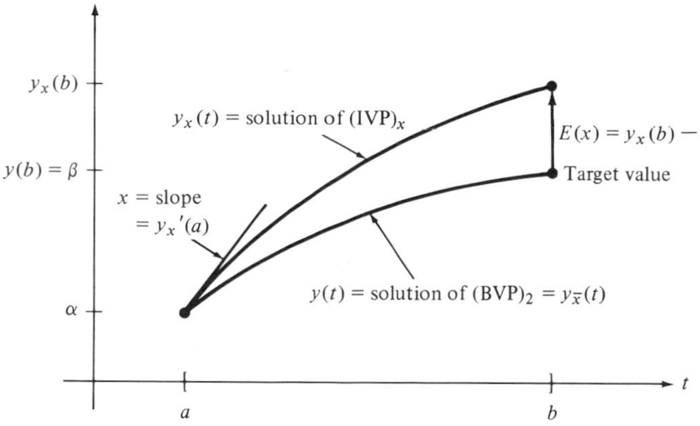

Figure 8.5-1 *Graphical representation of x, $y_x(t)$, y(t), and E(x).*

To carry out the Shooting Method, we introduce the error function

$$E(x) = y_x(b) - \beta \tag{6}$$

The value of $E(x)$ is the amount by which $y_x(b)$ misses the "target value" β. So *the problem of solving* $(\text{BVP})_2$ *can be viewed as that of finding a root* $\bar{x}$ *of* $E(x)$, that is, solving

$$E(x) = 0$$

Once $\bar{x} = y'(a)$ has been found, the desired $y(t)$ is $y_{\bar{x}}(t)$.

Since each evaluation of $E(x)$ requires a lot of work, namely integrating $(\text{IVP})_x$ from $t_0 = a$ to $t_N = b$, it is important to find the desired root of $E(x)$ using a method that converges rapidly. Although the Newton–Raphson (NR) method was the most rapidly convergent of Chapter 2, it is not suitable for this application because we generally have no analytic expression for $E(x)$ to differentiate. On the other hand, *the Secant (SEC) Method of Section 2.3D is ideally suited.* For this application, the iterative equation for the Secant Method becomes

$$x_{k+1} = x_k - \frac{E(x_k)(x_k - x_{k-1})}{E(x_k) - E(x_{k-1})} \quad \textbf{(7)}$$

EXAMPLE 8.5A Use the Shooting Method to solve the following nonlinear two-point BVP to about 5*s* accuracy.

$$y'' = y'\left[\frac{1}{t} + \frac{2y'}{y}\right], \qquad y(1) = 4, \qquad y(2) = 8 \quad \textbf{(8)}$$

SOLUTION: We wish to find a root $\bar{x}$ of

$$E(x) = y_x(b) - \beta = y_x(2) - 8 \quad \textbf{(9a)}$$

where $y_x(t)$ is the solution of the associated initial value problem

$$(\text{IVP})_x \quad y'' = y'\left[\frac{1}{t} + \frac{2y'}{y}\right], \qquad y(1) = 4, \qquad y'(1) = x \quad \textbf{(9b)}$$

The "natural" candidate for an initial guess of $y'(1)$ is

$$\frac{\beta - \alpha}{b - a} = \frac{8 - 4}{2 - 1} = 4 \qquad [\text{slope of the line from } (a, \alpha) \text{ to } (b, \beta)]$$

However, an attempt to solve $(\text{IVP})_x$ with $x = 4$ reveals that $y_x(t)$ has a singularity between $a = 1$ and $b = 2$. Consequently, we "aim low," taking say half this slope, namely $x = 2$, as an initial guess. This time

$$x = 2 \Leftrightarrow y_x(2) = 16 \Leftrightarrow E(x) = y_x(2) - 8 = 8 \quad \textbf{(9c)}$$

Since SEC requires two initial guesses and $E(2) > 0$, we try a smaller (more negative) slope, say $x = 1$. Using an accurate solver to solve $(\text{IVP})_x$ with $x = 1$, we find that

$$x = 1 \Leftrightarrow y_x(2) = 6.4 \Leftrightarrow E(x) = y_x(2) - 8 = -1.6$$

We can now begin the iteration. Taking $x_{-1} = 2$, $x_0 = 1$, and then

$$x_{k+1} = x_k + \Delta x_k, \qquad \text{where } \Delta x_k = \frac{-E(x_k)(x_k - x_{k-1})}{E(x_k) - E(x_{k-1})} \quad \textbf{(10)}$$

for $k = 0, 1, \ldots,$ we get the results shown in Table 8.5-1. The $y_x(2)$ values were obtained using an accurate $(IVP)_n$ solver.

TABLE 8.5-1 Shooting Method Iterations When $(BVP)_2$ Is Nonlinear

k	$x_k = x$	$y_x(2)$	$E(x) = y_x(2) - 8$	Δx_k	$x_{k+1} = x_k + \Delta x_k$
−1	2	16	8		
0	1	6.4	−1.6	0.1666667	1.1666667
1	1.1666667	7.1111111	−0.8888889	0.2083333	1.3750000
2	1.3750000	8.2580645	0.2580645	−0.0468750	1.3281250
3	1.3281250	7.9688716	−0.03112840	0.00504557	1.3331706
4	1.3331706	7.9902356	−0.00976443	0.00016339	1.3333334

A look at the Δx_k column of Table 8.5-1 reveals that

$$x_5 \doteq 1.333333 \text{ should approximate } \bar{x} = y'(1) \text{ to about } 6s \tag{11}$$

So the desired $y(t)$ can be obtained to about $5s$ by solving $(IVP)_x$ to about $6s$ on $[1, 2]$ with $x = 1.333333$.

As it turns out, this $(IVP)_x$ can be solved analytically. In fact,

$$y_x(t) = \frac{32}{8 + x(1 - t^2)} \qquad \text{so that } y_x(2) = \frac{32}{8 - 3x} \tag{12}$$

This formula for $y_x(2)$ makes it possible to find $\bar{x}$ analytically. We simply impose the boundary condition at $b = 2$:

$$y_{\bar{x}}(2) = \beta \quad \Leftrightarrow \quad \frac{32}{8 - 3\bar{x}} = 8 \quad \Leftrightarrow \quad \bar{x} = \frac{4}{3} \quad \Rightarrow \quad y(t) = y_{\bar{x}}(t) = \frac{24}{7 - t^2} \tag{13}$$

So the Shooting Method did indeed find $\bar{x}$ to $6s$. ■

In practice, when we do not have an analytic expression for $y_x(b)$, the Shooting Method must rely on an accurate $(IVP)_n$-solver such as that given in Section 8.4D to find $y_x(b)$ for a particular initial slope x.

Pseudocode for solving a nonlinear $(BVP)_2$ by the Shooting Method is shown in Algorithm 8.5A. Note that the initial *Xprev* is obtained by "aiming low" as we did in (9c), and the initial *X* is constrained to be close to *Xprev*.

8.5B The Importance of Linearity for the Shooting Method

The following example illustrates how the Shooting Method is particularly effective when $(BVP)_2$ is *linear*, that is, of the form

$$y'' = \phi_1(t)y + \phi_2(t)y' + F(t), \qquad y(a) = \alpha, \qquad y(b) = \beta \tag{14}$$

ALGORITHM 8.5A. THE SHOOTING METHOD

PURPOSE: To solve, to *NumSig* significant digits, the second-order BVP

$$y'' = f(t, y, y') \qquad y(a) = \alpha, \qquad y(b) = \beta$$

The method requires solving the associated $(\text{IVP})_x$, namely

$$y'' = f(t, y, y') \qquad y(a) = \alpha, \qquad y'(a) = x$$

for $y_x(t)$ in order to evaluate $E(x) = y_x(b) - \beta$. The Secant Method is used to get $\bar{x}$ (= slope of $y(t)$ at $t = a$) as a root of $E(x)$.

GET $a, b, \alpha, \beta,$ {parameters of BVP}
    NumSig, *MaxIt* {termination parameters}
RelTol $\leftarrow 0.5 \cdot 10^{-NumSig}$
Xprev $\leftarrow \frac{1}{4}(\beta - \alpha)/(b - a)$ {$\frac{1}{4}\cdot$slope from (a, α) to (b, β)}
Solve $(\text{IVP})_x$ with x = *Xprev* to get $y_x(b)$ to at least *NumSig* digits
Eprev $\leftarrow y_x(b) - \beta$
$X \leftarrow$ *Xprev* $- 0.2\cdot$sign(*Eprev*) {second initial guess of }

{**Iterate: Secant Method**}
DO FOR $k = 1$ TO *MaxIt*
  Solve $(\text{IVP})_x$ with $x = X$ to get $y_x(b)$ to at least *NumSig* digits
  $E \leftarrow y_x(b) - \beta$
  DeltaX $\leftarrow -E\cdot(X -$ *Xprev*$)/(E -$ *Eprev*$)$
  {***Update***} *Xprev* $\leftarrow X$; *Exprev* $\leftarrow E$; $X \leftarrow X +$ *DeltaX*
  UNTIL |*DeltaX*| $\leq$ *RelTol*$\cdot|X|$ **termination test**
IF **termination test** succeeded {$\bar{x} \doteq X$ to at least *NumSig* digits}
  THEN Solve $(\text{IVP})_x$ with $x = X$ to get $y_x(t) \approx y(t)$ on $[a, b]$
    OUTPUT ('Solution:' $(t_j, y_x(t_j))$ for $t_j = a$ to b)
  ELSE OUTPUT ('Convergence did not occur in' *MaxIt* 'iterations')
STOP

EXAMPLE 8.5B Use the Shooting Method to solve the linear BVP

$$y'' = -\frac{2}{t}(y' - 2), \qquad y\left(\frac{1}{2}\right) = 3, \qquad y(1) = 3 \tag{15}$$

SOLUTION: We wish to find a root $\bar{x}$ of

$$E(x) = y_x(b) - \beta = y_x(1) - 3 \tag{16a}$$

where $y_x(t)$ is the solution of the associated initial value problem

$$(\text{IVP})_x \quad y'' = -\frac{2}{t}(y' - 2), \qquad y\left(\frac{1}{2}\right) = 3, \qquad y'\left(\frac{1}{2}\right) = x \tag{16b}$$

Since $\alpha = \beta$ (both are 3), a reasonable initial guess for $y'(\frac{1}{2})$ is $x = 0$. Using an accurate $(\text{IVP})_n$ solver to solve $(\text{IVP})_x$ with $x = 0$ gives

$$x = 0 \Rightarrow y_x(1) = 3.5 \Rightarrow E(0) = 3.5 - 3 = \tfrac{1}{2} \tag{17a}$$

Since $E(0)$ is positive, we try a more negative initial slope, say $x = -1$, as our second initial guess. Solving $(\text{IVP})_x$ accurately gives

$$x = -1 \Rightarrow y_x(1) = 3.25 \Rightarrow E(-1) = 3.25 - 3 = \tfrac{1}{4} \tag{17b}$$

the two points

$$(x_{-1}, E(x_{-1})) = (0, \tfrac{1}{2}) \qquad \text{and} \qquad (x_0, E(x_0)) = (-1, \tfrac{1}{4})$$

are plotted on a set of x, $E(x)$-axes in Figure 8.5-2. The x-intercept of their secant line is

$$x_1 = x_0 - \frac{E(x_0)(x_0 - x_{-1})}{E(x_0) - E(x_{-1})} = -1 - \frac{\frac{1}{4}(-1 - 0)}{\frac{1}{4} - \frac{1}{2}} = -2$$

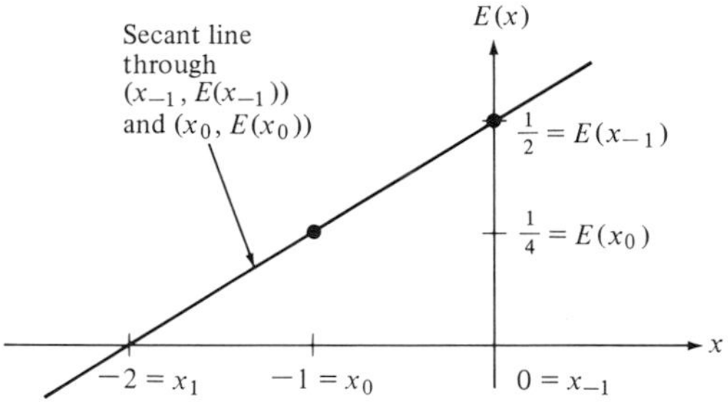

Figure 8.5-2 *Using SEC to get x_1 from $x_{-1} = 0$ and $x_0 = -1$.*

Remarkably, upon solving $(\text{IVP})_x$ with $x = x_1 = -2$, we find that

$$x = -2 \Rightarrow y_x(1) = 3 \Rightarrow E(-2) = 0, \qquad \text{that is, } \bar{x} = -2 \tag{17c}$$

We have thus found the desired root $\bar{x} = y'(\frac{1}{2})$ in one iteration! The desired solution is obtained by integrating $(\text{IVP})_x$ in (16b) from $a = \frac{1}{2}$ to $b = 1$ with $x = \bar{x} = -2$.

An analytic expression for the solution of $(\text{IVP})_x$ in (16b) is

$$y_x(t) = 2t + \frac{1}{t}\left(\frac{2 - x}{4}\right) + \left(\frac{2 + x}{2}\right) \tag{18}$$

as the reader can verify. Putting $t = b = 1$ gives

$$y_x(1) = 2 + \left(\frac{2 - x}{4}\right) + \left(\frac{2 + x}{2}\right) = \frac{7}{2} + \frac{x}{4} \tag{19}$$

Notice that $y_x(1)$ varies *linearly* with x. As a result, the graph of

$$E(x) = y_x(1) - 3 = \frac{1}{2} + \frac{x}{4} \tag{20}$$

is precisely the secant line drawn in Figure 8.5-2! This explains why the Secant Method found $\bar{x}$ in only one iteration. ■

In fact, *$E(x)$ will vary linearly with x whenever* (BVP)$_2$ *is linear*. In this case the first iterate x_1 determined by *any* two initial values

$$(x_{-1}, E(x_{-1})) \qquad \text{and} \qquad (x_0, E(x_0))$$

will be exactly $\bar{x} = y'(a)$ if $E(x_{-1})$ and $E(x_0)$ are exact. A proof of this important result is outlined in Problem M8-13.

8.5C The Finite Difference Method

An alternative strategy for solving

$$(\text{BVP})_2 \quad y'' = f(t, y, y'), \qquad y(a) = \alpha, \qquad y(b) = \beta \tag{21}$$

is to partition $[a, b]$ into n subintervals using $n + 1$ h-spaced t_j's

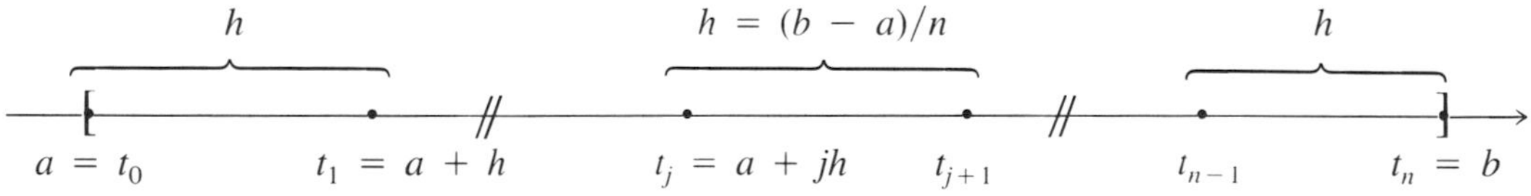

and then replace $y'(t)$ and $y'(t_j)$ by finite difference approximations such as the $O(h^2)$ central difference approximations

$$y''(t_j) \approx \frac{y_{j+1} - 2y_j + y_{j-1}}{h^2} \qquad \text{and} \qquad y'(t_j) \approx \frac{y_{j+1} - y_{j-1}}{2h} \tag{22}$$

for $j = 1, 2, \ldots, n - 1$ [see (27) of Section 7.2G]. This strategy is called the **Finite Difference Method**. It converts the *analytic* problem of solving (BVP)$_2$ to an approximating *algebraic* problem of solving $n - 1$ equations in the $n - 1$ unknowns

$$y_1, \ldots, y_{n-1}, \qquad \text{where } y_j \approx y(t_j) \quad \text{for } j = 1, 2, \ldots, n - 1 \tag{23}$$

Note that $y_0 = y(a) = \alpha$ and $y_n = y(b) = \beta$ are known from (BVP)$_2$.

EXAMPLE 8.5C Use the Finite Difference Method to solve the nonlinear $(BVP)_2$

$$y'' = y'\left[\frac{1}{t} + \frac{2y'}{y}\right], \qquad y(1) = 4, \qquad y(2) = 8 \tag{24}$$

SOLUTION: We first replace y'' and y' by their $O(h^2)$ central difference approximations at t_j for $j = 1, 2, \ldots, n-1$. The result is

$$\frac{y_{j+1} - 2y_j + y_{j-1}}{h^2} = \frac{y_{j+1} - y_{j-1}}{2h}\left[\frac{1}{t_j} + \frac{2(y_{j+1} - y_{j-1})}{2hy_j}\right], \qquad y_0 = 4, \qquad y_n = 8$$

Upon multiplying by $2h^2t_j$ and collecting the *linear* terms in y_{j-1}, y_j, and y_{j+1} on the lefthand side, this becomes

$$(2t_j + h)y_{j-1} - 4t_jy_j + (2t_j - h)y_{j+1} = d_j \tag{25a}$$

where $y_0 = 4$, $y_n = 8$, and

$$d_j = \frac{t_j}{y_j}(y_{j+1} - y_{j-1})^2, \qquad j = 1, 2, \ldots, n-1 \tag{25b}$$

The dominance (although not strict) of the $-4t_jy_j$ term on the lefthand side of (25a) suggests that we can solve the jth equation for

$$y_j^{(\text{new})} = \frac{1}{4t_j}\{(2t_j + h)y_{j-1} + (2t_j - h)y_{j+1} - d_j\}, \qquad j = 1, 2, \ldots, n-1 \tag{26}$$

and then use Gauss–Seidel iteration (Section 4.3A). Initially, we take the y_j's to be on the straight-line trajectory from (a, α) to (b, β), that is,

$$y_j = \alpha + \frac{\beta - \alpha}{b - a}(t_j - a), \qquad j = 0, 1, \ldots, n \tag{27}$$

The preceding approach, that is, applying Gauss–Seidel iteration to an approximating **discretized system**, is called the **Liebmann Process**. It is implemented in the Fortran program shown in Figure 8.5-3. The `FUNCTION` subprogram shown in Figure 8.5-4 calculates

$$\texttt{DELTAY} = y_j^{(\text{new})} - y_j$$

for use in the Gauss–Seidel iteration (lines 3600–4500 of Figure 8.5-3).

Two runs of the resulting program are shown in Figure 8.5-5. Since `TOL` = 1.0E−6, the y_j's shown approximate the solution of the discretized system (25) to about 6*s*. It can be shown that *if $O(h^n)$ formulas are used to get the approximating discretized system, then*

```
ØØ1ØØ            PROGRAM LIEBPR   !Liebmann Process calling SUBROUTINE LIEBMN
ØØ2ØØ            PARAMETER (MAXN=1ØØ, MAXIT=5ØØ, TOL=1.E-6, IW=5, IR=5)
ØØ3ØØ            DIMENSION T(Ø:MAXN), Y(Ø:MAXN)
ØØ4ØØ            LOGICAL CONVGD
ØØ5ØØ            DATA N, A, B, ALPHA, BETA /5, 1, 2, 4, 8/  !for Example 8.5C
ØØ6ØØ
ØØ7ØØ            CALL LIEBMN (N,A,B,ALPHA,BETA,T,Y,MAXIT,TOL,ITER,CONVGD)
ØØ8ØØ
ØØ9ØØ            IF (CONVGD) WRITE (IW,1) ITER, N, (T(J), Y(J), J=Ø,N)
Ø1ØØØ            IF (.NOT.CONVGD) WRITE (IW,2) N, MAXIT
Ø11ØØ            STOP
Ø12ØØ          1 FORMAT('ØAfter',I4,' iterations (n =',I3,'):'//5X,'tj',11X,'yj',
Ø13ØØ          &                                   1Ø1(/,G12.5,G14.6))
Ø14ØØ          2 FORMAT('ØConvergence not evident in',I4,' iterations (n =',I3,')')
Ø15ØØ            END
Ø16ØØ
Ø17ØØ            SUBROUTINE LIEBMN (N,A,B,ALPHA,BETA,T,Y,MAXIT,TOL,ITER,CONVGD)
Ø18ØØ            DIMENSION T(Ø:N), Y(Ø:N)
Ø19ØØ            LOGICAL CONVGD
Ø2ØØØ     C - - - - - - - - - - - - - - - - - - - - - - - - - - - - - - - - - C
Ø21ØØ     C  PURPOSE: To solve the second-order 2-point boundary value problem  C
Ø22ØØ     C           (BVP)2    y" = f(t,y,y'),   y(a) = alpha,   y(b) = beta   C
Ø23ØØ     C  by replacing  y'  and  y"  by their O(h*h) central difference      C
Ø24ØØ     C  approximations at tj = a + j*h (j = 1, ... , n-1), then solving     C
Ø25ØØ     C  the resulting system using Gauss-Seidel iteration.  The user must   C
Ø26ØØ     C  provide FUNCTION DELTAY(J, N, H, T, Y) = yj(new) - (current yj).     C
Ø27ØØ     C - - - - - - - - - - - - - - - - - - - - VERSION 2: 9/9/85 - - - - C
Ø28ØØ            H = (B - A)/N
Ø29ØØ     C      Initialize (tj,yj) on straight line from (a,alpha) to (b,beta)
Ø3ØØØ            SLOPE = (BETA - ALPHA)/(B - A)
Ø31ØØ            DO 1Ø J=Ø,N
Ø32ØØ               T(J) = A + J*H
Ø33ØØ               Y(J) = ALPHA + SLOPE*J*H
Ø34ØØ         1Ø CONTINUE
Ø35ØØ            ITER = Ø      !iteration counter
Ø36ØØ     C      Main Loop:  Perform Gauss-Seidel until CONVGD = .TRUE.
Ø37ØØ         2Ø CONVGD = .TRUE.
Ø38ØØ            DO 3Ø J = 1, N-1     !get dyj = yj(new) - yj(current) and test
Ø39ØØ               JJ = J
Ø4ØØØ               DYJ = DELTAY(JJ, N, H, T, Y)
Ø41ØØ               Y(J) = Y(J) + DYJ                                  !yj(new)
Ø42ØØ               IF (ABS(DYJ) .GT. TOL*ABS(Y(J))) CONVGD = .FALSE.   !test
Ø43ØØ         3Ø CONTINUE
Ø44ØØ            ITER = ITER + 1
Ø45ØØ            IF (.NOT.CONVGD .AND. ITER.LT.MAXIT) GOTO 2Ø  !end of loop
Ø46ØØ            END
```

Figure 8.5-3 *Fortran program for the Liebmann Process.*

```
ØØ1ØØ          FUNCTION DELTAY (J, N, H, T, Y)
ØØ2ØØ          DIMENSION T(Ø:N), Y(Ø:N)
ØØ3ØØ    C  - - - - - - - - - - - - - - - - - - - - - - - - - - - - - -  C
ØØ4ØØ    C          DELTAY = yj(new) - (current yj) for Example 8.5C      C
ØØ5ØØ    C  - - - - - - - - - - - - - - - - - - - - - - - - - - - - -    C
ØØ6ØØ          DJ = T(J)*(Y(J+1) - Y(J-1))**2/Y(J)
ØØ7ØØ          RESID = (2*T(J) + H)*Y(J-1) + (2*T(J) - H)*Y(J+1) - DJ
ØØ8ØØ          DELTAY = RESID/(4*T(J)) - Y(J)
ØØ9ØØ          RETURN
Ø1ØØØ          END
```

Figure 8.5-4 *User provided code for the program in Figure 8.5-3.*

the approximation $y(t_j) \approx y_j$ *is also* $O(h^n)$. To illustrate, since the exact $y(t)$ is $24/(7 - t^2)$ [see (13)], we see from Figure 8.5-5 that the truncation errors $\tau[h]$ at $t_j = 1.4$ are

$$\tau[0.2] = y(1.4) - y_j[0.2] = 4.76190 - 4.74349 \doteq 0.0184$$

$$\tau[0.1] = y(1.4) - y_j[0.1] = 4.76190 - 4.75752 \doteq 0.0044$$

```
After  28 iterations (n =  5):     After  96 iterations (n = 1Ø):

     tj          yj                     tj          yj
   1.ØØØØ     4.ØØØØØ                  1.ØØØØ     4.ØØØØØ
   1.2ØØØ     4.3Ø869                  1.1ØØØ     4.14421
   1.4ØØØ     4.74349                  1.2ØØØ     4.31467
   1.6ØØØ     5.37399                  1.3ØØØ     4.51672
   1.8ØØØ     6.34223                  1.4ØØØ     4.75752
   2.ØØØØ     8.ØØØØØ                  1.5ØØØ     5.Ø4676
                                       1.6ØØØ     5.39800
                                       1.7ØØØ     5.83Ø64
                                       1.8ØØØ     6.37359
                                       1.9ØØØ     7.Ø7188
                                       2.ØØØØ     8.ØØØØØ
```

Figure 8.5-5 *Runs of the program with h = 0.2 (n = 5) and h = 0.1 (n = 10).*

So $\tau[0.1] \approx \frac{1}{4}\tau[0.2]$. We can thus use Richardson's formula with $n = 2$ and $r = (0.2)/(0.1) = 2$ to get the improved estimate

$$y(1.4) \approx \frac{2^2(4.75752) - 4.74349}{4 - 1} = 4.76220 \qquad (28)$$

This is accurate to almost $5s$. ∎

8.5D The Finite Difference Method when (BVP)₂ Is Linear

Let us see what happens when $(BVP)_2$ is linear, that is, of the form

$$y'' = \phi_1(t)y + \phi_2(t)y' + F(t), \qquad y(a) = \alpha, \qquad y(b) = \beta \tag{29}$$

Replacing y'' by $\delta^2 y(t_j)/h^2$ and y' by $\delta y(t_j)/2h$, then multiplying by $2h^2$ and collecting y_{j-1}, y_j, and y_{j+1} terms on the left gives for $j = 1, 2, \ldots, n-1$

$$\underbrace{[2 + h\phi_2(t_j)]}_{a_j}y_{j-1} - \underbrace{[4 + 2h^2\phi_1(t_j)]}_{b_j}y_j + \underbrace{[2 - h\phi_2(t_j)]}_{c_j}y_{j+1} = \underbrace{2h^2F(t_j)}_{d_j} \tag{30}$$

Since $y_0 = y(t_0) = y(a) = \alpha$ and $y_n = y(t_n) = y(b) = \beta$, the matrix form of this *linear, tridiagonal* system is

$$\begin{bmatrix} b_1 & c_1 & & & \\ a_2 & b_2 & c_2 & & \\ & & \ddots & & \\ & & a_{n-2} & b_{n-2} & c_{n-2} \\ & & & a_{n-1} & b_{n-1} \end{bmatrix} \begin{bmatrix} y_1 \\ y_2 \\ \vdots \\ y_{n-2} \\ y_{n-1} \end{bmatrix} = \begin{bmatrix} d_1 - a_1\alpha \\ d_2 \\ \vdots \\ d_{n-2} \\ d_{n-1} - c_{n-1}\beta \end{bmatrix} \tag{31}$$

where a_j, b_j, c_j, and d_j are defined in (30), and

$$y_j \text{ approximates } y(t_j) = y(a + jh) \qquad \text{for } j = 1, 2, \ldots, n-1 \tag{32}$$

EXAMPLE 8.5D Use the Finite Difference Method to solve the linear $(\mathrm{BVP})_2$

$$y'' = \frac{-2}{t}(y' - 2) = 0, \qquad y(\tfrac{1}{2}) = 3, \qquad y(1) = 3 \tag{33}$$

Solution: Here $\phi_1(t) = 0$, $\phi_2(t) = -2/t$, and $F(t) = 4/t$; so (30) becomes

$$\left[2 - 2\frac{h}{t_j}\right] y_{j-1} - [4 + 2h^2 \cdot 0]y_j + \left[2 + 2\frac{h}{t_j}\right] y_{j+1} = \frac{8h^2}{t_j} \tag{34}$$

Multiplying both sides by $t_j/2$ and replacing $t_j \pm h$ by $t_{j\pm 1}$ gives

$$(\mathrm{E}_j) \quad t_{j-1}y_{j-1} - 2t_jy_j + t_{j+1}y_{j+1} = 4h^2, \qquad j = 1, \ldots, n-1 \tag{35a}$$

Imposing the boundary conditions in (E_1) and (E_{n-1}), we get

$$t_0y_0 = a\alpha = \tfrac{3}{2} \qquad \text{and} \qquad t_ny_n = b\beta = 3 \tag{35b}$$

For example, $n = 5$ $[h = (1 - \frac{1}{2})/5 = 0.1]$ gives the tridiagonal system

$$\begin{bmatrix} -1.2 & 0.7 & & \\ 0.6 & -1.4 & 0.8 & \\ & 0.7 & -1.6 & 0.9 \\ & & 0.8 & -1.8 \end{bmatrix} \begin{bmatrix} y_1 \\ y_2 \\ y_3 \\ y_4 \end{bmatrix} = \begin{bmatrix} 0.04 - \frac{3}{2} \\ 0.04 \\ 0.04 \\ 0.04 - 3 \end{bmatrix} = \begin{bmatrix} -1.46 \\ 0.04 \\ 0.04 \\ -2.96 \end{bmatrix} \tag{36}$$

Solving this tridiagonal system by Gaussian elimination shows that

$$y_1 = 2.8666667, \qquad y_2 = 2.8285714, \qquad y_3 = 2.8500000, \qquad y_4 = 2.9111111 \qquad \textbf{(37a)}$$

For this particular equation these values approximate

$$y(t_1) = y(0.6), \qquad y(t_2) = y(0.7), \qquad y(t_3) = y(0.8), \qquad y(t_4) = y(0.9) \qquad \textbf{(37b)}$$

accurately to 8*s*. In general, one should double n to check the accuracy and, if necessary, use Richardson's formula with $n = r = 2$, as we did in Section 8.5C. ■

8.5E Comparison of the Shooting Method and Finite Difference Method

Effect of Linearity. If $(\mathrm{BVP})_2$ is linear, the Shooting Method requires only two integrations of $(\mathrm{IVP})_x$ on $[a, b]$, whereas the Finite Difference Method yields a *linear* system with a diagonally dominant, banded coefficient matrix that can be solved quickly and accurately by Gaussian elimination even for large n. If $(\mathrm{BVP})_2$ is nonlinear, the Shooting Method becomes an iterative method (namely SEC), whereas the Finite Difference Method yields a nonlinear system that can be solved by Gauss–Seidel iteration (the Leibmann process).

Sources of Error. Whether $(\mathrm{BVP})_2$ is linear or nonlinear, the solution obtained by the Finite Difference Method reflects the truncation error of the finite difference approximations of y'' and y'; the accuracy should therefore be assessed by comparing the calculated y_j values to those obtained with n doubled. The accuracy obtained by the Shooting Method is limited only by the accuracy of the calculated $y_x(b)$ values.

Use of a Program. Whether $(\mathrm{BVP})_2$ is linear or nonlinear, the Shooting Method requires code for only $f(t, y, y')$. So this method should be tried whenever an $(\mathrm{IVP})_n$ solver (or a program for the Shooting Method itself) is available. However, the code for the Finite Difference Method is much easier to write; it is therefore recommended if a reliable solver is not around or if the BVP appears to be stiff.

More extensive comparisons can be found in [23 and 24].

PROBLEMS

Throughout this problem set, $\mathrm{IVP}_A - \mathrm{IVP}_F$ will refer to the following IVP's:

IVP$_A$	$y' = -t/y$,	$y(0) = 2$,	$h = 0.2$	$[y(t) = \sqrt{4 - t^2}]$
IVP$_B$	$y' = -2 + y/t$,	$y(1) = 2$,	$h = 0.2$	$[y(t) = 2t(1 - \ln t)]$
IVP$_C$	$y' = t(y + 1)$,	$y(0) = 0$,	$h = 0.4$	$[y(t) = e^{-t^2/2} - 1]$
IVP$_D$	$y' = t + 3y/t$,	$y(1) = 0$,	$h = 0.2$	$[y(t) = t^3 - t^2]$
IVP$_E$	$y' = \sqrt{y}$,	$y(2) = 4$,	$h = 0.5$	$[y(t) = (t + 2)^2/4]$
IVP$_F$	$y' = 2(y - 1)^2$,	$y(0) = 2$,	$h = 0.05$	$[y(t) = 2(1 - t)/(1 - 2t)]$

Section 8.1

8-1 For IVP_A–IVP_F of your choice, verify that the given $y(t)$ is the solution, and describe a t-interval over which this solution is unique.

8-2 For IVP_A–IVP_F of your choice, do (a) and (b) using Euler's method.
(a) Estimate $y(t_0 + h)$ by taking one h-step and by taking two $h/2$-steps, and confirm that at $t = t_0 + h$, $E[h/2] \approx E[h]/2^n$, where n is the order of the method.
(b) Use Richardson's formula and your results of part (a) to get an improved estimate of $y(t_0 + h)$. Did it actually improve the accuracy?

Section 8.2

8-3 Same as Problem 8-2 but for the Modified Euler method.

8-4 Same as Problem 8-2 but for Huen's method.

8-5 Same as Problem 8-2 but for the RK4 method.

8-6 Same as Problem 8-2 but for the second-order Taylor's method.

8-7 Same as Problem 8-2 but for the third-order Taylor's method.

8-8 For IVP_A–IVP_F of your choice, take one step of RKF4, using $Rmax = h^4$ and determine t_j and h for the second step.

8-9 ***(Integrating backwards)*** For IVP_D, take two (-0.4)-steps to estimate $y(0.2)$ using **(a)** Euler's method; **(b)** the Modified Euler method; **(c)** RK4.

8-10 ***(Integrating backwards)*** For IVP_B, take three (-0.2)-steps to estimate $y(0.4)$ using **(a)** Euler's method; **(b)** Huen's Method; **(c)** RK4.

8-11 Show that the solution of $y' = Ay$, $y_0 = 0$ is unstable for $At > 0$.

8-12 For IVP_A–IVP_F of your choice, suppose that Euler's method is used to get y_j.
(a) Should this y_j approximate $y(t_j)$ well for any *positive* h? Explain your reasoning (see Example 8.2E).
(b) Use (30) of Section 8.2E to determine the values of h for which y_j should approximate $y(t_j)$ well for five h-steps.

8-13 Are any of IVP_A–IVP_F stiff? Explain.

8-14 Verify that $y(t) = t + e^{-2t^2}$ is the solution of the linear IVP $y' = 1 - 4t(y - t)$, $y(0) = 1$. Why is this IVP stiff? Use (30) of Section 8.2E to find a formula for β in terms of t_N so that Euler's method will approximate $y(t_j)$ well for $0 \leq t \leq t_N$, provided that $0 < h < \beta$. Find β when t_N is 5; 10; 20; 100. Why are your results problematical?

8-15 Let $h = 0.4$ for the IVP $y' = 4(t - y) + 1$, $y(0) = 3$ $[y(t) = t + 3e^{-4t} \approx t$ for $t > 1]$.
(a) Explain why the IVP is stiff.
(b) Integrate the IVP from t_0 to $t_N - 10h$ by taking two h-steps followed by two $2h$-steps and one $4h$-step using Euler's method.
(c) Repeat part (b) using the Backward Euler method (33b). Compare the accuracy to that obtained in part (b).

8-16 Let $h = 0.1$ for the IVP $y' = -50(y - t^2) + 2t$, $y(0) = 1$ $[y(t) = t^2 + e^{-50t}]$. Do (a)–(c) of Problem 8-15 but use the $O(h^2)$ Modified Euler method in part (b).

8-17 Let $h = 0.8$ for the IVP $y' = 1/y - y$, $y(0) = 2$ $[y(t) = \sqrt{1 + 3e^{-2t}}]$. Do (a)–(c) of Problem 8-15 but use Huen's method in (b) and the Trapezoidal Rule (39) in part (c).

Section 8.3

8-18 Show that Huen's method can be viewed as a predictor–corrector method.

8-19 Explain why if $y(t)$ happens to be a polynomial of degee $\leqslant 4$ and exact $y(t_j)$ are used to start APC4, then both p_{j+1} and c_{j+1} will be exactly $y(t_{j+1})$ for $j \geqslant 3$.

8-20 For IVP_A–IVP_F of your choice, use APC4 of Section 8.3A, starting with the exact values $y_j = y(t_j)$, $j = 0, 1, 2, 3$, to get y_4 and y_5 three ways:
(a) With $y_{j+1} = c_{j+1}$ [two $f(t, y)$ evaluations per step].
(b) With y_{j+1} obtained by replacing p_{j+1} by c_{j+1} once in the corrector formula [three $f(t, y)$ evaluations per step].
(c) With y_{j+1} obtained by replacing p_{j+1} by c_{j+1} once in the corrector formula *only* if c_{j+1} does not appear to have $5s$ accuracy.

8-21 From your work in getting y_4 in (a) of Problem 8-20, determine t_5 and the h that would be used to get y_5 if the Halve/Double/Leave Stepsize Strategy [(5) of Section 8.2B] with *Rmax* = 0.0005, and δ_4 used as *ErrEst*.

8-22 Use your work in getting y_4 in (a) of Problem 8-20 to get y_5 with the stepsize (i) halved; (ii) doubled.

8-23 The **Second-Order Adams Predictor–Corrector Method (APC2)** uses

$$p_{j+1} = y_j + \frac{h}{2}[3f_j - f_{j-1}], \qquad e_{j+1}[h]_p = \frac{5}{12}y'''(\xi_p)h^2$$

$$c_{j+1} = y_j + \frac{h}{2}[f(t_{j+1}, p_{j+1}) + f_j], \quad e_{j+1}[h]_c = \frac{-1}{12}y'''(\xi_c)h^2$$

Show that $e_{j+1}[h]_c \approx \delta_{j+1} = \dfrac{-1}{12}[p_{j+1} - c_{j+1}]$

8-24 Same as Problem 8-20 but use APC2 of Problem 8-23, starting with the exact value $y_1 = y(t_1)$, to get y_2 and y_3.

8-25 The **Third-Order Adams Predictor–Corrector Method (APC3)** uses

$$p_{j+1} = y_j + \frac{h}{12}[23f_j - 16f_{j-1} + 5f_{j-2}], \qquad e_{j+1}[h]_p = \frac{3}{8}y^{(\text{iv})}(\xi_p)h^3$$

$$c_{j+1} = y_j + \frac{h}{12}[5f(t_{j+1}, p_{j+1}) + 8f_j - f_{j-1}], \quad e_{j+1}[h]_c = \frac{-1}{24}y^{(\text{iv})}(\xi_c)h^3$$

Show that $e_{j+1}[h]_c \approx \delta_{j+1} = \dfrac{-1}{24}[p_{j+1} - c_{j+1}]$.

8-26 Same as Problem 8-20 but use APC3 of Problem 8-25 to get y_3 and y_4, starting with the exact values $y_1 = y(t_1)$ and $y_2 = y(t_2)$

8-27 Suppose that the nth order APC method is used with exact starting values and no roundoff error. For which (if any) of IVP_A–IVP_F will y_j be exactly $y(t_j)$ for $j \geqslant 0$ when n is 2? 3? 4?

Section 8.4

In what follows, IVP_G and IVP_H refer to the following $(\text{IVP})_2$'s:

IVP_G	$x' = y + t,$	$x(0) = 1$	$[x(t) = 1 + \sin 2t]$; $h = 0.1$
	$y' = 3 - 4x,$	$y(0) = 2$	$[y = -t + 2\cos 2t]$
IVP_H	$y_1' = 1 + 2e^{-y_2},$	$y_1(1) = -1$	$[y_1(t) = t - 2/t]$; $h = 0.2$
	$y_2' t = -y_1,$	$y_2(1) = 0$	$[y_2(t) = 2 \ln t]$

8-28 For IVP_G–IVP_H of your choice, state whether the $(IVP)_2$ is linear and verify that the given $\mathbf{y}(t)$ is the solution.

8-29 For IVP_G–IVP_H of your choice, do (a) and (b) using Euler's method.
(a) Estimate $\mathbf{y}(t_0 + h)$ by taking one h-step and by taking two $h/2$-steps. At $t = t_0 + h$, confirm that $E[h/2] \approx E[h]/2^n$ in *each* component, where n is the order of the method.
(b) Use Richardson's formula and your results of part (a) to get an improved estimate.

8-30 Same as Problem 8-29 but use the Modified Euler method.

8-31 Same as Problem 8-29 but use Huen's method.

8-32 For IVP_G–IVP_H of your choice, take two h-steps using $RK4_n$.

8-33 Use a method of your choice to solve the $(IVP)_3$ in (1) to 4*s* on [1, 1.4].

8-34 In parts (a)–(d), change to a first order system in vector form. If the given nth-order IVP is linear, write the system as $\mathbf{y}' = A\mathbf{y} + \mathbf{F}(t)$.
(a) $y''' - \dfrac{y''}{y^2} - 2ty' = 0,\ y(0) = 1,\ y'(0) = -1,\ y''(0) = 2$
(b) $e^{-t}y^{(iv)} + e^t y'' + 2ty = 4,\ y(0) = y''(0) = 1,\ y'''(0) = y'(0) = 0$
(c) $ty'' + t^2y' - \dfrac{y}{t} = e^t,\ y(1) = 3,\ y'(1) = 0$
(d) $y'' - y'e^y + ty = t^2,\ y(1) = 0,\ y'(1) = 1$

8-35 **(Satellite Motion)** The equations of motion of a satellite fired from an orbiting space station are

$$x'' = -G\frac{x}{r^3}, \quad x(0) = x_0, \quad x'(0) = v_{0x}$$

$$y'' = -G\frac{y}{r^3}, \quad y(0) = 0, \quad y'(0) = v_{0y}$$

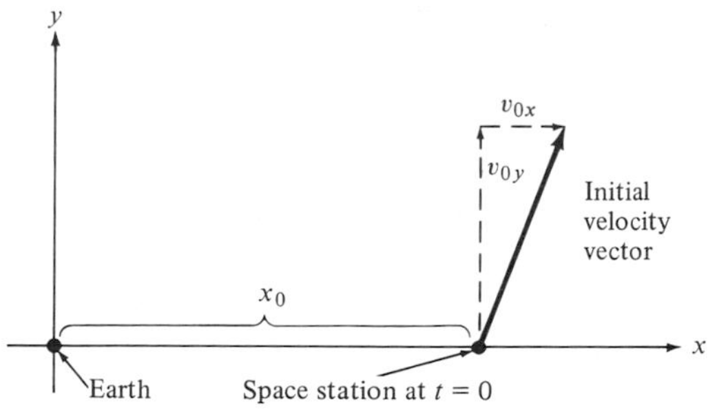

where $r = \sqrt{x^2 + y^2}$ is the distance to the earth as shown. Convert this second-order system to an equivalent $(IVP)_4$ using

$$y_1 = x, \quad y_2 = x', \quad y_3 = y, \quad y_4 = y'$$

Section 8.5

8-36 **(a)** Verify that $y_x(t)$ in (12) is the solution of (9b) in Section 8.5A.
(b) Verify that $y_x(t)$ in (18) is the solution of (16b) in Section 8.5B.

8-37 Suppose you know that $y_x(t) = e^{tx} - 2$ is the solution of

$$(IVP)_x \ y'' = f(t, y, y'), \qquad y(0) = -1,\ y'(0) = x$$

(a) Is the differential equation $y'' = f(t, y, y')$ linear? Explain.
(b) Find $y(t)$ for the BVP: $y'' = f(t, y, y')$, $y(0) = -1$, $y(1) = 0$

8-38 In (a) and (b), verify that the given $y_x(t)$ is the solution of the associated $(\text{IVP})_x$. Then use the Shooting Method with

$$x_{-1} = \frac{1}{4}\frac{\beta - \alpha}{b - a} \qquad \text{and} \qquad x_0 = x_{-1} - 0.2 \cdot \text{sign}(E(x_{-1}))$$

[and with the given $y_x(t)$ used to evaluate $y_x(\text{b})$] to find $\bar{x}$ to $6s$. Verify your answer by using $y_x(t)$ to find $\bar{x}$ analytically.

(a) $y'' = \dfrac{-2y'}{t}$, $y(1) = -1$, $y(2) = 0$ $\quad \left[y_x(t) = \dfrac{t - x}{t} \right]$

(b) $y'' = (y')^2$, $y(1) = 0$, $y(2) = \ln \dfrac{1}{2}$ $\quad \left[y_x(t) = -\ln\left(t - 1 - \dfrac{1}{x} \right) \right]$

In Problems 8-39–8-41, BVP_I–BVP_L refer to the following BVP's:

BVP_I: $y'' = 2y' - y - 3$, $y(0) = -3$, $y(2) = 1$ $\quad [y(t) = 2te^{t-2} - 3]$

BVP_J: $y'' = 3ty' - 9y + 3t^2 + 2$, $y(1) = 1$, $y(2) = -2$ $\quad [y(t) = t + t^2 - t^3]$

BVP_K: $y'' = \dfrac{2}{y' + 1}$, $y(1) = \dfrac{1}{3}$, $y(4) = \dfrac{20}{3}$ $\quad \left[y(t) = \dfrac{4}{3} t^{3/2} - t \right]$

BVP_L: $y'' = t(y')^2$, $y(0) = \dfrac{\pi}{2}$, $y(2) = \dfrac{\pi}{4}$ $\quad \left[y(t) = \cot^{-1}\left(\dfrac{t}{2} \right) \right]$

8-39 Apply the Finite Difference Method with n subintervals to BVP_I–BVP_L of your choice. If linear, find the jth equation of the tridiagonal system $T = \mathbf{d}$ [see (30) of Section 8.5D]; if nonlinear, solve the jth equation for y_j (to use the Liebmann Process).

8-40 For $\text{BVP}_I - \text{BVP}_L$ of your choice, use the Finite Difference Method with $n = 3$ subintervals to get $y_1[\approx y(a + h)]$ and $y_2[\approx y(a + 2h)]$ to $4s$.

8-41 For $\text{BVP}_I - \text{BVP}_L$ of your choice, use the Finite Difference Method, first with $n = 2$ and then with $n = 4$ subintervals. If nonlinear, get $4s$ accuracy. Apply Richardson's formula to your values at $t = \frac{1}{2}(a + b)$.

MISCELLANEOUS PROBLEMS

M8-1 Show that $y(t)$ in (6) of Section 8.1A is differentiable at $t = 0$ and is a solution of the given for any choice of C.

M8-2 **(a)** In the absence of roundoff error, why will y_j obtained by the nth order Taylor's method be exactly $y(t_j)$ whenever $f(t, y)$ is a polynomial (in t and y) of degree $\leq n$?
(b) We saw in Section 8.2D that $h\phi$ for an nth order Runge–Kutta method is an $O(h^n)$ approximation $h\phi_{T,n}$. For which (if any) of IVP_A–IVP_F might you expect to get especially good results using the Modified Euler method? RK4? Explain your answer.

M8-3 Consider the IVP $y' = e^y$, $y(0) = y_0$ for which $y(t) = -y \ln(e^{y_0} - t)$.

(a) For what values of y_0 is $y(t)$ stable? For what values is the IVP stiff?

(b) Take two steps of the Backward Euler method (33b) of Section 8.2F using $h = 0.1$. For this IVP, is there any reason to prefer the Backward Euler method to Euler's method? Explain.

M8-4 **(a)** Show that for the Test Equation $y' = Ay$, $y(0) = y_0$, the solution using both the Modified Euler and Huen methods will be $y_j = y_0(1 + hA + h^2A^2/2)^j$ (see Example 8.2E).

(b) Find a condition of the form $\alpha < hA < \beta$ under which y_j in part (a) will approximate $y(t_j) = y_0e^{Ahj}$ well (see Example 8.2E).

(c) Guess what the answer in part (a) would be if carried out for the RK4 algorithm? Explain your reasoning.

M8-5 Assuming that you had a computer program for solving IVP, describe a trial and error procedure for determining an h (roughly) as large as possible but small enough so that the y_j's are accurate to a desired number of significant digits. What assumptions about the IVP are you making?

M8-6 Which of **(a)** Huen's method; **(b)** $y_{j+1} = y_{j-1} + h[f_{j-1} + 4f_j + f(t_{j+1}, y_{j+1})]3$; **(c)** $y_{j+1} = y_{j-1} + hf_j$ can be used as a predictor? as a corrector? to solve stiff IVP's?

M8-7 Show that the Midpoint Predictor (26a) of Section 8.2E results from replacing the exact IVP condition $y'(t_j) = f(t_j, y(t_j))$ by the approximate finite difference condition

$$\frac{y_{j+1} - y_{j-1}}{2h} \approx f_j$$

M8-8 Use the values obtained in getting y_4 in Example 8.3A to get p_5 **(a)** with h halved to 0.05 (see Section 8.3B); **(b)** with h doubled to 0.2.

M8-9 Explain how the given $(\text{IVP})_3$ is related to a third-order IVP.

(a) $x' = y, \quad x(1) = 3$
$y' = z, \quad y(1) = 0$
$z' = tx^2 + yz, \quad z(1) = 0$

(b) $q' = r/s + t, \quad q(0) = 0$
$r' = q; \quad r(0) = 1$
$s' = r, \quad s(0) = 2$

M8-10 For $A = \begin{bmatrix} 0 & 1 \\ 1 & 0 \end{bmatrix}$ and $At = \begin{bmatrix} 0 & t \\ t & 0 \end{bmatrix}$, show the following:

(a) If n is even, then $A^n = I$; if n is odd, then $A^n = A$.

(b) Deduce from part (a) and (24) of Section 8.4F that

$$e^{At} = (\cosh t)I_2 + (\sinh t)A = \begin{bmatrix} \cosh t & \sinh t \\ \sinh t & \cosh t \end{bmatrix}$$

(c) Deduce from part (b) that $de^{At}/dt = Ae^{At}$ and $e^{-At} = e^{A(-t)} = [e^{At}]^{-1}$

(d) Use formula (26) of Section 8.4E with $M(t) = At$ to show that $y(t) = \cosh t - t$ is the solution of $y'' = y + t$, $y(0) = 1$, $y'(0) = -1$ [see (21) of Section 8.4E].
NOTE: You must solve $\mathbf{y}' = A\mathbf{y} + [0 \quad t]^T$, $\mathbf{y}(0) = [1 \quad -1]^T$. Why?

M8-11 In parts (a)–(d), consider the constant matrix

$$A = \begin{bmatrix} 0 & 1 & 0 \\ 0 & 0 & 1 \\ 0 & 0 & 0 \end{bmatrix}, \qquad \text{for which } At = \begin{bmatrix} 0 & t & 0 \\ 0 & 0 & t \\ 0 & 0 & 0 \end{bmatrix}$$

(a) A square matrix A for which $A^k = 0$ for some k is called **idempotent.** Show that A is idempotent.

(b) Deduce from your matrices $A, A^2, A^3, \ldots$ in part (a) that

$$e^{At} = \begin{bmatrix} 1 & t & \frac{1}{2}t^2 \\ 0 & 1 & t \\ 0 & 0 & 1 \end{bmatrix}, \qquad \text{hence that } \frac{d}{dt} e^{At} = Ae^{At} = \begin{bmatrix} 0 & 1 & t \\ 0 & 0 & 1 \\ 0 & 0 & 0 \end{bmatrix}$$

(c) Verify that $e^{-At} = e^{A(-t)}$ is $[e^{At}]^{-1}$ (i.e., $e^{-At}e^{At} = I_3$).
(d) Use formula (26) of Section 8.4E, with $M(t) = At$, to solve the $(\text{IVP})_3$

$$\begin{aligned} y_1' &= y_2 & y_1(0) &= 1 \\ y_2' &= y_3, & y_2(0) &= 0 \\ y_3' &= 6, & y_3(0) &= 0 \end{aligned} \qquad \left(\text{exact solution: } \mathbf{y}(t) = \begin{bmatrix} t^3 + 1 \\ 3t^2 \\ 6t \end{bmatrix}\right)$$

M8-12 Let $\int A(t)\,dt$ and $e^{\pm M(t)}$ be defined as in (23) and (24) of Section 8.4E, and perform differentiation and series summation of matrices componentwise. Prove (a)–(d) for any differentiable $n \times n$ matrices $A(t)$ and $B(t)$.

(a) Product Rule: $\dfrac{d}{dt} A(t)B(t) = A(t)\dfrac{dB(t)}{dt} + B(t)\dfrac{dA(t)}{dt}$.

(b) Chain Rule: $\dfrac{d}{dt} e^{A(t)} = A'(t)e^{A(t)}$, hence $\dfrac{d}{dt}(e^{\int A(t)\,dt}) = A(t)e^{\int A(t)\,dt}$

(c) Exponent Rule: $e^{A(t)+B(t)} = e^{A(t)}e^{B(t)}$.

HINT: Show that both sides have the same derivative, hence must differ by a constant matrix C. Take $t = 0$ to show that $C = \text{O}$.)

(d) Inverse Rule: $e^{-A(t)} = [e^{A(t)}]^{-1}$

(e) Using parts (a)–(d), verify that the formula (26) of Section 8.4E does indeed give a solution of $\mathbf{y}' = A(t)\mathbf{y} + \mathbf{F}(t)$, $\mathbf{y}(t_0) = \mathbf{y}_0$.

M8-13 **(a)** ***(Superposition Principle)*** Show that if $y_1(t)$ and $y_2(t)$ satisfy

$$y'' = \phi_1(t)y + \phi_2(t)y' + F(t)$$

then so does $ry_1(t) + sy_2(t)$ for any r and s such that $r + s = 1$.

(b) Deduce from part (a) that for any x_0, x_1, and x, the solution of $(\text{IVP})_x$ is

$$\circledast \quad y_x(t) = y_{x0}(t)\,L_0(x) + y_{x_{-1}}(t)\,L_{-1}(x)$$

where $L_0(x)$ and $L_{-1}(x)$ are the Lagrange polynomials

$$L_0(x) = \frac{x - x_{-1}}{x_0 - x_{-1}} \qquad \text{and} \qquad L_{-1}(x) = \frac{x - x_0}{x_{-1} - x_0}$$

(see Problem 6-5). Clearly, $\circledast$ is linear in x.

M8-14 Current i flowing through a resistor causes an increase in temperature, and this in turn increases its resistance R according to the formula $R = R_0 + bi^2$. When an E volt (V) battery is applied across such a resistor in series with an L henry (h) inductor at time $t = 0$, the circuit current $i = i(t)$ in milliamperes (ma) after t milliseconds (ms) satisfies

$$L\frac{di}{dt} = E - R_0 i - bi^3, \quad i(0) = 0$$

Suppose that $L = 4$ h, $R_0 = 1.2$ kilohms [KΩ], and $b = 0.4 \cdot 10^{-7}$ KΩ/ma^2. Use a method of your choice to find $i(t)$ to about 4s for $0 \leqslant t \leqslant 1$ if E is **(a)** 1.2 V; **(b)** 12 V; **(c)** 120 V.

M8-15 A three-loop *RLC* circuit has two capacitors with voltages v_1 and v_2 and one inductor with current i. Find v_1, v_2, and i at $t = \pi/2$ if

$$dv_1/dt = -v_1 - i - 2\sin t, \quad v_1(0) = 0 \text{ V}$$
$$dv_2/dt = -v_2 + i, \quad v_2(0) = 2 \text{ V}$$
$$di/dt = (v_1 - v_2)/2, \quad i(0) = 1 \text{ ma}$$

COMPUTER PROBLEMS

C8-1 Write and test a subprogram as indicated to integrate (IVP): $y' = f(t, y)$, $y(t_0) = y_0$ (the initial `Y`) from t_0 (the initial `T`) to t_N in *Nsteps* steps. If $Nprint > 0$, it should output j, and y_j every *Nprint* steps.

(a) `RK2(METHOD, F, T, TN, Y, NSTEPS, NPRINT)` that uses Huen's method if `METHOD` = 1 and the Modified Euler method if `METHOD` = 2. Test using Example 8.1D(b).

(b) `TAYLOR(N, F, TO, TN, YO, NSTEPS, NPRINT)` that uses the nth order Taylor's method, where n (= `N`) can be 1, 2, or 3. Test using Examples 8.1B and 8.2C.

C8-2 Write a subprogram `APC4(F, T, TN, Y, NSTEPS, NPRINT)` that solves (IVP) as described in Problem C8-1 using the Fourth-Order Adams Predictor–Corrector method (Algorithm 8.3A). It should call a subprogram such as `RK4` of Figure 8.2-1 to implement the **Start** phase. Test using Example 8.3A with $MaxIt = 1$.

NOTE: In the remaining problems, IVP_A–IVP_F refer to the IVP's given (with h) before Problem 8-1, and IVP_G and IVP_H refer to the IVP_2's given (with h) before Problem 8-28.

C8-3 For IVP_A–IVP_F of your choice and any available *fixed-stepsize* IVP-solver:

(a) Estimate $y(t_0 + 10h)$ using 10 h-steps, then 20 $(h/2)$-steps.

(b) If $E[h]$ and $E[h/2]$ at $t_0 + 10h$, confirm the order of the method, use Richardson's formula, and discuss the improvement obtained; if not, try to explain what went wrong.

C8-4 Assuming you did not know that $y(t)$ for IVP_F had a singularity at $t_s = \frac{1}{2}$, use any available computer program to try to estimate t_s to 4s.

C8-5 Write a calling program for the Fortran 77 `SUBROUTINE RKF4` of Figure 8.4-1 or an equivalent subprogram in another language of your choice. Test as in Figures 8.2-2 and 8.2-3.

C8-6 **(a)** For IVP_A–IVP_H of your choice, describe the input data and write a subprogram `EFALF` in order to use your calling program in Problem C8-5 to integrate from t_0 to $t_0 + 20h$.

(b) Execute your code in part (a). Are there significant differences between the numerical solution and the exact solution (if given)? If so, try to explain them. Does taking 40 $(h/2)$-steps help?

C8-7 Same as Problem C8-6 but for the $(IVP)_3$ in (1) of Section 8.4A with $h = 0.1$.

C8-8 Use any available $(IVP)_n$-solver to solve to 6s on the indicated interval.

(a) The third-order IVP in Problem 8-34(a) on [0, 2].

(b) The fourth-order IVP in Problem 8-34(b) on [0, 1].

(c) The second-order IVP in Problem 8-34(c) on [1, 4].

(d) The $(IVP)_4$ obtained in Problem 8-35 on $[0, \pi/2]$. Use the "normalized" values $G = 1$, $x_0 = 1$, $y_0 = 0$, $v_{0x} = 0$, $v_{0y} = 1$ (for which the exact solution is $x = \cos t$, $y = \sin t$).

C8-9 **(a)** Use any available $(IVP)_n$-solver to solve BVP_I–BVP_L of Problem 8-39 to 6s by the Shooting Method.

(b) Use any necessary available software to solve BVP_I–BVP_L of Problem 8-39 to 6s by the Finite Difference Method.

C8-10 Write a SUBROUTINE BKEULR(F, T, TN, Y, NSTEPS, NPRINT, NUMSIG) for the Backward Euler method, where F, T, TN, Y, NSTEPS, NPRINT, NUMSIG are as in Problem C8-1. For each j, use the Secant Method, with initial guesses $y_{j+1}^{(E)} = y_j + hf_j$ and $1.01y_{j+1}^{(E)}$, to solve (33b) of Section 8.2F for y_{j+1} to *NumSig* significant digits. Test using Problem M8-3 on $[0, 0.9e^{-y_0}]$ for $y_0 = -1$ and with $y_0 = 10$. Do your results confirm your answers to Problem M8-3?

C8-11 Create SUBROUTINE IMTRAP modifying SUBROUTINE BKEULR of Problem C8-10 so that it implements the implicit Trapezoidal Rule formula (39) of Section 8.2F. Test as in Problem C8-10.

C8-12 Create SUBROUTINE BKEUL2 by modifying SUBROUTINE BKEULR of Problem C8-10 so that it can solve a stiff IVP_2. Use NR_2 (see Problem C4-10) with $\mathbf{y}_{j+1}^{(E)}$ as an initial guess to get $\mathbf{y}_{j+1}$ to *NumSig* significant digits. Test using the stiff $(IVP)_2$

$$x' = -82x - 54y + 5, \quad x(0) = 5$$
$$y' = 108x + 71y + 4, \quad y(0) = 3$$

whose exact solution is $x(t) = 2e^{-t} + 3e^{-10t} + 5t$, $y(t) = -3e^{-t} - 4e^{-10t} + 4t + 10$.

C8-13 Create SUBROUTINE BKEULN by modifying SUBROUTINE BKEUL2 of Problem C8-12 (see Problem C4-11) so that it can solve a stiff IVP_n for $1 \leq n$ (= N) ≤ 10. Test using the stiff third-order IVP

$$y''' + 9.9y'' - y' = -1, \; y(0) = 0, \; y'(0) = -8.6, \; y''(0) = 100.04$$

whose exact solution is $y(t) = e^{-10t} + 4e^{.1t} + t - 5$.

9

Eigenvalues

9.0

Introduction

In this final chapter we present the basic properties of *eigenvalues* λ and *eigenvectors* $\mathbf{v}$ of an $n \times n$ matrix A, and we show how they help us gain a geometric understanding of the effect of multiplying $\mathbf{x}$ by A. We also show how they can be used to describe the solution of the linear first-order system

$$\frac{d\mathbf{y}}{dt} = A\mathbf{y}, \qquad \mathbf{y}(t_0) = \mathbf{y}_0, \qquad (A_{n \times n} = \text{constant})$$

and how they can help us find the *characteristic values* λ for which there exist nontrivial solutions of boundary value problems such as

$$\frac{d^2y}{dt^2} + \lambda y = 0, \qquad y(0) = y(L) = 0$$

which arise in studying axially loaded beams.

It will be assumed that all entries of matrices A are real numbers. However, eigenvalues λ may be complex.

The reader is expected to be thoroughly familiar with the basic algebraic properties of matrices (Section 3.1) and determinants (Section 3.5). In addition, the reader will need to know the following basic results about linear independence in n-space [2]:

1. The vectors $\mathbf{v}_1, \ldots, \mathbf{v}_k$ are called **linearly independent** if

$$\alpha_1 \mathbf{v}_1 + \alpha_2 \mathbf{v}_2 + \cdots + \alpha_k \mathbf{v}_k = \mathbf{0}$$

can hold only if $\alpha_1 = \cdots = \alpha_k = 0$; otherwise, they are **linearly dependent**.

2. Any n linearly independent vectors $\mathbf{v}_1, \ldots, \mathbf{v}_n$ are a **basis** for n-space, by which we mean that every $\mathbf{x}$ in n-space can be expressed in only *one way* as the linear combination

$$\mathbf{x} = \alpha_1 \mathbf{v}_1 + \alpha_2 \mathbf{v}_2 + \cdots + \alpha_n \mathbf{v}_n$$

These *unique scalars* $\alpha_1, \alpha_2, \ldots, \alpha_n$ are called the **components of x** with respect to the basis $\{\mathbf{v}_1, \ldots, \mathbf{v}_n\}$.

3. An $n \times n$ matrix V is nonsingular if and only if its columns $\mathbf{v}_1, \ldots, \mathbf{v}_n$ are linearly independent, hence form a basis for n-space.

9.1

Basic Properties of Eigenvalues and Eigenvectors

9.1A The Characteristic Polynomial $p_A(\lambda)$

We wish to find scalars λ for which there exist *nonzero* vectors $\mathbf{v}$ such that

$$A\mathbf{v} = \lambda \mathbf{v} \qquad (A \text{ multiplies } \mathbf{v} \text{ into a multiple of itself}) \tag{1}$$

When this occurs, we call λ an **eigenvalue**† for A, $\mathbf{v}$ an **eigenvector** for λ, and $(\lambda, \mathbf{v})$ an **eigenpair** for A. It is immediate from (1) that

$$\text{if } \mathbf{v} \text{ is an eigenvector for } \lambda, \text{ so is } \alpha\mathbf{v} \text{ for any } \alpha \neq 0$$

Eigenpairs $(\lambda, \mathbf{v})$ and $(\lambda, \alpha\mathbf{v})$ will not be considered as different.

Clearly, $\mathrm{O}\mathbf{x} = 0\mathbf{x}$ and $I_n\mathbf{x} = 1\mathbf{x}$ for *any* $\mathbf{x}$; so $(0, \mathbf{x})$ is an eigenpair for $\mathrm{O}_{n\times n}$ and $(1, \mathbf{x})$ is an eigenpair for I_n for *any* nonzero $\mathbf{x}$. To see that eigenpairs must exist for any $n \times n$ matrix A, note that

$$A\mathbf{v} = \lambda\mathbf{v}, \quad \mathbf{v} \neq \mathbf{0} \quad \Leftrightarrow \quad (A - \lambda I_n)\mathbf{v} = \mathbf{0}, \quad \mathbf{v} \neq \mathbf{0} \tag{2a}$$

So *eigenpairs are possible if and only if* $(A - \lambda I_n)$ *is singular*, that is,

$$p_A(\lambda) = \det(A - \lambda I_n) = 0 \tag{2b}$$

†The prefix "eigen" is German for "self." Some authors use **characteristic value** for eigenvalue, and **characteristic vector** for eigenvector.

It is immediate from the definition of det that $p_A(\lambda)$ is a polynomial in λ whose leading term, $(-1)^n\lambda^n$, comes from the diagonal product $\Pi_{k=1}^n (a_{kk} - \lambda)$ [see (2) of Section 3.5A]. We call $p_A(\lambda)$ the **characteristic polynomial** of A. Since $p_A(\lambda)$ has degree n, it follows from (2b) and the Fundamental Theorem of Algebra that *every $n \times n$ matrix A has exactly n (possibly repeated and possibly complex) eigenvalues, namely the roots of $p_A(\lambda)$*. If we denote them by $\lambda_1, \ldots, \lambda_n$, then $p_A(\lambda)$ can be factored as

$$p_A(\lambda) = (-1)^n(\lambda - \lambda_1)(\lambda - \lambda_2) \cdots (\lambda - \lambda_n) = \prod_{j=1}^{n} (\lambda_j - \lambda) \tag{3a}$$

Upon taking $\lambda = 0$ in (2b) and (3a), we see that

$$\det A = \lambda_1\lambda_2 \cdots \lambda_n = \text{constant term of } p_A(\lambda) \tag{3b}$$

So *A is invertible if and only if all eigenvalues are nonzero,* in which case multiplying (1) by $\lambda^{-1}A^{-1}$ gives $A^{-1}\mathbf{v} = (1/\lambda)\mathbf{v}$. Thus

$$(\lambda, \mathbf{v}) \text{ is an eigenpair for } A \Leftrightarrow \left(\frac{1}{\lambda}, \mathbf{v}\right) \text{ is an eigenpair for } A^{-1} \tag{4}$$

In theory, all eigenpairs for A can be found by first solving $p_A(\lambda) = 0$ and then finding nontrivial solutions of $(A - \lambda I_n)\mathbf{v} = \mathbf{0}$.

EXAMPLE 9.1A Find all eigenpairs for

$$\textbf{(a) } A = \begin{bmatrix} 3 & 0 & 1 \\ 0 & -3 & 0 \\ 1 & 0 & 3 \end{bmatrix} \qquad \textbf{(b) } A^{-1} = \begin{bmatrix} \frac{3}{8} & 0 & -\frac{1}{8} \\ 0 & -\frac{1}{3} & 0 \\ -\frac{1}{8} & 0 & \frac{3}{8} \end{bmatrix} \tag{5}$$

SOLUTION (a): We first use the "arrow rule" to get

$$p_A(\lambda) = \det(A - \lambda I) = \begin{vmatrix} 3-\lambda & 0 & 1 \\ 0 & -3-\lambda & 0 \\ 1 & 0 & 3-\lambda \end{vmatrix}$$
$$= (3-\lambda)^2(-3-\lambda) + (3+\lambda) = -(3+\lambda)[(3-\lambda)^2 - 1]$$
$$= -(3+\lambda)[\lambda^2 - 6\lambda + 8] = -(\lambda+3)(\lambda-2)(\lambda-4)$$

So the eigenvalues of A, in order of decreasing magnitude, are

$$\lambda_1 = 4, \qquad \lambda_2 = -3, \qquad \lambda_3 = 2 \tag{6a}$$

To find an eigenvector for $\lambda_1 = 4$, we must solve $(A - 4I)\mathbf{v} = \mathbf{0}$ for a *nonzero* $\mathbf{v} = [v_1, \ v_2 \ \ v_3]^T$. Using Basic Gaussian elimination, we get

$$[A - 4I \vdots \mathbf{0}] = \begin{bmatrix} -1 & 0 & 1 & : & 0 \\ 0 & -7 & 0 & : & 0 \\ 1 & 0 & -1 & : & 0 \end{bmatrix} \to \begin{bmatrix} -1 & 0 & 1 & : & 0 \\ 0 & -7 & 0 & : & 0 \\ 0 & 0 & 0 & : & 0 \end{bmatrix} \Leftrightarrow \begin{matrix} v_1 = v_3 \\ v_2 = 0 \end{matrix}$$

Solutions of $A\mathbf{v} = 4\mathbf{v}$ are therefore of the form $\mathbf{v} = [\alpha \quad 0 \quad \alpha]^T = \alpha[1 \quad 0 \quad 1]^T$, where $\alpha \neq 0$ is arbitrary. Similarly, the eigenvectors for $\lambda_2 = -3$ and $\lambda_3 = 2$ are scalar multiples of $\mathbf{v}_2 = [0 \quad 1 \quad 0]^T$ and $\mathbf{v}_3 = [1 \quad 0 \quad -1]^T$, respectively. So a set of eigenvectors for the eigenvalues in (6a) are, respectively,

$$\mathbf{v}_1 = \begin{bmatrix} 1 \\ 0 \\ 1 \end{bmatrix}, \qquad \mathbf{v}_2 = \begin{bmatrix} 0 \\ 1 \\ 0 \end{bmatrix}, \qquad \mathbf{v}_3 = \begin{bmatrix} 1 \\ 0 \\ -1 \end{bmatrix} \tag{6b}$$

Solution (b): Rather than proceed as in part (a), we simply use (4) and (6) to get *by inspection*:

$$(\tfrac{1}{4}, \mathbf{v}_1), \quad (-\tfrac{1}{3}, \mathbf{v}_2), \quad \text{and} \quad (\tfrac{1}{2}, \mathbf{v}_3) \quad \text{are the eigenpairs of } A^{-1}$$

The reader is urged to verify that $A^{-1}\mathbf{v}_i = (1/\lambda_i)\mathbf{v}_i$ for $i = 1, 2, 3$. ■

The method of finding eigenvalues directly as roots of $p_A(\lambda)$ is impractical for large n because of the combined effects of the ill-conditioned nature of root finding (Section 2.4A) and the roundoff error that is inevitable in calculating $p_A(\lambda)$. Alternative methods will be described in Sections 9.2 and 9.3. To help understand these methods, we first describe the role of *similarity transformations* in finding eigenpairs.

9.1B Similar Matrices and Diagonalizability

Of special importance for the study of eigenvalues are **diagonal matrices**

$$D = \operatorname{diag}(\lambda_1, \lambda_2, \ldots, \lambda_n) = \begin{bmatrix} \lambda_1 & & & \\ & \lambda_2 & & \\ & & \ddots & \\ & & & \lambda_n \end{bmatrix} \quad \left(\begin{matrix}\text{off-diagonal} \\ \text{entries are } 0\end{matrix}\right) \tag{7a}$$

It is easy to see that for $j = 1, 2, \ldots, n$,

$$D\mathbf{e}_j = \lambda_j \mathbf{e}_j, \qquad \text{where } \mathbf{e}_j = \operatorname{col}_j I_n \tag{7b}$$

Thus, by *inspection*, the eigenpairs of $D = \operatorname{diag}(\lambda_1, \ldots, \lambda_n)$ are $(\lambda_1, \mathbf{e}_1), \ldots, (\lambda_n, \mathbf{e}_n)$. This simple but useful result makes it desirable to find ways to transform a general $n \times n$ matrix A into a diagonal matrix having the same eigenvalues. Unfortunately, the 0_{ij}-subtracts that can be used to reduce $A \to D$ are *not* suitable because they alter eigenvalues (Problem 9-5). What is needed are *similarity transformations* $A \to V^{-1}AV$, described next.

Two $n \times n$ matrices A and B are called **similar** if there exists a nonsingular matrix V such that $B = V^{-1}AV$ (or, equivalently, $A = VBV^{-1}$). In this case,

$$A\mathbf{v} = \lambda\mathbf{v} \quad \Leftrightarrow \quad VBV^{-1}\mathbf{v} = \lambda\mathbf{v} \quad \Leftrightarrow \quad B(V^{-1}\mathbf{v}) = \lambda(V^{-1}\mathbf{v})$$

So *similar matrices have identical eigenvalues*; in fact,

$$(\lambda, \mathbf{v}) \text{ is an eigenpair for } A \Leftrightarrow (\lambda, V^{-1}\mathbf{v}) \text{ is an eigenpair for } V^{-1}AV \tag{8}$$

An $n \times n$ matrix A is called **diagonalizable** if it is similar to a diagonal matrix. The fundamental diagonalizability result is the following:

DIAGONALIZATION THEOREM. *Let A be an $n \times n$ matrix. Then*

$$V^{-1}AV = D = \text{diag}(\lambda_1, \ldots, \lambda_n), \qquad \textit{where } V = [\mathbf{v}_1 \;:\; \mathbf{v}_2 \;:\; \cdots \;:\; \mathbf{v}_n] \tag{9a}$$

*can hold if and only if the columns $\mathbf{v}_1, \ldots, \mathbf{v}_n$ of V are eigenvectors (for $\lambda_1, \ldots, \lambda_n$, respectively), which form a basis for n-space. This means that every n-vector **x** can be expressed uniquely as the linear combination*

$$\mathbf{x} = \alpha_1\mathbf{v}_1 + \alpha_2\mathbf{v}_2 + \cdots + \alpha_n\mathbf{v}_n, \qquad \textit{where } A\mathbf{v}_j = \lambda_j\mathbf{v}_j \;\; \textit{for } j = 1, \ldots, n \tag{9b}$$

If (9a) holds, then we know from linear algebra that the columns of the nonsingular matrix V are linearly independent, and hence are a basis. Also, since $VDV^{-1} = A$ and $D\mathbf{e}_j = \lambda_j\mathbf{e}_j$, we know from (8) that

$$A(V\mathbf{e}_j) = \lambda_j(V\mathbf{e}_j)$$

But $V\mathbf{e}_j = \text{col}_jV = \mathbf{v}_j$; so (9b) $\Rightarrow$ (9a). Conversely, if (9b) holds, we use $\mathbf{v}_1, \ldots, \mathbf{v}_n$ to form $V = [\mathbf{v}_1 \;:\; \cdots \;:\; \mathbf{v}_n]$; then

$$V^{-1}AV\mathbf{e}_j = V^{-1}A\mathbf{v}_j = V^{-1}(\lambda_j\mathbf{v}_j) = \lambda_jV^{-1}(V\mathbf{e}_j) = \lambda_j\mathbf{e}_j = \text{col}_jD$$

Since $V^{-1}AV\mathbf{e}_j = \text{col}_j(V^{-1}AV)$, we see that (9b) implies (9a), completing the proof of the Diagonalization Theorem.

It follows from the Diagonalization Theorem that *every matrix having distinct, real eigenvalues is diagonalizable* [Problem M9-2(b)]. However, there are matrices with repeated eigenvalues that nevertheless are diagonalizable (Problem M9-3). To illustrate, let us use the eigenvectors (6b) to form

$$V = [\mathbf{v}_1 \;:\; \mathbf{v}_2 \;:\; \mathbf{v}_3] = \begin{bmatrix} 1 & 0 & 1 \\ 0 & 1 & 0 \\ 1 & 0 & -1 \end{bmatrix} \qquad \text{for which } V^{-1} = \begin{bmatrix} \frac{1}{2} & 0 & \frac{1}{2} \\ 0 & 1 & 0 \\ \frac{1}{2} & 0 & -\frac{1}{2} \end{bmatrix} \tag{10a}$$

The reader should verify that $VV^{-1} = I_3$ and that

$$\underbrace{\begin{bmatrix} \frac{1}{2} & 0 & \frac{1}{2} \\ 0 & 1 & 0 \\ \frac{1}{2} & 0 & -\frac{1}{2} \end{bmatrix}}_{V^{-1}} \underbrace{\begin{bmatrix} 3 & 0 & 1 \\ 0 & -3 & 0 \\ 1 & 0 & 3 \end{bmatrix}}_{A} \underbrace{\begin{bmatrix} 1 & 0 & 1 \\ 0 & 1 & 0 \\ 1 & 0 & -1 \end{bmatrix}}_{V} = \underbrace{\begin{bmatrix} 4 & 0 & 0 \\ 0 & -3 & 0 \\ 0 & 0 & 2 \end{bmatrix}}_{\text{diag}(4,\; -3,\; 2)} \tag{10b}$$

Had we arrived at (10b) without a priori knowledge of the eigenpairs of A, we could have used the Diagonalization Theorem to deduce that

$$A\mathbf{v}_1 = 4\mathbf{v}_1, \quad A\mathbf{v}_2 = -3\mathbf{v}_2, \quad \text{and} \quad A\mathbf{v}_3 = 2\mathbf{v}_3 \tag{10c}$$

EXAMPLE 9.1B Show that the matrix $A = \begin{bmatrix} 5 & 4 \\ -1 & 1 \end{bmatrix}$ is not diagonalizable.

SOLUTION: We first get $p_A(\lambda) = \det(A - \lambda I_2)$.

$$p_A(\lambda) = \begin{bmatrix} 5-\lambda & 4 \\ -1 & 1-\lambda \end{bmatrix} = (5-\lambda)(1-\lambda) + 4 = \lambda^2 - 6\lambda + 9 = (\lambda - 3)^2$$

So $\mathbf{v} = [v_1 \quad v_2]^T \neq \mathbf{0}$ is an eigenvector of A only if $(A - 3I_2)\mathbf{v} = \mathbf{0}$, i.e.

$$\begin{bmatrix} 2 & 4 \\ -1 & -2 \end{bmatrix}\begin{bmatrix} v_1 \\ v_2 \end{bmatrix} = \begin{bmatrix} 0 \\ 0 \end{bmatrix} \quad \Leftrightarrow \quad v_1 = -2v_2 \quad \Leftrightarrow \quad \mathbf{v} = \alpha\begin{bmatrix} 2 \\ -1 \end{bmatrix}$$

Thus all eigenvectors for A are nonzero scalar multiples of $[2 \quad -1]^T$. Since *two* linearly independent vectors are needed for a basis, there cannot be a basis of eigenvectors for A. By the Diagonalization Theorem, A is not diagonalizable. ■

9.1C Using Eigenvectors to Uncouple Linear IVP's

Of all $(\text{IVP})_n$'s, the most frequently encountered is the *linear* first-order system

$$\begin{aligned} \frac{dy_1}{dt} &= a_{11}y_1 + a_{12}y_2 + \cdots + a_{1n}y_n, \qquad y_1(t_0) = y_{1,0} \\ &\ \ \vdots \qquad\qquad\qquad\qquad\qquad\qquad\qquad\qquad \vdots \\ \frac{dy_n}{dt} &= a_{n1}y_1 + a_{n2}y_2 + \cdots + a_{nn}y_n, \qquad y_n(t_0) = y_{n,0} \end{aligned} \tag{11a}$$

in which the a_{ij}'s are constant. By letting

$$\mathbf{y} = \begin{bmatrix} y_1 \\ \vdots \\ y_n \end{bmatrix}, \quad \mathbf{y}' = \frac{d\mathbf{y}}{dt} = \begin{bmatrix} y_1' \\ \vdots \\ y_n' \end{bmatrix}, \quad \text{and} \quad \mathbf{y}_0 = \begin{bmatrix} y_{1,0} \\ \vdots \\ y_{n,0} \end{bmatrix} \tag{11b}$$

this system can be expressed in matrix form as

$$\frac{d\mathbf{y}}{dt} = A\mathbf{y}, \quad \mathbf{y}(t_0) = \mathbf{y}_0, \qquad \text{where } A = (a_{ij})_{n\times n} \text{ is constant} \tag{11c}$$

Suppose that $V^{-1}AV = D = \operatorname{diag}(\lambda_1, \ldots, \lambda_n)$. Then the change of variables

$$\mathbf{y} = V\mathbf{z} \quad \text{satisfies} \quad \frac{d\mathbf{y}}{dt} = V\frac{d\mathbf{z}}{dt} \tag{12}$$

because the V entries are constant. Hence (11c) can be written as

$$V\frac{d\mathbf{z}}{dt} = A(V\mathbf{z}), \qquad V\mathbf{z}(t_0) = \mathbf{y}_0$$

But $V^{-1}AV = D$; so upon premultiplying by V^{-1}, we get

$$\frac{d\mathbf{z}}{dt} = D\mathbf{z}, \quad \mathbf{z}(t_0) = \mathbf{z}_0, \qquad \text{where } \mathbf{z}_0 = V^{-1}\mathbf{y}_0 = [z_{1,0} \cdots z_{n,0}]^T \tag{13a}$$

The equations represented by this *diagonal* system are

$$\begin{aligned} \frac{dz_1}{dt} &= \lambda_1 z_1, & z_1(t_0) &= z_{1,0} \\ &\vdots & &\vdots \\ \frac{dz_n}{dt} &= \lambda_n z_n, & z(t_0) &= z_{n,0} \end{aligned} \tag{13b}$$

Since dz_i/dt depends *only* on z_i, the solution of this "uncoupled" system is

$$\begin{aligned} z_1(t) &= z_{1,0}e^{\lambda_1(t-t_0)} \\ &\vdots \\ z_n(t) &= z_{n,0}e^{\lambda_n(t-t_0)} \end{aligned} \qquad \text{that is, } \mathbf{z}(t) = \begin{bmatrix} z_{1,0}e^{\lambda_1(t-t_0)} \\ \vdots \\ z_{n,0}e^{\lambda_n(t-t_0)} \end{bmatrix} \tag{14}$$

It follows from this and (12) that the solution of (11) can be expressed in terms of the columns $\mathbf{v}_1, \ldots, \mathbf{v}_n$ of V. Specifically,

$$\mathbf{y}(t) = V\mathbf{z}(t) = z_{1,0}e^{\lambda_1(t-t_0)}\mathbf{v}_1 + \cdots + z_{n,0}e^{\lambda_n(t-t_0)}\mathbf{v}_n, \qquad \text{where } \mathbf{z}_0 = V^{-1}\mathbf{y}_0 \tag{15}$$

[see Problem 3-5(b)]. In words, if A in (11) is diagonalizable, then the solution of (11) is a weighted sum of the eigenvectors of A in which the "weight" of $\mathbf{v}_i$ is the exponential $z_{i,0}e^{(t-t_0)}$, where $A\mathbf{v}_i = \lambda_i\mathbf{v}_i$ and $\mathbf{z}_0 = V^{-1}\mathbf{y}_0$. For example, to find the solution of the linear $(\text{IVP})_3$

$$\frac{d}{dt}\begin{bmatrix} y_1 \\ y_2 \\ y_3 \end{bmatrix} = \underbrace{\begin{bmatrix} 3 & 0 & 1 \\ 0 & -3 & 0 \\ 1 & 0 & 3 \end{bmatrix}}_{A}\begin{bmatrix} y_1 \\ y_2 \\ y_3 \end{bmatrix}, \qquad \mathbf{y}(0) = \begin{bmatrix} 0 \\ -2 \\ 2 \end{bmatrix} = \mathbf{y}_0 \tag{16a}$$

we can refer to (10b) where we saw that $V^{-1}AV = \text{diag}\,(4, -3, 2)$, where

$$V = [\mathbf{v}_1 \ : \ \mathbf{v}_2 \ : \ \mathbf{v}_3] = \begin{bmatrix} 1 & 0 & 1 \\ 0 & 1 & 0 \\ 1 & 0 & -1 \end{bmatrix} \quad \text{and} \quad V^{-1} = \begin{bmatrix} \frac{1}{2} & 0 & \frac{1}{2} \\ 0 & 1 & 0 \\ \frac{1}{2} & 0 & -\frac{1}{2} \end{bmatrix} \tag{16b}$$

Since $V^{-1}\mathbf{y}_0 = V^{-1}[0 \quad -2 \quad 2]^T = [1 \quad -2 \quad -1]^T$, we can use (15) with $t_0 = 0$ to express the solution of (16a) as

$$\mathbf{y}(t) = (1)e^{4t}\begin{bmatrix} 1 \\ 0 \\ 1 \end{bmatrix} + (-2)e^{-3t}\begin{bmatrix} 0 \\ 1 \\ 0 \end{bmatrix} + (-1)e^{2t}\begin{bmatrix} 1 \\ 0 \\ -1 \end{bmatrix} = \begin{bmatrix} e^{4t} - e^{2t} \\ -2e^{-3t} \\ e^{4t} + e^{2t} \end{bmatrix} \tag{16c}$$

Notice that for large t, $\mathbf{y}(t) \approx e^{4t}\mathbf{v}_1$. More generally, we see from (15) that if $\lambda_1 > |\lambda_i|$ for $i > 1$, then $\mathbf{y}(t) \approx z_{1,0}e^{\lambda_1 t}\mathbf{v}_1$ for large t, provided that $z_{1,0} \neq 0$. This is why it is often important to know only the *dominant* eigenvalue of A (i.e., the one of largest magnitude) and an associated eigenvector. ■

NOTE: Recall that a differential equation is *stiff* if its solution has terms $e^{\lambda_1 t}$ and $e^{\lambda_n t}$, where $|\lambda_1| >> |\lambda_n|$ [see Section 8.2F]. In view of (15), we see that the *linear* $(\text{IVP})_n$

$$\mathbf{y}' = A\mathbf{y}, \qquad \mathbf{y}(t_0) = \mathbf{y}_0$$

will be stiff whenever it has eigenvalues of *both* large *and* small magnitude. Thus (16a) is *not* stiff. For the general nonlinear $(\text{IVP})_n$,

$$\mathbf{y}' = \mathbf{f}(t, \mathbf{y}), \qquad \mathbf{y}(t_0) = \mathbf{y}_0$$

stiffness can be determined by examining the Jacobian matrix $\partial\mathbf{f}(t, \mathbf{y})/\partial\mathbf{y} = (\partial f_i(t, \mathbf{y})/\partial y_j)_{n\times n}$. Further details can be found in [22].

It should be apparent from (15) that the eigenvalues and eigenvectors of A are intimately related to the "natural responses" of linear systems described by $\mathbf{y}' = A\mathbf{y}$, where A is constant. A specific illustration is given next.

EXAMPLE 9.1C For $t \geq 0$, the current i in the "lumped parameter" RLC circuit shown in Figure 9.1-1 satisfies the second-order linear IVP

$$L\frac{d^2i}{dt^2} + R\frac{di}{dt} + \frac{1}{C}i = 0, \qquad i(0) = \frac{E}{R}, \quad i'(0) = 0 \tag{17a}$$

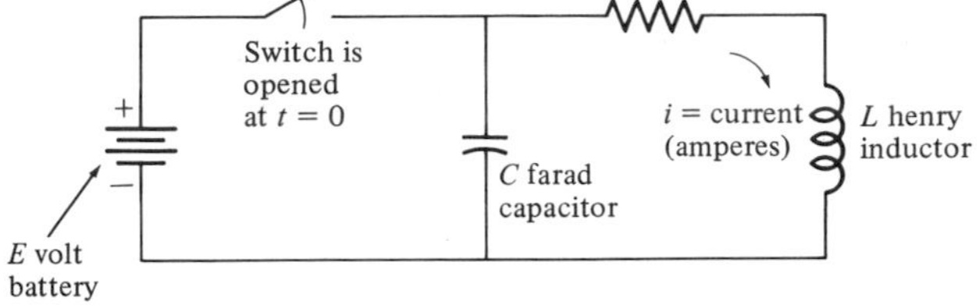

Figure 9.1-1 *RLC-circuit.*

Dividing by L, we can rewrite the equation as

$$\frac{d^2i}{dt^2} + 2\beta\frac{di}{dt} + \gamma i = 0, \qquad \text{where } \beta = \frac{R}{2L} \text{ and } \gamma = \frac{1}{LC} \tag{17b}$$

If we introduce the variables y_1 and y_2, where

$$y_1 = i(t), \qquad y_2 = \frac{di}{dt}, \qquad \text{that is, } \mathbf{y} = \begin{bmatrix} i \\ \dfrac{di}{dt} \end{bmatrix} \tag{18a}$$

then the equation (17b) can be written as the linear $(\text{IVP})_2$

$$\left.\begin{aligned} \frac{dy_1}{dt} &= y_2, & y_1(0) &= \frac{E}{R} \\ \frac{dy_2}{dt} &= -\gamma y_1 - 2\beta y_2, & y_2(0) &= 0 \end{aligned}\right\} \Leftrightarrow \frac{d\mathbf{y}}{dt} = \begin{bmatrix} 0 & 1 \\ -\gamma & -2\beta \end{bmatrix}\mathbf{y}, \quad \mathbf{y}(0) = \begin{bmatrix} \dfrac{E}{R} \\ 0 \end{bmatrix} \tag{18b}$$

The characteristic polynomial of the coefficient matrix A is

$$p_A(\lambda) = \begin{vmatrix} -\lambda & 1 \\ -\gamma & -2\beta - \lambda \end{vmatrix} = \lambda(\lambda + 2\beta) + \gamma = \lambda^2 + 2\beta\lambda + \gamma \tag{19}$$

Hence by the quadratic formula, the eigenvalues of A are

$$\lambda_1 = -\beta + \sqrt{\beta^2 - \gamma} = \frac{-R}{2L} + \sqrt{\left(\frac{R}{2L}\right)^2 - \frac{1}{LC}} \tag{20a}$$

$$\lambda_2 = -\beta - \sqrt{\beta^2 - \gamma} = \frac{-R}{2L} - \sqrt{\left(\frac{R}{2L}\right)^2 - \frac{1}{LC}} \tag{20b}$$

These will be real if what is under the radical in nonnegative, that is, if

$$\beta^2 - \gamma \geqslant 0 \Leftrightarrow \left(\frac{R}{2L}\right)^2 \geqslant \frac{1}{RC} \tag{21}$$

and a complex-conjugate pair otherwise. The circuit is called **overdamped, critically damped**, or **underdamped** according as $\beta^2 > \gamma$, $\beta^2 = \gamma$, or $\beta^2 > \gamma$, as shown in Table 9.1-1.

When the circuit is underdamped, $i(t)$ will "ring" (i.e., oscillate) with a **natural frequency** of $\omega = \sqrt{\gamma - \beta^2}$ hertz while decaying like $e^{-\beta t}$. In the limiting case when $R = 0$,

$$i(t) = \frac{E}{R}\cos\frac{t}{\sqrt{LC}}$$

TABLE 9.1-1 Possible Responses of Current in an *RLC* Circuit

Condition on $\beta = \frac{R}{2L}$ and $\gamma = \frac{1}{LC}$	Eigenvalues λ_1 and λ_2	Solution $i(t)$ = current in circuit
$\beta^2 > \gamma$ (overdamped)	$\lambda_1 = -\beta + \sqrt{\beta^2 - \gamma}$ $\lambda_2 = -\beta - \sqrt{\beta^2 - \gamma}$ (real, distinct, negative)	$i(t) = \frac{E}{R}\left(\frac{\lambda_1 e^{\lambda_2 t} - \lambda_2 e^{\lambda_1 t}}{\lambda_1 - \lambda_2}\right)$
$\beta^2 = \gamma$ (critically damped)	$\lambda_1 = \lambda_2 = -\beta$ (real, equal, negative)	$i(t) = \frac{E}{R}(1 + \beta t)e^{-\beta t}$
$\beta^2 < \gamma$ (underdamped)	$\lambda_1 = -\beta + i\omega$ $\lambda_2 = -\beta - i\omega$, $\omega = \sqrt{\beta - \beta^2}$ [complex conjugates, $Re(\lambda) < 0$]	$i(t) = \frac{E}{R}e^{-\beta t}\left(\cos \omega t + \frac{\beta}{\omega}\sin \omega t\right)$

that is, the circuit will ring indefinitely with a natural frequency of $\omega = 1/\sqrt{LC}$ hertz. Finally, given L and C, if R is adjusted for critical damping, then $|i(t)|$ will shrink to zero as rapidly as possible and without ringing (see Problem 9-9).

9.2 The Power Method

The **dominant** member of a set of numbers (e.g., eigenvalues of A, or components of $\mathbf{x}$) is the one of largest magnitude. For example, both $+4$ and -4 are dominant components of the vector $\mathbf{x} = [2 \quad -4 \quad 4 \quad -\frac{1}{2}]^T$; and $-\frac{1}{2}$ is the **least dominant**.

For the remainder of this chapter, eigenvalues will be indexed in order of decreasing magnitude, that is,

$$|\lambda_1| \geq |\lambda_2| \geq \cdots \geq |\lambda_n| \tag{1}$$

so that λ_1 will be dominant. We now describe a procedure for obtaining the **dominant eigenpair** $(\lambda_1, \mathbf{v}_1)$ when the following assumptions hold:

Assumption 1: *$|\lambda_1|$ is strictly greater than $|\lambda_i|$ for $i = 2, \ldots, n$.*

Assumption 2: *A has n eigenvectors $\mathbf{v}_1, \ldots, \mathbf{v}_n$ (where $A\mathbf{v}_i = \lambda_i\mathbf{v}_i$ for all i), which are a basis for n-space.*

9.2A The Power Method for Finding Dominant Eigenvalues

In its simplest form, the **Power Method** is the iterative algorithm defined by multiplying an initial guess $\mathbf{x}_0$ by successively higher powers of A:

$$\text{Choose } \mathbf{x}_0, \text{ then take } \mathbf{x}_{k+1} = A\mathbf{x}_k \quad \text{for } k = 0, 1, 2, \ldots \tag{2a}$$

Thus $\mathbf{x}_1 = A\mathbf{x}_0$, $\mathbf{x}_2 = A\mathbf{x}_1 = A^2\mathbf{x}_0$, $\mathbf{x}_3 = A\mathbf{x}_2 = A^3\mathbf{x}_0$, $\ldots$, so that

$$\mathbf{x}_k = A^k\mathbf{x}_0 \quad \text{for } k = 1, 2, 3, \ldots \tag{2b}$$

In view of Assumption 2, $\mathbf{x}_0$ can be represented uniquely as (hence we can choose any $n\times 1$ x_0) with real elements

$$\mathbf{x}_0 = \alpha_1\mathbf{v}_1 + \alpha_2\mathbf{v}_2 + \cdots + \alpha_n\mathbf{v}_n$$

But $A\mathbf{v}_i = \lambda_i\mathbf{v}_i$, hence $A^k\mathbf{v}_i = \lambda_i^k\mathbf{v}_i$ for $k \geqslant 1$. So in view of Assumption 1,

$$A^k\mathbf{x}_0 = \alpha_1\lambda_1^k\mathbf{v}_1 + \alpha_2\lambda_2^k\mathbf{v}_2 + \cdots + \alpha_n\lambda_n^k\mathbf{v}_n \tag{3a}$$

$$= \lambda_1^k\left[\alpha_1\mathbf{v}_1 + \alpha_2\left(\frac{\lambda_2}{\lambda_1}\right)^k\mathbf{v}_2 + \cdots + \alpha_n\left(\frac{\lambda_n}{\lambda_1}\right)^k\mathbf{v}_n\right] \tag{3b}$$

$$\approx \lambda_1^k\alpha_1\mathbf{v}_1 \quad \text{for large } k, \text{ provided that } \alpha_1 \neq 0 \tag{3c}$$

Since $\lambda_1^k\alpha_1\mathbf{v}_1$ is a scalar multiple of $\mathbf{v}_1$, $\mathbf{x}_k = A^k\mathbf{x}_0$ *will approach an eigenvector for the dominant eigenvalue* λ_1 (i.e., $A\mathbf{x}_k \approx \lambda_1\mathbf{x}_k$); so if $\mathbf{x}_k$ is scaled so that its dominant component is 1, then

$A^k x_0 = \lambda_1^k \alpha_1 v_1 = x_k \Rightarrow A x_k = \lambda_1(\lambda_1^k \alpha_1 v_1)$

$$(\text{dominant component of } A\mathbf{x}_k) \approx \lambda_1 \cdot (\text{dominant component of } \mathbf{x}_k) = \lambda_1 \tag{3d}$$

$\approx A v_1 = \lambda_1 v_1$

The **Scaled Power Method** described in Algorithm 9.2A is based on (3).

The ***Scale*** step sets the dominant component of the current $\mathbf{x}$ to 1. So when **termination test** is satisfied, it follows from (3d) that the current *BigXi* approximates the dominant eigenvalue λ_1, and from (3c) that the current $\mathbf{x}$ approximates an eigenvector for λ_1.

EXAMPLE 9.2A Perform four iterations of the Scaled Power Method for the matrix A of Example 9.1A. Start with $\mathbf{x}_0 = [0 \quad 1 \quad 2]^T$.

SOLUTION: Using the notation $A\mathbf{x}_k = BigXi\,\mathbf{x}_{k+1}$ we get

$$\mathbf{k = 0:}\quad A\mathbf{x}_0 = \begin{bmatrix} 3 & 0 & 1 \\ 0 & -3 & 0 \\ 1 & 0 & 3 \end{bmatrix}\begin{bmatrix} 0 \\ 1 \\ 2 \end{bmatrix} = \begin{bmatrix} 2 \\ -3 \\ 6 \end{bmatrix} = 6\begin{bmatrix} \frac{1}{3} \\ -\frac{1}{2} \\ 1 \end{bmatrix} \quad (= BigXi\,\mathbf{x}_1)$$

$$k = 1:\quad A\mathbf{x}_1 = \begin{bmatrix} 3 & 0 & 1 \\ 0 & -3 & 0 \\ 1 & 0 & 3 \end{bmatrix}\begin{bmatrix} \frac{1}{3} \\ -\frac{1}{2} \\ 1 \end{bmatrix} = \begin{bmatrix} 2 \\ \frac{3}{2} \\ \frac{10}{3} \end{bmatrix} = \frac{10}{3}\begin{bmatrix} \frac{3}{5} \\ \frac{9}{20} \\ 1 \end{bmatrix} \quad (= BigXi\,\mathbf{x}_2)$$

ALGORITHM 9.2A. SCALED POWER METHOD

PURPOSE: To find the dominant eigenpair $(\lambda_1, \mathbf{v}_1)$ of a given $n \times n$ matrix A to *NumSig* significant digits.

{**Initialize**}
GET n, A, $\mathbf{x}_0$, {$\mathbf{x}_0$ is an initial nonzero guess of $\mathbf{v}_1$}
NumSig, *MaxIt* {termination parameters}
$Tol \leftarrow 10^{-NumSig}$; $\mathbf{x} \leftarrow \mathbf{x}_0$ {$\mathbf{x} = [x_1 \ \ x_2 \ \ \cdots \ \ x_n]^T$ is the current $\mathbf{x}_k$}

DO FOR $k = 1$ TO *MaxIt*
$\mathbf{x}_{\text{new}} \leftarrow A\mathbf{x}$; $BigXi \leftarrow \max\limits_{1 \le i \le n}$ (|ith component of $\mathbf{x}_{\text{new}}$|)
{*Scale*} $\mathbf{x}_{\text{new}} \leftarrow \dfrac{1}{BigXi}\mathbf{x}_{\text{new}}$ {dominant component of $\mathbf{x}_{\text{new}}$ is now 1}
$\mathbf{dx} \leftarrow \mathbf{x}_{\text{new}} - \mathbf{x}$
{*Update*} $\mathbf{x} \leftarrow \mathbf{x}_{\text{new}}$
UNTIL $|dx_i| \le Tol \cdot \max(1, |x_i|)$, $i = 1, 2, \ldots, n$ {**termination test**}

IF **termination test** succeeded
THEN OUTPUT ('Dominant eigenpair, to '*NumSig*' digits, is ('*BigXi*, $\mathbf{x}$')')
ELSE OUTPUT ('Convergence did not occur in '*MaxIt*' iterations')
STOP

Two more iterations give $A\mathbf{x}_2 = BigXi\mathbf{x}_3$ and $A\mathbf{x}_3 = BigXi\mathbf{x}_4$:

$$A\mathbf{x}_2 = \begin{bmatrix} \frac{14}{5} \\ -\frac{27}{20} \\ \frac{18}{5} \end{bmatrix} = \frac{18}{5}\begin{bmatrix} \frac{7}{9} \\ -\frac{3}{8} \\ 1 \end{bmatrix}, \qquad A\mathbf{x}_3 = \begin{bmatrix} \frac{10}{3} \\ \frac{9}{8} \\ \frac{34}{9} \end{bmatrix} = \frac{34}{9}\begin{bmatrix} \frac{15}{17} \\ \frac{81}{272} \\ 1 \end{bmatrix}$$

Evidently, $BigXi \to 4 = \lambda_1$ and $\mathbf{x}_k \to [1 \ \ 0 \ \ 1]^T = \mathbf{v}_1$ [see (6) of Section 9.1]. ■

Notice that the second component of $\mathbf{x}_k$ alternates sign while decreasing in magnitude; *this indicates convergence to zero*. A component that alternates sign with *constant* magnitude indicates that Assumption 1 fails (i.e., $|\lambda_1| = |\lambda_2|$); in this case the method may still yield λ_1 but will not yield $\mathbf{v}_1$.

Practical Consideration 9.2A (Using the *Scale* step). Notice how the ***Scale*** step makes it easy to see if $\mathbf{x}_{\text{new}} \approx \mathbf{x}$. More important, it avoids the possibility of overflow (if $|\lambda_1| > 1$) or underflow (if $|\lambda_1| < 1$) in calculating the components of $\mathbf{x}_{\text{new}}$ when k is large [see the λ_1^k scalar in (3c)]. Consequently, the Power Method should always be carried out "scaled" as in Algorithm 9.2A. ■

9.2B Convergence Considerations

It follows from (3b) that the convergence of $\mathbf{x}_k$ to a scalar multiple of $\mathbf{v}_1$ is most rapid when $|\lambda_1| >> |\lambda_i|$ (i.e., $\lambda_i/\lambda_1 \approx 0$) for $i > 1$. The slow convergence of the second component of the preceding example results from the fact that $\lambda_2/\lambda_1 = (-3)/4 = 0.75 \approx 0$.

If an approximation of $\mathbf{v}_1$ is not known, $\mathbf{x}_0 = [1 \quad 1 \quad \cdots \quad 1]^T$ is usually tried. It may turn out that the $\mathbf{x}_0$ chosen has $\alpha_1 = 0$. When this occurs, the Power Method (in the absence of roundoff error) should yield λ_2 and $\mathbf{v}_2$ [see (2a) and Problem 9-14]. If there is any doubt that the dominant eigenvalue has been obtained, try a different $\mathbf{x}_0$.

If $|\lambda_1| = |\lambda_2|$ (i.e., either $\lambda_1 = \pm\lambda_2$ or λ_1 and λ_2 are a complex-conjugate pair), then the Power Method will fail whether Scaled or not. This will be indicated by irregular behavior of some component of $\mathbf{x}_k$ (see Problem 9-12). In such situations the method must be modified (see [51]), or a more sophisticated method such as the QR algorithm (Section 9.3D) can be used.

9.2C The Inverse Power Method for Finding Smallest Eigenvalues

We saw in (4) of Section 9.1A that for a nonsingular matrix A,

$$(\lambda, \mathbf{v}) \text{ is an eigenpair for } A \quad \Leftrightarrow \quad \left(\frac{1}{\lambda}, \mathbf{v}\right) \text{ is an eigenpair for } A^{-1} \tag{4}$$

It follows that *the dominant eigenvalue of A^{-1} is λ_n^{-1}, that is, the reciprocal of the least dominant eigenvalue of A*. Hence if $\lambda_n \neq 0$ and $|\lambda_n| < |\lambda_i|$ for $i \neq n$, we can find $1/\lambda_n$ and an associated eigenvector $\mathbf{v}_n$ by applying the Scaled Power Method to A^{-1}, that is, by repeating the iterative step

$$\mathbf{x}_{k+1} = \frac{1}{BigXi} A^{-1}\mathbf{x}_k \qquad (BigXi = \text{dominant component of } A^{-1}x_k) \tag{5}$$

until $\mathbf{x}_{k+1} \approx \mathbf{v}_n$ and $BigXi \approx 1/\lambda_n$. If A is singular, we shall be unable to find A^{-1}, indicating that $\lambda_n = 0$.

The application of the Scaled Power Method to A^{-1} is called the **Inverse Power Method** for finding the least dominant eigenpair $(\lambda_n, \mathbf{v}_n)$ for A.

EXAMPLE 9.2C Find the least dominant eigenpair for A if

$$A = \begin{bmatrix} 3 & 0 & 1 \\ 0 & -3 & 0 \\ 1 & 0 & 3 \end{bmatrix} \qquad \text{for which } A^{-1} = \begin{bmatrix} \frac{3}{8} & 0 & -\frac{1}{8} \\ 0 & -\frac{1}{3} & 0 \\ -\frac{1}{8} & 0 & \frac{3}{8} \end{bmatrix} \tag{6}$$

using the Inverse Power Method with $\mathbf{x}_0 = [0 \quad 1 \quad 2]^T$.

Solution: Using the notation $A^{-1}\mathbf{x}_k = BigXi\mathbf{x}_{k+1}$ for $k = 0, 1, 2$, we get

$$A^{-1}\mathbf{x}_0 = \begin{bmatrix} \frac{3}{8} & 0 & -\frac{1}{8} \\ 0 & -\frac{1}{3} & 0 \\ -\frac{1}{8} & \frac{3}{8} & \frac{3}{8} \end{bmatrix} \begin{bmatrix} 0 \\ 1 \\ 2 \end{bmatrix} = \begin{bmatrix} -\frac{1}{4} \\ -\frac{1}{3} \\ \frac{3}{4} \end{bmatrix} = \left(\frac{3}{4}\right) \begin{bmatrix} -\frac{1}{3} \\ -\frac{4}{9} \\ 1 \end{bmatrix} = BigXi\mathbf{x}_1$$

$$A^{-1}\mathbf{x}_1 = \begin{bmatrix} \frac{3}{8} & 0 & -\frac{1}{8} \\ 0 & -\frac{1}{3} & 0 \\ -\frac{1}{8} & 0 & \frac{3}{8} \end{bmatrix} \begin{bmatrix} -\frac{1}{3} \\ -\frac{4}{9} \\ 1 \end{bmatrix} = \begin{bmatrix} -\frac{1}{4} \\ \frac{4}{27} \\ \frac{5}{12} \end{bmatrix} = \left(\frac{5}{12}\right)\begin{bmatrix} -\frac{3}{5} \\ \frac{16}{45} \\ 1 \end{bmatrix} = BigXi\,\mathbf{x}_2$$

$$A^{-1}\mathbf{x}_2 = \begin{bmatrix} \frac{3}{8} & 0 & -\frac{1}{8} \\ 0 & -\frac{1}{3} & 0 \\ -\frac{1}{8} & 0 & \frac{3}{8} \end{bmatrix} \begin{bmatrix} -\frac{3}{5} \\ \frac{16}{45} \\ 1 \end{bmatrix} = \begin{bmatrix} -\frac{7}{20} \\ -\frac{16}{135} \\ \frac{9}{20} \end{bmatrix} = \left(\frac{9}{20}\right)\begin{bmatrix} -\frac{7}{9} \\ -\frac{64}{243} \\ 1 \end{bmatrix} = BigXi\,\mathbf{x}_3$$

The *BigXi*'s $\frac{3}{4}$, $\frac{5}{12}$, $\frac{9}{20}$, are approaching $\frac{1}{2}$ (the reciprocal if $\lambda_3 = 2$), and $\mathbf{x}_k$ is approaching a scalar multiple of the eigenvector $\mathbf{v}_3 = \begin{bmatrix} 1 & 0 & -1 \end{bmatrix}^T$ [see (6) of Example 9.1A].   ■

9.2D Shifting Eigenvalues   *gives the eigenvalue nearest s*

For any scalar s,

$$A\mathbf{v} = \lambda\mathbf{v} \quad \Leftrightarrow \quad (A - sI)\mathbf{v} = (\lambda - s)\mathbf{v} \tag{7a}$$

Thus, as illustrated graphically in Figure 9.2-1,

$$(\lambda, \mathbf{v}) \text{ is an eigenpair for } A \quad \Leftrightarrow \quad (\lambda - s, \mathbf{v}) \text{ is an eigenpair for } A - sI \tag{7b}$$

$(\lambda_s - s)^{-1}$ = dominant eigenvalue of $(A - sI)^{-1}$

Let λ_s be the eigenvalue of A nearest s and let $\mathbf{v}_s$ be an associated eigenvector. Then $(\lambda_s - s, \mathbf{v}_s)$ will be the least dominant eigenpair for $A - sI$, hence $((\lambda_s - s)^{-1}, \mathbf{v}_s)$ is the dominant eigenpair of $(A - sI)^{-1}$. Solving for λ_s, we get

$$\lambda_s = \frac{1}{\text{dominant eigenvalue of } (A - sI)^{-1}} + s \tag{8}$$

We will refer to this strategy for finding $(\lambda_s, \mathbf{v}_s)$ as the **Shifted Inverse Power Method**.

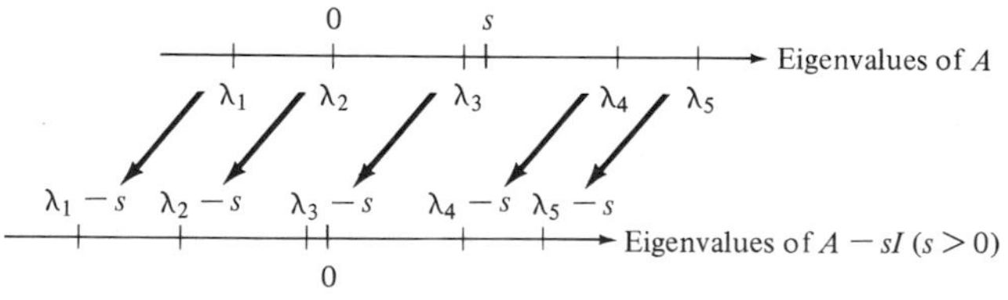

Figure 9.2-1   *Eigenvalues of A and A − sI (s > 0).*

EXAMPLE 9.2D   Use the Shifted Inverse Power Method with $\mathbf{x}_0 = \begin{bmatrix} 1 & 1 & 1 \end{bmatrix}^T$ to find the eigenvalue nearest $s = -\frac{5}{2}$ of

$$A = \begin{bmatrix} 3 & 0 & 1 \\ 0 & -3 & 0 \\ 1 & 0 & 3 \end{bmatrix} \quad \text{for which } A - sI = A + \frac{5}{2}I = \begin{bmatrix} \frac{11}{2} & 0 & 1 \\ 0 & -\frac{1}{2} & 0 \\ 0 & 0 & \frac{11}{2} \end{bmatrix} \tag{9}$$

SOLUTION: Applying the Scaled Inverse Power Method to $A - sI$ gives

$$(A + \tfrac{5}{2}I)^{-1}\mathbf{x}_0 = (-2)[-7.69231\text{E}-2 \quad 1 \quad -7.69231\text{E}-2]^T = BigXi\,\mathbf{x}_1$$
$$(A + \tfrac{5}{2}I)^{-1}\mathbf{x}_1 = (-2)[5.91716\text{E}-3 \quad 1 \quad 5.91716\text{E}-3]^T = BigXi\,\mathbf{x}_2$$
$$(A + \tfrac{5}{2}I)^{-1}\mathbf{x}_2 = (-2)[-4.55166\text{E}-4 \quad 1 \quad -4.55166\text{E}-4]^T = BigXi\,\mathbf{x}_3$$

In view of the alternating signs of the (shrinking) first and last components of $\mathbf{x}_k$, we see that $\mathbf{v}_s = [0 \quad 1 \quad 0]^T$; and -2 is the dominant eigenvalue of $(A + \frac{5}{2}I)^{-1}$. So, by (8), the eigenvalue of A nearest $s = -\frac{5}{2}$ is

$$\lambda_s = \frac{1}{(-2)} + \left(-\frac{5}{2}\right) = -3 \qquad \blacksquare$$

9.2E Practical Considerations When Using the Power Method

The following result, whose proof is given in [37], shows that *the diagonal entries of A are reasonable values to try for s in using the Shifted Inverse Power Method.*

used to get initial guesses on eigenvalues

GERSCHGORIN'S DISK THEOREM *For $A = (a_{ij})_{n\times n}$ and $i = 1, \ldots, n$ let D_i denote the circular disk consisting of all points z in the complex plane such that $|z - a_{ii}| \leq r_i$, where*

$|\lambda - a_{ii}| \leq \sum_{j\neq i} |a_{ij}|$

$$r_i = \sum_{j\neq i} |a_{ij}| \qquad \left(\begin{array}{l}\textit{the sum of the magnitudes of the}\\ \textit{off-diagonal entries of } \mathrm{row}_i A\end{array}\right)$$

Then each D_i contains at least one eigenvalue of A, and each eigenvalue of A lies in at least one D_i.

For the matrix A in (9), this theorem assures us that two eigenvalues satisfy $|\lambda - 3| \leq 1$ (rows 1 and 3) and the other one satisfies $|\lambda - (-3)| \leq 0$ (row 2).

The Shifted Inverse Power Method can be used to find any eigenpair $(\lambda_i, \mathbf{v}_i)$ for which we can find an s such that

$$|\lambda_i - s| < |\lambda_j - s| \qquad \text{for all } j \neq i \tag{10}$$

(i.e., s is closer to λ_i than any other λ_j). When this cannot be done, the method must be modified [51] or another method used (see Section 9.3D).

When $(A - sI)^{-1}$ is easy to find accurately (e.g., when $n = 2$ or perhaps 3), the iteration

$$\mathbf{x}_{\text{new}} \leftarrow \frac{1}{BigXi}\,\bar{\mathbf{x}}, \qquad \text{where } \bar{\mathbf{x}} = (A - sI)^{-1}\mathbf{x}_{\text{current}} \tag{11}$$

can be carried out using $(A - sI)^{-1}$, as we have done here. Otherwise, one should find an LU-decomposition for $A - sI$ and then find $\bar{\mathbf{x}}$ in (11) by completing the forback matrix

$$[\hat{L}\backslash\hat{U} : \hat{\mathbf{x}}_{\text{current}} : \bar{\mathbf{c}} : \bar{\mathbf{x}}] \qquad \text{(see Section 3.3C)} \tag{12}$$

at each iteration. This procedure requires essentially the same number of arithmetic operations as multiplying $\mathbf{x}_{\text{current}}$ by $(A - sI)^{-1}$ but without the extra roundoff introduced by actually finding this inverse.

In most situations the Shifted Inverse Power Method will find any desired eigenpairs of a given matrix A. However, if there is frequent need for *all* eigenpairs of A, then the methods of the next section should be used.

9.3

Methods for Finding All Eigenpairs of a Matrix

For any $m \times n$ matrix A, the **transpose** of A, written A^T, is the $n \times m$ matrix whose ith row is $(\text{col}_i A)^T$ for $i = 1, \ldots, n$. A straightforward verification shows that for any A and B we have $(A^T)^T = A$, and

$$(AB)^T = B^T A^T \tag{1}$$

that is, *the transpose of a product is the product of the transposes in the reverse order*.

9.3A Orthogonal Matrices

Recall that a square matrix A is **symmetric** if $a_{ij} = a_{ji}$ for all $i \neq j$, that is,

$$A \text{ is symmetric} \quad \Leftrightarrow \quad A^T = A \quad (A \text{ is its own transpose}) \tag{2}$$

The simplest symmetric matrices are diagonal matrices. In particular, I_n and $O_{n \times n}$ are symmetric. It follows from (1) that $(A^TA)^T = A^T(A^T)^T = A^TA$, and similarly that $(AA^T)^T = AA^T$; thus

$$A^TA \quad \text{and} \quad AA^T \qquad \text{are symmetric for any } m \times n \text{ matrix } A \tag{3}$$

Most numerical procedures for finding eigenvalues of a *symmetric* matrix A rely on the ability to diagonalize A as $U^{-1}AU$, where U^{-1} is especially easy to get, as described next.

A real $n \times n$ matrix U is called **orthogonal** if its transpose is its inverse, that is,

$$UU^T = U^TU = I_n \tag{4a}$$

In view of the definition of matrix multiplication, this means that

$$\text{row}_i U (\text{row}_j U)^T = (\text{col}_i U)^T \text{col}_j U = \begin{cases} 0 & \text{if } i \neq j \\ 1 & \text{if } i = j \end{cases} \tag{4b}$$

Notice that if U and V are orthogonal, then by (1) and (4a),

$$(UV)^T(UV) = (V^TU^T)(UV) = V^T(U^TU)V = V^TV = I_n$$

This shows that *the product of orthogonal matrices is orthogonal.*

Orthogonal matrices have certain desirable geometric properties. Specifically, if $\mathbf{x} = [x_1 \cdots x_n]^T$ and $\mathbf{y} = [y_1 \cdots y_n]^T$ are viewed as points in n-space and we define

$$\mathbf{x} \cdot \mathbf{y} = \mathbf{x}^T\mathbf{y} = x_1y_1 + \cdots + x_ny_n, \quad \text{the **dot product** of } \mathbf{x} \text{ and } \mathbf{y} \tag{5a}$$

$$\|\mathbf{x}\| = \sqrt{\mathbf{x}\cdot\mathbf{x}} = \sqrt{x_1^2 + \cdots + x_n^2}, \quad \text{the **Euclidean length** of } \mathbf{x} \tag{5b}$$

$$\angle(\mathbf{x}, \mathbf{y}) = \cos^{-1}\left(\frac{\mathbf{x}\cdot\mathbf{y}}{\|\mathbf{x}\|\|\mathbf{y}\|}\right), \quad \text{the **angle between x and y** (if } \mathbf{x}, \mathbf{y} \neq \mathbf{0}) \tag{5c}$$

then *an orthogonal matrix U preserves dot products, lengths, and angles*, that is,

$$U\mathbf{x}\cdot U\mathbf{y} = \mathbf{x}\cdot\mathbf{y} \quad \text{for all } \mathbf{x}, \mathbf{y} \tag{6a}$$

$$\|U\mathbf{x}\| = \|\mathbf{x}\| \quad \text{for all } \mathbf{x} \tag{6b}$$

$$\angle(U\mathbf{x}, U\mathbf{y}) = \angle(\mathbf{x}, \mathbf{y}) \quad \text{for all nonzero } \mathbf{x} \text{ and } \mathbf{y} \tag{6c}$$

Indeed, using (5a), (1), and (4a), we see that for any $\mathbf{x}$ and $\mathbf{y}$,

$$U\mathbf{x}\cdot U\mathbf{y} = (U\mathbf{x})^TU\mathbf{y} = \mathbf{x}^TU^TU\mathbf{y} = \mathbf{x}^T\mathbf{y} = \mathbf{x}\cdot\mathbf{y}$$

proving (6a); (6b) and (6c) now follow directly from (5b) and (5c).

In view of the definitions (5), (4b) asserts that *the rows (or columns) of an orthogonal matrix are* **unit vectors** (i.e., of length 1) *that are mutually perpendicular.*

EXAMPLE 9.3A Rotation Matrices The **rotation matrix** R_θ defined by

$$R_\theta = \begin{bmatrix} \cos\theta & -\sin\theta \\ \sin\theta & \cos\theta \end{bmatrix} \tag{7}$$

satisfies $R_{(-\theta)} = R_\theta^T = R_\theta^{-1}$, and hence is orthogonal for any angle θ (see Problem M3-2). Geometrically, multiplying $\mathbf{x}$ by R_θ has the effect of rotating $\mathbf{x}$ counterclockwise by θ radians about the origin in 2-space, as shown in Figure 9.3-1. More generally, the ***i, j*-rotation matrix** defined for $i < j$ as the $n \times n$ matrix

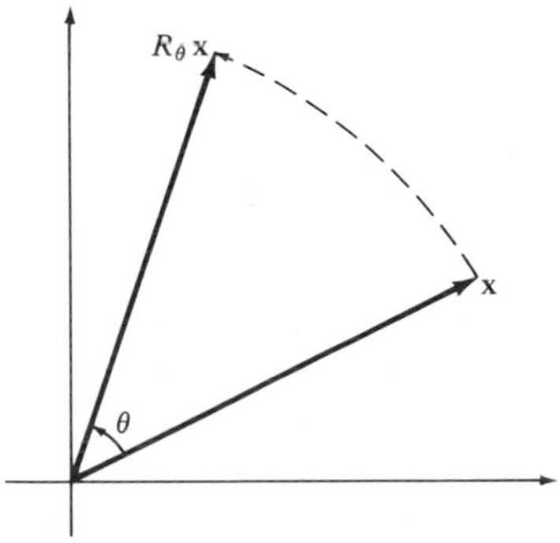

Figure 9.3-1 *Multiplying* **x** *by* R_θ in 2-space.

$$
R_{ij}(\theta) = \begin{bmatrix} 1 & 0 & \cdots & 0 & \cdots & 0 & \cdots & 0 \\ \vdots & \vdots & & \vdots & & \vdots & & \vdots \\ 0 & 0 & \cdots & \cos\theta & & -\sin\theta & \cdots & 0 \\ \vdots & \vdots & & \vdots & & \vdots & & \vdots \\ 0 & 0 & \cdots & \sin\theta & & \cos\theta & \cdots & 0 \\ \vdots & \vdots & & \vdots & & \vdots & & \vdots \\ 0 & 0 & \cdots & 0 & \cdots & 0 & \cdots & 1 \end{bmatrix} \begin{matrix} \\ \\ \leftarrow \text{row}_i \\ \\ \leftarrow \text{row}_j \\ \\ \\ \end{matrix} \qquad \textbf{(8)}
$$

$$\underbrace{}_{\text{col}_i} \qquad \underbrace{}_{\text{col}_j}$$

(in which the entries not shown are the same as I_n) is orthogonal for any real θ and any $i < j$. An example when $n = 3$ is

$$
R_{13}\left(\frac{\pi}{6}\right) = \begin{bmatrix} \cos\frac{\pi}{6} & 0 & -\sin\frac{\pi}{6} \\ 0 & 1 & 0 \\ \sin\frac{\pi}{6} & 0 & \cos\frac{\pi}{6} \end{bmatrix} = \begin{bmatrix} \frac{\sqrt{3}}{2} & 0 & \frac{-1}{2} \\ 0 & 1 & 0 \\ \frac{1}{2} & 0 & \frac{\sqrt{3}}{2} \end{bmatrix} \qquad \textbf{(9)}
$$

Multiplying $\mathbf{x} = [x \quad y \quad z]^T$ by $R_{13}(\pi/6)$ has the effect of rotating $\mathbf{x}$ counterclockwise by $\pi/6$ radians about the y-axis in 3-space.

The inverse of an orthogonal matrix U is U^T; so it can be found *by inspection* for *any* n. When such matrices appear in a similarity transformation, $B = U^{-1}AU$ becomes

$$
B = U^T A U \qquad \textbf{(10)}
$$

in which case B is said to be **orthogonally similar** to A.

The following important result, which is proved in linear algebra, assures us that *every symmetric matrix is orthogonally similar to a diagonal matrix.*

SYMMETRIC DIAGONALIZATION THEOREM. *If A is symmetric, then there is an orthogonal matrix U such that $U^TAU = \text{diag}(\lambda_1, \ldots, \lambda_n)$. Hence the eigenvalues of A are all real, and the columns $\mathbf{u}_1, \ldots, \mathbf{u}_n$ of U form a basis of mutually perpendicular unit eigenvectors (often referred to as an* **orthonormal basis of eigenvectors***).*

For example, the reader can verify that for the symmetric 3×3 matrix A in Example 9.2D, $U = R_{13}(\pi/4)$ satisfies

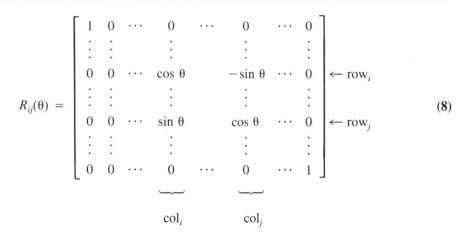

$$\underbrace{\begin{bmatrix} \frac{1}{\sqrt{2}} & 0 & \frac{1}{\sqrt{2}} \\ 0 & 1 & 0 \\ -\frac{1}{\sqrt{2}} & 0 & \frac{1}{\sqrt{2}} \end{bmatrix}}_{U^T = U^{-1}} \underbrace{\begin{bmatrix} 3 & 0 & 1 \\ 0 & -3 & 0 \\ 1 & 0 & 3 \end{bmatrix}}_{A = A^T} \underbrace{\begin{bmatrix} \frac{1}{\sqrt{2}} & 0 & -\frac{1}{\sqrt{2}} \\ 0 & 1 & 0 \\ \frac{1}{\sqrt{2}} & 0 & \frac{1}{\sqrt{2}} \end{bmatrix}}_{U = R_{13}\left(\frac{\pi}{4}\right)} = \underbrace{\begin{bmatrix} 4 & 0 & 0 \\ 0 & -3 & 0 \\ 0 & 0 & 2 \end{bmatrix}}_{\text{diag}(4,\ -3,\ 2)}$$

If A is symmetric, U is orthogonal, and $B = U^TAU$, then

$$B^T = U^TA^T(U^T)^T = U^TAU = B$$

So *any matrix orthogonally similar to a symmetric matrix is also symmetric*. In particular, if $A = (a_{ij})$ is symmetric and

$$B = R_{ij}(\theta)^TAR_{ij}(\theta) \tag{11a}$$

then B is symmetric, and a straightforward multiplication shows that

$$b_{ij} = b_{ji} = a_{ij}[\cos^2\theta - \sin^2\theta] + (a_{jj} - a_{ii})\sin\theta\cos\theta \tag{11b}$$

$$b_{ik} = b_{ki} = a_{ik}\cos\theta + a_{jk}\sin\theta \qquad \text{for } k \neq i,j \text{ (in row}_i\, B,\ \text{col}_i\, B) \tag{11c}$$

$$b_{jk} = b_{kj} = -a_{ik}\sin\theta + a_{jk}\cos\theta \qquad \text{for } k \neq i,j \text{ (in row}_j\, B,\ \text{col}_j\, B) \tag{11d}$$

$$b_{ii} = a_{ii}\cos^2\theta + a_{jj}\sin^2\theta + 2a_{ij}\sin\theta\cos\theta \tag{11e}$$

$$b_{jj} = a_{ii}\sin^2\theta + a_{jj}\cos^2\theta - 2a_{ij}\sin\theta\cos\theta \tag{11f}$$

with the remaining entries of B identical to those of A.

9.3B Jacobi's Method for Symmetric Matrices

If θ is chosen so that

$$a_{ij}\cos(2\theta) + \tfrac{1}{2}(a_{jj} - a_{ii})\sin(2\theta) = 0 \tag{12a}$$

then, in view of (11b), the matrix $B = R_{ij}(\theta)^TAR_{ij}(\theta)$ will satisfy

$$b_{ij} = b_{ji} = 0 \tag{12b}$$

The θ (in radians) that satisfies (12a) can be expressed as

$$\theta = \begin{cases} \dfrac{1}{2}\tan^{-1}\dfrac{2a_{ij}}{a_{ii} - a_{jj}} & \text{if } a_{ii} \neq a_{jj} \quad \textbf{(13a)} \\[2ex] \dfrac{\pi}{4} & \text{if } a_{ii} = a_{jj} \quad \textbf{(13b)} \end{cases}$$

If A is symmetric and i and j are the indices of the dominant superdiagonal entry of A, then in view of (12b), the matrix B should be "closer" to being diagonal than A. The iterative procedure incorporating this strategy is called **Jacobi's method**. If we start with $A_0 = A$, the method generates rotation matrices

$$R_k = R_{ij}(\theta_k), \qquad k = 1, 2, \ldots \qquad \textbf{(14)}$$

ALGORITHM 9.3B. JACOBI'S METHOD

PURPOSE: To find all eigenpairs of a given $n \times n$ matrix A to a prescribed accuracy. The method uses the rotation matrix $R_{ij}(\theta)$ defined in (8) iteratively to form an orthogonal matrix U such that $U^TAU = \text{diag}(\lambda_1, \ldots, \lambda_n)$ to the desired accuracy.

```
{Initialize}
GET n, A,              {matrix parameters}
    MaxIt, NumDec {termination parameters}
U ← I_n;     AbsTol ← 10^(−NumDec)

{Iterate}
DO FOR k = 1 TO MaxIt UNTIL termination test is satisfied
|   Get i, j of the dominant superdiagonal entry, a_ij {i < j}
|   MaxOffDiag ← |a_ij|
|   {Rotate i, j}
|   |   {Get θ} IF a_ii = a_jj THEN Theta ← π/4
|   |           |_________ELSE Theta ← ½tan^(−1)[2a_ij/(a_ii − a_jj)]
|   |   R ← R_ij(Theta)                                    {R is R_k}
|   |   A ← R^T A R                      {A is A_k;   a_ij is now zero}
|   |___U ← UR                                             {U is U_k}
|___UNTIL MaxOffDiag < AbsTol                       {termination test}

IF termination test succeeded
|   THEN OUTPUT ('The eigenpairs of A are ('a_jj, col_j U')', j = 1, ..., n)
|___ELSE OUTPUT ('Convergence did not occur in 'MaxIt' iterations')
STOP
```

such that $A_k = R_k^T A_{k-1} R_k$ has zeros where A_{k-1} had its dominant off-diagonal entries. It can be shown [51] that this selection of i, j, and θ_k for each k ensures that

$$A_k = R_k^T A_{k-1} R_k \to D = \operatorname{diag}(\lambda_1, \ldots, \lambda_n) \quad \text{as } k \to \infty \tag{15}$$

Since $A_0 = A$, we can express A_k in terms of A as follows:

$$A_k = (U_k)^T A U_k, \qquad \text{where } U_k = R_1 R_2 \cdots R_k = U_{k-1} R_k \tag{16}$$

The matrix U_k in (16) is a product of orthogonal matrices, and hence is orthogonal. As $k \to \infty$ it will approach the U guaranteed by the Symmetric Diagonalization Theorem given in Section 9.3A. Notice that Jacobi's method yields approximations of *all* eigenvalues (the diagonal entries of A_k) and *all* eigenvectors (the corresponding columns of U_k).

A pseudoprogram for Jacobi's method is given in Algorithm 9.3B. If A is 2×2, then only one rotation is needed to diagonalize A. This might also be the case for $n > 2$ [see (10)]. In general, however, A_k in (15) never actually equals D.

In discussing $R_k = R_{ij}(\theta_k)$, it will be convenient to use the abbreviations

$$S_k = \sin \theta_k \quad \text{and} \quad C_k = \cos \theta_k, \qquad k = 1, 2, \ldots \tag{17}$$

EXAMPLE 9.3B Perform two iterations of Jacobi's method for the symmetric matrix

$$A = \begin{bmatrix} 3 & 0.01 & 0.02 \\ 0.01 & 2 & 0.1 \\ 0.02 & 0.1 & 1 \end{bmatrix} \tag{18}$$

SOLUTION: The largest superdiagonal entry of A is $0.1 = a_{23}$. By (13a)

$$\theta_1 = \frac{1}{2}\tan^{-1}\left(\frac{2a_{23}}{a_{22} - a_{33}}\right) = \frac{1}{2}\tan^{-1}\left(\frac{2(0.1)}{2 - 1}\right) = 0.0986978 \text{ (radians)}$$

$$C_1 = \cos\theta_1 = 0.995133 \quad \text{and} \quad S_1 = \sin\theta_1 = 0.0985376$$

Putting these values in $R_1 = R_{23}(\theta_1)$, we get A_1 as

$$A_1 = R_1^T A_0 R_1 = \begin{bmatrix} 1 & 0 & 0 \\ 0 & C_1 & S_1 \\ 0 & -S_1 & C_1 \end{bmatrix} \begin{bmatrix} 3 & 0.01 & 0.02 \\ 0.01 & 2 & 0.1 \\ 0.02 & 0.1 & 1 \end{bmatrix} \begin{bmatrix} 1 & 0 & 0 \\ 0 & C_1 & -S_1 \\ 0 & S_1 & C_1 \end{bmatrix}$$

$$= \begin{bmatrix} 3 & 0.011922 & 0.018917 \\ 0.011922 & 2.009902 & 0 \\ 0.018917 & 0 & 0.990098 \end{bmatrix} \tag{19}$$

For this first rotation, U_1 is simply R_1 itself.

Notice that the rotation which "zeros" the 2,3-entry makes the 1,2-entry slightly larger. However, the dominant superdiagonal entry of A_1 (namely $a_{13} = 0.018917$) is much smaller than that of A_0 (namely $a_{23} = 0.1$). Continuing, we obtain

$$\theta_2 = \frac{1}{2}\tan^{-1}\left(\frac{2a_{13}}{a_{11} - a_{33}}\right) = \frac{1}{2}\tan^{-1}\left(\frac{2(0.018917)}{3 - 0.990098}\right) \doteq 0.00941079$$

$$C_2 = \cos\theta_2 = 0.999956 \qquad \text{and} \qquad S_2 = \sin\theta_2 = 0.00941065$$

Putting these values in $R_2 = R_{13}(\theta_2)$, we get A_2 as $R_2^T A_1 R_2$, that is, as

$$A_2 = \begin{bmatrix} C_2 & 0 & S_2 \\ 0 & 1 & 0 \\ -S_2 & 0 & C_2 \end{bmatrix} \begin{bmatrix} 3 & 0.011922 & 0.018917 \\ 0.011922 & 2.009902 & 0 \\ 0.018917 & 0 & 0.990098 \end{bmatrix} \begin{bmatrix} C_2 & 0 & -S_2 \\ 0 & 1 & 0 \\ S_2 & 0 & C_2 \end{bmatrix}$$

$$\doteq \begin{bmatrix} 3.000178 & 0.011922 & 0 \\ 0.011922 & 2.009902 & -0.000112 \\ 0 & -0.000112 & -0.989920 \end{bmatrix} \tag{20a}$$

and we get U_2 as $U_1 R_2 = R_{23}(\theta_1) R_{13}(\theta_2)$, that is, as

$$U_2 = \begin{bmatrix} 1 & 0 & 0 \\ 0 & 0.995133 & -0.098538 \\ 0 & 0.098538 & 0.995133 \end{bmatrix} \begin{bmatrix} 0.999956 & 0 & -0.009411 \\ 0 & 1 & 0 \\ 0.009411 & 0 & 0.999956 \end{bmatrix}$$

$$\doteq \begin{bmatrix} 0.999956 & 0 & -0.009411 \\ -0.000927 & 0.995133 & -0.098533 \\ 0.009365 & 0.098538 & 0.995089 \end{bmatrix} \tag{20b}$$

A third iteration, using $R_3 = R_{12}(\theta_3)$, where $\theta_3 = 0.012035$ (radians), yields the following values of $A_3 = R_3^T A_2 R_3$ and $U_3 = U_2 R_3$.

$$A_3 \doteq \begin{bmatrix} 3.00032 & 0 & -\underline{1.35\text{E}-6} \\ 0 & 2.00976 & -\underline{1.122\text{E}-4} \\ -\underline{1.35\text{E}-6} & -\underline{1.122\text{E}-4} & 0.989920 \end{bmatrix} \tag{21a}$$

$$U_3 \doteq \begin{bmatrix} 0.999883 & 0.0120\underline{355} & -0.009410\underline{8} \\ 0.0110502 & 0.9950\underline{72} & -0.0985\underline{332} \\ 0.01055\underline{03} & 0.0984\underline{177} & 0.9950\underline{89} \end{bmatrix} \tag{21b}$$

The underlined digits are those that would be wrong *after* rounding. Even had we not known the exact eigenvalues, we would have known from Gerschgorin's disk theorem and inspection of A_3 that

$$|\lambda_1 - 3.00032| < 1.4\text{E}-6; \quad |\lambda_2 - 2.00976| < 1.13\text{E}-4; \quad |\lambda_3 - 0.98992| < 1.13\text{E}-4$$

■

The Fortran subroutine ROTATE shown in Figure 9.3-2 performs one rotation of Jacobi's method. After selecting THETA in lines 1000 to 1400 [see (13)], it uses (11) to replace A by R^TAR and U by UR [without actually forming $R = R_{ij}$(THETA)] in lines 1700 to 3200.

```
ØØ1ØØ         SUBROUTINE ROTATE (I,J)
ØØ2ØØ         DIMENSION A(6,6), U(6,6)
ØØ3ØØ         COMMON N, NROTAT, A, U
ØØ4ØØ   C - - - - - - - - - - - - - - - - - - - - - - - - - - - - - - C
ØØ5ØØ   C   PURPOSE:  To perform the two rotation replacements    C
ØØ6ØØ   C               A <-- Rtranspose*A*R   and   U <-- U*R      C
ØØ7ØØ   C   where  R = Rij(THETA)  makes  A(i,j) = A(j,i) = Ø.     C
ØØ8ØØ   C - - - - - - - - - - - - - - - - - - -  VERSION 2:  9/9/85 - C
ØØ9ØØ         NROTAT = NROTAT + 1
Ø1ØØØ         IF (ABS(A(I,I)-A(J,J)) .LT. 1.E-6*ABS(A(I,I))) THEN
Ø11ØØ            THETA = ATAN(1.Ø)  != pi/4
Ø12ØØ         ELSE
Ø13ØØ            THETA = .5*ATAN(2.*A(I,J)/(A(I,I)-A(J,J)))
Ø14ØØ         ENDIF
Ø15ØØ         SI = SIN(THETA)          !critical entries
Ø16ØØ         CO = COS(THETA)          !  of desired Rij
Ø17ØØ         DO 1Ø K=1,N              !rotate U and A
Ø18ØØ            UKI = U(K,I)
Ø19ØØ            U(K,I) =  UKI*CO + U(K,J)*SI
Ø2ØØØ            U(K,J) = -UKI*SI + U(K,J)*CO
Ø21ØØ            IF (K.NE.I .AND. K.NE.J) THEN
Ø22ØØ               A(K,I) =  A(I,K)*CO + A(J,K)*SI
Ø23ØØ               A(K,J) = -A(I,K)*SI + A(J,K)*CO
Ø24ØØ               A(J,K) = A(K,J)
Ø25ØØ               A(I,K) = A(K,I)
Ø26ØØ            ENDIF
Ø27ØØ      1Ø CONTINUE                 !the k = i, j entries are
Ø28ØØ         AII = A(I,I)
Ø29ØØ         A(I,I) = AII*CO**2 + A(J,J)*SI**2 + 2.*A(I,J)*SI*CO
Ø3ØØØ         A(J,J) = AII*SI**2 + A(J,J)*CO**2 - 2.*A(I,J)*SI*CO
Ø31ØØ         A(I,J) = Ø.Ø
Ø32ØØ         A(J,I) = Ø.Ø
Ø33ØØ         RETURN
Ø34ØØ         END
```

Figure 9.3-2 *Fortran subroutine for one rotation of Jacobi's Method.*

9.3C Jacobi's Method with Thresholds

For large n, it is time consuming (i.e., expensive) to have a computer scan the $n(n-1)/2$ superdiagonal (or subdiagonal) entries of the current A to find the maximum off-diagonal $|a_{ij}|$. What is often done in practice is to set a "threshold" magnitude *Thresh* for a scan, and perform a rotation whenever $|a_{ij}| \geqslant$ *Thresh*, as shown in Algorithm 9.3C. The value of *Thresh* for any scan is set to

ALGORITHM 9.3C. JACOBI'S METHOD WITH THRESHOLDS

PURPOSE: To perform Jacobi's method more efficiently by using fewer scans and more rotations per scan. The parameter *Fraction* controls the frequency of rotations, with the number of rotations per scan increasing as *Fraction* $\to 0$.

GET n, A, *NumDec*, *MaxIt*, *Fraction* {as in Jacobi's method algorithm} $\{0 < Fraction < 1\}$

$U \leftarrow I_n$; $AbsTol \leftarrow 10^{-NumDec}$

$MaxOffDiag \leftarrow$ (largest superdiagonal $|a_{ij}|$) $\{i < j\}$

DO FOR $k = 1$ TO *MaxIt* UNTIL **termination test** is satisfied
- *Thresh* $\leftarrow$ *Fraction*$\cdot$*MaxOffDiag*; *MaxOffDiag* $\leftarrow 0$
- DO FOR $i = 1$ TO $n = 1$ {scan ith row of A}
  - DO FOR $j = i + 1$ TO n
    - IF $|a_{ij}| >$ *Thresh* THEN **rotate *i*, *j*** {as in Jacobi's method}
    - IF $|a_{ij}| >$ *MaxOffDiag* THEN *MaxOffDiag* $\leftarrow |a_{ij}|$
- {Now *MaxOffDiag* = largest unrotated $|a_{ij}|$ scanned}
- UNTIL *MaxOffDiag* < *AbsTol* {**termination test**}

IF **termination test** succeeded
- THEN OUTPUT ('The eigenpairs of A are ('a_{jj}, $\text{col}_j U$')', $j = 1, \ldots, n$)
- ELSE OUTPUT ('Convergence did not occur in '*MaxIt*' iterations'

STOP

```
Jacobi's Method With Thresholds (FRACT = 0.300) for the matrix

                    3.00000      0.01000      0.02000
                    0.01000      2.00000      0.10000
                    0.02000      0.10000      1.00000

Scan # 1 (THRESH = 0.030000):  1 Rotations.  [ A : U ] is:

 3.0000000  0.0119221  0.0189173      1.0000000  0.0000000  0.0000000
 0.0119221  2.0099020  0.0000000      0.0000000  0.9951333 -0.0985376
 0.0189173  0.0000000  0.9900981      0.0000000  0.0985376  0.9951333

Scan # 2 (THRESH = 0.006000):  2 Rotations.  [ A : U ] is:

 3.0003215 -0.0000021  0.0000000      0.9998833 -0.0120387 -0.0094088
-0.0000021  2.0097584 -0.0002277      0.0110524  0.9950612 -0.0986460
 0.0000000 -0.0002277  0.9899201      0.0105499  0.0985305  0.9950781

Scan # 3 (THRESH = 0.000068):  1 Rotations.  [ A : U ] is:

 3.0003215 -0.0000021  0.0000000      0.9998833 -0.0120366 -0.0094114
-0.0000021  2.0097585  0.0000000      0.0110524  0.9950832 -0.0984238
 0.0000000  0.0000000  0.9899200      0.0105499  0.0983083  0.9951001
```

Figure 9.3-3 *Three scans of Jacobi's method with thresholds.*

$$\textit{Fraction} \cdot \textit{MaxOffdiag}$$

where *Fraction* is a number between 0 and 1 and *MaxOffdiag* is the magnitude of the largest off-diagonal term that was *not* "zeroed" during the preceding scan. The **rotate *i, j*** step is performed in a subroutine such as that given in Figure 9.3-2.

If Algorithm 9.3C is applied to $A_{3\times 3}$ of Example 9.3B with *Fraction* taken to be 0.3, we get the results shown in Figure 9.3-3. All eigenpairs appear to be accurate to 6*s*.

It should be noted that Jacobi-type methods have been developed for nonsymmetric matrices. These are discussed in [40].

9.3D Factorization Methods

If all eigenpairs of a nonsymmetric matrix A are needed, then Jacobi's method cannot be used. The methods developed for this situation utilize a factorization much like the LU-factorization of Section 4.2B.

One of the more effective methods for finding all eigenpairs of any matrix A is the ***QR*-method**. It is usually performed in two steps.

Step 1: Replace A by a similar matrix $P^{-1}AP$ whose entries more than one row below the main diagonal are all zero. Such matrices are said to be in **upper-Hessenberg** form. A nonsingular matrix P that does this is usually obtained in one of two ways:

a. *Householder's method:* This method uses $n - 2$ **Householder transformations**

$$A_k = P_k^T A_{k-1} P_k, \qquad \text{where } P_k = I - 2\mathbf{u}_k\mathbf{u}_k^T \tag{22}$$

for some suitably chosen unit vector $\mathbf{u}_k$, $k = 1, \ldots, n - 2$. The P_k's are easily seen to be orthogonal *and* symmetric. If $A = A_0$ is symmetric, A_{n-2} will be tridiagonal and symmetric.

b. *Elementary Transformation Method*: This method uses Gaussian elimination in the form of $n - 2$ transformations (see Problems M3-8–M3-9)

$$A_k = E_k A_{k-1} E_k^{-1}, \qquad \text{where } E_k \text{ is an } \textbf{elementary matrix} \tag{23}$$

that is, E_k is a matrix such that E_kA has the effect of one of the elementary row operations.

The Elementary Transformation Method does not yield a tridiagonal matrix if A is symmetric; and unlike the Householder transformations (for which $P = P_1P_2 \cdots P_{n-2}$ is orthogonal, hence well conditioned), the use of $P = (E_{n-1} \cdots E_2E_1)^{-1}$ may cause $P^{-1}AP$ to be more ill conditioned than A (although this possibility can be minimized by the use of a suitable pivoting strategy). However, it requires only about half as many arithmetic operations as Householder's method.

Step 2: Starting with $A_0 = A$, form a sequence of decompositions

$$A_k - s_kI = Q_kR_k, \qquad A_{k+1} = R_kQ_k + s_kI \tag{24}$$

where Q_k is orthogonal, R_k is upper (i.e., right)-triangular, and s_k is a shifting scalar [see (7) of Section 9.2C] chosen so that

$$A_{k+1} \to \text{diag}(\lambda_1, \ldots, \lambda_n) \text{ rapidly as } k \to \infty \tag{25}$$

Although step 2 can be performed without step 1, the substantial savings in iterations needed for convergence makes step 1 worth incorporating. Details of the *QR*-method and of a related factorization method called the ***LR*-method** can be found in [40 and 51].

Well-tested software implementing these methods can be found in the EISPAK package developed at the Argonne National Laboratories.

9.4

Characteristic Values and Solutions of Homogeneous BVP's

A linear, second-order BVP is called **homogeneous** if both the differential equation and the boundary conditions are homogeneous, that is, if it is of the form

$$\phi_2(t)y'' + \phi_1(t)y' + \phi_0(t)y = 0, \qquad y(a) = y(b) = 0 \tag{1}$$

Such BVP's always have the **trivial solution** $y(t) \equiv 0$.

Our attention will be confined to homogeneous equations of the form

$$a_2(t)y'' + a_1(t)y' + [a_0(t) - \lambda]y = 0, \qquad y(a) = y(b) = 0 \tag{2}$$

Physical systems governed by (2) are found to have nontrivial solutions $y(t)$ (called **characteristic solutions**) for *certain* values of the parameter λ (called **characteristic values** or **eigenvalues**). The terminology follows from the fact that these $(\lambda, y(t))$ pairs satisfy

$$\mathcal{D}y = \lambda y, \qquad \text{where } \mathcal{D} = a_2(t)\frac{d^2}{dt^2} + a_1(t)\frac{d}{dt} + a_0(t)$$

As with eigenvectors of matrices, *if $y(t)$ is a characteristic solution of a BVP, so is $\alpha y(t)$ for any $\alpha \neq 0$.*

The following example illustrates the importance of knowing the characteristic values and solutions of homogeneous BVP's.

9.4A Buckling of Axially Loaded Beams

Suppose that a thin, uniform beam L units long is subjected to a constant compressing force P along its axis. Such a force is called an **axial load**. Let $y = y(x)$ denote the horizontal deflection x units up from the base, as shown in Figure 9.4-1. It can be shown that the shape of the beam is governed by the linear, homogeneous BVP

$$\frac{d^2y}{dx^2} + \frac{P}{EI}y = 0, \qquad y(0) = y(L) = 0 \tag{3a}$$

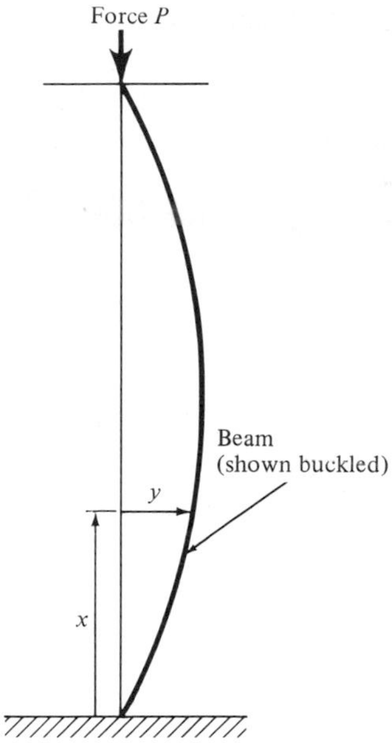

Figure 9.4-1 *Axially loaded beam.*

where E and I are constants that reflect the elasticity of the beam's material and the geometry of its cross section, respectively. Since the load P is assumed fixed, we can rewrite this as

$$y'' - \lambda y = 0, \qquad y(0) = y(L) = 0 \tag{3b}$$

where $\lambda = -P/EI$ is negative. It is easy to see that

$$y'' = \lambda y, \quad \text{where } \lambda < 0 \quad \Leftrightarrow \quad y(x) = A \cos \sqrt{-\lambda}x + B \sin \sqrt{-\lambda}x \tag{4}$$

Since $y(0) = 0$, A must be zero; hence the homogeneous boundary condition at the right endpoint can be written as

$$B \sin \sqrt{-\lambda}L = 0, \qquad \text{where } -\lambda = \frac{P}{EI} \tag{5}$$

As long as $\sqrt{-\lambda}L$ is not a multiple of π, (5) forces B to be zero, and the trivial solution $y(x) \equiv 0$ is the only possible one. However, if

$$\sqrt{-\lambda}L = k\pi, \qquad \text{that is, } \lambda = -\left(\frac{k\pi}{L}\right)^2 \tag{6a}$$

then $\sin \sqrt{-\lambda}L = 0$, leaving B unrestricted. For such λ's, $y(x)$ can be nonzero for $0 < x < L$ (i.e., *the beam can buckle*). Axial loads

$$P = -EI\lambda, \qquad \lambda \text{ an eigenvalue of the BVP} \tag{6b}$$

are called **buckling loads** or **critical loads** and should be avoided. The buckling load of primary interest is the *smallest* one because this load can cause the beam to break before larger loads can be applied!

9.4B Numerical Procedure for Estimating Characteristic Values

When E and or I are not constant but vary with x, we may not be able to find an analytic solution to use to obtain formulas like (6). The critical loads for such beams must therefore be found from a numerical solution. The general procedure when

$$a_2(x)y'' + a_1(x)y' + a_0(x)y = \lambda y, \qquad y(a) = y(b) = 0 \tag{7}$$

is to subdivide $[a, b]$ using the $n + 1$ h-spaced points

$$a = x_0 < x_1 < \cdots < x_j = a + jh < \cdots < x_{n-1} < x_n = b \tag{8}$$

and then use central difference approximations such as the $O(h^2)$ formulas

$$y'(x_j) \approx \frac{y_{j+1} - y_j}{2h} \quad \text{and} \quad y''(x_j) \approx \frac{y_{j+1} - 2y_j + y_{j-1}}{h^2}, \quad h = \frac{b-a}{n} \tag{9}$$

as we did in Section 8.5D. Putting (9) in (7) with y_0 and $y_n = 0$ gives

$$\frac{1}{h^2}\{\alpha_j y_{j-1} + \beta_j y_j + \gamma_j y_{j+1}\} = \lambda y_j \tag{10a}$$

where

$$\alpha_j = a_2(x_j) - \frac{h}{2}a_1(x_j), \quad \beta_j = -2a_2(x_j) + h^2 a_0(x_j), \quad \gamma_j = a_2(x_j) + \frac{h}{2}a_1(x_j) \tag{10b}$$

for $j = 1, \ldots, n - 1$. This is a tridiagonal system whose matrix form is

$$T\mathbf{y} = \lambda\mathbf{y} \qquad \text{where } T = \frac{1}{h^2}\begin{bmatrix} \beta_1 & \gamma_1 & & & \\ \alpha_2 & \beta_2 & & & \\ & & \ddots & & \\ & & \alpha_{n-2} & \beta_{n-2} & \gamma_{n-2} \\ & & & \alpha_{n-1} & \beta_{n-1} \end{bmatrix} \tag{10c}$$

Since $T\mathbf{y} = \lambda\mathbf{y}$ approximates the BVP in (7), the eigenpairs $(\lambda, \mathbf{y})$ of T should approximate the eigenpairs $(\lambda, y(x))$ of the BVP, with $\mathbf{y} = [y_1 \ \cdots \ y_{n-1}]^T$ and $y(x)$ related as follows

$$y_j \approx y(x_j), \qquad \text{where } x_j = a + jh \quad \text{for } h = 1, \ldots, n - 1 \tag{11}$$

EXAMPLE 9.4B Set up and solve the matrix equation $T\mathbf{y} = \lambda\mathbf{y}$ that approximates

$$y'' = \lambda y \ (\lambda < 0), \qquad y(0) = y(L) = 0 \tag{12}$$

using $n = 2, 3, 4, 5$, and discuss how well the eigenvalues of T approximate the characteristic values $\lambda = -(k\pi/L)^2$ obtained in (6a).

SOLUTION: For this simple BVP, $a_0(x) = a_1(x) \equiv 0$ and $a_2(x) \equiv 1$; hence

$$\alpha_j = 1,\ \beta_j = -2, \qquad \text{and} \qquad \gamma_j = 1 \qquad \text{for } j = 1, \ldots, n-1 \tag{13}$$

[see (10a)]. Thus for any n, the $(n-1) \times (n-1)$ matrix T in (10e) is

$$T = \frac{1}{h^2}\begin{bmatrix} -2 & 1 & & & \\ 1 & -2 & 1 & \ddots & \\ & & & & \\ & & & -2 & 1 \\ & & 1 & 1 & -2 \end{bmatrix}, \qquad \text{where } h = \frac{L}{n} \tag{14}$$

For small values of n the eigenvalues of this simple, symmetric T can be found analytically as roots of its characteristic polynomial. This is done for $n = 2, 3, 4, 5$ in Table 9.4-1. ■

Notice that reasonable estimates of the *least* dominant eigenvalue $\lambda_{\min} = -(\pi/L)^2$ (corresponding to the smallest buckling load P) can be obtained with smaller values of n than is required to get comparable estimates of the larger λ's. Even for more complicated differential equations, the Inverse Power Method can be applied to T to obtain reasonable approximations of $\lambda_{\min}$ and an associated characteristic solution $y(x)$ using small n's, especially if the following discussion is utilized.

9.4C Improving the Accuracy of Estimates of λ

Let $E[h]$ denote the error of approximating $\lambda_{\min} = -(\pi/L)^2$ by the least dominant eigenvalue of T. From Table 9.4-1,

$$h = \frac{L}{2} \Rightarrow E\left[\frac{L}{2}\right] = \lambda_{\min} - \left(-\frac{8}{L^2}\right) \doteq -\frac{1.87}{L^2}$$

$$h = \frac{L}{4} \Rightarrow E\left[\frac{L}{4}\right] = \lambda_{\min} - \left(-\frac{9.37}{L^2}\right) \doteq -\frac{0.50}{L^2} \approx \left(\frac{L/4}{L/2}\right)^2 E\left[\frac{L}{2}\right]$$

$$h = \frac{L}{5} \Rightarrow E\left[\frac{L}{5}\right] = \lambda_{\min} - \left(-\frac{9.55}{L^2}\right) \doteq -\frac{0.27}{L^2} \approx \left(\frac{L/5}{L/2}\right)^2 E\left[\frac{L}{2}\right]$$

Evidently, $E[L/n] \approx (2/n)^2 E[L/2]$, suggesting that the least dominant eigenvalue of T is an $O(h^2)$ approximation of $\lambda_{\min}$. In fact, the following more general result can be proved. *If $O(h^m)$ approximations are used to get the system $T\mathbf{y} = \lambda\mathbf{y}$ from the linear, homogeneous BVP*

$$a_2(x)y'' + a_1(x)y' + [a_0(x) - \lambda]y = 0, \qquad y(a) = y(b) = 0 \tag{15}$$

then the eigenvalues and eigenvectors of T are *$O(h^m)$ approximations of the corresponding characteristic values and solutions of the BVP.*

TABLE 9.4-1 Approximating Characteristic Values by Eigenvalues of *T*

n	h	Characteristic Polynomial of T $p_T(\lambda) = det(T - \lambda I_{n-1})$	Eigenvalues of T (roots of $p_T(\lambda)$)	Characteristic Values $\lambda = -\left(\frac{k\pi}{L}\right)^2$
2	$\frac{L}{2}$	$\frac{1}{h^2}\{-2 - \lambda h^2\}$	$-\frac{2}{h^2} = -\frac{8}{L^2}$	$-\left(\frac{\pi}{L}\right)^2 \doteq -\frac{9.87}{L^2}$
3	$\frac{L}{3}$	$\frac{1}{h^2}\{(-2 - \lambda h^2)^2 - 1\}$	$-\frac{1}{h^2} = -\frac{9}{L^2}$	$-\left(\frac{\pi}{L}\right)^2 \doteq -\frac{9.87}{L^2}$
			$-\frac{3}{h^2} = -\frac{27}{L^2}$	$-\left(\frac{2\pi}{L}\right)^2 \doteq -\frac{39.48}{L^2}$
4	$\frac{L}{4}$	$\frac{1}{h^2}\{(-2 - \lambda h^2)[(-2 - \lambda h^2)^2 - 2]\}$	$\frac{1}{h^2}(-2 + \sqrt{2}) \doteq -\frac{9.37}{L^2}$	$-\left(\frac{\pi}{L}\right)^2 \doteq -\frac{9.87}{L^2}$
			$-\frac{2}{h^2} = -\frac{32}{L^2}$	$-\left(\frac{2\pi}{L}\right)^2 \doteq -\frac{39.48}{L^2}$
			$\frac{1}{h^2}(-2 - \sqrt{2}) \doteq -\frac{54.63}{L^2}$	$-\left(\frac{3\pi}{L}\right)^2 \doteq -\frac{88.83}{L^2}$
5	$\frac{L}{5}$	$\frac{1}{h^2}\{(-2 + \lambda h^2)^4 - 3(-2 - \lambda h^2)^2 + 1\}$	$\frac{1}{h^2}\left(-2 + \sqrt{\frac{1}{2}(3 + \sqrt{5})}\right) \doteq -\frac{9.55}{L^2}$	$-\left(\frac{\pi}{L}\right)^2 \doteq -\frac{9.87}{L^2}$
			$\frac{1}{h^2}\left(-2 - \sqrt{\frac{1}{2}(3 - \sqrt{5})}\right) \doteq -\frac{34.55}{L^2}$	$-\left(\frac{2\pi}{L}\right)^2 \doteq -\frac{39.48}{L^2}$
			$\frac{1}{h^2}\left(-2 - \sqrt{\frac{1}{2}(3 - \sqrt{5})}\right) \doteq -\frac{65.45}{L^2}$	$-\left(\frac{3\pi}{L}\right)^2 \doteq -\frac{88.83}{L^2}$
			$\frac{1}{h^2}\left(-2 - \sqrt{\frac{1}{2}(3 + \sqrt{5})}\right) \doteq -\frac{90.45}{L^2}$	$-\left(\frac{4\pi}{L}\right)^2 \doteq -\frac{157.91}{L^2}$

We can therefore use Richardson's improvement formula of Section 7.2E. From the $h = L/4$ and $h = L/5$ estimates of $\lambda_{\min} = -(\pi/L)^2$, we get

$$(\lambda_{\min})_{\text{improved}} = \frac{(\frac{5}{4})^2(-9.55/L^2) - (-9.37/L^2)}{(\frac{5}{4})^2 - 1} = -\frac{9.87}{L^2} \qquad \textbf{(16)}$$

This is accurate to the 3d carried in the intermediate computations.

The reader is urged to use Richardson's formula whenever more than one approximation of λ and $y(t)$ are obtained.

9.4D The Sturm–Liouville Equation

A frequently occurring linear BVP is

$$\frac{d}{dx}[f(x)y'] + [r(x) - \lambda w(x)]y = 0, \qquad y(a) = y(b) = 0 \qquad \textbf{(17)}$$

in which $f(x)$, $r(x)$, and $w(x)$ are continuous on $[a, b]$ and $w(x) > 0$ on (a, b). This BVP, called the **Sturm–Liouville equation**, has some very desirable theoretical properties [5], and lends itself to the following numerical procedure. First, replace $d[f(x)y']/dx$ at x_j by the $O(h^2)$ central difference formula but *using a stepsize $h/2$*, getting

$$\frac{f(x_j + \frac{1}{2}h)y'(x_j + \frac{1}{2}h) - f(x_j - \frac{1}{2}h)y'(x_j - \frac{1}{2}h)}{2(\frac{1}{2}h)} + [r(x_j) - \lambda w(x_j)]y_j = 0$$

Then replace $y'(x_j \pm \frac{1}{2}h)$ in the same way, getting

$$\frac{1}{h}\left\{f(x_j + \tfrac{1}{2}h)\left(\frac{y_{j+1} - y_j}{h}\right) - f(x_j - \tfrac{1}{2}h)\left(\frac{y_j - y_{j-1}}{h}\right)\right\} + r(x_j)y_j = \lambda w(x_j)y_j$$

Finally, divide by $w(x_j)$ and collect y_{j-1}, y_j, and y_{j+1} terms to get

$$\alpha_j y_{j-1} + \beta_j y_j + \gamma_j y_{j+1} = \lambda y_j \qquad \textbf{(18a)}$$

where $y_0 = y_n = 0$, and for $j = 1, \ldots, n - 1$ $\left(\text{with } h = \dfrac{b - a}{n}\right)$

$$\alpha_j = \frac{f(x_j - \frac{1}{2}h)}{h^2 w(x_j)}, \quad \gamma_j = \frac{f(x_j + \frac{1}{2}h)}{h^2 w(x_j)}, \quad \text{and} \quad \beta_j = \frac{r(x_j)}{w(x_j)} - \alpha_j - \gamma_j \qquad \textbf{(18b)}$$

In matrix form, the $(n - 1) \times (n - 1)$ tridiagonal system (18a) is

$$T\mathbf{y} = \lambda\mathbf{y} \qquad \text{where } \mathbf{y} = [y_1 \ \cdots \ y_{n-1}]^T \text{ and } T = \text{trid}\,(\boldsymbol{\alpha}, \boldsymbol{\beta}, \boldsymbol{\gamma}) \qquad \textbf{(18c)}$$

If $w(x) \equiv$ constant, then it is immediate from (18b) that $\alpha_{j+1} = \gamma_j$, $j = 1, \ldots, n-2$, and hence that the tridiagonal matrix T is symmetric. Its n eigenpairs can therefore be found simultaneously using Jacobi's method.

EXAMPLE 9.4D Use (18) with $n = 3$, 5, and 7 to get an accurate approximation of the least dominant eigenvalue of

$$\frac{d}{dx}[(1 + x^2)y'] + 2\lambda y = 0, \qquad y(0) = y(2) = 0 \tag{19}$$

Solution: This is a Sturm–Liouville equation with

$$f(x) = 1 + x^2, \quad r(x) \equiv 0, \quad \text{and} \quad w(x) \equiv w = -2$$

Hence by (18),

$$\alpha_j = \frac{1 + (x_j - \frac{1}{2}h)^2}{-2h^2}, \qquad \gamma_j = \frac{1 + (x_j + \frac{1}{2}h)^2}{-2h^2}, \qquad \beta_j = -(\alpha_j + \gamma_j) \tag{20}$$

Since $w(x) =$ constant, T will be symmetric. We can therefore use either the Power Method or Jacobi's method to get the following results:

$n = 3$ ($h = \frac{2}{3}$):

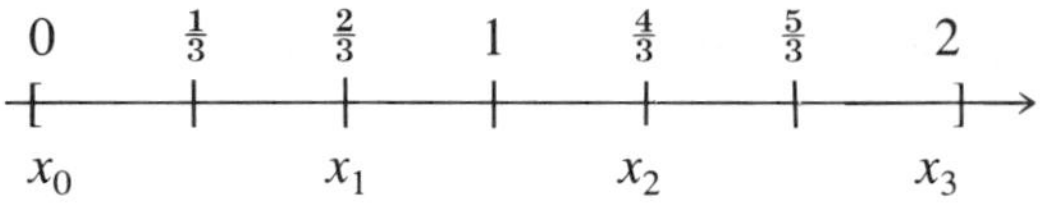

By (20),

$$T = \begin{bmatrix} \beta_1 & \gamma_1 \\ \alpha_2 & \beta_2 \end{bmatrix} = \begin{bmatrix} \frac{9}{8}(\frac{10}{9} + 2) & -\frac{9}{8}(2) \\ -\frac{9}{8}(2) & \frac{9}{8}(2 + \frac{34}{9}) \end{bmatrix} = \begin{bmatrix} 3.5 & -2.25 \\ -2.25 & 6.5 \end{bmatrix}$$

The eigenpairs of this $T_{2\times 2}$ are (to $5d$):

$$\left(2.29584, \begin{bmatrix} 0.88167 \\ 0.47186 \end{bmatrix}\right), \qquad \left(7.70416, \begin{bmatrix} -0.47186 \\ 0.88167 \end{bmatrix}\right)$$

$n = 5$ ($h = \frac{2}{5}$):

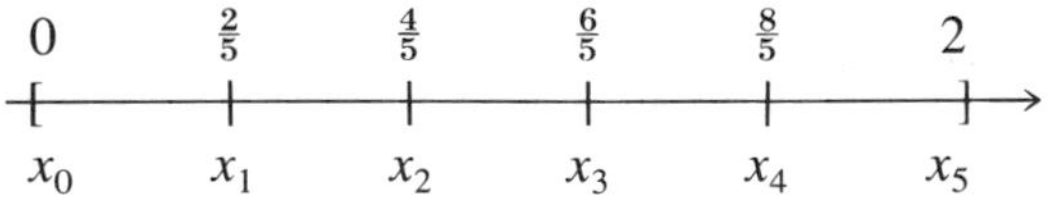

By (20),

$$T = \begin{bmatrix} 7.5 & -4.25 & & \\ -4.25 & 10.5 & -6.25 & \\ & -6.25 & 15.5 & -9.25 \\ & & -9.25 & 22.5 \end{bmatrix}$$

The two least dominant eigenpairs of this $T_{4\times 4}$ are

$$(2.46098, [\ \ 0.55283\ \ 0.65547\ \ 0.46716\ \ 0.21564]^T)$$
$$(8.17772, [-0.74538\ \ 0.11886\ \ 0.55102\ \ 0.35588]^T)$$

$n = 7$ ($h = \frac{2}{7}$): The two least dominant eigenpairs of the $T_{6\times 6}$ are

$$(2.51448, [\ \ 0.36022\ \ 0.54583\ \ 0.54293\ \ 0.42929\ \ 0.27720\ \ 0.12810]^T)$$
$$(8.89002, [-0.61134\ \ -0.38873\ \ 0.15935\ \ 0.45757\ \ 0.43205\ \ 0.23179]^T)$$

Since large values of h were used, we should use Richardson's formula to improve the three available estimates of $\lambda_{\min}$ (see Table 9.4-2).

TABLE 9.4-2 Richardson Table for $\lambda_{\min}$ Values

	h	$F[h]$, $O(h^2)$	$F_1[h]$, $O(h^4)$	$F_2[h]$
	$\frac{2}{3}$	2.29584		
$r = \frac{5}{3}$				
	$\frac{2}{5}$	2.46098	2.55387	
$r = \frac{7}{5}$				
	$\frac{2}{7}$	2.51448	2.57021	2.57902

It appears from this Richardson table that

$$\lambda_{\min} \doteq F_2[\tfrac{2}{7}] = 2.579$$

That $F_1[h]$ is $O(h^4)$ follows from the fact that the truncation error of the approximation $f'(x) \approx \delta f(x)/2h$ is of the form $Ch^2 + Dh^4 + \cdots$ [see (13b) of Section 7.2C]. ■

9.5 Using Eigenvalues to Uncover the Structure of A

In this final section we use the eigenpairs of A to help provide insight into the effect of multiplying $\mathbf{x}$ by A and to get computable formulas for $\|A\|$ and cond A (see Section 4.1D).

9.5A The Principal Axis Theorem

Let A be a symmetric $n \times n$ matrix. In this case $\|\cdot\|$ will denote the Euclidean norm $\|\cdot\|_2$. The Symmetric Diagonalization Theorem of Section 9.3A assures us of the existence of vectors $\mathbf{u}_1, \ldots, \mathbf{u}_n$ such that

$$A\mathbf{u}_i = \lambda_i \mathbf{u}_i, \quad i = 1, \ldots, n \text{ (the } \mathbf{u}_i\text{'s are eigenvectors } A) \tag{1a}$$

$$\mathbf{u}_i^T \mathbf{u}_j = \begin{cases} 1 & \text{if } i = j, \\ 0 & \text{if } i \neq j \end{cases} \quad \text{(the } \mathbf{u}_i\text{'s are orthonormal)} \tag{1b}$$

principal axes = orthonormal eigenvectors

It follows that every $\mathbf{x}$ can be represented uniquely as

$$\mathbf{x} = \alpha_1\mathbf{u}_1 + \alpha_2\mathbf{u}_2 + \cdots + \alpha_n\mathbf{u}_n = \sum_{i=1}^{n} \alpha_i\mathbf{u}_i \tag{1c}$$

Such orthonormal bases of eigenvectors are often called **principal axes** for A, and the components $\alpha_1, \ldots, \alpha_n$ along these axes the **principal components** of $\mathbf{x}$. It is immediate from (1) that

$$\|\mathbf{x}\|^2 = \mathbf{x}^T\mathbf{x} = (\Sigma_i\ \alpha_i\mathbf{u}_i)^T(\Sigma_j\ \alpha_j\mathbf{u}_j) = \Sigma_{i,j}\ \alpha_i\alpha_j\mathbf{u}_i^T\mathbf{u}_j = \Sigma_i\ \alpha_i^2 \tag{2}$$

Thus $\|\mathbf{x}\|^2$ *can be obtained as the sum of the squares of the principal components of* $\mathbf{x}$. And since $\mathbf{u}_i$ is an eigenvector for λ_i for each i,

$$A\mathbf{x} = \lambda_1\alpha_1\mathbf{u}_1 + \lambda_2\alpha_2\mathbf{u}_2 + \cdots + \lambda_n\alpha_n\mathbf{u}_n = \Sigma_i\ \lambda_i\alpha_i\mathbf{u}_i \tag{3}$$

In words, *the effect of multiplying* $\mathbf{x}$ *by* A *is to multiply the* ith *principal component of* $\mathbf{x}$ *by* λ_i. Hence, if we denote the ith principal component of $A\mathbf{x}$ by β_i, then by (3) and (2),

$$\left(\frac{\beta_1}{\lambda_1}\right)^2 + \left(\frac{\beta_2}{\lambda_2}\right)^2 + \cdots + \left(\frac{\beta_n}{\lambda_n}\right)^2 = \sum_{i=1}^{n}\left(\frac{\beta_i}{\lambda_i}\right)^2 = \sum_{i=1}^{n} \alpha_i^2 = \|\mathbf{x}\|^2$$

provided that all λ_i's are nonzero. Otherwise, $\Sigma_{i=1}^{n}$ should be replaced by $\Sigma_{\lambda_i \neq 0}$, that is, the summation over all i for which $\lambda_i \neq 0$. This proves:

THE PRINCIPAL AXIS THEOREM. *Let A be a symmetric matrix with principal axes $\mathbf{u}_1, \ldots, \mathbf{u}_n$ and let $\mathbf{y}$ be a vector with principal components $\beta_1 \ldots, \beta_n$. Then*

$$\mathbf{y} = A\mathbf{x}, \text{ where } \|\mathbf{x}\| = 1 \Leftrightarrow \sum_{\lambda_i \neq 0}\left(\frac{\beta_i}{\lambda_i}\right)^2 = 1 \tag{4}$$

In words, A multiplies the unit ball onto the hyperellipsoid with semiaxes of length $|\lambda_1|, \ldots, |\lambda_n|$ directed along the principal axes.

EXAMPLE 9.5A Sketch the set of all $A\mathbf{x}$, $\|\mathbf{x}\| = 1$, for $A = \begin{bmatrix} 0 & \sqrt{3} \\ \sqrt{3} & 2 \end{bmatrix}$.

SOLUTION: It is easy to verify that $\lambda_1, = 3$ and $\lambda_2 = -1$ are eigenvalues of A. In fact,

$$\begin{bmatrix} 0 & \sqrt{3} \\ \sqrt{3} & 2 \end{bmatrix}\begin{bmatrix} 1 \\ \sqrt{3} \end{bmatrix} = 3\begin{bmatrix} 1 \\ \sqrt{3} \end{bmatrix} \quad \text{and} \quad \begin{bmatrix} 0 & \sqrt{3} \\ \sqrt{3} & 2 \end{bmatrix}\begin{bmatrix} \sqrt{3} \\ -1 \end{bmatrix} = -1\begin{bmatrix} \sqrt{3} \\ -1 \end{bmatrix} \tag{5a}$$

hence a pair of principal axes for A is

$$\mathbf{u}_1 = \begin{bmatrix} \frac{1}{2} \\ \sqrt{\frac{3}{2}} \end{bmatrix} \quad \text{and} \quad \mathbf{u}_2 = \begin{bmatrix} \sqrt{\frac{3}{2}} \\ -\frac{1}{2} \end{bmatrix} \tag{5b}$$

So if $\mathbf{x} = \alpha_1\mathbf{u}_1 + \alpha_2\mathbf{u}_2$ and $A\mathbf{x} = \beta_1\mathbf{u}_1 + \beta_2\mathbf{u}_2$, then by (4),

$$\|\mathbf{x}\| = \alpha_1^2 + \alpha_2^2 = 1 \quad \Leftrightarrow \quad \left(\frac{\beta_1}{3}\right)^2 + \left(\frac{\beta_2}{-1}\right)^2 = 1 \tag{6}$$

Hence $A\mathbf{x}$ lies on the ellipse sketched in Figure 9.5-1.

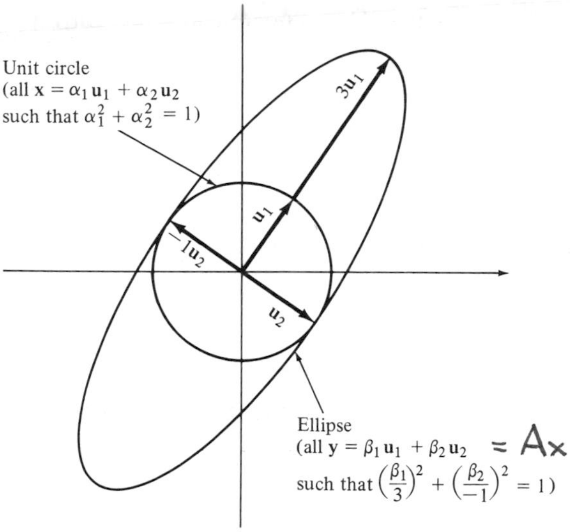

Figure 9.5-1 *Principal Axis Theorem when* $\gamma_1 = 3$, $\gamma_2 = -1$.

■

9.5B Describing $\|A\|$ and cond A When A Is Symmetric

Let λ_1 be the dominant eigenvalue of a symmetric matrix A. The Principal Axis Theorem suggests that $A\mathbf{x}$ is at most $|\lambda_1|$ times as long as $\mathbf{x}$, that is, that $\|A\| = |\lambda_1|$. To prove this, note from (2) and (3) that

$$\|A\mathbf{x}\|^2 = \Sigma\, \alpha_i^2\lambda_i^2 \leq \lambda_1^2\, \Sigma\, \alpha_i^2 = \lambda_1^2\|\mathbf{x}\|^2$$

So by the definition of matrix norm [see (25a) of Section 4.1D],

$$\|A\| = \max_{\mathbf{x}\neq 0} \frac{\|A\mathbf{x}\|}{\|\mathbf{x}\|} \leq \sqrt{\lambda_1^2} = |\lambda_1| \tag{7a}$$

Equality holds when $\mathbf{x} = \mathbf{u}_1$ (the eigenvector for λ_1) because

$$\mathbf{x} = \mathbf{u}_1 \quad \Rightarrow \quad \|A\mathbf{x}\| = \|\lambda_1\mathbf{x}\| = |\lambda_1|\,\|\mathbf{x}\| \tag{7b}$$

It follows that $\|A\| = |\lambda_1|$. Since A^{-1} is symmetric whenever A is (Problem 9-22), we have the following important result.

$\Rightarrow \|A\| = |\lambda_1|$

SPECTRAL RADIUS THEOREM *Let A be symmetric. Then with* $\|\cdot\| = \|\cdot\|_2$:

(a) $\|A\| = \max\{|\lambda_1|, \ldots, |\lambda_n|\}$, *the* **spectral radius** *of A.*

(b) *If A is also invertible, then* $\|A^{-1}\| = 1/\min\{|\lambda_1|, \ldots, |\lambda_n|\}$

(c)
$$\text{cond } A = \frac{\textit{largest } |\lambda_i|}{\textit{smallest } |\lambda_i|} \tag{8}$$

Thus, *for a symmetric matrix A, ill conditioning corresponds to A having eigenvalues of both large and small magnitude* (see Section 4.1D). Geometrically, this corresponds to the hyperellipsoid (4) being very eccentric in at least one cross section.

For example, the 2×2 matrix A in Example 9.5A satisfies

$$\|A\| = |\lambda_1| = |3| = 3, \quad \|A^{-1}\| = \frac{1}{|\lambda_2|} = \frac{1}{|-1|} = 1, \quad \text{and} \quad \text{cond } A = \frac{|3|}{|-1|} = 3$$

Since 3 is of the order of magnitude of 1, A is well conditioned.

9.5C Positive Definite Matrices

We have just seen how multiplication by a symmetric matrix A can "distort" the Euclidean unit sphere by stretching the components along its principal axes $\mathbf{u}_i$ by different amounts λ_i. One way to compensate for this distortion is to use

$$\|\mathbf{x}\|_A = \sqrt{\mathbf{x}^T A \mathbf{x}}, \qquad \text{the } \textbf{norm induced by } \boldsymbol{A} \tag{9}$$

rather than $\|\mathbf{x}\|$ to measure distance from $\mathbf{0}$. In order for $\|\mathbf{x}\|_A$ to be defined for all $\mathbf{x}$ and nonzero if $\mathbf{x} \neq \mathbf{0}$, A must have the property that

$$\mathbf{x}^T A \mathbf{x} > 0 \qquad \text{for all } \mathbf{x} \neq \mathbf{0} \tag{10}$$

Symmetric matrices having this property are called **positive definite**.

If $\mathbf{u}_1, \ldots, \mathbf{u}_n$ is a set of principal axes for the symmetric matrix A and $\alpha_1, \ldots, \alpha_n$ are the principal components of $\mathbf{x}$, then by (1) and (3),

$$\mathbf{x}^T A \mathbf{x} = \left(\sum_i \alpha_i \mathbf{u}_i\right)^T \left(\sum_j \lambda_j \alpha_j \mathbf{u}_j\right) = \sum_i \alpha_i^2 \lambda_i \tag{11}$$

Since $\alpha_i^2 > 0$, (10) is possible if and only if $\lambda_i > 0$ for all i. Thus

> A symmetric matrix is positive definite if and only if all of its eigenvalues are positive. **(12)**

The determination of whether a symmetric matrix A is positive definite can therefore be made *by inspection* whenever all its eigenvalues are known. Otherwise, this determination can be made by carrying out the Choleski Factorization Algorithm 4.2E to get

$$A = LL^T \qquad \text{where } L \text{ is lower triangular and invertible} \tag{13}$$

This factorization is possible if and only if A is positive definite.

Suppose that A is a nonsingular but not necessarily symmetric matrix. Then A^TA is symmetric [see (3) of Section 9.3A], and with $\| \ \|$ still $\|\cdot\|_2$,

$$\mathbf{x}^T(A^TA)\mathbf{x} = (\mathbf{x}^TA^T)(A\mathbf{x}) = (A\mathbf{x})^T(A\mathbf{x}) = \|A\mathbf{x}\|^2 \geq 0 \tag{14}$$

But $\|A\mathbf{x}\|^2 = 0 \Leftrightarrow A\mathbf{x} = \mathbf{0} \Leftrightarrow \mathbf{x} = \mathbf{0}$ because A is nonsingular. Similarly, $\mathbf{x}^T(AA^T)\mathbf{x} > 0$ all for $\mathbf{x} \neq \mathbf{0}$. Thus

$$A^TA \text{ and } AA^T \text{ are positive definite for any nonsingular } A \tag{15}$$

This result can be generalized to any $m \times n$ matrix A for which A^TA (or AA^T) turns out to be nonsingular. In this more general form, (15) plays an important role in statistics and optimization theory.

9.5D Relating cond A to Eigenvalues When A Is Not Symmetric

Let A be an invertible matrix that is not symmetric. If λ_1 is its dominant eigenvalue and $\mathbf{v}_1$ is an associated eigenvector, then

$$\|A\| = \max_{\mathbf{x}\neq 0} \frac{\|A\mathbf{x}\|}{\|\mathbf{x}\|} \geq \frac{\|A\mathbf{v}_1\|}{\|\mathbf{v}_1\|} = \frac{\|\lambda_1\mathbf{v}_1\|}{\|\mathbf{v}_1\|} = |\lambda_1| \tag{16}$$

no matter what $\|\cdot\|$ is used on n-space. However, $\|A\|$ *may be strictly larger than* $|\lambda_1|$. To get a more precise formula, note from (14) that when $\|\cdot\|$ is $\|\cdot\|_2$,

$$\|A\|^2 = \max_{\mathbf{x}\neq 0} \frac{\|A\mathbf{x}\|^2}{\|\mathbf{x}\|^2} = \max_{\mathbf{x}\neq 0} \frac{\mathbf{x}^T(A^TA)\mathbf{x}}{\|\mathbf{x}\|^2}$$

Since A^TA is positive definite, the expression on the right gives the largest eigenvalue of A^TA (Problem 9-34). So for a nonsymmetric matrix A, parts (a), (b), and (c) of the Spectral Radius Theorem of Section 9.5B get replaced by

$$\|A\| = \sqrt{\text{largest eigenvalue of } A^TA} \tag{17a}$$

$$\|A^{-1}\| = \sqrt{\text{smallest eigenvalue of } A^TA} \tag{17b}$$

$$\text{cond } A = \left(\frac{\text{largest eigenvalue of } A^TA}{\text{smallest eigenvalue of } A^TA}\right)^{1/2} \tag{17c}$$

where $\|A\|$ and $\|A^{-1}\|$ are with respect to $\|\cdot\|_2$ on n-space.

application in multitrait variance component estimation! (Canonical transformation)

9.5E The Generalized Eigenvalue Problem

There are times when it is necessary to find λ and $\mathbf{x} = \mathbf{0}$ such that

$$B\mathbf{x} = \lambda C\mathbf{x}, \qquad \text{where } B \text{ and } C \text{ are given } n \times n \text{ matrices} \tag{18}$$

If C is nonsingular, this can be rewritten as

$$(C^{-1}B)\mathbf{x} = \lambda\mathbf{x} \qquad \text{(find an eigenvalue of } C^{-1}B\text{)} \tag{19}$$

If B and C are both symmetric, then $C^{-1}B$ need not be symmetric (see Exercise 9-19); in this case the use of (19) destroys some of the geometric structure of (18). However, if C is positive definite, then the Choleski factorization can be used to write C as

$$C = LL^T, \qquad \text{where } L \text{ is lower triangular and invertible} \tag{20a}$$

This in turn can be used to transform (18) as follows:

$$B\mathbf{x} = \lambda C\mathbf{x} \quad \Leftrightarrow \quad B(L^T)^{-1}L^T\mathbf{x} = \lambda LL^T\mathbf{x} \quad \Leftrightarrow \quad L^{-1}B(L^T)^{-1}L^Tx = \lambda L^T\mathbf{x} \tag{20b}$$

This last equality can be written as

$$A\mathbf{v} = \lambda\mathbf{v}, \qquad \text{where } A = L^{-1}B(L^T)^{-1} \qquad \text{and} \qquad \mathbf{v} = L^T\mathbf{x} \tag{21}$$

The desired λ's and $\mathbf{x}$'s in (18) can now be obtained from the eigenpairs $(\lambda, \mathbf{v})$ of this matrix A. Specifically,

$$A\mathbf{v}_i = \lambda_i\mathbf{v}_i \quad \Leftrightarrow \quad B\mathbf{x}_i = \lambda_i C\mathbf{x}_i, \qquad \text{where } \mathbf{x}_i = (L^T)^{-1}\mathbf{v}_i \tag{22}$$

Since $(L^T)^{-1} = (L^{-1})^T$ (Problem 9-18), we see from (21) that $A^T = L^{-1}B^T(L^T)^{-1}$, and hence that A is symmetric whenever B is.

PROBLEMS

Section 9.1

9-1 In part (a) of the Example 9.1A, use Gaussian elimination to show that the *only* eigenvectors for $\lambda_2 = -3$ and $\lambda_3 = 2$ are scalar multiples of $\mathbf{v}_2 = [0 \quad 1 \quad 0]^T$ and $\mathbf{v}_3 = [1 \quad 0 \quad -1]^T$, respectively.

9-2 Find the characteristic polynomial and all eigenpairs of A; then deduce the eigenpairs of A^{-1}.

(a) $A = \begin{bmatrix} 0 & 1 \\ 1 & 0 \end{bmatrix}$ **(b)** $A = \begin{bmatrix} 4 & 5 \\ 3 & 2 \end{bmatrix}$ **(c)** $A = \begin{bmatrix} 0 & 1 \\ -1 & 2 \end{bmatrix}$ **(d)** $A = \begin{bmatrix} 0 & -1 \\ 1 & 0 \end{bmatrix}$

9-3 How are the eigenpairs of αA related to those of A? Justify.

9-4 Find all eigenpairs of D and state whether D is invertible; justify your answers.
(a) $D = \text{diag}\,(8, -1, 3)$ **(b)** $D = \text{diag}\,(-2, 3, 0)$

9-5 Which elementary row operations are needed to reduce $A \to B \to C$ where

$$A = \begin{bmatrix} 1 & 1 \\ 3 & 1 \end{bmatrix}, \quad B = \begin{bmatrix} 1 & 0 \\ 0 & -2 \end{bmatrix}, \quad \text{and} \quad C = \begin{bmatrix} 2 & 0 \\ 0 & -2 \end{bmatrix}$$

Do these operations alter eigenvalues? Explain.

9-6 For A in parts (a)–(d) of Problem 9-2: Find a matrix V such that $V^{-1}AV = \text{diag}(\lambda_1, \lambda_2)$ if such a V exists. If not, explain why not.

9-7 Use your answers to parts (a) and (b) of Problem 9-6 and (15) of Section 9.1C to solve

(a) $y_1' = y_2,\ y_1(1) = 2$ **(b)** $y_1' = 4y_1 + 5y_2,\ y_1(0) = 1$
$y_2' = y_1,\ y_2(1) = -1$ $y_2' = 3y_1 + 2y_2,\ y_2(0) = 0$

Describe the approximate behavior of $y(t)$ for large t.

9-8 **(a)** Apply (15) of Section 9.1C and your answer to Problem 9-6(c) to the $(\text{IVP})_2$

$$\begin{aligned} y_1' &= y_2, & y_1(0) &= 1 \\ y_2' &= -y_1 + 2y_2, & y_2(0) &= 0 \end{aligned}$$

Can you explain why the formula does not work?
(b) Verify that $\mathbf{y}(t) = e^t[\alpha_1 + \beta_1 t \quad \alpha_2 + \beta_2 t]^T$. Find $\alpha_1, \alpha_2, \beta_1, \beta_2$.

9-9 Suppose that $L = 50 \cdot 10^{-3}$ henries and $C = 0.5 \cdot 10^{-6}$ farads in the RLC circuit of Figure 9.1-1.
(a) What is the natural frequency when $R = 0$?
(b) Find R_{crit} (in ohms) needed for critical damping.
(c) Find $i(t)$ when (i) $R = \frac{1}{2}R_{\text{crit}}$; (ii) $R = R_{\text{crit}}$; (iii) $R = 2R_{\text{crit}}$.
(d) Sketch $i(t)$ for your answers to (i)–(iii) of part (c).

Section 9.2

For A and $\mathbf{x}_0$ given in Problems 9-10–9-13, do parts (a)–(d) if possible. If not, explain what is wrong.

(a) Find $\mathbf{x}_1, \mathbf{x}_2, \mathbf{x}_3, \mathbf{x}_4$ by the Scaled Power Method.
(b) Find $\mathbf{x}_1, \mathbf{x}_2, \mathbf{x}_3, \mathbf{x}_4$ by the Inverse Power Method. Use the formula for A^{-1}.
(c) Find $\mathbf{x}_1, \mathbf{x}_2, \mathbf{x}_3, \mathbf{x}_4$ using the Shifted Inverse Power Method, shifting by the given scalar s. Use the formula for $(A - sI)^{-1}$.
(d) Determine whether the iterations in parts (a)–(c) appear to be converging to the desired eigenpair $(\lambda, \mathbf{v})$. If not, try to explain why.

9-10 $A = \begin{bmatrix} 4 & -2 \\ -3 & 5 \end{bmatrix}$, $\mathbf{x}_0 = \begin{bmatrix} 1 \\ -1 \end{bmatrix}$, $s = 3$ Eigenpairs: $\left(2, \begin{bmatrix} 1 \\ 1 \end{bmatrix}\right), \left(7, \begin{bmatrix} -2 \\ 3 \end{bmatrix}\right)$

9-11 $A = \begin{bmatrix} 4 & -2 \\ -3 & 5 \end{bmatrix}$, $\mathbf{x}_0 = \begin{bmatrix} 1 \\ 1 \end{bmatrix}$, $s = 5$ Eigenpairs: $\left(2, \begin{bmatrix} 1 \\ 1 \end{bmatrix}\right), \left(7, \begin{bmatrix} -2 \\ 3 \end{bmatrix}\right)$

9-12 $A = \begin{bmatrix} 4 & -1 \\ 1 & 2 \end{bmatrix}$, $\mathbf{x}_0 = \begin{bmatrix} -1 \\ 1 \end{bmatrix}$, $s = -1$ Eigenpair: $\left(3, \begin{bmatrix} 1 \\ 1 \end{bmatrix}\right)$

9-13 $A = \begin{bmatrix} 4 & 5 \\ 3 & 2 \end{bmatrix}$, $\mathbf{x}_0 = \begin{bmatrix} 2 \\ 1 \end{bmatrix}$, $s = 3$ Eigenpairs: $\left(-1, \begin{bmatrix} 1 \\ -1 \end{bmatrix}\right), \left(7, \begin{bmatrix} 5 \\ 3 \end{bmatrix}\right)$

9.14 For A of Example 9.2C, do three iterations of the Inverse Power Method using $\mathbf{x}_0 = [1 \quad 1 \quad 1]^T$. Explain why it is converging to $(\lambda_2, \mathbf{v}_2)$ and not $(\lambda_3, \mathbf{v}_3)$ as intended (see Problem M9-7).

9-15 Use Gerschgorin's disk theorem to describe the regions of the complex plane that contain the eigenvalues of A:

$$\text{(a) } A = \begin{bmatrix} 2 & 0.1 & -0.5 \\ 0.2 & -4 & 0.6 \\ 0.1 & 0.1 & 3 \end{bmatrix} \qquad \text{(b) } A = \begin{bmatrix} 0 & 0.1 & -0.1 \\ -2 & 8 & 1 \\ 1 & 3 & 10 \end{bmatrix}$$

9-16 Use the methods of Section 9.2 to find all eigenpairs of the matrices in parts (a) and (b) of Problem 9-15 to 3*s*.

Section 9.3

9-17 Verify that $(AB)^T = B^TA^T$ when $A = \begin{bmatrix} 1 & 2 \\ 3 & 4 \end{bmatrix}$ and $B = \begin{bmatrix} 5 & 6 & 7 \\ 8 & 9 & 0 \end{bmatrix}$

9-18 Prove: If A is invertible, so is A^T; in fact, $(A^T)^{-1} = (A^{-1})^T$. Deduce that A^{-1} is symmetric if A is.

9-19 **(a)** Let $A = \begin{bmatrix} a & b \\ b & c \end{bmatrix}$ and $B = \begin{bmatrix} d & e \\ e & f \end{bmatrix}$ be symmetric 2×2 matrices. Find a condition on a, b, c, d, e, f which assures that both AB and BA are symmetric.
(b) Find symmetric 2×2 matrices A and B such that AB is *not* symmetric.

9-20 If $A_{n \times n}$ is symmetric and B is similar to A, is B symmetric? Justify.

9-21 For any $A_{n \times n}$, show that $\mathbf{e}_j = \text{col}_j\, I_n$ satisfies $A\mathbf{e}_j = \text{col}_j A$ for $j = 1, \ldots, n$.

9-22 For $\mathbf{x} = [1 \quad -1 \quad 1]^T$ and $\mathbf{y} = [2 \quad 3 \quad -1]^T$, find $\cos \angle(\mathbf{x}, \mathbf{y})$.

9-23 Let R_θ be the 2×2 rotation matrix in (7) of Section 9.3B.
(a) Prove that multiplying $\mathbf{x} = [x_1 \quad x_2]^T$ by R_θ rotates $\mathbf{x}$ by θ radians counterclockwise (Figure 9.3-1).
(b) Using trigonometric identities, show that $R_\theta R_\phi = R_{(\theta+\phi)}$. Interpret this result geometrically.
(c) Find a 2×2 orthogonal matrix that is *not* a rotation matrix.

9-24 Use Jacobi's method to find a rotation matrix U and a diagonal matrix $D = \text{diag}(\lambda_1, \lambda_2)$ such that $U^TAU = D$.

$$\text{(a) } A = \begin{bmatrix} 5 & -2 \\ -2 & 8 \end{bmatrix} \qquad \text{(b) } A = \begin{bmatrix} 2 & 1 \\ 1 & 2 \end{bmatrix} \qquad \text{(c) } A = \begin{bmatrix} -23 & 36 \\ 36 & -2 \end{bmatrix}$$

9-25 Perform two iterations of Jacobi's method:

$$\text{(a) } A = \begin{bmatrix} 1 & -1 & 2 \\ -1 & 1 & 0 \\ 2 & 0 & 2 \end{bmatrix} \qquad \text{(b) } A = \begin{bmatrix} 0 & 3 & 2 \\ 3 & 0 & 1 \\ 2 & 1 & 0 \end{bmatrix}$$

9-26 Repeat Problem 9-25 using Jacobi's method with thresholds (Algorithm 9.3C) with *Fraction* = 0.6.

Section 9.4

9-27 Suppose that $E = 30 \cdot 10^6$ psi, $I = 100$ in.^{3}, and $L = 100$ in. for the beam considered in Example 9.4B. Find the smallest critical load $P_{\min}$ and the approximation of $P_{\min}$ using $n = 4$ subintervals (see Table 9.4-1).

9-28 Consider the homogeneous BVP $xy'' + y' + [x^2 - \lambda]y = 0$, $y(0) = y(1) = 0$.
(a) Approximate this BVP by the linear system $T\mathbf{y} = \lambda\mathbf{y}$ as in Section 9.4B using (i) $n = 2$; (ii) $n = 3$; (iii) $n = 4$ subintervals.
(b) Using any method, find to 6*s* the least dominant eigenvalue of the matrices T obtained in (i)–(iii) of part (a).
(c) For the eigenvalues $\lambda_{(i)}$, $\lambda_{(ii)}$, and $\lambda_{(iii)}$ obtained in part (b), use Richardson's formula first

to use $\lambda_{(i)}$ to improve $\lambda_{(ii)}$, and then to use $\lambda_{(ii)}$ to improve $\lambda_{(iii)}$. Assuming that these improved approximations are $O(h^4)$, use $\lambda_{(ii),\text{improved}}$ to improve $\lambda_{(iii),\text{improved}}$.

9-29 Do parts (a)–(c) of Problem 9-28 for the BVP $(1 + x^2)y'' + 2xy' - \lambda y = 0$, $y(-1) = y(1) = 0$.

9-30 Consider the BVP $y'' - \lambda y = 0$, $y(0) = y(L) = 0$.
(a) Show that it is a Sturm–Liouville equation.
(b) Set up the T matrix in (18c) of Section 9.4C. How does it compare to that given in (11) of Section 9.4B?

9-31 Show that the BVP in Problem 9-28 is a Sturm–Liouville equation. Then do (a)–(c) of Problem 9-28 using the matrix T in (18c) of Section 9.4C. Compare the accuracy obtained to that of Problem 9-28.

9-32 Repeat Problem 9-31 for the BVP given in Problem 9-29.

Section 9.5

9-33 Use the definition of $\|A\|$ given in (25a) of Section 9.1D to prove that:
(a) If $D = \text{diag}(\lambda_1, \lambda_2, \ldots, \lambda_n)$, then $\|D\| = \max\{|\lambda_1|, \ldots, |\lambda_n|\}$.
(b) If $\|\cdot\| = \|\cdot\|_2$ on n-space, then $\|U\| = 1$ for any orthogonal matrix U.

9-34 Prove: If A is positive definite, then [see (11) of Section 9.5C]

$$\frac{\mathbf{x}^T A \mathbf{x}}{\|\mathbf{x}\|^2} = \text{largest (most positive) eigenvalue of } A$$

9-35 For the matrices A in parts (a)–(c) of Problem 9-24, and with $\|\cdot\| = \|\cdot\|_2$, sketch the set of all $A\mathbf{x}$ ($\|\mathbf{x}\| = 1$). Then use the Spectral Radius Theorem to find $\|A\|$, $\|A^{-1}\|$, and cond A.

9-36 Find $p_{AA^T}(\lambda)$, then use the results of Section 9.5D to find cond A.

(a) $A = \begin{bmatrix} 2 & 0 \\ -1 & \sqrt{3} \end{bmatrix}$ **(b)** $A = \begin{bmatrix} 100 & 1 \\ 0.1 & 0 \end{bmatrix}$

9-37 Which of the matrices in Problem 9-24 are positive definite? Justify.

9-38 Suppose that $B = \begin{bmatrix} 2 & 4 \\ 4 & 14 \end{bmatrix}$, $C = \begin{bmatrix} 1 & 3 \\ 3 & 13 \end{bmatrix}$, and $L = \begin{bmatrix} 1 & 0 \\ 3 & 2 \end{bmatrix}$.
(a) Verify that $C = LL^T$ and use this to transform $B\mathbf{x} = \lambda C\mathbf{x}$ into $A\mathbf{v} = \lambda\mathbf{v}$.
(b) Find the eigenpairs of A in part (a) and use them to get all solutions of $B\mathbf{x} = \lambda C\mathbf{x}$.

MISCELLANEOUS PROBLEMS

M9-1 Prove:
(a) If A is similar to B, then det A = det B and $p_A(\lambda) = p_B(\lambda)$.
(b) If A and B are diagonalizable with $p_A(\lambda) = p_B(\lambda)$, then A is similar to B.

M9-2 **(a)** *Prove*: If $\lambda_1, \ldots, \lambda_n$ are distinct eigenvalues of A (i.e., $\lambda_i \neq \lambda_j$ for $i \neq j$) with corresponding eigenvectors $\mathbf{v}_1, \ldots, \mathbf{v}_n$, then $\mathbf{v}_1, \ldots, \mathbf{v}_n$ are linearly independent. OUTLINE: Assume $\mathbf{v}_j = \Sigma_{i \neq j} \alpha_i \mathbf{v}_i$ where $\alpha_i \neq 0$ for all i, and reach a contradiction.
(b) Use the result in part (a) to prove that every matrix having n distinct eigenvalues is diagonalizable.

M9-3 Let $D = \text{diag}(1, 1, 3)$. Show that for *any* nonsingular matrix V, $A = VDV^{-1}$ is diagonalizable although $\lambda = 1$ is a repeated eigenvalue.

M9-4 Prove: $A\mathbf{v} = \lambda\mathbf{v} \Rightarrow A^k\mathbf{v} = \lambda^k\mathbf{v}$ for $k = 1, 2, \ldots$.

M9-5 (***Polar Decomposition***) Show that any nonzero vector $\mathbf{x}$ in n-space can be written as $\rho\mathbf{u}$, where $\rho = \|\mathbf{x}\| > 0$ and $\mathbf{u}$ is a unit vector.

M9-6 Prove:
(a) If A preserves length (e.g., if A is orthogonal), then all the eigenvalues of A satisfy $|\lambda| = 1$.
(b) For any unit vector $\mathbf{u}$, $P = I - 2\mathbf{u}\mathbf{u}^T$ is symmetric and orthogonal.

M9-7 (***Formula for Components Relative to an Orthogonal Basis***) Suppose that $\mathbf{v}_1, \ldots, \mathbf{v}_n$ are nonzero and orthogonal (i.e., $\mathbf{v}_i^T\mathbf{v}_j = 0$ for $i \neq j$). Then, for any $\mathbf{x}$,

$$\mathbf{x} = \sum_{j=1}^{n} \alpha_j \mathbf{v}_j \Rightarrow \alpha_j = \frac{\mathbf{x}^T\mathbf{v}_j}{\|\mathbf{v}_j\|^2}, \qquad j = 1, \ldots, n$$

Deduce that $\mathbf{v}_1, \ldots, \mathbf{v}_n$ are linearly independent (see Section 5.3D).

M9-8 **(a)** Find conditions on $a_2(x)$ and $a_1(x)$ that imply that

$$(*)\ a_2(x)y'' + a_1(x)y' + [a_0(x) - \lambda]y = 0, \qquad y(a) = y(b) = 0$$

is a Sturm–Liouville equation [see (17) of Section 9.4C].
(b) Find an integrating factor $\mu(x)$ such that multiplying $(*)$ by $\mu(x)$ converts it to the Sturm–Liouville equation

$$\frac{d}{dx}[a_2(x)\mu(x)y'] + \mu(x)[a_0(x) - \lambda]y = 0, \qquad y(a) = y(b) = 0$$

M9-9 It can be shown [37] that $A_{n\times n}$ is positive definite if and only if the determinants of all upper left-hand submatrices are positive, that is,

$$\det(a_{ij})_{k\times k} > 0 \qquad \text{for } k = 1, \ldots, n$$

Use this result to determine those A matrices of Problem 4-15 that are positive definite.

COMPUTER PROBLEMS

C9-1 **(a)** Write a subprogram `SCALE(N, X, BIGXI, XSCALED)` that finds *BigXi*, the dominant component of $\mathbf{x}$, and forms $\mathbf{x}_{\text{scaled}} = (1/BigXi)\mathbf{x}$.
(b) Use `MULT` of Problem C3-1(b) and `SCALE` of part (a) in a subprogram `POWER(N, A, X, BIGXI, MAXIT,` termination test parameters`)` that implements the **iterate** steps of the Scaled Power Method Algorithm 9.2A. Test it using Example 9.2A. Be sure that the subprogram returns A unchanged to the calling program.

C9-2 Modify `POWER` of Problem C9-1(b) by adding the parameters `SHIFT` and `S` ($= s$) to the parameter list. The modified program should be as in Problem C9-1(c) if `SHIFT = FALSE` but should do the Shifted Inverse Power Method of Section 9.2D if `SHIFT = TRUE`. Test it using Examples 9.2C ($s = 0$) and 9.2D ($s = -\frac{5}{2}$).

C9-3 Use any computer program(s) available to find to 5s all eigenpairs of the following matrices A.
(a) Problem 9-15(a) **(b)** Probelm 9-15(b) **(c)** Problem 9-25(a)
(d) Problem 9-25(b) **(e)** Problem 4-15(a) **(f)** Problem 4-15(b)

APPENDIX I

Using Pseudoprograms to Describe Algorithms

Pseudoprograms describe the essential steps of an algorithm. In this book, they will begin with

> **ALGORITHM.** Name (optional comments).
>
> **Purpose:** General description, introducing the important variables used.

This will be followed by the pseudocode itself, using structures and conventions described in the following paragraphs. Suggestions for translating pseudocode into a particular programming language are given in Sections 2.2C, 2.4E, and 5.2C.

I.1 Identifiers (Names of Variables)

Most variables assume a single value that is either a **scalar** (i.e., real number), an **integer**, or a **boolean** (or **logical**) value (TRUE or FALSE). Such variables will be denoted by a single Greek or italic letter (e.g., X, i, α, ϵ) or a descriptive names in italics (e.g., *Xprev, Numsig, DxMax, Tol, OK, Convgd*). Variables that store **vectors** (**one-dimensional arrays**) will be denoted by lowercase boldface letters as follows:

Vector	b	x_0	row
*i*th entry	b_i or **b**(*i*)	$x_{0,i}$ or x_{0i}	**row**(*i*)

Matrices (two-dimensional arrays) will be denoted by uppercase italic letters such as A, B, DD, with *ij*th entries a_{ij}, b_{ij}, and $DD(i, j)$, respectively. Finally **functions** will be denoted in the customary way using notation such as $f(x)$, $\phi(x)$, $f(t, y)$, $y(t)$, and so on.

I.2

Input/Output

The input values needed for the execution of an algorithm will be described using a statement of the form

GET input data list

For example, an algorithm to scale an m by n matrix A by a scalar α would begin with

GET *Alpha*, *m*, *n*, *A*

It is up to the person implementing the algorithm to decide whether the entries of the input data list are entered interactively from a terminal, read from a file, passed as parameters (or in a COMMON area) to a subroutine, or simply defined as constants within the program itself. These options are discussed in Sections 2.2C, 2.4E, and 5.2C.

Intermediate and final values and status messages that may be of interest to the user are indicated by a statement of the form

OUTPUT (values and/or message describing values or current status)

with the descriptive message be put in single quotes. For example,

OUTPUT (*X*' appears to have '*NumSig*' significant digit accuracy.')

OUTPUT statements are included to indicate when certain values are available. It is left to the person implementing the algorithm to decide whether to omit, include, or even augment these statements, depending on the nature of the implementation.

I.3

Assign Statements (←)

The back-arrow (←) will be used to assign a value to a variable. For example, the statement

$$Det \leftarrow Det \cdot a_{jj}$$

has the effect of multiplying the current value of *Det* by the current value of a_{jj} and storing the result as the new current value of *Det*. Generally, each line of pseudocode contains only one statement, either beginning at the left margin or indented as described below. Sometimes several assign statements will be made on a single line. When this is done, they will be separated by semicolons (;).

I.4

Conditional Statements [IF ··· THEN (··· ELSE)]

The structure used to take an action only if a certain condition holds is

If (logical variable or expression for the condition) THEN action

If the action consists of several statements, we shall use the indented format

```
IF (logical variable or expression for the condition) THEN
└─statements requiring one or several lines describing the action
```

For example, the conditional statement

IF $Val > MaxVal$ THEN $Val \leftarrow MaxVal$

ensures that *Val* never exceeds *MaxVal*. To count the number of times the assign statement $Val \leftarrow MaxVal$ was executed, we could use the alternative conditional statement

```
IF Val > MaxVal THEN
└─Val ← MaxVal; n ← n + 1
```

after initializing n to 0 earlier in the pseudoprogram.

The structure used to take one Action 1 if a certain condition holds, and Action 2 if it fails is

```
IF (logical variable or expression for the condition)
│ THEN statements describing Action 1
└─ELSE statements describing Action 2
```

For example, the conditional statement

```
IF L · Y > 0
│ THEN Y ← −Y
└─ELSE L ← 2L; Y ← Y/2
```

changes the sign of Y if L and Y have the same sign, and doubles L and halves Y otherwise.

I.5

Repetition Statements (Iteration)

If one or several statements are to be repeated a specified number of times, we shall use the **count controlled loop** structure

```
DO FOR Index = Istart TO Istop STEP k
└──statement(s) to be repeated
```

where $k > 0$. This causes the statement(s) to be executed with *Index* = *Istart*, then *Istart* + k, then *Istart* + $2k$, ..., until *Index* > *Istop*. No statements are executed if *Istart* > *Istop*†. We will omit the "STEP k" in the frequently occurring case when $k = 1$. To count backwards, we will use the analogous loop structure

```
DO FOR Index = Istart DOWNTO Istop STEP k
└──statement(s) to be repeated
```

where $k < 0$. Thus, for example, the two pseudoprogram fragments

```
Sum1 ← 0; Sign ← +1
DO FOR i = 2 TO 5
│   Sum1 < Sum1 + Sign · a_i
└───Sign ← -Sign
```

```
Sum2 ← 0; Sign ← -1
DO FOR i = 7 DOWNTO 2 STEP -2
│   Sum2 < Sum2 + Sign · a_i
└───Sign ← -Sign
```

form *Sum1* $= a_2 - a_3 + a_4 - a_5$ and *Sum2* $= -a_7 + a_5 - a_3$, respectively.

When the number of iterations is not known in advance, then a *logical condition* must be used to control the repetition. This will be done using one of the two **conditionally controlled loop** structures:

```
DO WHILE (condition)             DO UNTIL (condition)
└──statement(s) to repeat        └──statement(s) to repeat
```

These result in the following statements, respectively, being repeated indefinitely

```
IF (condition holds)                Execute statement(s)
│ THEN execute statement(s)         IF (condition fails)
└─ELSE exit loop                    └──THEN exit loop
```

Notice that the DO WHILE structure performs *no* iterations if (condition) fails upon entering the loop; on the other hand, the DO UNTIL structure executes the statement(s) *at least once*. To illustrate this difference, suppose that we want *GreatInt* to be the greatest integer that lies at or to the left of a given positive real number x on the number line. Both pseudoprogram fragments

†In Fortran 77, DO-loops are executed at least once, even if *Istart* > *Istop*. If this is a possibility, then a conditional statement of the form "IF (ISTART.LE.ISTOP) THEN execute the DO-loop" should be used to implement the psuedocode faithfully.

$GreatInt \leftarrow 0$
DO WHILE $GreatInt \leqslant x - 1$
└── $GreatInt \leftarrow GreatInt + 1$

$GreatInt \leftarrow 0$
DO UNTIL $GreatInt > x - 1$
└── $GreatInt \leftarrow GreatInt + 1$

will calculate *GreatInt* correctly for $x \geqslant 1$. However, if $0 < x < 1$, then the DO WHILE fragment will (correctly) yield $GreatInt = 0$, whereas the DO UNTIL fragment will (incorrectly) yield $GreatInt = 1$.

Numerical methods often require performing a calculation repeatedly until a desired accuracy is achieved. However, to guard against the possibility of an "infinite loop" if this accuracy is never achieved, it is prudent to put an upper limit, say *MaxIt*, on the number of iterations that can be performed. This will be done using a "hybrid" conditionally *and* count controlled loop structure of the form

DO FOR $k = 1$ TO *MaxIt*
│ statement(s) to be repeated
└── UNTIL a **termination test** is satisfied

Exit from such a loop will occur after *MaxIt* iterations, or sooner if it takes fewer iterations for the **termination test** to be satisfied.

I.6

Descriptive Statements, Comments, and Labels

Ordinary English prose, set in ordinary type, will be used to describe actions to be taken in *general terms*. For example,

GET termination test parameters

indicates the need to specify these parameters without specifying their precise nature. Similarly, a statement such as

Solve $A\mathbf{x} = \mathbf{b}$ for $\bar{\mathbf{x}}$

leaves the precise method of solution up to the user and suggests that it should be implemented by invoking a subprogram.

Clarifying comments will be inserted in braces ({ ··· }). For example,

If $L \cdot Y > 0$ {L and Y have the same sign}

leaves no doubt about the purpose of the logical expression $L \cdot Y > 0$. Also enclosed in braces will be **labels**. These are brief phrases, set in **boldface** or ***boldface italic***, that are placed immediately before or after the statement(s) they describe. Their purpose is to provide an outline of the major steps of the algorithm. For example, an iterative numerical method for getting *Val* to a prescribed accuracy proceeds as follows:

```
{Initialize}
GET Maxit, InitVal, termination test parameters
statements preparing for the iteration
Val ← InitVal                                        {initial guess of Val}

{Iterate}
DO FOR k = 1 TO MaxIt
|     statement(s) to be repeated                    {making Val more accurate}
|____UNTIL termination test
OUTPUT (results of the iteration)
STOP
```

Notice that the comments and labels (in braces) help clarify the pseudocode but are not actually part of it.

I.7

The STOP Statement

The STOP statement will be used as in the preceding pseudoprogram to indicate the successful end of an algorithm. It will also be used to terminate the algorithm if for some reason the algorithm fails before reaching its desired termination point. The person implementing such a STOP statement must decide how to record this failure (e.g., set a boolean flag and/or OUTPUT an error message) and must terminate the execution in an appropriate way.

APPENDIX II

Review of the Basic Results of Calculus

II.1

Continuity

A Definition

Let f be a function that is defined in a neighborhood of the point a. Then f is **continuous at a** if

$$\lim_{x \to a} f(x) = f(a)$$

Geometrically, the graph of f is "unbroken" at $(a, f(a))$. Thus, if f is continuous on a whole interval I, its graph is an unbroken curve over I. Two important results that say this more precisely are given next.

B Extreme Value Theorem (EVT)

If f is continuous on a closed interval $[a, b]$, then f assumes a maximum value M and a minimum value m; that is, there exist points x_M and x_m in $[a, b]$ such that

$$m = f(x_m) \leq f(x) \leq f(x_M) = M \qquad \text{for any } x \text{ in } [a, b]$$

C Intermediate Value Theorem (IVT)

If f is continuous on a closed interval $[a, b]$ and $m < c < M$, where m and M are as in the preceding theorem, then there exists at least one point ξ_c in $[a, b]$ such that

$$f(\xi_c) = c$$

It follows from IVT that if f is continuous on $[a, b]$ and $f(a)$ and $f(b)$ have opposite sign, then f has at least one root between a and b.

Taken together, EVT and IVT say that if f *is continuous on a closed interval* $[a, b]$, *then its range is also a closed interval*, namely $[m, M]$; moreover if $\xi_1, \ldots, \xi_n$ lie in $[a, b]$ then, since $nm \leq \sum_{k=1}^{n} f(\xi_k) \leq nM$, there must be a point ξ in $[a, b]$ such that

$$\frac{1}{n}\{f(\xi_1) + \cdots + f(\xi_n)\} = f(\xi)$$

II.2 Derivatives

A Definition

Let f be defined in a neighborhood of the point a. The **derivative of *f* at *a*** is the number

$$f'(a) = \lim_{x \to a} \frac{\Delta f(a)}{h}, \qquad \text{where } \frac{\Delta f(a)}{h} = \frac{f(x) - f(a)}{x - a} = \frac{f(a + h) - f(a)}{h}$$

Recall that *differentiability at a guarantees continuity at a.*

Geometrically, $f'(a)$ gives the slope of the *tangent line* at $(a, f(a))$ on the graph of f. The number $\Delta f(a)/h$, called the **difference quotient of *f*** at $x = a$, gives the slope of the *secant line* from $(a, f(a))$ to the nearby point $(x, f(x))$ on the graph of f. Similarly, $f''(a) = df'(x)/dx]_{x=a}$ indicates the *concavity* of the graph of f near $(a, f(a))$.

B Mean Value Theorem (MVT)

If f is continuous on $[a, b]$ and differentiable on (a, b), then there exists at least one point ξ between a and b such that

$$f(b) = f(a) + f'(\xi)(b - a)$$

Suppose that f is continuous on $[a, b]$ and $f^{(n)}$ exists on (a, b) and it is known that f has $n + 1$ distinct roots, say $x_0, x_1, \ldots, x_n$, on $[a, b]$. Since $f(x_i) = 0$ for $i = 0, 1, \ldots, n$, applying MVT to the n subintervals

$$[x_0, x_1], \quad [x_1, x_2], \quad \ldots, \quad [x_{n-1}, x_n]$$

shows that there are points ξ_i in (x_{i-1}, x_i) such that $f'(\xi_i) = 0$ for $i = 1, 2, \ldots n$. In other words, f' has n distinct roots on (a, b). Continuing inductively, we see that f'', f''', $\ldots$, $f^{(n)}$ have, respectively, $n - 1, n - 2, \ldots, 1$ roots. This proves the following result.

C Generalized Rolle's Theorem

If f is continuous and has $n + 1$ distinct roots on $[a, b]$ and $f^{(n)}$ exists on (a, b), then there exists a point ξ in (a, b) such that $f^{(n)}(\xi) = 0$.

D Taylor's Theorem with Lagrange's Form of the Remainder

If $f(a), f'(a), \ldots, f^{(n)}(a)$ exist and x is near a, then the ***n*th Taylor polynomial** (based at $x = a$), namely

$$P_n(x) = f(a) + \frac{f'(a)}{1!}(x - a) + \cdots + \frac{f^{(n)}(a)}{n!}(x - a)^n$$

can be used to approximate $f(x)$. Define the ***n*th remainder** $R_n(x)$ by

$$R_n(x) = f(x) - P_n(x) = \text{the error of approximating } f(x) \text{ by } P_n(x)$$

If $f^{(n)}$ is continuous on $[a, x]$ and $f^{(n+1)}$ exists on (a, x), then there exists at least one point c between a and x such that

$$R_n(x) = \frac{f^{(n+1)}(c)}{(n + 1)!}(x - a)^{n+1} \qquad \textbf{(Lagrange's form)}$$

Some frequently used Taylor Polynomials are shown, along with their remainders in Lagrange's form, in Table II.2-1.

TABLE II.2-1 Some Familiar Taylor Polynomial Approximations

Base Point	*Taylor Approximation*	*Remainder (c is between a and x)*
$a = 0$	$e^x \approx 1 + \frac{x}{1!} + \frac{x^2}{2!} + \cdots + \frac{x^n}{n!}$	$\frac{e^c}{(n+1)!}x^{n+1}$
$a = 1$	$\ln x \approx \frac{x-1}{1} - \frac{(x-1)^2}{2} + \cdots + (-1)^{n-1}\frac{(x-1)^n}{n}$	$\frac{(-1)^n(x-1)^{n+1}}{(n+1)c^{n+1}}$
$a = 0$	$\sin x \approx x - \frac{x^3}{3!} + \frac{x^5}{5!} - \cdots + (-1)^n\frac{x^{2n+1}}{(2n+1)!}$	$\frac{(-1)^{n+1}\cos c}{(2n+3)!}x^{2n+3}$
$a = 0$	$\cos x \approx 1 - \frac{x^2}{2!} + \frac{x^4}{4!} - \cdots + (-1)^n\frac{x^{2n}}{(2n)!}$	$\frac{(-1)^{n+1}\cos c}{(2n+2)!}x^{2n+2}$
$a = 1$	$x^p \approx 1 + p(x-1) + \frac{p(p-1)}{2!}(x-1)^2 + \cdots + \frac{p(p-1)\cdots(p-n+1)}{n!}(x-1)^n$	$\frac{p(p-1)\cdots(p-n)(1-x)^{n+1}}{(n+1)!\,c^{n+1-p}}$

E Taylor Series Representation

If f has a Taylor series representation

$$\textbf{(TS)}\ f(x) = \sum_{n=0}^{\infty} \frac{f^{(n)}(a)}{n!}(x-a)^n = f(a) + \frac{f'(a)}{1!}(x-a) + \frac{f''(a)}{2!}(x-a)^2 + \cdots$$

with radius of convergence R, then f has continuous derivatives of all orders for $|x - a| < R$; in fact, on the interval of convergence, $f^{(k)}(x)$ is represented by the series obtained by differentiating the Taylor series (TS) termwise k times, for $k = 1, 2, \ldots$.

II.3

h-Increment Notation

If we fix x instead of a, then a point near x will be denoted by $x + h$, where h is a small, nonzero increment. In this notation,

$$f \text{ is continuous at } x \quad \Leftrightarrow \quad \lim_{h\to 0} f(x+h) = f(x)$$

and the definition of the derivative of f at x can be written as

$$f'(x) = \lim_{h\to 0} \frac{\Delta f(x)}{h}, \qquad \text{where } \frac{\Delta f(x)}{h} = \frac{f(x+h) - f(x)}{h}$$

Finally, the nth Taylor approximation (based at x) can be described as

$$f(x+h) \approx P_n(x+h) = f(x) + \frac{f'(x)}{1!}h + \cdots + \frac{f^{(n)}(x)}{n!}h^n$$

$$R_n(x+h) = f(x+h) - P_n(x+h) = \frac{f^{(n+1)}(x+\theta h)}{(n+1)!}h^{n+1}, \qquad \text{where } 0 < \theta < 1$$

Notice that $x + \theta h$ lies between x and $x + h$ whether $h > 0$ or $h < 0$.

II.4

Definite Integrals

A Definitions

If f is defined on $[a, b]$ $(a < b)$, let

$$a = x_0 < x_1 < \cdots < x_k < \cdots < x_{n-1} < x_n = b$$

be a partition of $[a, b]$. A **Riemann sum** for this partition is

$$f(c_1)\Delta x_1 + f(c_2)\Delta x_2 + \cdots + f(c_n)\Delta x_n = \sum_{k=1}^{n} f(c_k)\Delta x_k$$

where

$$\Delta x_k = x_k - x_{k-1} \text{ (the length of } [x_{k-1}, x_k]) \qquad \text{and} \qquad x_{k-1} \leqslant c_k \leqslant x_k$$

The **definite integral of f on $[a, b]$** is defined as

$$\int_a^b f(x)\,dx = \lim_{\max\Delta x_k \to 0} \left(\sum_{k=1}^{n} f(c_k)\,\Delta x_k\right)$$

This means that for any $\epsilon > 0$ you can find a $\delta > 0$ such that

$$\max \Delta x_k < \delta \;\Rightarrow\; \left|\int_a^b f(x)\,dx - \sum_{k=1}^{n} f(c_k)\Delta x_k\right| < \epsilon$$

no matter what values are used for c_k, $k = 1, 2, \ldots, n$. When this limit exists, f is **(Riemann) integrable over $[a, b]$**.

B Fundamental Theorem of Calculus

If f is continuous on $[a, b]$, then f is integrable on $[a, b]$ and

$$\int_a^b f(x)\,dx = F(x)\Big]_a^b = F(b) - F(a)$$

where $F(x)$ is any antiderivative of $f(x)$, that is, $F'(x) = f(x)$ on $[a, b]$.

C Generalized Mean Value Theorem for Integrals

If f is continuous on $[a, b]$ and g is a *nonnegative* integrable function on $[a, b]$, then there exists a point c in (a, b) such that

$$\int_a^b f(x)g(x)\,dx = f(c)\int_a^b g(x)\,dx$$

When $\int_a^b f(x)\,dx$ is defined as in A above (i.e., as a limit of approximating Riemann sums), $\int_a^b f(x)\,dx$ is called a **proper** integral. It is easy to see that $\int_a^b f(x)\,dx$ *cannot* be proper if either $a = -\infty$ or $b = +\infty$, or if f is unbounded on $[a, b]$; **improper** integrals, defined for these situations, are described in Section 7.4.

II.5

Partial Derivatives

Suppose that f is a function of several variables, say $x_1, x_2, \ldots, x_n$. It will be convenient to use the vector $\mathbf{x}$ to denote the n-tuple $(x_1, \ldots, x_n)$. Then

$$f(x_1, x_2, \ldots, x_n) \text{ will be abbreviated as } f(\mathbf{x})$$

A Definition

The **partial derivative of f with respect to x_j** at $\mathbf{x} = (x_1, \ldots, x_n)$ is the number

$$\frac{\partial f}{\partial x_j}(\mathbf{x}) = \lim_{h \to 0} \frac{f(\mathbf{x} + h\mathbf{e}_j) - f(\mathbf{x})}{h}$$

where

$$\mathbf{x} + h\mathbf{e}_j = (x_1, \ldots, x_{j-1}, x_j + h, x_{j+1}, \ldots, x_n)$$

The number $\partial f(\mathbf{x})/\partial x_j$ is also denoted by $f_{x_j}(\mathbf{x})$

B Theorem

If the second mixed partials

$$f_{x_i x_j} = \frac{\partial}{\partial x_j}(f_{x_i}) \qquad \text{and} \qquad f_{x_j x_i} = \frac{\partial}{\partial x_i}(f_{x_j})$$

both exist in an open region (of n-space) on which one of them is known to be continuous, then $f_{x_i x_j}(\mathbf{x}) = f_{x_j x_i}(\mathbf{x})$ for all $\mathbf{x}$ in this region.

This result allows us to find higher-order partials of f in any convenient order as long as the result is continuous.

C Definition

If $f_{x_1}(\mathbf{x}), \ldots, f_{x_n}(\mathbf{x})$ all exist, then the **total differential of f at $\mathbf{x}$** is defined by

$$df(\mathbf{x}) = f_{x_1}(\mathbf{x})\,dx_1 + f_{x_2}(\mathbf{x})\,dx_2 + \cdots + f_{x_n}(\mathbf{x})\,dx_n$$

where $dx_1, \ldots, dx_n$ can be viewed as increments from $x_1, \ldots, x_n$, respectively.

The following result justifies the use of the **linear approximation**

$$f(\mathbf{x} + \mathbf{dx}) = f(x_1 + dx_1, \ldots, x_n + dx_n) \approx f(\mathbf{x}) + df(\mathbf{x})$$

when the increment $\mathbf{dx} = (dx_1, \ldots, dx_n)$ is "small" in the sense that

$$\|\mathbf{dx}\| = \max\{|dx_1|, \ldots, |dx_n|\} \approx 0$$

D Theorem

If all second partials of f are continuous in a neighborhood of $\mathbf{x}$, then the error of approximating $f(\mathbf{x} + \mathbf{dx})$ by $f(\mathbf{x}) + df(\mathbf{x})$ goes to zero faster than $\mathbf{dx}$ in the sense that

$$\lim_{\|\mathbf{dx}\| \to 0} \frac{f(\mathbf{x} + \mathbf{dx}) - [f(\mathbf{x}) + df(\mathbf{x})]}{\|\mathbf{dx}\|} = 0$$

An important consequence of this result is the following.

E Theorem (Chain Rule)

If f is a differentiable function of $x_1, x_2, \ldots, x_n$ and each of these is a function of $u_1, u_2, \ldots, u_m$, then for any $j = 1, 2, \ldots, m$,

$$\frac{\partial f}{\partial u_j} = \frac{\partial f}{\partial x_1} \cdot \frac{\partial x_1}{\partial u_j} + \frac{\partial f}{\partial x_2} \cdot \frac{\partial x_2}{\partial u_j} + \cdots + \frac{\partial f}{\partial x_n} \cdot \frac{\partial x_n}{\partial u_j}$$

where the partials with respect to u_j are at a fixed $\mathbf{u} = (u_1, \ldots, u_m)$ and the partials with respect to $x_1, \ldots, x_n$ are at the $\mathbf{x} = (x_1, \ldots, x_n)$ corresponding to this fixed $\mathbf{u}$. As a special case, if y is a twice differentiable function of t and $f(t, y)$ has continuous mixed second partial derivatives, then

$$\frac{d}{dt}[f(t, y)] = f_t + f_y y'$$

and

$$\frac{d^2}{dt^2}[f(t, y)] = f_{tt} + 2f_{ty}y' + f_{yy}f^2 + f_y[f_t + f_y f]$$

Bibliography

1. Acton, F. S. (1970), *Numerical Methods That (Usually) Work.* Harper & Row, Publishers, New York.
2. Anton, H. (1977), *Elementary Linear Algebra*, 2nd ed. John Wiley & Sons, Inc., New York.
3. Bailey, P. B., L. F. Shampine, and P. E. Waltman (1968). *Nonlinear Two-Point Boundary-Value Problems.* Academic Press, Inc., New York.
4. Bartle, R. G. (1976), *The Elements of Real Analysis*, 2nd ed. John Wiley & Sons, Inc., New York.
5. Birkhoff, G., and G. Rota (1962), *Ordinary Differential Equations.* John Wiley & Sons, Inc., New York.
6. Brent, R. (1973), *Algorithms for Minimization Without Derivatives.* Prentice-Hall, Inc., Englewood Cliffs, N.J.
7. Brent, R. P. (1976) ''Fast multiple precision evaluation of elementary functions.'' *Journal of the Association for Computing Machinery*, **23**, 242–251.
8. Brown, K. M. (1969), ''A quadratically convergent Newton-like method based upon Gaussian elimination.'' *SIAM Journal of Numerical Analysis*, **6**, No. 4, 560–569.
9. Broyden, C. G. (1965), ''A class of methods for solving nonlinear simultaneous equations.'' *Mathematics of Computation*, **19**, 577–593.
10. Bulirsch, R., and J. Stoer (1966), ''Numerical treatment of ordinary differential equations by extrapolation methods.'' *Numerische Mathematik*, **8**, 1–13.
11. Butcher, J. C. (1965), ''On the attainable order of Runge–Kutta methods.'' *Mathematics of Computation*, **19**, 408–417.
12. Dahlquist, G., and Å, Björk (1974), *Numerical Methods.* Prentice-Hall, Inc., Englewood Cliffs, N.J.
13. De Boor, C. (1978), *A Practical Guide to Splines.* Springer Verlag New York, Inc., New York.
14. Draper, N. R., and H. Smith (1966), *Applied Regression Analysis.* John Wiley & Sons, Inc., New York.
15. Enright, W. H. (1974), ''Optimal second derivative methods for stiff systems,'' R. A. Willoughby, ed., *Stiff Differential Equations.* Plenum Publishing Corporation, New York.
16. Enright, W. H., T. E. Hull, and B. Lindberg (1975), ''Comparing numerical methods for stiff systems of O.D.E.'s.'' *BIT*, **15**, 10–48.

17. Fehlberg, E. (1970), "Klassiche Runge–Kutta Formeln vierter und niedrigerer Ordnung mit Schrittweiten-Kontrolle und ihre Anwendung auf Wärmeleitungs-probleme." *Computing*, **6**, 61–71.
18. Forsythe, G. E. (1957), "Generation and use of orthogonal polynomials for data-fitting with a digital computer." *Society for Industrial and Applied Mathematics Journal*, **5**, 74–88.
19. Forsythe, G. E., and C. B. Moler (1967), *Computer Solution of Linear Algebraic Systems*. Prentice-Hall, Inc., Englewood Cliffs, N.J.
20. Forsythe, G. E., M. A. Malcolm, and C. B. Moler (1977), *Computer Methods for Mathematical Computation*. Prentice-Hall, Inc., Englewood Cliffs, N.J.
21. Gragg, W. B. (1965), "On extrapolation algorithms for ordinary initial-value problems." *SIAM Journal on Numerical Analysis*, **2**, 384–403.
22. Gear, C. W. (1971), *Numerical Initial-Value Problems in Ordinary Differential Equations*. Prentice-Hall, Inc., Englewood Cliffs, N.J. Hamming, R. W. (1973), *Numerical Methods for Engineers and Scientists*, 2nd ed. McGraw-Hill Book Company, New York.
23. Hamming, R. W. (1973), *Numerical Methods for Engineers and Scientists*, 2nd ed. McGraw-Hill Book Company, New York.
24. Hearn, A. (1976), "Scientific applications of symbolic computation." *Computer Science and Scientific Computing*, J. Ortega, ed. Academic Press, Inc., New York.
25. Henrici, P. (1963), *Error Propagation for Difference Methods*. John Wiley & Sons, Inc., New York.
26. Hildebrand, F. B. (1974), *Introduction to Numerical Analysis*, 2nd ed. McGraw-Hill Book Company, New York.
27. Hull, T. E., and W. H. Enright (1976), "Test results on initial-value methods for nonstiff ordinary differential equations." *SIAM Journal of Numerical Analysis*, **13**, No. 6, 944–961.
28. Hull, T. E., W. H. Enright, B. M. Fellen, and A. E. Sedgewick (1972), "Comparing numerical methods for ordinary differential equations." *SIAM Journal of Numerical Analysis*, **9**, No. 4, 603–637.
29. Isaacson, E., and H. B. Keller (1966), *Analysis of Numerical Methods*. John Wiley & Sons, Inc., New York.
30. Johnson, G. W., and N. H. Austria (1983), "A quasi-Newton method employing direct secant updates of matrix factorizations." *SIAM Journal on Numerical Analysis*, **20**, No. 2, 315–325.
31. Joyce, D. C. (1971), "Survey of extrapolation processes in numerical analysis." *SIAM Review*, **13**, No. 4, 435–490.
32. Keller, H. B. (1968), *Numerical Methods for Two-Point Boundary-Value Problems*. John Wiley & Sons, Inc., New York.
33. Larson, H. J. (1982), *Introduction to Probability Theory and Statistical Inference*, 3rd ed. John Wiley & Sons, Inc., New York.
34. Lawson, C. L., and R. J. Hanson (1974), *Solving Least Squares Problems*. Prentice-Hall, Inc., Englewood Cliffs, N.J.
35. Luke, Y. L. (1975), *Mathematical Functions and Their Approximations*. Academic Press, Inc., New York.
36. Moré, J. J., and M. Y. Consard (1979), "Numerical solution of nonlinear equations." *ACM Transactions on Mathematical Software*, **5**, No. 1, 64–85.
37. Noble, B., and J. W. Daniel (1977), *Applied Linear Algebra*, 2nd ed. Prentice-Hall, Inc., Englewood Cliffs, N.J.
38. Ortega, J. M. (1972), *Numerical Analysis–A Second Course*. Academic Press, Inc., New York.
39. Ortega, J. M., and W. C. Rheinboldt (1970), *Iterative Solution of Nonlinear Equations in Several Variables*. Academic Press, Inc., New York.
40. Ralston, A., and P. Rabinowitz (1978), *A First Course in Numerical Analysis*, 2nd ed. McGraw-Hill Book Company, New York.
41. Ralston, A., and H. S. Wilf, ed. (1967), *Numerical Methods for Digital Computers*, Vols. 1 and 2. John Wiley & Sons, Inc., New York.
42. Reklaitis, G. V. (1983), *Introduction to Material and Energy Balances*. John Wiley & Sons, Inc., New York.

43. Schultz, M. H. (1966), *Spline Analysis*. Prentice-Hall, Inc., Englewood Cliffs, N.J.
44. Shampine, L. F., and C. W. Gear (1979), "A user's view of solving stiff ordinary differential equations." *SIAM Review*, **21**, No. 1, 1–17.
45. Strang, W. G. (1980), *Linear Algebra and Its Applications*. Academic Press, Inc., New York.
46. Stroud, A. H., and D. Secrest (1966), *Gaussian Quadrature Formulas*. Prentice-Hall, Inc., Englewood Cliffs, N.J.
47. Varga, R. S. (1962), *Matrix Iterative Analysis*. Prentice-Hall, Inc., Englewood Cliffs, N.J.
48. Watkins, D. S. (1982), "Understanding the QR algorithm." *SIAM Review*, **24**, No. 4, 427–440.
49. Wendroff, B. (1966), *Theoretical Numerical Analysis*. Academic Press, Inc., New York.
50. Wilkinson, J. H. (1963), *Rounding Errors in Algebraic Processes*. H. M. Stationery Office, London.
51. Wilkinson, J. H. (1965), *The Algebraic Eigenvalue Problem*. Clarendon Press, Oxford, England.
52. Wilkinson, J. H., and C. Reinsch (1971), *Handbook for Automatic Computation*, Vols. 1 and 2. Springer-Verlag New York, Inc., New York.
53. Young, D. M., and R. T. Gregory (1973), *A Survey of Numerical Mathematics*, Vols, 1 and 2. Addison-Wesley Publishing Co., Inc., Reading, Mass.

Answers to Selected Problems

Chapter 1

1-1 (a) 90.91, 90.909 (c) 0.06, 0.062 (e) 1.00, 1.000

1-2

	$h = h_0$	$h = h_0/r$	$h = h_0/r^2$	$h = h_0/r^3$
(a)	3	3.9	3.99	3.999
(d)	1.38629	1.05361	1.01014	1.00201
(e)	0.894182	0.987894	0.998651	0.999850

1-3 [WARNING: Answers will vary from device to device.]

1-4 (a) 91., 90.9 (c) 0.062, 0.0625 (e) 1.0, 1.00

1-6 (a) (i) $\frac{1}{16}\cdot 2^{-64}$ (ii) $[1 - (\frac{1}{16})^6]\cdot 2^{63}$ (iii) $2[128\cdot 15\cdot 16^5] + 1$ (iv) 2^{-24} (v) 7 (vi) 32 bits (vii) $2^{31} - 1$

1-8 s and L of the text will be multiplied by 10.

1-10 (a) $z(e^{xy})$ makes it less likely that $|xy| = |\text{exponent}| > 39$.

1-11 (a) $M = .9091$ (rounded), $.9090$ (chopped); $c = 2$ (b) $M = .2449$ (*not* $-.2449$); $c = 1$

1-12 (a) $-\frac{1}{1100}$ (b) about $-4.9E-4$

1-13 (a) $-1\text{E}-5$ (b) about $-2\text{E}-4$

1-14 (a) $\epsilon = -\frac{1}{110}$, $\rho = -\frac{1}{10}$, $\% = 10$ (b) $\epsilon = -\frac{4}{900}$, $\rho = -\frac{4}{500}$, $\% = -0.8$

1-15 $r = 0.5\text{E}-3\cdot|x|$ (a) Gets all X's (d) Gets almost all of them

1-16 (a) $0.7426 \neq 0.7424$ (b) $1.932 \neq 1.933$

1-18 (b) NA (+ if $|x| >> 1$ or $|x| << 1$); SC (− if $x >> 1$); EM (* if $|x| >> 1$)
(c) NA (+ if $|y| << x$ or $|x| << y$); SC (− if $|x| << y$, + if $x \approx y$)
(d) NA (− if $|\cos x| \approx 0$); SC (− if $\cos x \approx 1$)

1-20 (a) $C = n$, well conditioned for $n < 10$ (c) $C \approx 3$, well conditioned

1-22 (b) $I = (a, \infty)$, where $a > 10$; $\phi(x) = 1/(x + \sqrt{x^2 + 1})$
(c) Let $u = x/y$ so that $A = \ln(1 + u)$ for $u > -1$
(d) "x in I" is best described as "$|\cos x - 1| < \epsilon, \epsilon \approx 0$"; $\phi(x) = \sin^2 x/(1 + \cos x)$; even better, use $A = 2\cos^2(x/2)$ for all x.

1-23 (a) $\ln \pi = 1.144729886$ (10s) (b) One way is $\ln[4\cdot\tan^{-1}(1)]$.

1-24 (a) Use $u*u*u$, where $u = x*x$, or $u*u$, where $u = x*x*x$.
(c) Use $u*u*(u*\ln u + 1)$, where $u = x*y$ (d) $1/(2 - \sqrt{x})$

1-25 (b) $((2x^2 + 1)x - 5)x^2 - 10$; -39.8125

Chapter 2

2-1 (a) $C_L = \frac{1}{2}$, $x_5 = 3\frac{3}{8}$; $\bar{x} = (x_4)_{\text{improved}} = \frac{7}{2}$
(c) $C_L = \frac{1}{2}$, $x_3 = 2.85$; $\bar{x} = (x_2)_{\text{improved}} = 2.9$

2-2 (a) $C_Q = 1$, $x_5 = 3.3125$; $\bar{x} \doteq x_7 \doteq 3.3164215$
(c) $C_Q = 2.5$, $x_3 = 2.825$; $\bar{x} \doteq x_5 \doteq 2.8265686$

2-3 (a) $\Delta x_i/\Delta x_{i-1} \approx 0.66 = C_L$, $\epsilon_4 = -(\Delta x_3)^2/(\Delta x_4 - \Delta x_3) \doteq -0.156098$
(b) $\Delta x_i/(\Delta x_{i-1})^2 \approx 3.1 = C_Q$
(c) $\Delta x_i/\Delta x_{i-1} \approx 0.8 = C_L$, $\epsilon_4 = -(\Delta x_3)^2/(\Delta x_4 - \Delta x_3) = 0.56$
(d) $|\Delta x_i|/|\Delta x_{i-1}|^{1.618} \approx 1.7 = C_S$

2-4

	x_0	x_1	x_2	x_3	x_4	Convergence Rate	
(a)	0	0.25	0.140625	0.184631	0.166207	$C_L \approx -0.4$	
	6	6.25	6.890625	8.67487	14.7259	$C_L \approx 2.5$	$[\approx g'(6)]$
(b)	1	2.5	2.05	2.00061	2.00000	$C_Q \approx -0.25$	
	−3	−2.16667	−2.00641	−2.00001	−2.00000	$C_Q \approx +0.25$	
(d)	1.8	1.58	1.0718	−0.287178	−5.18995	$C_L \approx 2.3$	$[\approx g'(1.8)]$
	3.5	3.875	3.99219	3.99997	4.00000	$C_Q \approx 0.6$	
(e)	2.5	3.71333	2.54318	3.63076	2.57571	$C_L \approx -0.97$	$[\approx -1]$
	3.5	2.63605	3.46907	2.65207	3.44322	$C_L \approx -0.97$	$[\approx -1]$

2-7 (a) $x_0 = 0$: $(x_2)_{\text{improved}} = 0.173913$ (new x_0), $x_1 = 0.170605$, $x_2 = 0.9171974$,
$(x_2)_{\text{improved}} = 0.171573$ ($\doteq \bar{x}$)
$x_0 = 6$: $(x_2)_{\text{improved}} = 5.84000$ (new x_0), $x_1 = 5.8564$, $x_2 = 5.89616$,
$(x_2)_{\text{improved}} = 5.82848$
(b) Not applicable; convergence is quadratic (not linear) for both $x_0 = 1$ and $x_0 = -3$
(d) $x_0 = 1.8$: $(x_2)_{\text{improved}} = 1.96794$ (new x_0), $x_1 = 1.93536$, $x_2 = 1.86864$,
$(x_2)_{\text{improved}} = 1.99901$
(e) $x_0 = 2.5$: $(x_2)_{\text{improved}} = 2.11766$ (new x_0), $x_1 = 2.89111$, $x_2 = 3.11719$,
$(x_2)_{\text{improved}} = 3.00427$
$x_0 = 3.5$: $(x_2)_{\text{improved}} = 3.06015$ (new x_0), $x_1 = 2.94220$, $x_2 = 3.06009$,
$(x_2)_{\text{improved}} = 3.00116$

2-8 (a) Linear for all a (c) quadratic for $a = -2$; linear for all other a in $(-3, -1)$

2-9 (a) $x_k = 9$ for all even k (oscillatory divergence); Theorem RS does not apply $[|g'(\sqrt{78.8})| = 1]$

2-11 Using the Absolute Difference Test $|\Delta x_{k+1}| < 0.00005$: k is (a) 5; (c) 41; (d) 4; (g) 5

2-14 (a) Terminates if x seems accurate to $3d$ when $|x| \leq 1$, and to $3s$ when $|x| > 1$

2-17 (a) $f(x) = x^5 - c$, where c is given

2-18 (a) $x^3 + 1 = 9/x^2$, $\bar{x} \approx 1.5$ (b) $3v^4 = 2(v + 1)$, $\bar{v} \approx 1.1, -0.7$
(e) $e^x = x^2 + 2$, $\bar{x} \approx 1.3$ (f) $\tan\theta = \theta^{-2}$, $\bar{\theta} \approx \pi/4, -7\pi/8$, and $(n\pi)^+$ for $n = 1, 2, \ldots$ and $-2, -3, \ldots$

2-19 (a) 1.4690 (b) 1.0859 (e)1.3191 (f) 0.89521

2-25 (a) 1.05, 0.925, 0.09875, 1.01875 (b) 1.1122, 1.0328, 1.0087, 1.0022 (c) Neither

2-26 The smallest integer greater than $\ln[(b - a)/\epsilon]/\ln 2$

2-27 Use $f(x) = g(x) - x$ to get (to $5s$) (a) 8.87694 (b) 3.35669

2-29 (a) Iterates are (i) −1.2248, 1.4702, −12.336; (ii) −27, −18.058, −12.125

2-30 For $x = -0.5$: $m = 2$, $\phi(x) = (x - 2.5)(x^2 - 3x + 2.3)$

2-32 Use $(x_8)_{\text{improved}} \doteq 0.0001$ for NR, and $(x_{12})_{\text{improved}} \doteq 0.0000138$ for SEC. (Why?)

2-33 Not if stored as REAL; storing as INTEGER will not help. (Why?)

2-34 (a) $r_1 = \frac{4}{3}$, $s_1 = -\frac{1}{3}$; $r_2 = -1.12397$, $s_2 = -0.332415$
(b) $r_1 = -2.41379$, $s_1 = 0.729064$; $r_2 = -3.11803$, $s_2 = 0.412618$

2-36 $\bar{x} = 1, 2$; $Q(x) = p(x)/(x^2 - 3x + 2) = x^3 - 0.5x^2 - 0.3x + 0.1$; $Q(x)$ has 0 or 2 positive roots and 1 or 3 negative roots, all in the interval $(-1.5, 1.5)$ (by the Root Bound Theorem). They are -0.5 and $0.5 \pm \sqrt{0.05}$.

Chapter 3

3-2 $AB = \begin{bmatrix} 4 & 3 \\ 10 & 7 \end{bmatrix}$, $BA = \begin{bmatrix} 5 & 8 \\ 4 & 6 \end{bmatrix}$, $\mathbf{c}B = [9 \quad 5]$; (e), (i), and (j) are not defined

3-4 $A^{-1} = \dfrac{1}{-2}\begin{bmatrix} 4 & -2 \\ -3 & 1 \end{bmatrix}$, $B^{-1} = \dfrac{1}{1}\begin{bmatrix} 1 & -1 \\ -1 & 2 \end{bmatrix}$, $(AB)^{-1} = \dfrac{1}{-2}\begin{bmatrix} 7 & -3 \\ -10 & 4 \end{bmatrix} =$
$B^{-1}A^{-1} \neq A^{-1}B^{-1}$

3-6 $\text{row}_i(DA) = d_{ii}(\text{row}_i A)$ and $\text{col}_j(AD) = d_{jj}(\text{col}_j A)$ for $1 \leq i, j \leq n$

3-8 (b) $\begin{bmatrix} -2 & 1 \\ 3 & -1 \end{bmatrix}\begin{bmatrix} s \\ t \end{bmatrix} = \begin{bmatrix} -1 \\ 3 \end{bmatrix}$, so $\begin{bmatrix} \bar{s} \\ \bar{t} \end{bmatrix} = \dfrac{1}{-1}\begin{bmatrix} -1 & -1 \\ -3 & -2 \end{bmatrix}\begin{bmatrix} -1 \\ 3 \end{bmatrix} = \begin{bmatrix} 2 \\ 3 \end{bmatrix} \begin{matrix} (\bar{s} = 2) \\ (\bar{t} = 3) \end{matrix}$

3-9 6 multiplications, 3 additions, 2 divisions (using $\det A = a_{11}a_{22} - a_{12}a_{21}$)

3-10 $\bar{x}_1 = (\text{row}_1 A^{-1})\mathbf{b}/(\det A)$, $\bar{x}_2 = (\text{row}_2 A^{-1})\mathbf{b}/(\det A)$

3-11 (a) $[3 \quad -2 \quad \tfrac{1}{3}]^T$ (b) $[\tfrac{1}{3} \quad -2 \quad 3]^T$ (d) $[0 \quad 0 \quad -1 \quad -1]^T$ (e) $[-1 \quad 2 \quad 0 \quad -\tfrac{7}{4}]^T$

3-12 $ad = a_{11}$, $ae = a_{12}$, $bd = a_{21}$, $be + cf = a_{22}$. These are $n^2 = 4$ equations in the $n^2 + 2 = 6$ unknowns a, b, c, d, e, f.

3-13 (a) (i) $\begin{bmatrix} 4 & 0 \\ 2 & \tfrac{1}{2} \end{bmatrix}\begin{bmatrix} 1 & \tfrac{3}{4} \\ 0 & 1 \end{bmatrix}$; (ii) $\begin{bmatrix} 1 & 0 \\ \tfrac{1}{2} & 1 \end{bmatrix}\begin{bmatrix} 4 & 3 \\ 0 & \tfrac{1}{2} \end{bmatrix}$; (iii) $\begin{bmatrix} 2 & 0 \\ 1 & \sqrt{\tfrac{1}{2}} \end{bmatrix}\begin{bmatrix} 2 & \tfrac{3}{2} \\ 0 & \sqrt{\tfrac{1}{2}} \end{bmatrix}$

(b) (i) $\begin{bmatrix} 2 & 0 \\ 1 & -\tfrac{1}{2} \end{bmatrix}\begin{bmatrix} 1 & \tfrac{1}{2} \\ 0 & 1 \end{bmatrix}$; (ii) $\begin{bmatrix} 1 & 0 \\ \tfrac{1}{2} & 1 \end{bmatrix}\begin{bmatrix} 2 & 1 \\ 0 & -\tfrac{1}{2} \end{bmatrix}$; (iii) not possible

3-14 (a) $[14 \quad -4 \quad 1]^T$ (b) $\tilde{a}_{22} = 0$ (c) $[-1 \quad 1 \quad 0]^T$ (d) $\tilde{a}_{11} = 0$ (e) $\tilde{a}_{11} = 0$
(f) $[-11 \quad -7 \quad 8 \quad 1]^T$

3-17 (a) $[1 \quad 1 \quad -1]^T$ (c) $[2 \quad -2 \quad 2 \quad -1]^T$

NOTE: Answers to 3-21 and 3-22 are given as $[\hat{L}\backslash\hat{U} : \hat{\mathbf{b}} : \bar{\mathbf{c}} : \bar{\mathbf{x}}]$.

3-21 (b) $\begin{bmatrix} \textcircled{2} & 2 & -2 & : & -4 & : & -4 & : & -15 = \bar{x} \\ \underline{-\tfrac{1}{2}} & \textcircled{2} & 2 & : & 4 & : & 2 & : & 7 = \bar{y} \\ \underline{1} & \underline{0} & \textcircled{-2} & : & 8 & : & 12 & : & -6 = \bar{z} \end{bmatrix}$

(d) $\begin{bmatrix} \textcircled{-1} & 1 & 0 & : & 2 & : & 2 & : & 0 = \bar{x}_1 \\ \underline{0} & \textcircled{2} & 1 & : & 3 & : & 3 & : & 2 = \bar{x}_2 \\ \underline{-2} & \underline{\tfrac{1}{2}} & \textcircled{\tfrac{5}{2}} & : & -5 & : & -\tfrac{5}{2} & : & -1 = \bar{x}_3 \end{bmatrix}$

3-22 Parts (b) and (d) turn out to have solutions identical to 3-21.

(a) $\begin{bmatrix} \textcircled{1} & 3 & 1 & : & 3 & : & 3 & : & 14 = \bar{x} \\ \underline{0} & \textcircled{2} & 2 & : & -6 & : & -6 & : & -4 = \bar{y} \\ \underline{2} & \underline{1} & \textcircled{-2} & : & -2 & : & -2 & : & 1 = \bar{z} \end{bmatrix}$

(e) $\begin{bmatrix} \textcircled{1} & -1 & -2 & 0 & : & 0 & : & 0 & : & 2 = \bar{s} \\ \underline{-1} & \textcircled{2} & 0 & 0 & : & 0 & : & 0 & : & 0 = \bar{t} \\ \underline{2} & \underline{1} & \textcircled{4} & 6 & : & -2 & : & -2 & : & 1 = \bar{u} \\ \underline{0} & \underline{1} & \underline{\tfrac{1}{4}} & \textcircled{\tfrac{5}{2}} & : & -3 & : & -\tfrac{5}{2} & : & -1 = \bar{v} \end{bmatrix}$

3-25 Find $\hat{A}$ as $\hat{L} \cdot \hat{U}$, then perform the *mp*-pivots *in reverse order* to get A:

(a) $\begin{bmatrix} 1 & 2 & 1 \\ 2 & -1 & 2 \\ 0 & 1 & 3 \end{bmatrix}$ (b) $\begin{bmatrix} 2 & 2 & -2 \\ 2 & 2 & -4 \\ -1 & 1 & 3 \end{bmatrix}$ (c) $\begin{bmatrix} 0 & 2 & 4 \\ 2 & 2 & 9 \\ 1 & -1 & 0 \end{bmatrix}$ (d) $\begin{bmatrix} 1 & 3 & 1 \\ 2 & 8 & 2 \\ 0 & 2 & 2 \end{bmatrix}$

(e) $\begin{bmatrix} 0 & 2 & 1 & 4 \\ 1 & -1 & -2 & 0 \\ 2 & 0 & 0 & 6 \\ -1 & 3 & 2 & 0 \end{bmatrix}$ (f) $\begin{bmatrix} 1 & 0 & -2 & 1 \\ 0 & 2 & 0 & 4 \\ 3 & 0 & -1 & 0 \\ 0 & -1 & -1 & 0 \end{bmatrix}$

3-26 (a) $\hat{\mathbf{b}} = [-4 \quad 2 \quad -3]^T$, $\bar{\mathbf{c}} = [-4 \quad 2 \quad 1]^T$, $\bar{\mathbf{x}} = [-5 \quad -1 \quad 1]^T$
(d) $\hat{\mathbf{b}} = [3 \quad -1 \quad -1 \quad 8]^T$, $\bar{\mathbf{c}} = [3 \quad -1 \quad -2 \quad \frac{42}{5}]^T$, $\bar{\mathbf{x}} = [2 \quad -2 \quad 3 \quad 3]^T$

3-27 NOTE: Lower-triangular entries of $\text{col}_m \tilde{A}$ used in the mth **pivot** step can be obtained from $\hat{L}\backslash\hat{U}$ as

$$[\hat{u}_{mm}, \quad (\hat{l}_{m+1,m} \cdot \hat{u}_{mm}), \quad \ldots, \quad (\hat{l}_{n,m} \cdot \hat{u}_{mm})$$

(b) From $\hat{A}$, $\hat{\mathbf{s}} = [2 \quad 4 \quad 3]^T$. $m = 1$ ratios: $\frac{2}{2} = 1$, $(\frac{1}{2} \cdot 2)/4 = \frac{1}{4}$, $(1 \cdot 2)/3 = \frac{2}{3}$; $m = 2$ ratios: $\frac{2}{4} = \frac{1}{2}$, $(0 \cdot 2)/3 = 0$. Since circled pivot entries of $\hat{L}\backslash\hat{U}$ have largest ratios, Scaled Partial Pivoting could have been used.

3-28 (a) $[1 \quad 2 \quad 3]^T$ (c) $[3 \quad 1 \quad 2]^T$ (e) $[3 \quad 4 \quad 2 \quad 1]^T$

3-30 (a) After reducing $\text{col}_1 A$, we see that $\rho_3 - 7\rho_1 = 2(\rho_2 - 4\rho_1)$ so that $\rho_3 = 2\rho_2 - \rho_1$, where ρ_i denotes $\text{row}_i A$.

3-31 (b) $\hat{\mathbf{b}}_1 = [1 \quad 2 \quad 7]^T$, $\bar{\mathbf{x}}_1 = [0 \quad -1 \quad 1]^T$ (c) $\hat{\mathbf{b}}_1 = [4 \quad 2 \quad 1]^T$, $\bar{\mathbf{x}}_1 = [-2 \quad 1 \quad 0]^T$

3-32 (a) $\dfrac{1}{4}\begin{bmatrix} 5 & -4 & -2 \\ -1 & 2 & 2 \\ 2 & -2 & 0 \end{bmatrix}$ (c) $\dfrac{1}{10}\begin{bmatrix} -12 & 3 & 8 & 9 \\ 0 & 5 & 0 & 5 \\ -6 & -6 & 4 & 2 \\ 4 & -1 & -1 & -3 \end{bmatrix}$

(d) $\dfrac{1}{209}\begin{bmatrix} 56 & -15 & 4 & 1 \\ -15 & 60 & -16 & 4 \\ 4 & -16 & 60 & -15 \\ 1 & 4 & -15 & 56 \end{bmatrix}$

3-37 (a) $(-1)^n a$ (d) a/b (e) $2^n a$ (see Problem 3-36)

3-38 The values of det A, det A^{-1} are (a) -15, $-\frac{1}{15}$ (b) 8, $\frac{1}{8}$ (e) 20, $\frac{1}{20}$

Chapter 4

4-1 (a) $[\hat{L}\backslash\hat{U} : \hat{\mathbf{b}} : \bar{\mathbf{c}} : \bar{\mathbf{x}}_{\text{calc}}]$ (with entries entered *then used* rounded to 3s) is

$$\left[\begin{array}{cc:c:c:c} \textcircled{0.13} & -0.536 & -5.23 & -5.23 & 1.41 \\ \underline{-1.85} & \textcircled{0.0084} & 9.76 & 0.0845 & 10.1 \end{array}\right]$$

[NOTE: A 1, 2-pivot is not necessary because $0.13/0.536 > 0.24/1.00$]

$[\hat{L}\backslash\hat{U} : \hat{\mathbf{r}} : \bar{\mathbf{c}} : \bar{\mathbf{e}}_{\text{calc}}]$ (with entries rounded to 3s as before) is

$$\left[\begin{array}{cc:c:c:c} \textcircled{0.13} & -0.536 & 0.0003 & 0.0003 & -0.511 \\ \underline{-1.85} & \textcircled{0.0084} & -0.0016 & -0.00104 & -0.124 \end{array}\right];$$

$$\bar{\mathbf{x}}_{\text{improved}} = \begin{bmatrix} 1.41 \\ 9.76 \end{bmatrix} + \begin{bmatrix} -0.511 \\ -0.124 \end{bmatrix} = \begin{bmatrix} 0.901 \\ 10.0 \end{bmatrix}$$

4-3 (a) $C_p(A) = 1.00/.0084 \doteq 119$; possible error in second significant digit.

4-7 $\text{row}_n\hat{A} = (\text{row}_n\hat{L}) \cdot \hat{U}$, so $C_p(A)$ is: (a) $s_3/u_3 = \frac{3}{3} = 1$ (d) $\hat{s}_3/\hat{u}_{33} = \frac{3}{1} = 3$ (e) $\hat{s}_4/\hat{u}_{44} = 4/(\frac{5}{2}) = \frac{8}{5}$

4-8 About (a) 5s (c) 4s *plus* roundoff error due to $\approx$ 0.7E6 arithmetic operations

4-9 (c) $\|A\|_1 = \max\{4, 6, 5, 10\} = 10$, $\|A^{-1}\|_1 = \max\{2.2, 1.5, 1.3, 1.9\} = 2.2$, cond $A = 10 \cdot 2.2 = 22$

4-10 (c) $\|A\|_\infty = \max\{7, 4, 8, 6\} = 8$, $\|A^{-1}\|_\infty = \max\{3.2, 1.0, 1.8, 0.9\} = 3.2$, cond $A = 8 \cdot 3.2 = 25.6$

4-15 Only $A_{(a)}$ and $A_{(c)}$ are positive definite. For $A_{(a)}$, $L = \dfrac{1}{\sqrt{5}}\begin{bmatrix} 5 & & \\ 3 & 1 & \\ -2 & -4 & \sqrt{5} \end{bmatrix} = U^T$

4-16 (a) Solving (E_1) for y, then (E_2) for x, then (E_3) for z gives $\mathbf{x}_1 = [\frac{11}{16} \quad \frac{1}{2} \quad -\frac{61}{48}]^T$, $\mathbf{x}_2 = [0.69401 \quad 0.94097 \quad -1.1217]^T$

4-17 Convergence should be expected in parts (a) and (c) (A is positive definite) and part (b) (A is strictly diagonally dominant). For part (b), $\mathbf{x}_1 = [-2 \quad -3 \quad 2]^T$, $\mathbf{x}_2 = [-1 \quad 0 \quad 1]^T$, $\mathbf{x}_3 = [-\frac{1}{2} \quad \frac{1}{8} \quad \frac{7}{4}]^T$

4-18 (a) $\mathbf{f}'(\mathbf{x}) = \begin{bmatrix} 2x & 2y \\ 2-2x & 1 \end{bmatrix}$; $\mathbf{x}_1 = \begin{bmatrix} \frac{15}{8} \\ \frac{3}{4} \end{bmatrix}$, $\mathbf{x}_2 = \begin{bmatrix} \frac{1517}{816} \\ \frac{301}{408} \end{bmatrix}$

(b) $\mathbf{f}'(\mathbf{x}) = \begin{bmatrix} y & x \\ 2x & 1 \end{bmatrix}$; $\mathbf{x}_1 = \begin{bmatrix} 2 \\ 0 \end{bmatrix}$, $\mathbf{x}_2 = \begin{bmatrix} \frac{13}{8} \\ \frac{1}{2} \end{bmatrix}$

4-19 (a) $\bar{\mathbf{x}} \doteq [1.47766 \quad 2.09348]^T$ (c) $\bar{\mathbf{x}} \doteq [-2.71241 \quad -1.83226]^T$ and $[2.12068 \quad 0.705174]^T$

4-20 (a) $\mathbf{x}_1 = [2.7 \quad 1.3 \quad -3.8]^T$ (b) $\mathbf{x}_1 = [-1 \quad 0 \quad -1]^T$

4-21 Solve (E_3) for y, (E_2) for x, then (E_1) for z; $\bar{\mathbf{x}} = [2 \quad 1 \quad -2]^T$

Chapter 5

5-1 (a) $\hat{\alpha} = [\Sigma(y_k - 1)x_k^2]/(\Sigma x_k^4)$ (b) $\hat{\alpha} = \frac{15}{17}$, $E(\hat{g}) = \frac{1088}{289}$ (c) $\frac{87}{98}$

5-2 (a) $\Sigma x_k e^{2\alpha x_k} = \Sigma x_k y_k e^{\alpha x_k}$ (nonlinear) (b) $\hat{\alpha} = \ln(3/\sqrt{2})$, $E(\hat{g}) = \frac{19}{4}$
(c) Need a computer. You can get an α_0 by passing $y = g(x)$ through P_2. (Why?)

5-4 $\hat{c} = (\Sigma y_k)/M$ and $E(\hat{g}) = \Sigma(\hat{c} - y_k)^2$ are the mean and variance of $y_1, \ldots, y_M$.

5-5 (a) $\hat{\alpha} = [\Sigma \phi(x_k)y_k]/[\Sigma \phi(x_k)^2]$ (b) $\hat{\alpha} = 2[\hat{g}(x) = 2 \sin(\pi x/3)]$

5-6 (a) $\hat{L} \doteq 0.85714 + 1.2143x$, $E(\hat{L}) \doteq 1.786$, $R(\hat{L}) \doteq 0.7940$, $T = \sqrt{3.854} < 12.706$ (poor fit)
(c) $\hat{L}(x) \doteq 5 - 1.6x$, $E(\hat{L}) = 1.2$, $R(\hat{L}) \doteq 0.91429$, $T = \sqrt{21.36} > 4.303$ (good fit)

5-7 $v_0 \approx \hat{L}(f) \doteq -2.2243 + 0.045031 \cdot f \cdot 10^{-13}$; so $h \approx 0.045031 \cdot e \cdot 10^{-13} \doteq 7.2\text{E}-34$ Joule · sec and f_t (when $v_0 = 0$) $\approx 49.4\text{E}+13$ Hz

5-8

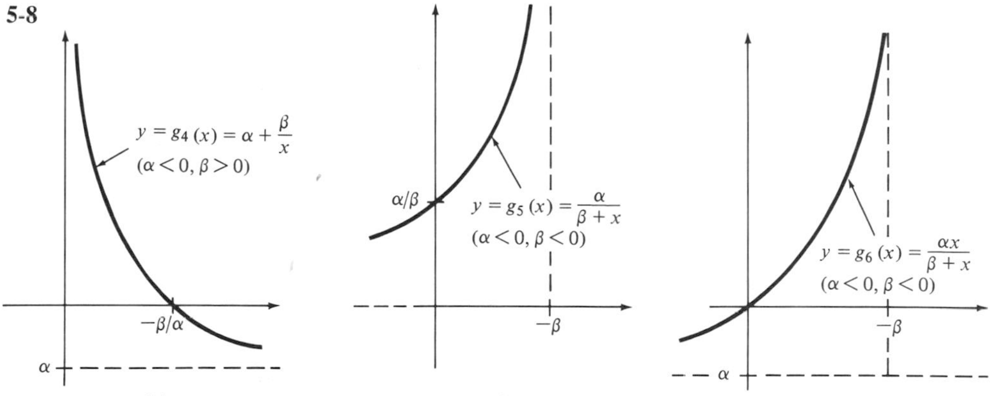

(a) (b) (c)

5-10 (a) $g(x) = -0.51852/(x - 2.1111)$ (c) $g(x) = -\frac{7}{3} + 6.4/x$

5-11 (a) $(\Sigma t_k)\alpha + (\Sigma 1/\sqrt{t_k})\beta = \Sigma y_k\sqrt{t_k}$, $(\Sigma 1/\sqrt{t_k})\alpha + (\Sigma 1/t_k^2)\beta = \Sigma y_k/t_k$ (linear)
(f) $\alpha(\Sigma x_k^{2\beta}) = \Sigma y_k x_k^{\beta}$, $\alpha(\Sigma x_k^{2\beta} \ln x_k) = \Sigma y_k x_k^{\beta} \ln x_k$ (nonlinear)

5-13 (b) For $g(x)$: $\alpha \doteq 1.7035$, $\beta \doteq 0.54715$, $R(\hat{L}) \doteq 0.94685$, $E(g) \doteq 24.808$
For $h(x)$: $\alpha \doteq 2.3039$, $\beta \doteq 1.3990$, $R(\hat{L}) \doteq 0.99999$, $E(h) \doteq 0.00137$

5-14 (a) $h(x)$ should give better fit [data appear to go through (0, 0)]

5-15 (a) $g(x)$ should give better fit (data do not appear to have a horizontal asymptote)

5-16 (a) On semilog paper: $g_1(x)$ and $g_3(x)$; on log-log paper: $g_2(x)$

5-17 $n = 4$ (integer nearest $\beta = 3.9818$), $K = 4.1142$

5-18 Using the Linearization Algorithm 5.2B to fit $\alpha e^{\beta t}$ to $(t_i, 10 - v_i)$ gives $\alpha \doteq 4.2321$, $\beta \doteq -0.27835$. So $v_0 \approx 5.77$ and $\tau \approx 3.59$.

5-22 $g_3(x)$ and $g_4(x)$; compare normal equations of least square and linearized fit.

5-23 (a) $\alpha = \frac{4}{11}$, $\beta = \frac{19}{11}$, $R(\hat{g}) \doteq 0.72328$
(c) $\alpha \doteq -0.344883$, $\beta \doteq 6.04994$, $R(\hat{g}) \doteq 0.99989$

5-24 When $n = 3$, M should be at least 6, preferably as much as 8–13.

5-25 (a) $\hat{g}(x) = -\frac{1}{3} + x/2 + 5x^2/6$; $R(\hat{g}) = 1$ (by inspection)
(c) $\hat{g}(x) = 7.5 - 4.1x + 0.5x^2$; $R(\hat{g}) \doteq 0.9864$

5-28 (a) $y \approx g_3(x) \doteq 6.3768 - 13.372x + 1.9994x^2$; $R(\hat{g}_3) = 0.99999$ $[R(\hat{g}_2) \doteq 0.57198]$

5-30 (a) $C_p(A_{2\times 2}) = 17 << 0.3\text{E}{+}7 = C_p(A_{3\times 3})$. Yes!
(c) (X, y) data, with $X = x/75 - \frac{70}{3}$: $(-2, 0)$, $(-\frac{2}{3}, 1)$, $(\frac{2}{3}, 2)$, $(2, 9)$

Normal equations: $\begin{bmatrix} 4 & 0 & \frac{80}{9} \\ 0 & \frac{80}{9} & 0 \\ \frac{80}{9} & 0 & \frac{2624}{81} \end{bmatrix} \begin{bmatrix} a \\ b \\ c \end{bmatrix} = \begin{bmatrix} 14 \\ 20 \\ \frac{344}{9} \end{bmatrix}$; $\begin{bmatrix} \hat{a} \\ \hat{b} \\ \hat{c} \end{bmatrix} = \begin{bmatrix} \frac{9}{4} \\ \frac{9}{4} \\ \frac{9}{16} \end{bmatrix}$

So $\hat{G}(X) = \frac{9}{4} + \frac{9}{4}X + \frac{9}{16}X^2 = \frac{9}{16}(2 + X)^2$ [for (X, y)] and hence the desired $\hat{g}(x)$ is $\hat{G}(x/75 - \frac{70}{3}) = (0.01x - 16)^2$

5-32 (a) Third- or fourth-degree polynomial; (c) second- or third-degree polynomial in $1/x$

5-35 (a) $\hat{b} = \frac{1}{3}$, the average of $f(x)$ on $[0, 1]$; $\epsilon_{\max} = \max\{\hat{\delta}(0), \hat{\delta}(1)\} = \frac{2}{3}$

(b) Normal equations: $\begin{bmatrix} \frac{1}{3} & \frac{1}{2} \\ \frac{1}{2} & 1 \end{bmatrix} \begin{bmatrix} m \\ b \end{bmatrix} = \begin{bmatrix} \frac{1}{4} \\ \frac{1}{3} \end{bmatrix}$; $\begin{bmatrix} \hat{m} \\ \hat{b} \end{bmatrix} = \begin{bmatrix} 1 \\ -\frac{1}{16} \end{bmatrix}$

$\hat{g}(x) = x - \frac{1}{6}$; $\epsilon_{\max} = \max\{\hat{\delta}(0), \hat{\delta}(\frac{1}{2}), \hat{\delta}(1)\} = \max\{\frac{1}{6}, \frac{1}{12}, \frac{1}{6}\} = \frac{1}{6}$

5-38 Normal equations are $\int_0^2 xe^{2\beta x}\,dx = \int_0^2 x^2 e^{\beta x}\,dx$. Now use a reduction formula for integrating $x^n e^{ax}$.

5-41 (a) $r(x) = (1 - x/2 + x^2/12)/(1 + x/2 + x^2/12)$
(b) $r(x) = (x - 7x^3/60)/(1 + x^2/20)$

5-42 (a) $p_{\text{econ}}(x) = 191/192 - x + 13x^2/24 - x^3/6$
(b) $p_{\text{econ}}(x) = (383x - 60x^3)/384$

5-43 (a) $p_{\text{econ}}(x) = -\frac{1}{2} - 3x/2 + 2x^2/3$
(b) $p_{\text{econ}}(x) = (23x - 3x^3)/24$

5-44 (a) $R_4(x) = -e^{-c}x^5/5!$ (i) $|R_4(x)| \leq e^{1/2}(\frac{1}{2})^5/5!$; (ii) $|R_4(x)| \leq e^1(1)^5/5!$
(b) $R_5(x) = R_6(x) = x^7(\cos c)/7!$ (i) $|R_6(x)| \leq 1(\frac{1}{2})^7/7!$ (ii) $|R_6(x)| \leq 1(1)^7/7!$

Chapter 6

6-1 (a) $A + B + C = 1$, $A + 2B + 4C = 8$, $A + 3B + 9C = 27$; $A = C = 6$, $B = -11$
(b) $p_{1,3}(x) = 1\,\dfrac{(x-2)(x-3)}{(-1)(-2)} + 8\,\dfrac{(x-1)(x-3)}{(1)(-1)} + 27\,\dfrac{(x-1)(x-2)}{(2)(1)}$

6-3 (a) $y = 3 + (-1)(x + 2)$ and $y = 3(x - 1)/(-3) + 0(x + 2)/3$; both are $y = 1 - x$

6-4 (a) $p_{2,3}(x) = y_2L_2(x) + y_3L_3(x) = -\frac{1}{2}(x - 2) + 2x$; $p_{2,3}(1) = \frac{5}{2}$ (b) $p_{1,3}(1) = 3$
(c) $p_{0,3}(x) = \frac{15}{8}(x + 1)x(x - 2) - \frac{2}{3}(x + 2)x(x - 2) - \frac{1}{4}(x + 2)(x + 1)(x - 2) + \frac{1}{6}(x + 2)(x + 1)x$; $\frac{3}{4}$

6-5 $p_{0,2}(0) = p_{0,2}(x_2) = y_2 = 1$, $p_{2,3}(2) = p_{2,3}(x_3) = y_3 = 4$, $p_{2,3}(0) = p_{2,3}(x_2) = y_2 = 1$

6-6 (a) – (c) $L_2(0) = L_2(x_2) = 1$, $L_2(2) = L_2(x_3) = 0$, $L_3(0) = 0$, $L_3(2) = 1$ (no others)
(d) All but $L_3(4)$ can be found (4 is not an interpolated x_i).

6-11 (a) With $s = \dfrac{z - x_3}{h} = \dfrac{1 - 0}{\frac{1}{2}} = 2$: $p_{1,3}(1) = 21\,\dfrac{(2+1)(2)}{(-1)(-2)} + 7\,\dfrac{(2+2)(2)}{(1)(-1)} +$

$1\,\dfrac{(2+2)(2+1)}{(0+2)(0+1)} = 13$

(c) With $s = \dfrac{z - x_1}{h} = \dfrac{-2 - (-1)}{\frac{1}{2}} = -2$: $p_{1,4}(-2) = 21(10) + 7(-20) + 1(15) - 3(-4) = 7$

6-12 (a) $p_{0,0}(-3)$ (c) $p_{1,3}(2)$ (e) $p_{-1,2}(-6)$

6-14 (a)

n	$\hat{p}_n(7) \approx \ln 7 \doteq 1.9459$	$\hat{\epsilon}_n(7)$	$\hat{\delta}_n(7)$ [for Problem 6-16(b)]	
0	$\hat{p}_0(7) = p_{3,3}(7) \doteq 2.1972$	-0.2513		
1	$\hat{p}_1(7) = p_{2,3}(7) \doteq 1.8728$	0.0731	-0.3244	
2	$\hat{p}_2(7) = p_{2,4}(7) \doteq 1.9125$	-0.0334	0.0397	$\leftarrow \hat{\delta}_n(7)$ stops decreasing
3	$\hat{p}_3(7) = p_{1,4}(7) \doteq 2.0238$	-0.0779	0.1113	(best approximation for
4	$\hat{p}_4(7) = p_{1,5}(7) \doteq 1.2851$	0.6608	-0.7387	this problem is $\hat{p}_1(7)$)

6-16 (a) With $n = 4$ in (25): $\hat{\xi}_0 = -0.951057$, $\hat{x}_0 = -0.453456$, $\hat{\xi}_2 = 0$, $\hat{x}_2 = 8.125$, etc.

(c)

n	$\hat{p}_n(7) \approx \ln 7 \doteq 1.9459$	$\hat{\epsilon}_n(7)$	$\hat{\delta}_n(7)$ (see Problem 6-16)	
0	$\hat{p}_0(7) = p_{2,2}(7) \doteq 2.0949$	-0.1490		
1	$\hat{p}_1(7) = p_{1,2}(7) \doteq 1.8900$	0.0559	-0.2049	
2	$\hat{p}_2(7) = p_{1,3}(7) \doteq 1.9261$	0.0198	0.0361	$\leftarrow$ smallest $\hat{\delta}_n(7)$
3	$\hat{p}_3(7) = p_{0,3}(7) \doteq 2.0124$	-0.0665	0.0863	(best approximation for
4	$\hat{p}_4(7) = p_{0,4}(7) \doteq 1.9802$	-0.0342	-0.0322	this problem is $\hat{p}_2(7)$)

6-17 $\Delta y_0 = 11$, $\Delta y_4 = 317$, $\Delta^2 y_0 = -22$, $\Delta^3 y_1 = -1$

6-18 $p_{1,3}(x) = \Delta^2 y_1 x^2 + \cdots = 2x^2 + \cdots$; $p_{0,4}(x) = \Delta^4 y_0 x^4 + \cdots = -2x^4 + \cdots$

6-19 (a) $[[[[1(x-2)-2](x-1)+9]x-34](x+2)+92](x+3)-71$; $p_{0,5}(-1) = 7$

(b) $[[[[1(x-3)+4](x-2)-1](x-1)+2](x+2)-10]x+1$; $p_{0,5}(-1) = 7$

6-20 $p''_{2,4}(x) = 2!\Delta^2 y_2 = 2(-2) = -4$; $p'''_{2,5}(x) = 3!\Delta^3 y_2 = 6(19) = 114$

6-21 (a) An examination of the forward difference table for $(x, f(x))$ [Problem 6-23(c)] shows that for $0 \leq x \leq 30$, $f(x)$ behaves like a fourth-degree polynomial whose forward-difference form (48b), when nested, becomes $[[[212(x-15)/20 - 176](x-10)/15 + 116](x-5)/10 + 5]x/5 + 2$. Solving $f(x) - 100 = 0$ using SEC with $x_0 = 9$, $x_{-1} = 10$, gives $\bar{x} = x_3 = 8.74485$ $(6s)$.

(b) The best interpolant of the knots $(f(x), x)$ is $\hat{p}_1(100) \doteq 8.59924$ [linear interpolation of $(7, 5)$ and $(128, 10)$]. This is only accurate to about $2s$.

6-23 $f(x)$ lies on a polynomial curve of degree (a) 3 (b) 5 (c) 4 (d) 4

6-26 (b) Taking $k = 2$, so that $s = (x - x_2)/h = (x - 12)/0.4$, gives

$$\left[\left[\left[\left[60\,\frac{(s+2)}{2.0} + 108\right]\frac{(s-2)}{1.6} - 3\right]\frac{(s-1)}{1.2} - 32\right]\frac{(s+1)}{0.8} - 16\right]\frac{s}{0.4} + 144$$

6-28 (a) For parts (I), (III), and (IV), $s(x) = x^3$ is the unique interpolating spline.

(b) (I) $\begin{bmatrix} 6 & 1 \\ 1 & 4 \end{bmatrix}\begin{bmatrix} \sigma_1 \\ \sigma_2 \end{bmatrix} = \begin{bmatrix} 6 \\ 24 \end{bmatrix}$; (II) $\begin{bmatrix} 8 & 1 \\ 1 & 5 \end{bmatrix}\begin{bmatrix} \sigma_1 \\ \sigma_2 \end{bmatrix} = \begin{bmatrix} -18 \\ 36 \end{bmatrix}$;

(III) $\begin{bmatrix} 4 & -1 \\ 0 & 3 \end{bmatrix}\begin{bmatrix} \sigma_1 \\ \sigma_2 \end{bmatrix} = \begin{bmatrix} -6 \\ 18 \end{bmatrix}$; (IV) $\begin{bmatrix} 5 & 1 \\ 1 & \frac{7}{2} \end{bmatrix}\begin{bmatrix} \sigma_1 \\ \sigma_2 \end{bmatrix} = \begin{bmatrix} 6 \\ 21 \end{bmatrix}$

(c) For (I), (III), and (IV) $\sigma_k = d^2(x^3)/dx^2$ at x_k for $k = 0, 1, 2, 3$; that is, $\sigma_0 = -12$, $\sigma_1 = 0$, $\sigma_2 = 6$, $\sigma_3 = 12$. For (II), $\sigma_0 = \sigma_1 \doteq -3.2308$, $\sigma_2 = \sigma_3 \doteq 7.8462$

6-29 (a) $\sigma_0 = \sigma_4 = 0$, $\sigma_1 \doteq -0.52091$, $\sigma_2 \doteq 0.055766$, $\sigma_3 \doteq -0.025904$

6-31 (a) $\sigma_0 = -0.59838$, $\sigma_1 = -0.43935$, $\sigma_2 = 0.037737$, $\sigma_3 = -0.017147 = \sigma_4$

6-32 (a) $c_{11} = c_{21} = 0$, $c_{31} = 1$ (c) $c_{12} \doteq 0.12864$, $c_{22} \doteq 0.027883$, $c_{32} \doteq -0.0027223$

Chapter 7

7-1

	Q	w_0	w_1	w_2	$f(x) = x^2$	$f(x) = x^3$	$f(x) = e^x$
(a)	$f'(0)$	$-\frac{1}{2}$	$\frac{1}{2}$	0	0 (exact)	1	1.1752 $(5s)$
(c)	$\int_0^1 f(x)\,dx$	$\frac{5}{36}$	$\frac{13}{12}$	$-\frac{2}{9}$	$\frac{1}{3}$ (exact)	$-\frac{5}{6}$	1.3539 $(5s)$

7-2 (a) $C = -1$ (b) Not applicable (c) $\frac{13}{12}$

7-6 (a) $F[0.1] \doteq 4.559019$, $F[0.05] = 4.781493$, $\tau[0.1]/\tau[0.05] \approx 2 = (0.1/0.05)^1$
(b) $F[\frac{1}{10}] = 0.9531018$, $F[\frac{1}{40}] = 0.9877045$, $\tau[\frac{1}{10}]/\tau[\frac{1}{40}] \approx 4^1$

7-7 (a) $F[0.1] \doteq 4.991671$, $F[0.05] = 4.997917$, $\tau[0.1]/\tau[0.05] \approx 4 = (0.1/0.05)^2$
(b) $F[\frac{1}{10}] \doteq 1.003354$, $F[\frac{1}{40}] \doteq 1.000208$, $\tau[\frac{1}{10}]/\tau[\frac{1}{40}] \approx 4^2$

7-9 (a) $F_1[0.05] \doteq 0.500397$ (b) $F_1[\frac{1}{40}] \doteq 0.999238$

7-10 (a) $F_1[0.05] \doteq 0.500000$ (b) $F_1[\frac{1}{40}] \doteq 0.999998$

7-12 (a) $F_1[0.4] = \frac{7}{15}$, $F_1[0.1] = \frac{1}{2}$, $F_2[0.1] \doteq 0.466471$
(c) $F_1[0.4] \doteq 0$, $F_1[0.1] = \frac{1}{3}$, $F_2[0.1] \doteq 0.33862$

7-14 (a) $F[1] = -1$, $F[0.5] = -\frac{3}{2}$, $F_1[0.5] = -2$ (exact for quadratics)
(c) $F[0.5] \doteq -1.042191$, $F[0.1] \doteq -1.001667$, $F[0.05] \doteq -1.000417$, $F_1[0.1] \doteq -0.9999790$, $F_1[0.05] \doteq -0.9999998$, $F_2[0.1] \doteq -1.000000$
(g) $F[-0.2] \doteq -1.10701$, $F[-0.1] \doteq -1.05171$, $F[-0.04] \doteq -1.02027$, $F_1[-0.1] \doteq -0.996408$, $F_1[-0.04] \doteq -0.999311$, $F_2[-0.04] \doteq -0.100028$
(h) $F[0.6] \doteq -1.23969$, $F[0.2] \doteq -1.02055$, $F[0.1] \doteq -1.00503$, $F_1[0.2] \doteq -0.993158$, $F_1[0.1] \doteq 0.999861$, $F_2[0.1] \doteq -1.00031$

7-16 (a) Using (26a): $F[1] = 38.5$, $F[2] = 35.5$, $F_1[1] = 39.5 = f'(-6)$ [(26a) is *not* cubic exact; see (13b)]
(c) Using *either* (26c) with $h = -1$ or (27c) with $h = 1$ gives $f'''(5) = 3$ [cubic exact]

7-17 (a) From a Richardson table for (26a) with $h = 0.4, 0.8, 1.2, 1.6$: $f'(0.4) \approx F_3[0.4] \doteq 2280$
(c) Using (27b) as $F[h]$: $F[0.4] = 2500$, $F[0.8] = 2143.75$, $F_1[0.4] = 2025 = f''(4.0)$ (exact for degree 5)

7-18 (a) Using (27b) as $F[h]$: $F[1] = 76$, $F[2] = 79$, $F_1[1] = 75 \approx f''(-3)$
(c) Using (26b) as $F[h]$: $F[-1] = -8$, $F[-2] = -41$, $F_1[1] = +3 \approx f''(0^-)$
[NOTE: (27b) will *not* give more accurate results for this $f(x)$.]
(e) [SUGGESTION: Use Problem 7-20(a) and a Richardson table if necessary.]

7-19 (a) 0 (b) 2 (c) 7.8 In choosing h_0, recall that 1 radian is about 57°.

7-24 All but $M]_a^b$ are cubic exact, hence exact for the tabulated $f(x)$.
(a) $M]_{-1}^1 = 2(-3) = -6$ (b) $CC]_1^2 = \frac{1}{24}[-1(-3) + 13(-4 + 0) - 1(12)] = -\frac{61}{24}$
(c) $AC]_{-6}^{-4}$ cannot be found directly (i.e., with $h > 0$). However,

$$\int_{-6}^{-4} = -\int_{-4}^{-6} \approx AC_{h=-2}]_{-4}^{-6} = -\frac{2}{24}[-3 - 5(2) + 19(-9) + 9(-60)] = \frac{181}{3}$$

(f) $AP_{h=2}]_6^8 = \frac{2}{24}[-9(-3) + 37(0) - 59(35) + 55(126)] = \frac{1223}{3}$

7-25 (a) $M]_{-1}^1 = 2(2) = 4$ is not accurate because $x = 0$ is interior to $[-1, 1]$.
Better: $AC_{h=1}]_{-1}^0 + -AC_{h=-1}]_1^0 = \frac{25}{6} + \frac{11}{12} = \frac{61}{12}$
Parts (b)–(f) can be evaluated as in Problem 7-24.

7-26 NC_5 is exact for fifth-degree polynomials, hence should be most accurate.

7-29 (a) $T[1] = 225.5$, $T[\frac{1}{2}] = \frac{1}{2}\{T[1] + 1[(\frac{3}{2})^4 + (\frac{5}{2})^4 + (\frac{7}{2})^4]\} = 209.8435$, $T_1[\frac{1}{2}] = S[\frac{1}{2}] = 204.62467$
(c) $T[2] = 100.513$, $T[1] = \frac{1}{2}\{T[2] + 2[f(1) + f(3)]\} = 51.1870$, $T_1[1] = S[1] = 34.7450$
(e) $T[4] = 4$, $T[2] = \frac{1}{2}\{T[4] + 4[\sqrt{2}]\} = 4.82843$, $T_1[2] = S[2] = 5.10457$
(f) $T[\pi/4] = 0.392699$, $T[\pi/8] = \frac{1}{2}\{T[\pi/4] + (\pi/4)[\tan(\pi/8)]\} = 0.359010$, $T_1[\pi/8] = S[\pi/8] = 0.347781$

7-30 (a) $S[\frac{3}{2}] = 206.525$, $S[\frac{3}{4}] = 204.7266$, $S_1[\frac{3}{4}] = 204.6$ (exact)
(c) $S[2] = 67.3110$, $S[1] = 34.7450$, $S_1[1] = 32.5739$
(e) $S[2] = 5.10457$, $S[1] = 5.25221$, $S_1[1] = 5.26205$
(f) $S[\pi/8] = 0.3477814$, $S[\pi/16] = 0.3466742$, $S_1[\pi/16] = 0.3466004$

7-33 (a) $T_2[\frac{1}{2}] = 206.4$ (exact) (c) $T_3[\frac{1}{4}] \doteq 12.17668$ has no accurate digits [see (33) of Section 7.4]. (e) $T_3[\frac{1}{2}] \doteq 5.27004$ has no accurate digits (f) $T_2[\pi/32] \doteq 0.346574$ has 5*s*

7-34 (a) $S_{1/3}[2]_{-3}^3$ or $S_{1/3}[1]_{-3}^3$ (cubic exact) will give $\int_{-3}^3 = 0$.
(d) $\int_{-7}^{-6} = -\int_{-6}^{-7} = -AC_{h=-1}]_{-6}^{-7} \doteq -82.54$

(g) $\int_8^9 = \int_7^9 - \int_7^8$; now use suitable Adams formulas.

7-35 (a) Using $h = 0.4$, 0.8, and 1.6 gives 202.34 (exact)

(b) $S_{1/3}]_{2.0}^{2.8}$ will not be exact. One way to get exactness is $\int_{2.0}^{2.8} = \int_{2.0}^{4.4} - \int_{2.8}^{4.4}$; now use $S_{3/8}]_{2.0}^{4.4}$, $S_{1/3}]_{2.8}^{4.4}$, and Richardson.

7-36 (a) $\int_{-1}^{1} = AC_{h=1}]_{-1}^{0} - AC_{h=-1}]_{1}^{0}$ (b) $\int_{-3}^{3} = S_{3/8}]_{-3}^{0} - S_{3/8}]_{3}^{0}$ [NOTE: $S_{1/3}[1]_{-3}^{3}$ is a poor strategy. Why?]

7-40 Solve $2n + 1 = 7$ to get $n = 3$.

7-41 (a) 20.25 (exact) (b) 1.3307 (c) 0 (exact)

7-44 Use 5-point Gauss quadrature from 0 to 60 to get ninth-degree exactness.

7-45 Singularities at $x = 0$ in (a)–(d); also at $x = 1$ in (c). Aside from isolating the singularities, we could change variables to get:

(a) $2\int_0^{\sqrt{\pi}} \dfrac{\sin u^2}{u^2}\, du$ (removable singularity at $u = 0$); $u = \sqrt{x}$

(b) $\int_1^{\infty} e^{1/u}\, du$ (diverges because $e^{1/u} \to 1 \neq 0$ as $u \to \infty$); $u = 1/x$

(c) $\int_0^{\pi/2} \ln(\sin\theta)\, d\theta$ (singularity only at $\theta = 0$); $x = \sin\theta$

(d) $\int_0^1 3ue^u\, du$ (proper); $u = x^{1/3}$

All of (e)–(h) are on unbounded intervals; and (g) has a singularity at $x = 0$. Aside from isolating $\pm\infty$ [and the singularity in (g)], one can let $u = e^x$ in (e) and $u = \sqrt{x}$ in (f).

7-47 (b) $(e^{1/\sqrt{3}} + e^{-1/\sqrt{3}})(\pi/3)\tan(\pi/6) \doteq 1.41639$

(c) $[g(2.42265) + g(3.57735)] \approx [2.631785 + 5.951549] \doteq 8.58333$ (exact)

(d) $x_1 = 2 - 2\sqrt{0.6}$, $x_2 = 2$, $x_3 = 2 + 2\sqrt{0.6}$;
$I \approx \frac{2}{9}[5(0.85437) + 8(8.77811) + 5(22.46949)] \doteq 41.5209$

Chapter 8

	IVP	$y(t_j)$	$y_1[h]$	$y_2[h/2]$	Improved	Exact
8-2	A	$y(0.2)$	2	1.995	1.99	1.989975
	C	$y(0.4)$	0	0.04	0.08	0.08328706
	E	$y(2.5)$	5	5.03033	5.06066	5.0625
8-3	A	$y(0.2)$	1.99	1.989981	1.98997	1.989975
	C	$y(0.4)$	0.08	0.082424	0.0832320	0.08328706
	E	$y(2.5)$	5.060660	5.062026	5.06248	5.0625
8-4	A	$y(0.2)$	1.99	1.989978	1.98997	1.989975
	C	$y(0.4)$	0.08	0.082832	0.0837760	0.08328706
	E	$y(2.5)$	5.06	5.061579	5.06243	5.0625
8-5	A	$y(0.2)$	1.99	1.989975	1.98997	1.989975
	C	$y(0.4)$	0.08	0.0832853	0.0832870	0.08328706
	E	$y(2.5)$	5.062495	5.06500	5.06250	5.0625
8-6	A	$y(0.2)$	1.99	1.98998	1.98997	1.989975
	C	$y(0.4)$	0.08	0.0820160	0.0826880	0.08328706
	E	$y(2.5)$	5.0625	5.0625	5.0625	5.0625
8-7	A	$y(0.2)$	1.99	1.98998	1.98997	1.989975
	C	$y(0.4)$	0.08	0.0828429	0.0823249	0.08328706
	E	$y(2.5)$	5.0625	5.0625	5.0625	5.0625

[NOTE: The solutions to Problems 8-6 and 8-7 used the following formulas.]
For IVP$_A$: $y' = -t/y$; $y'' = (ty' - y)/y^2$; $y''' = y^{-2}ty'' - 2y^{-1}y'y'' = -3y'y''/y$

For IVP_C: $y' = t(y + 1)$; $y'' = ty' + y + 1$; $y''' = ty'' + 2y'$
For IVP_E: $y' = \sqrt{y}$; $y'' = y'/(2\sqrt{y}) = \frac{1}{2}$ (constant); $y''' = 0$

8-10 (a) $\frac{5}{3}$ (b) 1.497222 (b) 1.532581 [$y(0.4) \doteq 1.533033$]

8-12 For IVP_E, $hf_y = \frac{1}{2}h/\sqrt{y}$. So instability is to be expected if $\frac{1}{2}h/\sqrt{y} < -2$, or $h < -4\sqrt{y} < 0$. However, when $t = t_j$ lies between $t_0 + h_{\min} = -1$ and $t_0 = 2$, $y(t) = (t + 2)/4$ will be no less than $\frac{1}{4}$. So $\sqrt{y} < \frac{1}{2}$, from which $hf_y > h \geqslant h_{\min} = -2$. We therefore do not expect instability for $|h| \leqslant 2$.

8-13 None is stiff. Examine $y(t)$ to justify.

8-14 $0 < h < 1/(2t_N) = \beta$. Note that $y(t) \approx t$ for $t > 1$.

8-15

	$t_1 = 0.4$	$t_2 = 0.8$	$t_3 = 1.6$	$t_4 = 2.4$	$t_5 = 4 = 10h$
(b)	-1.4	1.88	-0.776001	7.62720	-24.2269
(c)	1.55385	1.24379	1.70566	2.42516	4.00340
$y(t)$	1.00569	0.922287	1.60498	2.40020	4.00000

8-20

IVP	$y(t_j)$	(a) p_{j+1}	$c_{j+1} = y_{j+1}$	(b) δ_{j+1}	y_{j+1}	(c) y_{j+1}
A	$y(0.8)$	1.833112	1.833021	0.2384E−5	1.833020	1.833021
	$y(1.0)$	1.732209	1.732024	0.4878E−5	1.732017	1.732024
C	$y(1.6)$	2.498180	2.590094	−0.2426E−2	2.612154	2.612154
	$y(2.0)$	6.045721	6.335710	−0.7524E−2	6.470812	6.470812
D	$y(4.0)$	9.	9.	0	9.	9.
	$y(4.5)$	10.5625	10.5625	0	10.5625	10.5625

8-21 For IVP_A: $Ratio = |\delta_4/h| \doteq 0.96\text{E}{-5} < 10^{-3}/64 = 0.5Rmax/2^n$ (double). So $t_5 = t_4 + 0.4 = 1.2$.
For IVP_C: $Ratio = |\delta_4/h| \doteq 0.06 > Rmax$ (halve). So $t_5 = t_3 + 0.2 = 1.4$ (reject y_4)

8-27 When $n = 2$, only IVP_E.

8-28 Only IVP_G is linear.

8-29 (a) For IVP_H: $\mathbf{y}(1.2) \approx \mathbf{y}_1[0.2] = [-0.4 \quad 0.4]^T$ and $\mathbf{y}_2[0.1] = [-0.43625 \quad 0.38]^T$
(b) $[-0.4725 \quad 0.36]^T$

8-30 (a) For IVP_H: $\mathbf{y}(1.2) \approx \mathbf{y}_1[0.2] = [-0.43251 \quad 36]^T$ and $\mathbf{y}_2[0.1] = [-0.46802 \quad 0.36363]^T$
(b) $[-0.479857 \quad 0.364840]^T$

8-32 For IVP_H: $\mathbf{y}(1.2) \approx \mathbf{y}_1[0.2] = [-0.46662 \quad 0.36461]^T$

8-34 (a) $$\frac{d}{dt}\begin{bmatrix} y_1 \\ y_2 \\ y_3 \end{bmatrix} = \begin{bmatrix} y_2 \\ y_3 \\ \dfrac{y_3}{y_1^2} + 2ty_2 \end{bmatrix}, \qquad \mathbf{y}(0) = \begin{bmatrix} 1 \\ -1 \\ 2 \end{bmatrix}$$

(c) $$\frac{d}{dt}\begin{bmatrix} y_1 \\ y_2 \end{bmatrix} = \begin{bmatrix} 0 & 1 \\ \dfrac{1}{t^2} & -t \end{bmatrix}\begin{bmatrix} y_1 \\ y_2 \end{bmatrix} + \begin{bmatrix} 0 \\ \dfrac{e^t}{t} \end{bmatrix}, \quad \mathbf{y}(1) = \begin{bmatrix} 3 \\ 0 \end{bmatrix}$$

8-35 $y_1' = y_2$, $y_2' = Gy_1/r^3$; $y_3' = y_4$, $y_4' = -Gy_3/r^3$; $r = \sqrt{y_1^2 + y_3^2}$;
$\mathbf{y}_0 = [x_0 \quad v_{0x} \quad 0 \quad v_{0y}]^T$

8-37 (a) No (b) $\bar{x} = \ln 2$, so $y(t) = 2^t - 2$

8-38 The exact $\bar{x}$'s are (a) 2 (b) $-\frac{1}{2}$.

8-39 For BVP_I: $(1 + h)y_{j-1} + (-2 + h^2)y_j + (1 - h)y_{j+1} = -3h^2$, $y_0 = -3$, $y_n = 1$
For BVP_K $y_j = \frac{1}{2}\{y_{j-1} + y_{j+1} - 4h^3/(y_{j+1} - y_{j-1} + 2h)\}$, $y_0 = \frac{1}{3}$, $y_n = \frac{20}{3}$

8-40

Approx. of	BVP_I	BVP_J	BVP_K	BVP_L
$y(a + h)$	−2.762	0.9206	1.775	1.247
$y(a + 2h)$	−1.887	0.2063	3.930	0.9806

8-41 For BVP_K: $y_1 = 2.777$ ($n = 2$); $y_2 = 2.772$ ($n = 4$); $(4y_2 - y_1)/3 = 2.770$
For BVP_L: $y_1 = 1.101$ ($n = 2$); $y_2 = 1.106$ ($n = 4$); $(4y_2 - y_1)/3 \doteq 1.106$

Chapter 9

9-2 (a) $p_A(\lambda) = \lambda^2 - 1$; eigenpairs are $(1, [1 \quad 1]^T)$ and $(-1, [1 \quad -1]^T)$.
(b) $p_A(\lambda) = \lambda^2 - 6\lambda - 7$; eigenpairs are $(7, [5 \quad 3]^T)$ and $(-1, [1 \quad -1]^T)$

9-6 (a) $V = \begin{bmatrix} 1 & 1 \\ 1 & -1 \end{bmatrix}$, $V^{-1}AV = \text{diag}(1, \quad -1)$

(b) $V = \begin{bmatrix} 5 & 1 \\ 3 & 1 \end{bmatrix}$, $V^{-1}AV = \text{diag}(7, \quad -1)$

9-7 (a) $\mathbf{y}(t) = [y_1(t) \quad y_2(t)]^T = [\tfrac{1}{2}e^{t-1} + \tfrac{3}{2}e^{1-t} \quad \tfrac{1}{2}e^{t-1} - \tfrac{3}{2}e^{1-t}]^T$

	$\mathbf{x}_1^T$	$\mathbf{x}_2^T$	$\mathbf{x}_3^T$	$\mathbf{x}_4^T$	$\lim_k \mathbf{x}_k^T$
9-10 (a)	$[-0.7500 \quad 1]$	$[-0.6897 \quad 1]$	$[-0.6732 \quad 1]$	$[-0.6685 \quad 1]$	$[-\frac{2}{3} \quad 1]$
(b)	$[1 \quad -0.3333]$	$[1 \quad 0.3846]$	$[1 \quad 0.7867]$	$[1 \quad 0.9351]$	$[1 \quad 1]$
(c)	$[0 \quad 1]$	$[1 \quad \frac{1}{2}]$	$[0.8572 \quad 1]$	$[1 \quad 0.9615]$	$[1 \quad 1]$
9-13 (a)	$[1 \quad 0.6154]$	$[1 \quad 0.5978]$	$[1 \quad 0.6003]$	$[1 \quad 0.6000]$	$[0 \quad 0.6]$
(b)	$[\frac{1}{2} \quad 1]$	$[1 \quad -0.6250]$	$[-0.9318 \quad 1]$	$[1 \quad -0.9901]$	$\pm[1 \quad -1]$
(c)	$[1 \quad 0.7143]$	$[1 \quad 0.5]$	$[1 \quad 0.7143]$	$[1 \quad 0.5]$	None

9-14 Notice that $\mathbf{x}_0 = [1 \quad 1 \quad 1]^T = 1\mathbf{v}_1 + 1\mathbf{v}_2 + 0\mathbf{v}_3$
9-15 (a) $|\lambda_1 - (-4)| < 0.6$; $|\lambda_2 - 3| < 0.1$; $|\lambda_3 - 2| < 0.5$
9-16 (a) $\lambda_1 = -4.01$, $\mathbf{v}_1 = [-0.0178 \quad 1 \quad -0.0140]^T$
$\lambda_2 = 2.96$, $\mathbf{v}_2 = [-0.515 \quad 0.0714 \quad 1]^T$
$\lambda_3 = 2.06$, $\mathbf{v}_3 = [1 \quad 0.223 \quad -0.108]^T$
9-19 (a) $ae + bf = bd + ce$
9-22 $\cos \angle\mathbf{x}, \mathbf{y} = \mathbf{x}^T\mathbf{y}/(\|\mathbf{x}\|\,\|\mathbf{y}\|) = (-2)/(\sqrt{3}\sqrt{14}) = -2/\sqrt{42}$

9-24 (b) $U = \dfrac{1}{\sqrt{2}}\begin{bmatrix} 1 & -1 \\ 1 & 1 \end{bmatrix}$, $D = \begin{bmatrix} 3 & 0 \\ 0 & 1 \end{bmatrix}$ (c) $U = \dfrac{1}{5}\begin{bmatrix} 4 & 3 \\ -3 & 4 \end{bmatrix}$, $D = \begin{bmatrix} -50 & 0 \\ 0 & 25 \end{bmatrix}$

9-25 (a) Using first $R_1 = R_{13}(-0.662909)$, then $R_2 = R_{12}(0.395664)$ yields

$$A_1 = \begin{bmatrix} -0.561553 & -0.788205 & 0 \\ -0.788205 & 1 & -0.615412 \\ 0 & -0.615412 & 3.561553 \end{bmatrix}$$

$$A_2 = \begin{bmatrix} -0.890226 & 0 & -0.236853 \\ 0 & 1.328674 & -0.568008 \\ -10.236853 & -0.568008 & 3.561553 \end{bmatrix}$$

9-28 (a) $\alpha_j = x_j - h/2$, $\beta_j = x_j(-2 - h^2x_j)$, $\gamma_j = x_j + h/2$
(c)

	n	h	$F[h] = \lambda_{\min}$	$F_1[h]$	$F_2[h]$
	2	$\frac{1}{2}$	-3.75		
$r = \frac{3}{2}$					
	3	$\frac{1}{3}$	-3.40453	3.12815	
$r = \frac{4}{3}$					
	4	$\frac{1}{4}$	-3.20341	2.94483	2.85998

9-30 (a) $f(x) \equiv 1$, $r(x) \equiv 0$, $w(x) \equiv 1$ (b) The T matrix is the same.
9-31 (a) $f(x) = x$, $r(x) = x^2$, $w(x) = 1$ (b) and (c): Same as Exercise 9-28.
9-35 (b) $\|A\| = 3$, $\|A^{-1}\| = 1$, cond $A = 3$. Not ill conditioned.
(c) $\|A\| = 50$, $\|A^{-1}\| = \frac{1}{25}$, cond $A = 2$. Not ill conditioned.
9-36 (a) $P_{AA^T}(\lambda) = (\lambda - 4)^2 - 4$; cond $A = 3$

9-38 (a) $A = \begin{bmatrix} 2 & 4 \\ 4 & 14 \end{bmatrix}$ (b) $\lambda = 3$, $\mathbf{x} = \alpha\begin{bmatrix} 5 \\ -1 \end{bmatrix}$ and $\lambda = 1$, $\mathbf{x} = \alpha\begin{bmatrix} -1 \\ 1 \end{bmatrix}$

Index

Note: Citations are given in three ways, as follows:
By Page (e.g., 56–57, 181)
By Section (e.g., 1.3A, 1.4A–D)
By Problem (e.g., **8-23**, **C1-7–C1-8**).

F

G

S

Abbreviations of Names

		Section(s)
APC	Adams predictor–corrector method	8.3A
BIS	Bisection Method	2.3G
BVP	Boundary value problem	8.5
BP	Basic Pivoting strategy	3.3C
DD	Divided difference	6.2B
FP	False Position Method	2.3G
IVP	Initial value problem	8.1
NC	Newton–Cotes	7.3A
NR	Newton–Raphson	2.3C, 4.4B
PC	Predictor–corrector	8.3A
PEP	Partial Extended Precision	1.4C
PP	Partial Pivoting strategy	3.3C
RK4	Fourth-order Runge–Kutta method	8.2A
RKF4	Runge–Kutta–Fehlberg method	8.3B
RS	Repeated Substitution	2.1B
SEC	Secant Method	2.3D
SPP	Scaled Partial Pivoting strategy	3.3C